Handbook of Milkfat Fractionation Technology and Applications

Handbook of Milkfat Fractionation Technology and Applications

Kerry E. Kaylegian
Center for Dairy Research
University of Wisconsin—Madison
Madison, Wisconsin

Robert C. Lindsay
Department of Food Science
University of Wisconsin—Madison
Madison, Wisconsin

Champaign, Illinois

AOCS Mission Statement
To be a forum for the exchange of ideas, information, and experience among those with a professional interest in the science and technology of fats, oils, and related substances in ways that promote personal excellence and provide high standards of quality.

AOCS Books and Special Publications Committee

E. Perkins, chairperson, University of Illinois, Urbana, Illinois
T. Applewhite, Austin, Texas
J. Bauer, Texas A&M University, College Station, Texas
T. Foglia, USDA–ERRC, Philadelphia, Pennsylvania
M. Mossoba, Food and Drug Administration, Washington, D.C.
Y.-S. Huang, Ross Laboratories, Columbus, Ohio
L. Johnson, Iowa State University, Ames, Iowa
J. Lynn, Lever Brothers Co., Edgewater, New Jersey
G. Maerker, Oreland, Pennsylvania
G. Nelson, Western Regional Research Center, San Francisco, California
F. Orthoefer, Riceland Foods Inc., Stuttgart, Arkansas
J. Rattray, University of Guelph, Guelph, Ontario
A. Sinclair, Royal Melbourne Institute of Technology, Melbourne, Australia
T. Smouse, Archer Daniels Midland Co., Decatur, Illinois
G. Szajer, Akzo Chemicals, Dobbs Ferry, New York
L. Witting, State College, Pennsylvania

Copyright © 1995 by AOCS Press. All rights reserved. No part of this book may be reproduced or transmitted in any form or by any means without written permission of the publisher.

The paper used in this book is acid-free and falls within the guidelines established to ensure permanence and durability.

Library of Congress Cataloging-in-Publication Data
Kaylegian, Kerry E.
    Handbook of milkfat fractionation technology and applications/
Kerry E. Kaylegian, Robert C. Lindsay.
      p. cm.
    ISBN 0-935315-57-8
    1. Milkfat fractionation—Handbooks, manuals, etc.   I. Lindsay, Robert C. (Robert Clarence), 1936- . II. Title.
SF275.M56K38   1994
637' . 14—dc20   94-37325
                                                CIP

Printed in the United States of America with vegetable oil–based inks.

Preface

The purpose of this monograph is to organize and summarize the widely distributed literature on milkfat fractionation technology and applications. This monograph is intended for use by

1. Producers of milkfat fractions for specialty food ingredients
2. Users of milkfat fractions in food applications
3. Other interested groups, including research scientists and technologists

While work on milkfat fractionation dates back to the late 1950s, in recent years an accelerated interest in this technology has generated a significant body of literature. This monograph brings this information together so that producers, users, and researchers can easily access the earlier work on milkfat fractionation as a foundation for decision-making.

It must be emphasized that the monograph contains only information available in the public domain, and thus encompasses data on milkfat fractionation technology and applications of milkfat fractions obtained from research journals, books, conference proceedings, graduate theses, patents, and available company literature. Milkfat fractionation is practiced commercially in Europe and New Zealand, and as is found in any industry, some process- and product-specific information is proprietary. Access to this information requires development of confidential arrangements between interested parties.

Chapters in this book are dedicated to the major areas of subject matter relating to milkfat fractionation, and include the following:

1. Milkfat Usage and Modification
2. Raw Materials for Milkfat Fractionation
3. Milkfat Fractionation Technologies
4. Methods for the Characterization of Milkfat Fractions
5. Technical Data for Experimental and Commercial Milkfat Fractions
6. Overview of Milkfat Fractionation
7. The Functionality of Milkfat and Milkfat Fractions in Foods
8. Application of Milkfat Fractions in Foods
9. Future Directions for Milkfat Fractionation

Chapter 1 provides a general perspective of the current usage of milkfat and technologies available to modify milkfat. Chapters 2, 3, and 4 review the raw materials used for milkfat fractionation, fractionation methods, and methods used to characterize milkfat fractions. Chapter 5 contains detailed descriptions of the characteristics of experimental and commercially available milkfat fractions. Technical data are organized primarily by the general melting behavior of the individual fractions and secondarily by the fractionation method employed. In cases where overlap of processing conditions occur, such as fractions produced in a similar manner from both melted milkfat and from solvent solution, the data are compared and interpreted. Chapter 6 summarizes Chapter 5 and provides a broad overview of the characteristics of different categories of milkfat fractions. Chapter 6 also summarizes the effects of fractionation variables on milkfat

Chapter 7 discusses the inherent functionalities of milkfat and milkfat fractions and reviews functionality attributes that are important in food applications. Chapter 8 provides information on the production of experimental milkfat fraction-based ingredients, and the use of experimental and commercially available milkfat ingredients in a variety of food applications.

Chapter 9 summarizes the concepts addressed in the earlier chapters and identifies current applications for milkfat fractions. Chapter 9 also highlights potential new uses of milkfat fractions in food and nonfood products. Basic research needs that have been identified throughout the book are summarized in Chapter 9.

This book has been prepared so that sections of interest to a particular user are understandable without the need to fully comprehend all aspects of milkfat fractionation. Interaction and communication between the researcher, producer, and user are encouraged, because this should lead to an increased understanding of the functionality and behavior of milkfat fractions and the development of specially tailored milkfat ingredients designed to meet the specialized needs of the food industry.

Kerry E. Kaylegian
Robert C. Lindsay

Acknowledgments

The authors thank the University of Wisconsin Center for Dairy Research and the Wisconsin Milk Marketing Board for their continued support in the area of milkfat fractionation. The authors also thank W.J. Harper for his review of the manuscript and suggestions for improvement, and M. Lohman for her assistance with the figures.

Contents

	Preface	v
	List of Tables	xiii
	List of Figures	xxiii

Chapter 1 **Milkfat Usage and Modification** ... 1
    Trends in Milkfat Usage ... 1
        Retail Consumption; Commercial Consumption; Milkfat Surplus
    Milkfat as a Food Ingredient ... 2
        Milkfat Sources; Milkfat Functionality; Factors Affecting Milkfat Functionality
    Milkfat Fractionation ... 4
        The Literature on Milkfat Fractionation; Fractionation Methods; Commercial Fractionation of Milkfat
    Other Technologies Available for Milkfat Modification ... 12
        Physical Processes; Chemical Processes; Biological Processes;
    Cholesterol Removal from Milkfat ... 15
    Blends of Milkfat With Other Fats and Oils ... 17
        Milkfat Blends with Vegetable Oils and Animal Fats; Milkfat and Cocoa Butter

Chapter 2 **Raw Materials for Milkfat Fractionation** ... 19
    Milkfat Biosynthesis ... 19
        Fatty Acid Synthesis; Triglyceride Synthesis; Milkfat Globule Formation
    Factors Affecting the Uniformity of Milkfat ... 20
        Stage of Lactation; Seasonal Variation; Feeding Regimens; Administration of Bovine Somatotropin
    Manufacture of Anhydrous Milkfat ... 21
    Chemical Composition of Milkfat ... 22
        Fatty Acid Composition of Milkfat; Triglyceride Composition and Structure of Milkfat; Other Selected Chemical Properties of Milkfat; Flavor Constituents of Milkfat
    Physical Properties of Milkfat ... 29
        Crystallization Behavior of Milkfat; Melting Behavior of Milkfat
    Other Important Properties of Milkfat ... 33
        Deterioration of Milkfat; Nutritional Properties of Milkfat

Chapter 3 **Milkfat Fractionation Technologies** ... 39
    Fractionation Based on the Crystallization and Solubility Behaviors of Milkfat ... 39
        Milkfat Crystallization; Fractionation Procedures; Fractionation of Milkfat by Crystallization from Melted Milkfat; Fractionation of Milkfat by Crystallization from Solvent Solution
    Fractionation Based on the Solubility and Volatility of Milkfat Components ... 51

Fractionation of Milkfat by Supercritical Fluid Extraction; Fractionation by Short-Path Distillation

Chapter 4 **Methods for the Characterization of Milkfat Fractions**   81
Methods Used to Characterize Chemical Properties of Milkfat Fractions   81
Fatty Acid Composition; Triglyceride Composition; Other Selected Chemical and Physical Properties; Flavor Profile
Methods Used to Characterize Physical Properties of Milkfat Fractions   84
Crystal Characteristics; Melting Behavior; Textural Characteristics
Methods Used to Characterize Other Important Properties of Milkfat Fractions   89
Deterioration of Milkfat Fractions; Nutritional Properties

Chapter 5 **Technical Data for Experimental and Commercial Milkfat Fractions**   141
Nomenclature and Categorization of Milkfat Fractions   141
Fraction Designation; General Fraction Property
Experimentally Prepared Milkfat Fractions   142
Intact Anhydrous Milkfats Used to Produce Experimental Milkfat Fractions; Very-High-Melting Milkfat Fractions; High-Melting Milkfat Fractions; Middle-Melting Milkfat Fractions; Low-Melting Milkfat Fractions; Very-Low-Melting Milkfat Fractions; Unknown-Melting Milkfat Fractions; Oxidative Stability of Experimental Milkfat Fractions; Nutritional Properties of Experimental Milkfat Fractions; Blends of Experimental Milkfat Fractions with Other Fats and Oils
Commercially Prepared Milkfat Ingredients   211
Chemical Characteristics of Commercially Prepared Milkfat Ingredients; Physical Characteristics of Commercially Prepared Milkfat Ingredients; Oxidative Stability of Commercially Prepared Milkfat Ingredients; Nutritional Properties of Commercially Prepared Milkfat Ingredients

Chapter 6 **Overview of Milkfat Fractionation**   489
Conventions Used to Describe Milkfat Fractions   489
Fraction Designation; General Fraction Property
Technologies Used to Fractionate Milkfat   490
Crystallization from Melted Milkfat; Crystallization from Solvent Solution; Supercritical Fluid Extraction
Experimental Milkfat Fractions   491
Fractionation Conditions and Yields; Effects of Fractionation on Selected Characteristics of Milkfat Fractions
Commercially/Prepared Milkfat Ingredients   508

Chapter 7 **The Functionality of Milkfat and Milkfat Fractions in Foods**   509
Interactions of Milkfat With Other Food Components   509
Milkfat-Fat Interactions; Milkfat-(Non-Fat)-Solid Interactions; Milkfat-Gas Interactions; Milkfat-Water Interactions
Basic Functionality of Milkfat Ingredients   513
Chemical Functionality; Physical Functionality
Functionality Attributes of Milkfat Ingredients   516
Firmness, Hardness and Softness; Structure Formation; Viscosity

	and Flow Properties; Solution Properties; Dispersion Properties; Spreadability; Plasticity; Aeration Properties; Layering Properties; Shortening Properties; Coating Properties and Lubricity; Heat Transfer	
	Factors Affecting the Functionality of Milkfat Fractions	521
	Chemical Composition; Crystallization Behavior; Melting Behavior	
	Summary of the Functionality of Milkfat Ingredients in Food Products	522
Chapter 8	**Application of Milkfat Fractions in Foods**	525
	Milkfat Ingredients	526
	Fractionation of Milkfat to Produce Stock Fractions; Blending of Milkfat and Milkfat Fractions; Incorporation of Functional Additives to Milkfat Ingredients; Texturization of Milkfat Ingredients; Packaging of Milkfat Ingredients; Tempering of Milkfat Ingredients	
	The Use of Milkfat Ingredients in Food Products	530
	Cold-Spreadable Butter; Dairy-Based Spreads; Bakery Products; Chocolate; Dairy Products; Frying Oils	
Chapter 9	**Future Directions for Milkfat Fractionation**	631
	Milkfat Ingredients	631
	Applications for Specialty Milkfat Ingredients	632
	Current Applications for Specialty Milkfat Ingredients; Potential Applications for Specialty Milkfat Ingredients	
	Research Needs Identified for Milkfat Fractions	634
	Basic Research Needs Identified for Milkfat Fractions; Applications-Based Research Needs Identified for Milkfat Fractions	
	Commercialization of Milkfat Fractionation in the U.S.	636
	References	637
	Index	657

List of Tables

Chapter 1
1.1	Composition of some milkfat ingredients	3
1.2	Milkfat fractionation studies listed in chronological order	5
1.3	A brief summary of advantages and disadvantages of milkfat fractionation methods	9
1.4	Properties of some commercially available milkfat fraction products	10

Chapter 2
2.1	Fatty acids identified in milkfat	23
2.2	Composition of the major fatty acids in milkfat	24
2.3	Average triglyceride composition of milkfat, expressed as total carbon number	24
2.4	Typical ranges for selected chemical and physical properties of milkfat	26

Chapter 3
3.1	Fractionation methods used to produce milkfat fractions by crystallization from melted milkfat	55
3.2	Fractionation methods used to produce milkfat fractions by crystallization from solvent solution	67
3.3	Fractionation methods used to produce milkfat fractions by supercritical fluid extraction	74
3.4	Fractionation method used to produce milkfat fraction by short-path distillation	79

Chapter 4
4.1	Methods used for the determination of fatty acid and triglyceride compositions of milkfat fractions	91
4.2	Methods used for the determination of selected chemical indices and flavor profile of milkfat fractions	111
4.3	Methods used to determine the crystal characteristics of milkfat fractions	117
4.4	Methods used to determine the melting behavior of milkfat fractions	121
4.5	Methods used to determine the textural properties of milkfat fractions and finished products containing milkfat fractions	132
4.6	Methods for characterization of the quality of milkfat fractions	135
4.7	Methods for the determination of selected nutritional characteristics of milkfat fractions	137

Chapter 5
5.1	Experimental milkfat fractions obtained by crystallization from melted milkfat	228
5.2	Experimental milkfat fractions obtained by crystallization from solvent solution	247
5.3	Experimental milkfat fractions obtained by supercritical fluid extraction	253
5.4	Experimental milkfat fractions obtained by short-path distillation	258
5.5	Experimental milkfat fractions obtained by unspecified fractionation methods	259
5.6	Overall summary of experimental milkfat fractions produced by various methods	260
5.7	Selected melting behavior of anhydrous milkfats used to produce milkfat fractions	262
5.8	Melting behavior of intact anhydrous milkfats used to produce milkfat fractions from melted milkfat	263
5.9	Melting behavior of intact anhydrous milkfats used to produce milkfat fractions from solvent solution	265

5.10	Melting behavior of intact milkfats used to produce milkfat fractions by supercritical fluid extraction	265
5.11	Melting behavior of intact anhdyrous milkfats used to produce milkfat fractions by short-path distillation	266
5.12	Yield and melting behavior of very-high-melting milkfat fractions obtained from melted milkfat	267
5.13	Yield and melting behavior of very-high-melting milkfat fractions obtained from solvent solution	269
5.14	Yield and melting behavior of very-high-melting milkfat fractions obtained by supercritical fluid extraction	270
5.15	Yield and melting behavior of high-melting milkfat fractions obtained from melted milkfat	271
5.16	Yield and melting behavior of high-melting milkfat fractions obtained from solvent solution	276
5.17	Yield and melting behavior of high-melting milkfat fractions obtained by supercritical fluid extraction	277
5.18	Yield and melting behavior of high-melting milkfat fractions obtained by short-path distillation	277
5.19	Yield and melting behavior of middle-melting milkfat fractions obtained from melted milkfat	278
5.20	Yield and melting behavior of middle-melting milkfat fractions obtained from solvent solution	281
5.21	Yield and melting behavior of middle-melting milkfat fractions obtained by supercritical fluid extraction	282
5.22	Yield and melting behavior of middle-melting milkfat fractions obtained by short-path distillation	282
5.23	Yield and melting behavior of low-melting milkfat fractions obtained from melted milkfat	283
5.24	Yield and melting behavior of low-melting milkfat fractions obtained from solvent solution	287
5.25	Yield and melting behavior of low-melting milkfat fractions obtained by supercritical fluid extraction	288
5.26	Yield and melting behavior of low-melting milkfat fractions obtained by short-path distillation	288
5.27	Yield and melting behavior of very-low-melting milkfat fractions obtained from melted milkfat	289
5.28	Yield and melting behavior of very-low-melting milkfat fractions obtained from solvent solution	290
5.29	Yield and melting behavior of unknown-melting milkfat fractions obtained from melted milkfat	291
5.30	Yield and melting behavior of unknown-melting milkfat fractions obtained from solvent solution	297
5.31	Yield and melting behavior of unknown-melting milkfat fractions obtained by supercritical fluid extraction	300
5.32	Yield and melting behavior of unknown-melting milkfat fractions obtained by short-path distillation	301
5.33	Individual fatty acid composition of intact anhydrous milkfats used to produce milkfat fractions from melted milkfat	302
5.34	Individual fatty acid composition of intact anhydrous milkfats used to produce milkfat fractions from solvent solution	306

5.35	Individual fatty acid composition of intact anhydrous milkfats used to produce milkfat fractions by supercritical fluid extraction	308
5.36	Individual fatty acid composition of intact anhydrous milkfats used to produce milkfat fractions by short-path distillation	309
5.37	Individual fatty acid composition of very-high-melting milkfat fractions obtained from melted milkfat	310
5.38	Individual fatty acid composition of very-high-melting milkfat fractions obtained from solvent solution	312
5.39	Individual fatty acid composition of very-high-melting milkfat fractions obtained by supercritical fluid extraction	313
5.40	Individual fatty acid composition of high-melting milkfat fractions obtained from melted milkfat	314
5.41	Individual fatty acid composition of high-melting milkfat fractions obtained from solvent solution	318
5.42	Individual fatty acid composition of high-melting milkfat fractions obtained by supercritical fluid extraction	319
5.43	Individual fatty acid composition of high-melting milkfat fractions obtained by short-path distillation	320
5.44	Individual fatty acid composition of middle-melting milkfat fractions obtained from melted milkfat	320
5.45	Individual fatty acid composition of middle-melting milkfat fractions obtained from solvent solution	323
5.46	Individual fatty acid composition of middle-melting milkfat fractions obtained by supercritical fluid extraction	324
5.47	Individual fatty acid composition of middle-melting milkfat fractions obtained by short-path distillation	324
5.48	Individual fatty acid composition of low-melting milkfat fractions obtained from melted milkfat	325
5.49	Individual fatty acid composition of low-melting milkfat fractions obtained from solvent solution	331
5.50	Individual fatty acid composition of low-melting milkfat fractions obtained by supercritical fluid extraction	332
5.51	Individual fatty acid composition of low-melting milkfat fractions obtained by short-path distillation	333
5.52	Individual fatty acid composition of very-low-melting milkfat fractions obtained from melted milkfat	333
5.53	Individual fatty acid composition of very-low-melting milkfat fractions obtained from solvent solution	334
5.54	Individual fatty acid composition of unknown-melting milkfat fractions obtained from melted milkfat	334
5.55	Individual fatty acid composition of unknown-melting milkfat fractions obtained from solvent solution	336
5.56	Individual fatty acid composition of unknown-melting milkfat fractions obained by supercritical fluid extraction	340
5.57	Individual fatty acid composition of unknown-melting milkfat fractions obtained by short-path distillation	344
5.58	Fatty acid composition (selected fatty acid groupings) of intact anhydrous milkfats used to produce milkfat fractions from melted milkfat	345
5.59	Fatty acid composition (selected fatty acid groupings) of intact anhydrous milkfats used to produce milkfat fractions from solvent solution	347

5.60	Fatty acid composition (selected fatty acid groupings) of intact anhydrous milkfats used to produce milkfat fractions by supercritical fluid extraction	348
5.61	Fatty acid composition (selected fatty acid groupings) of intact anhydrous milkfats used to produce milkfat fractions by short-path distillation	348
5.62	Fatty acid composition (selected fatty acid groupings) of very-high-melting milkfat fractions obtained from melted milkfat	349
5.63	Fatty acid composition (selected fatty acid groupings) of very-high-melting milkfat fractions obtained from solvent solution	351
5.64	Fatty acid composition (selected fatty acid groupings) of very-high-melting milkfat fractions obtained by supercritical fluid extraction	352
5.65	Fatty acid composition (selected fatty acid groupings) of high-melting milkfat fractions obtained from melted milkfat	353
5.66	Fatty acid composition (selected fatty acid groupings) of high-melting milkfat fractions obtained from solvent solution	355
5.67	Fatty acid composition (selected fatty acid groupings) of high-melting milkfat fractions obtained by supercritical fluid extraction	356
5.68	Fatty acid composition (selected fatty acid groupings) of high-melting milkfat fractions obtained by short-path distillation	356
5.69	Fatty acid composition (selected fatty acid groupings) of middle-melting milkfat fractions obtained from melted milkfat	357
5.70	Fatty acid composition (selected fatty acid groupings) of middle-melting milkfat fractions obtained from solvent solution	358
5.71	Fatty acid composition (selected fatty acid groupings) of middle-melting milkfat fractions obtained by supercritical fluid extraction	359
5.72	Fatty acid composition (selected fatty acid groupings) of middle-melting milkfat fractions obtained by short-path distillation	359
5.73	Fatty acid composition (selected fatty acid groupings) of low-melting milkfat fractions obtained from melted milkfat	360
5.74	Fatty acid composition (selected fatty acid groupings) of low-melting milkfat fractions obtained from solvent solution	363
5.75	Fatty acid composition (selected fatty acid groupings) of low-melting milkfat fractions obtained by supercritical fluid extraction	363
5.76	Fatty acid composition (selected fatty acid groupings) of low-melting milkfat fractions obtained by short-path distillation	364
5.77	Fatty acid composition (selected fatty acid groupings) of very-low-melting milkfat fractions obtained from melted milkfat	364
5.78	Fatty acid composition (selected fatty acid groupings) of very-low-melting milkfat fractions obtained from solvent solution	364
5.79	Fatty acid composition (selected fatty acid groupings) of unknown-melting milkfat fractions obtained from melted milkfat	365
5.80	Fatty acid composition (selected fatty acid groupings) of unknown-melting milkfat fractions obtained by short-path distillation	366
5.81	Fatty acid composition (selected fatty acid groupings) of unknown-melting milkfat fractions obtained from solvent solution	367
5.82	Fatty acid composition (selected fatty acid groupings) of unknown-melting milkfat fractions obtained by supercritical fluid extraction	369
5.83	Triglyceride composition (individual triglyceride) of intact anhydrous milkfats used to produce milkfat fractions from melted milkfat	370
5.84	Triglyceride composition (individual triglyceride) of intact anhydrous milkfats used to produce milkfat fractions from solvent solution	371

List of Tables   xvii

5.85	Triglyceride composition (individual triglyceride) of intact anhydrous milkfats used to produce milkfat fractions by supercritical fluid extraction	372
5.86	Triglyceride composition (individual triglyceride) of intact anhydrous milkfats used to produce milkfat fractions by short-path distillation	373
5.87	Triglyceride composition (individual triglyceride) of very-high-melting milkfat fractions obtained from melted milkfat	373
5.88	Triglyceride composition (individual triglyceride) of very-high-melting milkfat fractions obtained from solvent solution	374
5.89	Triglyceride composition (individual triglyceride) of high-melting milkfat fractions obtained from melted milkfat	375
5.90	Triglyceride composition (individual triglyceride) of high-melting milkfat fractions obtained from solvent solution	377
5.91	Triglyceride composition (individual triglyceride) of high-melting milkfat fractions obtained by supercritical fluid extraction	378
5.92	Triglyceride composition (individual triglyceride) of high-melting milkfat fractions obtained by short-path distillation	379
5.93	Triglyceride composition (individual triglyceride) of middle-melting milkfat fractions obtained from melted milkfat	379
5.94	Triglyceride composition (individual triglyceride) of middle-melting milkfat fractions obtained from solvent solution	380
5.95	Triglyceride composition (individual triglyceride) of middle-melting milkfat fractions obtained by supercritical fluid extraction	381
5.96	Triglyceride composition (individual triglyceride) of middle-melting milkfat fractions obtained by short-path distillation	381
5.97	Triglyceride composition (individual triglyceride) of low-melting milkfat fractions obtained from melted milkfat	382
5.98	Triglyceride composition (individual triglyceride) of low-melting milkfat fractions obtained from solvent solution	383
5.99	Triglyceride composition (individual triglyceride) of low-melting milkfat fractions obtained by supercritical fluid extraction	383
5.100	Triglyceride composition (individual triglyceride) of low-melting milkfat fractions obtained by short-path distillation	384
5.101	Triglyceride composition (individual triglyceride) of very-low-melting milkfat fractions obtained from solvent solution	384
5.102	Triglyceride composition (individual triglyceride) of unknown-melting milkfat fractions obtained by short-path distillation	385
5.103	Triglyceride composition (individual triglyceride) of unknown-melting milkfat fractions obtained from melted milkfat	385
5.104	Triglyceride composition (individual triglyceride) of unknown-melting milkfat fractions obtained from solvent solution	387
5.105	Triglyceride composition (individual triglyceride) of unknown-melting milkfat fractions obtained by supercritical fluid extraction	388
5.106	Triglyceride composition (selected triglyceride groupings) of intact anhydrous milkfats used to produce milkfat fractions from melted milkfat	389
5.107	Triglyceride composition (selected triglyceride groupings) of intact anhydrous milkfats used to produce milkfat fractions from solvent solution	391
5.108	Triglyceride composition (selected triglyceride groupings) of intact anhydrous milkfats used to produce milkfat fractions by supercritical fluid extraction	392
5.109	Triglyceride composition (selected triglyceride groupings) of intact anhydrous milkfats used to produce milkfat fractions by short-path distillation	392

5.110	Triglyceride composition (selected triglyceride groupings) of very-high-melting milkfat fractions obtained from melted milkfat	393
5.111	Triglyceride composition (selected triglyceride groupings) of very-high-melting milkfat fractions obtained from solvent solution	394
5.112	Triglyceride composition (selected triglyceride groupings) of very-high-melting milkfat fractions obtained supercritical fluid extraction	394
5.113	Triglyceride composition (selected triglyceride groupings) of high-melting milkfat fractions obtained from melted milkfat	395
5.114	Triglyceride composition (selected triglyceride groupings) of high-melting milkfat fractions obtained from solvent solution	397
5.115	Triglyceride composition (selected triglyceride groupings) of high-melting milkfat fractions obtained by supercritical fluid extraction	397
5.116	Triglyceride composition (selected triglyceride groupings) of high-melting milkfat fractions obtained by short-path distillation	398
5.117	Triglyceride composition (selected triglyceride groupings) of middle-melting milkfat fractions obtained from melted milkfat	399
5.118	Triglyceride composition (selected triglyceride groupings) of middle-melting milkfat fractions obtained from solvent solution	401
5.119	Triglyceride composition (selected triglyceride groupings) of middle-melting milkfat fractions obtained by supercritical fluid extraction	401
5.120	Triglyceride composition (selected triglyceride groupings) of middle-melting milkfat fractions obtained by short-path distillation	401
5.121	Triglyceride composition (selected triglyceride groupings) of low-melting milkfat fractions obtained from melted milkfat	402
5.122	Triglyceride composition (selected triglyceride groupings) of low-melting milkfat fractions obtained from solvent solution	403
5.123	Triglyceride composition (selected triglyceride groupings) of low-melting milkfat fractions obtained by supercritical fluid extraction	404
5.124	Triglyceride composition (selected triglyceride groupings) of low-melting milkfat fractions obtained by short-path distillation	404
5.125	Triglyceride composition (selected triglyceride groupings) of very-low-melting milkfat fractions obtained from melted milkfat	404
5.126	Triglyceride composition (selected triglyceride groupings) of very-low-melting milkfat fractions obtained from solvent solution	405
5.127	Triglyceride composition (selected triglyceride groupings) of unknown-melting milkfat fractions obtained from melted milkfat	405
5.128	Triglyceride composition (selected triglyceride groupings) of unknown-melting milkfat fractions obtained from solvent solution	406
5.129	Triglyceride composition (selected triglyceride groupings) of unknown-melting milkfat fractions obtained by supercritical fluid extraction	407
5.130	Triglyceride composition (selected triglyceride groupings) of unknown-melting milkfat fractions obtained by short-path distillation	408
5.131	Selected chemical characteristics of the intact anhydrous milkfats used to produce milkfat fractions from melted milkfat	408
5.132	Selected chemical characteristics of the intact anhydrous milkfats used to produce milkfat fractions from solvent solution	410
5.133	Selected chemical characteristics of the intact anhydrous milkfats used to produce milkfat fractions by supercritical fluid extraction	410
5.134	Selected chemical characteristics of very-high-melting milkfat fractions obtained from melted milkfat	410

5.135	Selected chemical characteristics of very-high-melting milkfat fractions obtained from solvent solution	410
5.136	Selected chemical characteristics of very-high-melting milkfat fractions obtained by supercritical fluid extraction	411
5.137	Selected chemical characteristics of high-melting milkfat fractions obtained from melted milkfat	411
5.138	Selected chemical characteristics of high-melting milkfat fractions obtained from solvent solution	413
5.139	Selected chemical characteristics of high-melting milkfat fractions obtained by supercritical fluid extraction	414
5.140	Selected chemical characteristics of middle-melting milkfat fractions obtained from melted milkfat	414
5.141	Selected chemical characteristics of middle-melting milkfat fractions obtained from solvent solution	415
5.142	Selected chemical characteristics of middle-melting ilkfat fractions obtained by supercritical fluid extraction	416
5.143	Selected chemical characteristics of low-melting milkfat fractions obtained from melted milkfat	416
5.144	Selected chemical characteristics of low-melting milkfat fractions obtained from solvent solution	418
5.145	Selected chemical characteristics of low-melting milkfat fractions obtained by supercritical fluid extraction	419
5.146	Selected chemical characteristics of very-low-melting milkfat fractions obtained from melted milkfat	419
5.147	Selected chemical characteristics of very-low-melting milkfat fractions obtained from solvent solution	419
5.148	Selected chemical characteristics of unknown-melting milkfat fractions obtained from melted milkfat	420
5.149	Selected chemical characteristics of unknown-melting milkfat fractions obtained from solvent solution	421
5.150	Lactone concentration of anhydrous milkfat used to produce milkfat fractions	422
5.151	Lactone concentration of very-high-melting milkfat fractions	423
5.152	Lactone concentration of high-melting milkfat fractions	423
5.153	Lactone concentration of middle-melting milkfat fractions	424
5.154	Lactone concentration of low-melting milkfat fractions	424
5.155	Lactone concentration of unknown-melting milkfat fractions	425
5.156	Methyl ketone concentration of anhydrous milkfat used to produce milkfat fractions	427
5.157	Methyl ketone concentration of very-high-melting milkfat fractions	427
5.158	Methyl ketone concentration of high-melting milkfat fractions	428
5.159	Methyl ketone concentration of middle-melting milkfat fractions	428
5.160	Methyl ketone concentration of low-melting milkfat fractions	428
5.161	Aldehyde concentration of anhydrous milkfat used to produce milkfat fractions	429
5.162	Aldehyde concentration of very-high-melting milkfat fractions	429
5.163	Aldehyde concentration of high-melting milkfat fractions	429
5.164	Aldehyde concentration of middle-melting milkfat fractions	429
5.165	Aldchydc concentration of low-melting milkfat fractions	430
5.166	Crystal morphology for anydrous milkfats used to produce milkfat fractions	430
5.167	Crystal morphology of very-high-melting milkfat fractions	431
5.168	Crystal morphology of high-melting milkfat fractions	432
5.169	Crystal morphology of middle-melting milkfat fractions	432

5.170	Crystal morphology of low-melting milkfat fractions	433
5.171	Crystal morphology of unknown-melting milkfat fractions	433
5.172	Crystal size of high-melting milkfat fractions	434
5.173	Crystal size of unknown-melting milkfat fractions	435
5.174	Solid fat content of intact anhydrous milkfats used to produce milkfat fractions from melted milkfat	436
5.175	Solid fat content of intact anhydrous milkfats used to produce milkfat fractions from solvent solution	437
5.176	Solid fat content of intact anhydrous milkfats used to produce milkfat fractions by supercritical fluid extraction	437
5.177	Solid fat content of intact anhydrous milkfats used to produce milkfat fractions by short-path distillation	438
5.178	Solid fat content of very-high-melting milkfat fractions obtained from melted milkfat	438
5.179	Solid fat content of very-high-melting milkfat fractions obtained from solvent solution	439
5.180	Solid fat content of very-high-melting milkfat fractions obtained by supercritical fluid extraction	439
5.181	Solid fat content of high-melting milkfat fractions obtained from melted milkfat	440
5.182	Solid fat content of high-melting milkfat fractions obtained from solvent solution	442
5.183	Solid fat content of high-melting milkfat fractions obtained by supercritical fluid extraction	442
5.184	Solid fat content of high-melting milkfat fractions obtained by short-path distillation	442
5.185	Solid fat content of middle-melting milkfat fractions obtained from melted milkfat	443
5.186	Solid fat content of middle-melting milkfat fractions obtained from solvent solution	444
5.187	Solid fat content of middle-melting milkfat fractions obtained by supercritical fluid extraction	444
5.188	Solid fat content of middle-melting milkfat fractions obtained by short-path distillation	444
5.189	Solid fat content of low-melting milkfat fractions obtained from melted milkfat	445
5.190	Solid fat content of low-melting milkfat fractions obtained from solvent solution	447
5.191	Solid fat content of low-melting milkfat fractions obtained by supercritical fluid extraction	447
5.192	Solid fat content of low-melting milkfat fractions obtained by short-path distillation	448
5.193	Solid fat content of very-low-melting milkfat fractions obtained from melted milkfat	448
5.194	Solid fat content of very-low-melting milkfat fractions obtained from solvent solution	448
5.195	Textural characteristics of anhydrous milkfats used to produce milkfat fractions	449
5.196	Textural characteristics of high-melting milkfat fractions	449
5.197	Textural characteristics of middle-melting milkfat fractions	450
5.198	Textural characteristics of low-melting milkfat fractions	450
5.199	Textural characteristics of unknown-melting milkfat fractions	451
5.200	Oxidative and lipolytic stability characterization of intact milkfat and milkfat fractions	452
5.201	Oxidation products formed in milkfat and milkfat fractions	461
5.202	Nutritional composition of intact milkfat and milkfat fractions	468
5.203	Formulas for mixtures of milkfat fractions with other fats and oils	473
5.204	Solid fat content profiles for mixtures of milkfat fractions with other fats and oils	477
5.205	List of some commercially-available milkfat fraction-based products	480
5.206	Fatty acid composition of commercially-available milkfat fraction products	482
5.207	Triglyceride composition of commercially-available milkfat fraction products	484
5.208	Selected chemical characteristics of commercially-available milkfat fraction products	486
5.209	Solid fat content commercially-available milkfat fraction products	487

Chapter 6

6.1	Effects of fractionation on the chemical composition of milkfat fractions relative to intact milkfat	502

Chapter 7

7.1	Functionality categories of milkfat ingredients	510

Chapter 8

8.1	Formulations and processing information for experimental cold-spreadable butters	564
8.2	Solid fat content of experimental cold-spreadable butters	570
8.3	Textural characteristics of experimental cold-spreadable butters	572
8.4	Lactone concentration of cold-spreadable butters	576
8.5	Formulations and processing information for dairy-based spreads	577
8.6	Solid fat content profiles of dairy-based spreads	580
8.7	Textural characteristics of dairy-based spreads	581
8.8	Formulations and processing information for experimental bakery milkfat ingredients	582
8.9	Chemical and physical properties of the milkfat phase of experimental and commercial milkfat ingredients for bakery products	586
8.10	Solid fat content profiles of the milkfat portion of experimental and commercial milkfat ingredients used in bakery products	591
8.11	Formulations and evaluation of milkfat ingredients in bakery products	594
8.12	Formulations and processing information for milkfat fractions in chocolate	599
8.13	Commercial milkfat ingredients for use in chocolate	605
8.14	Solid fat content of the experimental milkfat ingredients used in chocolates	606
8.15	Solid fat content of the fat phase used in experimental chocolates	607
8.16	Evaluations of milkfat ingredients in experimental chocolate	609
8.17	Formulations and processing information for milkfat ingredients used in experimental dairy products	614
8.18	Information for commercial milkfat ingredients in dairy products	623
8.19	Solid fat content profiles of the commercial milkfat ingredients for use in dairy products	624
8.20	Evaluation of experimental milkfat ingredients in dairy products	625
8.21	Evaluation of milkfat fractions in deep-fat frying	629

List of Figures

Chapter 2
2.1	Typical thermal profile of anhydrous milkfat	32
2.2	Typical solid fat content profile of anhydrous milkfat	32

Chapter 3
3.1	Typical schematic diagram of fractionation equipment used for experimental fractionation from melted milkfat	44
3.2	Schematic diagram of Tirtiaux fractionation equipment	45
3.3	Typical schematic diagram of equipment used for milkfat fractionation by supercritical fluid extraction	53
3.4	Schematic diagram of equipment used for milkfat fractionation by short-path distillation	54

Chapter 5
5.1	Thermal profiles (DSC curves) for intact anhydrous milkfats used to produce milkfat fractions from melted milkfat	213
5.2	Thermal profiles (DSC curves) for intact anhydrous milkfats and other starting materials used to produce milkfat fractions from solvent solutions	213
5.3	Thermal profiles (DSC curves) of intact anhydrous milkfats used to produce milkfat fractions by supercritical fluid extraction	214
5.4	Thermal profile (DSC curve) of intact anhydrous milkfats used to produce milkfat fractions by short-path distillation	214
5.5	Thermal profiles (DSC curves) of very-high-melting milkfat fractions obtained from melted milkfat	215
5.6	Thermal profiles (DSC curves) of very-high-melting milkfat fractions obtained from solvent solution	215
5.7	Thermal profile (DSC curve) of very-high-melting milkfat fractions obtained by supercritical fluid extraction	216
5.8	Thermal profiles (DSC curves) for high-melting milkfat fractions obtained from melted milkfat	216
5.9	Thermal profiles (DSC curves) of high-melting milkfat fractions obtained from solvent solution	217
5.10	Thermal profiles (DSC curves) of high-melting milkfat fractions obtained by supercritical fluid extraction	217
5.11	Thermal profile (DSC curve) of high-melting milkfat fractions obtained by short-path distillation	218
5.12	Thermal profiles (DSC curves) of middle-melting milkfat fractions obtained from melted milkfat	218
5.13	Thermal profiles (DSC curves) of middle-melting milkfat fractions obtained from solvent solution	219
5.14	Thermal profiles (DSC curves) of middle-melting milkfat fractions obtained by supercritical fluid extraction	219
5.15	Thermal profile (DSC curve) of middle-melting milkfat fractions obtained by short-path distillation	220
5.16	Thermal profiles (DSC curves) of low-melting milkfat fractions obtained from melted milkfat	220

5.17	Thermal profiles (DSC curves) of low-melting milkfat fractions obtained from solvent solution	221
5.18	Thermal profiles (DSC curves) of low-melting milkfat fractions obtained by supercritical fluid extraction	221
5.19	Thermal profiles (DSC curves) of low-melting milkfat fractions obtained by short-path distillation	222
5.20	Thermal profiles (DSC curves) of very-low-melting milkfat fractions obtained from melted milkfat	222
5.21	Thermal profiles (DSC curves) of very-low-melting milkfat fractions obtained from solvent solution	222
5.22	Solid fat content profiles of very-high-melting milkfat fractions	223
5.23	Solid fat content profiles of high-melting milkfat fractions	223
5.24	Solid fat content profiles of middle-melting milkfat fractions	224
5.25	Thermal profiles (DSC curves) for blends of milkfat fractions with other fats and oils	224
5.26	Isosolid diagrams of blends of milkfat fractions and other fats and oils	225
5.27	Phase diagrams of milkfat fraction and cocoa butter mixtures	226
5.28	Thermal profiles (DSC curves) of commercially-available milkfat fraction products	227

Chapter 6

6.1	Thermal profiles (DSC curves) of milkfat fractions obtained by multiple-step crystallization from melted milkfat	502
6.2	Thermal profiles (DSC curves) of milkfat fractions obtained by multiple-step crystallization from acetone solution	503
6.3	Solid fat content profiles of milkfat fractions obtained by single-step crystallization from melted milkfat	504
6.4	Solid fat content profiles of milkfat fractions obtained by multiple-step crystallization from melted milkfat	505
6.5	Solid fat content profiles of milkfat fractions obtained by multiple-step crystallization from acetone solution	506

Chapter 8

8.1	Thermal profiles (DSC curves) of dairy-based spreads	562
8.2	Isosolid diagrams of milkfat-cocoa butter mixtures used for experimental chocolates	562
8.3	Thermal profiles (DSC) for experimental chocolates	563

Chapter 1
Milkfat Usage and Modification

Trends in Milkfat Usage

Milkfat has been a traditional part of human diets—historically, through the consumption of fluid milk, butter, and cheese, and more recently, through commercially prepared foods that contain milkfat-based ingredients as well. In recent years milkfat has experienced a worldwide decline in usage from both retail and commercial markets. Some of the reasons behind this decline include health concerns, price, and restricted functionality.

Retail Consumption

At the retail level, consumption patterns have shifted away from traditional dairy foods toward the low-fat varieties. In the U.S., per capita consumption of fluid milk declined 13% between 1975 and 1990. In 1975 whole milk had 68% of the market but declined to 39% in 1990, while low-fat and skim milk went from 26% of the market in 1975 to 56% in 1990 (Milk Industry Foundation, 1991). Health-related trends toward low-fat diets, particularly low-cholesterol diets, have resulted in a shift away from butter toward margarine and butter–margarine spreads. While per capita spread consumption has seen only a slight overall decline, butter went from 64% of the market share in 1950 to 30% in 1980, while margarine's share increased from 36% in 1950 to 71% of the market in 1980 (Anon., *Dairy Field,* 1982). In a survey of 18 countries conducted by the International Dairy Federation (1984), the following barriers to sales of retail butterfat were given in order of perceived importance:

1. Price of butter compared to margarine
2. Consumer attitudes that butter can be dangerous for health
3. Poor spreadability of butter from the refrigerator
4. Inadequate promotional spending for butter
5. Product innovation in margarine and spreads
6. Inadequate retail margin for butter compared to other yellow fats
7. Legislative barriers (e.g., regulations hampering the introduction of new types of butter or dairy spreads)
8. Other barriers (e.g., poor display of butter in shops)

Commercial Consumption

Current consumer trends influence the commercial markets for fats, particularly in the area of saturated fats. This can be illustrated by recent U.S. trends away from the use of animal fats and tropical oils in processed foods and by a large increase in the number of low-fat foods available in the marketplace. Price and functionality are also important factors of milkfat usage at the commercial level; milkfat has the disadvantage of both high price and limited functionality compared to tailored vegetable fats and oils. In the vegetable fats and oils industry a variety of techniques such as fractionation, hydrogenation, and interesterification are commonly used to modify fats and oils for use as specialized bakery fats, confectionery fats, frying oils, and other liquid shortenings, but milkfat is only churned into butter or processed into butter oil before it is used in the food industry.

Milkfat Surplus

Historically, U.S. farmers have been paid for milk on the basis of the milkfat content and therefore have been rewarded for the production of large quantities of milkfat. The continued production of large quantities of milkfat coupled with the decrease in usage has led to increasing surplus in the U.S. Jesse (1989) has stated several alternatives to help decrease current and future milkfat surplus in the U.S.:

- Give butterfat away domestically or subsidize exports.
- Produce less milk.
- Change the pricing structure of milk and its components.
- Genetically modify the cow to change the composition of milk.
- Modify milkfat in such a way as to increase its value and thus its usage in the food industry.

Subsidization of milkfat exports and production quotas for milkfat are associated with economic and political considerations (Jesse, 1989). The restructuring of milk pricing in the U.S. is currently under consideration (Jesse, 1989). Genetic modifications of the cow's production of milkfat are being evaluated at the research stage (Gibson, 1991). Of these options, the modification of milkfat for use as specialized food ingredients appears to be the most feasible and holds the most promise for increasing the value of milkfat and reducing the surplus.

Milkfat as a Food Ingredient

The trend away from home-prepared foods toward commercially prepared foods has created many opportunities for dairy ingredients, particularly milkfat-based ingredients. Milkfat ingredients can be found in a wide range of food products, including prepared baked goods and bakery mixes, confectionery items, frozen dinners and entrée mixes, soups, and snacks. Milkfat-based ingredients contribute in several ways to food products:

- Flavor
- Mouthfeel
- Texture
- Structure
- Pseudomoistness
- Appearance characteristics

Characteristics relevant to milkfat's performance as a food ingredient (Kirkpatrick, 1982) include the following:

- The flavor of milkfat is inherently desirable to the human palate.
- Milkfat is viewed by consumers as less processed than other fats, and thus is perceived as pure and unadulterated.
- Compared to other fats, milkfat contains a higher proportion of short-chain fatty acids, which are thought to contribute to its ease of digestibility.
- Milkfat is or can be rendered relatively stable in storage in a wide variety of forms without the need for additives.
- The physical properties of milkfat, especially the melting and plastic range, often do not suit it to application in some important areas of edible fats usages.

Milkfat Sources

Milkfat can be incorporated into commercially prepared foods through the use of anhydrous milkfat, butter, cream, and other traditional dairy products. Typical analyses of these products are presented in Table 1.1. Alternate sources of milkfat include cheese, yogurt, buttermilk, and a number of low-fat dairy products, which are found in fluid, condensed, and dried forms. These milkfat sources vary in chemical composition and physical properties. The nonfat milk components (protein and lactose) and the presence of an aqueous phase can contribute substantially to the functionality of these ingredients in foods. The choice of milkfat-based ingredients usually depends on the needs of the final application.

TABLE 1.1 Composition of Some Milkfat Ingredients (Bassette and Acosta, 1988)

Ingredient	Milkfat (%)	Water (%)	Solids not fat (%)
Butter	81.1	15.9	3.0
Butter oil	99.3	0.5	0.2
Anhydrous milkfat	99.8	0.1	0.1
Heavy whipping cream	37.0	57.7	5.3
Light whipping cream	30.9	63.5	5.6
Light cream	19.3	73.7	7.0
Half and half	11.5	80.6	7.9
Fluid whole milk	3.3	88.5	8.2
Dried whole milk	26.7	2.5	70.8
Evaporated milk	7.6	74.0	18.4
Sweetened condensed milk	8.7	27.1	64.2

Milkfat Functionality

Milkfat functionalities in foods, which are related to its chemical and physical properties, are discussed in detail in Chapter 7. Chemical functionality of milkfat is expressed through its unique flavor and, perhaps more importantly, through the formation of secondary flavor compounds derived directly from milkfat and through subsequent reactions with other constituents of foods.

Physical properties of milkfat are a function of the solid and liquid phase interactions, combined with the effects of temperature. These properties are generally described in terms of crystallization and melting behaviors. Milkfat is composed of many triglyceride species that melt over a relatively wide temperature range ($-40°C$ to $+40°C$), and it exhibits complex melting behavior. For instance, butter containing intact milkfat is too hard at refrigerator temperature to spread easily, yet it is too soft for optimal use as a pastry butter. The opportunity to manipulate the melting behavior of milkfat through fractionation allows the creation of butter with milkfat fractions that spread easily at refrigerator temperature, as well as another milkfat fraction blend that is sufficiently firm for use in pastries (Chapter 8). The crystallization behavior of milkfat is also important to its functionality. Milkfat is one of the few fats that is somewhat compatible with cocoa butter, and it has the added benefit of inhibiting fat bloom in chocolate (Chapter 8).

The expanded use of milkfat in foods, where it can contribute substantially to flavor and quality, is frequently inhibited by functional incompatibilities with other ingredients. The complexity of milkfat is often viewed as a negative aspect of milkfat functionality. However, it is the complexity of milkfat that allows diversity in the tailoring of milkfat to meet specific functional needs of the food industry.

Factors Affecting Milkfat Functionality

The functionality of milkfat ingredients is influenced by chemical composition and processing conditions. The functionality of milkfat in foods is often closely related to the physical properties of the milkfat, which are largely determined by the chemical composition of the native milkfat. The chemical composition of milkfat varies according to environmental and genetic factors such as season, feeding regimen, and breed of cow (Chapter 2). Processing conditions used for fractionation, texturization, and tempering can greatly influence the melting and crystallization behavior of milkfat ingredients (Chapter 3).

Milkfat Fractionation

Fractionation of anhydrous milkfat has been shown to improve the functional properties of milkfat. Milkfat fractionation, selective blending, and texturization provide opportunities to overcome functional variability in the raw material and produce consistent finished products that are tailored to the desired application. Opportunities for increasing the use of milkfat by the incorporation of milkfat fractions include pastry and bakery fats; cocoa butter extenders; confectionery fats; cold-spreadable butter and other dairy-based spreads; dairy foods; and convenience foods. Several applications in which fractionated milkfats perform better than unmodified milkfat have been studied:

- Improved flavor and performance by high-melting milkfat fractions when used as roll-in pastry fats for croissants and Danish (Baker, 1970b; Deffense, 1989; Humphries, 1971; Munro and Illingworth, 1986; Pedersen, 1988; Schaap, 1982; Tolboe, 1984)
- The inhibition of bloom by high-melting milkfat fractions when used in chocolate manufacture (Baker, 1970b; Timms and Parekh, 1980; Yi, 1993)
- Cold-spreadability provided by combinations of low-melting and high-melting fractions in the manufacture of butter (Bumbalough, 1989; Deffense, 1987; Dolby, 1970b; Jamotte and Guyot, 1980; Kaylegian, 1991; Kaylegian and Lindsay, 1992; Makhlouf et al.,1987; Munro et al., 1978)
- Increased foam stability by the addition of high-melting fractions to whipped creams (Bratland, 1983; Tucker, 1974)

The Literature on Milkfat Fractionation

Table 1.2 presents a historical perspective of milkfat fractionation research that has been conducted since the late 1950s in a wide variety of locations, including the United States, Canada, Belgium, The Netherlands, France, Italy, Germany, Finland, Denmark, Sweden, Ireland, Australia, New Zealand, India, Egypt, Japan, and Russia. These studies cover a range of topics including the following:

- Design of fractions for specific applications (e.g., cold-spreadable butter, croissant pastry fats)
- Studies of the feasibilities of new methodologies
- Determination of optimum processing conditions
- Comparison of fractions produced by different fractionation methods
- Pilot-scale fractionation studies
- Chemical and physical characterization of fractions
- Comparison of fractions obtained from summer and winter milkfats

- Study of compatibility of milkfat fractions with other fats (e.g., cocoa butter)
- Oxidation stability studies
- Determination of the distribution of cholesterol and flavoring components in milkfat

TABLE 1.2 Milkfat Fractionation Studies Listed in Chronological Order

		Fractionation method				
Author	Year	Crystallization from melted milkfat	Crystallization from solvent solution	Supercritical fluid extraction	Short-path distillation	Fractionation method not specified
Baker et al.	1959	X	X			
Bhalerao et al.	1959		X			
McCarthy, et al.	1962				X	
Sherbon	1963	X	X			
Chen and deMan	1966		X			
Rolland and Riel	1966		X			
Schultz and Timmen	1966	X				
Jensen et al.	1967		X			
deMan	1968	X				
Richardson	1968	X				
Woodrow and deMan	1968		X			
Mattsson et al.	1969		X			
Antila and Antila	1970	X				
Baker	1970a	X				
Baker	1970b					X
Dolby	1970a	X				
Dolby	1970b	X				
Fjaervoll	1970a	X				
Fjaervoll	1970b	X				
Jebson	1970	X				
Schaap and van Beresteyn	1970	X				
Norris et al.	1971	X				
Voss et al.	1971	X				
Riel and Paquet	1972	X				
Sherbon et al.	1972	X				
Walker	1972	X				
Black	1973	X				
Kehagias and Radema	1973					X
Sherbon and Dolby	1973		X			
Thomas	1973a	X				
Vovan and Riel	1973	X				
Avvakumov	1974		X			
Black	1974b	X				
Deroanne and Guyot	1974	X				
Dixon and Black	1974	X				
Jebson	1974a	X				
Jebson	1974b					X

Continued

TABLE 1.2 (Continued)

Author	Year	Crystallization from melted milkfat	Crystallization from solvent solution	Supercritical fluid extraction	Short-path distillation	Fractionation method not specified
Kankare and Antila	1974a	X				
Parodi	1974a		X			
Stepanenko and Tverdokhleb	1974	X				
Timmen	1974	X				
Tucker	1974					X
Walker	1974					X
Black	1975b	X				
Lechat et al.	1975	X	X			
Schaap et al.	1975	X	X			
Deroanne	1976	X				
Doležálek et al.	1976	X				
El-Ghandour et al.	1976	X				
Evans	1976	X				
Helal et al.	1977	X				
Norris	1977		X			
Walker et al.	1977		X			
Youssef et al.	1977		X			
Munro et al.	1978		X			
Timms	1978b	X				
Walker et al.	1978		X			
Larsen and Samuelsson	1979		X			
deMan and Finoro	1980	X				
Frede et al.	1980	X				
Jamotte and Guyot	1980	X				
Timms	1980a		X			
Timms	1980b		X			
Timms and Parekh	1980	X				
Sreebhashyam et al.	1981	X				
El-Ghandour and El-Nimr	1982	X				
Guyot	1982	X				
Hayakawa and deMan	1982	X				
Kaufmann et al.	1982				X	
Martine	1982	X				
Badings et al.	1983b	X				
Bhat and Rama Murthy	1983	X				
Bratland	1983	X				
Lambelet	1983		X			
Helal et al.	1984	X				
Lansbergen and Kemps	1984		X			
Timmen et al.	1984			X		
Verhagen and Bodor	1984	X				
Verhagen and Waarnar	1984	X				

Continued

TABLE 1.2 (Continued)

		Fractionation method				
Author	Year	Crystallization from melted milkfat	Crystallization from solvent solution	Supercritical fluid extraction	Short-path distillation	Fractionation method not specified
Amer et al.	1985	X				
Banks et al.	1985	X				
Biernoth and Merk	1985			X		
Lakshminarayana and Rama Murthy	1985	X				
Muuse and van der Kamp	1985		X			
Jordan	1986	X				
Kankare and Antila	1986	X				
Keogh and Higgins	1986b	X				
Kuprancyz et al.	1986	X				
Lakshminarayana and Rama Murthy	1986	X				
Shishikura et al.	1986			X		
Arul et al.	1987			X		
Deffense	1987	X				
Keshava Prasad and Bhat	1987a	X				
Keshava Prasad and Bhat	1987b	X				
Kulkarni and Rama Murthy	1987	X				
Lund and Danmark	1987	X				
Makhlouf et al.	1987	X				
Ramesh and Bindal	1987b	X				
Arul et al.	1988a				X	
Kankare and Antila	1988a	X		X		
Kankare and Antila	1988b	X				
Khalifa and Mansour	1988	X				
Pedersen	1988					X
Bradley	1989			X		
Bumbalough	1989	X				
Büning-Pfaue et al.	1989			X		
Deffense	1989	X				
Ensign	1989			X		
Keogh	1989	X				
Kankare et al.	1989			X		
Fouad et al.	1990	X				
Keogh and Morrissey	1990	X				
Rizvi et al.	1990			X		
Chen and Schwartz	1991			X		
Hamman et al.	1991			X		
Kaylegian	1991	X	X			
Kuwabara et al.	1991	X				
Versteeg	1991	X				
Barna et al.	1992	X				
Chen et al.	1992			X		

Continued

TABLE 1.2 (Continued)

		Fractionation method				
Author	Year	Crystallization from melted milkfat	Crystallization from solvent solution	Supercritical fluid extraction	Short-path distillation	Fractionation method not specified
Grall and Hartel	1992	X				
Kaylegian and Lindsay	1992	X	X			
Laakso et al.	1992	X				
Mayhill and Newstead	1992					X
Bhaskar et al.	1993a			X		
Bhaskar et al.	1993b			X		
Branger	1993	X				
Deffense	1993b	X				
Rizvi et al.	1993				X	
Yi	1993		X			
Shukla et al.	1994			X		

Fractionation Methods

Although a large number of fractionation studies have been performed with a variety of different objectives, only three major fractionation methods have been employed:

- Crystallization from melted milkfat
- Crystallization of milkfat dissolved in a solvent solution
- Supercritical fluid extraction

The most common method employed in these studies has been crystallization from melted milkfat. This method uses no additives, so it is a relatively simple, unadulterated process. Crystallization from melted milkfat is also relatively inexpensive compared with fractionation from solvent solution and extraction by supercritical carbon dioxide. For these reasons, fractionation from melted milkfat has been the method of choice for commercialization of milkfat fractionation.

The following brief summary of these fractionation methods is given to facilitate subsequent discussions; detailed information on fractionation methods is presented in Chapter 3. Some of the advantages and disadvantages associated with these methods are presented in Table 1.3.

Molecular (short-path) distillation and chromatographic techniques are also used to fractionate milkfat, but are usually employed on a small scale in the preparation of samples for detailed chemical compositional and structural analyses.

Crystallization from Melted Milkfat

Crystallization of milkfat from melted milkfat is also known as *fractionation from the melt* or *dry fractionation,* because it does not employ solvents or additives. It is a temperature-based process in which the milkfat is held at a given temperature to allow a portion of the milkfat to crystallize, and then the crystals are physically separated from the liquid fraction. The process can be completed at this point for a single-stage process, producing one liquid and one solid fraction. Alternatively, both the liquid and solid fractions can be subsequently refractionated at a different temperature for a multistage process, yielding more than two fractions from the initial batch of milkfat.

Several methods are available to separate the solid and liquid fractions, including vacuum filtration, pressure filtration using a membrane filter press, and centrifugation. Separation of the solid and liquid phases can also be accomplished by mixing the fat solution with an aqueous detergent solution. The solid crystals will partition into the aqueous detergent phase and can then be separated from the liquid fat phase. The aqueous detergent phase, containing the crystals, is heated, causing the crystals to melt, and then the liquid fat is separated from the aqueous phase and dried.

TABLE 1.3 A Brief Summary of the Advantages and Disadvantages of Milkfat Fractionation Methods

	Crystallization from melted milkfat	Crystallization from solvent solution	Supercritical carbon dioxide extraction
Advantages	No additives Simple physical process Successfully commercialized	More discrete fractions produced Can use low temperatures	No additives CO_2 is nontoxic More discrete fractions produced
Disadvantages	Less pure fractions Limited temperature range	Potential toxicity of solvent Flavor changes in milkfat Safety hazards High cost of operation and solvent recovery	High capital investment

Crystallization from Solvent Solution

Crystallization from solvent solution involves dissolving the melted milkfat in a solvent prior to crystallization, in a ratio ranging from one part milkfat to four to ten parts solvent. The solvent employed is generally acetone, although solvents such as ethanol, pentane, and hexane have been used. This is also a temperature-based process and is conducted in a manner similar to fractionation from melted milkfat. As the solid milkfat fraction crystallizes, it precipitates from the milkfat–solvent solution and is easy to separate from the liquid fraction using vacuum filtration. The milkfat fractions are heated to allow the solvent to evaporate, often using a vacuum rotary evaporator; then they are washed and dried to remove trace solvent and moisture.

Supercritical Carbon Dioxide Extraction

A gas in the supercritical state—that is, above its critical pressure and temperature—exhibits unique solvent properties. Supercritical fluid extraction of milkfat is generally performed with carbon dioxide, which is nontoxic and relatively inexpensive. Milkfat fractions are selectively dissolved in the supercritical carbon dioxide by changing the temperature and pressure of the system. The fractions are collected at the end of the process, where the temperature and pressure return to normal atmospheric conditions. Carbon dioxide evaporates at atmospheric conditions and leaves no residues in the milkfat fractions.

Short-Path Distillation

Short-path (molecular) distillation is a form of vacuum distillation used to separate compounds based on their molecular weight, melting temperature, volatility, and intermolecular interaction. Although this method has found other applications in the food, chemical, and pharmaceutical industries, it has received limited attention as a means for the fractionation of milkfat.

Commercial Fractionation of Milkfat

Fractionation of milkfat to create ingredients with specific functional properties has received considerable attention around the world. Fractionation from melted milkfat to produce specialty food ingredients has been in commercial practice in Europe since 1973 (Tirtiaux, undated). S.A.N. Corman and Aveve, based in Belgium, produce a wide range of milkfat fraction–based ingredients that are suited for the bakery, pastry, confectionery, and ice cream industries (Aveve, undated; Corman, undated). In addition to these ingredients, cold-spreadable butter and other dairy-based spreads containing milkfat fractions are sold directly into retail markets (Corman, undated). In the European Economic Community milkfat is available to bakers at lower prices than other bakery fats, and hard milkfat fractions have become serious competitors for tailor-made vegetable fats and margarines (Timmen et al., 1984). Specialty milkfats for the bakery, confectionery, and ice cream industries are also produced by fractionation from melted milkfat in New Zealand by Alaco (Alaco, undated). Some of the commercially available milkfat fractions from S.A.N. Corman, Aveve, and Alaco are summarized in Table 1.4. Information on the product lines of other companies fractionating milkfat (Tirtiaux, undated) was not available.

TABLE 1.4 Properties of Some Commercially Available Milkfat Fraction Products

Company Product name	General properties	Suggested applications
Alaco		
Bakery Butter	General purpose butter	General purpose bakery
	Standardized melting profile	
Butter Shortening	Anhydrous form of general purpose butter	General purpose bakery
	Standardized melting profile	
Butter Sheets	Butter in sheeted form	Croissants
		Danish pastry
		Puff pastry
Confectionery Butterfat 42	42°C melting point	Confectionery
		Chocolate
Soft Butteroil 21	21°C melting point	Ice cream
		Recombined dairy products
Soft Butteroil 28	28°C melting point	Ice cream
		Recombined dairy products
Pastry Butter	37°C melting point	Croissants
	Plasticized and work toughened	Danish pastry
		Puff pastry
Cookie Butter	Tailored for high butter biscuits and cookies	Shortbread
		High butter biscuits
	Inhibits bloom formation	High butter cookies
Aveve		
Anhydrous Milk Fat	99.8% milkfat	General purpose food and bakery
		Recombined milk products
		Ice cream
Pure Cow Anhydrous Milk Fat 13–15°C Melting Point	99.8% milkfat	—
	13–15°C melting point	
Concentrated Butter	99.8% milkfat	Ice cream

Continued

TABLE 1.4 (Continued)

Company Product name	General properties	Suggested applications
Formula B	28–32°C melting point	
Concentrated Butter "Creme au Beurre"	99.8% milkfat 28°C melting point Bright yellow or white color	Recombined cream Butter cookies
Concentrated Butter "4/4"	99.8% milkfat 22–32°C melting point Strong buttery flavor Bright yellow color	Doughs Biscuits Chocolate
Concentrated Butter "Patissier"	99.8% milkfat 28–32°C melting point Bland flavor Bright yellow color	Doughs Biscuits Chocolate
Concentrated Butter "Croissant"	99.8% milkfat 38°C melting point Bland flavor Yellow-orange color	Doughs
Concentrated Butter "Millefeuille"	99.8% butter 40–42°C melting point Bland flavor Yellow-orange color	All puff pastry applications
Concentrated Butter Feuilletage 2000	99.8% milkfat 24–26°C melting point Yellow-orange color	All puff pastry applications
S.A.N. Corman		
Standard Anhydrous Milk Fat	32°C melting point Neutral flavor	General purpose bakery Recombined milk products Ice cream Culinary dressings
Crème au Beurre	99.9% milkfat	Butter icings and fillings Petits fours Soft centers for chocolates
4/4 Concentrated Butter	99.9% milkfat Neutral flavor Ivory to pale yellow color	Doughs requiring the use of melted butter Liquid doughs (sponge cake)
Danish 4/4	99.9% milkfat 30°C melting point Neutral flavor Ivory to pale yellow color Suited for extrusion and rotary molding	Danish butter cookies Shortbread Finger biscuits
Patissier Concentrated Butter	99.9% milkfat	Self-raising pastries (Viennese shortcakes, bread, rolls)
Croissant Concentrated Butter	99.9% milkfat 38°C melting point	Short pastry dough Raised puff pastry

Continued

TABLE 1.4 (Continued)

Company Product name	General properties	Suggested applications
	Butter flavor	(croissants, Danish)
	Yellow color	Puff pastry
	Also comes in slice form	
Millefeuille Concentrated Butter	99.9% milkfat	Any puff pastry application
	41°C melting point	
	Butter flavor	
	Yellow color	
	Also comes in slice form	
Glacier Concentrated Butter	99.9% milkfat	Ice cream
	30–32°C melting point	
	Strong butter flavor	
	Yellow color	
Glacier Extra Concentrated Butter	99.9% milkfat	Ice cream
	30–32°C melting point	
	Neutral flavor	
	Ivory to pale yellow color	
Extra White Anhydrous Milk Fat	99.9% milkfat	Soft cheeses (goat, Feta)
	32°C melting point	
	Neutral flavor	
	White color	
Milk Extra Anhydrous Milk Fat	99.9% milkfat	Recombined milk products
	28°C melting point	Yogurt
	Strong butter flavor	Ice cream
	Yellow color	

Other Technologies Available for Milkfat Modification

Technologies available for milkfat modification can also be applied to milkfat sources other than anhydrous milkfat. For example, one of the goals of early milkfat research was to improve the spreadability of butter at refrigerator temperature. These research efforts concentrated on manipulating the processing conditions of the cream to change the properties of milkfat to yield a more spreadable butter. As technology advanced, other processes became available to modify the chemical composition of milkfat and subsequently change its physical properties. While this monograph concentrates on fractionation of milkfat, it should be noted that other physical, chemical, and biological technologies are available for the modification of milkfat to improve its functionality in foods, and these are summarized in this section.

Physical Processes

Cream Tempering

Cream tempering techniques such as the Alnarp process (Alfa-Laval, 1987; Dolby, 1959) provide controlled crystallization of the milkfat. The tempering results in a higher liquid-to-solid fat phase ratio in the butter and therefore yields a softer butter (Kleyn, 1992). Selection of the temperatures used for cream tempering is based on the iodine value of the milkfat (Alfa-Laval, 1987; Mogensen, 1984), which is a reflection of the unsaturation and melting point of the milkfat.

Butter Tempering and Reworking

The crystalline structure of butter can be influenced by reworking and temperature control of the fat crystallization process after churning (Alfa-Laval, 1987; Huebner and Thomsen, 1957; Mogensen, 1984; Tverdokhleb and Avvakumov, 1978). The greatest disadvantage to the reworking process is that the butter will regain much of its original firmness if it is subjected to fluctuations in temperature (Mogensen, 1984).

Chemical Processes

Hydrogenation

Hydrogenation is a common technique used to increase the saturated fat content of vegetable fats and oils and thereby increase their hardness, oxidative stability, and plastic range. Hydrogen is added to the double bonds of unsaturated fatty acids with the aid of metal catalysts, yielding saturated fatty acids. Vasishtha et al. (1970) and Mukherjea et al. (1966) have found that trace hydrogenation of anhydrous milkfat improved its keeping qualities. Smith and Vasconcellos (1974) have investigated the use of hydrogenation as an alternative to fractional crystallization to produce hard milkfats for specialty uses. Campbell et al. (1969) reported that hydrogenated milkfat was an effective inhibitor of bloom formation in dark chocolate. Banks (1991c) has recently stated that there appears to be little future for hydrogenated milkfat, because of current criticism over its saturated nature and the cost of production.

Interesterification

Interesterification is a chemically or enzymatically catalyzed process that results in a rearrangement of the fatty acids of milkfat triglycerides. An important concern with interesterification is flavor defects, which have resulted from free fatty acids and other side products formed during the interesterification reaction. Mickle (1960) found that conducting the interesterification reaction in a nitrogen atmosphere and refining the finished product diminished the off-flavors. Frede (1991) has stated that interesterification of milkfat has not been applied commercially, primarily because of the flavor problems associated with the reaction and refining processes.

Chemical interesterification of fats and oils is carried out in the presence of an inorganic catalyst, such as sodium methoxide or other sodium alkoxide, to promote the exchange of fatty acids on the glycerol molecule. If the reaction temperature is high enough for the triglycerides to be liquid when formed, the fatty acid exchange will proceed until an equilibrium composition, representing a completely random arrangement, is obtained. However, if the temperature of the reaction is controlled or directed so that some of the glycerides are in a solid form, these glycerides are effectively removed from the reaction and are not further interesterified (Mickle, 1960). Riel (1966) used a controlled interesterification reaction at 35°C to produce a milkfat with a solid fat index lower at temperatures less than 20°C, and higher at temperatures above 20°C, than that of unmodified milkfat. Milkfat from uncontrolled chemical interesterification has been found to be harder than unmodified milkfat (deMan, 1961a; McNeill, 1988), and milkfat chemically interesterified with temperature control was softer than unmodified milkfat (Mickle, 1960).

Enzymatic interesterification is based on the use of lipases: enzymes that effect the exchange of fatty acids on the glycerol molecule. Lipases used for interesterification reactions have been classified by their selectivity for fatty acids. The three main classes of lipases are nonspecific; 1,3-specific; and fatty acid–specific. Interesterification reactions performed with nonspecific lipases result in products with a randomized fatty acid arrangement, similar to that obtained with chemical interesterification. The 1,3-specific lipases cause a change in the fatty acid profile only at the sn-1 and sn-3 positions on the glycerol molecule. Fatty acid–specific lipases cause rearrangement of a specific fatty acid, regardless of its position on the glycerol molecule (Jensen et al., 1990; Macrae, 1983). The use of lipase-catalyzed interesterification allows more control of the reaction and the end products and can be used to yield products with a range of physical properties. Intact milkfat, interesterified with enzyme catalysis, has resulted in a softer milkfat than unmodified or chemically interesterified milkfat (Elliot, 1990; McNeill, 1988).

Biological Processes

Milkfat Composition Changes Through Diet

Many studies have shown that the chemical composition of milkfat can be affected by the diet of the dairy cow (Baer, 1991). Changes in the chemical composition of milkfat can yield changes in the physical and nutritional properties of milkfat (Grummer, 1991). The long-chain fatty acids in milkfat (C16 and C18 saturated and unsaturated) are derived from the cow's blood supply, and their relative composition in milkfat can be manipulated through the diet (Banks, 1991b). Grummer (1991) has estimated that 50% of milkfat synthesized is derived from plasma lipids.

Feeding studies have been conducted to evaluate the effects of seasonal diets and dietary energy level (Precht et al., 1984). Extensive studies have been conducted on the supplementation of the diet with a variety of fat sources, including saturated and unsaturated fats, oils and oilseeds, and protected and unprotected lipids (Baer, 1991; Grummer, 1991). For example, Precht et al. (1984) conducted a study in which cows were fed three diet regimens: normal, young pasture grass, and an energy-deficient diet. They found that the type of feeding regimen had a direct and marked influence on the composition of the milkfat. They also found that diet manipulation caused flavor defects and affected the size of the milkfat globule, resulting in a change in the physical behavior of the milkfat.

Supplementation of the cow's diet with a source of long-chain fatty acids has been shown to result in increased amounts of stearic (C18:0) and oleic (C18:1) acids in milkfat, with a concurrent decrease in the medium-chain fatty acid content (Baer, 1991; Grummer, 1991). Fat supplements fed to dairy cows have included sunflower oil, high-oleic sunflower oil, whole sunflower seeds, rolled sunflower seeds, soybean oil, soybean oil meal, safflower oil, canola oil, tallow, cottonseed oil, marine oils, coconut oil, menhaden oil, and oleic acid (Baer, 1991; Grummer, 1991). Grummer (1991) noted that the feeding of unprotected vegetable oils high in polyunsaturated fatty acid content has resulted in elevated *trans:cis* ratios in milkfat, whereas the feeding of fats and oils high in saturated and monounsaturated fatty acids (e.g., tallow or high-oleic sunflower or safflower oils) has resulted in low *trans:cis* ratios in milkfat.

Feeding studies also have evaluated the effect of the physical form of the fat supplement, including the use of free oil, whole seed, and protected lipids (Baer, 1991; Grummer, 1991). In studies where free oil was compared with whole oilseeds, both milkfats exhibited similar fatty acid composition, but the cows that were fed free oil exhibited a decrease in milkfat yield, while the cows that were fed oilseeds either maintained or increased their milkfat yield (Grummer, 1991).

The most effective means of diet manipulation has been to feed the cow protected (e.g., encapsulated or formaldehyde-treated) lipid supplements, which pass through the rumen and avoid the hydrogenation process (Baer, 1991; Banks, 1991b; Christie, 1979; Grummer, 1991; Hawke and Taylor, 1983; Mogensen, 1984). Banks (1991b) states that 20 to 25% of the daily intake of protected fat appeared in the milk. Thus, feeding protected unsaturated lipids to the cow produces milkfat with a higher ratio of unsaturated to saturated fat.

Although the scientific knowledge to manipulate the composition of milkfat is well established, Banks (1991b) observes that economic considerations have prevented its exploitation. Grummer (1991) predicts that feeding-based research will be limited until the dairy industry recognizes the potential to improve milkfat quality through feeding practices and provides incentives for producers to do so.

Genetic Manipulation of the Dairy Cow

Genetic manipulation of the dairy cow for the purpose of the modification of milkfat can be accomplished by natural genetic selection or through transgenic approaches.

Gibson (1991) stated that there is sufficient genetic variation among cattle breeds and species to effect gradual changes in milkfat composition through natural selection. Genetic improvement in cattle is slow (less than 3% per year) but cumulative; however, until breeders and the dairy industry can foresee an economic benefit of this approach, the use of natural selection to modify the composition of the milkfat remains unlikely (Gibson, 1991).

The transgenic approach to milkfat modification may yield more marked changes in the milkfat composition than natural selection can, which may result in a more nutritionally attractive product from the consumer's point of view (Gibson, 1991). In a recent review, Yom and Bremel (1993) discuss several transgenic approaches for the modification of milk components. They note that reduction or extinction of key enzymes (e.g., acetyl coenzyme A (CoA) carboxylase) would lead to a dramatic reduction in the fat content of the milk and that cows producing lower-fat milk would have lower dietary requirements, which would result in reduction of operating costs for the milk producers. They conclude that although more research needs to be conducted in several areas to make this technology commercially viable, the transgenic approach to milk modification will have a substantial effect on the dairy and related industries.

Cholesterol Removal from Milkfat

The removal of cholesterol from milkfat does not provide physical or functional benefits, but is practiced in a limited manner for nutritional and advertising claim reasons. Concerns about the relationship between cholesterol intake and heart disease have prompted the food and dairy industries to respond to the request for low-cholesterol products. Technologies available for reduction or removal of cholesterol from milkfat include the following:

- Steam-stripping/distillation
- Extraction by supercritical carbon dioxide
- Removal by complexing agents
- Enzyme processes
- Solvent extraction

Steam-Stripping

Steam-stripping of cholesterol from milkfat is accomplished by the use of vacuum steam distillation, in which cholesterol is flash-vaporized and then carried away in the steam phase, which separates it from the remaining milkfat (Sperber, 1989). Steam-stripping is currently practiced at the commercial level in the U.S. (The OmegaSource Corp., Burnsville, MN) and at two plants in Europe that use the Tirtiaux Turbo Deodorizer (Boudreau and Arul, 1993; Deffense, 1993a). Lanzani et al. (1994) modified a molecular distillation plant to improve efficiency and safety, and reported a 90% reduction in the cholesterol content of milkfat without the undesirable changes normally associated with steam distilled milkfat. Conte and Johnson (1992) also have obtained a patent for a multistage steam sparging process to remove cholesterol from milkfat.

A drawback to steam-stripping is that most of the volatile flavor components are also lost during the process. However, Schroeder and Baer (1990) recombined steam-stripped, cholesterol-reduced milkfat with skim milk to prepare cholesterol-reduced fluid milks for consumer evaluation, and they reported that an acceptable low-fat, cholesterol-reduced fluid milk could be produced.

Boudreau and Arul (1993) evaluated the efficiency of cholesterol removal from milkfat by short-path distillation, supercritical carbon dioxide extraction, and melt crystallization. They reported that short-path distillation was the most efficient method for the removal of cholesterol, and the process resulted in the least change in composition of the milkfat. Boudreau and Arul (1993) noted that the conditions used for the removal of cholesterol also remove low-molecular-weight triglycerides and flavor compounds.

Supercritical Carbon Dioxide Extraction

The technology used for extraction of cholesterol with supercritical carbon dioxide is similar to that used for fractionating milkfat, but the pressure and temperature are chosen to selectively dissolve and remove cholesterol. Studies have shown that 90% of the cholesterol of milkfat can be removed by extraction with supercritical carbon dioxide (Bradley, 1989; Yeh et al., 1991). The use of multiple staging and in-line adsorbents has been shown to improve the efficiency of cholesterol removal (Boudreau and Arul, 1993). The use of supercritical carbon dioxide extraction for cholesterol removal from milkfat as yet has not been commercialized; but supercritical fluid processing is used on a commercial scale for decaffeination of coffee; removal of bitter oil from hops; isolation of essential oils from spices; and purification of ethyl esters of ω-3 fatty acids from fish oils (Bradley, 1989).

Removal of Cholesterol by Complexing Agents

Cholesterol can also be removed from milkfat by complexing it with any of several chemical reagents and then separating the cholesterol–reagent complex from the milkfat.

The CSIRO Food Research Laboratory in Australia has developed a process (SIDOAK™) that is based on the specific affinity of ß-cyclodextrin for cholesterol (Oakenfull and Sidhu, 1991). The SIDOAK™ process, which removes 80 to 90% of cholesterol from cream in a single operation, is performed at low temperature, thus minimizing microbial spoilage and flavor deterioration; it also has low capital and production costs (Oakenfull and Sidhu, 1991). Oakenfull and Sidhu (1991) reported that cyclodextrins have been approved for food use in Australia, Japan, and some European countries.

Micich and coworkers have reported the successful removal of cholesterol from butter oil with the use of digitonin (Micich, 1990; Micich et al., 1992) and tomatine (Micich, 1991; Micich et al., 1992) as complexing agents. Tomatine gave more incentives for further research on polymer-supported systems because of its lower toxicity compared to digitonin.

In a recent review on cholesterol removal technologies, Boudreau and Arul (1993) noted that promising work was ongoing at the Department of Food Science, University of California (Davis, CA) and referenced a patent application (Richardson and JiminezFlores, 1991) for the use of food-grade saponins as adsorbents for the removal of cholesterol from milkfat.

Enzyme-Based Processes

Enzyme processes used to convert cholesterol into other compounds have been reviewed by Sperber (1989), Versteeg (1991) and Boudreau and Arul (1993); these have stated that the use of enzyme technology for cholesterol reduction and removal from milkfat is still in the research stage. These reviewers have concluded that the lack of affordable enzymes and the requirement for further studies make the commercial use of enzymatic processes unlikely in the short term. Sperber (1989) pointed out that one of the enzymes evaluated, cholesterol reductase, is derived from fecal sources, and this could pose regulatory problems.

Solvent Extraction

Johnson and Conte (1991) have patented a process of preferentially extracting cholesterol from the milkfat globule membrane using an organic polar solvent, such as ethanol or acetone, without substantial loss of milkfat. The resulting low-cholesterol milkfat was reported to have a composition and sensory properties similar to the native milkfat.

Blends of Milkfat with Other Fats and Oils

Milkfat is blended with other fats and oils for several reasons:

- Improved flavor
- Increased functionality
- Decreased cost

Blends of milkfat with other fats and oils are common in butter-margarine spreads, where the milkfat is blended with vegetable oil, and in chocolate, where it is blended with cocoa butter. The blending of two fats or oils of different composition influences all aspects of the functionality of fats. Chemical functionality relies in part on the ratio of saturated and unsaturated fatty acids, which has nutritional as well as flavor implications. Physical functionality includes the crystallization properties and melting properties. One of the most important concerns in the blending of fats is that the crystallization properties of the glycerides in the mixture remain compatible. Incompatible glycerides in fats can cause retardation or complete lack of crystallization, due to intersolubility effects and formation of eutectic mixtures.

A detailed discussion of the effects of intact milkfat combined with other fats and oils is beyond the scope of this monograph, but it is discussed briefly in the following paragraphs.

Milkfat Blends with Vegetable Oils and Animal Fats

The blending of milkfat with vegetable oils in butter-margarine blends is well established in the edible oil industry. In this case, the motive for the blend is to add butter flavor to margarine, to provide a more spreadable product and a lower cost than butter, and to potentially enhance the nutritive value of the product. However, at this time there is much debate about the notion that margarine is healthier than butter from the perspective of saturated and *trans*-fatty acids.

Evaluation of mixtures of milkfat and beef tallow have been reported in the literature by Kalo et al. (1986) and Timms (1979). The purpose of these blends was to provide a shortening at a lower cost than butter but with some of the flavor of butter.

Milkfat and Cocoa Butter

The use of milkfat in chocolate is also well established. Milkfat can be used as a bloom inhibitor in dark chocolate, as a flavor and texture agent in milk chocolate, and in general as a cocoa butter replacer. Milkfat is one of the few fats that is both reasonably compatible with cocoa butter and legally permitted in chocolate. These factors, combined with the cost savings of milkfat compared with cocoa butter, make it a highly desirable ingredient in chocolate. However, as the ratio of milkfat to cocoa butter increases, blends exhibit an undesirable softening effect. Consequently, there has been much incentive from the confectionery industry to evaluate the use of modified milkfat in chocolate.

The physical effects (e.g., crystallization behavior, tempering effects, melting behavior) of intact milkfat and cocoa butter blends have been investigated by Petersson (1986), Timms (1980b), and Timms and Parekh (1980). Milkfats modified by interesterification (Timms and Parekh, 1980) and fractionation (Chapter 8) have been evaluated in chocolate.

Chapter 2
Raw Materials for Milkfat Fractionation

Milkfat used for fractionation is generally of bovine origin, but milkfat fractions also have been obtained from buffalo milkfat and ghee (Bhat and Rama Murthy, 1983; El-Ghandour and El-Nimr, 1982; El-Ghandour et al., 1976; Helal et al., 1984; Lakshminarayana and Rama Murthy, 1985; Prasad and Gupta, 1983; Ramesh and Bindal, 1987b; Rifaat et al., 1973; Youssef et al., 1977), goat ghee (Ramesh and Bindal, 1987b), and camel milkfat (Abu-Lehia, 1989) in Egypt, India, and Saudia Arabia. Hydrogenated milkfat (Banks et al., 1989a, 1989b; Lansbergen and Kemps, 1983) and lipase-modified butter oil (Kemppinen and Kalo, 1993) have also been used as raw materials for fractionation. However, the data presented and discussed in this handbook are for milkfat fractions obtained from unmodified bovine milkfat.

Milkfat Biosynthesis

Fatty Acid Synthesis

Milkfat contains a unique array of fatty acids that result from microbial activity in the rumen of the cow (Jensen and Clark, 1988). Microorganisms ferment vegetable matter in the diet to form acetic, propionic, and butyric acids (Banks, 1991d; Jensen and Clark, 1988). Acetate and β-hydroxybutyrate, produced from ruminal fermentation, serve as the building blocks for the synthesis of fatty acids in the epithelial cells of the mammary gland (Baer, 1991; Banks, 1991d; Christie, 1983; Grummer, 1991; Hawke and Taylor, 1983; Jensen and Clark, 1988; Mulder and Walstra, 1974). All of the C4 to C14:0 fatty acids and half of the C16:0 in milkfat are synthesized *de novo* in the mammary gland; the remaining half of the C16:0 and the C18 fatty acids in milkfat are obtained directly from the blood supply to the mammary gland (Baer, 1991; Banks, 1991d; Grummer, 1991; Hawke and Taylor, 1983). These fatty acids can be either of dietary origin or liberated from depot fat. The high content of C18:1 results from the actions of desaturase enzymes in the gut wall and mammary gland, which converts C18:0 to 9-*cis*-18:1 (Banks, 1991b; Hawke and Taylor, 1983), and the hydrogenation of C18:2 to 11-*trans*-18:1 in the rumen (Grummer, 1991).

Triglyceride Synthesis

Triglyceride synthesis (i.e., esterification of fatty acids onto the glycerol molecule) in bovine mammary tissue occurs primarily via the α-glycerol phosphate pathway, which uses a glycerol phosphate as the fatty acid acceptor (Hawke and Taylor, 1983; Jensen and Clark, 1988). Jensen and Clark (1988) report observations that the molecular weights of bovine milkfat triglycerides exhibit a bimodal distribution, with one maximum at 38 fatty acid carbons and a second maximum at 48 carbons. It has been postulated that the lower-molecular-weight triglycerides may be synthesized via a monoglyceride pathway that utilizes an sn-2-monoglyceride as the fatty acid acceptor (Jensen and Clark, 1988). However, although the monoglyceride pathway is operational in bovine intestinal mucosa and adipose tissue, the presence of this pathway in mammary tissue has been questioned (Hawke and Taylor, 1983). In any event, fatty acids are not randomly distributed among the three positions on the glycerol molecule, but the control of biosynthetic esterification is not well understood.

Milkfat Globule Formation

Milkfat secreted in the mammary gland exists in the form of globules. The globules are formed throughout the cell, but particularly at the basal membrane, and they grow as they move through the cell toward the apical membrane. Globules are then extruded through the apical membrane into the lumen, being enveloped in a cell membrane in the process (Banks, 1991d; Christie, 1983; Jensen et al., 1991; Keenan et al., 1983 and 1988; Mulder and Walstra, 1974; Precht, 1988).

Milkfat globules have an inner core, composed of 98% triglycerides, which is surrounded by a lipid membrane composed primarily of phospholipids. The milkfat globule membrane is a layered structure that has an inner layer of high-melting triglycerides, followed by a layer of polypeptides, which are interspersed in a lipid continuum of phospholipids and cholesterol (Jensen et al., 1991; Keenan et al., 1983; King, 1955). The membrane contains a variety of compounds including phospholipids; cerebrosides; cholesterol and other sterols; carotenoids; vitamins A and D; tocopherols; xanthophyll; squalene; waxes; enzymes; copper; and iron (Christie, 1983; Keenan et al., 1983; King, 1955; Mulder and Walstra, 1974). The surface properties of the phospholipid membrane are largely responsible for the high emulsion stability of the milkfat phase in fluid milk (King, 1955; Precht, 1988).

Milkfat globules range in diameter from 0.2 µm to 20 µm or more (Jensen, 1988; Jensen et al., 1991; Keenan et al., 1988; King, 1955; Precht, 1988). Small globules (less than 1 µm) are the most abundant, but they account for only a small percentage of the total lipid; intermediate globules (1 to 8 µm) comprise 90% of the total lipid but only 10% to 30% of the number of globules (Jensen and Clark, 1988; Jensen et al., 1991; Keenan et al., 1988; Precht, 1988; Walstra, 1983). The size distribution of milkfat globules varies among species and breeds of cattle (Jenness, 1988).

Factors Affecting the Uniformity of Milkfat

The complex process involved in the biosynthesis of milkfat provides an opportunity for many factors to influence the final chemical composition:

- Stage of lactation
- Seasonal variation and influences of temperature
- Feeding regimen
- Breed and genetic variations
- Age of the cow
- Nutrition of the cow
- Infections of the udder
- Administration of bovine somatotropin

(Baer, 1991; Christie, 1979; Grummer, 1991; Hawke and Taylor, 1983; Hinrichs et al., 1992; Jenness, 1988; Muuse and van der Kamp, 1985; Norris et al., 1973; Palmquist et al., 1993; Parodi, 1974b; Sonntag, 1979b). Many of these factors, particularly seasonal variations and feeding regimens, have been well studied, but a detailed discussion is beyond the scope of this handbook.

Stage of Lactation

In the dairy cow the absolute amount of milkfat secreted is greatest in early lactation and then decreases to a minimum around the 10th week before it again begins to increase (Christie, 1979). During early lactation there is a relatively constant rate of synthesis of short-chain fatty acids in the mammary gland and extensive mobilization of fatty acids from adipose tissues

(Christie, 1979). These biological phenomena result in a concentration of the C4 fatty acid that is highest during the first month of lactation, while the concentrations of C6 to C14:0 fatty acids all increase during the first eight weeks of lactation and then decrease toward the end of lactation (Baer, 1991; Christie, 1979; Hawke and Taylor, 1983; Palmquist et al., 1993; Parodi, 1974b). The concentration of C16:0 remains relatively constant throughout lactation (Christie, 1979; Hawke and Taylor, 1983; Parodi, 1974b). The concentrations of C18:0 and C18:1 fatty acids are highest in early lactation (Christie, 1979; Hawke and Taylor, 1983); decrease until mid-lactation; then increase toward the end of lactation (Baer, 1991; Parodi, 1974b). Changes in fatty acid composition after the 10th week of lactation are relatively minor (Baer, 1991; Christie, 1979).

Seasonal Variation

Seasonal variations in the composition of milkfat are largely a result in the change in feed of the animal, although it has been suggested that temperature fluctuations may contribute to these variations (Baer, 1991; Sonntag, 1979b). In general, summer milkfat contains increased concentrations of C4:0 to C10:0 and unsaturated fatty acids (primarily C18:1), and decreased concentrations of C16:0 and C18:0 fatty acids, compared with winter milkfat. The reverse trends are observed in the fatty acid composition of winter milkfat. Other chemical and physical indices show similar seasonal trends. For example, the iodine number and softening point of summer milkfat are generally greater than those of winter milkfat because of the increase in the unsaturated fatty acid content (Norris et al., 1973; Sonntag, 1979b).

Feeding Regimens

The long chain fatty acids in milkfat are derived from the cow's blood supply, and their composition of milkfat can be directly affected by diet (Banks, 1991b). Dietary changes can occur through normal seasonal variation in the feed (i.e., pasture and stall periods) or by the addition of oil supplements to the feed. In general, a diet relatively high in long-chain fatty acids results in milkfat with increased concentrations of C18:0 and C18:1 fatty acids and a concurrent decrease in medium-chain fatty acid content (Baer, 1991; Grummer, 1991). The effects of diet manipulation were also discussed in Chapter 1.

Administration of Bovine Somatotropin

Lynch et al. (1988) reported that the relative percentages of individual fatty acids of milkfat obtained from cows injected with bovine somatotropin were similar to those of the control milkfat, although the percentages and output of some fatty acids cycled relative to the day of injection. In a review, Baer (1991) noted that in some lactation studies no effects were observed on milkfat composition but that in other studies the milkfat from cows treated with bovine somatotropin was more unsaturated due to higher amounts of oleic acid; that these results may indicate that high-producing cows produce milkfat with a higher percentage of unsaturated fatty acids, or may indicate a direct effect of bovine somatotropin; and that further research is needed to clarify these effects.

Manufacture of Anhydrous Milkfat

Anhydrous milkfat or anhydrous butter oil are the preferred raw materials for milkfat fractionation. Anhydrous milkfat and anhydrous butter oil are defined as containing 99.8% milkfat and

0.1% water, and butter oil is defined as containing 99.3% milkfat and 0.5% water (Bassette and Acosta, 1988; Rajah, 1991). The goal of anhydrous milkfat production is to disrupt the milkfat globule membrane and liberate the free milkfat to form a bulk phase (Rajah, 1991).

Anhydrous milkfat is generally obtained from pasteurized cream (40% fat) that is first concentrated to 70% to 80% milkfat. The high-fat cream is processed through a specialized phase inversion unit or separator, such as those produced by Alfa-Laval or Westfalia. After phase inversion is complete, the milkfat is concentrated up to 99.6% milkfat, and the residual moisture is removed by vacuum drying (Rajah, 1991; Sadler and Wong, 1970). Butter oil and anhydrous butter oil are obtained by heating fresh or stored butter to 80°C to melt the milkfat and disrupt the butter emulsion. The milkfat is then concentrated in separators and vacuum dried in a manner similar to that used for anhydrous milkfat production (Banks, 1991d; Rajah, 1991; Sadler and Wong, 1970). Sometimes the butter oil is washed with water prior to the final drying stage to remove trace impurities (Banks, 1991d).

Chemical Composition of Milkfat

The lipid composition of bulk milkfat is much different from the total lipid composition of milkfat globules, because of the presence of the milkfat globule membrane. The bulk milkfat, or inner core of the globule, is composed of approximately 98% triglycerides, while the milkfat globule membrane is composed primarily of phospholipids, as well as glycerides (tri-, di-, mono-), cholesterol and other sterols, carotenoids, vitamin A, and other minor components (Banks, 1991d; Jensen et al., 1991; Mulder and Walstra, 1974; Sonntag, 1979b). Subsequent discussions on the chemical composition of milkfat in this monograph refer to bulk anhydrous milkfat without traces of milkfat globule membrane components.

The precise chemical composition of any given milkfat is quite complex and varies with respect to the factors discussed previously (stage of lactation, season, feeding regimen, etc.). Despite these differences, general trends relating to chemical composition and physical properties are observed for milkfat and are discussed in this section.

Fatty Acid Composition of Milkfat

The fatty acid composition is one of the most basic inherent chemical features of milkfat and is determined by genetic and environmental factors. The combination of the fatty acid composition and fatty acid position on triglyceride molecules affects the entire scope of milkfat functionality. Certain fatty acids are responsible for much of the unique flavor of milkfat. The saturated and unsaturated fatty acid content of milkfat affects its melting behavior and oxidative stability. The manner in which the fatty acids are arranged on the triglyceride molecule affects flavor release, crystallization behavior, and nutritional aspects of milkfat. The relationship between fatty acid composition and other functionalities of milkfat (flavor, melting behavior, nutritional aspects) are discussed in their respective sections later in this chapter.

Sonntag (1979b) has noted that the fatty acid composition of milkfat is one of its most distinguishing characteristics, because of its complexity and relatively high proportion of C4 to C10 fatty acids compared with other fats and oils. Over 400 fatty acids have been identified in milkfat (Table 2.1), but only 10 to 20 of these account for 90% of the total fatty acids in milkfat (Banks, 1991d; Jensen and Clark, 1988; Sonntag, 1979b). The average composition of the major fatty acids found in milkfat is presented in Table 2.2.

The two most abundant fatty acids in milkfat are C16:0 and C18:1, approximately 20 to 30% of the total fatty acids each. The C4 to C10 fatty acids account for about 10%, and the C12 and C14 fatty acids contribute 10 to 20% of the fatty acid composition. Milkfat is composed of approximately 70% saturated fatty acids and 30% unsaturated fatty acids, using the strict chemical definition of *unsaturated* (i.e., presence of double bonds). In contrast, the nutrition and medical communities sometimes use a definition of *saturated* that refers only to the combined content of C12, C14, and C16:0 fatty acids, because of their hypercholesterolemic properties (Banks, 1991d).

Triglyceride Composition and Structure of Milkfat

The triglyceride composition is expressed by total fatty acid carbon number or triglyceride class (e.g., trisaturated), and the triglyceride structure is defined as the stereochemical position of the fatty acids on the glycerol backbone. Because of the large number of fatty acids present in milkfat (at least 400), the possible number of triglycerides is at least 400^3 (Jensen et al., 1991); consequently, detailed analysis of triglyceride composition and structure is difficult. Technologies available for triglyceride analysis have been limiting, but recent advances in technologies such as high-performance liquid chromatography and gas chromatography using polar columns, have expanded our capabilities to explore triglyceride composition and structure in more detail. Techniques used to determine triglyceride compostion and structure are discussed in Chapter 4. Many studies on triglyceride composition have separated milkfat triglycerides into fractions based on molecular weight or level of saturation to facilitate the determination of triglyceride composition and structure (Banks et al., 1987; Barbano and Sherbon, 1974; Breckenridge and Kuksis, 1968a, 1968b; Itabashi et al., 1993; Kerkhoven and deMan, 1966; Kuksis and Breckenridge, 1968; Kuksis et al., 1973; Laakso and Kallio, 1993a, 1993b; Lund, 1981; McCarthy et al., 1962; McCarthy et al., 1988; Myher et al., 1993; Parodi, 1973, 1979, 1982; Taylor and Hawke, 1975a). The following data refer only to the triglyceride composition of intact milkfat.

TABLE 2.1 Fatty Acids Identified in Milkfat[1]

Fatty acid class	Fatty acid carbon number
Saturates	
Normal	C2 to C28
Monobranched	C9 to C26; C13 to C19; three or more positional isomers; C5 to C11; Me, Et
Multibranched	C16 to C28
Monoenes	
cis-	C9 to C26; positional isomers of C12:1, C14:1, C16:1 to C18:1, and C23:1 to C25:1
trans-	C14, C16 to C24; positional isomers of C14:1, C16:1, C18:1, and C23:1 to C25:1
Dienes	C14 to C26 evens only: *cis, cis; cis, trans;* or *trans, cis* and *trans, trans* geometric isomers unconjugated and conjugated and positional isomers
Polyenes	
Tri-	C18, C20, C22; geometric positional conjugated and unconjugated isomers
Tetra-	C18, C20, C22; positional isomers
Penta-	C20, C22
Hexa-	C22
Keto (oxo)-	
Saturated	C10, C12, C14, C15 to C20, C22, C24; positional isomers
Unsaturated	C14, C16, C18; positional isomers of carbonyl and double bond
Hydroxy-	
2-position	C14:0, C16:0 to C26:0, C16:1, C18:1, C21:1, C24:1, C25:1
4- and 5-positions	C10:0 to C16:0, C14:ω-6 and C12:1-9
Cyclic	
Hexyl	C11; terminal cyclohexyl
Furan fatty acids	12,15-epoxy-13,14-dimethyleicosa-12,14-dienoic acid and 10,13-epoxy-11,12-dimethyloctadeca-10,12-dienoic acid

[1]Compiled from Guth and Grosch (1992), Jensen et al. (1991), and Sonntag (1979b).

TABLE 2.2 Composition of the Major Fatty Acids in Milkfat[1]

Fatty acid carbon number	Fatty acid common name	Average range (wt%)
4:0	Butyric	2–5
6:0	Caproic	1–5
8:0	Caprylic	1–3
10:0	Capric	2–4
12:0	Lauric	2–5
14:0	Myristic	8–14
15:0	— (Pentadecanoic)	1–2
16:0	Palmitic	22–35
16:1	Palmitoleic	1–3
17:0	Margaric	0.5–1.5
18:0	Stearic	9–14
18:1	Oleic	20–30
18:2	Linoleic	1–3
18:3	Linolenic	0.5–2

[1]Data compiled from data presented in Chapter 5 for intact milkfats used for fractionation (Tables 5.34 to 5.37), and from Banks (1991d), Jensen et al. (1991), and Mulder and Walstra (1974).

Triglyceride Composition Expressed by Fatty Acid Carbon Number

The average triglyceride composition of milkfat as expressed by total fatty acid carbon number is presented in Table 2.3. The C36 to C40 and the C48 to C52 triglycerides are the most abundant in milkfat.

Hinrichs et al. (1992) reported that summer milkfat generally had a higher concentration of the C30 to C42 triglycerides and higher proportions of unsaturated fatty acids, while winter

TABLE 2.3 Average Triglyceride Composition of Milkfat, Expressed as Total Carbon Number[1]

Total fatty acid carbon number	Average range (wt%)
C26	0.1–1.0
C28	0.3–1.3
C30	0.7–1.5
C32	1.8–4.0
C34	4–8
C36	9–14
C38	10–15
C40	9–13
C42	6–7
C44	5–7.5
C46	5–7
C48	7–11
C50	8–12
C52	7–11
C54	1–5

[1]Data compiled from triglyceride composition data presented in Chapter 5 for intact milkfats used for fractionation (Tables 5.84 to 5.86).

milkfat had more saturated C26 to C42 and fewer unsaturated C44 to C54 triglycerides. Hinrichs et al. (1992) also noted that two triglyceride fractions with a fatty acid carbon number of 48 appeared only in summer milkfat samples. Bornaz et al. (1992) identified 34 triglyceride peaks (fractions) and reported that most of the seasonal variation of milkfat involved only seven peaks, representing the C36, C38, C44, C46, and C48 triglycerides. They also reported that the C46 and C48 triglyceride concentrations were greater in summer milkfat than in winter fat, that the C36 and C38 triglyceride concentrations were greater in winter milkfat, and that the other peaks did not display defined or significant patterns.

Triglyceride Composition Expressed by Triglyceride Class

Triglycerides classes of milkfat are determined by the saturation of the constituent fatty acids and stereoisomer: saturated (S), monoene (M), diene (D), triene (T), polyunsaturated (U), cis (^{c}), $trans$ (^{t}). Trisaturated (SSS), cis-monoene (SSM^c), $trans$-monoene (SSM^t), cis-$trans$-dimonoene (SM^{c}M^t), and cis-cis-dimonoene (SM^{c}M^c) triglycerides represent approximately 90% of the triglyceride classes of milkfat (Parodi, 1982). The trisaturated triglycerides comprised approximately 38% of intact milkfat (Chen and deMan, 1966; Kerkhoven and deMan, 1966). Small amounts of SSD, MMM, SMD, MMD, SDD, SST triglyceride classes are also present in milkfat (Parodi, 1982).

Triglyceride Structure of Milkfat

The importance of the stereospecific position of fatty acids on the triglyceride molecule is recognized, but the precise relationship among triglyceride structure, fatty acid composition, and other functionalities is not well understood. Simple fatty acid composition fails to provide information about the distribution of fatty acids in triglycerides which also affects functionality. For example, a physical equimolar mixture of tributyrin, tripalmitin, and triolein exhibit markedly different properties from a single triglyceride composed of equimolar concentrations of butyric, palmitic and oleic acids. Triglyceride structure also affects the release of flavor constituents from milkfat, influences the melting and crystallization behaviors of milkfat, and has been shown to be important to the hypercholesterolemic properties of milkfat (Jensen et al., 1991; Kermasha et al., 1993).

Fatty acids are esterified onto the glycerol backbone in the mammary gland in an orderly (not random) fashion and show a high degree of asymmetry (Jensen et al., 1991; Parodi, 1982). Sonntag (1979a) has classified three basic types of triglycerides: type I triglycerides have 48 to 54 carbon atoms and are made up of long-chain 1, 2-diglycerides esterified with long-chain fatty acids, such as oleic, linoleic and stearic acids; type II triglycerides have 36 to 46 carbon atoms and are made up of long-chain 1, 2-diglycerides esterified to short-chain fatty acids in the 3-position, such as butyric, hexanoic, and octanoic acids; and type III triglycerides have 26 to 34 carbon atoms and are made up of medium-chain 1, 2-diglycerides combined with short- and medium-chain fatty acids in the 3-position.

The following generalizations can be made regarding the stereospecific position of the major fatty acids in milkfat on the glycerol backbone:

- C4 and C6 are in the sn-3 position.
- C8 is located in the sn-2 and sn-3 positions.
- C10 is located preferentially in the sn-2 position.
- C12 is located preferentially in the sn-2 position.
- C14:0 is in the sn-2 position.
- C16:0 is almost equally distributed in the sn-1 and sn-2 positions.
- C18:0 is in the sn-1 or sn-3 position.
- C18:1 is in the sn-1 or sn-3 position.

(Bornaz et al., 1992; Breckenridge and Kuksis, 1968; Christie, 1983; Jensen et al., 1991; Kermasha et al., 1993; Parodi, 1982; Taylor and Hawke, 1975b). Taylor and Hawke (1975b) reported that the position of C18:1 varied but that it was preferentially esterified at position sn-1, with lesser but comparable amounts esterified at the sn-2 and sn-3 positions. The percentage of a fatty acid at the sn-3 position generally decreased with an increase in the chain length of the acid, and the reverse occurred at the sn-1 position. Fatty acid variation was greater at positions sn-1 and sn-3 than at position sn-2, and when the proportion of an acid varied at position sn-1, it usually varied at position sn-3 in the reverse manner (Parodi, 1979). Although the relative proportions of fatty acids vary with seasonal changes in the milkfat, no major variations in the stereospecific distribution of fatty acids through the year have been observed (Parodi, 1979a; Taylor and Hawke, 1975b).

Other Selected Chemical Properties of Milkfat

Historically, chemical properties such as iodine value, Reichert–Meissl, Polenske, and saponification numbers have been reported for milkfat to indicate general trends in the fatty acid composition. Use of these indicators has to a large extent been replaced by the use of the actual fatty acid and triglyceride compositions, which have been determined by gas chromatography or high-performance liquid chromatography. However, indicators such as iodine value are still used as ingredient specifications in the edible oils industry. Other selected properties that have been reported in the literature include refractive index and density, although these relate to the physical, not chemical, properties of milkfat. Typical values for these chemical and physical characteristics of milkfat are presented in Table 2.4, and the characterization methods are described in Chapter 4.

Flavor Constituents of Milkfat

Flavor is one of the most important functionalities of milkfat in food products. Milkfat flavor is unique because of its complexity and because many of the constituents of milkfat flavor are found in precursor form. Milkfat flavor functionality is often discussed in terms of milkfat flavor and flavor potential because of the influence of the precursor compounds. The precursor compounds are certain fatty acids, many of which are minor constituents of milkfat esterified to the triglyceride. These flavor precursor fatty acids are unstable and can be released from the triglyceride structure in the presence of heat, water, acid, or lipases (Sonntag, 1979a; Urbach, 1991). Such released fatty acids can undergo further reactions, including lactonization, to form flavor compounds. Flavor compounds may also be formed as the result of lipid oxidation reactions.

TABLE 2.4 Typical Ranges for Selected Chemical and Physical Properties of Milkfat[1]

Milkfat	Iodine value	Reichert-Meissl number	Polenske number	Saponification number	Refractive index	Density
Average	29–40	20–34	2–4	210–250	1.4530–1.4550	~0.8936 (25°C) ~0.9624 (30°C)
Summer milkfat	> 35	—[2]	—	—	1.4550	—
Winter milkfat	< 35	—	—	—	1.4534	—

[1]Data compiled from Tables 5.131 to 5.133; Newlander and Atherton, 1964; Sonntag, 1979b.
[2]Data not available.

A large number of compounds have been shown to contribute to milkfat flavor; major classes of compounds important to milkfat flavor include

- Lactones
- Methyl ketones
- Low-molecular-weight and branched-chain fatty acids
- Aldehydes

Most of these compounds impart desirable flavors to milkfat at low concentration levels, but at higher concentrations, are responsible for off-flavors of milkfat. Because much of the flavor of milkfat is derived from its fatty acid content, it is subject to the same genetic and environmental influences as fatty acid composition (Dimick et al., 1969; Forss, 1971; Urbach, 1990). An important influence on milkfat flavor is the feed of the cow, which can impart off-flavors to the milk. These off-flavors can result from the presence of compounds in the feed that are directly transferred to the milkfat or from the production of relatively high proportions of precursor compounds that yield high concentrations of normal flavor components, such as indole and skatole (Forss, 1971; Urbach, 1990).

Lactones

Lactones are considered to be one of the most important classes of flavor compounds in milkfat and are a major contributor of heated milkfat flavor. Small quantities of free lactones (5 to 30 ppm) are naturally present in milkfat, but most of the total lactone concentration (110 to 150 ppm) is generated when milkfat is heated (Hammond, 1989; Kinsella, 1975; Sonntag, 1979b). Lactone production can vary with time and temperature of heating (Ellis and Wong, 1975; Sadler and Wong, 1970).

Lactones are spontaneously formed from β-hydroxy acids that have been released from triglycerides upon heating, prolonged storage, or lipolysis or as a result of oxidation (Boldingh and Taylor, 1962; Dimick et al., 1969; Sadler and Wong, 1970; Sonntag, 1979b; Urbach, 1991). Lactones occur in two homologous series, including the 4-hydroxy acids, which form the four-carbon ring γ-lactones, and the 5-hydroxy acids, which form five-carbon ring δ-lactones (Kinsella, 1975). Lactones are produced from saturated and unsaturated, odd- and even-numbered C6 to C16 hydroxy acids (Boldingh and Taylor, 1962; Kinsella et al., 1967; Sonntag, 1979b). The most important lactones in milkfat flavor are δ-C10, δ-C12, and δ-C14 lactones (Dimick et al., 1969; Ellis and Wong, 1975; Sonntag, 1979b). The δ-lactones impart coconut, apricot, and peach flavors, and the γ-lactones impart sweet, raspberrylike flavors to milkfat (Kinsella, 1975; Urbach, 1991).

Methyl Ketones

Methyl ketones are also produced in milkfat by heating, lipolysis, or, in some cases, oxidation reactions (Boldingh and Taylor, 1962; Kinsella, 1975; Sadler and Wong, 1970; Urbach, 1991). The precursors of native methyl ketones are β-keto acids attached to glycerides, which are decarboxylated by heating after hydrolysis to form free methyl ketones (Kinsella, 1975). Methyl ketones have been identified as a homologous series with odd numbers of carbon atoms from C3 to C15 (Kinsella et al., 1967). Methyl ketones in high concentrations have a characteristic flavor and aroma of blue cheese (Urbach, 1991), but lower concentrations contribute to the overall flavor of butter and milkfat. Diketones are also important to butter flavor, particularly 2,3-butadienone, which is commonly known as *diacetyl* (Kinsella, 1975). Diacetyl is produced abundantly by lactic fermentations, although low concentrations can arise from carbonyl–amino degradations.

Low-Molecular-Weight and Branched-Chain Fatty Acids

Milkfat is the only edible fat that contains appreciable amounts of butyric acid (C4), and it has a relatively high concentration of low-molecular-weight fatty acids (C4 to C10) compared with other edible oils (Sonntag, 1979b). These fatty acids greatly contribute to the unique flavors derived from milkfat (Sonntag, 1979b; Urbach, 1991). The C4 to C8 fatty acids are preferentially esterified to the sn-3 position of the triglyceride molecule and are relatively easy to release from the structure by heat, hydrolysis, and lipolysis reactions (Hammond, 1989; Jensen et al., 1991). In high concentrations these compounds generally have a flavor of lipase-treated Italian cheeses (Urbach, 1991).

The presence of branched-chain fatty acids in milkfat has been known since the 1950s (Ha and Lindsay, 1991b; Sonntag, 1979b), but the correlation between branched-chain fatty acids and flavor in dairy products has been recognized only recently (Ha and Lindsay, 1991b). Although branched-chain fatty acids are present in small quantities in milkfat, their threshold values have been found to be low compared to corresponding straight-chain fatty acids (Ha and Lindsay, 1991b). Brennand et al. (1989) evaluated the threshold values and aroma properties of some branched-chain and minor volatile fatty acids occurring in milkfat and reported a wide range of threshold values; they noted that branched-chain fatty acids exhibited unique aromas, and the shorter branched-chain fatty acids were related to cheeselike flavors. The concentrations of volatile branched-chain fatty acids in aged Italian and varietal cheeses and butter have been quantified by Ha and Lindsay (1991a, 1991b), who reported (1991a) that the minor branched-chain fatty acids did not appear to contribute to the flavor of fresh, sweet cream butter, but they did contribute to the flavor of some cheeses when limited hydrolysis of the fatty acids occurred at some stage during processing or storage.

Aldehydes

Aldehydes from C2 to C18 have been identified in milkfat and include odd and even carbon numbers, saturated and unsaturated, and straight and branched-chain aldehydes (Kinsella et al., 1967; Parks et al., 1961). Parks et al. (1961) noted that the C11 to C16 saturated aldehydes occurred as plasmalogens where they were bound to the glycerol molecule in milkfat, and that aldehydes with less than 12 carbons were a result of oxidation. Aldehydes and some ketones are common end products of oxidation reactions in milkfat. Slight oxidation has been shown to enhance the flavor of milkfat, but more pronounced oxidation is one of the primary sources of off-flavors in milkfat (Kinsella et al., 1967; Kinsella, 1975; Urbach, 1991).

Other Compounds

Minor but important trace constituents of milkfat flavor include

- Alcohols
- Phenol
- *o*-methoxyphenol
- *m*- and *p*-cresol
- Skatole
- Indole
- Hydrocarbons
- Dimethyl sulfide

(Kinsella, 1975; Sonntag, 1979b; Stark et al., 1976; Urbach et al., 1970).

Physical Properties of Milkfat

The physical properties of milkfat are expressed in crystallization and melting behaviors. The physical properties of milkfat are dictated by the inherent chemical composition and structure of milkfat and are influenced by the conditions employed during processing. Consequently, the physical properties of milkfat are influenced by the same factors that affect the chemical composition of milkfat (e.g., season). Parameters, such as crystallization rate and temperature employed during processing, can have a marked effect on the physical properties of milkfat.

Crystallization Behavior of Milkfat

The crystallization behavior of milkfat affects its melting properties and other functionalities in food products, such as aeration properties and compatibility with other fats. Consequences of crystallization encompass the effects of the crystallization process (nucleation and growth) and crystal characteristics (morphology and size).

The crystallization behavior of milkfat is very complex because of the large number of triglyceride species involved. The fatty acid composition and triglyceride structure are recognized as affecting the crystallization behavior of milkfat, although the details of these relationships are largely unknown. The crystallization properties of milkfat can be altered by variations in processing conditions such as cooling rate, agitation rate, and final temperature.

The Crystallization Process

The crystallization process consists of two basic steps: nucleus formation and crystal growth (Deffense, 1993b). Alternatively, Schaap and Rutten (1976) have divided the crystallization process of milkfat into three steps: (a) spontaneous seeding, (b) growth at a high cooling rate, and (c) growth at a low cooling rate. The course of the crystal growth curve is influenced by spontaneous seeding and growth of crystals at a high cooling rate; crystal growth at a low cooling rate occurs at the end of the crystallization process (Schaap and Rutten, 1976).

The initial number and size distribution of nuclei that are formed greatly influence crystal distribution and growth, and this has a direct effect on the melting properties of milkfat (Tverdokhleb and Vergelesov, 1970).

Crystal nucleation Crystallization is initiated from molten milkfat by supercooling below its final melting point, causing spontaneous crystal formation (Antila, 1979; Banks 1991a; Deffense, 1993b; Mulder and Walstra, 1974; van Beresteyn and Walstra, 1971). The deeper the supercooling, the greater the probability that any impurity present can catalyze nucleation (Mulder and Walstra, 1974). The nucleation rate can be increased by decreasing the temperature until a maximum is reached, and the viscosity of the molten fat reduces the rate of diffusion (Mortensen, 1983).

The first nuclei formed are the smallest crystals that can exist in molten milkfat without dissolving; they are composed of high-melting triglycerides (Grishchenko and Ivantsova, 1974; Tverdokhleb and Vergelesov, 1970; van Beresteyn and Walstra, 1971). The nuclei are formed in the shape of a sphere composed of fine radial monocrystals (α-form) (Mortensen, 1983; Tverdokhleb et al., 1974).

The number and size of milkfat crystals at the end of nucleation are governed by the cooling rate, the speed of the stirrer, and thermal history of the molten milkfat (Antila, 1979; Beyerlein and Voss, 1973; Grall and Hartel, 1992; Schaap and Rutten, 1976; Tverdokhleb and Vergelesov, 1970). Rapid cooling results in the formation of many small crystal nuclei that are unstable and will recrystallize to more stable forms when tempered. Slow cooling results in the formation of fewer, large nuclei. The use of agitation may accelerate cooling, which accelerates crystallization and results in the formation of a fine crystalline structure (Grall and Hartel, 1992; Schaap and Rutten, 1976; Tverdokhleb and Vergelesov, 1970).

Crystal growth Growth of the spherulite crystal nuclei occurs by the deposition of successive layers of molecules on an already ordered crystal surface (Mortensen, 1983). Spherulite crystal formation is rhythmical in nature, and subsequent crystal layers form concentric surfaces around the nuclei (Mortensen, 1983; Tverdokhleb et al., 1974). Spherulite growth is unique to β and β' crystal forms (Mortensen, 1983: Tverdokhleb et al., 1974).

Foley and Brady (1984) stated that rapid cooling results in the incorporation of more high-melting triglycerides than result from slow cooling, and this leads to a milkfat fraction with a higher solid fat content profile and increased firmness compared to fractions obtained with slow cooling. If cooling occurs very rapidly, a considerable amount of low-melting triglycerides is incorporated in the higher-melting crystal lattice, resulting in the formation of mixed crystals (Foley and Brady, 1984; Mortensen, 1983).

The kinetics of milkfat crystallization are similar to those of chemical reactions, that is, there is a free energy barrier opposing the aggregation of milkfat crystals (Mortensen, 1983). For practical applications the crystallization process can be considered a first-order reaction (Mortensen, 1983; Mulder and Walstra, 1974). The growth rate depends on the degree of supersaturation and on the time needed for a molecule to attain a perfect fit into a vacant site on the crystal lattice (Mortensen, 1983; Mulder and Walstra, 1974). Crystallization of fats is an exothermal process. This property can interrupt the growth of the crystals due to localized increases in temperature, but crystal growth will resume after the excess heat is dissipated (Andrianov and Kornelyuk, 1974; Mortensen, 1983; Mulder and Walstra, 1974; Tverdokhleb et al., 1974).

In addition to regular crystal growth, milkfat held at a constant temperature can undergo three phenomena (van Beresteyn, 1972):

1. Additional crystallization of triglycerides that did not crystallize immediately (liquid-to-solid transition)
2. Polymorphic transition that involves alteration of the crystal lattice structure (solid-to-solid transition)
3. Recrystallization of impure and mixed crystals without alteration of the crystal lattice structure (solid-to-solid transition)

Crystal Characteristics

The crystal characteristics are described by crystal morphology and size and depend on the conditions employed during the crystallization process.

Crystal morphology Milkfat, like most other fats, exhibits polymorphism, which results from a change in the crystal structure of the triglyceride. Crystals of the γ-, α-, β'- and β-forms have been identified in milkfat (Banks, 1991a; Belousov and Vergelesov, 1962; Deffense, 1989; deMan, 1963; Foley and Brady, 1984; Kuzdzal-Savoie and Saada, 1978b; Tverdokhleb and Vergelesov, 1970). The γ-form is highly unstable and has been observed only during photomicrographic studies (Belousov and Vergelesov, 1962). The α-form has little spatial arrangement and a low melting point; β' crystals have a tighter arrangement and higher melting point; and β crystals have a very dense arrangement and a very high melting point (Deffense, 1989; Mortensen, 1983). The β'-form is generally the most stable form for milkfat crystals (Deroanne, 1976; Mulder and Walstra, 1974).

Rapid and deep cooling promotes the formation of the low-melting unstable crystal forms γ and α, but raising the temperature and reducing the cooling rate promote the formation of more stable (β') crystal forms (Belousov and Vergelesov, 1962; deMan, 1961b, 1963;

Deroanne, 1976; Foley and Brady, 1984; Mortensen, 1983; Mulder and Walstra, 1974; Sherbon and Dolby, 1973; Tverdokhleb and Vergelesov, 1970). The β-form of milkfat crystals is not very common and has been identified only for some high-melting fractions of milkfat (Belousov and Vergelesov, 1962; Sherbon and Dolby, 1973).

Crystal size The size of milkfat crystals varies considerably, depending on the rate of crystallization. If milkfat is cooled rapidly, numerous small crystals, of needle or platelet shape with a diameter of 0.1 to 3 μm, are formed (deMan, 1961b; Foley and Brady, 1984; Mortensen, 1983; Mulder and Walstra, 1974). Slow cooling of milkfat results in the formation of a few large, spherulite-shaped crystals with diameters up to 40 μm (deMan, 1961b; Foley and Brady, 1984; Mortensen, 1983). Spherulite crystals easily form aggregates that can range from 100 to 1000 μm in diameter (Mortensen, 1983). Crystal size is also affected by agitation rate; an increased agitation rate will produce a greater quantity of small crystals than will slower agitation rates (Chapter 3).

Melting Behavior of Milkfat

Melting behavior is important to many applications of milkfat in food products. Unfortunately, the melting behavior is often the limiting factor in milkfat usage in food products that would otherwise benefit from other milkfat functionalities (e.g., flavor, antiblooming properties in chocolate). The melting behavior of milkfat is characterized by melting point, thermal profile, and solid fat content profile.

Melting Point of Milkfat

The melting point of milkfat is defined as the temperature at which milkfat becomes visually clear and free of crystals; it is approximately 32 to 36°C. Other measurements of the melting properties of milkfat that are commonly employed are dropping point, softening point, and slip point. Milkfat will begin to soften, slip, and drop prior to the final clear point, so the temperatures reported for these measurements are lower than the actual clear point of milkfat.

Thermal Profile of Milkfat

Milkfat is completely solid at –40°C, completely liquid at +40°C, and a mixture of solid and liquid fat at intermediate temperatures. The thermal profile, measured using differential scanning calorimetry (DSC), provides a macroscopic view of the melting behavior of milkfat. A typical DSC profile of intact milkfat is shown in Figure 2.1. Milkfat consists of three major melting species:

- Low-melting glycerides, which melt below 10°C
- Middle-melting glycerides, which melt between 10 and 20°C
- High-melting glycerides, which melt above 20°C

Solid Fat Content of Milkfat

The solid fat content (SFC) of milkfat is a measurement of the percentage of solid milkfat in a sample at a selected temperature. When SFC is measured over a selected temperature range, a solid fat content *profile* for the milkfat sample results. Solid fat content can be measured using pulsed nuclear magnetic resonance, differential scanning calorimetry, or dilatometric techniques. The most accurate and increasingly most popular method for SFC determination is nuclear magnetic resonance. A detailed discussion of solid fat content methodologies is presented in Chapter 4.

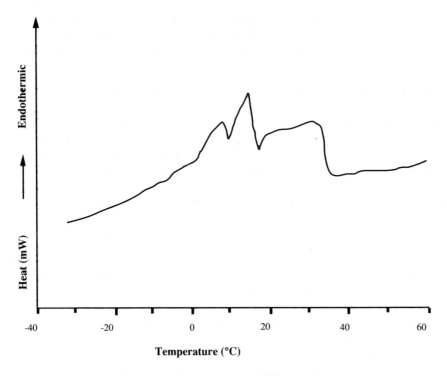

Figure 2.1. Typical thermal profile of anhydrous milkfat (Kaylegian, 1991).

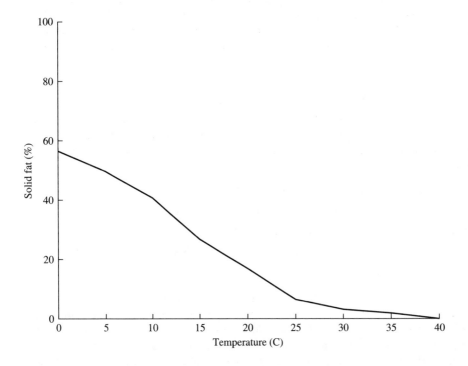

Figure 2.2. Typical solid fat content profile of anhydrous milkfat (Kaylegian and Lindsay, 1992).

A typical SFC profile of intact milkfat is presented below and shown graphically in Figure 2.2 (Kaylegian and Lindsay, 1992):

0°C = 57.1% solid fat

5°C = 50.3

10°C = 41.1

15°C = 26.8

20°C = 16.7

25°C = 6.5

30°C = 2.8

35°C = 1.7

40°C = 0.0

Milkfat exhibits a gradual melting profile from approximately 50% solid fat at 5°C to a final melting point of 35 to 40°C.

Influence of Chemical Composition on Melting Behavior and Texture

Since the melting behavior is responsible for the texture of fatty foods, butter texture has been traditionally indexed by iodine values and fatty acid contents. However, Parodi (1971, 1976, 1981) has established a relationship between the chemical composition and structure and softening point of milkfat. More recently, advances in instrumentation has permitted the accurate determination of the triglyceride compostion, and firmness has now been related to four triglyceride fractions in butter (Bornaz et al., 1993). As a result, measurement of these prevalent triglycerides has been proposed as better indices of butter texture than traditional methods. Futher research on the relationship between the structure of triglyceride and physical properties of milkfat and its fractions is necessary and important to the understanding of the functionality of milkfat.

Increases in short-chain (C4 to C10) and long-chain unsaturated (C16:1 and C18:1), with concurrent decreases in long-chain saturated (C16:0 and C18:0), fatty acids result in milkfats with lowered melting points, a greater proportion of low-melting glyceride species, and a lower SFC profile. Conversely, an increase in C16:0 and C18:0, with a concurrent decrease in the C4 to C10 and C18:1, fatty acids increases the melting point of milkfat and results in increased proportions of high-melting glyceride species that cause a higher SFC profile (Chapter 5).

Other Important Properties of Milkfat

The chemical and physical properties are most often referred to when discussing milkfat functionality. However, other aspects of milkfat are important to its use in foods—namely, its potential for deterioration and nutritional properties. The potential for, and prevention of, deterioration of all fat ingredients is of great concern from a quality perspective in the food industry. Recent recommendations from the medical community to decrease consumption of dietary fats, particularly saturated fats, have put more emphasis on the nutritional properties of fat ingredients used in food products.

Deterioration of Milkfat

The deterioration of lipids can occur through oxidative and hydrolytic reactions, and such deterioration is often detected by the presence of rancid off-flavors. Oxidative reactions involve the uptake of oxygen by unsaturated fatty acids in milkfat and yield off-flavor compounds consisting primarily of aldehydes, and leads to the development of oxidative rancidity off-flavors. Hydrolytic reactions, often involving lipolytic enzymes, yield free fatty acids and lead to hydrolytic rancidity off-flavors.

Oxidation of Milkfat

Factors affecting the oxidative stability of milkfat   Factors affecting the oxidative stability of milkfat include

- Saturated fatty acid content
- Availability of oxygen
- Storage temperature
- Light
- Metallic contamination (especially copper and iron)
- Naturally occurring α-tocopherol level

(Hamm et al., 1968; Kaylegian, 1991; Kehagias and Radema, 1973; Keogh and Higgins, 1986a; Sadler and Wong, 1970; Weihrauch, 1988). Milkfat has been shown to be more stable to oxidation than some vegetable oils because of its high saturated fatty acid content (Kupranycz et al., 1986).

Oxidation mechanisms   Oxidation mechanisms commonly encountered in lipid oxidation are free radical chain reactions initiated by light exposure or the presence of active oxygen species (Richardson and Korycka-Dahl, 1983; Weihrauch, 1988; Wong, 1989). Exposure to light readily initiates free radical oxidation via proton abstraction from unsaturated fatty acids and subsequent reaction of radicals with oxygen (Wong, 1989). Several active oxygen species have been found to initiate milkfat oxidation; they include singlet oxygen, superoxide anion, hydrogen peroxide, and hydroxyl radical (Richardson and Korycka-Dahl, 1983; Wong, 1989). Lipid oxidation mechanisms are complex, and a detailed discussion of the topic is beyond the scope of this monograph. However, many excellent reviews are available (Frankel, 1985; Nawar, 1985a, 1985b; Richardson and Korycka-Dahl, 1983; Weihrauch, 1988; Wong, 1989).

Oxidative off-flavors   One of the most important consequences of milkfat oxidation is the formation of off-flavors. Oxidation of milkfat results in a variety of complex chemical changes that yield the formation of numerous volatile and nonvolatile degradation products (Keogh and Higgins, 1986a; Kupranycz et al., 1986; Richardson and Korycka-Dahl, 1983). Although these compounds may be present at low concentrations, many still impart off-flavors to milkfat. Aldehydes formed as products during oxidation are responsible for many of the characteristic oxidative off-flavors associated with milkfat (Forss, 1971; Keogh and Higgins, 1986a; Weihrauch, 1988). Other oxidation products include ketones, alcohols, lactones, saturated and unsaturated hydrocarbons, and semialdehydes (Forss, 1971; Hammond, 1989; Keogh and Higgins, 1986a; Sadler and Wong, 1970; Weihrauch, 1988). Oxidized off-flavors in milkfat have been described by terms such as *painty, grassy, tallowy, oily, cardboardy, fishy, metallic, mushroomy,* and *cucumber-* and *melon-like* (Hammond, 1989; Sadler and Wong, 1970; Weihrauch, 1988).

Measurement of oxidative stability   Many methods provide indices of the oxidative stability of milkfat, but this property is often measured by peroxide value (PV). The maximum PV permitted in milkfat ingredients by International Dairy Federation (IDF) standards is 0.2 meq oxygen/kg fat (Keogh and Higgins, 1986a).

Use of antioxidants Because of strict standards of identity for butter, antioxidants generally are not used in milkfat ingredients. However, experimental evaluations have shown that the presence of many antioxidants in milkfat prolongs the onset of oxidation or lessens the level of deterioration. Antioxidants that have been evaluated include α-tocopherol, phospholipids, propyl gallate, octyl gallate, dodecyl gallate, butylated hydroxyanisole, and ascorbic acid (Chen and Nawar, 1991; Keogh and Higgins, 1986a; Kaylegian, 1991; Sokolov, 1974; Timmen, 1978; Weihrauch, 1988). At low levels (less than 500 ppm) α-tocopherol has antioxidant properties, but at higher levels (1000 ppm) it may behave as a prooxidant (Kaylegian, 1991; Keogh and Higgins, 1986a). Most antioxidants have been shown to exhibit synergistic effects when used in combination; milkfats containing combined antioxidants have improved oxidative stability compared with milkfats that only contained a single antioxidant (Sokolov, 1974; Timmen, 1978).

The removal of oxygen from the system by nitrogen blanketing or purging is also an effective means of retarding oxidation (Higgins and Keogh, 1986; Keogh and Higgins, 1986a; Sadler and Wong, 1970).

Hydrolysis of Milkfat

The hydrolysis of milkfat usually occurs through the action of lipolytic enzymes on the triglyceride molecule and yields diglycerides, monoglycerides, and free fatty acids (Higgins and Keogh, 1986; Keogh and Higgins, 1986a). Hydrolysis of fats also can be effected by chemical means, especially under alkaline conditions. Generally, chemical hydrolysis of milkfat does not occur or contribute to deterioration of fats in foods, however.

Lipolytic enzymes Two general types or sources of enzymes are involved in milkfat lipolysis: those that occur naturally in milk and those that result from microbial metabolism (Deeth and Fitz-Gerald, 1983; Higgins and Keogh, 1986). Native lipolytic enzymes are inactivated by adequate pasteurization prior to anhydrous milkfat manufacture and generally do not play a part the deterioration of milkfat during further processing, such as fractionation (Higgins and Keogh, 1986). However, lipolysis that occurs during storage of anhydrous milkfat may be caused by either type of lipase (Higgins and Keogh, 1986). A detailed discussion of the causes of hydrolytic rancidity and specific lipases is beyond the scope of this monograph, but this information can be found in reviews (Deeth and Fitz-Gerald 1983; Weihrauch, 1988).

Lipolytic off-flavors Free fatty acids produced in low concentration by lipolysis often contribute to the desirable flavor of milkfat and other dairy products. However, at higher levels free fatty acids can contribute to off-flavor in milkfat. Lipolytic off-flavors have been described as *rancid, butyric, bitter, goaty, unclean, soapy,* and *astringent* (Deeth and Fitz-Gerald, 1983; McNeill et al., 1986; Woo and Lindsay, 1983).

Measurement of lipolytic quality The lipolytic quality of milkfat can be measured by determination of the total free fatty acid (FFA) content or acid degree value (ADV), which is a titration of carboxyl groups or individual fatty acids. The maximum FFA content permitted by the USDA (1975) and the International Dairy Federation (Keogh and Higgins, 1986a) standard is 0.3% expressed as oleic acid.

The various specificities of lipases result in various proportions of short- and long-chain faaty acids from milkfat, and this causes difficulties in relating ADV to sensory properties. However, the measurement gives an approximation of the amount of free fatty acids present in a milkfat sample. In butter normal acid degree values are below 1, and as lipolysis occurs the values may rise to 4 or above. However, the presence of a rancid flavor usually occurs at about 1.5, and the flavor becomes objectionable above 2.0 (Woo and Lindsay, 1983).

Nutritional Properties of Milkfat

Milkfat and Hypercholesterolemia

Nutritional concerns regarding fats and oils has recently focused on the saturated fatty acid and cholesterol contents of fats. Cholesterol has several important functions in the body, where it serves as a structural element of cell membranes and as a precursor of bile acids, steroid hormones, and vitamin D (McBean and Speckmann, 1988). However, epidemiological studies have repeatedly demonstrated a clear association between elevated serum cholesterol levels and an increased risk of mortality from coronary heart disease (Ney, 1991). Milkfat has been identified as a hypercholesterolemic fat because it contains cholesterol and is primarily saturated (70%), although appreciable amounts of monounsaturated (25 to 35%) and small amounts of polyunsaturated (4%) fatty acids are present (Banks, 1991d; Jensen et al., 1991; McBean and Speckmann, 1988; Ney, 1991). However, recent research has suggested that different types of saturated fats do not have equivalent effects on plasma cholesterol levels (Gurr, 1992; Ney, 1991).

Milkfat contains a relatively high proportion (10%) of C4 to C12 fatty acids (Ney, 1991). Saturated fatty acids with chain lengths up to C10 do not influence plasma cholesterol, because they are absorbed directly from the stomach into the portal vein, where they are carried to the liver and rapidly metabolized (Gurr, 1992; Jensen and Clark, 1988; Jensen et al., 1991). Thus, the metabolic handling of the shorter-chain fatty acids is different from that of the longer-chain fatty acids, which are transported as chylomicrons (Gurr, 1992; Jensen and Clark, 1988; Jensen et al., 1991).

The hypercholesterolemic effect of saturated fats is largely due to 12-, 14-, and 16-carbon fatty acids, which represent about 40% to 50% of the total fatty acids in milkfat (Gurr, 1992; Jensen et al., 1991; Ney, 1991). Gurr (1992) reported that myristic acid (14:0) is more effective than lauric acid (12:0) in raising cholesterol.

Approximately 35% of the total fatty acids in milkfat consist of stearic (C18:0) and oleic acids (C18:1). Recent studies suggest that stearic acid and oleic acid lowered plasma cholesterol levels when either fatty acid replaced palmitic acid in the diet of men (Ney, 1991).

The stereospecific distribution of the fatty acids on the triglyceride affects its digestibility and influence on plasma cholesterol, irrespective of the overall fatty acid composition (Gurr, 1992; McBean and Speckmann, 1988). The 4:0, 6:0, and 8:0 fatty acids are preferentially located at the sn-3 position on the triglyceride and are easily hydrolyzed by pancreatic and lingual lipases and then transported via the portal vein to the liver (Jensen and Clark, 1988; McBean and Speckmann, 1988). Animal and human studies have suggested that the capability of a saturated fat to increase serum cholesterol depends on the presence of a long-chain saturated fatty acid in the sn-2 position on the triglyceride molecule (Kermasha et al., 1993). Thus, palmitic, myristic, and lauric acids, which are preferentially esterified to the sn-2 position, appear to provide hypercholesterolemic properties, whereas similar fatty acids at sn-1 or sn-3 positions may not contribute such properties (Kermasha et al., 1993). Further, linoleic acid has been found to be more hypocholesterolemic when present at the sn-2 position than in positions sn-1 and sn-3 (Gurr, 1992). The fact that stearic acid is normally esterified in the sn-1 position, and rarely at position sn-2, may in part explain the neutral effect of this fatty acid on blood cholesterol (Gurr, 1992).

Vitamin Content of Milkfat

Milkfat contains small amounts of the fat-soluble vitamins:

- Vitamin A (19.8 to 40.7 IU/g milkfat)

- Vitamin D (0.34 to 0.84 IU/g milkfat)
- Vitamin E (0.024 mg/g of milkfat)
- Vitamin K (0.001 mg/g of milkfat)

(McBean and Speckmann, 1988; Newlander and Atherton, 1964; Sonntag, 1979b). The vitamin content, like all other properties of milkfat, varies with breed, season, and other factors affecting the chemical composition of milkfat (Sonntag, 1979b).

Vitamin A exists in two forms in milkfat: preformed vitamin A (retinol and retinyl esters) and carotene (McBean and Speckmann, 1988; Sonntag, 1979b). Vitamin A is necessary for growth and reproduction, resistance to infection, maintenance and differentiation of epithelial tissues, stability and integrity of membrane structure, and vision (McBean and Speckmann, 1988).

Vitamin D plays a central role in calcium and phosphorus homeostasis in the body by stimulating the absorption of these materials in the intestine and promoting normal mineralization of newly forming bone (McBean and Speckmann, 1988).

Vitamin E is a generic term for the tocopherols and tocotrienols, but α-tocopherol has the greatest biological activity. It is generally accepted that vitamin E functions in cellular and subcellular membranes as a biological antioxidant or free radical scavenger, thus stabilizing cellular membranes (McBean and Speckmann, 1988).

Vitamin K is required for the synthesis of blood-clotting factors: prothrombin (factor II) and factors VII, IX, and X (McBean and Speckmann, 1988).

Conjugated Linoleic Acid in Milkfat

Ruminant dairy and meat products have been identified as principal dietary sources of conjugated linoleic acid (CLA). Chin et al. (1992) and Chang et al. (1994) surveyed a variety of foods and reported that milk, butter, and natural cheeses contained considerable amounts of natural CLA. CLA isomers recently have been found to be anticarcinogenic in animal models (Chin et al., 1992; Ip and Scimeca, 1994), and they also have been found to have antioxidant properties (Chin et al., 1992; Pariza et al., 1993). The role of CLA in health maintenance and its use in food preservation is currently under investigation (Pariza et al., 1993).

Chapter 3
Milkfat Fractionation Technologies

Milkfat can be separated into fractions based on some of its inherent properties, including crystallization behavior, solubility, and volatility. Fractionation methods and conditions employed by investigators have varied, but they can be loosely placed into two general groups:

1. Fractionation methods based on the crystallization and solubility behaviors of milkfat triglyceride species, including

   - Crystallization from melted milkfat
   - Crystallization from milkfat dissolved in solvent

2. Fractionation methods based on the solubility and volatility of milkfat components, including

   - Supercritical fluid extraction
   - Short-path distillation

The oldest, and still most common, method of milkfat fractionation is crystallization, either from melted milkfat or from solvent solution. Specific conditions employed for experimental milkfat fractionation are summarized in Tables 3.1 to 3.4, respectively, and each method is discussed further in this chapter.

Fractionation Based on the Crystallization and Solubility Behaviors of Milkfat

Fractionation methods in this group are generally temperature-driven crystallization processes. Milkfat is composed of many triglyceride species that have different melting points, and intact milkfat melts over the range –40°C to +40°C. This broad melting range provides many opportunities to create numerous fractions by crystallization of milkfat at selected temperatures and then separating the solid and liquid fractions. Additionally, individual fractions can be recrystallized, providing subfractions with more homogeneous composition than parent fractions. Also, milkfat fractions can be prepared by crystallization either from pure, melted milkfat or from milkfat dissolved in a solvent solution.

Milkfat Crystallization

The inherent crystallization behavior of milkfat has been discussed in Chapter 2. The basic phenomena of the crystallization process (nucleation and growth) and crystal characteristics (morphology and size) remain valid for milkfat under the various circumstances encountered. These instances include the fundamental crystallization of milkfat (Chapter 2), crystallization during the fractionation of milkfat (Chapter 5), and crystallization during texturization of milkfat ingredients (Chapter 8). However, the specific conditions employed during processing greatly affect crystallization parameters, and the physical and chemical properties of the milkfat fractions are likewise affected.

The Crystallization Process

The processes employed for fractionation have been studied extensively (Black, 1974b; Deffense, 1993b; Kuzdzal-Savoie and Saada, 1978a; Rajah, 1991; Schaap and Rutten, 1976; Schaap and van Beresteyn, 1970; Voss et al., 1971), and can be divided into several general steps:

1. Preparation of the milkfat starting material
2. Cooling to the desired temperature
3. Crystallization
4. Separation of the solid phase from the liquid phase.

Milkfat is initially heated to approximately 60°C, and held for 30 min to remove any crystal history that may influence the current crystallization process. If appropriate, milkfat may also be dissolved in a solvent to form a homogeneous solution. Cooling of milkfat to the desired fractionation temperature is usually done under controlled conditions. The milkfat is held at the fractionation temperature for a specified period of time to allow crystallization. Agitation of the milkfat solution is common during fractionation from melted milkfat, but quiescent conditions are generally preferred for fractionation from solvent solution. Many techniques have been employed to separate the solid and liquid milkfat phases, and these will be discussed subsequently.

Factors Affecting the Crystallization of Milkfat During Fractionation

The fractionation process is a result of complex interactions between the inherent chemical composition and crystallization behavior of milkfat combined with processing-controlled parameters such as cooling rate, agitation rate, fractionation temperature, and separation method (Antila, 1984; Deffense, 1987, 1989, 1993b; Doležálek et al., 1976; Grall and Hartel, 1992; Keogh and Higgins, 1986b; Rajah, 1991; Schaap and Rutten, 1976; Tverdokhleb and Vergelesov, 1970). Because these parameters are interrelated, processing conditions can be manipulated to achieve the desired properties in milkfat fractions. Several factors affect crystallization during fractionation, including:

Seeding The crystallization of milkfat is considered a naturally slow process, but it has been accelerated by the use of seeding (Doležálek et al., 1976; Voss et al., 1971). The seeding process involves adding high-melting seed crystals to the milkfat to serve as nucleation sites. Seeding accelerates the crystallization process by providing nuclei quicker than would naturally occur, and consequently it accelerates the entire crystallization process. Doležálek et al. (1976) reported a crystallization time of 2 h for milkfat that had been seeded, compared with a crystallization time of 4 h for the control milkfat.

Supercooling Milkfat must be supercooled to initiate spontaneous nucleation of the high-melting triglycerides. The temperature at which crystallization occurs spontaneously has often been defined as the initial temperature of milkfat crystallization (Black, 1973; Keogh and Higgins, 1986b). Keogh and Higgins (1986b) have evaluated three initial temperatures (31.0, 34.0, and 37.0°C) for the crystallization of slowly cooled milkfat, and found little influence of the initial temperature on the crystallization process. However, Black (1973) and Voss et al. (1971) found that extreme supercooling resulted in the formation of a large number of nuclei, causing an acceleration in the crystallization process. The accelerated crystallization process inhibited crystal growth and resulted in a large number of small crystals in the solution (Black, 1973; Voss et al., 1971).

Cooling rate The cooling rate has a major effect on the separation efficiency of the milkfat fractions, and programmed, slow cooling has been found to be superior to linear, rapid cooling (Antila, 1979; Black, 1975b; Doležálek et al., 1976; Keogh and Higgins, 1986b; Schaap and van Beresteyn, 1970). Slow cooling promotes the formation of larger, more differentiated crystals which are easier to separate from the liquid phase (Black, 1975b; Deffense, 1987,

1989; Tverdokhleb and Vergelesov, 1970). Rapid cooling promotes the formation of smaller crystals; this in turn results in the formation of mixed crystals and increased viscosity, both of which contribute to separation difficulties (deMan, 1968; Deffense, 1987, 1989; Doležálek et al., 1976; Schaap and van Beresteyn, 1970; Tverdokhleb and Vergelesov, 1970).

Keogh and Higgins (1986b) tested three cooling rates (0.55, 0.71, and 1.0°C/h), and found little difference in crystallization characteristics. This effect could be attributed to the fact that the three cooling rates tested are all relatively slow compared to other cooling rates used for milkfat fractionation (i.e., 0.005 to 2°C/min; Table 3.1).

Kankare and Antila (1974a) investigated the effect of cooling rates on crystal shape. They reported that if the milkfat was cooled slowly, spherulite crystals were formed in which the individual crystal needles were arranged radially. Rapid cooling of the milkfat produced a supersaturated solution, which resulted in crystal needles that were markedly bent and forked and had little radial arrangement.

Agitation rate Milkfat can be fractionated by crystallization processes with or without agitation. The rate of agitation is important to crystallization rate, crystal growth, and separation efficiency during fractionation (Doležálek et al., 1976; Schaap and van Beresteyn, 1970). Crystallization of milkfat without agitation produces open-textured spherulite crystals, which entrap some of the liquid phase, increase the viscosity of the mixture, and decrease the separation efficiency of the solid from the liquid milkfat (Black, 1973; Keogh and Higgins, 1986b).

Agitation accelerates cooling, which in turn accelerates the crystallization rate (Doležálek et al., 1976; Grall and Hartel, 1992; Tverdokhleb and Vergelesov, 1970). Crystallization is an exothermic reaction; the temperature of the milkfat can rise several degrees during crystallization, and this results in a retardation of the crystallization process (Antila, 1984; Deffense, 1989). Agitation has been found to promote the formation of crystals by increasing the dissipation of the heat of crystallization.

Slow stirring during crystallization produces crystals with less incorporated liquid fat, and this results in more distinct crystals that allow a better separation of the solid phase from the liquid phase (Black, 1975; Keogh and Higgins, 1986b). Doležálek et al. (1976) reported that low agitation rates (13.8 rpm) produced crystals of nonuniform size, while medium agitation rates (60 rpm) produced crystals of more uniform size with most crystals larger than 100 µm. Fast agitation rates (120 rpm) produced crystals that were relatively uniform, but under these conditions most crystals were less than 100 µm in size (Doležálek et al., 1976). At very low and very high agitation speeds very fine crystals were formed, and it was difficult to separate the crystal mass from the liquid milkfat (Schaap and Rutten, 1976). High agitation rates result in smaller crystals because of shear damage (Schaap and van Beresteyn, 1970).

Grall and Hartel (1992) reported that the final mean crystal diameter decreased with agitator velocity at 30 and 20°C but increased at 15°C, indicating that the crystals at 15°C were not easily sheared by the higher agitation rates. They also reported that crystal growth rates increased with agitator velocity at 20 and 15°C but decreased with agitator velocity at 30°C, indicating a different growth mechanism. At 20 and 30°C aggregation was the primary mechanism of crystal growth, but little aggregation was observed at 15°C.

Fractionation temperature The fractionation temperature influences the crystallization behavior, melting behavior, chemical composition, and yields of milkfat fractions (Black, 1973; Doležálek et al., 1976; Grall and Hartel, 1992; Jebson, 1970; Keogh and Higgins, 1986b).

Doležálek et al. (1976) reported that at fractionation temperatures of 27 to 30°C crystal growth was uniform, but at 25°C large clusters were formed along with regular small particles. Grall and Hartel (1992) have studied fractionation conditions and also reported a change in crystal appearance as fractionation temperatures decreased. Crystals formed at 30°C consisted of a conglomeration of needle-like structures, while crystals formed at 20°C appeared to be a conglomeration of smaller, tighter aggregates (Grall and Hartel, 1992). Crystals formed at 15°C were uniform spheres with a single birefringent cross in polarized light (Grall and Hartel, 1992).

Crystallization time Because the inherent crystallization process of milkfat is relatively slow, the time period allotted for crystallization can affect crystal size and yield. Keogh and Higgins (1986b) reported an increase in solid fraction yield as crystallization time increased from 0 to 3 h. Black (1973) employed either 16 or 21 h crystallization periods, and reported no significant difference in crystal size. However, longer crystallization times resulted in increased liquid fat content at separation. Antila (1979) noted that 15 h was usually sufficient for crystallization. Many investigators have employed a 24 h crystallization period (Tables 3.1 and 3.2) to ensure that the crystallization process was completed.

Crystallization Kinetics

Grall and Hartel (1992) noted that there is little quantitative data on the kinetics of milkfat crystal formation and growth; they studied the effects of fractionation temperature (30, 20, and 15°C) and agitation rate (50, 75, 100, 125, and 150 rpm) on crystal nucleation, mass deposition and growth. For the agitator velocities tested, the nucleation rates were highest for crystals formed at 15°C and lowest at 20°C. The increase in nucleation rate with agitator velocity for all temperatures could be explained by the increase in impeller shear, where the impeller blades caused tiny crystals to form from disruption of existing crystals. Lag times for crystallization were longest at 15°C, shortest at 30°C, and decreased as agitation increased.

Grall and Hartel (1992) reported that mass deposition rates increased with agitator velocity for all temperatures evaluated. Crystal growth rates at 15°C and 20°C increased with increasing agitator velocities, while growth rates at 30°C initially increased between 50 and 100 rpm but then decreased with agitator velocity. This trend suggested that the crystals were held together in loose aggregates at 30°C.

Fractionation Procedures

Milkfat can be fractionated using either single-step or multiple-step fractionation procedures, and both solid and liquid fractions can be refractionated.

Single-Step Fractionation

Single-step fractionation yields one solid and one liquid fraction from a single batch of milkfat at a selected temperature. As fractionation temperature decreases, the yield of the solid fraction increases. The increase in yield is caused by increases in viscosity and results in decreases in separation efficiency at lower temperatures. Consequently, there are some temperature limitations with single-step fractionation. Fractions obtained by single-step fractionation have shown gradual changes in physical and chemical characteristics with decreasing fractionation temperatures (Chapter 5).

Multiple-Step Fractionation

Multiple-step fractionation yields two or more solid fractions and one liquid fraction from each batch of milkfat. Usually multiple-step fractionation procedures are initiated at the highest desired temperature, and the liquid fraction is refractionated at subsequently lower temperatures (Tables 3.1 and 3.2). However, Rolland and Riel (1966) employed an increasing temperature profile and refractionated the solid fraction for their multiple-step fractionation procedure. For this process, milkfat samples were placed in tubes in a temperature-controlled centrifuge, and the temperature was gradually increased. In each case the liquid fraction was removed after centrifugation at 30 min intervals, and the solid fraction remained in the centrifuge tube for refractionation at the next highest temperature.

Multiple-step fractionation removes a portion of the milkfat at each fractionation step. Consequently, the starting materials for each successive step have different chemical compositions. The removal of the high-melting fractions as the fractionation temperature profile decreases results in the remaining liquid portions having lower viscosities at lower temperatures than do comparable fractions obtained by single-step fractionation at the same temperatures. Therefore, milkfat fractions can be obtained at lower temperatures using multiple-step fractionation procedures than for single-step fractionation procedures. The yield of the second solid fraction, and subsequent solid fractions, similarly will be lower than yields for solid fractions obtained by single-step fractionations at similar temperatures.

A series of solid fractions produced by multiple-step fractionation showed more diversity in physical properties than a series of solid fractions produced by single-step fractionation at corresponding temperatures using a decreasing temperature profile (Kaylegian, 1991). Multiple-step fractionation produced solid milkfat fractions that contained less liquid oil than solid fractions obtained at corresponding temperatures by single-step fractionation (Kaylegian, 1991).

Kankare and Antila (1974a) observed that milkfat fractions obtained from multiple-step fractionation exhibited different crystallization behaviors, particularly with respect to cooling rates. Milkfat crystals and agglomerates also decreased in size at subsequent fractionation steps. Kankare and Antila (1974a) reported spherulite crystals and agglomerates of 0.1 to 1.0 mm diameter at the first fractionation step, and of 0.05 to 0.2 mm diameter for the third fractionation step.

Refractionation

Refractionation of liquid fractions is employed as a matter of course during traditional multiple-step fractionation procedures. Refractionation of solid fractions is generally employed for the purpose of increasing melting points and purity of the high-melting triglycerides. Liquid fractions are generally refractionated at lower temperatures, whereas solid fractions are often recrystallized at the same or higher temperatures.

Fractionation of Milkfat by Crystallization from Melted Milkfat

Crystallization from melted milkfat is the most commonly employed fractionation method, because it is a relatively simple physical process. Fractionation by crystallization from melted milkfat has also been referred to as *fractionation from the melt*, *dry fractionation*, and *simple fractionation*, because it does not use additives or solvents. An exception to this occurs when an aqueous detergent solution is used as a processing aid during the separation of the solid and liquid phases.

Fractionation Methods

Experimental fractionation of milkfat by crystallization from melted milkfat has been performed using laboratory equipment and processing parameters ranging from small-scale (500 g) laboratory experiments to pilot plant–scale operations of 45 L and larger sizes (Table 3.1). Commercial milkfat fractionation operations also employ crystallization from melted milkfat, but details of these processes are not generally available (Alaco, undated; Aveve, undated; Corman, undated).

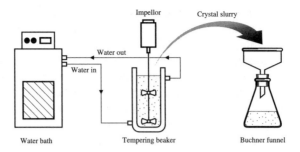

Figure 3.1 Typical schematic diagram of fractionation equipment used for experimental fractionation from melted milkfat.

General method for fractionation by crystallization from melted milkfat Although many conditions have been employed for fractional crystallization processes, the basic crystallization process consists of four steps:

1. Heat milkfat to 60°C and melt completely.
2. Cool milkfat to the fractionation temperature.
3. Crystallize milkfat.
4. Separate the solid and liquid fractions.

The crystallization of melted milkfat generally is conducted in a temperature-controlled crystallization vessel, such as a jacketed vessel or a beaker placed in a water bath or incubator (Figure 3.1). The temperature of the milkfat is then adjusted to the fractionation temperature, either by programmed cooling or simply by allowing the temperature to reach equilibrium unassisted. Variables employed during fractionation from melted milkfat (Table 3.1) include the following:

- Cooling rates = 0.005 to 2°C/min.
- Agitation rates = 0 to 120 rpm.
- Crystallization periods = 0 to 48 h.
- Fractionation temperatures:
  - Single-step fractionation = 34 to 17°C.
  - Multiple-step fractionation = 34 to 4.5°C.

The information presented in Table 3.1 illustrates the range of conditions used, but the reasoning behind certain choices of conditions may not always be apparent. Sometimes processing equipment and parameters have been chosen for specific reasons, such as pilot-scale studies of fractionation or the investigation of the effects of agitation rate. Sometimes parameters appear to have been chosen out of convenience, such as the availability and cost of small-scale equipment compared with pilot-scale fractionation equipment. The use of a 24 h crystallization period appears to ensure that the crystallization process is completed, and it provides a convenient work schedule for the investigator.

The Tirtiaux fractionation process The Tirtiaux fractionation process is a commercially available technology that involves controlled crystallization and separation of the fractions by vacuum filtration (Tirtiaux, undated). The accurate crystallization control of this process allows the choice of crystallization conditions and separation temperatures for the purpose of obtaining milkfat fractions of specific quality (Deffense, 1989, 1993b).

Milkfat is heated, then cooled, in double-walled stainless steel crystallizers equipped with a cooling coil and a fixed- or variable-speed agitator (Figure 3.2). The crystallization processes is controlled so that suitable crystal seeds are formed, and crystal growth rates regulate the heat transfer from the milkfat to the coolant (Deffense, 1987, 1989, 1993b; Tirtiaux, 1976; Tirtiaux, 1983). Crystals of the β' form are preferred, because they stay in suspension in the liquid and are easy to separate using the Tirtiaux Florentine filter (Deffense, 1987, 1989).

The Tirtiaux Florentine continuous vacuum filter (Figure 3.2) is equipped with a stainless steel perforated belt as the filtration support (Deffense, 1987, 1989; Tirtiaux, 1976; Tirtiaux, 1983). The coarse mesh of the belt and the large size of the crystals obtained allow easy filtration with low vacuum, even if viscosity of the milkfat is high. A recycling device for any crystals sucked through the belt ensures filtration on a preformed cake. The filtration surface is air-conditioned to provide filtration at the same temperature as the milkfat. The crystals are scraped off the filtration belt, and the filter is heated to clean the belt as it rotates.

The De Smet Process The De Smet fractionation process is a commercially available technology that was designed for palm oil fractionation and is being used to fractionate milkfat (Aveue, undated; De Smet, undated). The De Smet process uses specially designed, patented coolers-crystallizers combined with program cooling to produce medium sized, easily filterable crystals in approximately five hours (De Smet, undated).

The De Smet process employs a membrane filter press with a polypropylene membrane to separate the solid and liquid fractions (De Smet, undated). Filtration using a membrane filter press is described in the next section.

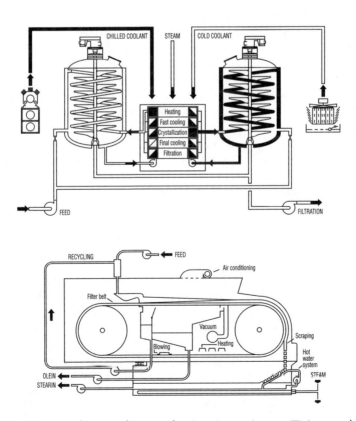

Figure 3.2 Schematic diagram of Tirtiaux fractionation equipment (Tirtiaux, undated).

The Alfa-Laval process The Alfa-Laval process is also a commercial fractionation process that involves crystallization from melted milkfat, but separation of the fractions is accomplished by centrifugation with the aid of an aqueous detergent solution (Sherbon, 1974). The process can be divided into three main steps: crystallization, separation, and washing and drying. Several crystallizers in a row are used to provide a continuous flow of crystallized fat to the separators (Fjaervoll, 1970a).

After crystallization is complete, the solid and liquid milkfat mixture is covered with water containing sodium lauryl sulfate (0.2 to 1%) and an electrolyte, usually sodium sulfate (2 to 5%). This crystallized mass passes through two successive mixers; the first mixer splits large crystals and the second mixer disperses both phases thoroughly. The "light" liquid fraction is isolated, and the aqueous phase–solid fat complex is sent to a plate heat exchanger. The "heavy" solid milkfat phase is collected by heating the mixture to approximately 60°C to melt the crystals, and then the milkfat phase is separated from the aqueous phase using a hermetic separator. The melted milkfat fractions are mixed with warm water, reseparated, and finally vacuum-dried to reduce the water content to 0.1% (Antila, 1979; Fjaervoll, 1970a; Kuzdzal-Savoie and Saada, 1978a).

Jebson (1970, 1974a) studied several processing parameters and found that the size distribution of the crystal aggregates was the most significant variable affecting the overall fractionation efficiency. Fractionation efficiency at optimum size distribution (crystal size of 200 μm) was affected by the concentration of surfactant and to a lesser extent by surfactant flow rate. The speed of the mixer was not important, but the effect of large crystals was reduced by increasing the intensity of mixing of detergent and fat.

Other fractionation processes Esdaile and Wilson (1974) reported that a method of milkfat fractionation named *liquation* produced milkfat fractions with good qualities, but they did not give descriptions or characteristics of these fractions. They performed pilot-scale fractionation by *liquation* experiments (45 kg) using trays with perforated bases. The perforations were temporarily sealed, and liquid butter oil was cooled quickly to 30°C and poured into the trays to a depth of 10 cm. High-melting seed crystals were added, and crystallization was continued by slowly cooling at the base of the trays with a stream of nitrogen. Within a period of 20 to 24 h a matrix of heterogeneous crystals was formed that was sufficiently porous to allow liquid oil to penetrate easily. The perforations in the trays were uncovered, and the top layer of the solid mass was liquefied by passing hot nitrogen gas across the surface. A temperature gradient of up to 13°C was maintained between the upper and lower surfaces. Hot liquid milkfat from the top surface percolated down through the mass and melted crystals of lower melting point, which dripped through the perforations in the bottom of the tray. The softening point of the fractions obtained gradually increased as the operation proceeded.

Separation Techniques

Several techniques have been employed to separate fractions obtained by crystallization from melted milkfat (Table 3.1). Efficient separation of the solid phase from the liquid phase has been somewhat challenging, and considerable research has been focused on the crystallization parameters that result in good separation. Factors affecting separation efficiency include the quantity and size of the milkfat crystals; these are strongly influenced by the chemical composition of the milkfat and crystallization parameters employed (Amer et al., 1985; Beyerlein and Voss, 1973). Well-formed, distinct spherulite crystals are preferred, and the optimal size depends on the separation technique employed. Crystal purity is also of concern, because the formation of mixed crystals yields a solid fraction that has a greater proportion of liquid fraction than distinct spherulites. The solid fraction usually contains a small quantity of liquid milkfat, but it may contain substantial amounts of liquid milkfat because the liquid tends to adsorb to the surface of the spherulites. Liquid milkfat may also be entrapped within solid milkfat crystals. The solid and liquid fractions are very difficult to separate completely without the aid of detergents or solvent.

Vacuum filtration Vacuum filtration is the most commonly employed technique for the separation of fractions produced by crystallization from melted milkfat. Vacuum filtration has been performed using Buchner funnels, Melitta porcelain filters, rotating drum filters, and the Tirtiaux Florentine filter (Table 3.1). Filter supports that have been employed include filter paper, nylon cloth, and stainless steel mesh. The basic vacuum filtration process is as follows:

- Set up filtration equipment and adjust to filtration temperature.
- Turn vacuum on to decrease pressure under the filter bed.
- Transfer the melted milkfat slurry to the filter bed while spreading milkfat evenly and sufficiently thickly to seal the filter.
- Reduce the pressure under the filter, and filter until the liquid is removed and the filter cake appears dry.
- Release vacuum from filter.
- Scrape filter cake from filter support.

The filtration equipment should be maintained at the fractionation temperature to prevent further crystallization of the liquid milkfat or melting of solid milkfat. The methods of liquid-solid milkfat slurry transfer and the amount of slurry to process depend on the filter. For example, a small Buchner funnel is generally filled in small batches to avoid overloading the filter, but the Tirtiaux Florentine filter has a continuous filter bed that allows for continuous loading of the filter.

Black (1974b, 1975b) investigated several factors affecting separation efficiency using vacuum filtration and reported that crystal size and degree of aggregation had a large effect on the filtration rate. Milkfat crystallized without agitation produced crystal clusters that trapped too much liquid milkfat, and this resulted in high viscosity of the milkfat and poor filtration. Black (1975b) observed that filtration rates were best when programmed cooling was employed, agitator speed was relatively low (10 rpm), and crystal size was 200 to 350 µm.

Pressure filtration Separation of milkfat fractions has been performed by placing liquid-solid milkfat slurries over a polymer, cloth, or wire gauze filter and then applying pressure from above the slurry (Table 3.1). Pressure filtration can also be performed with a membrane filter press, which is a plate and frame assembly with nylon or polypropylene filter membranes. The milkfat slurries are pumped between membranes until the chambers are full. Air is then pumped into the spaces between the membranes and the plates, causing the membranes to squeeze the crystal slurry toward the center of the plate chamber, forming a filter cake. As the system is pressurized, the liquid fraction is expelled from the filter cake, collected in channels in the plates, and pumped to a holding vessel. The pressure is then reduced, the plates are opened, and the filter cake is gravity-discharged into melting trays at the base of the filter. The filter cake is melted and pumped to a storage facility. Membrane filter presses are available commercially from De Smet (De Smet, undated) and Tirtiaux (Tirtiaux, undated) in Belgium, and from Hoesch (Rajah, 1991) in Germany, and from Duriron Co. Inc., Filtration System Div. (Angola, NY) in the U.S. Kuwabara et al. (1991) have obtained a U.S. patent for a pressure filtration process for nonlauric fats, including milkfat.

Pressure filtration may allow for a more efficient separation of the liquid phase from the solid phase, and this results in a more pure crystalline fraction with a higher melting point than results from vacuum filtration (Deffense, 1993b). A disadvantage associated with pressure filtration is the potential for the temperature of the filter cake to rise upon pressurizing, which may result in the melting of some crystals (Schaap and van Beresteyn, 1970). Deffense (1989) noted that the membrane filter press may not be suitable for delicate crystals and for separation at low temperatures, because of the increased viscosity encountered in milkfat slurries at such temperatures.

Centrifugation Milkfat fractions have also been separated using centrifugation (Table 3.1). In a batch process, the liquid fraction is decanted after centrifugation, and then the solid fraction remaining in the tube is melted and decanted. Alternatively, a filter or basket-type centrifuge tube can be used, in which the solid fraction remains on the top and the liquid drains through the perforated filter into the bottom of the centrifuge tube. Branger (1993) reported that efficient separation of solid and liquid milkfat fractions was achieved in pilot scale studies using a Westfalia separator with a desludging feature.

Black (1974b) investigated some of the factors affecting separation by a filter centrifuge and reported that the most important variables were centrifuge speed and feed rate (optimal = 500 to 700 kg/h). Filter screen type and size were less important. If the centrifuge speed was too high, crystals were sheared, causing contamination of the liquid phase with solid crystals. Crystals produced during crystallization with or without agitation were satisfactory for centrifugal separation (Black, 1974b). Crystals of 100 to 150 μm are preferred for centrifugation methods, compared with 200 to 350 μm for vacuum filtration (Antila, 1984; Black, 1975b).

Separation with the aid of an aqueous detergent solution The most commonly used method for fraction separation with the aid of an aqueous detergent solution is the Alfa-Laval process, which was described earlier. Other detergent solutions (Table 3.1) have been employed, but the general processes are similar to the Alfa-Laval process. Separation with the aid of an aqueous detergent solution yields solid fractions of much greater crystalline purity than those from other methods associated with fractionation from melted milkfat, because the solid milkfat crystals disperse into the aqueous detergent solution phase while the liquid milkfat remains in a separate lipid phase. The solid fat–aqueous phase complex is heated to melt the milkfat crystals, which effects a separation of the milkfat and aqueous detergent phases. The fractions are washed to remove residual detergent solution, and then are dried.

Jebson (1974a) investigated some of the factors affecting the separation efficiency and found, as is the case in most separation techniques, that crystal size was one of the most important factors. Jebson suggested that a crystal size of approximately 200 μm was optimal for separation in the Alfa-Laval process, and this was consistent with centrifugation methods. Banks et al. (1985) evaluated the separation of milkfat fractions with the aid of a proprietary aqueous detergent solution (Byprox), and found that a 10-fold increase in the detergent concentration yielded three layers (a semisolid fraction, an aqueous layer, and a crystalline solid) instead of the two layers normally obtained during detergent separation.

The use of processing aids in the separation of milkfat fractions is often viewed as a disadvantage, despite the improvement in crystal purity. Jordan (1986) noted that detergent fractionation required a final deodorization step, which may result in loss of flavor, but did not state why this step was required.

Other separation techniques Separations of milkfat fractions by filtration through muslin cloth, a milk filter, a cheese press, and a casein-dewatering press have been reported (Table 3.1). Although these methods were less common, they appeared to provide satisfactory separation and were influenced by processing parameters similar to other separation techniques. Black (1973) investigated some of the processing parameters for separation using a casein-dewatering press and reported that controlled cooling was essential for good separation.

Fractionation of Milkfat by Crystallization from Solvent Solution

Crystallization of milkfat from solvent solutions involves the same basic principles as crystallization from melted milkfat. Most of the factors affecting crystallization from melted milkfat (i.e., chemical composition, fractionation temperature) are also important for crystallization from solvent solution. However, the presence of the solvent has a significant effect on the crystallization behavior of milkfat, largely resulting from the lower viscosity of the milkfat–solvent solution compared with melted milkfat.

Crystal formation of milkfat dissolved in solvent occurs without the interference of liquid milkfat, because the liquid milkfat remains dissolved in the solvent. This enables the high-melting triglycerides to form relatively pure nuclei, and growth of crystals proceeds by the layering of progressively lower-melting triglycerides on the nuclei without entrapping liquid milkfat. The formation of mixed crystals is not of as great a concern for fractionation from solvent solution, because of the low interaction between the liquid and solid milkfat phases. Larsen and Samuelsson (1979) observed that crystals formed from melted milkfat yield spherulite crystals, which have a porous structure that allows liquid milkfat to become entrapped and adsorbed onto the crystal, whereas milkfat crystals formed from polar solvents yield rod-shaped crystals, which form relatively compact aggregates and thus accumulate smaller amounts of liquid milkfat. They also reported that milkfat crystals formed from nonpolar solvents (e.g., hexane, benzene) may not yield well-defined crystals but instead may form a gel-like structure. The time required for crystallization of milkfat from solvent solution is generally shorter than for fractionation from melted milkfat (Schaap and van Beresteyn, 1970).

One of the primary advantages of fractionation from solvent solution is that the fractions have more distinctive, and sometimes more desirable, properties than fractions obtained from melted milkfat. Thus, the approach is valuable from a research standpoint because it provides target functional fractions, which then can be sought by more food-acceptable fractionation techniques. From a commercial point of view, many concerns regarding the use of solvent fractionation for milkfat exist, including

- Loss of aroma and flavor compounds
- Cost of processing facilities and solvent recovery
- Potential toxicological concerns about solvent residues
- Loss of fat-soluble vitamins

(Amer et al., 1985; Antila, 1979; Jordan, 1986; Kaylegian, 1991). For these reasons, techniques employing crystallization of milkfat from solvent solutions have been employed only in laboratory studies and have not been used in commercial ventures (Kuzdzal-Savoie and Saada, 1978a; Sherbon, 1974).

Fractionation Methods

Equipment employed for fractionation from solvent solution is similar to that used for fractionation from melted milkfat. Fractionation from solvent solution consists of the following steps:

1. Heat milkfat to 60°C and melt completely.
2. Blend milkfat with solvent to form solution.
3. Crystallize milkfat.
4. Separate the solid and liquid fractions.
5. Remove solvent from fractions.

Milkfat is heated to 60°C to melt the crystals fully; then it is blended with solvent, which is generally at or slightly above room temperature to avoid solvent loss through boiling. The solution is cooled to fractionation temperature, usually without agitation. As the solution cools, the milkfat triglycerides crystallize and precipitate, leaving a solution of liquid milkfat dissolved in solvent. Because of the low liquid viscosity, the crystals are easily filtered; then the solvent is removed from fractions using techniques discussed in the next section.

Crystallization parameters employed during fractionation of milkfat from solvent solutions (Table 3.2) have included

- Milkfat:solvent ratio = 1:3 to 1:10
- Agitation rate = 0 rpm
- Crystallization time = 30 min to 24 h
- Fractionation temperatures:
  - Single-step fractionation = 28 to 0°C
  - Multiple-step fractionation = 25 to –70°C.

Because crystal formation in solvent solution occurs virtually without the interference of liquid milkfat, agitation is generally not required for good crystal formation. The cooling rate of the milkfat is also not as important for fractionation from solvent solution as it is for fractionation from melted milkfat. Crystallization time periods employed for fractionation from solvent solution are generally shorter than those employed for fractionation from melted milkfat, because the crystallization process occurs more rapidly. Crystallization of milkfat from solvent solution also requires lower temperatures to effect crystallization than does crystallization from melted milkfat, because of solubility effects. In addition, crystallization from solvent solution allows fractionation at much lower temperatures than are practical for fractionation from melted milkfat, and this can be advantageous.

Separation Techniques

Milkfat fractions obtained from solvent solution have been separated using vacuum filtration and centrifugation (Table 3.2). Separation of the solid phase from the liquid phase for fractionation from solvent solution is much easier than for fractionation from melted milkfat, because the liquid milkfat rapidly drains from the crystals and remains in the solvent solution. Consequently, parameters such as crystal size and quantity are not as crucial for efficient separation as they are with fractionation from melted milkfat.

Once the fractions have been separated, the solvent must be removed. Solvent removal has been accomplished by simple evaporation, vacuum distillation, and steam distillation (Table 3.2). Acetone has also been removed from milkfat fractions by washing several times with saturated sodium chloride solutions and then drying over sodium sulfate (Kaylegian, 1991; Yi, 1993). The most common method selected for solvent removal has been vacuum distillation, often using a rotary evaporator.

Conditions of solvent removal (e.g., amount of vacuum, time, temperature) generally have not been reported, and data regarding the efficiency of solvent removal has also been lacking. However, trace amounts of acetone have been found as normal constituents of milkfat (Sonntag, 1979b), and ethanol is a component of many beverages. Extraction of oilseed with solvents such as hexane is commonplace in the edible oils industry, and solvent residues are removed by steam distillation (Norris, 1982).

Solvents Employed

Milkfat has been fractionated by crystallization from acetone, ethanol, isopropanol, pentane, and hexane solutions (Table 3.2). Acetone has been the most commonly used solvent for milkfat fractionation. Fractionation temperatures have been dependent upon the solvent employed, because of differences in the solubility of milkfat in the solvents. For example, lower temperatures were required to effect crystallization in pentane than were necessary for crystallization in acetone.

Fractionation Based on the Solubility and Volatility of Milkfat Components

These techniques use changes in temperature and pressure to separate fractions based on the inherent chemical solubility and volatility functionalities of milkfat. Milkfat has been fractionated by two techniques in this category, namely supercritical fluid extraction and short-path distillation.

Fractionation of Milkfat by Supercritical Fluid Extraction

Supercritical Fluid Extraction Principles

A *supercritical fluid* is a substance at a temperature and pressure above its *critical value*—a temperature above which a gas cannot be liquefied by the application of pressure alone. A fluid in a supercritical condition exhibits physicochemical properties intermediate between those of liquid and gas, and this usually enhances its ability to act as a solvent. The relatively high density of a supercritical fluid provides good solvent power, while its relatively low viscosity and diffusivity values provide appreciable penetrating power into the solute matrix. Supercritical fluid extraction is similar to both distillation and solvent extraction and may be considered to be a combination of the two (Arul et al., 1987; Bradley, 1989; Rizvi, 1991; Rizvi et al., 1990; Shishikura et al., 1986; Timmen, 1974).

The advantages of supercritical carbon dioxide extraction include

- Pollution-free solvent
- Low-temperature operation
- Low energy cost
- High-purity end products
- Better extraction and separation of hard-to-fractionate systems

(Bhaskar et al., 1993a; Rizvi, 1991; Rizvi et al., 1990). Although operating costs of a supercritical fluid extraction system are relatively low, the initial capital costs are high. However, in a recent economic analysis of supercritical fluid–processing equipment, Singh and Rizvi (1994) have reported that the investments for continuous systems are lower than for batch or semicontinuous systems.

The solvent used for food applications is carbon dioxide, which is inexpensive, nontoxic, nonflammable, and nonpolluting (Arul et al., 1987; Bhaskar et al., 1993a; Rizvi, 1991; Rizvi et al., 1990; Shishikura et al., 1986; Timmen et al., 1984). Kankare et al. (1989) noted that the solubility of milkfat in carbon dioxide is poor, which makes the extraction less effective and more costly. Therefore, they suggested that supercritical carbon dioxide extraction be used to obtain fractions containing maximum amounts of flavor and aroma substances.

Fractionation Methods

As with other fractionation methods, fractionation equipment and processing conditions have varied between investigators, but the basic fractionation process (Figure 3.3) remains constant:

1. Milkfat is placed into the extraction vessel.
2. CO_2 is heated and pressurized to the desired supercritical conditions.
3. Supercritical CO_2 is allowed into the extraction vessel.
4. Supercritical CO_2 is kept in contact with the milkfat for a selected period of time for extraction.
5. Supercritical CO_2, with the solubilized milkfat components, passes through a pressure reduction valve and into a collection vessel.
6. Temperature and pressure are returned to atmospheric conditions, and the CO_2 evaporates, leaving the milkfat components in the collection vessel.

Milkfat fractions have been obtained by supercritical fluid extraction using both batch and continuous processes. Fractionation methods are summarized in Table 3.3, and fractionation conditions that have been employed include

- Fractionation temperatures = 40 to 80°C
- Fractionation pressures = 100 to 400 bar; 2100 to 6000 psig; 10 to 28 MPa
- Extraction time = 12 min to 15 h.

Milkfat fractions have been extracted using single-step and multiple-step procedures. Single-step conditions yield one extract and one residual fraction at each set of temperature and pressure conditions. Batch multiple-step procedures yield two or more extracted fractions and one residual fraction, and can be obtained in two ways:

1. At constant temperature and pressure conditions, but at different time intervals
2. By varying the conditions of fractionation during the extraction

Factors Affecting Fractionation by Supercritical Fluid Extraction

Efficiency of fractionation by supercritical fluid extraction is affected primarily by the temperature and pressures employed. The solubility of lipids is generally proportional to the density of the supercritical fluid at constant temperature, and solubility increases rapidly when the pressure increases (Chen et al., 1992). At low gas densities (low pressures), the low-molecular-weight triglycerides are dissolved into the supercritical fluid phase while the higher-molecular-weight triglycerides remain undissolved in the extraction chamber. As the pressure of the gas increases at a constant temperature, the intermediate and higher-molecular-weight triglycerides become soluble in the supercritical fluid and can be extracted (Arul et al., 1987; Büning-Pfaue et al., 1989; Hamman et al., 1991; Kankare and Antila, 1988a; Kankare et al., 1989). Under high enough pressures and temperatures, all triglycerides are solubilized in the supercritical fluid, and fractionation cannot be achieved.

Fractionation by Short-Path Distillation

Short-path molecular distillation is a process used for some food applications (e.g., stripping of vitamins and volatiles from oils and separation of mono- and diglycerides), but is more commonly employed in the chemical and pharmaceutical industries (Arul et al., 1988a). Molecular distillation was first employed for fractionation of milkfat triglycerides used in detailed chemical and structural analyses of milkfat (McCarthy et al., 1962). Subsequently, this technique has been further evaluated as a means of fractionating milkfat for food ingredient applications (Arul et al., 1988a). Short-path distillation processes require relatively high temperatures (245 to 265°C), and this may result in thermal damage to the milkfat and loss in flavor (Arul et al.,

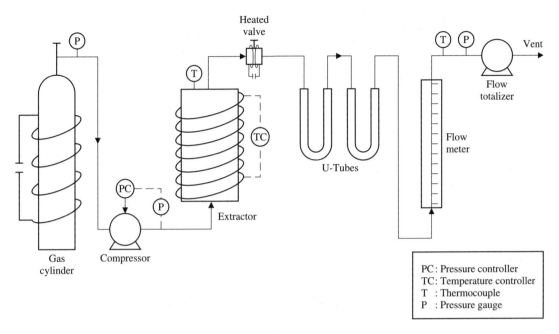

Figure 3.3. Typical schematic diagram of equipment used for milkfat fractionation by supercritical fluid extraction (Arul et al., 1987; reprinted with permission).

1988a). Arul et al. (1988a) noted that thermal damage can be minimized by reducing the time of thermal exposure, which can be achieved in wiped-film evaporators or centrifugal stills.

Fractionation Method

Fractionation of milkfat by short-path distillation has been investigated only on an experimental basis (Table 3.4). Milkfat fractionation has been carried out in a high-vacuum wiped-film pilot-scale evaporator (Figure 3.4). This evaporator is an agitated thin-film evaporator that uses rotating slotted wiper blades, which spread the feed material into a uniform thin film on the wall of the chamber. The slots in the wipers provide a pumping action to move the film down the heated wall with constant agitation, which can be adjusted to yield a residence time of about 1 s. Vaporized material passes through an entrainment separator and condenses on an internal condenser. The noncondensable gases flow out through the vapor outlet (Arul et al., 1988a). Conditions that have been employed during fractionation are

- Fractionation temperatures = 150°C to 265°C
- Fractionation pressures = 220 and 100 mm Hg; 5 to 15 µ
- Residence time = 1 s.

Milkfat has been fractionated by short-path distillation using multiple-step processes.

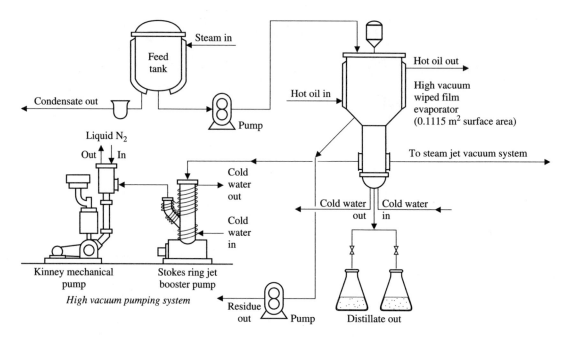

Figure 3.4. Schematic diagram of equipment used for milkfat fractionation by short-path distillation (Arul et al., 1988a; reprinted with permission).

TABLE 3.1 Fractionation Methods Used to Produce Milkfat Fractions by Crystallization from Melted Milkfat

Fraction separation method by author[1]	Fractionation method[2]	Fractionation parameters[3]	Fractionation temperatures (°C)[4]	Separation method
Vacuum filtration				
Amer et al., 1985	MF was heated to 60°C, and 1.5 L was transferred to a Hobart mixer bowl (12 L) equipped with a water jacket. Temperature control of the bowl was accomplished by using a Haake D1 circulating water bath and an EK 12 cooling unit. The bowl was preheated to 60°C. The MF was then added, the cooling unit turned on, and the MF cooled slowly to the fractionation temperature.	AG = 20 rpm (Hobart flat beater) t = 4 h (total cooling + crystallization)	SS = 19, 23, 26, 29	Vacuum filtration (50–100 bar below atm), 14 mesh stainless filter (Tirtiaux, Belgium).
Barna et al., 1992	MF was cool under agitation in a 2 L crystallizer and allowed to equilibrate.	CR = 0.45°C/min AG = 100 rpm t = 24 h	MS = 30, 20	The crystal slurry was separated by vacuum filtration in a Buchner funnel through a Whatman filter.
Black, 1975b	Crystallization was accomplished in a 45 L jacketed stainless steel vat with a variable-speed, two-blade agitator fitted with vertical rods to sweep the walls of the vessel free of crystals. The configuration of the crystallizer was: ID 374 mm, height 500 mm, top-driven, centrally mounted agitator shaft fitted with two pitched impellers, dia. 335 mm, height 45 mm, the top impeller 20 mm below the surface of the liquid and the bottom impeller 20 mm above the bottom of the vat. The vertical rods joining the impellers were fixed at the outside of the impellers and swept within 20 mm of the vat wall.	AG = 10, 20 rpm t = 8, 20 h (cooling)	SS = 25	The partially crystallized MF was filtered on a laboratory test leaf filter of the type used by a manufacturer of commercial vacuum filters. Commercial filter cloths were examined for suitability and a fine nylon cloth was selected.
Deffense, 1987	Tirtiaux process.	—[5]	—	Tirtiaux Florentine filter
Deffense, 1989	Tirtiaux process.	—	MS = unknown	Tirtiaux Florentine filter
Deffense, 1993b	Tirtiaux process.	—	MS = unknown	Tirtiaux Florentine filter
deMan and Finoro, 1980	MF (500 g) was melted at 60°C in beakers (1 L) which were kept in a thermostated bath. The fat was stirred (100 rpm; Fisher Stedi-Speed stirrer).	AG = 100 rpm t = 6 h	SS = 32, 31, 30, 29, 28 27, 26, 25	The fat was filtered in a Buchner funnel.
Frede et al., 1980	MF was melted at 65°C and held for 6 h, then cooled to 28°C and held for 45 h.	t = 45 h	SS = 28	A sieve of 80 µm mesh and 42.5 cm diameter was used to separate the fractions. The sieve was mounted vacuum-tight on a milk can. During filtration there was a maximum pressure decrease of 0.2 bar.

Continued

Milkfat Fractionation Technologies   55

TABLE 3.1 (Continued)

Fraction separation method by author[1]	Fractionation method[2]	Fractionation parameters[3]	Fractionation temperatures (°C)[4]	Separation method
Grall and Hartel, 1992	Approx. 2000 g of MF was placed in a stainless steel jacketed vessel (2.3 L) with a 30 cm height and a 10 cm inner dia. The fat was held at 60°C by a circulating water bath for 0.5 h. The crystallization vessel was equipped with 4 baffles of 8.5 mm width, which extended the length of the tank. The impeller consisted of a 6 mm dia stainless steel shaft with 4 to 5 cm dia, 45° pitched-blade impellers and 2 radial fins (16 cm x 2 cm)	CR = 0.4°C/min AG = 75, 100, 125 rpm t = 20–24 h	MS = 30, 20, 15	Filtration through Buchner funnels housed in a temperature-controlled box. Filter cake: 30 & 20°C = 1 cm, 15°C = 0.5 cm
Guyot, 1982	Milkfat was fractionated by successive crystallizations.	—	MS = 23, 15, 10, 5	Filtration under vacuum with a pilot scale rotating drum filter, with a nylon band.
Hayakawa and deMan, 1982	MF (500 g) was melted at 60°C in beakers (1 L), which were kept in a thermostated bath. The fat was stirred (100 rpm; Fisher Stedi-Speed stirrer).	AG = 100 rpm t = 6 h	SS = 32, 31, 30, 29, 28, 27, 26, 25	The fat was filtered in a Buchner funnel.
Jamotte and Guyot, 1980	Crystallization was carried out in a specially designed vessel (150 L) with a programmed system for chilling, controlled by a Hetofrig unit. An inclined paddle stirrer was used.	CR = 0.005–2°C/min AG = 0 at initial crystallization = 10 rpm after nucleation = 40 rpm at filtration	unknown	Pilot plant rotating drum filter designed by De Smet Engineering (Belgium) and built by Stockdale (Cheshire). A perforated drum which could be regulated between 0.1–3.5 rpm, with a nylon mesh filter 1 m long x 0.12 m wide. The drum rotates in a trough, which receives the mixture to be separated. The interior of the drum consists of 12 sections with 3 vacuum lines and a line for removal of air.
Kaylegian, 1991	MF was heated to 60°C and filtered through Whatman #4 filter paper under vacuum (<50 mm Hg) to remove water and proteinaceous material. MF was heated to 60°C and transferred to a stainless steel tempering beaker (2 L), controlled with a Lauda refrigerated, circulating water bath.	AG = 50 rpm t = 24 h	SS = 34, 30, 25, 20 MS = 34, 30, 25, 20, 16, 13	Vacuum filtration (< 50 mm Hg) in a Buchner funnel through Whatman #1 filter paper in a chamber that was maintained at crystallization temperature.
Kaylegian and Lindsay, 1992	MF was heated to 60°C and transferred to a stainless steel tempering beaker (2 L), controlled with a Lauda refrigerated, circulating water bath.	AG = 50 rpm t = 24 h	SS = 34, 30, 25, 20 MS = 34, 30, 25, 20, 16, 13	Vacuum filtration (< 50 mm Hg) in a Buchner funnel through Whatman #1 filter paper in a chamber that was maintained at crystallization temperature.
Kupranycz et al., 1986	MF was heated to 60°C, and 1.5 L was transferred to a Hobart mixer bowl (12 L) equipped with a water jacket. Temperature control of the bowl was accomplished by using a Haake D1 circulating water bath and an EK 12 cooling unit. The bowl was preheated to 60°C. The MF was then added, the cooling unit turned on, and the MF cooled slowly to the fractionation temperature.	AG = 20 rpm (Hobart flat beater) t = 4 h (total cooling + crystallization)	SS = 29, 19	Vacuum filtration (50–100 bar below atm), 14 mesh stainless filter (Tirtiaux, Belgium).

Reference	Description	Conditions	Separation
Laakso et al., 1992	Tirtiaux process.	—	Tirtiaux Florentine filter
Lund and Danmark, 1987	The fractionation were performed by the butter factory in Rødkjærsbro on a commercial plant.	—	—
Makhlouf et al., 1987	Crystallization of MF (9–10 kg) was done in beakers (4 L) placed in a forced air incubator fitted with a refrigeration unit that was maintained with a temperature controller.	AG = 0 rpm t = 8–15 h	SS = 26, 18
Schaap and van Beresteyn, 1970	MF is heated to 70–80°C; then it is cooled very slowly, with stirring, to the crystallization temperature.	—	MS = 26, 21, 15, 9
Schultz and Timmen, 1966	By cooling molten MF first down to 30 or 25°C and then by steps further down, the least meltable components, in the form of fat granules, can be separated out. After removal of the solids, the remaining oil was cooled down to the next lower temperature, and after sitting for several hours, the newly precipitated fat crystals were removed by filtration.	—	MS = 27, 17, 7, 2
Sherbon, 1963	MF was melted and subjected to progressive crystallization. The temperatures were achieved as follows: 23°C, allowed to sit in a sheltered spot on the laboratory desk; 17 and 10°C, submersion of the sample container in a properly adjusted refrigerated water bath.	—	MS = 30, 25, 20, 15, 10, 7, 4.5
Timms, 1978b	Crystallization was accomplished in a jacketed stainless steel vat (45 L) with a variable-speed, two-blade agitator fitted with vertical rods to sweep the walls of the vessels free of crystals.	—	MS = 23, 17, 10
Timms and Parekh, 1980	Crystallization was accomplished in a jacketed stainless steel vat (45 L) with a variable-speed, two-blade agitator fitted with vertical rods to sweep the walls of the vessels free of crystals.	—	SS = 25

Continued

TABLE 3.1 (Continued)

Fraction separation method by author[1]	Fractionation method[2]	Fractionation parameters[3]	Fractionation temperatures (°C)[4]	Separation method
Versteeg, 1991	Tirtiaux process.	—	SS = 28, 24, 22, 18 MS = 28, 26, 24, 22, 20, 18, 15, 14, 13, 12, 11.5	Tirtiaux Florentine filter.
Voss et al., 1971	MF was melted at 72°C and cooled to crystallization temperature with careful stirring.	—	SS = 30	A sieve of 80 μm mesh and 42.5 cm diameter was used to separate the fractions. The sieve was mounted vacuum-tight on a milk can. During filtration there was a maximum pressure decrease of 0.2 bar.
Vovan and Riel, 1973	MF was fractionated using simple crystallization at the appropriate temperature.	t = 48 h	MS = 21, 10	The crystals were separated by vacuum filtration.
Pressure filtration				
Badings et al., 1983a	MF was melted at 65°C and cooled to the first crystallization temperature in beakers (2 L) placed in a water bath. Stepwise crystallization was used.	AG = 0 rpm t = 32 h	MS = 23.0, 18.5, 15.0, 11.3, 5.5 (stall lot) MS = 21.5, 15.3, 13.0, 9.0, 7.1 (pasture lot)	The solid fraction was separated from the liquid by filtration over a filter cloth with openings of 150 μm under a pressure of 540 Pa. The filtration was stopped as soon as the pressure on the filter cake dropped.
Deffense, 1993b	Tirtiaux process.		MS = unknown	Membrane filter press.
deMan, 1968	Melted MF was cooled slowly or rapidly, left to crystallize, and separated.	CR = slow (leaving a beaker of melted fat in a room at the required temp) CR = rapid (placing the beaker in a water bath maintained at the required temp) t = 24 h	SS = 30, 27, 24	Pressure filtration was carried out in a controlled-temperature room.
El-Ghandour et al., 1976	The fractions were obtained by cooling melted pure MF to a set temperature at a specific rate, followed by separation of the resultant crystallized mass from the noncrystallized liquid fat.	CR = 0.5°C/h	MS = 25, 15	The fat was pressed through a double layer of cloth to obtain the liquid portion of the fat.

Reference	Description	Parameters	Additional
Helal et al., 1977	The fractions were obtained by cooling melted pure MF to a set temperature at a specific rate, followed by separation of the resultant crystallized mass from the noncrystallized liquid fat.	CR = 0.5°C/h	MS = 25, 15	The fat was pressed through a double layer of cloth to obtain the liquid portion of the fat.
Helal et al., 1984	The fractions were obtained by cooling melted pure MF to a set temperature at a specific rate, followed by separation of the resultant crystallized mass from the noncrystallized liquid fat.	CR = 0.5°C/h	MS = 25, 15	—
Kuwabara et al., 1991	MF was fractionated in a vertical crystallization tank equipped with a stirrer. The MF was slowly stirred and cooled with air at fractionation temperature.	CR = 0.3°C/min t = 4 h	SS = 23, 20	The MF cake was crushed, and the resulting crystalline mass or agglomerates were pressed at pressure of 10 kg/cm^2 for 10 min by using a small filter press having a frame thickness of 20 mm.
Riel and Paquet, 1972	A continuous process for fractionation was used. The crystallized fat mixture is thinly spread over a moving wire gauze. This gauze band carries the fat through successive thermic zones provided with a temperature gradient. Corresponding to each zone, a receiver under vacuum is adapted under the moving gauze creating the pressure necessary for filtration.	t = 24 h	MS = 38, 4	Pressure filtration through a wire gauze.
Stepanenko and Tverdokhleb, 1974	The fractions were obtained by cooling melted pure MF to a set temperature at a specific rate, followed by separation of the resultant crystallized mass from the noncrystallized liquid fat.	—	MS = 20, 11	—
Centrifugation			
Antila and Antila, 1970	Method of Antila, 1966.	—	—	—
Baker et al., 1959	MF (8000 g) was melted and was then maintained at a temperature below the melting point of the fat for the desired period of time. When a lower-melting fraction was desired, the liquid fat was again cooled to a lower temperature than was employed in the first crystallization.	t = 18 h (first step, 28°C) t = 5 h (second step, 22°C)	MS = 28, 22	The solid fat was separated from the liquid fraction by centrifugation.
Bhat and Rama Murthy, 1983	MF was melted to 100°C and filtered through Whatman #4 filter paper at 37°C. The MF samples were kept overnight in 50 mL centrifuge tubes.	AG = 0 rpm t = overnight	SS = 28	The MF was centrifuged at 1500 g for 15 min. The liquid fraction was decanted.

Continued

TABLE 3.1 (Continued)

Fraction separation method by author[1]	Fractionation method[2]	Fractionation parameters[3]	Fractionation temperatures (°C)[4]	Separation method
Black, 1974b	—	—	—	Filtering centrifuge
Branger, 1993 (Bench-top)	1 kg of MF was placed in a 2 L beaker. The samples were totally melted in a water bath at 60°C and placed in an Imperial IV convection oven (Lab-line Instruments Inc., Melrose Park, IL), which was previously heated to the crystallization temperature of choice. Samples were allowed to crystallize for 24 h at constant temperature without agitation. After the crystallization period the samples were removed from the oven and were manually agitated with a glass rod to yield a homogeneous distribution of crystals within the oil phase.	AG = 0 rpm t = 24 h	Trial 1:MS = 32, 22 Trial 2:MS = 32, 20 Trial 3:MS = 32, 22, 12	The samples were poured into 200 mL centrifuge bottles, placed in an International Centrifuge, Model K (International Equipment Co., Needham Hts, MA) and centrifuged at 2500 rpm for 5 min. The samples were separated by decanting the liquid fraction from the centrifuge bottles into 2 L beakers. The centrifuge bottles were placed in a water bath at 60°C until the fat melted. The melted solid fractions were decanted, frozen, and stored.
(Pilot-scale)	MF was melted at 55°C in a 380 L tank. Once the sample was melted completely, it was cooled to 32°C by circulating cold water through the jacket of the tank while using high agitation speed (60 rpm). Once the MF reached crystallization temperature, water at the same temperature was circulated through the jacket. MF was allowed to crystallize at the determined temp for 6 or 12 h. The speed of agitation during crystallization was controlled by a connecting the drive motor for the agitator to a voltage regulator. Once the crystallization process was finished, the speed of agitation was increased to 60 rpm to obtain an even distribution of crystals in the oil phase for 10–15 min immediately before centrifugation.	AG = 6 rpm for 2 h, 30 rpm for 4 h t = 6 h	MS = 32, 25, 20, 17	Separation of MF into fractions was accomplished with a Westfalia separator, model NA 7-03-576, with desludging feature and peripheral nozzles. The centrifuge was temperature-conditioned with water for 10–15 min prior to separation. To begin separation, the positive rotary pump was operated at maximum speed to fill the system, and as soon as the AMF reached the minimum to avoid excessive overflow. When product appeared in the outlet window of the centrifuge, the bowl was full. At that moment the pressure in the manometer was increased to a level high enough that a clear, crystal-free oil was discharged, but low enough to avoid excessive overflow. The centrifuge was flushed using the manual valve, not by using the control panel timer. The interval between flushes was determined by the time required for crystals to emerge in the oil outlet after the last flush. In order to flush, the handle of the manual valve was opened until the opening sound of the bowl was heard. At that instant the valve was closed rapidly to procure a minimum opening time. The flushed fraction was collected in a 19 L bucket. The overflow of the centrifuge was collected in another 19 L bucket and recycled.
Bratland, 1983	MF was held at fractionation temperature until partial crystallization occurred.	—	SS = 28, 25, 24, 20	MF was separated in a centrifuge.

Milkfat Fractionation Technologies

Reference	Description	Parameters	Separation
Evans, 1976	MF was crystallized in a vessel (900 L) under nitrogen atmosphere.	—	Separation of the crystals from the liquid was done by centrifugation for 2 min at 25,000 m/s². The liquid fraction was removed with a syringe.
Jordan, 1986	MF was melted in a beaker (1 L) at 60°C. The beaker was stood in a water bath set at crystallization temperature, and the oil was stirred with an electric paddle under a nitrogen atmosphere.	AG = 60 rpm t = 24 h	SS = 25
Kankare and Antila, 1974a	The temperature of MF in a totally liquid state was lowered to the fractionation temperature. The fat was allowed to crystallize and then filtered.	AG = none, or gentle stirring t = 24 h	SS = 28, 24, 15
Kankare and Antila, 1986	The temperature of MF in a totally liquid state was lowered to the fractionation temperature. The fat was allowed to crystallize and then filtered.	AG = none, or gentle stirring t = 24 h	MS = 24, 18, 12
Kankare and Antila, 1988a	MF in a completely melted state was fractionated by crystallization where the fat was cooled slowly with careful stirring.	AG = slow t = 24 h	MS = 24, 18, 12
Kankare and Antila, 1988b	The temperature of MF in a totally liquid state was lowered to the fractionation temperature. The fat was allowed to crystallize and then filtered.	AG = none, or gentle stirring t = 24 h	MS = 24, 12
Keogh, 1989	A soft fraction (SLP = 20°C) was obtained by a single thermal fractionation step (Keogh and Higgins, 1986b).	—	MS = 24, 18, 12
Keogh and Higgins, 1986b	MF was completely melted by heating to 55°C for 0.5 h and then cooling to the initial temperature (IT). Programmed cooling at a given rate was initiated after the final temperature was reached. Cooling of the MF took place in a 400 mL beaker placed in a CB8 Heto bath fitted with an 04PG623 programmed cooler capable of a variable cooling gradient. On reaching the final temperature, the MF was stirred by means of a paddle stirrer, which swept within 6 mm of the beaker wall.	CR = 0.55, 0.71, 1.00 °C/h IT = 31.0, 34.0, 37.0°C agitation t = 0.0, 1.0, 3.0 h	SS = 28.0, 23.5, 19.0
Keogh and Morrissey, 1990	MF fractions (SLP = 35, 38, and 41°C) were obtained by dry fractionation (Keogh and Higgins, 1986b).	—	Approx. 20 g of stirred crystallized fat sample was poured into a 100 mL glass centrifuge tube fitted with a No. 1 sintered filter disk positioned at one-third of the tube height. After centrifuging for 5 min at 2500 g, the hard fraction was removed from the filter disk and the soft fraction poured from the lower part of the centrifuge tube by releasing a rubber stopper from a hole in the base.

Continued

TABLE 3.1 (Continued)

Fraction separation method by author[1]	Fractionation method[2]	Fractionation parameters[3]	Fractionation temperatures (°C)[4]	Separation method
Keshava Prasad and Bhat, 1987a	Butter was heated over flame with continuous stirring to 115°C, and the molten fat was filtered through Whatman #4 filter paper to obtain clear fat. MF samples were kept in 150 mL centrifuge tubes at fractionation temperature overnight and were centrifuged at 1500 g for 30 min.	t = overnight	SS = 28, 23	The low melting fraction obtained was decanted.
Keshava Prasad and Bhat, 1987b	Butter was heated to 115°C and the molten fat was filtered through Whatman #4 filter paper to obtain clear fat. MF samples were kept in 50 mL centrifuge tubes at fractionation temperature overnight and were centrifuged at 1500 g for 30 min.	t = overnight	SS = 28, 23	The low melting fraction obtained was decanted.
Lakshminarayana and Rama Murthy, 1985	Butter was clarified at 100°C and filtered through Whatman #4 filter paper. MF was melted to 70–80°C and then slowly cooled in about 1 h to 31°C and held for 12 h at fractionation temperature in an incubator. The crystals were separated from the liquid portion; the liquid portion was further cooled to the next lowest temperature; and the separation process repeated.	AG = 0 rpm t = 12 h	MS = 31, 23, 15	The crystals were separated from the liquid fraction by decanting.
Lakshminarayana and Rama Murthy, 1986	Butter was clarified at 100°C and filtered through Whatman #4 filter paper. MF was melted to 70–80°C and then slowly cooled in about 1 h to 31°C and held for 12 h at fractionation temperature in an incubator. The crystals were separated from the liquid portion; the liquid portion was further cooled to the next lowest temperature; and the separation process repeated.	AG = 0 rpm t = 12 h	MS = 31, 23, 15	The crystals were separated from the liquid fraction by decanting.
Ramesh and Bindal, 1987b	MF samples were fractionated into solid and liquid fractions by keeping the melted fat at 28°C for 20–24 h followed by centrifugation.	AG = 0 rpm t = 20–24 h	SS = 28	Solid and liquid fractions were separated by centrifugation for 5 min at 1500 g.
Richardson, 1968	MF was crystallized under nitrogen, using 72 h incubation periods. Crystals were packed by centrifuging, and the liquid fat was decanted off and placed at the next incubation temperature.	AG = 0 rpm t = 72 h	MS = 30, 25, 20, 15	Crystals were separated by centrifugation.

Reference	Process	Parameter 1	Parameter 2	Separation
Riel and Paquet, 1972	A continuous process for fractionation was used.	$t = 24$ h	MS = 38, 4	Centrifugation.
Separation with the aid of an aqueous detergent solution				
Banks et al., 1985	Butter (ca. 20 g) was mixed with glass beads in a proprietary detergent solution (50 mL, Byprox). Liquid was drained from the beads, which were washed with aqueous ammonia (50 mL, 25% ammonia) and the washings added to the fat-rich liquid. The dispersion was centrifuged (25,000 g, 15 min).	—	SS = 2, 16	The oil layer was removed by suction, the aqueous layer decanted, and the residual solid material washed with detergent solution and the separation procedure repeated. All of the above operations were carried out at fractionation temperature. The fractions were washed repeatedly with water and ethanol and dried.
Dolby, 1970a	Alfa-Laval process.	—	—	Alfa-Laval process.
Doležálek et al., 1976	MF (130 g) was crystallized under constant stirring in a beaker (250 mL) placed in a water bath at crystallization temperature. The stirrer rod was made of Z-shaped glass rod 5 mm dia., 60 mm height, 30 mm width.	AG = 13.8, 60, 120 rpm	SS = 30, 27, 24	Centrifugation (663 g) for 15 min
Fjaervoll, 1970a	MF entering the crystallizer is in a fluid state. Cooling is programmed to the fractionation temperature.	—	—	Alfa-Laval process: Separation involves the addition of a diluted aqueous surfactant to the fat and conditioning of the fat before separation. The mixture is pumped through two mixers. The first mixer disaggregates large crystals, and the second mixes the phases thoroughly. A special type of separator is used to separate the low-melting fraction, which is further washed and dried. The high-melting phase (crystals and aqueous solution) is heated in a plate heat exchanger to melt the fat. The aqueous phase is recirculated into the process. The high-melting fraction is washed and vacuum-dried.
Fjaervoll, 1970b	MF in fluid form is partly crystallized by programmed cooling, then conditioned by means of an aqueous detergent solution prior to separation.	—	—	From the separator a fluid oil is discharged through the light-phase outlet, and a suspension of fat crystals in a water phase leaves the heavy-phase outlet. The fluid oil is mixed with water, and a further separation takes place. The residual water is removed by passing through a vacuum chamber. Finally the oil is cooled to about crystallizing temperature in a plate heat exchanger.
Jebson, 1970	Alfa-Laval process.	—	SS = 27, 24	Alfa-Laval process.

Continued

TABLE 3.1 (Continued)

Fraction separation method by author[1]	Fractionation method[2]	Fractionation parameters[3]	Fractionation temperatures (°C)[4]	Separation method
Jebson, 1974a	Alfa-Laval pilot fractionation plant.	AG = 1100–2200 rpm surfactant flow rate = 65–100 L/h surfactant conc = 0.2–0.3% (w/v) separator pressure = optimum ± 3 kPa	SS = 25°C	Alfa-Laval process.
Norris et al., 1971	MF was crystallized by holding overnight.	t = overnight	SS = 26, 25	Partly crystallized MF was mixed with water, containing a wetting agent and an electrolyte, at the same temperature as the fat, and passed through a centrifugal separator, which discharged the liquid fat fraction and an aqueous phase containing suspended crystals of solid fat. The aqueous mixture was heated to liquefy the fat and passed through a second separator. The fat fractions were washed and dried.
Schaap et al., 1975	MF was heated to 60°C and held 1 h. The MF was transferred to a beaker (800 mL) and cooled 40°C. It was subsequently cooled under programmed cooling to crystallization temperature.	CR = 0.1°C/min t = 24 h	SS = 28	The crystal mass was separated by filtration through filter cloth (150 μm mesh), by applying pressure at 540 N/m². The filter cake was washed with detergent solution (5% dioctyl sodium succinate in water). A separation into crystals and oil by adding ammonia was achieved. The crystal mass was washed twice with water, dried, and stored over silica gel in a desiccator.
Sherbon et al., 1972	MF was separated using the Alfa-Laval process. Solid fractions were refractionated.	—	SS = 28 (lot A) MS = 25.5, 28, 31.7 (lot B) MS = 28, 30.6 (lot C)	Alfa-Laval process.
Walker, 1972	Alfa-Laval process.	—	SS = 26.5	Alfa-Laval process.

Filtration through muslin cloth

Khalifa and Mansour, 1988	MF was melted at 70°C, transferred to a separating funnel, and kept in an incubator for 24 h. Crystallized fat was obtained from the funnel by straining through muslin cloth. Upper retained portion represents the solid fat at 30°C or higher. The same process was repeated after incubation of the liquid portion at 25°C and 20°C.	AG = 0 rpm t = 24 h	MS = 30, 25, 20	Crystallized fat was obtained from the funnel by straining through muslin cloth. Upper retained portion represents the solid fat at 30°C or higher.

Reference	Procedure	Parameters	Separation
Kulkarni and Rama Murthy, 1985	MF was melted at 80°C and slowly cooled to 23°C and held for 24 h in an incubator.	$t = 24$ h	SS = 23	The solid fraction was separated from the liquid fraction by filtering through a muslin cloth.
Sreebhashyam et al., 1981	MF was heated to 60°C. The MF was held undisturbed at 30°C for 24 h. The fractions were separated and the process was repeated for the filtrate.	AG = 0 rpm $t = 24$ h	MS = 30, 25, 18	The crystallized fat was removed by filtering the mixture through muslin cloth.

Filtration through milk filter

Reference	Procedure	Parameters	Separation
Fouad et al., 1990	MF was fractionated using the method of Amer et al. (1985): MF was heated to 60°C, and 1.5 L was transferred to a Hobart mixer bowl (12 L) equipped with a water jacket. Temperature control of the bowl was accomplished by using a Haake D1 circulating water bath and an EK 12 cooling unit. The bowl was preheated to 60°C. The MF was then added, the cooling unit turned on, and the MF cooled slowly to the fractionation temperature.	AG = 0 rpm $t = 4$ h	SS = 29, 25, 21, 17	The precipitated MF was filtered at the selected temperature using milk filters (Kendall), which allowed complete separation of the crystals from the mother liquor within 5–15 min.

Filtration through cheese press

Reference	Procedure	Parameters	Separation
Baker, 1970a	MF was crystallized for 48 h at 13°C and pressed. The hard fraction was then remelted, recrystallized at 30°C, and again pressed.	$t = 48$ h	MS = 13, 30	Partially crystallized fat was separated using a pilot-scale cheese press. Cheese hoops and cloths were used as the filter medium.

Filtration through casein-dewatering press

Reference	Procedure	Parameters	Separation
Black, 1973	Quick cooling of MF was achieved by pumping it through a plate heat exchanger into the crystallization vessel. A cylindrical jacketed vessel (365 L) was fitted with circular immersion cooling coils.	$t = 21$ or 16 h	SS = 21, 18.5	A pilot-scale casein-dewatering press was adapted by the inclusion of extra seals to reduce contamination of one fraction with the other. Separation was achieved by drainage of liquid through the bottom perforated roller, assisted by pressure from the top roller on the crystals to produce dryness. The crystals were then scraped from the top roller.

Separation technique not reported

Reference	Procedure	Parameters	Separation
Deroanne, 1976	MF was melted at 60°C and slowly cooled to 21°C and then separated.	—	SS = 21	—
Deroanne and Guyot, 1974	Fractions were obtained after heating MF to 70°C with subsequent slow cooling and filtration.	—	SS = 21	—

Continued

TABLE 3.1 (Continued)

Fraction separation method by author[1]	Fractionation method[2]	Fractionation parameters[3]	Fractionation temperatures (°C)[4]	Separation method
Dixon and Black, 1974	MF was split into 3 components using two-stage fractionation.	—	—	—
Dolby, 1970b	Fractions were produced by twice fractionating a quantity of MF.	—	—	—
El-Ghandour and El-Nimr, 1982	Method of El-Nimr & El-Ghandour, 1980.	—	MS = 25, 15	—
Lechat et al., 1975	Direct crystallization, with one to four successive crystallizations.	—	—	—
Martine, 1982	Dry fractionation.	—	—	—
Thomas, 1973a	Commercial MF was fractionated in the pilot plant at the Gilbert Chandler Institute of Dairy Technology, Werribee, Victoria.	—	SS = 20	—
Timmen, 1974	MF fractions were obtained by allowing MF to stand at appropriate temperatures and subsequent filtration.	AG = 0 rpm	—	—
Verhagen and Bodor, 1984	MF was heated to 40°C and subsequently cooled to 25°C.	CR = 3°C/h	SS = 25	—
Verhagen and Warnaar, 1984	MF was heated to 40°C and subsequently brought to 25°C by programmed cooling.	CR = 3°C/h	—	—

[1]Sorted by fraction separation method and listed alphabetically by author.

[2]MF = milkfat; ID = inside diameter.

[3]AG = agitation rate; CR = cooling rate; IT = initial temperature; t = time.

[4]MS = multiple-step fractionation, SS = single-step fractionation.

[5]Data not available.

TABLE 3.2 Fractionation Methods Used to Produce Milkfat Fractions by Crystallization from Solvent Solution

Type of solvent solution[1]	Fractionation method[2]	Fractionation parameters[3]	Fractionation temperatures (°C)[4]	Separation method
Acetone				
Avvakumov, 1974	MF was fractionated by successive crystallization in acetone solution.	—[5]	MS = 20, 10, 0, −10, −20	—
Baker et al., 1959	MF (8000 g) was dissolved in acetone (30 L) at 37°C. The solution was maintained at 5°C for 12 h.	SR = 8000 g: 30 L t = 12 h	SS = 5	The insoluble fat was removed by centrifugation, and the acetone-soluble fraction was recovered by evaporation of the acetone under reduced pressure.
Bhalerao et al., 1959	MF (1 kg) was dissolved in acetone (10 L) at 50°C, and the solution was cooled to 0°C and held overnight.	SR = 1 kg: 10 L t = overnight	SS = 0	The fractions were separated by filtration and both fractions freed of acetone on a steam bath with the aid of a Rinco evaporator under vacuum.
Chen and deMan, 1966	MF was dissolved in acetone and left to crystallize. The crystals were filtered off, and the filtrate was left to crystallize at successively lower temperatures.	SR = 1:10	MS = 15, 5, −15, −25, −35, −45	The crystals were filtered off with a glass filter.
Jensen et al., 1967	MF (1058 g) was dissolved in acetone (10 L), allowed to stand overnight (−25°C) and filtered. The filtrate was reduced to 8 L and placed overnight in a refrigerated bath at −70°C.	SR = 1058 g MF: 10 L acetone (initial ratio) t = overnight	MS = −25, −70	The crystals were removed by filtration and washed with 4 L cold acetone (at crystallization temperature). The solvent was removed.
Kaylegian, 1991	MF was heated to 60°C and filtered through Whatman #4 filter paper under vacuum (<50 mm Hg) to remove water and proteinaceous material. MF was heated to 60°C and dissolved in acetone (21°C). The solution was adjusted to fractionation temperature and held to crystallize.	SR = 1:4 (v/v) AG = 0 rpm t = 24 h	SS = 25, 21, 15, 11, 5, 0 MS = 25, 21, 15, 11, 5, 0	The solid crystals were separated from the liquid fractions by vacuum filtration (< 50 mm Hg) through Whatman #1 filter paper. Acetone was removed by vacuum evaporation with a Buchi Rotovap for 20 min. Fractions were melted, washed with 4–6 volumes of aqueous saturated sodium chloride solution to remove residual acetone, and then dried over excess sodium sulfate.
Kaylegian and Lindsay, 1992	MF was heated to 60°C and dissolved in acetone (21°C). The solution was adjusted to fractionation temperature and held to crystallize.	SR = 1:4 (v/v) AG = 0 rpm t = 24 h	MS = 25, 21, 15, 11, 5, 0	The solid crystals were separated from the liquid fractions by vacuum filtration (< 50 mm Hg) through Whatman #1 filter paper. Acetone was removed by vacuum evaporation with a Buchi Rotovap for 20 min. Fractions were melted and washed with 4 to 6 volumes of aqueous saturated sodium chloride solution to remove residual acetone, and then dried over excess sodium sulfate.

Continued

TABLE 3.2 (Continued)

Type of solvent solution[1]	Fractionation method[2]	Fractionation parameters[3]	Fractionation temperatures (°C)[4]	Separation method
Lambelet, 1983	Acetone partitioning fractions of MF were obtained using a two-step fractionation scheme described by Luddy et al., 1980.	—	—	—
Lansbergen and Kemps, 1984	All experiments were carried out in a solution of butterfat in acetone, of which the ratio of acetone: butterfat (or butterfat fractions) was fixed at 5 (by weight). The fractionations were carried out in a 600 L crystallization vessel consisting of a cylindrical vessel and a cylindrical rotor (dia 0.6 and 0.3 m, respectively, and a height of about 3 m), with a rotor speed of 1.5 rev/sec. The solution of butterfat and acetone was made externally and the crystallization vessel was filled. If necessary, the solution was heated to e.g. 35°C to obtain a clear solution (without crystal nuclei).	SR = 1:5 (w/w) AG = 1.5 rps t = 2 h	MS = 12, 6, 0	The contents of the crystallizer were filtered in a Seitz filter (dia 0.5 m), which was preconditioned at the crystallization temperature; a polyester filter cloth with a mean pore size of 80 μm was used. Batches of 100–150 L were filtered, and the filter cake was washed 3 times with fresh acetone of about −8°C. The wash acetone was added to the first filtrate. Thereafter the acetone was distilled off.
Larsen and Samuelsson, 1979	—	—	SS = 21.4, 15.0, 9.6, 5.0, −5.0, −8.0, −16.4, −21.0	Using 8 times the amount of acetone to butterfat after filtration gave a nearly complete fractionation of the crystalline phase from the liquid fat phase.
Lechat et al., 1975	MF was dissolved in acetone, crystallized, and filtered.	SR = 1:3 (v/v) t = 24 h	MS = −9, 0, 10	Filtration was done on a fritted glass number 3 under reduced pressure. The crystals were washed with acetone.
Mattsson et al., 1969	MF (5 g) was dissolved in acetone (40 mL) and kept at 15°C for 3 h.	SR = 5 g/40 mL t = 3 h	SS = 15	The precipitate was collected on a glass filter and washed with a cold mixture of ethanol and ethyl ether (2:3, v/v). The precipitate was dried under vacuum at room temperature.
Munro et al., 1978	In a pilot-scale plant, a solution of acetone and MF was crystallized initially at 10°C to yield a precipitate and a filtrate, which was crystallized subsequently at −15°C.	—	MS = 10, −15	—

Norris, 1977 (Example 1)	MF was dissolved in acetone in a 500 mL flat bottomed flask. The solution was heated to 26.5°C and placed in a water bath for 2 h without stirring.	SR = 1:4 (w/w) AG = 0 rpm t = 2 h	MS = 22, 12, −14	The slurry was filtered in a jacketed sintered glass funnel. The precipitate was washed on the filter with 8 g acetone at 22°C and the wash liquor combined with the filtrate. The bulk of the solvent was removed from the precipitate in a rotary vacuum evaporator and the last traces removed by stripping with nitrogen gas.
(Example 2)	MF was dissolved in acetone and placed in a jacketed stainless steel vessel. This vessel consists of an upper and lower chamber. The upper chamber is fitted with a "near wall" paddle and the lower chamber contains a filter medium as its floor. Refrigerant can be circulated through the jacket from a controlled temperature bath. The solution was heated to 33°C cooled at 24°C/h to 22°C and held for 70 min before being dropped to the lower chamber where the slurry was filtered. The filtrate was returned to the top chamber, reheated to 30°C, cooled at 17°C/h to 11.5°C and held 1 hr before filtration. The filtrate from the 12°C crystallization was returned to the vessel heated to 12°C, cooled at 16.2°C/h to −15°C and held for 110 min before filtration.	SR = 1:4 (w/w) CR: 1st step: 24°C 2nd step: 17°C/h to 11.5°C 3rd step: 16.2°C/h to −15°C t = 1st step: 70 min 2nd step: 15 min 3rd step: 110 min	MS = 22, 11.5, −15	The crystals were washed on the filter with 0.4 kg of acetone at 18°C. The precipitate was removed from the vessel by heating the jacket to 50°C and dissolving the fat in 4 kg of propan-2-ol at 50°C. The solvent was then removed from the fat. The second filtrate was washed with 2 kg propan-2-ol at −7°C.
Parodi, 1974a	MF (100 g) was dissolved in acetone (900 mL) in a stoppered Erlenmeyer flask (1 L). The solution was held at 20°C for 3 h.	SR = 1:9 t = 3 h	SS = 20	The precipitate was collected on a glass filter and washed with a cold mixture of ethanol and diethyl ether (2:3). The precipitate was removed from the filter by dissolving in chloroform. The solvent was removed and the fraction dried.

Continued

TABLE 3.2 (Continued)

Type of solvent solution[1]	Fractionation method[2]	Fractionation parameters[3]	Fractionation temperatures (°C)[4]	Separation method
Rolland and Riel, 1966 (fractional crystallization method)	A solution (1 L) containing MF (40 g) was allowed to stand in a temperature controlled room. The crystals were filtered and the filtrate was held at the next lowest temperature.	SR = 40 g:1 L (total) t = 24 h	MS = 18, 10, 2, −6, −15	The crystals were vacuum filtered through Whatman #1 filter paper in the cold room at the same temperature.
(centrifugation method)	MF (200 g) was mixed with acetone (40 mL) and allowed to crystallize at 0°C for 24 h. The mixture of liquid fat and fat crystals was poured into the centrifuge at 0°C and the first liquid fraction collected immediately. The temperature of the room was raised rapidly to 8°C and held for 30 min while a second liquid fraction was collected. The same procedure was followed at higher temperatures, until all of the MF was liquefied and collected as fractions.	SR = 200 g:40 mL t = 24 h (1st step) t = 30 min (2nd–6th steps)	MS = 0, 8, 16, 24, 32, 40	The separation of liquid fractions from the unmelted crystals was done in an International Chemical Centrifuge (model CH) equipped with a 5 in. perforated basket lined with 80-mesh Dacron textile. The centrifuge speed was 4000 rpm throughout fractionation.
Schaap et al., 1975	MF was dissolved in acetone in a stoppered Erlenmeyer flask (1 L). The solution was crystallized and filtered.	SR = 1:9 (v/v) t = 3 h	SS = 28	The precipitated crystals were collected on a glass filter and washed with a cold mixture of ethanol and diethyl ether (2:3). The crystal mass was dried in a rotating vacuum evaporator and stored over silica gel in a desiccator.
Sherbon, 1963	Initially MF was dissolved in 5 vol acetone. No attempt was made to keep this ratio constant throughout fractionation.	SR = 1:5 (v/v) t = 24 h	MS = 23, 3, −20	The precipitates formed were removed by vacuum filtration, the Buchner funnel was incubated with the fat to reduce dissolving of the precipitate during filtration. Each fraction was placed in a laboratory evaporator, and the remaining traces of acetone were removed at 54°C and a pressure of 28 mm Hg below atm.
Sherbon and Dolby, 1973	MF was fractionated by progressive crystallization in acetone solution.	—	MS = 32, 25	Vacuum filtration removed the crystallized fat from the filtrate.
Timms, 1980a	MF was fractionated by crystallization from acetone solution.	—	—	—
Timms, 1980b	MF was dissolved in acetone (5 mL/g of fat) at 40–45°C, cooled to 20°C, and held 2 h, and the crystals were filtered off. After the first separation, the acetone was evaporated from the wash and mother liquors. The fat was redissolved in acetone (5 mL/g fat) at 40–45°C and cooled to 3°C and held overnight.	CR = 0.5–1°C/min SR = 1 g: 5mL t = 2 h (1st step) t = overnight (2nd step)	MS = 20, 3	The crystals were filtered off, washed 3 times with acetone (1 mL/g fat for each wash) and the acetone removed in a rotary evaporator.

Reference	Process	Parameters	Notes
Walker et al., 1977	MF (400 kg) was dissolved in acetone and crystallized.	—	MS = 14, 2, −13	The solid fraction was filtered from the liquid fraction. The bulk of the acetone was removed from each fraction in an evaporator and recovered in a surface condenser. Residual acetone was removed from fat fractions by steam stripping in a vacreator.
Walker et al., 1978	MF was crystallized from acetone solution.	—	MS = 14, 2, −13	—
Woodrow and deMan, 1968	The high melting fraction was obtained by crystallization of a 10% solution in acetone.	SR: 1:10	SS = 15	—
Yi, 1993	MF was melted to 60°C and vacuum filtered through Whatman #4 filter paper to remove residual water and protein. MF was then melted to 60°C and dissolved in acetone, and the solution placed in a 25°C CH/P Forma Scientific water bath (Forma Scientific, Inc., Marietta, OH) for 24 h, without agitation to induce formation of the crystalline fractions.	SR = 1:4; AG = 0 rpm; t = 24 h	MS = 25, 20, 15, 10, 5, 0	The crystalline fraction was obtained by vacuum filtration through Whatman #1 filter paper. To remove the acetone, the MF fractions were vacuum evaporated with a Rinco Rotovap (Valley Electromagnetics Corp., Spring Valley, IL) at 60°C. Residual acetone was removed with 4–6 volumes of aqueous sodium chloride solution, and residual water was removed with excess anhydrous sodium sulfate.
Youssef et al., 1977	MF dissolved in acetone (5 cc/g fat) was held overnight (18°C) and the fractions were separated. The filtrate was refractionated at 0°C overnight and the fractions were separated. The solid fraction in acetone (3 cc/g fat) was held at 18°C overnight and separated.	AG = 0 rpm; t = overnight	MS = 18, 0, 18	The separated fractions were transferred to suitable flasks, the solvent was removed completely by evaporation.

Ethanol

Reference	Process	Parameters	Notes
Bhalerao et al., 1959	MF (1 kg) was dissolved in absolute ethyl alcohol (10 L) at 70°C and the solution was cooled to 20°C and held overnight.	AG = 0 rpm; t = overnight	SS = 20	The solid fraction was removed by filtration and both fractions freed of the solvent on a steam bath under vacuum.
Rolland and Riel, 1966	A solution (1 L) containing MF (40 g) was allowed to stand in a temperature-controlled room. The crystals were filtered, and the filtrate was held at the next lowest temperature.	SR = 40 g:1 L(total); t = 24 h	MS = 18, 10, 2, −6, −15	The crystals were vacuum-filtered through Whatman #1 filter paper in the cold room at the same temperature.

Continued

TABLE 3.2 (Continued)

Type of solvent solution[1]	Fractionation method[2]	Fractionation parameters[3]	Fractionation temperatures (°C)[4]	Separation method
Hexane				
Muuse and van der Kamp, 1985	Dissolve 25.0 g of melted MF in 50 mL of hexane in a conical flask. Put the conical flask in a water bath at 12.5°C ± 0.2°C and stir with a magnetic stir apparatus till the first crystals are visible. Stop stirring and allow to crystallize for 30 min. (Note: also stop stirring when no crystals are formed within 5 min). Filter off the crystals and collect the two 25 mL-fractions and crystallize the fat again under the same conditions.	SR = 25 g fat: 50 mL hexane	SS = 12.5	Filter off the crystals with a Buchner funnel, under slight vacuum. Quantitatively transfer the crystals from the conical flask to the filter by rinsing the flask with the obtained filtrate. Put the isolated mass of crystals with the filter paper in 25 mL of warm hexane in the conical flask. Take the filter paper out of the solution and wash it with 25 mL of warm hexane. Collect the two 25 mL fractions and crystallize the fat again under the same conditions. Filter off as described and finally wash the crystals on the filter paper with 50 mL of cold hexane (4°C).
Isopropanol				
Norris, 1977	MF was dissolved in anhydrous propan-2-ol and placed in a jacketed stainless steel vessel. This vessel consists of an upper and lower chamber. The upper chamber is fitted with a "near wall" paddle, and the lower chamber contains a filter medium as its floor. Refrigerant can be circulated through the jacket from a controlled temperature bath. The solution was heated to 36°C cooled at 6°C/h to 20°C and held for 1 h before being dropped to the lower chamber, where the slurry was filtered. The filtrate was returned to the top chamber, reheated to 33°C, cooled at 20°C/h to 14°C and held 15 min, cooled at 1°C/h to 11°C, and then cooled at 6°C/h to −5°C and held for 1 h before filtration.	CR: 1st step: 6°C/h to 20°C 2nd step: 20°C/h to 14°C 1°C/h to 11°C 6°C/h to −5°C	MS = 20, −5	The crystals were washed on the filter with 1 kg of propan-2-ol at 18°C. The precipitate was removed from the vessel by heating the jacket to 50°C and dissolving the fat in 4 kg of propan-2-ol at 50°C. The solvent was then removed from the fat. The second filtrate was washed with 2 kg propan-2-ol at −7°C.

Pentane				
Kaylegian, 1991	MF was heated to 60°C and filtered through Whatman #4 filter paper under vacuum (<50 mm Hg) to remove water and proteinaceous material. MF was heated to 60°C and dissolved in pentane (21°C). The solution was adjusted to fractionation temperature and held to crystallize.	SR = 1:4 (v/v) AG = 0 rpm t = 24 h	SS = 11, 5, 0 MS = 11, 5, 0	The solid crystals were separated from the liquid fractions by vacuum filtration (< 50 mm Hg) through Whatman #1 filter paper. Pentane was removed by vacuum evaporation with a Buchi Rotovap for 20 min.

[1] Sorted by type of solvent solution and listed alphabetically by author.

[2] MF = milkfat.

[3] AG = agitation rate, CR = cooling rate, SR = solvent ratio (milkfat:solvent), t = time.

[4] MS = multiple-step fractionation, SS = single-step fractionation.

[5] Data not available.

TABLE 3.3 Fractionation Methods Used to Produce Milkfat Fractions by Supercritical Fluid Extraction

		Fractionation conditions	
Type of Solvent[1]	Fractionation and separation methods[2]	Temperature (°C)	Pressure
Carbon dioxide			
Arul et al., 1987	The system consists of a double-ended diaphragm compressor (Superpressure, Inc.), a temperature-controlled extraction vessel (2 cm ID x 30 cm long standard high-pressure tubing), a heated micrometering valve (flow-regulating and pressure-reducing), two glass U-tubes (Kimax, Inc.) in series, and in-line flowmeter (Fischer-Porter, Inc.) and a flow totalizer (Singer, Inc.). The inlet and outlet of the extractor were packed with glass wool, and 6.05 g MF was placed into the reactor vessel. The extractor was sealed, and CO_2 (commercial grade) was supplied at about 87 bar and 40°C to the suction side of the compressor and was compressed to the required pressure. The high-pressure gas passing downstream of the compressor was heated in a tube preheater to the required temperature prior to entering the temperature-controlled extraction vessel. The SC-CO_2, containing dissolved solute, was passed through the heated micrometering valve and expanded to ambient pressure. The dissolved solute precipitated in the first U-tube in the system. The second U-tube was employed to ensure trapping of all solute. Both tubes were packed with glass wool at the outlet. A gas flow rate of about 10 SLM was maintained.	50 70	100–250 bar 250–350
Bhaskar et al., 1993a	The basic system consists of a packed column, feed and solvent metering controls and a sampling vessel. The packed column is 1.8 m long and 4.9 cm ID with 6 inlet/outlet ports. The column was packed with SS 204 Goodloe knitted mesh packing (surface area: 19.2 cm^2/cm^3; void fraction: 0.95). The continuous system can be operated in either cocurrent or countercurrent mode. A Milton Roy Type B reciprocating pump (maximum flow rate: 8.0 L/h) is used for feeding AMF. The feed can be introduced from the top (Z = 1.22 m) or center (Z = 0.61 m) of the column. The packed column and sampling vessels are equipped with drain valves at the bottom and heating jackets for temperature control. A positive-displacement, reciprocating pump (max flow rate 113L/h) was used to compress the gas to the desired operating pressure. Pressure in the system was maintained with the help of back pressure regulators. The AMF and CO_2 flow rates were monitored by rotameters. The packed column was operated as a stripping column with SC-CO_2 as the continuous phase and AMF as the dispersed phase. The flow of the column was countercurrent with AMF entering at the center and SC-CO_2 entering at the bottom. SC-CO_2 with the dissolved fat then was passed into the separation vessel through a pressure reduction valve, where the soluble triglycerides were precipitated and collected.	40	24.1 MPa

Continued

TABLE 3.3 (Continued)

Type of Solvent[1]	Fractionation and separation methods[2]	Fractionation conditions	
		Temperature (°C)	Pressure
Bhaskar et al., 1993b	A continuous/batch, pilot-scale supercritical fluid extraction system was used. This unit could be adapted for liquid-liquid as well as solid-liquid extraction. The system consisted of a packed column and four separation vessels. The column was packed with SS 304 Goodloe knitted mesh packing and was 1.8 m long and 4.9 cm ID with six inlet/outlet ports. The continuous system could be operated in either cocurrent or countercurrent mode. A Milton Roy Type B reciprocating pump (max flow rate: 8.0 L/hr) was used for feeding AMF. The feed could be introduced from the top or from the center of the column. The packed column and the separators were equipeed with sampling valves and heateing jackets. A positive- displacement reciprocating pump (max flow rate:113 L/hr) was used to compress the gas to the desired operating pressure. System pressure was maintained by back pressure regulators. The AMF and CO_2 flow were monitored by rotameters. Commercial–grade butter was converted into AMF by melting at 60°C, decanting the top layer, and filtering through Whatman No. 1 filter paper.	40–75	24.1–3.4 MPa
Biernoth and Merk, 1985	MF was extracted with supercritical CO_2 with an apparatus that consisted of a 4 L extraction vessel, a 2 L separator vessel, compressor, 2 heat exchangers. Extraction time = 15 h, amount of CO_2 in circuit = 5 kg, flow rate = 6 kg/h. Separation conditions were: pressure = 30 bar, temperature = 30°C.	80	200 bar
Bradley, 1989	The sample is held in the extractor at a preset temperature while the supercritical fluid adjusted to the correct pressure with a diaphragm pump is allowed to pass through.	80	2300, 2500, 2800, 3050, 3350, 3400 3750, 6000 psig
	Fractions are collected in the U-tubes, which are at atmospheric temperature and pressure. The rotameter monitors gaseous flow per unit of time while the test meter monitors total flow.	60	2150, 2300, 2400, 2500, 2650, 2850 2900 psig
Büning-Pfaue et al., 1989	A continuous counter current setup was used. The CO_2 was raised to extraction pressure by the circulation pump (P1). The temperature is adjusted in the heat exchanger (W1). In column K the CO_2 becomes charged with the fat fraction, which is different from that of the original material. The charged CO_2 is depressurized in the throttle valve (DV), changed to gas in the evaporator (W2) such that it loses its solvent capacity, and then the extracted fat fraction is drawn off in precipitator A. At least two fat fractions can be obtained with this process. Further fractions can be obtained with additional stages of throttling and precipitation.	50 30	25,000 kPa (injection t = 12 min, contact t = 15 min) 20,000 kPa (injection t = 47 min, contact t = 30 min)

Continued

TABLE 3.3 (Continued)

		Fractionation conditions	
Type of Solvent[1]	Fractionation and separation methods[2]	Temperature (°C)	Pressure
Chen et al., 1992	Fractionation was carried out in a continuous supercritical fluid extraction system (Superpressure Division of Newport Scientific, Inc.). Commercial-grade liquid CO_2 (5.3 MPa; 25°C) was fed into the double end, diaphragm-type variable speed compressor. The SC-CO_2 was pumped through the temperature-controlled extraction vessel (6.5 cm ID and 25.4 cm long), containing the MF sample, at a rate of 12–17 L/min (1 atm, 25°C). The extractor inlet and outlet were packed with glass wool. The internal temperature of the extractor was monitored by an inserted thermocouple. The depressurized stream of SC-CO_2 discharged the extracted lipids in a U-shaped tube connected at the outlet of the extraction valve. The other end of the U-shaped tube was packed with glass wool to retain condensed particles. The accurate volume of CO_2 passed in each measurement was recorded by a flow totalizer and corrected for standard temperature and pressure.	40	10.3, 13.8, 17.2, 20.7, 24.1, 27.6 MPa
Ensign, 1989	The sample is held in the extractor at a preset temperature while the supercritical fluid adjusted to the correct pressure with a diaphragm pump is allowed to pass through.	60	2150, 2300, 2400, 2500, 2650, 2850 2900 psig
	Fractions are collected in the U-tubes, which are at atmospheric temperature and pressure. The rotameter monitors gaseous flow per unit of time, while the test meter monitors total flow.	80	2300, 2500, 2800, 3050, 3350, 3400 3750, 6000 psig
Hamman et al., 1991	Experiments carried out using a homemade laboratory pilot plant described in Hamman and Sivik (1991): The gas was cooled to get it in a liquid state, and a HPLC pump was used to increase the pressure. A 7 μm filter was placed before the pump to avoid contamination. The pressure was regulated with a relief valve, and a return loop took care of the overpressure. The same type of valve was used as a safety relief valve. The extraction vessel, a cylinder with a volume of 7.5 mL, was placed in a water bath, where the temperature was controlled by a thermostat. In both ends of the cylinder there were sintered steel plates. The separation of the extract from the CO_2 was achieved by expansion to atmospheric pressure on the micrometering valve. The extract was collected at room temperature in a plastic tube. The CO_2 passed through a filter to remove the entrained oils and then through a HI-TEC thermal mass flow meter. The amount of solvent consumed could be registered. The pressure in the system was measured with a pressure transducer and registered on a digital pressure indicator.	40	125, 350 bar
Kankare and Antila, 1988a	A cyclically operated supercritical extraction apparatus was used. The main elements of the extraction device are the extraction container, the pressure-reduction valve, the separator, and the pump.	50–80	100–400 bar

Continued

TABLE 3.3 (Continued)

Type of Solvent[1]	Fractionation and separation methods[2]	Fractionation conditions	
		Temperature (°C)	Pressure
Kankare et al., 1989	The extraction experiments were carried out with the Swiss Nova Werke AG pilot plant at the Chemical Laboratory of the Technical Research Centre of Finland.	50 48	100, 200, 300, 400 bar 120 bar (3 h 12 min) + 150 bar (4 h), 200 bar
Kaufmann et al., 1982	MF was dissolved in SC-CO_2 in a one-step apparatus, piped out of the extraction vessel, and separated, now subcritical, in a precipitator tank. The CO_2 was then directed back to the extractor after pressure and temperature were raised.	80	200 bar (15 h)
Rizvi et al., 1990	MF is charged into the extraction vessel and oxygen is purged from the system by passing CO_2 and venting the gases. The process is repeated 3 or 4 times and then SC-CO_2 is admitted again. As the extraction vessel and preheater are heated to the desired temperature, the pressure is increased to the desired pressure. The extraction is begun by opening the pressure reduction valve and permitting SC-CO_2 to flow through the MF in the extraction vessel. The components of MF that are soluble in the SC-CO_2 pass through the pressure reduction valve and into the separation vessel. Pressure reduction causes the soluble glycerides to precipitate and collect in the separation vessel. The volume of CO_2 used is determined and vented. Alternatively, the CO_2 can be reused after recompression.	40 48	241.3 bar 275.8 bar
Rizvi et al., 1993	A continuous pilot-scale supercritical fluid system for extraction and fractionation of AMF was used as described (Bhaskar et al., 1993b). Both AMF and SC-CO_2 were continuously fed countercurrently into a packed column at 24.1 MPA and 40°C and separated into extract and raffinate. The column was packed with SS 304 Goodloe knitted mesh packing. The extract was fractionated sequentially in four separation vessels at different temperatures and pressures. A cold trap, 316-stainless steel sampling vessel submerged in an ice bath, collected the remaining solutes prior to release of CO_2 to the atmosphere.	40 60	24.1, 6.9, 3.4, MPa 17.2, 13.8 MPa
Shishikura et al., 1986	CO_2 (commercial grade) was cooled and filtered, then compressed by a compressor to the desired pressure, which was regulated by a variable back pressure regulator and checked by a monitor. The compressed gas was passed through a coil to adjust the temperature and allowed to flow through the vertically mounted extractor from the bottom. The extractor was partially immersed in a thermally regulated bath. The extractor is a stainless steel cylinder (4 cm ID and 20 cm height). The bottom, which contains a gas inlet, rupture disk and gas filter, is threaded into the cylinder. The	40	300 kg/cm^2

Continued

TABLE 3.3 (Continued)

Type of Solvent[1]	Fractionation and separation methods[2]	Fractionation conditions	
		Temperature (°C)	Pressure
	contents of the extractor were constantly agitated by a propeller to mix the CO_2 well with the sample. In the case of the silicic acid absorption method, the oil-laden gas was passed through a column of Wakogel C-100 (Wako Pure Chem. Inc.) set up in the extractor. Silicic acid was activated at 120°C for 4 h, and both sides of the silicic gel column were fixed with cotton. In this method, the extractor was not agitated by a propeller. The oil-laden gas from the extractor was passed through a metering valve, where the CO_2 was depressurized, and the separated oil was collected in a receiver whose temperature was controlled at 30°C. The volume of gas flowing in the extractor was determined with a dry test meter. The CO_2 flow rate was regulated at 7–10 NL/min, and the extractions were run on 19–20 g MF.		
Shukla et al., 1994	MF was fractionated using a continuous pilot-scale SC-CO_2 system using the process developed by Bhaskar et al., 1993b.	40	24.1 MPa
Timmen et al., 1984	MF was dissolved in SC-CO_2 in a one-step apparatus, piped out of the extraction vessel and separated, now subcritical, in a precipitator tank. The CO_2 was then directed back to the extractor after pressure and temperature were raised.	80	200 bar (15 h)
Propane			
Biernoth and Merk, 1985	MF was extracted with supercritical propane with an apparatus that consisted of a 4L extraction vessel, a 2 L separator vessel, compressor, 2 heat exchangers. Extraction time = 6 h, flow rate = 2 kg/h. Separation conditions were: pressure = 15 bar, temperature = 90°C.	125	85 bar

[1]Sorted by type of solvent and listed alphabetically by author.

[2]AMF = anhydrous milkfat, ID = inside diameter, MF = milkfat, SC-CO_2 = supercritical carbon dioxide.

TABLE 3.4 Fractionation Methods Used to Produce Milkfat Fractions by Short-Path Distillation

Author	Fractionation and separation methods[1]	Fractionation conditions	
		Temperature (°C)	Pressure
Arul et al., 1988a	Short-path distillation was carried out in a high-vacuum wiped-film pilot-scale evaporator (The Pfaudler Co.). The evaporator is an agitated thin-film evaporator, which uses rotating slotted wiper blades that spread the feed material into a uniform thin film. The slots in the wipers provide a pumping action to move the film down the heated wall with constant agitation. The rapid and positive action of the wiper blades in moving the residue down and off the heated wall controls the residence time to about 1 s. Vaporized material passes through the entrainment separator and condenses on the internal U-bundle condenser, and noncondensables flow out through the vapor outlet. The pilot unit had 0.1115 m² evaporation surface and 0.15 m² of condensing surface and was fitted with a Chevron–type entrainment separator between the evaporator and condenser, which were 3.5 cm apart. MF was melted under nitrogen blanket and then vacuum-distilled (93°C and 700 μm Hg) to dehydrate and degas the fat. The MF was distilled in 3 steps. The desired pressures were reached by varying the feed rates.	245 265	220, 100 mm Hg 100 mm Hg
McCarthy et al., 1962 (initial distillation)	The molecular distillates of butter oil were obtained through the courtesy of Distillation Products Industries (Rochester, NY). MF (777 pounds) were initially distilled into the most volatile 10% fraction, the next most volatile 40% cut, and the 50% residue.	146–148 177–182	11–12μ 5–7μ
(second distillation)	The most volatile 10% fraction was redistilled into four approximately equal fractions.	150 160 180 185	15μ 15μ 11–12μ 11–12μ

[1]MF = milkfat.

Chapter 4
Methods for the Characterization of Milkfat Fractions

Milkfat fractions are characterized to identify them, determine their quality, and describe their functionality. Fractions can be identified by chemical and physical characteristics, such as fatty acid composition and solid fat content. Parameters used to indicate the quality of milkfat fractions include peroxide value and free fatty acid content; sensory analysis also can be used to evaluate the quality of a fraction. Nutritional quality fats has become increasingly important, particularly the degree of saturation and cholesterol content.

The link between some parameters and functionality is well known, such as the use of solid fat content in predicting the spreadability of butter at refrigerator temperature. However, the relationships between many parameters and functionalities are not clearly defined, such as the effect of triglyceride composition on the crystallization behavior of milkfat. An improved understanding of the relationships between milkfat characteristics that can be easily quantified and the expressed functional attributes of milkfat would have great benefit for predicting the functionality of milkfat fractions in food products.

Most characteristics of milkfat fractions can be measured by several different methods. The choice of methods to characterize milkfat fractions depends on the availability of equipment and technology. For example, triglyceride analyses were not common before the 1980s because of inadequate technology, but now gas chromatography (GC) and high-performance liquid chromatography (HPLC) have become relatively commonplace in research and commercial laboratories. The availability of new technologies has also changed the standard methods for some analyses. For instance, the availability of nuclear magnetic resonance (NMR) technology has led to its preference for the determination of solid fat content in lipids, whereas previously, dilatometry was the primary method for solid fat index. Improvements in technology also have allowed more precise determination of some characteristics, such as the exact solid fat content instead of an approximate index of solid fat content.

In some analyses the same basic methodology may be applied (e.g., gas chromatography), but the sample preparation, equipment, and experimental conditions employed can be quite different. For other analyses of milkfat fractions, such as melting point and iodine values, standard methods have evolved, and they are published by organizations such as the American Oil Chemists' Society (AOCS) and the Association of Official Analytical Chemists (AOAC).

Methods employed for characterization of milkfat fractions are presented in Tables 4.1 to 4.7. These methods also have been used to characterize blends of milkfat (Chapter 5) and finished milkfat ingredients (Chapter 8). General descriptions of these methods are presented in the following section, and typical values for intact milkfat were presented in corresponding sections in Chapter 2.

Methods Used to Characterize Chemical Properties of Milkfat Fractions

Fatty Acid Composition

Historically, measurements such as iodine value have been used to indicate general trends in the saturated and unsaturated fatty acid content of fats. Advances in gas chromatography (GC) technology have eliminated the need for these more tedious analytical methods for the determination of fatty acid composition (Sonntag, 1982). Chromatographic methods are continually being refined for convenience and precision (Apps and Willemse, 1991; Sonntag, 1982).

The fatty acid composition of milkfat fractions has commonly been determined by conversion of fatty acids in triglycerides to methyl or butyl esters, followed by separation of these esters with gas or gas–liquid chromatography (Table 4.1). Milkfat contains a relatively high amount of low-molecular weight fatty acids that are prone to loss during analysis (Sonntag, 1979b). Methyl esters of short chain fatty acids are particularily vulnerable to losses during analysis because of their volatility. Care should be taken to prevent losses of fatty acids during analysis, and butyl esters usually provide a more reliable means of analysis than methyl esters (Iverson and Sheppard, 1986). High-performance liquid chromatography (HPLC) has also been used to determine the fatty acid composition of milkfat fractions (Chen et al., 1992; Laakso et al., 1992).

Milkfat has the most complex fatty acid composition of all edible fats, making the determination of its fatty acid composition more difficult than for plant-derived and other animal fats. However, about 90% of milkfat is composed of approximately 10 major fatty acids that are relatively easy to analyze. The most abundant fatty acids are the n-chain 4:0, 6:0, 8:0, 10:0, 12:0, 14:0, 16:0, 18:0, 18:1, and 18:2 acids. More abundant minor fatty acids include the 14:1, 15:0, 16:1, 17:0, and 18:3 acids and the *cis* and *trans* isomers of 18:1.

Triglyceride Composition

Older methods for triglyceride separations were generally accomplished by triglyceride class separation based on unsaturation (trisaturated, monounsaturated, etc.) using column or thin-layer chromatography (Chen and deMan, 1966). The use of gas–liquid chromatography was also being evaluated in the early 1960s (Kuksis and McCarthy, 1962; McCarthy et al., 1962). The use of column and thin-layer chromatography for the broad separation of triglyceride classes has since been coupled with GC or HPLC analyses, yielding two-step separation processes that result in further definition of the triglyceride composition (Breckenridge and Kuksis, 1986b; Kuksis et al., 1973; Lund, 1988; Shehata et al., 1971, 1972). Considerable advances in the determination of triglyceride composition of milkfat have been made in recent years, especially in the improvement of columns and on-column injectors (Lund, 1988).

The use of newer and combined chromatography techniques has allowed a more precise view of the triglyceride composition of intact milkfat and milkfat fractions. Some of these techniques include

- HPLC (Frede and Thiele, 1987)
- HPLC coupled with GC (Bornaz et al., 1992)
- Silver ion HPLC, reversed-phase chromatography, and GC (Laakso et al., 1992; Laakso and Kallio, 1993a and 1993b)
- Reversed-phase liquid chromatography and GC (Gresti et al., 1993; Maniongui et al., 1991)
- Reverse-phase HPLC and mass spectometry (Kuksis et al., 1991; Marai et al., 1983)
- GC and GLC coupled with mass spectrometry (Myher et al., 1988 and 1993; Kalo and Kemppinen, 1993)

However, an understanding of the complex triglyceride composition of milkfat is just beginning to emerge from detailed studies. Bornaz et al. (1992) and Gresti et al. (1993) have reported extensive experimental data on the triglyceride composition and structure of milkfat compared with theoretical values calculated using a random distribution hypothesis. High temperature capillary gas chromatography is used extensively to separate triglycerides according to the sum of the number of carbon atoms in the component fatty acids (McCarthy et al., 1962; Kuksis and McCarthy, 1962; Bornaz et al., 1992; Gresti et al., 1993). Since the component fatty acids of a triglyceride along with positional siting of these fatty acids determine the chemical and physical properties of that triglyceride, considerable analytical attention has been given to positional analysis of fatty acids in milkfat (Barbaro and Sherbon, 1975; Itabashi et al., 1993; Kuksis et al., 1973; Parodi, 1979, 1982).

Gas chromatography and high-performance liquid chromatography methods employed for the determination of triglyceride composition of milkfat fractions are summarized in Table 4.1.

Other Selected Chemical and Physical Properties

The measurement of selected chemical and physical properties of milkfat and other edible fats and oils has been standard practice in industry (Newlander and Atherton, 1964; Sonntag, 1982). Many of these methods have been used to determine properties that relate to fatty acid composition, but their power of discrimination is limited, and many of them are being replaced by the determination of fatty acid composition using chromatographic methods. The methods used to characterize selected properties of milkfat fractions are presented in Table 4.2, and a brief summary of some of these methods follows.

Iodine Value

Iodine value is a measurement of the unsaturated fatty acid content of a lipid and is expressed in terms of the amount of iodine (%) absorbed by double bonds per gram of sample. Iodine values increase with increasing unsaturation of the milkfat sample. Iodine value can be determined using the Wijs method, Method Cd 1-25 (AOCS, 1973) or Method 920.159 (AOAC, 1990), or the Hanus method, Method 920.158 (AOAC, 1990). The Wijs method is the preferred AOCS method and uses iodine monochloride as the reagent, whereas the Hanus method is the preferred AOAC method and uses iodine monobromide as the iodine reagent (Sonntag, 1982).

Reichert–Meissl Value and Polenske Value

The Reichert–Meissl value is a measure of water-soluble volatile fatty acids: butyric (C4) and caproic (C6). The Polenske value is a measure of the water-insoluble volatile fatty acids: caprylic (C8), capric (C10), and lauric (C12). The Reichert-Meissl value increases at the C4 and C6 fatty acid content increases, likewise the Polenske value increases as the C8, C10, and C12 content increases. These properties are measured in terms of the amount of sodium hydroxide required to neutralize the volatile fatty acids in a fat sample after recovery by distillation (AOCS, 1973, Method Cd 5-40; AOAC, 1990, Method 925.41).

Milkfat has a high percentage of volatile fatty acids and exhibits a distinguishably high range of Reichert–Meissl numbers (27 to 31) compared with other animal and vegetable oils (Newlander and Atherton, 1964).

Refractive Index

The refractive index is practically defined as the ratio of the speed of light in air to the speed of light in the fat, and is characteristic for oils within certain limits (AOCS, 1973, Method Cc 7-25; AOAC, 1990, Method 921.08). Refractive index increases with increasing unsaturation of the milkfat sample and can be affected by heat and oxidation (AOCS, 1973). Refractive index is usually determined with an Abbé refractometer or a Zeiss Butyrorefractometer.

Specific Gravity

The specific gravity is a general characteristic of fats and oils, but not highly definitive for various lipids (Sonntag, 1982). The specific gravity is the ratio of the weight of a unit volume of the fat at 25°C to the weight of an equivalent volume of water at 25°C (AOCS, 1973, Method Cc 10a-25). The AOAC method (920.212) for specific gravity employs a pycnometer.

Saponification Number

The saponification number is defined as the amount of alkali required to saponify a fat sample (AOCS, 1973, Method Cd 3-25; AOAC, 1990, Method 920.160). The fat species that are saponified during analysis include triglycerides, diglycerides, monoglycerides, free fatty acids, and other reactive esterlike components such as lactones (Sonntag, 1982). The saponification value is a measure of the average molecular weight of fatty materials (Sonntag, 1982). Fats containing short-chain fatty acids exhibit higher values than those composed entirely of long-chain fatty acids.

Flavor Profile

Milkfat flavor involves complex interactions between many flavor compounds and precursors, combined with the effects of processing conditions, storage, and oxidation (Chapter 2). Although a qualitative evaluation of milkfat flavor readily can be measured using a sensory panel, the identification and quantification of the specific compounds contributing to milkfat flavor has proven difficult. Methods used to describe the flavor profile of milkfat fractions are presented in Table 4.2.

Some compounds that contribute to milkfat flavor (i.e., lactones, methyl ketones, and aldehydes) have been quantified in intact milkfat and milkfat fractions, but the minor contributors to milkfat flavor have not been routinely analyzed in milkfat fractions. Lactone and methyl ketone concentrations have been expressed in terms of potential concentration and total (free plus potential) concentration (Kankare et al., 1989; Walker, 1972; Walker et al., 1977). Potential and total lactone concentrations have been measured using gas or gas–liquid chromatography (Kankare et al., 1989; Walker, 1972; Walker et al., 1977). Potential and total methyl ketone concentrations and aldehyde concentrations have been determined by conversion to 2,4-dinitrophenylhydrazones, followed by separation with thin-layer chromatography and quantified spectrophotometrically (Bhat and Rama Murthy, 1983; Walker, 1972).

Sensory panels have been employed more frequently for the evaluation of flavor in finished products containing milkfat fractions (e.g., cold-spreadable butter; see Chapter 8) than for the evaluation of individual milkfat fractions. The flavor of milkfat fractions has been assessed qualitatively for attributes such as creaminess, butyric acid intensity, rancidity, diacetyl intensity, and overall pleasantness (Jordan, 1986; Khalifa and Mansour, 1988; Sreebhashyam et al., 1981).

Methods Used to Characterize Physical Properties of Milkfat Fractions

Crystal Characteristics

The crystal characteristics of milkfat influence many aspects of fractionation, including the efficient separation of the solid and liquid fractions (Chapter 3) and the production of milkfat ingredients with selected functionalities (Chapter 8). Crystal characteristics can be described by the morphology and size of the milkfat crystals.

Crystal Morphology

Crystal morphology of milkfat fractions is usually determined with X-ray diffraction or microscopic methods (Table 4.3). Crystal forms exhibit distinctive X-ray diffraction patterns and are easily identified. Microscopic methods have also been employed because the crystal forms have characteristic appearances; β' crystals are generally spherulite-shaped, and β crystals appear as individual needle-shaped crystals. Microscopic methods have the advantage of convenience and availability, whereas X-ray diffraction equipment is often unavailable. Scanning electron microscopy has emerged as a newer technique for the determination of microstructure in fat products, and this technique can provide information valuable to the understanding of the relationship between processing conditions and fat functionality (Heertje et al., 1987).

Crystal Size

The sizes of crystals in milkfat fractions have been determined via calibrated fields in traditional light, polarized light, and scanning electron microscopy (Table 4.3).

Melting Behavior

Milkfat melts over a fairly wide temperature range, and many measurements can be used to describe the melting behavior of milkfat fractions. The three most commonly used parameters for melting behavior are melting point, thermal profile, and solid fat content.

Melting Point

The melting point of a sample is defined as a temperature at which the milkfat sample has completely melted. Several methods are available, but these methods do not necessarily provide measures of the same degree of melting. Because of the range of melting points of milkfat triglycerides and the empirical nature of melting point determinations, the experimental conditions for melting point methods are well-defined, and the reported value is specific to those conditions (Sonntag, 1982).

Melting point Generally, the most accepted melting point of a fat is defined by the clear point of the fat, which can be determined using a capillary tube method (AOCS, 1973, Method Cc 1-25; AOAC, 1990, Method 920.157). This method involves filling capillary tubes with fat and allowing them to solidify for a specified period of time. The tubes are then placed next to a thermometer in an agitated water bath and heated. The melting point is recorded as the temperature at which the milkfat has become visually clear.

The Wiley method can also be used to measure the point at which the fat is melted. This method (AOCS, 1973, Method Cc 2-38; AOAC, 1990, Method 920.156), measures the temperature at which a cylindrical fat disc becomes spherical.

Softening point and dropping point The softening point is the temperature at which fat softens or becomes sufficiently fluid to slip or run (AOCS, 1973, Method Cc 3-25). To conduct this test, milkfat is solidified in capillary tubes and then heated according to the specified procedures. The softening point is recorded as the average temperature at which the fat column in the capillary tubes begins to rise.

Another method for determining the softening point of milkfat is the Barnicoat method. This method involves dropping a standard ball bearing (3 mm dia.) through fat placed in a test tube that has been melted and held under specified conditions, and then measuring the temperature at which the ball bearing has sunk halfway through the sample (Black, 1974a; Timms, 1978b). Other softening point methods have been developed based on modifications of the Barnicoat method (Black, 1974a; Dolby, 1961).

A modification of softening point determination is the dropping point determination, which is made using a Mettler automatic dropping point apparatus (Timms, 1978b). This method involves controlled melting of the fat sample until the fat melts and molten fat drops through a small hole in the apparatus. The apparatus contains a light beam which is directed to a photocell. As the melted fat drops and interrupts the light beam, the temperature is automatically recorded.

Slipping point The slipping point of milkfat (AOCS, 1973, Method Cc 4-25) determines the softening point of the sample as it currently exists, and is not necessarily a constant of the fat itself (AOCS, 1973). The slipping point is measured in a manner similar to the AOCS softening point, except that the fat is forced through a brass cylinder instead of first being melted and filled into capillary tubes. The slipping point is determined as the average temperature at which the fat column rises in the cylinders.

Because the fat sample is taken as is, the slipping point can be strongly influenced by the previous thermal history of the sample. Consequently, most melting determinations include sample tempering methods in order to ensure that any influence of previous crystal history is removed and that the true melting behavior of the fat is characterized.

Cloud point The cloud point of a fat is not truly a melting point determination, which measures the temperatures at which crystals are destroyed, but instead, it measures the temperature at which crystals are formed. The cloud point is the temperature at which a cloud is induced in the fat sample, caused by the first stage of crystallization (AOCS, 1973, Method Cc 6-25). The melting point determines the point at which the milkfat goes from a solid (cloudy) state to a liquid (clear) state. However, the cloud point differs from the melting point in that supercooling is required to induce crystal formation. Therefore, for a given fat, the temperature at which the fat passes from clear to cloudy (cloud point) is lower than the temperature at which it passes from cloudy to clear (melting point).

Thermal Profile

The thermal profile provides a generalized view of the melting and crystallization behavior of milkfat over its melting range (–40 to +40°C). Thermal profiles of milkfat fractions have been measured with differential scanning calorimetry (Table 4.4). Differential scanning calorimetry measures the amount of heat absorbed by the sample as it is heated or cooled compared to a reference sample, which is usually an empty pan. Milkfat is generally analyzed using heating and cooling rates of 5 to 10°C/min. Melting profiles of milkfat are more commonly reported than cooling profiles.

Milkfat exhibits three major melting peaks, representing the high-melting, middle-melting, and low-melting glycerides. The conditions employed during analysis (e.g., heating and cooling rates) can affect the resultant profiles. For example, faster heating rates shift peaks to higher temperatures, and peaks become more broad (Cebula and Smith, 1991).

Solid Fat Content

Solid fat content profiles are more informative measurements of the melting behavior of milkfat over its melting range, and they provide the actual percent of solid fat in the milkfat at a given temperature. Solid fat contents of milkfat can be measured using dilatometry, differential scanning calorimetry (DSC), or nuclear magnetic resonance (NMR) techniques, but NMR is by far the most desirable technique currently.

Several studies have been performed to determine the differences in data generated by these three methods. Walker and Bosin (1971) evaluated all three methods on tallow and hydrogenated soybean oil, but they tempered the samples only for the dilatometric and DSC methods. They reported that all three methods were suitable for solid fat determination, but the results from the methods were not directly comparable. Some of the differences were attributed to the untempered samples used for the NMR determinations.

More recently, Lambelet (1983) has compared NMR and DSC methods for the determination of solid fat content on milkfat samples and has reported differences between the SFC values determined by the two methods. In that study, DSC values were always greater than those determined by NMR. Correcting the DSC data to account for differences in melting enthalpies of the different triglyceride species in the milkfat sample lowered the DSC values, but the results were still not directly comparable with the NMR values. Lambelet (1983) related the differences between the values reported by DSC and NMR to the chemical composition of the sample: the more low-melting triglycerides in a milkfat sample, the greater the differences were between values determined by the two methods.

Dilatometry Dilatometric determination of solid fat index (SFI) is an empirical determination of the solid fat content in a fat sample instead of the actual percentage of solid fat content. Dilatometry relies on the volumetric changes caused by thermal expansion of the sample, and SFI is calculated from the specific volumes of a milkfat sample at various temperatures (AOCS, 1973, Method Cd 10-57). The use of correction factors to compensate for thermal expansion is necessary for the calculation of solid fat index (AOCS, 1973). Dilatometry is the oldest method available for the determination of SFI and has been widely used as the standard method in the edible oils industry. However, it is a laborious and time-consuming method compared to the more recently developed techniques (DSC and NMR).

Differential scanning calorimetry Differential scanning calorimetry is most commonly used for the determination of thermal profiles of milkfat, but it also has been employed to determine the solid fat content of milkfat fractions (Amer et al., 1985; Fouad et al., 1990; Kankare and Antila, 1988b; Norris et al., 1971; Sherbon et al., 1972). The principles of DSC were previously described in the section on thermal profile. The solid fat contents of milkfat samples at a given temperature have been determined by the ratio of the area under the curve at that temperature to the area under the entire melting curve (Lambelet, 1983). An advantage to using DSC is that the melting curve obtained for the thermal profile of milkfat also can be used to determine the solid fat content, thereby obtaining two measurements of melting behavior from one analysis.

Nuclear magnetic resonance The solid fat contents of milkfat samples can be determined by nuclear magnetic resonance, using an NMR spectrometer that is designed for measuring solid fat contents. Milkfat samples are placed in a constant magnetic field and a pulsed radio frequency is applied. Protons in milkfat absorb specific radio frequency energy, and following individual pulses, radio frequency energy is emitted. The emitted energies from liquid and solid milkfat components of the sample exhibit different decay rates, which permits them to be differentiated and measured (Bruker, 1993). The direct method for solid fat content measures both the solid and liquid content in a sample (Bruker, 1993). The indirect method measures only the signal from the liquid content of the sample, which is then compared to the signal of a fully melted reference sample (Bruker, 1993; AOCS, 1973, Method Cd 16-81). Tempering procedures are often employed for fats prior to analysis to obtain a true measurement of solid fat content that has not been influenced by the temperature history of the sample (Bruker, 1993; AOCS, 1973, Method Cd 16-81). Specific NMR instrumentation for SFC has become readily available in recent years and has rapidly become the preferred method for solid fat content determinations because of its speed, precision, and ease of analysis.

Textural Characteristics

The textural characteristics of milkfat are influenced by both the solid and liquid phases and therefore can be described by textural attributes commonly associated with solid foods (e.g., firmness) and by rheological attributes commonly associated with liquid foods (e.g., viscosity).

There is a wide variety of methods available to determine the textural properties of foods (Bourne, 1982). Empirical tests, such as cone penetrometry, are commonly used to describe the texture of foods and have been shown to be related to textural quality, but they are not based on well-defined parameters (Bourne, 1982). Rheological properties, such as elasticity and viscosity, describe the more fundamental stress and strain properties of foods under defined conditions. However, fundamental rheological determinations are based on materials science and often do not represent what is sensed in the mouth (Bourne, 1982). Sensory perception of food textures in the mouth at different stages of chewing is the standard by which all objective measurements are compared (Bourne, 1982).

The measurement of textural characteristics has been more commonly employed for finished products that incorporate milkfat fractions than for individual milkfat fractions. The textural characteristics of individual milkfat fractions can be quite similar to the finished product, as in the case of cold-spreadable butter, or they can be different, as in the case of chocolate and pastry applications.

The texture of milkfat fractions is often described by empirical terms such as *hardness*, *firmness*, or *softness*. Analytical methods used to evaluate the textural characteristics of milkfat fractions and similar products (i.e., experimental butters and spreads) containing milkfat fractions are described in the following paragraphs. Cone penetrometry has been the most common method employed for the evaluation of individual milkfat fractions (Table 4.5). Milkfat fractions and experimental butters have also been evaluated by sensory panels to describe textural attributes such as spreadability, firmness, and plasticity (Deffense, 1987; Kaylegian and Lindsay, 1992; Pedersen, 1988).

The textural characterization of other finished products containing milkfat fractions includes the measurement of hardness of chocolate chip cookies (Peck, 1990), milk chocolate (Bunting, 1991), and dark chocolate (Yi, 1993). These determinations were made with an Instron Universal Testing Machine.

Cone and Disc Penetrometry

Cone penetrometry has been used to measure spreadability (Deffense, 1987; Jamotte and Guyot, 1980; Makhlouf et al., 1987), hardness (deMan, 1968), softness (Sreebhashyam et al., 1981), and the yield value (Keogh and Morrissey, 1990) of milkfats. Cone penetrometry is an empirical measure of the firmness of fats, provided by the distance a given weight of defined shape will penetrate the fat in a certain period of time (AOCS, 1973, Method Cc 16-60). Although the AOCS method prescribes the use of plasticized fats, the methods used in the analysis of milkfat fractions have not stated whether the fats were plasticized prior to analysis (Table 4.5).

Disc penetrometry has been used to measure the firmness of cold-spreadable butter (Dixon and Black, 1974). Disc penetrometry is similar to cone penetrometry, except that the instrument penetrating the fat is of a different shape.

Constant Speed Penetrometry

Constant speed penetrometry has been employed to measure the hardness (Hayakawa and deMan, 1982) of milkfat fractions. Constant speed penetrometry involves measurement of the force required to push a cylindrical punch moving at constant speed through a milkfat sample.

Sectility

These methods are often referred to as *wire cutting* measurements and have been used for the evaluation of cutting resistance (Frede et al., 1980; Kankare and Antila, 1974b) and hardness (Norris, 1977) of experimental butters and spreads. These methods involve drawing a wire through a milkfat sample, and values have been reported in a variety of units, including arbitrary units (Dolby, 1970b), resistance in grams (Kankare and Antila, 1974b), and Newtons (Frede et al., 1980).

Apparent Viscosity

The yield value and apparent (Bingham) viscosity of experimental cold-spreadable butters have been measured, but the specific methods employed were not reported (Makhlouf et al., 1987).

Methods Used to Characterize Other Important Properties of Milkfat Fractions

Deterioration of Milkfat Fractions

The quality of milkfat fractions is related to its oxidative and lipolytic stability. Quality indicators such as peroxide value and free fatty acid level are commonly used as standard specifications for fat ingredients. A number of different methods have been employed to characterize the quality of milkfat fractions (Table 4.6).

Oxidative Stability

Oxidative stability of milkfat can be measured by predictive tests, such as the active oxygen method, and by methods that quantify the compounds that are formed during oxidation, such as the hydroperoxide content determined by peroxide value.

Active oxygen method The active oxygen method (AOCS, Method Cd 12-57, 1973) measures the amount of time required for a specified peroxide value to develop under continuous aeration (Sonntag, 1982). However, employment of the active oxygen method for the evaluation of milkfat fractions has not been reported.

Peroxide value (PV) The most common method employed to evaluate the oxidative stability of milkfat fractions has been peroxide value (Table 4.6). The peroxide value measures the amount of hydroperoxides present, which are formed at the initial stages of oxidation. There are several different methods available for the determination of peroxide value (Table 4.6), including standard methods published by AOCS and AOAC. The standard methods for peroxide value determine the amount of all substances, generally assumed to be peroxides, that oxidize potassium iodide under the specified test conditions (AOCS, 1973, Method Cc 8-53; AOAC, 1990, Method 965.33). These methods are highly empirical, and any variation in the procedure may result in variations in the results (AOCS, 1973).

Thiobarbituric acid (TBA) value The thiobarbituric acid value also has been used to measure oxidation stability of milkfat fractions (Table 4.6), and it quantifies the amount of malonaldehyde present, which is formed during the early stages of oxidation (Sonntag, 1982). The malonaldehyde present in the milkfat reacts with thiobarbituric acid to produce a red color, which is measured spectrophotometrically (Sonntag, 1982).

Measurements of aliphatic aldehyde content Aliphatic aldehydes are produced during the course of oxidation and, therefore, serve as indicators of oxidative stability. Other methods involving the quantification of aldehydes as a measure of oxidation of milkfat fractions include anisidine value (Jamotte and Guyot, 1980), determination of conjugated polyunsaturated fatty acids (Keshava Prasad and Bhat, 1987a), and determination of total and monocarbonyls (Keshava Prasad and Bhat, 1987b).

Lipolytic Stability

The lipolytic stability of milkfat is generally determined by the content of free fatty acids in the milkfat and is expressed by acid degree value or free fatty acid value. Both of these methods quantify the free fatty acid content in milkfat, but on slightly different premises (Sonntag, 1982).

Free fatty acid value The free fatty acid value determines the actual quantity of free fatty acids existing in the milkfat (AOCS, 1973, Method Ca 5a-40; AOAC, 1990, Method 940.28). The free fatty acid value is determined by titration with alkali and is expressed as the content equivalent to oleic acid on percentage basis (Sonntag, 1982). Usually the free fatty acid value is approximately one-half of the acid degree value for a fat (Sonntag, 1982).

Acid degree value The acid degree value is a measure of the equivalent weight of the free fatty acids present. The acid degree value is the amount of alkali required to neutralize the free acids in a fat sample (AOCS, 1973, Method Cd 3a-63). Thus, a given weight of free fatty acids containing short-chain fatty acids will have higher acid degree values than samples containing a greater weight of long-chain free fatty acids.

Nutritional Properties

Characterization of specific nutritional properties of milkfat fractions has been limited, but it has included the determination of cholesterol content and some fat-soluble vitamins (Table 4.7).

Cholesterol Content

The determination of the cholesterol content has been most commonly associated with fractions produced by supercritical fluid extraction (Bradley, 1989; Chen et al., 1992; Ensign, 1989; Kankare et al., 1989; Kaufmann et al., 1982; Rizvi et al., 1990; Shishikura et al., 1986; Timmen et al., 1984). This has probably resulted from recent concerns regarding the cholesterol content of fats and human health and from the use of supercritical fluid extraction for cholesterol reduction in milkfat (Chapter 1).

The cholesterol content of individual milkfat fractions has been determined using several methods (Table 4.7). The standard AOAC method for cholesterol determination in animal fats (Method 967.18) involves the removal of free 3-β-OH sterols from the fat by complexing with digitonin, followed by separation of the sterols from the digitonin using gas chromatography. The cholesterol contents of milkfat fractions have also been determined using an enzymatic assay kit (Chen et al., 1992). Cholesterol content also can be determined using direct-injection GC techniques, from which milkfat triglyceride profiles are also obtained (Kaufmann et al., 1982).

Fat-Soluble Vitamin Content

The methods used to determine the vitamin A (Antila and Antila, 1970; Norris et al., 1971), carotene (Norris et al., 1971), and tocopherol (Antila and Antila, 1970) contents of milkfat fractions were referenced to other authors, and details of the procedures were not given (Table 4.7). Sonntag (1982) has described numerous methods for the analyses of vitamins in fats, including bioassay, spectrophotometry, gas chromatography, thin-layer chromatography, and high-performance liquid chromatography.

TABLE 4.1 Methods Used to Determine Fatty Acid and Triglyceride Compositions of Milkfat Fractions

Fractionation method[1]	Fatty acid composition[2]	Triglyceride composition[3]
Crystallization from melted milkfat		
Vacuum filtration		
Amer et al., 1985	*Methyl ester preparation*. MF (200 mg) was placed in a vial (7 mL), and petroleum ether (2 mL) was added to dissolve the sample. Sodium methoxide solution (0.1 mL; 2 N NaOCH$_3$ in anhydrous methanol) was added, and the contents of the vial were mixed (1 min) using a vortex mixer. After sedimentation of the sodium glycerolate, a portion of the clear supernatant was dissolved in hexane (1 part supernatant to 100 parts hexane). *Sample analysis*. An aliquot (0.2 µL) was injected into a fused silica capillary column (30 m × 0.32 mm ID) coated with SP-2340 (Supelco Inc., Canada) (0.2 µm). The column was placed into a Varian 3700 gas chromatograph equipped with a flame ionization detector and a cold on-column injector. The oven temperature was programmed from 50 to 160°C (5°C/min) after a 1 min hold at 50°C. The carrier gas (helium) flow rate was 1.5 mL/min. The injector temperature was programmed from 70°C to 200°C (100°C/min), and the detector temperature was maintained at 210°C. Correction factors were determined by analysis of a standard mixture of fatty acid methyl esters (Nu Chek Prep, Elysian, MN) having a composition that resembled that of an average butterfat sample.	*Sample preparation*. Dissolve 15 mg of MF in 6 mL of hexane. *Sample analysis*. Inject an aliquot (0.2 µL) into a bonded-phase fused silica capillary column (DB-5, 0.1 µm; 15 m × 0.32 mm ID; J & W Scientific Inc., Rancho Cordova, CA). The same GC was used as for FA analysis. The oven temperature was programmed in two stages as follows: first, from 50 to 240°C at a rate of 25°C/min, and then from 240 to 350°C at a rate of 3°C/min. The injector temperature was programmed from 70 to 330°C (100°C/min), and the detector temperature was maintained at 350°C. The carrier gas was helium (1.5 mL/min). Identification of the major groups of triglycerides according to carbon number was made by comparison of retention times to those of standard mixtures of simple triglycerides from C18 to C54 (Nu Chek Prep, Elysian, MN).
Barna et al., 1992	Nr[4]	—[5]
Deffense, 1939	Nr	TG were analyzed by GC on a short capillary column. TG were separated according to their total carbon number. Identification of subgroups were done by comparison with standards.
Deffense, 1987	Nr	TG were also analyzed by HPLC, and separated by Partition Number (PN = CN −2 × number of double bonds).

Continued

TABLE 4.1 (Continued)

Fractionation method[1]	Fatty acid composition[2]	Triglyceride composition[3]
deMan and Finoro, 1980	*Methyl ester preparation.* Methyl ester preparation was performed as specified by Shehata et al. (1970). *Sample analysis.* FA comp was determined with a Hewlett-Packard 402 GC with dual column, 250 cm × 5 mm dia., packed with 15% DEGS on 80/100 mesh Chromosorb W, acid washed. Operating conditions and weight response factors for the correction of peak areas were as specified by Shehata et al. (1970).	*Sample analysis.* For GLC separation of TG a Beckman GC-5 was used. Temperature programming started at 160°C, and a final temperature of 320°C was reached in 16 min. Molar and weight response factors were determined as outlined by Shehata et al. (1971).
Grall and Hartel, 1992	*Butyl ester preparation.* Butyl esters were produced by a modified method of Iverson and Shepard (1986). *Sample analysis.* FA comp was determined by GC analysis. Samples were injected into the GC, and concentration of each ester was determined by peak area analysis.	—
Jamotte and Guyot, 1980	Nr	TG determination by GLC.
Kaylegian, 1991	*Butyl ester preparation.* Weigh 150 mg MF into screw cap test tubes (18 mm ID × 150 mm), add 2 mL of 0.5 N sodium hydroxide in butanol solution. Tightly cap and heat in boiling water 10 min; cool under tap water. Add 2 mL of 14% boron trifluoride butanol reagent (Supelco Inc., Bellefonte, PA), cap and boil 10 min. Cool under tap water, add 10 mL pentane and 15 mL methanol in distilled water (6.39 v/v), shake 2 min, centrifuge for 1 min. The bottom layer was removed and discarded. The washing procedure with pentane and methanol was repeated two more times. *Sample analysis.* An aliquot (6 μL) was injected into a Varian 2700 GC equipped with a glass column (3 m × 2 mm ID) packed with 10% Carbowax 20M on 80/100 Chromosorb WAW (Supelco Inc.). The oven temperature was programmed from 50 to 220°C at 8°C/min. The carrier gas was helium at 24 mL/min. Peak areas were integrated with a Spectra-Physics 4200 integrator (Spectra-Physics, San Jose, CA) and the concentrations of each ester were calculated: peak are of each fatty acid ester/0.9 × total peak area.	*Sample preparation.* 0.02 g of MF was dissolved in 10 mL isooctane. *Sample analysis.* 0.3 μL were injected in a Varian 3410 high-temperature GC equipped with a programmable injector and FID. The carrier and make-up gas were high-purity helium at 2 mL/min and 30 mL/min, respectively. High-purity hydrogen (30 mL/min) and compressed air (300 mL/min) were supplied to the detector. The injector and detector temperature were 350°C, and the oven was programmed from 200 to 340°C at 2°C/min, followed by a 20 min hold. The concentration of each group of triglycerides based on carbon number was calculated based on the total peak area of the chromatogram.

Kupranycz et al., 1986

Methyl ester preparation. The preparation of methyl esters was described by Amer et al. (1985).
Sample analysis. The determination of FA composition by capillary GLC was described by Amer et al. (1985).

Sample preparation. Intact TG were analyzed by GPC. The samples (30 to 50 mg) were dissolved in 1 mL THF. Methyl esters of the unheated and heated fats were prepared by IUPAC Method 2.301 (4.2) for acid oils and fats with minor modifications. The methyl esters were extracted with hexane and the resultant solution dried over anhydrous sodium sulfate. The hexane was evaporated using a stream of nitrogen, and the methyl esters were redissolved in THF (30 to 50 mg methyl esters/1 mL solvent). High-purity dimer and trimer acids were obtained from Emery Industries (Cincinnati, OH); they were converted to methyl esters by the procedure given previously. Standard mixtures of triglycerides, diglycerides, monoglycerides and free fatty acids were purchased from Sigma Chemical Co. (St. Louis, MO).
Sample analysis. Samples were analyzed by gel permeation chromatography: Samples were analyzed by HPLC using an instrument supplied by Waters Assoc. (Milford, MA): a Model 510 pump, a U6K Universal Injector, an R401 Differential Refractometer (attenuation 8 × 8) and a series of 3 columns (7.8 mm ID × 30 cm length; 10^3 Å, 500 Å, and 100 Å Ultrastyragel) operated at room temperature. The column packing material was highly crosslinked styrenedivinylbenzene copolymer (<10 μ). The total permeation vol was 36 mL (or 12 mL/column), and the total void volume was 18 mL. THF stabilized with 250 ppm BHT (Anachemia, Lachine, Quebec) was used as the solvent at a flow rate of 1.0 mL/min. Peak integration was performed using a Spectra-Physics (San Jose, CA) SP-4720 Integrator.

Sample analysis by HPLC. HPLC analyses were carried out with a Shimadzu (Kyoto, Japan) LC-9A pump together with a FCV-9AL low-pressure gradient elution unit and a Cunow (Cergy-Saint-Christophe, France) DDL21 light scattering detector. Chromatograms were recorded with a Shimadzu C-R5A Chromatopac integrator. For preparative purposes a stream splitter was inserted between column and detector.
Sample analysis by AG-HPLC. AG-HPLC was methoxide performed according to Christie (1987, 1988). The separation was achieved

Laakso et al., 1992

Sample preparation for TG analysis. TG (10 mg) were purified prior to GC and HPLC analyses on short Florosil columns by elution with 10 mL hexane–diethyl ether (4:1 v/v). All solvents were HPLC grade. Whole TG fractions (100–400 μg) plus internal standard (75 μg, methyl heneicosanoate, Serva 24576) in 100 μL hexane and 3 μL methyl acetate were methylated by adding 1 μL of 1 M methanolic sodium methoxide (freshly prepared from Fluka 71748 reagent) and vortexing for a few minutes. The turbid preparation was acidified with oxalic acid and

Continued

TABLE 4.1 (Continued)

Fractionation method[1]	Fatty acid composition[2]	Triglyceride composition[3]
	transferred to a 200 mL insert in a 2 mL vial for immediate GC analysis. *Sample analysis by GC.* Samples were analyzed with a fused silica column (0.32 mm ID × 30 m) coated with 0.2 μm SP-2340 phase (Supelco, Bellafonte, PA). A Carlo Erba (Milan, Italy) 5160 GC fitted with a Hewlett-Packard (Palo Alto, CA) 7673A injector was used. The injector and detector temperatures were 250°C, and the split ratio was about 25:1. Helium (99.9999%) was the carrier gas, at 50 kPa (flow rate 1.5 mL/min) until the emergence of methyl butyrate, and thereafter at 80 kPa (2.1 mL/min). The temperature was held at 70°C until the emergence of the methyl butyrate, followed by programming at 30°C/min to 120°C, 10°C/min to 160°C, and 2°C/min to 190°C. The response correction factor for each FAME, used to convert peak area % to weight %, was determined by analyzing a butter oil of known FA comp.	with a cation-exchange column (4.6 mm ID × 250 mm) of Nucleosil 5SA (HPLC Technology, Macclesfield, U.K.) loaded with silver ions. TG (1–2 mg) dissolved in 10 μL of 1,2-dichloroethane were applied to the column and then eluted at ambient temperature with a binary gradient of (A) dichloromethane–1,2-dichloroethane (4:1 v/v) and (B) acetone at a flow rate of 1.0 mL/min. The linear gradient was 100% A to 80% A–20% B in 20 min and then to 100% B in 15 min (held at 100% B for 5 min). The preparative collection was repeated several times to get sufficient material for further analyses. Total TG, as well as their fractions obtained by AG-HPLC, were separated by reverse-phase HPLC. Samples were dissolved in 1,2-dichloroethane, and volumes of 10 μL or less were injected onto the column. Two columns (4.6 mm ID × 250 mm) with ODS phase (5 μm particles, Zorbax, Du Pont, Willmington, DE; Spheri-5, Brownlee Labs, Santa Clara, CA) in series were utilized at ambient temperature with a binary gradient of (A) dichloromethane–1,2-dichloroethane (4:1 v/v) and (B) acetonitrile as the mobile phase at a flow rate of 0.8 mL/min. The linear gradient was 30% A–70% B to 55% A–45% B in 65 min, to 60% A–40% B in 10 min, and then to 65% A–35% B in 5 min. The final solvent composition was held for 2 min.
Lund and Danmark, 1987	Nr	Nr

Makhlouf et al., 1987	*Methyl ester preparation.* Fatty acid methyl esters were prepared according to Luddy et al. (1968). *Sample analysis.* The separation of the methyl esters was done using a Hewlett-Packard HP 5890 GC with an FID, equipped with a capillary column (fused silica) OV 225, and the liquid phase was bound (Durabond) to a 0.25 μm thickness, and dimensions were 0.25 μm ID × 30 m (J & W Scientific, Rancho Cordova, CA). The injection was done by the "split" method (yield 300:1). The temperature program was 60°C for 2 min, 10°C/min until 190°C, and then maintained isothermally for 8 min.	*Sample preparation.* MF was dissolved in *n*-heptane (1:100). *Sample analysis.* A 5 μL sample was injected using a cold on-column injection method on the HP 5890 GC used for FA analyses. Separation of the TG was done with a capillary column (fused silica) SE 54 with the liquid phase bound (Durabond) at a 0.5 μm thickness and the dimensions of 0.25 mm ID × 15 m (J & W Scientific). The hydrogen carrier gas was maintained at a constant pressure of 55 kPa at the head of the column. The temperature program was 90°C for 2 min, 10°C/min until 350°C, and then maintained isothermally for 10 min.
	The correction factors for each FA were determined with the aid of a quantitative mixture of methyl ester standards (Applied Science Laboratories, Inc., State College, PA) possessing a composition similar to MF.	The identification of the TG was done in comparison to retention times for some TG standards from 24 to 54 (Applied Science Laboratories).
Schaap and van Beresteyn, 1970	FA composition was determined by GC analysis.	—
Schultz and Timmen, 1966	FA composition determined by GC analysis.	—
Timms, 1978b	FA analyzed as their methyl esters, as outlined by Timms (1978a).	—
Timms and Parekh, 1980	FA composition by analysis of methyl esters was performed according to Timms (1978a).	—
Versteeg, 1991	Nr	
Vovan and Riel, 1973	*Sample analysis.* FA were determined by GC using a Microtek model MT 220 GC with a double hydrogen flame detector. The column was 2 m with Gas Chrom Q, 80/100 mesh as the adsorbent, with diethylene glycol succinate (5%). Methylation of the FA was done in the presence of potassium methanolate 0.4 N in methanol. The reaction was done under nitrogen atmosphere at 65°C for 2 min. A mixture of silica gel and calcium dichloride (50/50) was used to remove any undesirable traces of the reaction. All the analyses were done using a temperature program of 50 to 170°C.	

Continued

TABLE 4.1 (Continued)

Fractionation method[1]	Fatty acid composition[2]	Triglyceride composition[3]
Pressure filtration		
Badings et al., 1983a	*Methyl ester preparation.* 100 mg of MF were weighed into an 8 mL Sovierel culture tube with cap, and 6 mL heptane were added to dissolve the sample. 0.06 mL of 2 N sodium methoxide solution in methanol was added, the contents of the tube capped and shaken vigorously for 60 s with a Vortex mixer. *Sample analysis.* After sedimentation of the sodium glycerolate, a 1 μL sample of the clear supernatant was injected directly into a glass capillary column (40 m × 0.7 mm ID) coated with diethylene glycol succinate ($d_f \approx 0.9$ μm) and mounted in a Varian 1860 GC equipped with an FID. Oven temperature was programmed from 30 to 180°C at 6°C/min. Carrier gas was helium, and flow rate was 7 mL/min. The sample was injected directly into the capillary column at 30°C the detector temperature was 250°C. Quantitative analysis of FAME was done by determination of peak areas with a Sigma 10B (Perkin-Elmer) data station and by conversion of peak areas into mass percentages by means of response factors.	*Sample preparation.* 100 mg of MF was dissolved in 1.7 mL toluene. *Sample analysis.* A 1 μL sample was injected into a glass capillary column (11.7 m × 0.7 mm ID) coated with OV-1 ($d_f \approx 0.44$ μm). The column was mounted in a Varian 2100 GC using glass liners in the injector and detector parts. The carrier gas was helium at 13 mL/min. Oven temperature was programmed from 150 to 310°C at 2°C/min. The temperature of the injector and detector was 330°C.
deMan, 1968	FA composition determined by the method of deMan (1964).	—
Riel and Paquet, 1972	FAME were analyzed with an Aerograph 200 GC. Temperature was programmed from 50 to 180°C at 8°C/min. The peak areas were determined for each FA by triangulation.	—
Stepanenko and Tverdokhleb, 1974	FA composition determined by GLC analysis.	—

Centrifugation		
Antila and Antila, 1970	FA composition determined by GC per Antila (1966).	
Branger, 1993	—	*Sample preparation.* Weigh 0.5 g of AMF, dilute with choloroform to 10 mL. *Sample analysis.* TG were analyzed using GLC. The GC was a Hewlett-Packard Model 5890 (Hewlett-Packard, Avondale, PA) equipped with an auto sampler (Hewlett-Packed model 7673A) and a flame ionization detector. The GC was equipped with a 0.25 mm × 15 m DB-5 column with a 0.1 mm film thickness (J & W Scientific, Folsom, CA). The operating conditions were split ratio 20:1 with hydrogen carrier at head pressure of 35 kPa; injector temperature 300°C; detector temperature 350°C. The temperature program was initial temperature 200°C with a 1 min hold and then increasing at rate of 7.0 °C/min to 250°C. At this point temperature was increased at 30 °C/min to 350°C followed by a 10 min hold. A 1 mL volume of the prepared sample was injected into the GC. The percentage area under each peak was interpreted as the percentage of TG. *Sample analysis.* TG carbon number analysis was carried out by high-temperature GLC. A MF of known composition was used for calibration. The GC used was a Pye Unicam GCD with a glass column (0.5 m × 4 mm ID) packed with 3% OV-1 on 100/120 mesh Gas Chrom Q was used. The initial temperature was 240°C, and the final temperature was 355°C with a rate of 4°C/min. Injector and detector temperatures were 380°C, and nitrogen was the carrier gas at 50 mL/min.
Jordan, 1986	*Methyl ester preparation.* Weigh 1 g MF into a screw-capped vial and dissolve in *n*-pentane (10 mL). Add 0.5 mL methanolic potassium hydroxide (2M), shake for 30 s and leave to stand. Pipette off upper layer when clear, and dilute with redistilled *n*-pentane (3 drops to 2 mL). *Sample analysis.* Analysis performed with a Carlo Erba Mega Series GC, with on-column injection and secondary cooling. A fused silica WCOT column (50 m × 0.32 mm ID) with Silar 10C or CP Sil88 (0.2 μm) as a stationary phase was used. Temperature program was held at 46°C for 4 min and then increased at 15°C/min to 106°C, then increased at 1°C/min to 170°C, then increased at 2°C/min 200°C and held for 30 min. Detector temperature was 235°C, carrier gas was hydrogen, and linear flow rate was 35 cm/s. Data interpretation was by means of a Spectra-Physics SP4270 computing integrator. Response factors were determined for the even-numbered FA (C4 to C18) using a prepared FAME standard (similar in composition to MF FAME).	

Continued

TABLE 4.1 (Continued)

Fractionation method[1]	Fatty acid composition[2]	Triglyceride composition[3]
Kankare and Antila, 1974a	Nr	—
Kankare and Antila, 1986	FA composition was determined using the method of Antila and Kankare (1983).	TG composition was determined by GC, using a capillary column and on-column injector (Kalo et al., 1986). The determinations were carried out in the Department of General Chemistry, University of Helsinki.
Kankare and Antila, 1988a	FA profiles were measured using a GC with a capillary column.	TG profiles were obtained with the help of a capillary column and on-column injector. The work was done in the Institute for General Chemistry of the University of Helsinki.
Kankare and Antila, 1988b	FA composition was determined using the method of Antila and Kankare (1983).	TG composition was determined by GC, using a capillary column and on-column injector (Kalo et al., 1986). The determinations were carried out in the Department of General Chemistry, University of Helsinki.
Keogh and Higgins, 1986b	*Sample analysis.* FA were analyzed as methyl esters by GLC with a Pye Unicam 204 GC fitted with dual-flame ionization detectors. Separation was carried out on a 2.13 m long × 2 mm ID glass column packed with 10% EGA (ethylene glycol adipate) on Gas Chrom Q 100/120 mesh (Analabs, USA). The N_2 carrier gas and H_2 flow rates were 20 mL/min, and the air flow rate 300 mL/min; injector temperature 200°C; detector temperature 250°C. The chromatograph was programmed from 80°C (with an initial delay of 2 min) at 16°C min during each analysis. Peak areas were computed using an Infortronics 304 integrator, and response factors of the individual acids were calculated relative to C_{16}, which was assigned a response factor = 1.00.	*Sample analysis.* TG analysis was carried out on the same chromatograph using an 0.46 m long × 3 mm ID glass column packed with 3% JXR on Gas Chrom Q 100/120 mesh. The operating temperature of the injector and detector were 340 and 350°C, respectively. The chromatograph was programmed from 218 to 340°C at 6°C/min. The flow rates of the carrier gas nitrogen, hydrogen and air were adjusted to 100, 1000 and 1500 mL/min respectively.
Keshava Prasad and Bhat, 1987a	Conjugated polyunsaturated FA in MF fractions were determined by measuring the absorbance of the fat in iso-octane, as described in AOCS (1962) using Bausch and Lomb Spectrometer 2000.	—
Lakshminarayana and Rama Murthy, 1985	FA composition determined by a GLC method of Rama Murthy and Naryanan (1971).	—
Ramesh and Bindal, 1987b	FA method reported in Ramesh and Bindal (1987a).	—
Richardson, 1968	Nr	—

Separation with the aid of an aqueous detergent solution

Banks et al., 1985	FA were analyzed as methyl esters by GLC using methods previously described (Banks et al., 1976).
	Sample preparation. MF was dissolved in heptane (2.5 g fat/L).
	Sample analysis. TG were separated on a Pye Unicam PU 4500 GLC fitted with a Chrompack On-column injector (Chrompack Ltd., London) and an FID detector. 1 μL of the MF:heptane solution was injected directly on to the WCOT silica column (12 m × 0.25 mm ID) coated with 0.13 μm chemically-bonded CP-Sil-8 (Chrompack Ltd.). Hydrogen was used as the carrier gas at a flow rate of 13 mL/min. An injection temperature of 50°C was used, an oven temperature of 60°C and a detector temperature of 350°C. Immediately after injection, the oven was heated rapidly to 250°C, held for 4 min, then heating continued at 4°C/min to 340°C and held for 17 min.
Dolby, 1970a	FAME were analyzed by GLC.
Doležálek et al., 1976	—
	Sample preparation. The samples were prepared by pre-esterification of MF by methanol in the presence of sodium methylate and petroleum ether.
	Sample analysis. FA analysis was done with a Chrom 4 GC (Laboratorní přístorje, Prague), with an FID. A glass column (200 cm × 3 mm ID) was packed with 10% PEGA on Chromaton N-AW-DMCS. Column temperature was 180°C (isothermic) or 70-180°C (programmed, 4 °C/min). Injection chamber temperature was 240°C, nitrogen flow 20–40 mL/min, hydrogen flow 20 mL/min.
	Quantitative chromatogram evaluation was carried out by the triangulation method used a factor for calculation of FA content.
Fjaervoll, 1970a	Nr
Fjaervoll, 1970b	Nr
Norris et al., 1971	*Methyl ester preparation.* FAME containing 14% (w/v) or boron trifluoride were prepared according to van Wijngaarden's (1967) modification of the procedure of Metcalfe et al. (1966).

Continued

TABLE 4.1 (Continued)

Fractionation method[1]	Fatty acid composition[2]	Triglyceride composition[3]
	Fatty acid analysis. FAME were separated by GLC using a Varian aerograph 1520 Chromatograph fitted with an FID and a matrix temperature programmer. A column (8 ft × 1/8 in. ID) packed with 12% diethylene glycol succinate on acid-washed DMCS-treated Chromosorb W (60/80 mesh) was used. Nitrogen flow rate was 25 mL/min, and the temperature was programmed from 40 to 60°C at 4°C/min and then to 175°C at 6°C/min. The identity of the methyl esters was established from the log values of retention volumes, and the proportion of esters present was determined by measuring peak areas by height × width at half-height. *Trans-fatty acid analysis. trans*-FA content of the C18 monoene acids were estimated by GLC after the separation of the methyl esters of the FA into saturated, *cis*-monoene, *trans*-monoene, diene, and triene fractions, by TLC on Silica-gel G (E. Merk AG) impregnated with 20% silver nitrate. Chromoplates were prepared by the method of Lees and DeMuria (1960) and developed in benzene. Compounds were detected under UV light after spraying with 0.2% 2′,7′-dichlorofluorescein in 95% ethanol (Mangold and Mallins, 1960). The *trans*- and *cis*-monoenes bands were then scraped off the chromatoplates into centrifuge tubes and the esters extracted from the Silica-gel G three times with 5 mL portions of chloroform–methanol (2:1; w/v). This solvent was removed *in vacuo*. The methyl esters were dissolved in 0.2 mL hexane and subjected to GLC.	
Schaap et al., 1975	*Methyl ester preparation.* FA were converted to methyl esters according to deMan (1964). *Sample analysis.* Analysis of methyl esters was carried out by a programmed-temperature GC, with a wide-bore glass-capillary column (32 m × 0.7 mm) coated with diethylene glycol succinate. The preparation and operation for this column has been described by Badings et al. (1975).	—
Filtration through muslin cloth		
Kulkarni and Rama Murthy, 1987	FA composition determined by the method of Rama Murthy and Narayanan (1971).	—

Methods for the Characterization of Milkfat Fractions 101

Filtration through milk filter	
Fouad et al., 1990	*Methyl ester preparation.* FAME were prepared by a modified method described by Christopherson and Glass (1969), always using freshly prepared solutions of sodium methoxide. *Sample analysis.* 0.2 μL injections were done on a Varian 3700 GC equipped with an FID and on-column injector, a Supelco (Supelco Inc., Canada) fused capillary column (30 m × 0.32 mm) coated with SP-2340, and helium at 1.5 mL as the carrier gas. The temperature was programmed to start at 50°C (1 min), increasing to 160°C at 5°C/min, while the detector temperature was maintained at 210°C. The detector response to individual FA was quantified by the method of external standards using commercially available mixtures of FAME dissolved in *n*-hexane (Nu Check Prep, Elysian, MN). Using the constants derived for each FA, the integrator response was calibrated to provide the results of each component on a weight basis.	*Sample preparation.* 15 μg of MF was dissolved in 6 mL *n*-hexane. *Sample analysis.* 0.2 μL were injected onto the column. TG were separated using 0.1 μm DB-5 stationary phase coated onto a 15 m × 0.32 mm ID fused silica capillary column (J & W Scientific, Rancho Cordova, CA) and helium (1.5 mL/min) as the carrier gas. The injector temperature was programmed to increase from 70 to 330°C at 100°C/min, and the detector temperature was maintained at 350°C. Separation of TG was based on differences in carbon number and the identity of the MF components were inferred from co-chromatography with components of a standard commercial mixture of TG from C18 to C154 (Nu Check Prep).
Filtration through cheese press	
Baker, 1970a	FA composition determined by GC.	Nr
Filtration through casein-dewatering press	
Black, 1973	*Butyl ester preparation.* Butyl esters were prepared by the method of Parodi (1970). *Sample analysis.* A Beckman GC 5 fitted with dual FIDs was used, and columns were 1.8 m × 2 mm stainless steel packed with 15% DEGS (HIEFF-IBP). On-column injection was used. Carrier gas was nitrogen at a flow rate of 25 mL/min. Temperature program of the oven was 75°C for 2 min, 6°C/min to 155°C, then 2°C/min to 195°C and isothermal until the last component was eluted. Identification was by comparison of retention times with those of butyl esters of a mixture of pure FA. Area measurement was by triangulation, and correction factors for each FA were calculated from a quantitative mixture of pure butyl esters.	—
Separation technique not reported	
Deroanne, 1976	FAME were analyzed by GC.	—

Continued

TABLE 4.1 (Continued)

Fractionation method[1]	Fatty acid composition[2]	Triglyceride composition[3]
Deroanne and Guyot, 1974	*Sample analysis.* FAME were analyzed by GC. The esters were separated on a stainless steel column 3 m × 3 mm dia., with 20% DEGS phase on Chromosorb W (80–100), with a temperature program from 90 to 210°C. The peak areas were measured by triangulation. The total peak area for the C4 to C18 acids was taken to represent 100% of the fatty acids of the MF.	—
Lechat et al., 1975	*Sample analysis.* FA with at least 8 carbons were converted to methyl esters and analyzed by GC in isothermal condition at 190°C on butandial columns. Butyric and caproic acids were measured using barium soaps method (Kuzdzal-Savoie and Kuzdzal, 1968). This method recommends direct analysis with GC at 145°C on columns with diethylene glycol succinate mixed with phosphoric acid or on Carbowax 20M column reticulated with nitroterephthalic acid.	TG composition by analysis of GC in OV1 column (3% OV1 on Chromosorb W or Gas Chromosorb Q).
Martine, 1982	Nr	—
Timmen, 1974	FA composition determined with FAME by GC.	—
Crystallization from solvent solution		
Acetone		
Chen and deMan, 1966	FA composition was determined by dual column temperature-programmed GLC, as described by deMan (1964).	Determination of the trisaturated TG by a mercuric acetate adduct procedure, described by Kerkhoven and deMan (1966).
Jensen et al., 1967	Butyl esters were analyzed for FA composition by programmed GLC using the method of Sampugna et al. (1966).	—
Kaylegian, 1991	*Butyl ester preparation.* Weigh 150 mg MF into screw cap test tubes (18 mm ID × 150 mm), add 2 mL of 0.5 N sodium hydroxide in butanol solution. Tightly cap and heat in boiling water 10 min, cool under tap water. Add 2 mL of 14% boron trifluoride butanol reagent (Supelco Inc, Bellefonte, PA), cap and boil 10 min. Cool under tap water, add 10 mL pentane and 15 mL methanol in distilled water (6:39 v/v), shake 2 min, centrifuge for 1 min. The bottom layer was removed and discarded. The washing procedure with pentane and methanol was repeated two more times.	*Sample preparation.* 0.02 g of MF was dissolved in 10 mL isooctane. *Sample analysis.* 0.3 μL were injected in a Varian 3410 high temperature GC equipped with a programmable injector and FID. The carrier and make-up gas were high-purity helium at 2 mL/min and 30 mL/min, respectively. High-purity hydrogen (30 mL/min) and compressed air (300 mL/min) were supplied to the detector. The injector and detector temperature were 350°C, and the oven was programmed from 200 to 340°C at 2°C/min, followed by a 20 min hold.

	Sample analysis. An aliquot (6 μL) was injected into a Varian 2700 GC equipped with a glass column (3 m × 2 mm ID) packed with 10% Carbowax 20M on 80/100 Chromosorb WAW (Supelco Inc.). The oven temperature was programmed from 50 to 220°C at 8°C/min. The carrier gas was helium at 24 mL/min.	The concentration of each group of triglycerides based on carbon number was calculated based on the total peak area of the chromatogram.
	Peak areas were integrated with a Spectra-Physics 4200 integrator (Spectra-Physics, CA) and the concentrations of each ester were calculated as peak area of each fatty acid ester (0.9 × total peak area).	
Lambelet, 1983	—	*Sample analysis.* TG composition was determined by GLC using a Carlo Erba GLC (model 4160) equipped with FID detector and cold on-column injector. A 10 m glass capillary column coated with OV1 was used with the following temperature program: 80°C isothermal 1 min, 80–140°C 35°C/min, isothermal 1 min, 140–220°C/min, isothermal 1 min, 220–310°C 4°C/min, isothermal 32 min.
Larsen and Samuelsson, 1979	FA composition was determined using a Perkin-Elmer 900 GC.	—
Lechat et al., 1975	*Sample analysis.* FA with at least 8 carbons were converted to methyl esters and analyzed by GC in isothermal condition at 190°C on butandial columns. Butyric and caproic acids were measured using barium soaps method (Kuzdzal-Savoie and Kuzdzal, 1968). This method recommends direct analysis with GC at 145°C on columns with diethylene glycol succinate mixed with phosphoric acid or on Carbowax 20M column reticulated with nitroterephthalic acid.	TG composition by analysis of GC in OV1 column (3% OV1 on Chromosorb W or Gas Chromosorb Q).
Mattsson et al., 1969	*Methyl ester prepration.* FAME were prepared by the method of Smith (1961).	—
	Sample analysis. FAME were analyzed with a Perkin-Elmer vapor fractometer, model 116E, equipped with an FID, and 2 m columns of ethylene glycol succinate (Perkin-Elmer 2S 42.14) at 195°C and with helium as carrier gas. The units were calibrated with standard FAME (Hormel Foundation).	
	Values for butyric acid were too small, so they were analyzed by column chromatography on silicic acid by the method of Kishonti and Sjöström (1965).	

Continued

104   Milkfat Fractionation

TABLE 4.1 (Continued)

Fractionation method[1]	Fatty acid composition[2]	Triglyceride composition[3]
Parodi, 1974a	*Butyl ester preparation.* FA were converted to butyl esters using previously described methods (Parodi, 1970, 1972). *Sample analysis.* Samples were analyzed on 2.44 m × 2 mm ID stainless steel columns packed with 10% EGSS-X on 100/120 mesh Gas Chrom P (Applied Science Laboratories, Inc.) using previously described methods (Parodi, 1970, 1972).	GLC analysis of TG was carried out by the method of Parodi (1972).
Schaap et al., 1975	*Methyl ester preparation.* FA were converted to methyl esters according to deMan (1964). *Sample analysis.* Analysis of methyl esters was carried out by a programmed-temperature GC, with a wide-bore glass-capillary column (32 m × 0.7 mm) coated with diethylene glycol succinate. The preparation and operation for this column has been described by Badings et al. (1975).	—
Sherbon and Dolby, 1973	*Methyl ester preparation.* FAME were prepared according to van Wijngaarden (1967). *Sample analysis.* Samples were analyzed by GLC with a 2.44 m × 3.17 mm column of 11% DEGS on 60/80 mesh DMCS-treated acid-washed Chromosorb W, with FID. The column was programmed from 60 to 200°C at 6°C/min. Peak areas were determined by triangulation. Retention volume plots based on isothermal GLC at 172°C were used to tentatively identify minor methyl esters.	—
Timms, 1980b	FA composition by GLC analysis of FAME was determined by the method of Timms (1978a).	TG composition was determined by high-temperature GLC by the method of Padley and Timms (1978).
Yi, 1993	*Butyl ester preparation.* FA were converted to butyl esters using a modified method of Iverson and Sheppard (1986). Approx. 150 mg	*Sample preparation.* Sample preparation was done following a modified method of Lund (1988). 6 mg of fat was dissolved in 1 mL isooctane and an

	of fat was placed into a 45 mL screw top centrifuge tube with 4 mL of 0.5 N NaOH in butanol. 1 mL of pelargonic acid (C9) internal standard (10 mg C9/1 mL pentane) was added to the mixture and flushed with nitrogen gas, capped, and vortexed before being placed into a heated water bath. The mixture was boiled for 10 min and cooled to room temperature. 5 mL of 14% boron trifluoride–butanol reagent (Supelco, Inc., Bellefonte, PA) was then added to the cooled mixture and once again flushed with nitrogen gas, capped, vortexed, boiled for 20 min, and cooled to room temperature. 15 mL of pentane and 15 mL of methanol in water (6:39 v/v) wash solution was added to the cooled mixture, shaken by hand for 3 min and spun in a Sowall RT6000B refrigerated centrifuge (Dupont, Hoffman Estates, IL) for 10 min at 1000 rpm and 4°C. The lower phase was discarded, and 15 mL of the solution was added to the upper phase and centrifuged again. This procedure was repeated two more times. After the final lower phase was discarded, 6 g of anhydrous sodium sulfate was added to the final upper phase to absorb residual water. 1 µL of diluted upper phase was then injected into the GC. *Sample analysis.* FA composition was analyzed by GC. The column was Supelcowax 10 (Supelco, Inc. Canada) fused silica capillary column (30 m × 0.32 mm), and the carrier gas was helium with a flow rate of 2 mL/min. The detector temperature was 300°C, the GC was programmed for an initial hold of 2 min at 55°C, 55–160°C at 30°C/min, 160–190°C at 3°C/min, 190–227°C at 4°C/min with a 15 min hold.	internal standard of trinonanoin (164 µg in isooctane) was added. 1 µL of this solution was injected into the GC. *Sample analysis.* TG analysis was conducted on a Varian Model 3700 (Varian Association, Palo Alto, CA) GC equipped with a FID and on-column injector. The column was a 25 m × 0.25 mm WCOT fused silica stainless steel capillary column (Chrompack, Raritan, NJ) coated with TAP-CB. The carrier gas was helium with a flow rate of 2 mL/min. The detector temperature was held at 370°C, while the GC temperature was programmed according to the Chrompack protocol for this column: initial hold at 280°C for 1 min and increasing to 355°C at 3°C/min. Data was processed with a NEC PowerMate SX Plus computer (Boxborough, MA) with Baseline 810 software program (Dynamic Solutions, Ventura, CA).

Hexane

Muuse and van der Kamp, 1985	*Methyl ester preparation.* The FA were transesterified by a modified Christopherson and Glass (1969) methylation by shaking 50 mg fat in hexane for 20 s with 100 µL KOH 2 mol/L in methanol and afterwards neutralizing the hydroxide by 0.5 g crystalline sodium hydrogen sulfate monohydrate to prevent saponification of the methyl esters. The clear solution of methyl esters in hexane that was obtained was ready for analysis. *Sample analysis.* FA composition was determined on a packed column coated with 15% Silar 9 CP on Chromosorb WAW 100/120 mesh and standardized by means of a reference material.	*Sample analysis.* TG were analyzed by capillary GC on a Varian 3700 GC fitted with a fused silica column with chemically bound phase of CP Sil 5. Column length about 10 m, ID 0.23 mm. Detection by FID and integration by Spectra-Physics CPU-SP 4000. Operating conditions: Split ratio 1:20, carrier gas helium (constant pressure 69 kPa); injection volume 1 µL of a 2% solution of fat in hexane; injector temperature: 330°C; oven temperature programmed from 260 to 350°C at 4°C/min.

Continued

TABLE 4.1 (Continued)

Fractionation method[1]	Fatty acid composition[2]	Triglyceride composition[3]
Pentane		
Kaylegian, 1991	*Butyl ester preparation.* Weigh 150 mg MF into screw cap test tubes (18 mm ID × 150 mm), add 2 mL of 0.5 N sodium hydroxide in butanol solution. Tightly cap and heat in boiling water 10 min, cool under tap water. Add 2 mL of 14% boron trifluoride–butanol reagent (Supelco Inc., Bellefonte, PA), cap, and boil 10 min. Cool under tap water, add 10 mL pentane and 15 mL methanol in distilled water (6.39 v/v), shake 2 min, centrifuge for 1 min. The bottom layer was removed and discarded. The washing procedure with pentane and methanol was repeated two more times. *Sample analysis.* An aliquot (6 μL) was injected into a Varian 2700 GC equipped with a glass column (3 m × 2 mm ID) packed with 10% Carbowax 20M on 80/100 Chromosorb WAW (Supelco Inc.). The oven temperature was programmed from 50 to 220°C at 8°C/min. The carrier gas was helium at 24 mL/min. Peak areas were integrated with a Spectra Physics 4200 integrator (Spectra-Physics, CA) and the concentrations of each ester were calculated as peak area of each fatty acid ester (0.9 × total peak area).	*Sample preparation.* 0.02 g of MF was dissolved in 10 mL isooctane. *Sample analysis.* 0.3 μL were injected in a Varian 3410 high temperature GC equipped with a programmable injector and FID. The carrier and make-up gas were high-purity helium at 2 mL/min and 30 mL/min, respectively. High-purity hydrogen (30 mL/min) and compressed air (300 mL/min) were supplied to the detector. The injector and detector temperature were 350°C, and the oven was programmed from 200 to 340°C at 2°C/min, followed by a 20 min hold. The concentration of each group of triglycerides based on carbon number was calculated based on the total peak area of the chromatogram.
Supercritical fluid extraction		
Carbon dioxide		
Arul et al., 1987	*Methyl ester preparation.* FA were converted into their methyl esters by the method of Luddy et al. (1968). *Sample analysis.* The separation of the fatty acid esters was carried out in a fused silica capillary column (30 m × 0.25 mm ID) coated with OV-225 Durabond (0.25 μm) (J & W Scientific Inc., Rancho Cordova, CA). The same GC with hydrogen as carrier gas was used for TG analyses. The temperature program was, 60°C for 2 min, 10°C/min until 190°C and held for 8 min. Relative response factors were determined by analysis of a standard mixture of FAME having a composition similar to average MF (Applied Science Laboratories, Inc., State College, PA).	*Sample analysis.* TG composition was analyzed on a GC equipped with a FAD (HP 5890, Hewlett-Packard, Avondale, PA) using the cold on-column injection technique (Freeman, 1981; Traitler and Prévot, 1981). The separation was carried out in a fused silica capillary column (15 m × 0.25 mm ID) coated with SE-54 Durabond (0.5 μm; J&W Scientific, Inc.). TG from C24 to C54 were identified from retention times of a standard mixture of TG (Applied Science Laboratories). The fat samples were diluted (1:100) in *n*-heptane, and an aliquot of 0.5 μL was injected onto the column. The temperature program was held isothermally at 90°C for 2 min, then heated to 350°C at 10°C/min and held at 350°C for 10 min. The carrier gas was hydrogen held at a constant pressure head of 1.5 bar.

Methods for the Characterization of Milkfat Fractions 107

Reference	Description
Bhaskar et al., 1993a	—
	Sample analysis. TG were analyzed by a modified method of Amer et al. (1985). The feed, raffinate, and extract of AMF were directly analyzed for their TG compositions on a GC using a DB-5 capillary glass column (30 m × 0.25 mm; J & W Scientific, Folsom, CA) fitted with an FID (Hewlett-Packard Model 5890, Avondale, PA). The oven temperature was programmed in 3 stages: from 50 to 240°C at 25°C/min, from 240 to 245°C at 3°C/min, held for 25 min, and from 345 to 350°C at 0.1°C/min. The injector and detector temperatures were maintained at 330 and 345°C, respectively. The carrier gas was helium (1.5 mL/min).
Bhaskar et al., 1993b	*Sample analysis.* TG were analyzed by a modified method of Amer et al. (1985), using the GC described for FA analysis. The TG were directly analyzed using a Durabond-5 capillary glass column (30 m × 0.25 mm; J & W Scientific, Folsom, CA). The oven temperature was programmed in 3 stages: from 50 to 240°C at 25°C/min, from 240 to 245°C at 3°C/min, held for 25 min, and from 345 to 350°C at 0.1°C/min. The injector and detector temperatures were maintained at 330 and 345°C, respectively.
	Sample preparation. The fatty acids were converted into methyl esters (AOCS, 1989).
	Sample analysis. AMF and fractions were analyzed on a GC fitted with a flame ionization detector (HP 5890, Hewlett-Packard, Avondale, PA). The carrier gas was helium, 1.5 mL/min. A capillary glass column 30 m × 0.25 mm Durabond-225 (J & W Scientific, Folsom, CA) was used. The oven temperature was held at 60°C for 2 min, then increased at 4°C/min to 220°C and held for 10 min. The injector was at 200°C and detector at 260°C.
Bradley, 1989	MF was converted to FAME with boron trifluoride and analyzed by GC.
Büning-Pfaue et al., 1989	*Sample analysis.* GC determination of FAME was done with a Hewlett-Packard Research Chromatograph G with FID; packed column Chromosorb WAW, 10% DEGS, 80–100 mesh, 6′ × 1/8″; injector and detector temperature at 250°C; column temperature 50 to 190°C at 6°C/min and 190°C isothermally for 5 min; carrier gas nitrogen 20 mL/min; hot empty needle sampling method; Sigma FAME reference substance, purity > 99%.
	Sample analysis by GC. GC determination with a Carlo Erba GC 6000 Vega series with sample-feeding system PAS 1 (evaporating split injector); capillary column FS-SE-54-CB, 15 m × 0.25 mm ID, 0.11 μm; injection volume 1 μL (150 mg MF/10 mL *n*-hexane); injector temperature 380°C; split 1:10; column temperature 240–360°C at 5°C/min and 360°C isothermally for 9 min; carrier gas hydrogen 100 kPa; hot empty needle sampling.
	Sample analysis by HPLC. HPLC determination: LKB pump 2150; Rheodyne injector valve 7125; sample carrier 20 μL (1 g MF/20 mL hexane/mobile phase [1 + 3 vol]); Knauer high-temperature oven; LKB differential refractometer detector 2142; column combination Multisorb RP-18, 5 μm, 125 × 12 × 8 mm and Lichrosorb RP-18, 5 μm, 250 ¥ 8 ¥ 4 mm; mobile phase 1 mL/min acetone/acetonitrile (5 + 3, vol); column temperature 35°C; Sigma TG reference substance, purity > 99%.

Continued

TABLE 4.1 (Continued)

Fractionation method[1]	Fatty acid composition[2]	Triglyceride composition[3]
Chen et al., 1992	*Sample preparation.* 100 mg MF was hydrolyzed in potassium hydroxide following the method described by Christie (1987). To derivatize FFA into esters of PBPB, 100 μL of FFA were transferred to a 5 mL reaction vial and 400 μL of fresh solution of PBPB (mg per mL in acetonitrile), 40 μL of 1,4,7,10,13,16-hexaoxacyclooctodecane (mg/mL acetonitrile) and 20 mg of potassium carbonate were added to the vial. The vial was filled to 1 mL with acetonitrile and heated for 30 min in a water bath at 75–80°C. After cooling to room temperature, 10 μL of formic acid (4% w/v) were added, and the solution was heated for 5 min (75–80°C) and cooled (4°C) for at least 1 h. All samples were filtered (0.45 μm) before HPLC analysis. *Sample analysis.* HPLC analysis was carried out on a Waters HPLC model 510 and 501 pumps, U6K injector, R401 differential refractive index (RI) detector, and 900 photodiode array detector (Waters Assoc., Milford, MA). 15 μL of FA ester were loaded into the HPLC and separated on a ZORBAX C18 column (5 μ, 4.6 mm ID × 25 cm, DuPont Chromatography products, MacMod Analytical Inc.) and monitored using a photodiode array detector at 254 nm. Mobile phase consisted of H$_2$O:acetonitrile, 80:10 (solvent A) and methanol (solvent B). The elution program followed a linear gradient from 55% B to 91% B in 13 min, 91% B isocratic for 5 min, followed by a 91 to 100% B linear gradient in 2 min, and holding at 100% B for 10 min. Flow rate was 1 mL/min. FA species were identified by comparison with authentic standard FA esters.	*Sample analysis.* An HPLC method described by Robinson and Macrae (1984) was modified and adapted for the analysis of TG composition of MF. The analysis was carried out on two reverse-phase ZORBAX C18 columns (5 μ, 4.6 mm ID × 25 cm, DuPont Chromatography products, MacMod Analytical Inc.) connected in series. MF was dissolved in tetrahydrofuran (100 mg/mL). 15 μL samples (filtered through a 0.45 μm filter) were loaded and eluted with acetone–acetonitrile (65:35 v/v) isocratically at a flow rate of 1 mL/min. Attenuation of the RI detector was set at 16. The separation of the TG was completed within 70 min.
Ensign, 1989	*Methyl ester preparation.* TG were saponified and the FA were liberated with 4 mL of 0.5 N sodium hydroxide in methanol. Boron trifluoride (14% w/v, Alltech/Applied Science Labs) was used to esterify the FFA to methyl esters. Distilled water and petroleum ether were added to separate the aqueous layer from the organic layer, which contained the FAME. The organic layer was reduced by evaporation under nitrogen and then transferred to a 10 mL volumetric flask. Chloroform or petroleum ether was added to bring the flask contents to volume. Anhydrous granular sodium sulfate was added to the flask to remove trace water. FAME standards (C4–C18) from Alltech/Applied Science Labs were used to make standard curves to calculate the concentration of the unknowns.	—

Hamman et al., 1991	*Sample analysis.* A Hewlett-Packard model 5890 (Avondale, PA) GC equipped with an autosampler (model 7673A) and integrator (model 3393A) was used. A DB-5 capillary column, 0.25 mm ID × 15 m (J & W Scientific, Folsom, CA), coated internally with 5% diphenyl and 95% dimethyl polysilicone crosslinked gum with a thickness of 0.1 μm was used. Injector and FID temperatures were 280°C, oven temperature was programmed from 50 to 190°C at 5°C/min, then increased to 260°C at 10°C/min, for a total run time of 53 min. Flow rate of the helium carrier gas was 0.75 mL/min, with a split ratio of 1:20 and a head pressure of 10 psi. *Methyl ester preparation.* FA composition was determined by conversion to FAME. To a 0.1 mL sample was added 1.0 mL hexane, 1.0 mL dimethyl carbonate, and 1.0 mL 0.5 N potassium methylate. The mixture was shaken for some minutes. 0.5 mL distilled water was added and the mixture shaken for 10 s. The phases were separated, and the upper phase was used for analysis. *Sample analysis.* A Varian model 3700 GC with split injection and FID was used. The FA esters were separated in a fused silica capillary column DB-225 (30 m × 0.25 mm ID) with a film thickness of 0.25 μm (J &W Scientific, Inc. Rancho Cordova, CA). Nitrogen was used as a carrier gas. The injector and detector temperatures were 230 and 260°C, respectively. The temperature program was 80°C for 1 min, 4°C/min until 233°C, and held for 7 min. A standard mixture of FAME having a composition similar to MF was used to determine the relative response factors.	—
Kankare and Antila, 1988a	FA profiles were measured using a GC with a capillary column.	TG profiles were obtained with the help of a capillary column and on-column injector. The work was done in the Institute for General Chemistry of the University of Helsinki.
Kankare et al., 1989	FA composition was determined with a capillary column on a GC (Antila and Kankare, 1983).	TG analysis was determined with a capillary column on a GC with an on-column injector (Kalo et al., 1986).
Kaufmann et al., 1982	*Methyl ester preparation.* FA composition was determined after obtaining the methyl esters through direct interchange of ester radical (Christofferson and Glass, 1969). *Sample analysis.* Analysis took place under the following conditions: GC Packard 429; capillary column, 27 m, SP 1000; carrier gas was nitrogen 0.4 mL/min; multiple step temperature program from 50 to 225°C. The Spectra-Physics system SP 4100 served as injector. Calibration was carried out with a test mixture made from the FAME close to the composition of MF.	*Sample analysis.* TG analysis was done with the following: Packard GC 429; compensation switching with two packed columns (80 cm × 3 mm ID, 3% OV-1 on 100/200 mesh Gas Chrom Q); carrier gas was nitrogen 60 mL/min; temperature program from 230 to 350°C. Calibration was done with 6 saturated TG. In the evaluation the odd-numbered acyl-C-numbers are recorded together with neighboring even-numbered TG.

Continued

TABLE 4.1 (Continued)

Fractionation method[1]	Fatty acid composition[2]	Triglyceride composition[3]
Rizvi et al., 1990	—	GC analysis was performed on a capillary column (Tse, 1987).
Shishikura et al., 1986	The FA composition was determined by GC after conversion to the n-propyl ester (Fujikawa et al., 1971).	*Sample analysis.* TG composition was determined by GC with JOEL JGC-20K apparatus equipped with a glass column (0.3 × 80 cm) containing 1% OV-1 on Chromosorb W (80/100 mesh). The column temperature was programmed from 200 to 350°C at 2°C/min.
Shukla et al., 1994	*Sample analysis.* Samples were analyzed on a GC fitted with a flame ionization detector (HP 5890, Hewlett Packard Co., Avondale, PA). The carrier gas was helium at 1.5 mL/min.	*Sample preparation.* Triglycerides were saponified into free fatty acids, followed by the their subsequent derivitization of fatty acids into methyl esters. *Sample analysis.* Methyl esters were analyzed using a capillary column 30 m × 0.25 mm (Durabond-225; J & W Scientific, CO., Folsom, CA).
Timmen et al., 1984	Nr	Nr
Short-path distillation		
Arul et al., 1988a	FA composition determined by GC as described by Arul et al. (1987).	TG composition determined by GC as described by Arul et al. (1987).
McCarthy et al., 1962	*Methyl ester preparation.* The methyl esters of fatty acids were prepared by transesterification (James, 1960). *Sample Analysis.* Samples were chromatographed using a Beckman GC 2A gas chromatograph equipped with a filament cell, a Minneapolis-Honeywell Brown 1-mv recorder, and a Model K1-1 Disc Integrator. A stainless steel column (6ft × 1/4 in OD) packed with 20% (w/w) DEGS on acid-washed firebrick (60–80 mesh) was used. The long-chain fatty acids were determined using a column temperature of 222°C and a helium flow rate of 100mL per min. The short-chain fatty acids were estimated at 180°C and 80mL/min. Fatty acid compositions were calculated from peak areas given by the integrator record. Runs with known mixtures indicated that these estimates approximated the weight distribution of the fatty acid test mixture.	GLC of the triglycerides was performed as previously described (Kuksis and McCarthy, 1962). For the chromatography of the most volatile distillate these conditions were altered slightly to increase the retention time. In this case, the starting temperature was 190°C and the temperature increment 2.1°C/min.

[1]Sorted by fractionation method and listed alphabetically by author.

[2]FA = fatty acid, FAME = fatty acid methyl ester, MF = milkfat.

[3]GC = gas chromatograph, GLC = gas liquid chromatography, GPC = gel permeation chromatography, HPLC = high-performance liquid chromatography, TG = triglyceride.

[4]Nr = data was reported in the literature, but the method employed was not available.

[5]No method or data reported.

TABLE 4.2 Methods Used to Determine Selected Chemical Indices and Flavor Profile of Milkfat Fractions

Fractionation method[1]	Selected chemical indices[2]	Flavor profile[3]
Crystallization from melted milkfat		
Vacuum filtration		
Amer et al., 1985	*IV*. AOCS Method Cd 1-25.	—[4]
Deffense, 1989	*IV*. Nr[5]	—
Deffense, 1987	*IV*. Nr	—
deMan and Finoro, 1980	*IV*. AOCS Method Cd 1-25. *Reichert-Meissl value*. AOCS Method Cd 5-40. *Polenske value*. AOCS Method Cd 5-40. *Refractive index at 40°C*. Zeiss refractometer. *Specific gravity*. AOCS Method Cc 10a-25.	—
Frede et al., 1980	*IV*. Nr	—
Jamotte and Guyot, 1980	*IV*. Nr *Reichert value*. Nr	—
Lund and Danmark, 1987	*IV*. Nr	—
Voss et al., 1971	*IV*. Wijs method. *Refractive index:* Nr	—
Vovan and Riel, 1973	*IV*. AOAC Wijs method. *Refractive index*. AOAC Method at 40°C.	—
Pressure filtration		
Badings et al., 1983a	*IV*. Method of Wijs according to Netherlands Standard Specifications (NEN 1046). *Refractive index*. With Abbe refractometer as described in NEN 1046. *Density at 35°C*. Estimated with a pycnometer as described in NEN 1046.	—
Riel and Paquet, 1972	*IV*. AOAC Method for Wijs IV.	—
Stepanenko and Tverdokhleb, 1974	*IV*. Nr *Reichert-Meissl value*. Nr *Refraction at 40 and 60°C*. Nr *Saponification*. Nr	—
Centrifugation		
Antila and Antila, 1970	*IV*. According to Hanus.	—
Bhat and Rama Murthy, 1983	—	*Carbonyl, monocarbonyl, and ketoglyceride analysis*. Total carbonyls, monocarbonyls, and ketoglycerides were isolated and estimated as 2,4-dinitrophenyl (DNP) hydrazones by the method of Schwartz et al. (1963). Monocarbonyls were separated into classes by TLC on MgO:Celite plates. The DNP hydrazones of methyl ketones were separated into individual components on Kieselguhr-G plates impregnated with 2-phenoxy ethanol (Urbach, 1963). The DNP hydrazones of alkanals were separated into individual components on Carbowax-coated Kieselguhr-G plates (Badings and Wassink, 1963). Individual components were tentatively identified by simultaneously running authentic mixtures of DNP hydrazones and quantified spectrophotometrically.

Continued

TABLE 4.2 (Continued)

Fractionation method[1]	Selected chemical indices[2]	Flavor profile[3]
Jordan, 1986	—	*Sensory analysis.* MF samples were assessed organoleptically and scores were given for diacetyl, creamy, butyric, and rancid flavors.
Kankare and Antila, 1986	*IV.* Nr	—
Kankare and Antila, 1988b	*IV.* Nr	—
Lakshminarayana and Rama Murthy, 1985	*IV.* Nr *Reichert-Meissl value.* Nr *Polenske value.* Nr *Refractive index at 40°C.* Nr *Specific gravity.* Determined by the pycnometer method of Davis and Macdonald (1953).	—
Ramesh and Bindal, 1987b	*IV.* Method reported by Ramesh and Bindal (1987a).	—
Separation with the aid of an aqueous detergent solution		
Dolby, 1970a	*IV.* Nr *Refractive index.* Nr *Reichert value.* Nr *Saponification value.* Nr	—
Dolby, 1970b	—	Nr
Doležálek et al., 1976	*IV.* Determined according to Hanus. Refraction at 40°C. Abbe refractometer.	—
Fjaervoll, 1970a	*IV.* Nr *Refractive index.* Nr	—
Fjaervoll, 1970c	*IV.* Nr *Refraction value.* Nr	—
Norris et al., 1971	*IV.* Determined according to British Standards Institution (1961). *Reichert value.* Determined according to British Standards Institution (1961). *Refractive index.* Determined according to British Standards Institution (1961). *Saponification value.* Determined according to British Standards Institution (1961).	—
Sherbon et al., 1972	*IV.* According to the British Standards Institution (1961).	—

Continued

TABLE 4.2 (Continued)

Fractionation method[1]	Selected chemical indices[2]	Flavor profile[3]
Walker, 1972	*Analysis of lactone precursors.* 300 g MF were vacuum steam distilled for 5 h, according to Dimick and Walker (1967). The MF temperature was 160–170°C and vacuum at 1–1.5 mm Hg absolute. Steam was generated at a rate that produced 100 mL distillate/h. After complete deodorization the thawed distillate, with diethyl ether washings of the cold traps, were transferred to a 2 L separating funnel. The aqueous phase was saturated with sodium chloride and extracted twice with 500 mL diethyl ether. The combined extracts were dried over anhydrous sodium sulfate and evaporated to dryness on a hot water bath under nitrogen. The residue was refluxed for 10 min with 5 mL of freshly prepared 7.8% (w/v) potassium hydroxide in ethanol. The solution was diluted 1:1 with water and extracted 3 times with petroleum ether. The petroleum ether extract was placed on a column (18 mm dia) containing 20 g silicic acid (water washed and activated at 120°C for 48 h). The column was washed with 100 mL 1% (v/v) diethyl ether in petroleum ether, and 150 mL 10% (v/v) diethyl ether in petrol ether to elute the majority of the FA, and 150 mL diethyl ether to elute the lactones. This fraction was reduced to 1 mL, and individual lactones were separated by GLC on a 2.5 m × 3.175 mm stainless steel column packed with 10% (w/w) stabilized diethylene glycol adipate (Analabs Inc., North Haven, CT) and 2% (w/w) of (88%) phosphoric acid on 60/80 mesh Chromosorb W (acid-washed, DMCS-treated). The carrier gas was nitrogen at 25 mL/min. Column temperature was programmed from 90 to 220°C at 6°C/min. The injector and detector (FID) temperatures were 240°C. Peak areas were measured and compared with those of standard solutions of depolymerized δ- and γ-lactones.	*Analysis of methyl ketone precursors.* 10 g MF were placed in 30 mL constricted test tubes, sealed with flame, and put in an oven at 160°C for 16 h to generate the methyl ketones from their precursors. Tubes was cooled, broken open in a mortar under carbonyl-free hexane, the fat liberated being washed into a flask with a total of 200 mL carbonyl-free hexane. Monocarbonyl compounds were converted into 2,4-dinitrophenyl-hydrazones and subsequently isolated from the fat according to Schwartz and Parks (1961). The isolated monocarbonyl DNP-hydrazones were separated into classes using a column of Celite 545 and Sea Sorb 43 (1:1) according to Schwartz et al. (1966). Gradients of chloroform in hexane followed by methanol in chloroform were employed to sequentially elute methyl ketone, saturated aldehyde, and unsaturated aldehyde hydrazones from the column. Classes were identified and quantitatively estimated by absorption at characteristic wavelengths in chloroform, i.e., 363 nm for methyl ketones, 358 nm for *n*-alkanals, 373 nm for alk-2-enals, and 391 nm for alk-2,4-dienals. Individual methyl ketone DNP-hydrazones were then separated by reverse-phase TLC on Kieselguhr G impregnated with 2-phenoxy-ethanol, using petroleum ether as mobile phase. Methyl ketone DNP-hydrazones were scraped individually from the plate into matched glass spectrophotometer cells with Teflon stoppers, and 2.5 mL chloroform was added to each cell. The cells were stoppered, shaken vigorously, and allowed to stand 30 min to settle the Kieselguhr. Optical densities were measured at 363 against a blank.

Filtration through muslin cloth

Khalifa and Mansour, 1988	*IV.* AOAC method. *Acid value.* AOAC method. *Total acidity.* AOAC method. *Refractive index.* AOAC method.	*Sensory analysis.* Organoleptic quality was evaluated by a 10-score panel; 10-marks were assigned each for flavor and texture.
Kulkarni and Rama Murthy, 1987	*IV.* AOAC method (1975).	—
Sreebhashyam et al., 1981	*IV.* Hanus method, AOAC (1970).	*Sensory analysis.* Fractions were evaluated by a panel using a 9-point scale; attributes evaluated were spreadability and flavor (pleasant to unpleasant).

Continued

TABLE 4.2 (Continued)

Fractionation method[1]	Selected chemical indices[2]	Flavor profile[3]
Filtration through cheese press		
Baker, 1970a	*IV*. Nr	Nr
Filtration through casein-dewatering press		
Black, 1973	*Reichert-Meissl value*. Method of the Australian Standard N63-1968, Section 5. *Polenske value*. Method of the Australian Standard N63-1968, Section 5.	—
Separation technique not reported		
Deroanne, 1976	*IV*. Method of Wijs as reported by Wolff (1968).	—
Deroanne and Guyot, 1974	*IV*. Method of FIL 8 (1959). *Refractive index*. Zeiss butyrorefractometer.	—
Lechat et al., 1975	*IV*. Nr	—
Crystallization from solvent solution		
Acetone		
Avvakumov, 1974	*IV*. Nr	—
Bhalerao et al., 1959	*IV*. Nr *Refractive index at 40°C*. NR	—
Larsen and Samuelsson, 1979	*IV*. Nr	—
Lechat et al., 1975	*IV*. Nr	—
Rolland and Riel, 1966	*IV*. Wijs Method.	—

Continued

TABLE 4.2 (Continued)

Fractionation method[1]	Selected chemical indices[2]	Flavor profile[3]
Walker et al., 1977	*IV*. Nr	*Lactone analysis.* 10 g MF was ground onto 35 g Celite 545 (Johns-Manville Products Corp., Lompoc, CA) in a mortar and packed into a chromatographic column above 20 g anhydrous sodium sulfate and 6 g deactivated acidic alumina. The column was eluted with redistilled acetonitrile until 20 mL was collected. The acetonitrile extract was reduced to 0.5–1 mL under nitrogen and cooled in ice. Acetonitrile was removed from above the small quantity of fat in the tube, and the fat was successively extracted with two 1 mL portions of fresh acetonitrile. The combined acetonitrile extracts were evaporated under nitrogen to 0.5–1 mL, and acetonitrile was removed and transferred to a small column (made from a disposable pipette) containing 1 g deactivated acidic alumina. The trace amount of fat remaining in the tube was extracted twice with 1 mL fresh acetonitrile, which was also transferred to the small alumina column. The column was finally washed with a further 1 mL acetonitrile, and the combined eluate from the column was reduced to 1 mL for GC analysis on a 1.8 m × 2 mm ID glass column containing 7.5% (w/w) stabilized ethylene glycol adipate (Analabs, Inc., North Haven, CT) + 2% (w/w) phosphoric acid on 80/100 mesh H/P Chromosorb W (Varian Aerograph, Walnut Creek, CA). The carrier gas was nitrogen at 25 mL/min, and the column temperature was programmed from 130–210°C at 4°C/min. The instrument was a Varian 1700 GC with the injector and FID temperatures at 240°C. For quantitative analysis, peak areas were measured with a model 3380A reporting integrator (Hewlett-Packard, Palo Alto, CA) and compared with those obtained from standard solutions of authentic depolymerized δ- and γ-lactones. Peak identities were established by GC/MS.
Walker et al., 1978	—	*Lactone analysis.* Lactones were extracted in acetonitrile from a column packed with fat-impregnated Celite 545 analyzed by GC on a 1.8 m × 2 mm ID glass column containing 7.5% EGA (stabilized) + 2% H_3PO_4 on 80/100 mesh Chromosorb WHP programmed from 130 to 210°C at 4°C/min. Potential lactone concentrations (free lactone + precursor) were determined similarly.
Youssef et al., 1977	*IV*. Method of Jacobs (1964). *Saponification value.* Method of Jacobs (1964).	—
Ethanol		
Bhalerao et al., 1959	*IV*. Nr *Refractive index at 40°C.* Nr	—

Continued

TABLE 4.2 (Continued)

Fractionation method[1]	Selected chemical indices[2]	Flavor profile[3]
Rolland and Riel, 1966	*IV*. Wijs Method.	—
Supercritical fluid extraction		
Carbon dioxide		
Bhaskar et al., 1993b	*IV*. AOCS Method Cd 1-25. *Saponification value*. AOCS Method Cd 3-25.	—
Kankare et al., 1989	—	*Lactone analysis*. Lactones were extracted with the method presented by Ellis and Wong (1975) on a column chromatograph and determined on a GC.
Kaufmann et al., 1982	*IV*. Wijs Method according to IDF Standard No. 8.	—
Rizvi et al., 1990	*IV*.	—
Rizvi et al., 1993	—	*Lactone Analysis*. δ– lactones were analyzed following the method of Ellis and Wong (1975). Using a gas-liquid chromatograph equipped with a flame ionized detector (HP 5890, Hewlett Packard, Arondale, PA), samples (2µL) were injected through a split sleeve at an injection ratio of 1:50. The carrier gas was helium at 1.5mL/min. δ– lactones were separated on a capillary glass column, 30 mm × 0.25 mm, Durabound-5 (J&W Scientific Inc., CA, USA) with programmed temperature starting at 100°C for 1.0 min., increasing at 4.0°C/min to a final 250°C. γ-12 lactone was used as internal standard. Peak areas were quantified using a Hewlett Packard 3393A integrator. The standard error of the results in the analysis was < 5% based on duplicate samples.
Timmen et al., 1984	*IV*. Nr	—
Short-path distillation		
Mc Carthy et al., 1962	*Saponification value*. AOCS Method Cd 3-25.	—

[1]Sorted by fractionation method and listed alphabetically by author.

[2]*IV* = iodine value; MF = milkfat.

[3]GC = gas chromatography; MF = milkfat; MS = mass spectroscopy; TLC = thin-layer chromatography.

[4]No method or data reported.

[5]Nr = Data was reported in the literature, but the method was not available.

TABLE 4.3 Methods Used to Determine the Crystal Characteristics of Milkfat Fractions

Fractionation method[1]	Crystal morphology[2]	Crystal size[3]
Crystallization from melted milkfat		
Vacuum filtration		
Black, 1975	—[4]	*Microscopy.* Crystal size was measured microscopically using polarized light and a calibrated graticule. Six fields were examined for each sample and the number of crystals within a range of class sizes counted.
Grall and Hartel, 1992	*Microscopy.* Photomicrographs of crystals were taken. Polarized light was useful for crystal definition and observation of crystal uniformity.	*Microscopy.* Photomicrographs of crystals were taken. The perimeter of each crystal was manually traced by means of a digitizer board, and the results were automatically transferred to a microcomputer. Crystal area (μm^2) was collected by image analysis software, and equivalent circular diameters and crystal size distribution were calculated with a custom-made program written in BASIC.
Voss et al., 1971	*Microscopy.* Crystal morphology was determined by microscopy.	—
Pressure Filtration		
El-Ghandour et al., 1976	*Microscopy.* Microscopic examination was carried out at 8–10°C using the technique of deMan and Wood (1959). A few drops of melted MF were fixed on a covered glass slide and the temperature was gradually decreased to 8–10°C at 0.5°C/h. The microscope used was a polarizing intereference Microscope MPI-S. The photographs were prepared with photomicrograph equipment (MNF) and Universal photomicrographic reflex (UKF). Magnifying powers were 9 and 20.	—
Helal et al., 1977	*X-ray diffraction.* X-ray diffraction patterns were obtained with X-ray diffractometer ORS-50 E Y PC-50 U with automatic recording. The radiation used was Cu K α (λ = 1.5148 Å). For identification of the crystal forms, Lutton's (1955) nomenclature was used.	—
Helal et al., 1984	*X-ray diffraction.* Samples were mounted on an X-ray diffractometer ORS-50 E Y PC-50 U with automatic recording. The radiation used was Cu K α (λ = 1.5148 Å). For identification of the crystal forms, Lutton's (1955) nomenclature was used: α = 4.12 Å, B = 3.8 Å, B = 4.6 Å.	
Centrifugation		
Kankare and Antila, 1974a	*Microscopy.* The crystal form was determined by microscopy.	*Microscopy.* The crystal size was determined by microscopy.
Keogh and Higgins, 1986h	—	*Microscopy.* The mean size of the crystals was determined microscopically using 400X magnification under polarized light. Twelve crystals in each of three fields were measured.

Continued

TABLE 4.3 (Continued)

Fractionation method[1]	Crystal morphology[2]	Crystal size[3]
Lakshminarayana and Rama Murthy, 1985	—	*Microscopy.* Crystal size was measured using a microscope having a calibrated stage micrometer of 100 divisions per mm.
Separation with the aid of an aqueous detergent solution		
Doležálek et al., 1976	—	*Microscopy.* Crystal size determined with an Ultrathermostat microscope (Meopta).
Jebson, 1970	—	Nr[5]
Schaap et al., 1975	*X-ray diffraction.* X-ray diffraction patterns were made using a conventional X-ray counter diffractometer (Philips/Norelco PW 1050 with scintillation counter), using nickel-filtered copper Kα radiation. The specimen holder consisted of a flat piece of copper that could easily be connected to a perforated copper block, through which the coolant from a cryostat could circulate. The entire sample holder was housed in a vacuum chamber provided with a slit sealed with Mylar, through which the X-ray beams could reach and leave the specimen. The sample chamber could be evacuated to prevent the formation of water droplets or ice crystals on the specimen in case of work done at low temperature.	*Scanning electron microscopy.* Fat crystals were observed by scanning electron microscopy, under a JEOL JSM-U3 instrument. Preparations were made by the replica method described by Reimer & Pfefferkorn (1973). The crystals on an object glass were covered with a layer of Xantopren (Bayer) mixed with the matching hardener. After hardening for at least 2 h at 28°C, the elastic Xantopren mass was peeled from the glass and the fat washed away with toluene. Then the Xantopren matrix was filled with polystyrene dissolved in toluene, which was later evaporated. The Xantopren mass could easily be removed, after which the remaining polystyrene block was covered with a thin layer of carbon and gold and observed electron-microscopically.
Filtration through casein-dewatering press		
Black, 1973	—	*Microscopy.* An MF sample was prepared on a microscope slide by slurrying gently with a light oil. The slides were viewed in a microscope with polarized light, and the crystal aggregates were measured using a calibrated graticule.
Separation technique not reported		
Deroanne, 1976	*X-ray diffraction.* Crystal morphology was determined by X-ray diffraction with a Phillips 1120 Diffractometer.	—

Continued

TABLE 4.3 (Continued)

Fractionation method[1]	Crystal morphology[2]	Crystal size[3]
Crystallization from solvent solution		
Acetone		
Schaap et al., 1975	*X-ray diffraction.* X-ray diffraction patterns were made using a conventional X-ray counter diffractometer (Philips/Norelco PW 1050 with scintillation counter), using nickel filtered copper Kα radiation. The specimen holder consisted of a flat piece of copper that could easily be connected to a perforated copper block, through which the coolant from a cryostat could circulate. The entire sample holder was housed in a vacuum chamber provided with a slit sealed with Mylar, through which the X-ray beams could reach and leave the specimen. The sample chamber could be evacuated to prevent the formation of water droplets or ice crystals on the specimen in case of work done at low temperature.	*Scanning electron microscopy.* Fat crystals were observed by scanning electron microscopy, under a JEOL JSM-U3 instrument. Preparations were made by the replica method described by Reimer & Pfefferkorn (1973). The crystals on an object glass were covered with a layer of Xantopren (Bayer) mixed with the matching hardener. After hardening for at least 2 h at 28°C, the elastic Xantopren mass was peeled from the glass and the fat washed away with toluene. Then the Xantopren matrix was filled with polystyrene dissolved in toluene, which was later evaporated. The Xantopren mass could easily be removed, after which the remaining polystyrene block was covered with a thin layer of carbon and gold and observed electron-microscopically.
Timms, 1980a	*X-ray diffraction.* X-ray diffraction carried out by the method of Timms (1979).	—
Timms, 1980b	*X-ray diffraction.* A Philips diffractometer with proportional counter and chart recorder was used with Cu Kα radiation and a nickel filter. The instrument was calibrated with tripalmitin using data given by Lutton (1945). The sample holder was maintained at 5–10°C by a stream of cool nitrogen, generated by boiling liquid nitrogen with a small heater controlled by a variable transformer. The sample holder was enclosed in a cylindrical Perpex chamber fitted with a Mylar window transparent to the X-ray beam and the diffracted rays. The diffraction pattern was recorded for 2θ from 1.5 to 27°.	—

Continued

TABLE 4.3 (Continued)

Fractionation method[1]	Crystal morphology[2]	Crystal size[3]
Woodrow and deMan, 1968	*X-ray diffraction.* X-ray diffraction patterns were obtained with a Philips Geiger Counter Diffractometer, the radiation used was Cu, Kα (λ = 1.5405 Å). Infrared spectra were obtained with a Perkin-Elmer 21 infrared spectrophotometer. For measurements at controlled temperatures the instrument was fitted with a variable-temperature chamber, Model 104 of the Barnes Engr. Co. Infrared spectra of the MF were obtained by placing a few drops of the fat, liquefied at 40°C, on a sodium chloride disc; this was covered by a second disc and placed in a cell holder. The cell holder was mounted in the variable-temperature chamber and the spectrum of the melt obtained at 40°C. The temperature was then lowered at a rate of 4°C/h and a spectrum run every h until 0°C was reached. The fast cooling method involved cooling the melt as quickly as possible in the variable-temperature chamber from 40 to 0°C. Spectra were run at 0°C and after 10 h of tempering at 5°C. X-ray diffraction cell mounts filled with MF were similarly treated in the variable-temperature chamber before running X-ray diffraction patterns. The shield of the diffractometer was packed with dry ice to control the temperature of the sample at about 0°C.	—
Supercritical fluid extraction		
Carbon dioxide		
Hamman et al., 1991	*X-ray diffraction.* MF was melted at 65°C to avoid the influence of structural history prior to X-ray diffraction analysis. A DPT camera with equipment for precise temperature control was used. The X-ray measurements were performed under isothermal conditions obtained by cooling in steps of 5°C from 40°C to 10°C. To obtain longspacing data, the time of exposure was 2 h and for measurements in the shortspacing region the time of exposure was 1 h.	—

[1]Sorted by fractionation method and listed alphabetically by author

[2]MF = milkfat.

[3]MF = milkfat.

[4]No method or data reported.

[5]Nr = Data of this category was reported in the literature, but the method was not.

TABLE 4.4 Methods Used to Determine the Melting Behavior of Milkfat Fractions

Fractionation method[1]	Melting point[2]	Thermal analysis[3]	Solid fat content[4]
Crystallization from melted milkfat			
Vacuum filtration			
Amer et al., 1985	MP. AOCS Method Cc 1-25.	Crystallization and melting curves were recorded using a heat flow differential scanning calorimeter (Mettler TA 3000, Mettler Instruments Ltd., Switzerland), calibrated with indium. 20–30 mg of MF was placed in an aluminum crucible; the crucible was covered with a pierced lid and sealed. The measuring cell was purged with nitrogen gas (50 mL/min) during the analysis. The samples were treated as follows: (1) heated at 80°C for 10 min to destroy previous thermal history; (2) crystallization over the temperature range 80 to –40°C (cooling rate 10°C/min, using liquid air); (3) fusion of the crystals formed by heating over the temperature range of –40 to 80°C (heating rate 10°C/min).	DSC. The relative percentages of liquid fat as a function of temperature were determined by integration of the DSC melting curve using a Mettler TC 10 data processor.
Barna et al., 1992	DSC. Melting point taken at peak temperature.	A Perkin-Elmer DSC-7 was used to determine melting profiles. Heating rate was 10°C/min.	—[5]
Black, 1975b	SP. Method of Black, 1972.	—	—
Deffense, 1989	DP. Mettler.	—	NMR. Nr
Deffense, 1987	DP. Nr[6] MP. Nr	—	NMR. NMR using Bruker pc 20.
deMan and Finoro, 1980	DP. With Mettler FP3 Automatic Dropping Point Apparatus with furnace attachment FP53. Details of the procedure are given by Mertens and deMan (1972).	—	NMR. SFC was determined using wide-line NMR with a Newport Quantity Analyzer MK II with supplementary modulation and temperature controller as described by Mertens and deMan (1972).

Continued

TABLE 4.4 (Continued)

Fractionation method[1]	Melting point[2]	Thermal analysis[3]	Solid fat content[4]
El-Ghandour and El-Nimr, 1982	—	—	*Dilatometry.* MF was melted at 70°C and filled in dilatometers. Dilatometers were immersed in a water bath and subjected to regime 6–16–13°C and then to 0°C as stated by El-Nimr and El-Ghandour (1980). Solid fat percentage (%T) and melting coefficient (D) were calculated every 2 degrees from 0°C till almost complete melting as follows: $\%T$ = contraction at T (mm^3/mg)/$D_s + \Delta D_s \cdot T$ where D_s = melting dilatation (mm^3/mg); ΔD_s = difference in coefficient of expansion of solid and liquid milkfat (mm^3/mg); T = definite temperature.
Grall and Hartel, 1992	*MP.* Estimated using DSC peak temperature.	Thermal heating curves were determined using a Perkin Elmer DSC 7 (Norwalk, CT) with liquid nitrogen as a heat sink. Samples were cooled to –40°C. A constant temperature ramp of 10°C/min was used for all samples over the range of –40 to 80°C.	—
Jamotte and Guyot, 1980	*DP.* Nr	—	*NMR.* Nr
Kaylegian, 1991	*MP.* AOCS Method Cc 1-25.	Thermal profiles were recorded with a Perkin-Elmer 7 DSC, calibrated with indium and mercury. 5–10 mg samples were encapsulated in hermetically sealed, aluminum-coated DSC pans (DuPont Co., Wilmington, DE). The DSC was fitted with a glove box and purged with air. The sample cell was purged with helium and cooled with liquid nitrogen during analysis. Samples were held at 60°C for 3 min to destroy crystal history, cooled to –60°C at 10°C/min, held for 3 min, and then heated to 60°C at 10°C/min to obtain a melting profile.	*NMR.* AOCS Method Cd 16-81; SFC values were measured every 5°C from 0°C to 55°C or until the sample was melted. Analyses were performed using a Bruker AM400 wide-bore multinuclear spectrometer with Aspect 3000 off-line data processing station (Bruker Spectrospin, Canada) at the Nuclear Magnetic Resonance Facility at Madison (Univ. of WI—Madison).
Kaylegian and Lindsay, 1992	*MP.* AOCS Method Cc 1-25.	—	*NMR.* AOCS Method Cd 16-81; SFC values were measured every 5°C from 0°C to 55°C or until the sample was melted. Analyses were performed using a Bruker AM400 wide-bore multinuclear spectrometer with Aspect 3000 off-line data processing station (Bruker Spectrospin, Canada) at the Nuclear Magnetic Resonance Facility at Madison (Univ. of WI—Madison).
Lund and Danmark, 1987	*DP.* Nr	—	Nr

Reference		Description
Makhlouf et al., 1987	—	Differential scanning calorimetric analysis was done according to the method proposed by Timms (1980a). The calorimeter used was a DuPont model 990 Thermal Analyzer.
		Dilatometry. Dilatometric analysis was done as described by Jasperson and McKerrigan (1957) and 4.5 μL/5°C/g was used as the correction factor for thermal expansion.
Schaap and van Beresteyn, 1970	—	Melting and solidification patterns were characterized by a DuPont 900 Differential Thermal Analyzer. MF was heated or cooled at 5°C/min.
		—
Timms, 1978b	*SP.* Method of Timms (1978c).	Heats of fusion were determined using a Perkin-Elmer DSC-2 differential scanning calorimeter, calibrated with indium. Samples were 5–10 mg, and heating rate was 10°C/min.
		—
Timms and Parekh, 1980	*SP.* Using Mettler dropping point apparatus.	—
		NMR. Method previously described by Timms (date not given).
Versteeg, 1991	*SP.* Nr	—
		Nr
Vovan and Riel, 1973	*CP.* AOAC method.	—
		Dilatometry. SFI was determined with volumetric dilatometers with a 5 mL bulb and a 0.900 mL capillary graduated in 0.005 mL. To confine the MF in the dilatometer, distilled water colored with methyl red was used. 5 g of fat and 1.5 g of water were placed with precision in the dilatometer in order to exclude air bubbles. The dilatometers were placed in a 0°C water bath overnight to obtain a stable form. Dilatation was measured at 5°C intervals after 20 min at each interval. The SFI was calculated using the heats of fusion curves and extrapolation of the liquid curve at 1% solid (Jasperson and McKerrigan, 1957).
Pressure filtration		
deMan, 1968	—	Differential thermal analysis was performed with a DuPont 900 Differential Thermal Analyzer with Calorimeter cell.
		—
Riel and Paquet, 1972	*CP.* AOCS method.	—
		—

Continued

TABLE 4.4 (Continued)

Fractionation method[1]	Melting point[2]	Thermal analysis[3]	Solid fat content[4]
Stepanenko and Tverdokhleb, 1974	MP. Nr	Nr	Dilatometry. Nr
Centrifugation			
Antila and Antila, 1970a	—	DSC method according to Antila (1966).	—
Bhat and Rama Murthy, 1983	SP. Nr	—	—
Black, 1974b	SP. Nr	—	—
Branger, 1993	—	—	NMR. Method is based on a combination of published methods and the AOCS method for SFC used pulsed NMR, and was designed to meet the requirements of the equipment available at the University of Wisconsin NMR facility in the Biochemistry Department. The technique required a Bruker AM-400 wide-bore multinuclear spectrometer, 9.45T/8.9 cm (400 Hz), with Bruker Aspect 1000/3000 data processing station. The probe used is a proton probe ^{1}H 5mm.
Evans, 1976	SP. Method of Dolby (1961).	—	—
Jordan, 1986	SLP. By the method in BS684, section 1.3.	—	—
Kankare and Antila, 1974	—	Melting and crystallization characteristics determined in a Perkin-Elmer DSC-1 differential scanning calorimeter.	—
Kankare and Antila, 1986	—	Melting characteristics were determined with a Perkin-Elmer DSC-4 apparatus. The crystallization and melting speed was 8°C/min.	—
Kankare and Antila, 1988a	—	Melting and crystallization properties were determined using a differential scanning calorimeter, Perkin-Elmer DSC-4. The cooling and heating rates were 8°C/min.	—
Kankare and Antila, 1988b	SP. Nr	Melting characteristics were determined with Perkin-Elmer DSC-4 equipment. The scan rates for	DSC. DSC equipment was used to measure the proportions of solid fat (SFI).

Keogh and Higgins, 1986b	SLP. Determined by the open capillary tube method (AOCS, 1981) as modified by Black (1972).	solidification and melting were 8°C/min. Percent solid fat was measured as outlined by Higgins and Keogh (1986b).
Lakshminarayana and Rama Murthy, 1985	MP. Method in ISI 3508 (ISI, 1966).	—
Ramesh and Bindal, 1987b	MP. Reported in Ramesh and Bindal (1987a). SP. Reported in Ramesh and Bindal (1987a).	—
Richardson, 1968	SLP. Nr	—
Separation with the aid of an aqueous detergent solution		
Banks et al., 1985	—	DSC was carried as described by Banks et al. (1976), except that sealed pans were used. Calibration was performed using indium and stearic acid.
Dolby, 1970a	SP. Nr	—
Dolby, 1970b	SP. Nr	—
Doležálek et al., 1976	MP. MF kept 1 h at −4°C, then for 23 h at 15°C, and capillary MP was determined.	—
Fjaervoll, 1970a	MP. Nr	—
Fjaervoll, 1970c	MP. Nr	—
Jebson, 1970	SP. Nr	—
Norris et al., 1971	SP. By Dolby's modification (1961) of Barnicoat's method (1944).	Thermal analysis was performed using a Perkin-Elmer DSC-1 differential scanning calorimeter, calibrated with high-purity melting point standards. MF was completely melted at 60°C and 4–9 mg samples were sealed into hermetically sealed sample pans. The DSC was held isothermally for 15 min at 60°C, and a cooling thermogram was recorded down to −60°C at 8°C/min. The MF was held for 5 min, and then a heating thermogram was recorded at 8°C/min up to 60°C. DSC. SFC was determined using normalized integral curves constructed from each DSC melting thermogram. SFC at temperature T was given by the ratio of the partial area above temperature T to the total area.

Continued

TABLE 4.4 (Continued)

Fractionation method[1]	Melting point[2]	Thermal analysis[3]	Solid fat content[4]
Sherbon et al., 1972	SP. By Dolby's modification (1961) of Barnicoat's method (1944).	Thermal analysis was performed using a Perkin-Elmer DSC-1 differential scanning calorimeter, calibrated with high-purity melting point standards. MF was sampled and thermograms obtained as described by Norris et al. (1971). Thermograms were corrected for thermal-lag and heat-capacity effects using the equations presented by Heuvel and Lind (1970).	DSC. Liquid fat contents were determined by integration of the corrected DSC thermograms.
Filtration through muslin cloth			
Khalifa and Mansour, 1988	MP. AOCS method.	—	—
Sreebhashyam et al., 1981	MP. ISI method (1966).	—	—
Filtration through milk filter			
Fouad et al., 1990	MP. AOCS Method Cc 1-25.	Thermal melting profiles were obtained with a Mettler TA 3000 heat flow differential scanning calorimeter, calibrated with indium. A 20–30 mg sample was held at 80°C for 10 min to destroy crystal history, cooled to −40°C at 10°C/min to crystallize, and then heated to 80°C at 10°C/min.	DSC. The liquid fat content was determined over 0 to 30°C by integration of the DSC melting curve using the Mettler TC 10 data processing accessory.
Filtration through cheese press			
Baker, 1970a	SP. Nr	—	—
Filtration through casein-dewatering press			
Black, 1973	SP. Method of Black (1972).	—	—
Separation technique not reported			
Deroanne, 1976	DP. Using Mettler FP 5 + FP 53 method. MP. Using Mettler FP. 5 + FP 51 method.	A Perkin-Elmer DSC-1B differential scanning calorimeter was used, calibrated with indium and tripalmitin. 10 mg MF were heated to 70°C, then cooled rapidly to −53°C, then heated at 16°C/min.	—

Reference			
Deroanne and Guyot, 1974	*SP.* Mettler FP5 + FP53 apparatus.	Perkin-Elmer DSC-1B. 10 mg MF was used. Melting curves were obtained by heating MF to 70°C, then cooling rapidly to –43°C, then heating at 16°C/min. Solidification curves were recorded after the melting curve by cooling at 16°C/min.	—
Lechat et al., 1975	*MP.* Nr	—	
Martine, 1982	*MP.* Nr	Nr	Nr
Crystallization from solvent solution			
Acetone			
Avvakumov, 1974	*MP.* Nr	—	*Dilatometry.* Nr
Baker et al., 1959	*MP.* Nr	—	—
Kaylegian, 1991	*MP.* AOCS Method Cc 1-25.	Thermal profiles were recorded with a Perkin-Elmer 7 DSC, calibrated with indium and mercury. 5–10 mg samples were encapsulated in hermetically sealed, aluminum-coated DSC pans (DuPont Co., Wilmington, DE). The DSC was fitted with a glove box and purged with air. The sample cell was purged with helium and cooled with liquid nitrogen during analysis. Samples were held at 60°C for 3 min to destroy crystal history, cooled to –60°C at 10°C/min, held for 3 min, and then heated to 60°C at 10°C/min to obtain a melting profile.	*NMR.* AOCS Method Cd 16-81; SFC values were measured every 5°C from 0°C to 55°C or until the sample was melted. Analyses were performed using a Bruker AM400 wide-bore multinuclear spectrometer with Aspect 3000 off-line data processing station (Bruker Spectrospin, Canada) at the Nuclear Magnetic Resonance Facility at Madison (Univ. of WI—Madison).
Kaylegian and Lindsay, 1992	*MP.* AOCS Method Cc 1-25.	—	*NMR.* AOCS Method Cd 16-81; SFC values were measured every 5°C from 0°C to 55°C or until the sample was melted. Analyses were performed using a Bruker AM400 wide-bore multinuclear spectrometer with Aspect 3000 off-line data processing station (Bruker Spectrospin, Canada) at the Nuclear Magnetic Resonance Facility at Madison (Univ. of WI—Madison).

Continued

TABLE 4.4 (Continued)

Fractionation method[1]	Melting point[2]	Thermal analysis[3]	Solid fat content[4]
Lambelet, 1983	—	Melting curves were recorded with a heat flow differential scanning calorimeter (model TA 2000 B, Mettler Instrument, Ltd., Zürich, Switzerland), calibrated with 99.9999% indium (Presussag Metall Ltd., Göslar, Germany). The samples of about 20 mg fat in aluminum crucibles were heated for 2°C/min against air. For data sampling, the calorimeter was connected to a Hewlett-Packard 3356 Laboratory Automation Computer System according to a scheme already described (Lambelet and Raemy, in press). Semireduced data from the 3356 Computer System was transferred to a Hewlett-Packard 3000 computer for further treatment.	DSC. Relative SFI values as a function of the temperature were determined by sequentially integrating the DSC curve of the fat sample. NMR. The Bruker P20$_i$ spectrometer was used. The temperature in the sample holder was controlled by external fluid circulation. NMR signals were sampled 70 μs after the start of the pulse. The analogous calculation unit was adjusted to take the average of 3 successive measurements separation by 10 s. SFI at temperature T was determined by comparing the amplitude of the NMR signals A_T and A_F. A_T was the signal of the liquid part of the sample at temperature T and A_F the corresponding signal at the temperature T_F, where the fat was completely melted (T and T_F in °Kelvin). $$\text{SFI }(\%) = 100 - A_T/A_F \cdot T/T_F \cdot 100$$ The correcting term T/T_F was introduced to take into account the temperature dependence of the NMR signal.
Lechat et al., 1975	MP. Nr	—	—
Parodi, 1974a	SP. As described by Parodi and Dunstan (1971).	—	—
Rolland and Riel, 1966	CP. Nr	—	—
Sherbon and Dolby, 1973	SP. Method described by Sherbon et al. (1972).	DSC method described by Sherbon et al. (1972).	—
Timms, 1980a	—	DSC determinations were carried out according to Timms (1979).	NMR. SFC determinations were carried out according to Timms (1979b).
Timms, 1980b	—	The calorimeter used was a Perkin-Elmer DSC-2 calibrated with indium, dodecane, and octadecane. Sample weight was about 8 mg, with an empty sealed pan and two flat aluminum lids as a reference. Sample pans were transferred from storage at 13°C in a small insulated container that could pass through the airlock on the glove box of the DSC. The temperature program was 280°K (sample loading temperature), cooled at 5°K/min to 270°K and held for 5 min; heated at 10°K/min to 330°K.	NMR. SFC was determined with a Bruker Minispec P20 pulsed NMR instrument.

Reference	Method	Description
Yi, 1993	MP. AOCS Method Cc 1-25 (1973).	Thermal profiles were determined with a Perkin-Elmer DSC 7 Differential Scanning Calorimeter (Perkin-Elmer Corp., Norwalk, CT). The DSC was calibrated with indium and mercury. The temperature of the sample holder cells was controlled with liquid nitrogen, and dry helium gas was flushed through the cells at 20 cc/min to help expel heat. 8–15 mg of fat samples were placed into aluminum-coated DSC pans (TA Instruments, New Castle, DE) and hermetically sealed. The samples were heated to 60°C for 3 min to erase past crystalline structure, cooled to –60°C at 10°C/min and held for 3 min to induce crystallization, and reheated to 60°C at 10°C/min to record a melting curve.
		NMR. SFC were analyzed at the National Magnetic Resonance Facility (Madison, WI) with a Bruker AM400 wide-bore multinuclear spectrometer in conjuction with a Bruker Aspect 3000 off-line data processing station (Bruker Spectrospin, Burlington, Ontario, Canada). NMR tubes 5 mm dia. were filled to a level of 20–25 mm. A reference sample of olive oil was used. A modified AOCS Method Cd 16-81 (AOCS, 1989) was used.
Youssef et al., 1977	MP. Method of Jacobs (1964).	—
Ethanol		
Rolland and R:el, 1966	CP. Nr	—
Pentane		
Kaylegian, 1991	MP. AOCS Method Cc 1-25.	Thermal profiles were recorded with a Perkin-Elmer 7 DSC, calibrated with indium and mercury. 5-10 mg samples were encapsulated in hermetically sealed, aluminum-coated DSC pans (DuPont Co., Wilmington, DE). The DSC was fitted with a glove box and purged with air. The sample cell was purged with helium and cooled with liquid nitrogen during analysis. Samples were held at 60°C for 3 min to destroy crystal history, cooled to –60°C at 10°C/min, held for 3 min, and then heated to 60°C at 10°C/min to obtain a melting profile.
		NMR. AOCS Method Cd 16-81; SFC values were measured every 5°C from 0°C to 55°C or until the sample was melted. Analyses were performed using a Bruker AM400 wide-bore multinuclear spectrometer with Aspect 3000 off-line data processing station (Bruker Spectrospin, Canada) at the Nuclear Magnetic Resonance Facility at Madison (Univ. of WI—Madison).

Continued

TABLE 4.4 (Continued)

Fractionation method[1]	Melting point[2]	Thermal analysis[3]	Solid fat content[4]
Supercritical fluid extraction			
Carbon dioxide			
Arul et al., 1987	—	Melting curves were performed on a DuPont model 990 thermal analyzer (DuPont Instruments, Toronto, Ontario) by the method of Timms (1980a).	—
Bhaskar et al., 1993b	—	Melting and crystallization curves were determined using a DSC (DSC-1, Perkin-Elmer, Norwalk, CT) as outlined by Norris et al. (1971).	DSC. SFC was calculated from DSC melting thermograms.
Büning-Pfaue et al., 1989	—	Nr	—
Chen et al., 1992	—	Melting curves were obtained using a Perkin-Elmer DSC-7 (Perkin-Elmer Corp., Norwalk, CT) differential scanning calorimeter with an empty pan as reference. 10 mg samples were used in each measurement. The melting thermograms were recorded while scanning from −40 to 60°C at 10°C/min.	—
Hamman et al., 1991	—	DSC measurements were performed with a Perkin-Elmer DSC-2C calorimeter. 1–2 mg samples, stored in a refrigerator at 7°C, were weighed into Perkin-Elmer standard volatile sample aluminum pans, using an empty pan as reference. The scanning rate was 5°C/min from 0 to 50°C.	—
Kankare and Antila, 1988a	—	Melting and crystallization properties were determined using a differential scanning calorimeter, Perkin-Elmer DSC-4. The cooling and heating rates were 8°C/min.	—
Kankare et al., 1989	—	Melting and crystallization properties were determined using a differential scanning calorimeter, Perkin-Elmer DSC-4. The cooling and heating rates were 8°C/min.	—

Methods for the Characterization of Milkfat Fractions

Kaufmann et al. 1982	—	Melting and crystallization behavior was examined by differential scanning calorimetry using a Perkin-Elmer DSC-2. Sample size was 8 mg in aluminum pans. MF was held for 10 min at 60°C, cooled to −60°C at 5°C/min, warmed at 5°C/min to 60°C.	*NMR*. SFC determinations were carried out on a Bruker Minispec (NMR) PC 20, using the method of van den Embden et al. (1978).
Rizvi et al., 1990	*MP*. AOCS method.	DSC was done according to the procedure of Norris et al. (1971).	—
Shukla et al., 1994	—	Thermal analysis was performed using a DSC (Model DSC-1; Perkin-Elmer, Norwalk, CT) according to the procedure described by Norris et al. (1971).	*DSC*. SFC at any temperature was given by the ratio of the partial area above that temperature to the total area under the DSC curve.
Timmen et al., 1984	—	Nr	*NMR*. Nr
Short-path distillation			
Arul et al., 1988a	—	Melting curves were performed on a DuPont model 990 thermal analyzer (DuPont Instruments, Toronto, Ontario) by the method of Timms (1980a).	*Dilatometry*. SFI was determined by dilatometry as described by Jasperson and McKerrigan (1957). A correction factor for thermal expansion of 4.25 $\mu l/5°C/g$ fat was applied as proposed by Riel (1966).

[1]Sorted by fractionation method and listed alphabetically by author.

[2]CP = cloud point; DP = dropping point; DSC = differential scanning calorimetry; MP = melting point; SLP = slip point; SP = softening point.

[3]DSC = differential scanning calorimeter; MF = milkfat.

[4]DSC = differential scanning calorimeter; MF = milkfat; NMR = nuclear magnetic resonance; SFC = solid fat content; SFI = solid fat index.

[5]No method or data reported.

[6]Nr = Data of this category reported in the literature, but the method employed was not.

TABLE 4.5 Methods Used to Determine the Textural Properties of Milkfat Fractions and Some Finished Products Containing Milkfat Fractions

Fractionation method[1]	Product characterized	Textural properties measured and method employed[2]
Crystallization from melted milkfat		
Vacuum filtration		
Deffense, 1987	Cold-spreadable butter	*Spreadability measured by penetrometry and sensory evaluation.* Spreadability was determined by penetrometry at 5°C and reported in arbitrary units. Spreadability and consistency were measured by sensory evaluation with a knife.
Frede et al., 1980	Milkfat fractions Cold-spreadable butter	*Cutting resistance measured by sectility.* Cutting resistance was determined according to the official regulatory wire cutting method at 15°C, and results reported in N.
Guyot, 1982	Cold-spreadable butter	*Spreadability measured by cone penetrometry.* Spreadability was measured by cone penetrometry at 4 and 18°C.
Hayakawa and deMan, 1982	Milkfat fractions	*Hardness measured by constant speed penetrometry.* MF was melted at 60°C and filled into plastic cups (20 mm dia.). Samples were cooled rapidly and kept for 2 h at –20°C, then left overnight in a thermostatically controlled room at measurement temperature of 5, 10, 15, 20 and 25°C. Measurements were made with a constant speed penetrometer using a set of cylindrical penetration punches with flat ends having diameters of 2, 4, 6, and 8 mm. The speed was 0.0005 m/s. A punch was chosen arbitrarily depending on the expected hardness of the sample and sensitivity of the recorder. Force readings of the force-distance curves were taken at points where the force exhibited yield points. Results are hardness in kg/cm^2.
Jamotte and Guyot, 1980	Cold-spreadable butter	*Spreadability measured by cone penetrometry.* Spreadability was measured by cone penetrometry at 4, 8, and 14°C, and reported in units of 0.1 mm. A Hertzog penetrometer, carrying a load of 150 g, applied for 5 sec with a 30° cone, was used.
Kaylegian and Lindsay, 1992	Cold-spreadable butter	*Spreadability and firmness of body measured by sensory evaluation.* Samples were evaluated by descriptive sensory analysis panels composed of 25 to 30 panelists. Panelists were presented with a ballot, two individual-serving packages of saltine crackers, a plastic knife, and a cup of water for rinsing. Ballots contained descriptors selected to focus on appearance, flavor, physical, and spreadability attributes.
Makhlouf et al., 1987	Cold-spreadable butter	*Spreadability determined by cone penetrometry.* Tests were done with an Instron "Universal Testing Machine" (Burlington, Ontario). The apparatus was fitted with an isothermal sample chamber to maintain constant temperature of the samples during analysis. The conditions used were as follows: cone with a 40° angle, a penetration level of 1 cm, 4 speeds of penetration (5, 10, 20 and 20 cm/min) and 3 temperatures (7, 15 and 20°C). For each temperature the butter samples were tempered for 48 h before the test. The spreadability was expressed by the force of resistance to penetration as a function of the speed of the cone.
		Yield value and apparent (Bingham) viscosity. Yield value was measured at 7, 15, and 20°C and reported in kPa. Apparent viscosity was measured at 7, 15, and 20°C and reported in Pa.s.

Continued

TABLE 4.5 (Continued)

Fractionation method[1]	Product characterized	Textural properties measured and method employed[2]
Vovan and Riel, 1973	Cold-spreadable butter	*Spreadability.* Spreadability was measured with the apparatus described by Huebner and Thomsen (1957), and expressed as resistance in g.
Pressure filtration		
deMan, 1968	Milkfat fractions	*Hardness measured by cone penetrometry.* Hardness was determined after cooling the liquid fractions rapidly in a blast freezer, and measuring with a cone penetrometer: AOCS Method Cc 16-60.
Centrifugation		
Bratland, 1983	Cold-spreadable butter	*Spreadability.* Nr[3]
Kankare and Antila, 1974b	Cold-spreadable butter	*Cutting resistance:* Consistency was measured by the cutting resistance method according to MOHR, at 10°C, and reported in g.
Keogh and Morrissey, 1990	Baking fats	*Yield values measured by cone penetrometry.* The yield value was measured by cone penetrometry according to Haighton (1959) after storage at 15°C for 1 day. Data reported in g/cm^2.
Richardson, 1968	Milkfat fractions	*Penetration.* Micropenetration was determined according to Feuge and Bailey (1944).
Separation with the aid of an aqueous detergent solution		
Dolby, 1970b	Milkfat fractions Cold-spreadable butter	*Hardness by sectility.* Hardness was determined after 30 d.
Filtration through muslin cloth		
Sreebhashyam et al., 1981	Milkfat fractions	*Softness by cone penetrometer.* Cone penetrometer AOCS method (1963).
Separation technique not reported		
Dixon and Black, 1974	Cold-spreadable butter	*Firmness measured by disc penetrometry.* Samples were tested for firmness at 5, 13, and 20°C using a disc penetrometer.
El-Nimr, 1980	Cold-spreadable butter	*Softness measured using a penetrometer.* Penetration values were determined at 5 and 12.5°C using penetrometer type Metrimpex 0B-204 (0-630) for 5 sec intervals, following the method of El-Nimr and Bass (1979).
Verhagen and Bodor, 1984	Dairy-based spreads	*Hardness.* Hardness, expressed in C-values (g/cm^2 was determined according to Haighton (1959) at 5, 10 and 20°C.
Verhagen and Warnaar, 1984	Dairy-based spreads	*Hardness.* Hardness, expressed in C-values (g/cm^2 was determined according to Haighton et al. (1959) at 5, 10 and 20°C.
Crystallization from solvent solution		
Acetone		
Kaylegian and Lindsay, 1992	Cold-spreadable butter	*Spreadability and firmness of body measured by sensory evaluation.* Samples were evaluated by descriptive sensory analysis panels composed of 25 to 30 panelists. Panelists were presented with a ballot, two individual-serving packages of saltine crackers, a plastic knife, and a cup of water for rinsing. Ballots contained descriptors selected to focus on appearance, flavor, physical and spreadability attributes.

Continued

TABLE 4.5 (Continued)

Fractionation method[1]	Product characterized	Textural properties measured and method employed[2]
Munro et al., 1978	Cold-spreadable butter	*Hardness.* Hardness was measured at 5°C and reported in arbitrary units.
Norris, 1977	Cold-spreadable butter	*Hardness measured by sectility.* Hardness was measured by sectility at 5°C and reported in arbitrary units. A "stand-up" property was evaluated at 22°C and data reported in arbitrary units.
Supercritical fluid extraction		
Carbon dioxide		
Shukla et al., 1994	Dairy-based spreads	*Complex viscosity, storage modulus, loss modulus.* Small-amplitude oscillatory measurements were performed on a Bohlin VOR Rheometer (Bohlin Instruments, Cranbury, NJ) using a 90 g.cm torsion bar and parallel plate geometry (15 mm plate diameter and 4 mm plate gap). Four test temperatures were used in the range of 12 to 32°C. The complex viscosity (η^*), the storage modulus (G'), and the loss modulus (G'') were determined for frequencies in the range of .1 s^{-1} and 10 s^{-1} at strains < .1%, where the viscoelastic properties were linear. From the flow curves of η^* versus frequency obtained at differed temperatures, master curves were obtained by shifting curves to a reference temperature (22°C) using the method of reduced variables (Ferry, 1980).
Fractionation method not reported		
Pedersen, 1988	Puff pastry butter	*Plasticity measured by finger judgment and penetration.* Finger judgment of plasticity was made at 18°C. Penetration measurements were made before and after work softening at 22°C and data reported in terms of decrease in firmness.

[1]Sorted by fractionation method and listed alphabetically by author.

[2]MF = milkfat.

[3]Nr = Data of this category reported in the literature, but the method employed was not.

TABLE 4.6 Methods for Characterization of the Quality of Milkfat Fractions

Fractionation method[1]	Oxidative stability[2]	Lipolytic stability[3]
Crystallization from melted milkfat		
Vacuum filtration		
Amer et al., 1985	PV. AOCS Method Cd 8-53.	—[4]
Jamotte and Guyot, 1980	PV. The method in BS684, section 2.14. Anisidine value. The method in BS684, section 2.24.	FFA. Method in BS684, section 2.10.
Kaylegian, 1991	PV. Hydroperoxide concentrations were determined by the method of Buege and Aust (1978).	—
Centrifugation		
Bhat and Rama Murthy, 1983	PV. Method of Lea (1939).	—
Keshava Prasad and Bhat, 1987a	Conjugated polyunsaturated fatty acids. Conjugated polyunsaturated fatty acids were determined by measuring the absorbance of the fat in iso-octane, as described in AOCS (1962) using Bausch and Lomb Spectrometer 2000.	—
Keshava Prasad and Bhat, 1987b	PV. Method of Lea (1931). Monocarbonyls. Total and monocarbonyl were estimated by the method of Schwartz et al. (1963).	—
Lakshminarayana and Rama Murthy, 1986	PV. Method of Lea (1931).	Hydrolysis rate. Rate of hydrolysis was determined by the potentiometric method of Marchis-Mouren et al. (1959). The method involved incubating the lipase (Type II, Sigma, Co.) and the substrate mixture at 37°C at the optimum pH, and following the lipolysis by continuously titrating the fatty acids liberated and maintaining the pH. The substrate emulsion was prepared by dispersing 20 mL of MF in 180 mL of 10% gum arabic solution at 10°C, followed by homogenization for 10 min in a mixer. 25 mL of the emulsion were tempered at 37°C in a 100 mL beaker placed in a constant-temperature water bath. The electrodes of a pH meter were immersed in the substrate emulsion. The pH of the substrate was brought to 9.0 by adding a few drops of 0.1 N sodium hydroxide solution with stirring, and then 1 mL of enzyme preparation was added. The fatty acids liberated by the enzyme tended to lower the pH, but the optimum pH was maintained at 9.0 by adding 0.1 N NaOH solution every 2 min for 20 min.
Richardson, 1968	—	FFA. Method of Iyer et al. (1967).
Filtration through muslin cloth		
Khalifa and Mansour, 1988	PV. AOAC Method. TBA. AOAC Method.	—

Continued

TABLE 4.6 (Continued)

Fractionation method[1]	Oxidative stability[2]	Lipolytic stability[3]
Crystallization from solvent solution		
Acetone		
Kaylegian, 1991	*PV*. Hydroperoxide concentrations were determined by the method of Buege and Aust (1978).	—
Supercritical fluid extraction		
Carbon dioxide		
Kaufmann et al., 1982	*PV*. According to a modified Wheeler process reported by Timmen (1975b).	*FFA*. FFA content as oleic acid was obtained by multiplying the acid concentration (determined according to IDF Standard No. 8 1960 by a factor of 0.282).
Timmen et al., 1984	Nr[5]	Nr
Fractionation method not reported		
Kehagias and Radema, 1973	*PV*. The ferric thiocyanate method was described by Loftus Hills and Thiel (1945–1946) with one modification. The ferrous chloride solution was prepared by adding 2 mL of conc. HCl and 1 g of hydrated ferrous sulfate to 100 mL distilled water. A fresh solution was made each time the PV had to be estimated. PV was expressed as mEq of oxygen/kg MF. *TBA*. According to Koops (1960). TBA values were expressed as extinction values (as defined by Koops).	*FFA*. Estimated by dispersing approx 5 g melted MF in 50 mL of a 1:1 mixture of ethyl ether and ethyl alcohol and titrating with 0.05 N NaOH with phenolphthalein as indicator. FFA was expressed as mEq/100 g MF.

[1]Sorted by fractionation method and listed alphabetically by author.

[2]MF = milkfat; PV = peroxide value; TBA = thiobarbituric acid.

[3]FFA = free fatty acid; MF = milkfat.

[4]No method or data reported.

[5]Nr = Data of this category reported in the literature, but the method employed was not.

TABLE 4.7 Methods for the Determination of Selected Nutritional Characteristics of Milkfat Fractions

Fractionation method[1]	Cholesterol content[2]	Selected compounds
Crystallization from melted milkfat		
Centrifugation		
Antila and Antila, 1970	—[3]	*Vitamin A*. According to Kaufmann (1958). *Tocopherol*. According to Erickson and Dunkley (1964).
Branger, 1993	Weigh 0.5 g of MF in 15 × 150 mm screw cap glass tube. Add 10 mL of alcoholic KOH solution (8.56 M), cap tightly with Teflon-lined screw caps, and vortex mix for 20 sec. Hold 1 h at 70 ± 2°C water bath, agitate every 15 min. Cool to room temperature. Add 5 mL of a saturated NaCl solution, 5 mL water, 1 mL of internal standard solution (1 mg 5 α-cholestane/1 mL 2,2,4-trimethylpentane), and 10 mL of petroleum ether. Vortex mix for 20 sec. Centrifuge (1,571 g) for 10 min or until the emulsion breaks and two clear layers develop. Remove upper layer with Pasteur pipette and transfer to a 15 × 200 mm glass tube. Re-extract twice. Evaporate the combined extracts under nitrogen in a water bath at 50 ± 1°C to about 1 mL but not to dryness. Add 2 mL of choloroform–methanol (2+1), and mix well. Transfer quantitatively to a 10 mL volumetric flask. Rinse tube three times, combine extracts and then dilute to 10 mL volume with choloroform–methanol (2+1).

A volume of 5 mL of the prepared sample was injected into the GC. The GC was a Hewlett-Packard Model 5890 (Hewlett-Packard, Avondale, PA) equipped with an auto sample (Hewlett-Packed model 7673A) and a flame ionization detector. The GC was equipped with a 0.25 mm × 15 m DB-5 column with a 0.1 mm film thickness (J & W Scientific, Folsom, CA). The operating conditions were: split ratio 20:1 with hydrogen carrier at head pressure of 35 KPa; injector temperature 325°C; detector temperature 325°C. The temperature program was isothermal maintained at 275°C for 4 min. The solvent peak appeared at about 0.8 min, 5 α-cholestane at about 1.9 min, and cholesterol at about 2.8 min. Quantitation of cholesterol was done with internal standards (5 α-cholestane and cholesterol). | |
Richardson, 1968	Method of Moore et al. (1965).	—
Separation with the aid of an aqueous detergent solution		
Norris et al., 1971	—	*Vitamin A*. Estimated as described by Thompson et al. (1949). *Carotene*. Estimated as described by Thompson et al. (1949).

Continued

TABLE 4.7 (Continued)

Fractionation method[1]	Cholesterol content[2]	Selected compounds
Crystallization from solvent solution		
Acetone		
Chen and deMan, 1966	Free and ester cholesterol were estimated with tomatine as indicated by Kabara et al. (1961). The absorbance of the tomatinide solution was measured with a Bausch & Lomb Spectronic 20 at 625 μm. Readings were made 30 min after the Liebermann–Burchard reagent was added.	—
Hexane		
Muuse and van der Kamp, 1985	Cholesterol was determined according to NEN 6350 (1977). The sterols were gravimetrically determined after saponification of the fat and precipitation of the sterols as their sterol digitonides.	—
Supercritical fluid extraction		
Carbon dioxide		
Bhaskar et al., 1993b	Cholesterol was determined by a modified method of Lynch and Barbano (1988).	*Carotene content.* Carotenoids were determined as β-carotene by AOAC Method 938.04 (1990).
Bradley, 1989	Cholesterol was assayed according to the AOAC (1984) digitonin procedure for β-3-OH sterols.	—
Chen et al., 1992	Cholesterol content was determined by an enzymatic assay (Boehringer Mannheim Corp., Indianapolis, IN) according to the instructions provided with the enzymatic kit.	—
Ensign, 1989	The AOAC (1984) method for β-sitosterol in butter oil (28.104) was modified as indicated. An appropriate amount of MF was added to a small centrifuge tube, 1.5 × 12 cm, with a Teflon-lined screw cap. 3 mL *n*-hexane and 0.5 mL of a 1.0 mg/mL β-sitosterol internal standard solution were added to the tube and mixed thoroughly, then transferred to the digitonin–Celite column. The centrifuge tube was washed twice with 2 mL of *n*-hexane, and the washes were added to the column. The column was washed with five 2 mL volumes of *n*-hexane and five 2 mL volumes of benzene. When the last portion of benzene was about 1 cm above the top of the digitonin–Celite packing, the tubing was removed, and the column tip was rinsed thoroughly with benzene. 10 mL of DMSO were added to the column to elute the cholesterol and β-sitosterol. DMSO was drained and collected from the column in a large centrifuge tube, 2.5 × 25 cm. The column was washed with two 5 mL volumes of DMSO and the eluate collected. 9 mL of *n*-hexane were added to the combined eluates; the mixture was vortexed and then centrifuged at 3000 rpm for 10 min. The upper layer, hexane, containing the sterols was transferred to a second large centrifuge tube. The DMSO solution was washed with two 12 mL volumes of *n*-hexane–benzene (1/1 v/v), followed by vortexing and centrifuging. The upper layer was removed and transferred to the second centrifuge tube each time. The combined upper layers were evaporated to dryness under nitrogen on a warm water bath. The dry sample was redissolved in 1 mL chloroform, and a 1 μL sample was analyzed by GC. Either a DB-5 column, 0.25 mm ID	—

Methods for the Characterization of Milkfat Fractions 139

Reference	Method	Carotene
Ensign, 1989 (continued)	× 15 m (J & W Scientific, Folsom, CA) or an HP-1 column, 0.53 mm ID × 25 m (Hewlett-Packard Co., Avondale, PA) was used. Both columns were coated with 1 μm 5% diphenyl and 95% dimethyl polysilicone crosslinked gum. The oven temp was 250°C, and the injector and FID temps were 280°C. The heat pressure was 10 psi, the carrier gas was helium for the DB-5 column and hydrogen for the HP-1 column, at 0.75 mL/min.	
Kankare et al., 1989	The total cholesterol was determined from the unsaponified part, using GC.	—
Kaufmann et al., 1982	Cholesterol content was found in the course of GC analysis of TG composition.	—
Rizvi et al., 1990	Cholesterol content determined according to AOCS methods.	—
Shishikura et al., 1986	Cholesterol content was determined by GC method of Kaneda et al. (1980) using 5-α-cholestane as an internal standard. GC was carried out with JEOL JGC-20K apparatus equipped with a glass column (0.3 × 150 cm) containing 5% SE-30 on Chromosorb W (60/80 mesh).	—
Shukla et al., 1994	Cholesterol was analyzed by the method of Lynch and Barbano (1991), but modified to form a trimethylsilyl ether derivative (TMS). The analysis was performed using a capillary column coated with SE-30 (Chrompack Co., Middelburg, The Netherlands).	*Carotene content.* Carotenoids were determined as β-carotene using AOAC Method 938.04.
Timmen et al., 1984	Nr[4]	—

[1]Sorted by fractionation method and listed alphabetically by author.

[2]GC = gas chromatography; MF = milkfat; TG = triglyceride.

[3]No method or data reported.

[4]Nr = Data of this category reported in the literature, but the method employed was not.

Chapter 5
Technical Data for Experimental and Commercial Milkfat Fractions

In this chapter the physical and chemical characteristics of both experimental and commercial milkfat fractions are tabulated and evaluated with respect to fractionation conditions and to the potential performance of these fractions in food products. Other aspects important to the use of specific milkfat fractions in food products, including flavor potential and oxidative stability, are also discussed. The detailed technical data presented in this chapter are summarized in Chapter 6 to provide a broad overview of the effects of fractionation on milkfat.

Nomenclature and Categorization of Milkfat Fractions

Two sets of conventions have been developed to identify milkfat fractions: *fraction designation, general fraction property.*

Fraction Designation

The fraction designation is a term that identifies individual milkfat fractions. Milkfat fractions produced by crystallization from melted milkfat or from solvent solutions are designated by a number indicating their fractionation temperature, followed by a letter indicating physical state at separation (S = solid, L = liquid) and a subscript that indicates the fractionation procedure employed (S = single-step, M = multiple-step). Notations in brackets following the main designation refer to the parent milkfat. Designations in parentheses refer to conventions used by the original authors. For example, a $25S_M$ fraction was the solid fraction obtained at 25°C using a multiple-step fractionation procedure. A fraction designated as $20L_M$ [S] (III) is the liquid fraction obtained at 20°C using a multiple-step fractionation procedure using summer milkfat and was called fraction III by the original investigator.

In cases where information on fractionation conditions was not available, and for all fractions produced by supercritical carbon dioxide extraction and short-path distillation, the fraction designations used are those assigned by the original investigators. Designations commonly employed by investigators include *hard, soft, stearin* (solid fraction), *olein* (liquid fraction), *HMF* (high-melting fraction), *MMF* (middle-melting fraction), *LMF* (low-melting fraction), or numbers or letters (#1, #2, A, B, C). Designations employed for milkfat fractions obtained by supercritical carbon dioxide extraction and short-path distillation generally reflect the fractionation process (e.g., extract #1, residual, and distillate #3). Where possible, fractionation conditions (temperature and pressure) are also identified for these fractions.

General fractionation conditions, and hence fraction designations, have been duplicated among researchers. However, physical, chemical, and functional properties of milkfat fractions are influenced by factors other than fractionation conditions; hence, two or more fractions with the same designation may not exhibit the same functionality.

General Fraction Property

The general fraction property is used to reflect the overall functionality in terms of milkfat fractions of broad categories, based on melting point (MP):

- Very-high-melting milkfat fractions: MP greater than 45°C
- High-melting milkfat fractions: MP between 35 and 45°C

- Middle-melting milkfat fractions: MP between 25 and 35°C
- Low-melting milkfat fractions: MP between 10 and 25°C
- Very-low-melting milkfat fractions: MP less than 10°C
- Unknown-melting milkfat fractions; which could not be categorized by melting behavior because data were not reported

These categories have been selected to facilitate interpreting and discussing the physical and chemical characteristics of milkfat fractions presented in this chapter, and to form a framework within which to link these characteristics with the functional properties of milkfat fractions in food products. Generalizations of the physical and chemical characteristics of milkfat fractions provide a perspective of the functionality of milkfat fractions in foods, but specific characteristics of individual milkfat fractions within the categories can vary.

Many options are available for characterizing the melting behavior of fats:

- Capillary melting or clear point, AOCS (1989) Method Cc 1-25
- Softening point, AOCS (1989) Method Cc 3-25
- Dropping point, Mettler method (Mertens and deMan, 1972)
- Slipping point, AOCS (1989) Method Cc 4-25
- Cloud test, AOCS (1989) Method Cc 6-25
- Solid fat content, AOCS (1989) Method Cd 16-81
- Solid fat index, AOCS (1989) Method Cd 10-57
- Thermal profiles (DSC curves)

For specific information relating to each of these analyses, refer to Chapter 4.

The melting point is the temperature that yields a transparent fat (AOCS, 1989; Sonntag, 1982). Other melting indicators, such as softening and slip point, are obtained prior to the clear point and, therefore, are reported at lower temperatures than the clear point. Thus, if the melting behavior of a milkfat fraction was categorized by methods other than capillary melting point by the original investigators, the fraction may have been placed in a lower-melting category than if it had been categorized based on the true clear point.

The melting point of a fat can also be determined using DSC curves (Grall and Hartel, 1992), although this is not commonly employed. Fats melt over a range of temperatures and often do not exhibit a sharp melting peak. Consequently, difficulties are associated with the determination of the melting point of a fat using DSC curves. Grall and Hartel (1992) interpreted the melting point to be the peak temperature. However, the shape of melting peaks is often not symmetrical and has accompanying shoulders, so melting point determinations made by the authors of this monograph have been interpreted to be the temperature at which the DSC curve returns to its baseline. Some of the literature included in this monograph provided DSC curves but not specific melting point behavior, so melting point behavior has been determined using the original DSC curves, and these interpretations have been so indicated.

Experimentally Prepared Milkfat Fractions

Over 850 milkfat fractions have been produced experimentally by the following fractionation methods:

- Crystallization from melted milkfat (Table 5.1)
- Crystallization from solvent solution (Table 5.2)

- Supercritical carbon dioxide extraction (Table 5.3)
- Short-path distillation (Table 5.4)
- Unspecified fractionation method (Table 5.5)

All tables and figures referred to in this chapter, are found in the section "Tables and Figures" at the end of this chapter. Each of these methods offers a number of options, discussed in detail in Chapter 3, for the production of milkfat fractions. To allow organized discussion of milkfat fractions obtained by similar methods, subcategories are used in addition to the broad categories that indicate the general fractionation method, as already seen in Chapters 3 and 4:

1. Crystallization from melted milkfat
   - Separation by vacuum filtration
   - Separation by pressure filtration
   - Separation by centrifugation
   - Separation with the aid of an aqueous detergent solution
   - Filtration through muslin cloth
   - Filtration through milk filter
   - Filtration with cheese press
   - Filtration with casein-dewatering press
   - Filtration technique not reported

2. Crystallization from solvent solution
   - Acetone
   - Ethanol
   - Isopropanol
   - Pentane
   - Hexane

3. Supercritical fluid extraction
   - Carbon dioxide
   - Propane

4. Short-path distillation

More specific indicators of fractionation conditions for individual fractions are often incorporated into the fraction designation. An overall summary of the experimental milkfat fractions produced by the different fractionation methods is presented in Table 5.6.

Crystallization from melted milkfat has been the most common method employed to produce milkfat fractions (537 fractions; Table 5.1), followed by crystallization from solvent solution (189 fractions; Table 5.2). Crystallization methods have been widely used for many years by the commercial edible fats and oils industry to modify fats for specific applications (Thomas, 1985). Experimental studies on the crystallization of milkfat have been conducted since the late 1950s (Baker et al., 1959; Bhalerao, 1959). Crystallization from melted milkfat is used to fractionate milkfat on a commercial scale throughout the world. Supercritical fluid extraction, a much newer technology, has been used to fractionate milkfat experimentally since the early 1980s, and 110 milkfat fractions (Table 5.3) have been described in the literature. Short-path distillation has also been used experimentally, and four milkfat fractions have been reported in the literature (Table 5.4). Table 5.5 lists 14 fractions that were used in food applications (Chapter 8), but fractionation information was not reported.

Approximately 60% of the milkfat fractions reported in the literature have been categorized based on melting behavior (Table 5.6). The high-melting milkfat fractions (171 fractions) and the low-melting milkfat fractions (163 fractions) were the most abundant with respect to the total number of fractions produced. However, the quantitative abundance (percentage yield) of individual milkfat fractions varies with the type of fractionation method used as well as the type of fractionation procedure (single-step or multiple-step) employed. Yield values for specific fractions are discussed within their respective general fraction property categories.

Many milkfat fractions (331 fractions; Table 5.6) reported in the literature were categorized as unknown-melting milkfat fractions because the melting behavior data were not reported. Melting behavior data were not reported by some investigators because the project concentrated on other aspects of milkfat fractions, such as flavor potential or oxidative stability. Additionally, analyses performed on milkfat fractions reflect the time period when the experiment was conducted; for example, technologies for individual triglyceride analyses were not available until the 1970s. Sample size can also influence the choice of analyses performed, particularly in the case of milkfat fractionated using supercritical carbon dioxide extraction, which produces smaller experimental samples than do the crystallization methods.

Fractionation parameters and selected chemical and physical properties of milkfat fractions are discussed in the following sections and have been arranged into the following subgroups according to the general fraction properties:

- Intact anhydrous milkfats used to produce milkfat fractions
- Very-high-melting milkfat fractions
- High-melting milkfat fractions
- Middle-melting milkfat fractions
- Low-melting milkfat fractions
- Very-low-melting milkfat fractions
- Unknown-melting milkfat fractions

Tables 5.1 through 5.5 detail the overall fractionation schemes employed by individual investigators, and milkfat fractions are listed in fractionation sequence. Other relevent information, including location, season of production of the parent intact AMF, yield, and melting behavior of the fractions, is also included in these tables. Selected chemical and physical characteristics of milkfat fractions have been tabulated at the end of this chapter:

- Yield and melting points (Tables 5.8 to 5.32)
- Fatty acid compositions (individual fatty acids) (Tables 5.33 to 5.57)
- Fatty acid compositions (selected fatty acid groupings) (Tables 5.58 to 5.82)
- Triglyceride compositions (individual triglycerides) (Tables 5.83 to 5.105)
- Triglyceride compositions (selected triglyceride groupings) (Tables 5.106 to 5.130)
- Selected chemical characteristics (Tables 5.131 to 5.149)
- Lactone concentrations (Tables 5.150 to 5.155)
- Methyl ketone concentrations (Tables 5.156 to 5.160)
- Aldehyde concentrations (Tables 5.161 to 5.165)
- Crystal morphologies (Tables 5.166 to 5.171)
- Crystal sizes (Tables 5.172 to 5.173)

- Thermal profiles (Figures 5.1 to 5.21)
- Solid fat content profiles (Tables 5.174 to 5.194)
- Textural characteristics (Tables 5.195 to 5.199)

Within each of these tables the milkfat fractions have been presented primarily by fractionation and separation method and secondarily in order of descending fractionation temperature. Thus, the general fraction properties of milkfat fractions (e.g., very-high-melting milkfat fractions) can be readily compared with respect to the specific fractionation methods employed. For a detailed listing, refer to the lists of tables and figures at the beginning of this book.

Other important properties of milkfat fractions, namely deterioration properties (Tables 5.200 and 5.201) and nutritional properties (Table 5.202), are presented following the discussion of fractionation parameters and chemical and physical characteristics of milkfat fractions. Data for the deterioration and nutritional properties of milkfat fractions have been sorted primarily by fractionation method and secondarily by author instead of by fractionation temperature, which was the convention employed for the data presented in tables for the chemical and physical characteristics of milkfat fractions.

The last topic addressed for experimental milkfat fractions in this chapter is blends of milkfat fractions. Milkfat fractions have been blended with other milkfat fractions, intact AMF, cocoa butter, and other edible oils (Tables 5.203 and 5.204).

Intact Anhydrous Milkfats Used to Produce Experimental Milkfat Fractions

Milkfat fractionation has been studied experimentally worldwide, and this has generated a wealth of data on milkfat fractions produced from different intact anhydrous milkfats. Factors such as breed of cow, season of milkfat production, and location influence the inherent characteristics of milkfat (Chapter 2), and these factors are further influenced by fractionation conditions to yield milkfat fractions with unique characteristics.

Chemical Characteristics of Intact Anhydrous Milkfats Used to Produce Milkfat Fractions

Fatty acid composition Fatty acid composition data are presented by individual fatty acids in Tables 5.33 to 5.36, and by selected fatty acid groupings in Tables 5.58 to 5.61, for the intact AMFs used to produce milkfat fractions from melted milkfat, solvent solution, supercritical fluid extraction, and short-path distillation, respectively. The data presented in these tables exhibit a wide range of percent composition for individual fatty acids. The ranges and average values for selected fatty acids of intact milkfat are:

FATTY ACID	RANGE (wt %)	AVERAGE (wt %)
C4	0.9 to 11.8	3 to 4
C14	7.3 to 15.5	9 to 12
C16:0	21.9 to 38.2	25 to 30
C18:0	6.8 to 18.0	10 to 15
C18:1	16.0 to 30.6	23 to 28
C4–C10	4.6 to 22.8	8 to 13
C12–C15	9.5 to 19.9	13 to 15
C16 C18	31.9 to 49.1	35 to 43
C18:1/C18:0	1.68 to 3.48	2.0 to 2.5

Because of the range of fatty acid compositions encountered, consistent trends relating to season and location are not easily discerned. In general, winter AMF contained higher concentrations of the C16 to C18 saturated fatty acids, and lower concentrations of C4 to C10 fatty acids, than summer milkfats. This is supported by data from Amer et al. (1985), Deroanne and

Guyot (1974), and Badings et al. (1983b). Interestingly, Badings et al. and Deroanne and Guyot reported lower 18:0 concentrations for the winter milkfat than for summer milkfat, but the 16:0 was greater for the winter milkfats.

Laakso et al. (1992) reported the fatty acid composition of six triglyceride classes of AMF (Tables 5.33 and 5.58). As expected, the trisaturated (SSS) fraction contained the highest concentration of the 4:0, 6:0, 8:0, 10:0, 12:0, 14:0, 16:0, and 18:0 fatty acids. The 18:1 fatty acid was found in the highest concentration in the dimonounsaturated triglycerides (SM^{c}M^c and SM^{c}M^t), and was greater for the *cis, cis* triglyceride, than for the *cis, trans* triglyceride. This is consistent with data from Banks et al. (1985; Table 5.33) who reported that the *cis*-18:1 isomer is present in greater concentration than the *trans* isomer.

Triglyceride composition The triglyceride composition of intact anhydrous milkfats also showed considerable variation. Triglyceride compositions are presented by individual triglyceride in Table 5.83 to 5.86 and by selected triglyceride groupings in Tables 5.106 to 5.109 for intact anhydrous milkfats used to produce milkfat fractions from melted milkfat, solvent solution, supercritical fluid extraction, and short-path distillation, respectively. The ranges and average values for selected triglycerides of intact AMFs are:

TRIGLYCERIDES	RANGE (wt %)	AVERAGE (wt %)
C24–C34	2.5 to 15.2	9 to 12
C36–C40	29.2 to 43.7	30 to 38
C42–C54	39.4 to 74.1	50 to 60

Only Badings et al. (1983b; Table 5.106) reported triglyceride composition data for both winter and summer milkfats, and the winter AMF had a greater concentration of C24 to C34 and C36 to C40 and a lower concentration of the C42 to C54 compared with the summer AMF.

Laakso et al. (1992) and Deffense (1993b) reported the triglyceride concentration of milkfat in terms of the triglyceride classes (Table 5.106). The most abundant triglyceride class is the monounsaturated (SSU) at 40–50%, followed by the trisaturated (SSS) triglycerides, ranging from 33 to 45%. The combined diunsaturated and triunsaturated (SUU and UUU) classes comprised 14 to 18% of the total triglycerides.

Selected chemical characteristics Selected chemical characteristics are presented in Tables 5.131, 5.132, and 5.133 for intact anhydrous milkfats used to produce milkfat fractions from melted milkfat, solvent solution, and supercritical fluid extraction, respectively. Iodine values were often reported for AMF samples, and values ranged from 29 to 39.5. The summer milkfats had greater iodine values than winter milkfats (Amer et al., 1985; Deffense, 1989). The refractive indices ranged from 1.4495 to 1.4550 at 40°C. Reichert–Meissl values, Polenske values, and density were seldom reported for intact milkfat.

Flavor The concentrations of lactones, methyl ketones, and aldehydes in intact milkfats used to produce milkfat fractions are listed in Tables 5.150, 5.156, and 5.161, respectively. The range of concentrations of these flavor compounds in intact milkfat are:

Lactones	60 to 130 mg/kg
Methyl ketones	115 to 125 mg/kg
Aldehydes	3.7 to 6.3 μmoles/10 g

Physical Characteristics of Intact Anhydrous Milkfats Used to Produce Milkfat Fractions

Crystal characteristics Crystal morphologies of intact AMF were determined by X-ray diffraction (Table 5.166). Most intact AMFs were composed primarily of stable β' crystals. Milkfat samples that were cooled rapidly exhibited the presence of unstable α-crystals, which transformed into more stable β' crystals upon storage. Very stable β crystals were observed only in samples that had been stored prior to analysis, but these milkfats also contained β' crystals. Data on the size of intact milkfat crystals was not available.

Melting point The melting points for AMF used to produce milkfat fractions from melted milkfat, from solvent solution, by supercritical fluid extraction, and by short-path distillation are shown in Tables 5.8, 5.9, 5.10 and 5.11, respectively. Overall, melting points of AMF ranged from 28.2 to 42.8°C, and most melting points were in the 32 to 35°C range.

Melting point calculations by the authors of this monograph using DSC curves provided by the original investigators appeared to provide slightly greater values (38 to 42°C) than melting points determined by the capillary method (32 to 35°C). This poses questions as to the accuracy and validity of assigning the melting point to the temperature at which the DSC curve returns to its baseline. Melting points for pure compounds are taken at the temperature of the apex of the melting peak, but fat melts over a broad range and often exhibits a shoulder on the DSC curve that can distort the melting point. An interesting example of the differences between peak temperature, baseline temperature, and capillary melting point is illustrated by the data of Amer et al. (1985), who reported a capillary melting point of 34.6°C for the intact milkfat (Table 5.8), whereas the DSC curve (Figure 5.1) shows a peak temperature of approximately 40°C, and the temperature at which the curve returns to the baseline is approximately 50°C. These data exhibit a 5°C difference between melting point and peak temperature and a 10°C difference between peak and baseline temperature, both of which are considerable differences. Consequently, the use of DSC curves to determine a true melting point of a fat, and the temperature at which to read the melting point, remains questionable. Conversely, a melting point of 35°C was obtained from the baseline temperature of the DSC curves reported by Arul et al. (1987; Figure 5.3), which is in general agreement with melting points for other anhydrous milkfats from Canada (Table 5.7). Bhaskar et al. (1993b) also used the baseline temperature to assign melting behavior, because that was the temperature at which all of the milkfat was in the liquid state.

There is a wide range of melting points reported for intact anhydrous milkfats that exhibited no consistent trends between locations and seasons, although some seasonal differences within a location may be apparent. The range of melting point values reported for some AMFs obtained from the same locations and from the same season is shown in Table 5.7. For example, Amer et al. (1985) reported a melting point of 34.6°C and Baker et al. (1959) reported a melting point of 30°C for winter anhydrous milkfats produced in Canada, whereas Timms and Parekh (1980) reported a softening point of 34.5°C for summer milkfat produced in Australia.

Thermal profile Thermal profiles obtained using differential scanning calorimetry (DSC) provide a generalized view of the melting behavior of major melting triglyceride species of milkfat. Milkfat is a very heterogeneous fat that exhibits a wide melting range, generally accepted in the literature to be from −40°C to +40°C. Intact milkfat has three main triglyceride species, as illustrated by the three distinct peaks in the DSC curves presented in Figures 5.1 through 5.4:

- Low-melting triglyceride species, which melt below 10°C
- Middle-melting triglyceride species, which melt between 10 and 20°C
- High-melting triglyceride species, which melt above 20°C

These designations should not be confused with low-melting, middle-melting, and high-melting milkfat fractions, which have been defined by different melting ranges than the major triglyceride species with similar names.

Most DSC curves for intact anhydrous milkfat exhibit three main melting peaks, although the size and shape of each peak differ slightly. Several AMF samples exhibit only two broad peaks. Sherbon and Dolby (1973) used a high-melting glyceride fraction as the starting material for fractionation, which explains the single peak shown in Figure 5.2 (curve D).

Solid fat content profile The solid fat contents (SFC) of anhydrous milkfats used to produce milkfat fractions from melted milkfat, solvent solution, supercritical fluid extraction, and short-path distillation are shown in Tables 5.174, 5.175, 5.176, and 5.177, respectively. As seen with the melting point data, anhydrous milkfats used to produce milkfat fractions exhibit a range of SFC values over their melting ranges. However, they all melt gradually throughout the melting ranges and are usually fully melted between 35 and 40°C. A typical SFC curve for AMF was shown in Chapter 2 (Figure 2.2).

The SFC profiles of summer and winter intact milkfats show slight differences within a single production location, but these differences are often negligible when comparing fractions from different locations. As an illustration, Amer et al. (1985; Table 5.174) produced winter and summer AMFs in Canada that showed slight differences in SFC profiles, with the winter AMF being more solid than the summer AMF throughout the melting range, but Kankare and Antila (1988b; Table 5.174) produced a summer milkfat in Finland that exhibited a melting profile similar to that of the winter fat produced by Amer et al. in Canada.

Textural characteristics The texture of intact milkfat has been measured by constant-speed penetrometry and cone penetrometry (Table 5.195). Penetration values for the cone penetrometer ranged from 2.7 to 3.4 mm at 10°C, and 5.3 mm at 15°C. Hardness at 15°C was measured at 2.58 kg/cm^2 by constant-speed penetrometry and at 1.02 kg/cm^2 by cone penetrometry.

Very-High-Melting Milkfat Fractions

A total of 58 experimental very-high-melting milkfat fractions (melting point greater than 45°C) have been reported in the literature (Table 5.6), and this represents 12% of the number of fractions that were categorized by melting behavior. Very-high-melting milkfat fractions were produced by:

1. Crystallization from melted milkfat (Table 5.12; 25 fractions) using the following separation techniques:
   - Vacuum filtration (Kankare and Antila, 1988a; Kaylegian, 1991; Kaylegian and Lindsay, 1992; Makhlouf et al., 1987; Schaap and van Beresteyn, 1970; Sherbon, 1963)
   - Pressure filtration (Deffense, 1993b; deMan, 1968)
   - Centrifugation (Antila and Antila, 1970; Kankare and Antila, 1986, 1988b)
   - Separation with the aid of an aqueous detergent solution (Banks et al., 1985; Schaap et al., 1975; Walker, 1972),

2. Crystallization from solvent solutions (Table 5.13; 29 fractions),
   - Acetone (Kaylegian, 1991; Kaylegian and Lindsay, 1992; Schaap et al., 1975; Sherbon, 1963; Sherbon and Dolby, 1973; Timms, 1980a, 1980b; Yi, 1993; Youssef et al., 1977)
   - Pentane (Kaylegian, 1991),

3. Supercritical fluid extraction (Table 5.14; 2 fractions)
   - Carbon dioxide (Bhaskar et al., 1993b; Hamman et al., 1991)

4. Fractionation methods not reported (Table 5.5; 2 fractions) (Jebson, 1974; Mayhill and Newstead, 1992)

Fractionation Conditions for Very-High-Melting Milkfat Fractions

Crystallization from melted milkfat The very-high-melting milkfat fractions produced by crystallization from melted milkfat are the solid fractions obtained either from the first step of a multiple-step procedure or from a single-step procedure (Table 5.12). The average fractionation temperature was 24.8°C. The range of fractionation temperatures (23 to 34°C) used to produce very-high-melting milkfat fractions was much greater than the range of their melting points (45.1 to 49.2°C). Kaylegian (1991) reported a melting point of 49.0°C for a fraction (34S_M [W]) obtained at 34°C, and Sherbon (1963) reported a fraction (23S_M) with a melting point of 48.9°C but a fractionation temperature of only 23°C. There was an 11°C difference in fractionation temperature, but only a 0.1°C difference in melting point between these fractions. Therefore, higher fractionation temperatures do not necessarily yield very-high-melting milkfat fractions with higher melting points.

The solid fractions obtained from the first step of a multiple-step fractionation procedure had melting points similar (± 2.5°C) to those obtained by a single-step procedure at equivalent fractionation temperatures. Kaylegian in the U.S. (1991) produced very-high-melting milkfat fractions at 34°C, using both single-step and multiple-step procedures with vacuum filtration with the same intact winter AMF source, and reported melting points 46.6°C and 49.0°C, respectively. Kankare and Antila in Finland reported three very-high-melting milkfat fractions, obtained at 24°C in different studies from summer AMF using a multiple-step procedure with centrifugation that had melting points of 47°C (1988a), 48.0°C (1988b), and 46.3°C (1986). deMan in Canada (1968) produced a very-high-melting milkfat fraction at 24°C that had a 46.7°C melting point, but employed a single-step fractionation procedure and pressure filtration. Both Schaap and van Beresteyn in The Netherlands (multiple-step, vacuum filtration, 1970) and deMan (single-step, pressure filtration, 1968) produced very-high-melting milkfat fractions at 27°C and reported melting points of 49.2°C and 47.6°C, respectively. Very-high-melting milkfat fractions obtained from melted milkfat at similar temperatures exhibited similar melting points, regardless of AMF source and location, provided that the separation method used did not employ additives.

Sherbon (1963) refractionated a very-high-melting milkfat fraction (23S_M, MP = 48.9°C) obtained at 23°C and produced another very-high-melting milkfat fraction (23ppt$_M$ refractionated, MP = 51.8°C) with a higher melting point. Refractionation of very-high-melting milkfat fractions obtained from melted milkfat yields fractions with higher melting points.

Most of the very-high-melting milkfat fractions that were obtained from melted milkfat and separated with the aid of an aqueous detergent solution had melting points higher than those of milkfat fractions produced at similar temperatures using vacuum filtration, pressure filtration, or centrifugation. Schaap et al. (1975) reported a melting point of 54.0°C for the fraction produced at 28°C with the use of an aqueous detergent solution (28S_S), compared with melting points of 49.2 and 47.6°C for fractions obtained at 27°C by vacuum filtration (Schaap and van Beresteyn, 1970) and pressure filtration (deMan, 1968). This elevation of melting points results from the dispersion of the solid milkfat crystals into the aqueous phase, which separates them from the bulk liquid oil fraction. The crystals are relatively free of contamination from liquid oil and contain more pure high-melting triglycerides; consequently, they have higher melting points than fractions from other separation methods commonly employed for crystallization from melted milkfat.

Banks et al. (1985) studied the effect of detergent levels on the separation of milkfat fractions produced using a single-step fractionation procedure, and found that the use of high levels of detergent yielded three fractions—a solid, a semi-solid, and a liquid fraction—whereas low levels of detergent yielded only one solid and one liquid fraction. They reported that a very-high-melting milkfat fraction obtained at 16°C using a high level of detergent had a melting point of 52.3°C, but using a low level of detergent lowered the melting point to 45.4°C. They also reported two other very-high-melting milkfat fractions, produced with a high level of detergent at 21°C, that had melting points of 51.6°C (solid fraction) and 53.0°C (semi-solid fraction). Although the melting point of the semi-solid fraction obtained at 21°C is higher than that of the solid fraction obtained at 21°C, the melting point of the semi-solid fraction obtained at 16°C (MP = 44.7°C, a high-melting milkfat fraction) is lower than that of the solid fraction obtained at 16°C (MP = 45.4°C, a very-high-melting milkfat fraction). The reason for this difference in melting behavior between the solid and semi–solid fractions obtained at 21°C and 16°C using high levels of detergent was not reported.

Crystallization from solvent solution Very-high-melting milkfat fractions obtained from solvent solution are similar to those obtained from melted milkfat in that they are all solid fractions obtained from single-step fractionation or at the first step of a multiple-step procedure (Table 5.13). Very-high-melting milkfat fractions were obtained from solvent solution at lower temperatures than fractions obtained from melted milkfat. Fractionation temperatures ranged from 0 to 32°C for fractions obtained from solvent solution, compared to a range of 16 to 34°C for melted milkfat. Lower temperatures are required to effect crystallization because of the solubility of milkfat in nonaqueous solvents.

Multiple-step fractionation sequences that employ a decreasing temperature profile yield the highest-melting solids at the first fractionation step. Very-high-melting milkfat fractions can be obtained at the first through fourth steps of a multiple-step procedure from solvent solution. Because milkfat fractionated from the melt has the crystalline and liquid fractions in direct contact with each other, liquid fat is often entrapped within the solid crystal matrix or interspersed among the spherulite crystal needles, yielding a less pure crystalline fraction. Conversely, milkfat in solvent solution will remain dissolved in solution until it has crystallized, at which time it settles out of solution. Because the liquid oil stays in solution and apparently does not interact with the crystalline milkfat, solvent fractionation yields purer, higher-melting crystals. Therefore, it is possible to produce very-high-melting milkfat fractions from the second, third, and fourth fractionation steps from solvent solution, but not from melted milkfat.

Very-high-melting milkfat fractions obtained from solvent solution have higher melting points than do fractions obtained from melted milkfat at similar temperatures. Melting points have ranged from 45.0 to 60.3°C for fractions obtained from solvent solution, compared with a range of 45.1 to 54°C for fractions obtained from melted milkfat. Kaylegian (1991) produced two very-high-melting milkfat fractions from solvent solution at 25°C, one using a single-step procedure and the other at the first step of a multiple-step procedure, that had melting points of 53.0 and 53.7°C, respectively. Kaylegian (1991) also produced milkfat fractions at 25°C by crystallization from melted milkfat (vacuum filtration, Table 5.15), but these fractions had melting points of 40.6 and 42.0°C and were categorized as high-melting milkfat fractions. Schaap et al. (1975) produced very-high-melting milkfat fractions both from solvent solution and from melted milkfat (separation with the aid of an aqueous detergent solution), and also reported a higher melting point for the fraction obtained from solvent solution ($28S_S$ fraction, MP = 58.6°C) than for the fraction obtained from melted milkfat ($28S_S$ fraction, MP = 54.0°C).

Very-high-melting milkfat fractions obtained from solvent solution exhibited similar melting behavior for fractions obtained by single-step and multiple-step fractionation. The melting points decreased as fractionation temperature decreased. Kaylegian (1991) produced very-high-melting milkfat fractions with both single-step and multiple-step procedures from acetone solution at 25, 21 and 15°C, and from pentane solution at 11, 5 and 0°C. Fractions produced at equivalent temperatures from the same solvent had similar melting points (± 2.5°C), but trends were not consistent as to which method, single-step or multiple-step, produced fractions with higher melting points.

Very-high-melting milkfat fractions obtained by refractionation at the same temperature from solvent solution exhibited lower melting points than did fractions obtained by refractionation from melted milkfat, which exhibited an increase in melting point when refractionated at the same temperature. Sherbon and Dolby (1973) used a high-melting glyceride as the starting material for fractionation and produced three very-high-melting milkfat fractions using a multiple-step procedure ($32S_M$ (I) [HMG], $25S_M$ (II) [HMG], and $25S_M$ (III refractionated) [HMG]). These fractions exhibited decreasing melting points with subsequent steps (56, 54 and 48°C, respectively), even though the fractionation temperature for the second and third fraction was the same. Youssef et al. (1977) also obtained two very-high-melting milkfat fractions at the first and third steps of a multiple-step fractionation procedure at the same temperature, and the melting point decreased for the fraction obtained at the third step. Although these fractions were refractionated at 18°C, they were separated by a step at 0°C.

The use of high-melting glycerides did not raise the melting point of milkfat fractions obtained from solvent solution, compared with fractions obtained using intact AMF. Sherbon and Dolby (1973) reported a melting point of 56°C for the fraction obtained from high-melting glycerides at 32°C, and Schaap et al. (1975) produced a fraction with a higher melting point (58.6°C) at a lower fractionation temperature (28°C) from intact AMF. Sherbon and Dolby (1973) and Kaylegian (1991) both obtained very-high-melting milkfat fractions at 25°C, and the melting points were 54°C and 53.7°C, respectively, although the starting materials were different.

Fractionation from pentane solution required lower temperatures than from acetone solution to effect crystallization, because of the greater solubility of milkfat in pentane than in acetone. Fractionation temperatures ranged from 11 to 32°C for very-high-melting milkfat fractions obtained from acetone solution, and from 0 to 11°C for fractions obtained from pentane solution (Table 5.13). Kaylegian (1991) reported that very-high-melting milkfat fractions obtained from pentane had higher melting points than the fractions obtained at equivalent temperatures from acetone solution, and that the first three solid fractions obtained from pentane solution had similar melting points to the first three solid fractions obtained from acetone solution.

Supercritical fluid extraction Two very-high-melting milkfat fractions with melting points of 45°C (Bhaskar et al., 1993b) and 48°C (Hamman et al., 1991) were produced by supercritical carbon dioxide extraction. These were the residual fractions that remained in the vessel after the extraction process was completed (Table 5.14). The supercritical carbon dioxide extraction process extracts the lower-melting triglycerides first, followed by species with increased melting points. The fraction remaining in the extraction vessel after the process has been completed usually contains few low-melting triglyceride species.

Percentage Yields of Very-High-Melting Milkfat Fractions

The percentage yields for individual very-high-melting milkfat fractions ranged from 0.9 to 36.3%, with an average of 16% for fractions obtained by crystallization from melted milkfat (Table 5.12) and 8% for fractions obtained by crystallization from solvent solution (Table 5.13). The increase in yields for very-high-melting milkfat fractions obtained from melted milkfat, compared with those obtained from solvent solution, is most likely caused by the presence of liquid milkfat entrapped within the crystal matrix. The yield for a very-high-melting milkfat fraction obtained by supercritical carbon dioxide extraction was 21% (Table 5.14).

The most abundant very-high-melting milkfat fraction obtained from *melted milkfat* was the F.S. fraction (36.3%) produced by Makhlouf et al. (1987). However, this fraction was obtained by the combination of the $26S_M$ (S.1) and $21S_M$ (S.2) fractions, which had yields of 10.7% and 25.6%, respectively. Other very-high-melting milkfat fractions with yields greater than 10% were produced using vacuum filtration (the $21S_M$ (S.2) fraction, Makhlouf et al., 1987); pressure filtration (the $30S_S$ and $24S_S$ fractions, deMan, 1968), and separation with the aid of an aqueous detergent solution (the 21 semi-S_S fraction, Banks et al., 1985). Yield data for fractions obtained from melted milkfat did not exhibit consistent trends between separation methods.

Yields of very-high-melting milkfat fractions obtained from melted milkfat were similar between single-step and multiple-step fractionation procedures at equivalent temperatures. Kaylegian (1991) reported yields of 3.7% for the fraction obtained from the first step of a multiple-step procedure at 34°C, and 2.5% for the fraction obtained using a single-step procedure at 34°C. Kaylegian (1991) also reported similar values for the very-high-melting milkfat fractions obtained from the first step of a multiple-step procedure and from a single-step procedure at 25°C from acetone solution; these were 1.7 and 2.2%, respectively. However, very-high-melting milkfat fractions obtained at the second through fourth steps in a multiple-step fractionation procedure from solvent solution exhibited lower yields than did fractions obtained using a single-step procedure at equivalent temperatures. Kaylegian (1991) reported a yield of 4.5% for the fraction obtained at the third step of a multiple-step procedure at 11°C, compared with a yield of 11.1% for the single-step fraction obtained at 11°C.

The most abundant very-high-melting milkfat fraction obtained from *acetone* was the fraction obtained at 25°C (33.1%; 25S_M (II) [HMG] fraction) by Sherbon and Dolby (1973), who employed a high-melting glyceride as the starting material. Interestingly, this is the fraction produced in the second step of a multiple-step fractionation procedure. Sherbon and Dolby refractionated this fraction at 25°C and reported a marked decrease in yield of the refractionated fraction (2.7%; 25S_M (III refractionated) [HMG] fraction). However, Youssef et al. (1977) found a slight increase in yield upon refractionation at 18°C and reported a yield of 14.2% for this fraction (18S_M (E)), which was at the third step of their multiple-step fractionation scheme, compared with a yield of 11.5% for the first fraction obtained at 18°C (18S_M (A)).

Very-high-melting milkfat fractions obtained from acetone solution had greater yields than fractions obtained from *pentane* solution at equivalent temperatures. Kaylegian (1991) produced very-high-melting milkfat fractions by single-step fractionation at 11°C from both acetone and pentane and reported yields of 19.3% and 2.6%, respectively. The decrease in yield for the fraction from pentane was probably caused by the greater solubility of milkfat in pentane than in acetone.

In general, yields for very-high-melting milkfat fractions increased as the fractionation temperature decreased for fractions obtained by single-step fractionation from both melted milkfat and solvent solutions.

Chemical Characteristics of Very-High-Melting Milkfat Fractions

The chemical characteristics of very-high-melting milkfat fractions exhibit general trends as a broad category, but specific characteristics for individual fractions can vary greatly and often depend on the chemical composition of the intact milkfat. For example, most of the very-high-melting milkfat fractions obtained from melted milkfat (Table 5.37) contained 1 to 3% C4 fatty acid, except for the fractions produced by Sherbon (1963), who reported C4 contents of 9.2% and 4.2% for the 23S_M and 23ppt$_M$ refractionated fractions, respectively. Sherbon (1963) also reported unusually high C4 contents for two fractions obtained from acetone solution (Table 5.38), the 2B and 2AJAN [Jan] fractions, with 8.2 and 8.7% C4, respectively. In contrast, the C4 contents for most very-high-melting milkfat fractions obtained from solvent solution ranged from 0.2 to 2.1%. The causes for these differences may have been errors in fatty acid analysis or in the composition of the starting milkfat. However, Sherbon did not report the fatty acid content for the intact AMF used to produce the fractions, and the cause of the high values remains speculative.

Fatty acid composition The fatty acid composition of very-high-melting milkfat fractions obtained from melted milkfat (Table 5.37 and 5.62) and from solvent solution (Table 5.38 and 5.63) showed similar trends compared with intact AMF, but these trends were more pronounced for fractions obtained from solvent solution. In general,

- C4–C10 fatty acid content decreased.
- C12–C15 fatty acid content increased slightly.
- C16–C18 saturated fatty acid content increased.

A very-high-melting milkfat fraction obtained using supercritical carbon dioxide extraction (Table 5.39 and 5.64) had less C4–C10 and C12–C15 fatty acids and more C16–C18 saturated fatty acids than did the starting AMF.

Very-high-melting milkfat fractions obtained from melted milkfat using vacuum filtration or centrifugation had higher concentrations of the C4–C10 fatty acids and lower concentrations of C16–C18 saturated fatty acids than did fractions obtained using pressure filtration or an aqueous detergent solution. As was previously discussed, separation of milkfat fractions with the aid of an aqueous detergent solution yields more pure crystalline fractions, but it also appears that pressure filtration is an effective way of expelling liquid milkfat from the crystal matrix to yield more pure crystalline very-high-melting milkfat fractions without the use of additives. An exception is the 21 semi-S_S [A] fraction obtained by Banks et al. (1985), which had a higher concentration (7.3%) of the C4–C10 fatty acids than did the solid fraction (21S_S [A]; 5.5%) produced using the same fractionation conditions; this reflects the increased liquid milkfat content in the semi-solid fraction.

Sherbon (1963) reported an increase in melting point for the fraction refractionated at 23°C from melted milkfat (23ppt$_M$ fraction), and this fraction had a lower content of C4–C10 fatty acids (9.4%) than did the original 23S_M fraction (13.6%). However, the refractionated fraction also showed a decrease in the C16–C18 saturated fatty acid content, no change in the C12–C15 fatty acid content, and an increase in the C18:1/C18:0 ratio—all of which indicate that the refractionated fraction would be likely to melt at a lower temperature than the original fraction. This unexpected result suggests the need for further investigation into the effects of refractionation on the properties of milkfat fractions.

The greater crystalline purity of very-high-melting milkfat fractions obtained from solvent solution is illustrated by the low content of C4–C10 fatty acids (less than 5%; Tables 5.38 and 5.63), compared with fractions obtained from melted milkfat, which had C4–C10 contents of approximately 4 to 9% (Tables 5.37 and 5.62). For fractions obtained from solvent solution, the C4–C10 and the C12–C15 fatty acid contents increased, with a concurrent decrease in the saturated C16–C18 fatty acid content, as the fractionation temperature decreased and with successive steps in a multiple-step fractionation procedure. This trend is illustrated by the fractions produced by Kaylegian (1991) using single-step and multiple–step fractionation procedures from acetone and pentane solutions. The 25S_S and 21S_S fractions produced from acetone solution by Kaylegian exhibited similar fatty acid compositions, and the 15S_S and 11S_S fractions also exhibited similar fatty acid compositions compared with each other, but there was a slight difference in fatty acid composition between these two groups of fractions. The similarities in behavior between the 25S_S and the 21S_S and between the 15S_S and the11S_S fractions were also exhibited in their SFC and DSC profiles. The DSC profiles showed that the 15S_S and 11S_S fractions had increased amounts of the middle-melting glyceride species; this is consistent with the compositional data.

Refractionation of very-high-melting milkfat fractions obtained from solvent solution yields fractions with decreased concentrations of C4–C10 and C16–C18 saturated fatty acids and with an increase in C12–C15 fatty acids. Sherbon and Dolby (1973) refractionated a very-high-melting milkfat fraction obtained at 25°C and reported a decrease from 1.5 to 0.7% for C4–C10 fatty acids; an increase from 18.2 to 23.7% for C12–C15 fatty acids; and a decrease from 71.9 to 60.3% for the C16–C18 saturated fatty acids for the refractionated fraction compared with the original fraction obtained at 25°C. This change in fatty acid composition coincides with the lower melting point observed for this refractionated fraction (Sherbon and Dolby, 1973; Table 5.13).

The very-high melting milkfat fractions obtained from supercritical carbon dioxide extraction (Table 5.39) were the residual fractions obtained after the lower-melting fractions had been extracted. This is illustrated by comparing the low concentration of C4–C10 fatty acids present in the supercritical extracted fraction (1.2–1.3%; Table 5.64) with that in the parent intact milkfat (7.80–8.55%; Table 5.60). The very-high-melting milkfat fraction obtained from supercritical carbon dioxide extraction had a higher quantity of C18:1 (30%) compared with its parent milkfat (23.3%) and with other very-high-melting milkfat fractions obtained from melted milkfat (approximately 16–20%) and from solvent solution (approximately 9–11%). At the same time, the concentration of C16:0 and C18:0 fatty acids were lower in the very-high-melting milkfat fraction obtained by supercritical carbon dioxide extraction than in fractions obtained from melted milkfat or solvent solutions.

Triglyceride composition The triglyceride composition of very-high-melting milkfat fractions obtained from melted milkfat (Tables 5.87 and 5.110) and from solvent solution (Tables 5.88 and 5.111) showed similar trends compared with intact AMF, and these trends were more pronounced for fractions obtained from solvent solution. In general,

- C24–C34 triglyceride content decreased.
- C36–C40 triglyceride content decreased.
- C42–C54 triglyceride content increased.

The general trend for the C24–C34 and C42–C54 triglycerides for very-high-melting milkfat fractions follows the same pattern that was seen for the C4–C10 and C16–C18 saturated fatty acids. However, the C36–C40 triglyceride content decreased while the C12–C15 fatty acid content increased slightly for very-high-melting milkfat fractions. This implies that the C12–C15 fatty acids are found at higher concentrations in the C42–C54 triglycerides than in the C36–C40 triglycerides.

The decreased content of liquid milkfat in very-high-melting milkfat fractions obtained from melted milkfat and separated with the aid of an aqueous detergent solution is evident in the lower content of C24–C34 triglycerides in these fractions than in very-high-melting milkfat fractions separated using vacuum filtration or centrifugation (Tables 5.87 and 5.110). The semi-solid fraction produced by Banks et al. (1985) using an aqueous detergent solution has a higher concentration of C24–C34 triglycerides than does the solid fraction obtained using the same fractionation conditions (7.9 and 2.8%, respectively), and it has a lower concentration of the C42–C54 triglycerides (47.4 and 73.1%, respectively). These trends are consistent with the other physical and chemical characteristics of very-high-melting milkfat fractions produced using an aqueous detergent solution, compared to fractions obtained from melted milkfat using other separation techniques. Deffense (1993b) reported that the very-high-melting milkfat fraction was composed of 54% trisaturated, 37% monounsaturated, and 9% di- and polyunsaturated triglycerides (Table 5.110).

Sufficient triglyceride composition data were not reported (Deffense, 1993b) to permit a statement regarding trends in fractionation temperature for very-high-melting milkfat fractions obtained from melted milkfat, although it would be expected that the C24–C34 triglyceride contents would increase as fractionation temperature decreased, with a concurrent decrease in the C42–C54 triglycerides.

Very-high-melting milkfat fractions obtained from solvent solution (Table 5.111) showed a marked decrease in the C24–C34 content (2.5 to 4.6%), compared with intact AMF (9.6 to 11.4%; Table 5.107) and with very-high-melting milkfat fractions obtained from melted milkfat using vacuum filtration or centrifugation (3.0 to 12.0%; Table 5.110). However, the C24–C34 content for fractions obtained from solvent solution was similar to the $21S_S$ [A] fraction obtained by Banks et al. (1985) using crystallization from melted milkfat and separation with the aid of an aqueous detergent solution. The approximate content of the C42–C54 triglycerides for fractions obtained from solvent solution was in the 80% range, compared with the 60–70% range for very-high-melting milkfat fractions obtained from melted milkfat and with a range of 40–60% for intact AMF.

Triglyceride composition for very-high-melting milkfat fractions obtained from solvent solution exhibited a gradual increase in the C24–C34 and C36–C40 triglyceride contents and a decrease in the C42–C54 content, as the fractionation temperature decreased and with successive steps in a multiple-step fractionation sequence (Kaylegian, 1991; Yi, 1993). The content of individual triglycerides increased with decreasing fractionation temperature for triglycerides below C46, and decreased with decreasing fractionation temperature for triglycerides above C46. The C46 triglyceride composition did not show any distinct trends relating to fractionation temperature.

The very-high-melting milkfat fraction obtained from supercritical carbon dioxide extraction exhibited a marked decrease in triglyceride content below C40 (Table 5.108) compared with intact AMF: 17.07% and 67.57%, respectively (Bhaskar et al., 1993b). A concurrent increase was observed in the C42–C54 triglyceride content: 32.93% for intact AMF and 82.93% for the very-high-melting milkfat fraction (Bhaskar et al., 1993b). These data further illustrate the extraction of the lower-melting triglycerides in the first fractions obtained by supercritical carbon dioxide extraction, resulting in a higher-melting residual fraction.

Selected chemical characteristics Iodine values were reported for four very-high-melting milkfat fractions. Antila and Antila (1970) reported an iodine value of 10.5 for the very-high-melting milkfat fraction obtained from melted milkfat (Table 5.134). Youssef et al. (1977) reported iodine values of 22.5 and 25.6 for the first fraction obtained at 18°C from acetone solution and the refractionated fraction obtained at 18°C, respectively (Table 5.135). The iodine values for fractions obtained from solvent solution should be lower than those obtained from melted milkfat because of the increased content of liquid milkfat in the fractions obtained from melted milkfat. However, this was not the case for these fractions, and fatty acid compositions were not available for these fractions to further assess the level of unsaturated fatty acids in these fractions.

Bhaskar et al. (1993b) reported an iodine value of 42.90 for the very-high-melting milkfat fraction obtained from supercritical carbon dioxide extraction (Table 5.136), compared with 35.20 for intact AMF (Table 5.133). The increase in iodine value of the very-high-melting milkfat fraction over the intact milkfat is an indication of the increase in the C16–C18 unsaturated fatty acid content of the fraction (Table 5.64). The C16–C18 fatty acids, both saturated and unsaturated, are the last to be solubilized by a supercritical fluid and tend to remain in the residual fraction. On the other hand, the unsaturated fatty acids are lower-melting than saturated fatty acids and tend to be separated with the first liquid fractions using thermal-based crystallization methods.

Flavor Walker (1972) reported the potential concentrations of lactones and methyl ketones and concentrations of aldehydes in a very-high-melting milkfat fraction (26.5S_M (F3S)) obtained from melted milkfat and separated with the aid of an aqueous detergent solution. The concentration of the lactones (Table 5.151) and methyl ketones (Table 5.157) in the very-high-melting milkfat fraction were about 60% of their concentration in the intact milkfat (Tables 5.150 and 5.156, respectively). The aldehyde concentration of the very-high-melting milkfat fraction (Table 5.162) was greater than that of the AMF (Table 5.161); this was caused by slight oxidation of the fat during AMF production and fractionation (Walker, 1972).

Physical Properties of Very-High-Melting Milkfat Fractions

Crystal characteristics The crystal *morphologies* of very-high-melting milkfat fractions obtained by crystallization from melted milkfat, solvent solution, and supercritical fluid extraction are shown in Table 5.167. Very-high-melting milkfat fractions exhibited the α, β', and β polymorphs. The β' was the most common stable crystal form observed. β crystals were observed in very-high-melting milkfat fractions obtained from solvent solution

(Schaap et al., 1975) and supercritical carbon dioxide extraction (Hamman et al., 1991). Sherbon and Dolby (1973) determined the polymorphic changes during cooling using DSC curves; they reported the transition of $\alpha \rightarrow$ liquid $\rightarrow \beta_2'$ for the very-high-melting milkfat fraction ($32S_M$ (I) fraction).

Data on the crystal sizes of very-high-melting milkfat fractions were not reported.

Melting point The melting points of very-high-melting milkfat fractions are presented in Tables 5.12, 5.13, and 5.14 for fractions obtained from melted milkfat, from solvent solution, and by supercritical fluid extraction, respectively. The relationship between melting points and fractionation conditions was discussed in the previous section on fractionation conditions of very-high-melting milkfat fractions. The general melting point characteristics of very-high-melting milkfat fractions are:

- By definition, melting points were greater than 45°C; they ranged from 45.0 to 60.3°C.
- Very-high-melting milkfat fractions obtained from melted milkfat had similar melting points when obtained at similar temperatures using vacuum filtration, pressure filtration, or centrifugation, and these fractions had lower melting points than fractions obtained with the aid of an aqueous detergent solution.
- Very-high-melting milkfat fractions obtained from melted milkfat had lower melting points than fractions obtained from solvent solution at equivalent temperatures.
- Fractions obtained from acetone solution had lower melting points than those obtained from pentane solution at equivalent temperatures.
- Refractionation of very-high-melting milkfat fractions yielded fractions with higher melting points when crystallized from melted milkfat, but with lower melting points when crystallized from solvent solution.

Thermal profile Very-high-melting milkfat fractions obtained from melted milkfat (Figure 5.5) exhibited a broader peak in the high-melting glyceride range and a small peak in the low-melting glyceride species range than did very-high-melting milkfat fractions obtained from solvent solution (Figure 5.6). The very-high-melting milkfat fraction obtained by supercritical fluid extraction exhibited a small low-melting and middle-melting peak and a large, broad high-melting peak (Figure 5.7).

Fractions obtained from melted milkfat contain more liquid milkfat, which results in more complex melting behavior and, consequently, a more complex DSC curve. The broad high-melting peak of the very-high-melting milkfat fraction obtained by supercritical carbon dioxide extraction also indicates a complex melting behavior in the high-melting region, which may be caused by the increased content of C16–C18 unsaturated fatty acids compared with fractions obtained from melted milkfat and solvent solution.

Very-high-melting milkfat fractions obtained from solvent solution exhibit DSC curves that consist primarily of one relatively narrow peak, which is generally located at the high-temperature end of the high-melting glyceride species range. This indicates a greater purity of the crystalline fraction obtained from solvent solution compared with fractions obtained from melted milkfat. The greater purity of crystalline fractions obtained from melted milkfat using an aqueous detergent solution (Banks et al., 1985), compared with fractions obtained from melted milkfat using other separation methods, was also evident in the DSC curves. These fractions (Figure 5.5) had a more distinct high-melting glyceride peak, similar to fractions obtained from solvent solution, but they also had a smaller, but definite, low-melting glyceride peak, which is consistent with fractions obtained from melted milkfat.

As the fractionation temperature decreased for both melted milkfat and solvent solutions, the high-melting glyceride peak shifted to lower temperatures, became broader, and developed a leading shoulder. The shift of the high-melting glyceride peak to lower temperatures is indicative of the decrease in melting points for fractions obtained using lower fractionation temperatures. The increase in peak width and development of the leading shoulder is caused by the increased presence of the middle-melting glyceride species as fractionation temperature is decreased. For fractions obtained from melted milkfat, an increase in the low-melting glyceride species peak also occurred, and this is illustrated by the DSC curves reported by deMan (1968) for fractions obtained using pressure filtration at 30, 27, and 24°C.

Very-high-melting milkfat fractions obtained from solvent solution using multiple-step fractionation procedures exhibited a more pronounced shift to lower temperatures in the high-melting glyceride peak than did fractions obtained using single-step fractionation. This shift represents the decrease in melting points of fractions obtained at successive steps with a multiple-step fractionation procedure; it is illustrated by comparing the fractions obtained by Kaylegian (1991) from acetone solution at 25, 21, and 15°C using multiple-step fractionation, with those obtained at 25, 21, 15 and 11°C using single-step fractionation. Very-high-melting milkfat fractions obtained from pentane solution exhibited trends similar to fractions obtained from acetone solution with respect to peak shifting and broadening for multiple-step and single-step fractionation procedures.

Solid fat content   The solid fat contents of very-high-melting milkfat fractions have been tabulated and are presented in Table 5.178 for fractions obtained from melted milkfat; in Table 5.179 for fractions obtained from solvent solution; and in Table 5.180 for fractions obtained by supercritical fluid extraction.

Although tabular data provide a ready reference of exact SFC values of milkfat fractions at specific temperatures, it is often useful to view this information graphically, and SFC profiles of representative very-high-melting milkfat fractions are presented in Figure 5.22. This figure illustrates the utility of SFC profiles in the determination of the functionality of milkfat fractions, because even though these fractions have similar melting points (49.0°C for $34S_M$ fraction from melted milkfat, and 50.7°C for $21S_M$ fraction from acetone solution, Kaylegian, 1991; 45°C for S1 fraction from supercritical carbon dioxide extraction, Bhaskar et al., 1993b), the profiles of their melting behavior are quite different, and this behavior should be of significance in food applications.

Major differences in melting behavior between very-high-melting milkfat fractions obtained from solvent solution and those obtained from melted milkfat are:

- Very-high-melting milkfat fractions obtained from solvent solution tend to have greater solid fat contents at most temperatures reported, compared with fractions obtained from melted milkfat.

- Solid fat content profiles of fractions obtained from solvent solution tend to melt more slowly at first, followed by sharp melting as the fraction nears the final melting point, whereas fractions obtained from melted milkfat melt more gradually throughout the melting range.

These two differences can be attributed to the greater crystalline purity of fractions obtained from solvent solution than of fractions obtained from melted milkfat. The presence of liquid milkfat, or lower-melting triglycerides, in the crystalline matrix of fractions obtained from melted milkfat results in an overall decrease in the SFC of these fractions over their melting range. The increased heterogeneity of these fractions also results in a more gradual melting profile. In contrast, the increased purity of the high-melting triglyceride species in fractions obtained from solvent solution results in slow melting of the fraction until the melting points of these higher-melting triglyceride species is approached, at which time the fraction melts rather sharply.

Very-high-melting milkfat fractions obtained from melted milkfat using vacuum filtration exhibited decreased SFC profiles as the fractionation temperature decreased (Table 5.177). The two fractions produced by Kaylegian (1991) at 34°C were similar, although the $34S_S$ [W] fraction had a slightly higher SFC than the 34S_M [W] fraction throughout the melting range. The 26S_M (S.1) and 21S_M (S.2) fractions produced by Makhlouf et al. (1987) exhibited decreasing SFC profiles as the fractionation temperature decreased, both relative to each other and to the fractions produced by Kaylegian (1991). The F.S. fraction produced by Makhlouf et al. (1987) was a combination of the 26S_M and 21S_M fractions; it had an SFC profile that was closer to that of the 21S_M fraction than to the middle range of its two constituent fractions.

The very-high-melting milkfat fraction produced from melted milkfat using centrifugation (24S_M [S]; Kankare and Antila, 1988b) exhibited a profile that also melted gradually but was more solid than that for fractions obtained using vacuum filtration. Kankare and Antila (1988b) reported the solid fat index for this fraction, and SFI values often do not correlate well with SFC values (Walker and Bosin, 1971).

Very-high-melting milkfat fractions obtained from solvent solution exhibited a decrease in SFC profile as the fractionation temperature decreased. Kaylegian (1991) obtained four very-high-melting milkfat fractions by single-step fractionation from acetone solution at 25, 21, 15, and 11°C, and these fractions exhibited decreases in SFC profiles as the fractionation temperature decreased. However, the SFC profiles were similar between fractions obtained at 25 and 21°C and between those obtained at 15 and 11°C, but there was a more pronounced difference in melting profile between these groups. The DSC curves of these fractions (Figure 5.6) illustrate that the 25S_S [W] and 21S_S [W] fractions are composed of mostly high-melting triglycerides, and the curves for the 15S_S [W] and the 11S_S [W] fractions have a larger shoulder at the onset of the main melting peak, which is indicative of the increased presence of the middle-melting glyceride species. This supports the contention that increased proportions of lower-melting glycerides, or liquid milkfat, in a milkfat fraction will lower its SFC profile.

Very-high-melting milkfat fractions obtained by single-step fractionation from pentane solution (Table 5.178; Kaylegian, 1991) also exhibited decreasing SFC profiles as the fractionation temperature decreased, but to a lesser extent than the fractions obtained from acetone solution.

The decrease in SFC profile with decreasing fractionation temperature was more pronounced in fractions obtained using a multiple-step procedure compared with a single-step fractionation procedure. This effect can be seen in fractions obtained from both acetone and pentane solutions (Table 5.178; Kaylegian, 1991).

The solid fat content profile for the very-high-melting milkfat fraction obtained by supercritical carbon dioxide extraction exhibited melting behavior similar to the fractions obtained from solvent solution at temperatures below 20°C and similar to fractions obtained from melted milkfat at temperatures above 25°C (Tables 5.180, 5.179, and 5.178, respectively). The more solid behavior at lower temperatures may be explained by the increased amount of longer-chain fatty acids present (Table 5.64) compared with fractions obtained from melted milkfat (Table 5.62) and from solvent solution (Table 5.63). The more liquid behavior at lower temperatures may be explained by the higher amount of longer-chain unsaturated fatty acids compared with fractions obtained from solvent solution, as was indicated by iodine values (Table 5.136 and 5.135, respectively).

Textural characteristics Data on the textural characteristics of very-high-melting milkfat fractions were not reported.

High-Melting Milkfat Fractions

Approximately one-third of the milkfat fractions characterized by melting behavior in the literature were high-melting milkfat fractions (Table 5.6). High-melting milkfat fractions have a melting point greater than 35°C and less than 45°C, and 159 of these fractions have been produced by:

1. Crystallization from melted milkfat (Table 5.15; 139 fractions), using the following separation techniques:

- Vacuum filtration (Amer et al., 1985; Barna et al., 1992; Bumbalough, 1989; Deffense, 1987, 1989, 1993b; deMan and Finoro, 1980; Grall and Hartel, 1992; Jamotte and Guyot, 1980; Kaylegian, 1991; Kaylegian and Lindsay, 1992; Laakso et al., 1992; Lund and Danmark, 1987; Schaap and van Beresteyn, 1970; Timms, 1978b; Timms and Parekh, 1980; Versteeg, 1991)
- Pressure filtration (Stepanenko and Tverdokhleb, 1974)
- Centrifugation (Bhat and Rama Murthy, 1983; Black, 1974b; Branger, 1993; Evans, 1976; Jordan, 1986; Kankare and Antila, 1988a, 1988b; Keogh and Higgins, 1986b; Keogh and Morrissey, 1990; Lakshminarayana and Rama Murthy, 1985; Ramesh and Bindal, 1987b; Richardson, 1968; Riel and Paquet, 1972)
- Separation with the aid of an aqueous detergent solution (Banks et al., 1985; Dolby, 1970a, 1970b; Doležálek et al., 1976; Fjaervoll, 1970a, 1970b; Jebson, 1970; Norris et al., 1971; Sherbon et al., 1972; Walker, 1972)
- Filtration through muslin cloth (Khalifa and Mansour, 1988; Sreebhashyam et al., 1981)
- Filtration through milk filter (Fouad et al., 1990)
- Filtration through cheese press (Baker, 1970a)
- Filtration with a casein-dewatering press (Black, 1973)
- Filtration technique not reported (Deroanne, 1976; Deroanne and Guyot, 1974; Dixon and Black, 1974; El-Ghandour and El-Nimr, 1982; Martine, 1982; Verhagen and Bodor, 1984)

2. Crystallization from solvent solutions (Table 5.16, 13 fractions) using the following solvents:

- Acetone (Avvakumov, 1974; Kaylegian, 1991; Kaylegian and Lindsay, 1992; Lambelet, 1983; Rolland and Riel, 1966; Sherbon, 1963; Sherbon and Dolby, 1973; Timms, 1980a, 1980b; Yi, 1993)
- Ethanol (Rolland and Riel, 1966),

3. Supercritical fluid extraction (Table 5.17; 17 fractions)

- Carbon dioxide (Arul et al., 1987; Bhaskar et al., 1993b; Biernoth and Merk, 1985; Büning-Pfaue et al., 1989; Chen et al., 1992; Kankare and Antila, 1988a; Kankare et al., 1989; Kaufmann et al., 1982; Rizvi et al., 1990; Shukla et al., 1994; Timmen et al., 1984),

4. Short-path distillation (Table 5.18; 1 fraction) (Arul et al., 1988a)

5. Fractionation method not reported (Table 5.5; 2 fractions) (Pedersen, 1988; Tucker, 1974).

Fractionation Conditions of High-Melting Milkfat Fractions

Crystallization from melted milkfat High-melting milkfat fractions produced by crystallization from melted milkfat were predominantly solid fractions produced using both single-step and multiple-step fractionation procedures and a variety of separation techniques (Table 5.15). Two liquid high-melting milkfat fractions were produced by refractionation of a high-melting milkfat fraction and separated with the aid of an aqueous detergent solution (Sherbon et al., 1972). The majority of high-melting milkfat fractions were produced using a single-step fractionation procedure. High-melting milkfat fractions were also produced at the first through fourth steps of a multiple-step fractionation procedure, in contrast to very high-melting milkfat fractions, which could only be obtained at the first step. Fractionation temperatures for high-melting milkfat fractions obtained from melted milkfat ranged from 12 to 38°C, a wider range than that employed for very-high-melting milkfat fractions (24 to 34°C). The most commonly employed fractionation temperatures were between 20 and 30°C.

Vacuum filtration was the most commonly employed separation technique. Fractionation temperatures used to produce high-melting milkfat fractions ranged from 18 to 32°C, and melting points ranged from 37.9 to 44.5°C. A number of fractions were produced at equivalent temperatures using both single-step and multiple-step fractionation procedures, and the following points were observed:

- Melting points of high-melting milkfat fractions obtained at equivalent temperatures were similar (± 1.5°C), regardless of the fractionation procedure employed.
- High-melting milkfat fractions obtained at temperatures greater than 25°C generally had melting points greater than 40°C, although increased fractionation temperatures did not always yield high-melting milkfat fractions with increased melting points.
- Differences in location and season of AMF production were not a major influence on the melting points of fractions obtained at equivalent temperatures.

Kaylegian (1991) produced high-melting milkfat fractions using single-step and multiple-step fractionation procedures at 30, 25, and 20°C, which corresponded to the second, third, and fourth steps for the multiple-step procedure. The fractions produced at 30°C both had a melting point of 44.2°C. The fractions produced at 25 and 20°C from multiple-step fractionation had slightly higher melting points (42.0°C and 40.5°C) than the fractions obtained using the single-step fractionation procedure (40.6°C and 39.6°C, respectively). Versteeg (1991) also obtained high-melting milkfat fractions using both multiple-step and single-step fractionation and reported a melting point of 43°C for the first step in the multiple-step procedure at 18°C, along with a melting point of 41.2°C for the single-step fraction.

The difference between fractionation temperature and melting point was approximately 20°C for fractionation temperatures less than 20°C (Amer et al., 1985; Lund and Danmark, 1987; Schaap and van Beresteyn, 1970; Versteeg, 1991), but there was only a 10 to 14°C difference when the fractionation temperature was 30°C or greater (Barna et al., 1992; deMan and Finoro, 1980; Grall and Hartel, 1992; Kaylegian, 1991; Kaylegian and Lindsay, 1992). Versteeg (1991) produced a high-melting milkfat fraction at 18°C and reported a softening point of 43°C compared with a melting point of 43.0°C for high-melting milkfat fractions obtained at 30°C (Barna et al., 1992; Grall and Hartel, 1992). Thus, the melting point did not increase as fractionation temperatures increased; this was compensated by the decrease in the difference between fractionation temperatures and melting point as the fractionation temperature increased.

Separation with the aid of an aqueous detergent solution was the second most popular separation technique employed to produce high-melting milkfat fractions (Table 5.15). Comparison of high-melting milkfat fractions obtained with the aid of an aqueous detergent solution and those obtained with vacuum filtration revealed the following:

- High-melting milkfat fractions obtained using an aqueous detergent solution had equivalent or slightly lower melting points for fractions obtained at equivalent temperatures.
- As with vacuum filtration, increasing fractionation temperature did not necessarily result in an increase in melting point for fractions obtained using either separation technique.

Doležálek et al. (1976) produced four high-melting milkfat fractions with the aid of an aqueous detergent solution at 30°C and reported melting points of 40.5, 41.5 (2 fractions), and 43.5°C; in contrast fractions produced at 30°C using vacuum filtration had melting points of 43.0°C (Barna et al., 1992; Grall and Hartel, 1992) and 44.2°C (Kaylegian, 1991; Kaylegian and Lindsay, 1992). High-melting milkfat fractions produced at 25°C with the aid of an aqueous detergent solution also had similar or slightly lower melting points, 33.1 to 40.4°C (Doležálek et al., 1976; Norris et al., 1971; Sherbon et al., 1972), compared with fractions obtained using vacuum filtration, which had melting points from 39.2 to 42.0°C (deMan and Finoro, 1980; Kaylegian, 1991; Timms and Parekh, 1980).

The differences in melting points for fractions reported by Doležálek et al. (1976) were related to differences in the melting points for their respective parent AMFs (Table 5.8). AMF (4) was the highest-melting AMF and yielded the highest-melting fractions. However, although AMF (1) was the lowest-melting AMF, the fractions produced from AMF (1) did not always exhibit the lowest melting points of the fractions produced at 30, 27, or 25°C. Although the fractions obtained by Doležálek et al. (1976) using four different parent milkfats did not exhibit consistent trends with respect to their parent AMFs, similar melting points at equivalent fractionation temperatures were observed. Fractions produced by other investigators at temperatures equivalent to those used by Doležálek et al. (1976) also had similar melting points. Thus, high-melting milkfat fractions obtained by crystallization from melted milkfat and separated with the aid of an aqueous detergent solution at similar fractionation temperatures exhibited similar melting points, regardless of AMF source.

High-melting milkfat fractions obtained with the aid of an aqueous detergent solution did not exhibit an increase in melting points as the fractionation temperature increased; this trend was similar to that observed for fractions obtained using vacuum filtration. For example, Doležálek et al. (1976) reported a melting point of 40.4°C for a fraction obtained at 25°C, and Sherbon et al. (1972) reported fractions with softening points of 40.6 and 40.9°C at fractionation temperatures of 31.7°C and 28°C, respectively.

Sherbon et al. (1972) studied recrystallization of high-melting milkfat fractions and reported increased melting points for the refractionated solid fractions. They also reported the only two liquid high-melting milkfat fractions; these had lower melting points than the solid fraction obtained at the same temperature. Sherbon et al. (1972) employed an increasing temperature profile with successive steps (25.5, 28, and 31.7°C), and this could also contribute to the increase in melting point with successive refractionations. Very-high-melting milkfat fractions obtained by refractionation from melted milkfat also exhibited in increase in melting point with successive fractionation steps, except that decreasing temperature profiles were generally employed.

The highest-melting fraction obtained with the aid of an aqueous detergent solution was obtained at the lowest fractionation temperature reported (16°C; Banks et al., 1985). Banks et al. (1985) studied the effects of detergent levels on milkfat fractionation and reported that the use of high levels of detergent resulted in fractions with higher melting points than did separation with low levels of detergent solution. Banks et al. (1985) defined "high" levels of detergent as a 10-fold increase in detergent concentration over the low levels of detergent solution employed.

High-melting milkfat fractions produced using *other separation techniques* (pressure filtration, centrifugation, filtration through muslin cloth, filtration through a milk filter, filtration through a cheese press, filtration through a casein-dewatering press, and separation technique not reported) did not exhibit any unusual trends or major deviations compared to trends observed for fractions obtained by vacuum filtration or with the aid of an aqueous detergent solution. High-melting milkfat fractions produced at equivalent temperatures using the same separation method generally had similar melting points (± 1°C), and the difference was slightly greater, generally ± 3°C, for fractions obtained at equivalent temperature, regardless of the separation method employed.

Crystallization from solvent solution High-melting milkfat fractions were solid fractions produced using single-step and multiple-step fractionation from acetone solution, and by multiple-step from ethanol solution (Table 5.16). Fractionation temperatures ranged from 3 to 25°C for the multiple-step procedures and from 0 to 5°C for the single-step procedures. The following observations were made when these fractions were compared with high-melting milkfat fractions obtained from melted milkfat (Table 5.15) and from solvent solution (Table 5.16):

- High-melting milkfat fractions obtained from solvent solution were produced at lower temperatures (0 to 25°C), compared with fractionation temperatures of 12 to 38°C for high-melting milkfat fractions obtained from melted milkfat.

- High-melting milkfat fractions were obtained from solvent solution using multiple-step fractionation procedures, whereas high-melting milkfat fractions obtained from melted milkfat were generally obtained at the first step of a single-step fractionation procedure.
- High-melting milkfat fractions obtained from solvent solution and from melted milkfat had similar melting points when obtained at similar fractionation temperatures.

High-melting milkfat fractions produced using multiple-step fractionation were obtained at the first through fourth step of the fractionation procedure, like the high-melting milkfat fractions obtained from melted milkfat, except that the fractionations temperatures employed for crystallization from solvent solution were lower. Milkfat fractions obtained from solvent solution generally require lower temperatures to effect crystallization because of the dilution effects of the solvent solution.

Although most high-melting milkfat fractions from solvent solution were obtained at lower fractionation temperatures than those employed for crystallization from melted milkfat, fractions obtained at equivalent temperatures from solvent solution and from melted milkfat had similar melting points. Sherbon and Dolby (1973) obtained a high-melting milkfat fraction by refractionation of a fraction with a melting point of 41°C from acetone solution at 25°C. This fraction was similar to fractions obtained from melted milkfat at 25°C, which had melting points from 39.2 to 42.0°C (Table 5.15; deMan and Finoro, 1980; Kaylegian, 1991; Kaylegian and Lindsay, 1992; Timms and Parekh, 1980). The refractionation process contributed to the decreased melting point of the acetone fraction. Rolland and Riel (1966) produced a high-melting milkfat fraction from acetone solution at 18°C that also had a melting point (42°C) similar to the fractions obtained from melted milkfat at 18°C by Versteeg (1991), which had melting points of 42 and 41.2°C (Table 5.15). The fraction obtained from solvent solution was the first step in a multiple-step procedure; one of the fractions produced by Versteeg (1991) from melted milkfat was also obtained at the first step of a multiple-step procedure, and the other fraction was produced using single-step fractionation.

High-melting milkfat fractions obtained from solvent solution (Table 5.16) were obtained at temperatures that were similar to those employed for very-high-melting milkfat fractions (Table 5.13). Kaylegian (1991) produced both a very-high-melting milkfat fraction and a high-melting milkfat fraction from acetone solution at 11°C. However, the very-high-melting milkfat fraction was produced using single-step fractionation, whereas the high-melting milkfat fraction was from the fourth step in a multiple-step fractionation procedure. Youssef et al. (1977) reported two very-high-melting milkfat fractions, obtained at the first and third steps of a multiple-step fractionation procedure at 18°C, that had melting points of 51 and 46°C, respectively. Rolland and Riel (1966) produced a high-melting milkfat fraction at 18°C that was the first step of a multiple-step procedure; they reported a cloud point of 42°C. They did not report melting behavior data for the parent AMF (Table 5.9); therefore, influences by the parent AMF on the fraction melting point could not be determined.

Sherbon and Dolby (1973) obtained three fractions from acetone solution at 25°C. Two were very-high-melting milkfat fractions (first fraction MP = 54°C; refractionated fraction MP = 48°C), and the third fraction was refractionated twice at 25°C, yielding a high-melting milkfat fraction, MP = 41°C. These observations further support the trend that refractionation from solvent solution at equivalent temperatures yields fractions with decreased melting points.

High-melting milkfat fractions were produced from both acetone and ethanol solutions by Rolland and Riel (1966), and both of these fractions were obtained at the first step of a multiple-step fractionation procedure. The fraction obtained from acetone had a higher cloud point (42°C) than the fraction obtained from ethanol solution (35°C).

Supercritical fluid extraction High-melting milkfat fractions obtained using supercritical carbon dioxide extraction were either the second through eighth extracted fractions or the residual fraction, which remained after the extraction process was completed (Table 5.17). Melting points for high-melting milkfat fractions ranged from 36.0 to 44.6°C and showed a small increase (< 5°C) in melting point compared to their respective parent AMF (Table 5.10). High-melting milkfat fractions were produced using supercritical carbon dioxide extraction at temperatures between 30 and 80°C and at pressures of 200 to 400 bar or 20 to 27.6 MPa. Because of the variety of temperatures and pressures employed by different investigators, it is difficult to compare data directly for high-melting milkfat fractions obtained by the various investigators. However, Kaufmann et al. (1982) and Timmen et al. (1984) produced residual fractions at 80°C and 200 bar that exhibited similar melting points (36.0 and 36.2°C, respectively) even though there was a slight difference in the melting points of the parent AMFs (34.3 and 35.6°C, respectively).

Arul et al. (1987) obtained two high-melting milkfat fractions, which were the seventh and eighth samples collected at 70°C and 250–350 bar; and these fractions had melting points of 39.3 and 43.3°C. The data reported by Arul et al. (1987) further confirmed that supercritical carbon dioxide extraction extracts the lower-melting triglycerides first. Chen et al. (1992) collected high-melting milkfat fractions with increasing pressure, but these fractions did not exhibit a well-defined trend for melting points as the pressure increased. These high-melting milkfat fractions included the second, fourth, fifth, and sixth fractions collected, but the third fraction collected was a middle-melting milkfat fraction (Table 5.21).

High-melting milkfat fractions obtained using supercritical carbon dioxide extraction had melting points that generally were between 35 and 40°C. High-melting milkfat fractions obtained by crystallization from melted milkfat at fractionation temperatures below 25°C also had melting points that ranged between 35 and 40°C; in contrast, high-melting milkfat fractions obtained from melted milkfat at fractionation temperatures above 25°C exhibited melting points above 40°C. High-melting milkfat fractions obtained from solvent solution also had melting points greater than 40°C.

Short-path distillation   One high-melting milkfat fraction (MP = 42.9°C) has been produced by short-path distillation (Table 5.18); this was the residual fraction obtained after the distillation process was completed. The melting point for the residual high-melting milkfat fraction obtained from short-path distillation was higher (3.5°C) than the melting points reported for the residual high-melting milkfat fractions obtained using supercritical carbon dioxide extraction (average MP = 39.4°C; Table 5.17).

Percentage Yields of High-Melting Milkfat Fractions

The percentage yields for individual high-melting milkfat fractions ranged from 10 to 90% for fractions obtained by crystallization from melted milkfat (Table 5.15), from 2.7 to 47.9% for fractions obtained by crystallization from solvent solution (Table 5.16), and from 1.5 to 80% for fractions obtained using supercritical carbon dioxide extraction (Table 5.17); the yield was 56.4% for the high-melting milkfat fraction obtained by short-path distillation (Table.16). Despite the wide range of yields reported, certain trends within fractionation methods were observed:

- High-melting milkfat fractions obtained using single-step fractionation from melted milkfat and from solvent solutions exhibited increased yields as the fractionation temperature decreased.

- The yields for high-melting milkfat fractions obtained by multiple-step fractionation from melted milkfat and from solvent solutions decreased with successive fractionation steps.

- High-melting milkfat fractions obtained by refractionation exhibited the highest yields reported.

- Yields for high-melting milkfat fractions obtained from solvent solution were generally lower than for those obtained from melted milkfat.

- Residual high-melting milkfat fractions obtained using supercritical carbon dioxide extraction generally had greater yields than did the extracted fractions.

Crystallization from melted milkfat Percentage yields did not exhibit major differences between separation techniques used for crystallization from melted milkfat (Table 5.15). High-melting milkfat fractions obtained using single-step fractionation at temperatures above 30°C generally had yields under 20%, whereas yields were 30 to 50% for fractionation temperatures below 30°C. Multiple-step fractionation yielded fractions with decreased yields at successive fractionation steps; this was observed by Kaylegian (1991), who used vacuum filtration and obtained three high-melting milkfat fractions at 30, 35, and 20°C, which represented the second, third, and fourth steps of the multiple-step fractionation procedure, and yields were 17.7, 11.2, and 7.0%, respectively. On the other hand, Kaylegian (1991) also produced high-melting milkfat fractions at 30, 25, and 20°C using single-step fractionation and reported increased yields as the fractionation temperatures decreased (20.2, 46.5, and 70.0%, respectively). Thus, the fractionation procedure employed greatly influenced fraction yield.

Sherbon et al. (1972) also produced high-melting milkfat fractions using a multiple-step fractionation procedure, but the solid fractions obtained at each step were refractionated at elevated temperatures, whereas Kaylegian (1991) refractionated the liquid fractions and employed a decreasing temperature profile for the multiple-step fractionation procedure. Sherbon et al. (1972) reported yields of 31, 92, and 94% for the solid fractions obtained at 25.5°C and the refractionated fractions obtained at 28 and 31.7°C, respectively. Refractionation of high-melting milkfat fractions using an increasing temperature profile increased the yields.

Crystallization from solvent solution High-melting milkfat fractions obtained from solvent solution exhibited lower yields for fractions obtained from multiple-step fractionation than for those obtained using single-step fractionation. Fractions obtained using multiple-step fractionation had yields of 2.7 to 15% for fractions obtained between 10 and 18°C (Avvakumov, 1974; Kaylegian, 1991; Rolland and Riel, 1966), compared with yields of 30.7 and 47.9% for fractions obtained using single-step fractionation at 5 and 0°C, respectively (Kaylegian, 1991). Sherbon and Dolby (1973) produced a high-melting milkfat fraction with a yield of 35.2% that was obtained from the fourth step in a multiple-step procedure but also had been refractionated twice at 25°C. Sherbon and Dolby (1973) obtained three fractions at 25°C using multiple-step fractionation (second to fourth steps) and reported yields of 33.1% for the first fraction and 2.7% for the second fraction; both of these fractions were categorized as very-high-melting milkfat fractions (Table 5.13). The refractionation of a very-high-melting milkfat fraction at the same temperature produced a high-melting milkfat fraction in greater yield than the original very-high-melting milkfat fraction (35.2% and 2.7%, respectively).

Rolland and Riel (1966) produced two fractions at 18°C and reported a greater yield for the fraction obtained from ethanol solution (12.7%) than for the fraction obtained from acetone solution (3.0%). Rolland and Riel (1966) also reported a decrease in melting point for the ethanol fraction, and this could indicate a greater percentage of liquid in this fraction, which would also account for the increase in yield. However, DSC data for these fractions were not available to confirm this.

Supercritical fluid extraction Yields for residual fractions were generally greater than those reported for the extracted high-melting milkfat fractions obtained from supercritical carbon dioxide extraction (Table 5.17). Yields were 57.5 and 80% for two residual fractions; obtained by Kankare et al. (1989) and Timmen et al. (1984), respectively, and the yields ranged from 1.5 to 28.7% for extracted fractions. However, Rizvi et al. (1990) reported similar yields for the extract and residual fractions, 28.7 and 28.9%, respectively.

High-melting milkfat fractions obtained by supercritical carbon dioxide extraction did not exhibit consistent trends in yields with respect to changes in temperature or pressure. For example, Arul et al. (1987) reported yields of 17 to 20% for fractions obtained at 70°C, and Chen et al. (1992) reported yields of 1.5 to 30% for high-melting milkfat fractions obtained at 40°C. Chen et al. (1992) employed an increasing pressure profile for the extraction of fractions; the yield values increased from 10.3 MPa to a maximum yield at 20.7 MPa, and then the yields decreased as the pressure continued to increase to 27.6 MPa (Table 5.17). The increased yields in the middle of the pressure sequence employed by Chen et al. (1992) suggested that increased solubility of milkfat in supercritical carbon dioxide does not correlate with higher pressures or temperatures.

The lack of apparent trends in yields, with respect to fractionation temperature and pressure (Table 5.17), is probably a result of many factors, including differences in fractionation parameters employed by various investigators, such as extraction times and the number of fractions obtained. The relationship between milkfat composition and solubility in supercritical fluids at various temperatures and pressures is complex and not well understood and may also contribute to differences observed in fraction yields.

Short-path distillation The yield of the high-melting milkfat fraction obtained using short-path distillation (Table 5.18) was 56.4% and appeared to be similar to the yields reported for residual high-melting milkfat fractions obtained using supercritical carbon dioxide extraction (Table 5.17).

Chemical Composition of High-Melting Milkfat Fractions

Fatty acid composition The fatty acid composition of high-melting milkfat fractions produced from melted milkfat (Tables 5.40 and 5.65) and from solvent solution (Tables 5.41 and 5.66) exhibited the same trends as that of very-high-melting milkfat fractions when compared to the intact AMFs, except that the change in percent composition of individual fatty acids was not as great for the high-melting milkfat fractions:

- C4–C10 fatty acid content decreased.
- C12–C15 fatty acid content decreased.
- C16–C18 saturated fatty acid content increased.

High-melting milkfat fractions obtained from melted milkfat exhibited a wide range of values for specific fatty acid content. With a few exceptions, the data for fractions obtained from melted milkfat did not exhibit distinct trends relating to fractionation temperature, fractionation procedure employed (multiple-step or single-step), separation method, or source of intact AMF. However, Kaylegian (1991) used a single source of AMF to produce high-melting milkfat fractions using both multiple-step and single-step fractionation and reported a slight increase in the C4–C10 fatty acid content and a slight decrease in the C16–C18 saturated fatty acid content as the fractionation temperature decreased for fractions obtained using single-step fractionation. Kaylegian (1991) also reported that fractions obtained using multiple-step fractionations at equivalent temperatures exhibited lower amounts of C4–C10 fatty acids and higher amounts of C16–C18 fatty acids than did fractions obtained using single-step fractionation. The trends reported by Kaylegian (1991) appear to hold true for fractions obtained from a single AMF source, but these trends were not consistent when high-melting milkfat fractions were obtained using single-step and multiple-step fractionation from different AMF sources.

Laakso et al. (1992) reported the percent fatty acid composition of several triglyceride classes of the high-melting milkfat fraction obtained from melted milkfat (Fraction S1, Tables 5.40 and 5.65). The distribution of the fatty acids in the triglycerides followed expected trends: the saturated fatty acids were found in the tri- and disaturated triglyceride classes (SSS and SSU), whereas the unsaturated fatty acids were in the more unsaturated triglyceride classes (SUU and UUU). The highest concentration of C4–C10 fatty acids was found in the trisaturated class (S1 SSS fraction, Table 5.65). The lowest concentrations of the C12 and C14 fatty acids were found in the di-monounsaturated triglyceride classes (S1 SMcMt fraction and S1 SMcMc fraction, Table 5.65). The highest concentration of C18:1 also was in the di-monounsaturated triglyceride class (S1 SMcMc fraction, Table 5.40), and the highest concentration of C18:2 was in the triunsaturated class (S1 UUU fraction, Table 5.40). The preferential concentration of C18:1 in the SM^{c}M^c triglyceride fraction over the SM^{c}M^t triglyceride fraction was confirmed by the greater concentration of *cis*-unsaturated fatty acids compared with *trans*-unsaturated fatty acids observed in high-melting milkfat fractions (Table 5.65).

High-melting milkfat fractions obtained with the aid of an aqueous detergent solution generally had lower contents of C4–C10 fatty acids and higher C16–C18 saturated fatty acid contents than did high-melting milkfat fractions produced using other separation techniques. This difference in fatty acid content between fractions obtained using an aqueous detergent solution and those obtained using other separation methods was also seen for very-high-melting milkfat fractions and was consistent with trends expressed in the physical properties of these fractions.

High-melting milkfat fractions obtained from solvent solution (Tables 5.41 and 5.66) exhibited an increase in the C4–C10 fatty acid content and a decrease in the C16–C18 saturated fatty acid content as fractionation temperature decreased, regardless of the fractionation procedure employed. The fraction obtained by refractionation of a high-melting glyceride at 25°C (Sherbon and Dolby, 1973) had the lowest concentration of C4–C10 fatty acids and the highest concentration of C16–C18 saturated fatty acids for high-melting milkfat fractions obtained from solvent solution. The fraction produced by Sherbon (1963) had an unusually high C4–C10 fatty acid content (20%), and either analytical inaccuracies or a high C4–C10 fatty acid content in the parent AMF could have been responsible. However, data for the parent AMF were unavailable, and therefore the compositional hypothesis could not be tested.

High-melting milkfat fractions obtained using supercritical carbon dioxide extraction generally exhibited fatty acid concentrations (Tables 5.42 and 5.67) similar to those for fractions obtained from melted milkfat for the extracted fractions, and the residual fractions obtained by supercritical carbon dioxide extraction were similar to fractions obtained from solvent solution. Several investigators (Arul et al., 1987; Chen et al., 1992; Kaufmann et al., 1982) have reported C4–C10 fatty acid concentrations that were greater than expected, but this trend correlated with relatively high concentrations of C4–C10 fatty acids in the parent AMF (Table 5.60). Chen et al. (1992) studied the effects of different pressures on high-melting milkfat fractions and reported a decrease in the C4–C10 and C12–C15 fatty acids and an increase in the C16–C18 saturated fatty acid content as pressure increased. The C18:1 content of very-high-melting milkfat fractions obtained from supercritical carbon dioxide extraction (Table 5.42) was relatively high compared to milkfat fractions obtained from solvent solution and only slightly greater than the contents reported for fractions obtained from melted milkfat.

The high-melting milkfat fractions obtained from short-path distillation exhibited a fatty acid composition (Tables 5.43 and 5.68) that was similar to that of fractions obtained from solvent solution with respect to saturated fatty acids, but contained a greater amount of C18:1 than did fractions obtained from solvent solution.

Triglyceride composition The triglyceride composition of high-melting milkfat fractions obtained from melted milkfat (Tables 5.89 and 5.113) and from solvent solutions (Tables 5.90 and 5.114) exhibited trends similar to those reported for the very-high-melting milkfat fractions with respect to the parent AMF:

- C24–C34 triglyceride content decreased.
- C36–C40 triglyceride content decreased.
- C42–C54 triglyceride content increased.

The difference in individual triglyceride contents was greater for the very-high-melting milkfat fractions than for the high-melting milkfat fractions when each was compared to its parent AMFs (Tables 5.83 and 5.84). The triglycerides of high-melting milkfat fractions were composed of 50 to 60% trisaturated (SSS), 30 to 40% monounsaturated (SSM), and 10% di- and triunsaturated triglyceride species (SUU and UUU; Table 5.113).

The high-melting milkfat fractions obtained from melted milkfat (Tables 5.89 and 5.113) with the aid of an aqueous detergent solution (Banks et al., 1985) showed a marked decrease in the C24–C34 triglycerides compared with fractions obtained using other separation methods, and this further illustrated the decreased amount of liquid milkfat found in crystalline fractions obtained with the aid of an aqueous detergent solution.

Although high-melting milkfat fractions obtained from melted milkfat (Tables 5.89 and 5.113) did not show an overall trend in triglyceride composition between investigators, a consistent trend existed within data reported by individual investigators. Both deMan and Finoro (1980) and Kaylegian (1991) reported an increase in C24–C34 and C36–C40 triglyceride contents, with a concurrent decrease in the C42–C54 triglycerides, as the fractionation temperatures decreased. This trend was also seen in fractions obtained from solvent solutions (Tables 5.90 and 5.114; Kaylegian, 1991).

High-melting milkfat fractions obtained using supercritical fluid extraction (Tables 5.91 and 5.115) exhibited a decrease in C24–C34 triglyceride content and an increase in both the C36–C40 and C42–C54 triglycerides. High-melting milkfat fractions obtained by supercritical carbon dioxide extraction had similar contents of C24–C34 triglycerides, but had higher concentrations of the C36–C40 triglycerides than did fractions obtained from melted milkfat.

Kankare et al. (1989) and Kaufmann et al. (1982) both used parent AMFs with a C24–C34 content of 15%, but the residual fractions obtained exhibited markedly different C24–C34 concentrations (10.2% and 6.6%, respectively). Kaufmann et al. (1992) employed a higher temperature (80°C) and lower pressure (200 bar) than Kankare et al. (50°C, 400 bar), and this resulted in a decreased content of the C24–C34 and C36–C40 triglycerides and an increase in the C42–C54 triglycerides.

The high-melting milkfat fractions obtained using short-path distillation appeared to have lower concentrations of all triglyceride groups (Tables 5.92 and 5.116) than did fractions obtained by other methods. However, the percentages for the various triglycerides reported totaled only 70% (Arul et al., 1988a). This indicated that a considerable amount of the total GC peak area in the triglyceride analysis could not be assigned to triglycerides. Even though the peak area was used in the calculation, it was omitted from the reported data. Thus, these data should be viewed with caution.

Selected chemical characteristics Iodine values and other selected chemical characteristics for high-melting milkfat fractions obtained from melted milkfat are presented in Table 5.137; for solvent solution in Table 5.138; and for supercritical fluid extraction in Table 5.139. Consistent trends were not observed for iodine values, refractive indices, Reichert–Meissl numbers, Polenske numbers, or densities for all high-melting milkfat fractions, regardless of fractionation method, procedure, separation technique, or fractionation temperature. High-melting milkfat fractions that exhibited relatively high iodine values for fractions obtained from melted milkfat (Deroanne and Guyot, 1974; Norris et al., 1971) were obtained from parent AMFs with high iodine values (Table 5.131). Ranges of chemical characteristics for high-melting milkfat fractions are:

CHARACTERISTIC	RANGE	AVERAGE
Iodine value	14 to 37	26 to 33
Refractive index	1.4533 to 1.4596	1.4533 to 1.4548
Reichert–Meissl value	17.3 to 28.1	22 to 26
Polenske value	1.3 to 2.1	1.5 to 1.7
Density	0.8906 to 0.9390	0.8910 to 0.8920

Flavor Potential lactone concentrations, methyl ketone concentrations, and aldehyde concentrations of high-melting milkfat fractions are presented in Tables 5.152, 5.158, and 5.163, respectively. The lactone and methyl ketone concentrations of high-melting milkfat fractions were approximately 60 to 70% of the intact milkfat (Tables 5.150 and 5.156). The total aldehyde concentration of the high-melting milkfat fractions is greater than the total aldehyde concentration of the intact milkfat (Table 5.161). Walker (1972) reported that the aldehyde concentration increased during the production of AMF from cream and during fractionation, indicating that oxidation had occurred during processing. Jordan (1986) used sensory panels to evaluate the flavor of the high-melting milkfat fraction ($28S_s$), and reported that the flavor of the fractionated milkfat was not significantly different from that of the intact milkfat. Jordan (1986) also reported that the milkfat samples had a pleasant flavor.

Physical Properties of High-Melting Milkfat Fractions

Crystal characteristics The *crystal morphology* of high-melting milkfat fractions obtained from melted milkfat and solvent solution are shown in Table 5.168, and both fractions exhibited the α and β' forms. Grall and Hartel (1992) reported that the crystals of high-melting milkfat fractions obtained from melted milkfat by vacuum filtration were a conglomeration of needle-like structures.

The *crystal sizes* of high-melting milkfat fractions obtained from melted milkfat are shown in Table 5.172. Doležálek et al. (1976) produced high-melting milkfat fractions at 30, 27, and 25°C, using four different AMF sources, and reported ranges in crystal sizes in these fractions. Some differences in crystal size were observed for high-melting milkfat fractions obtained at similar temperatures from the different AMF sources. However, for all AMF sources the crystals obtained at 30°C were larger than those obtained at 27°C, and the crystals obtained at 25°C were the smallest. Black (1973) evaluated the effects of fractionation parameters on the crystal size of high-melting milkfat fractions. Elevated temperatures, both initial and final, and increased fractionation times yielded larger crystals than did lower values of these fractionation conditions.

Melting point The relationship between melting point and fractionation conditions was discussed in detail in the previous section on fractionation conditions of high-melting milkfat fractions. The melting points of high-melting milkfat fractions are presented in Tables 5.15, 5.16, 5.17, and 5.18 for fractions obtained from melted milkfat, from solvent solution, by supercritical fluid extraction, and by short-path distillation, respectively. The general melting point characteristics of high-melting milkfat fractions were:

- By definition, the melting points of high-melting milkfat fractions were between 35 and 45°C.
- High-melting milkfat fractions obtained at equivalent temperatures from melted milkfat and from solvent solution exhibited similar melting points.
- Increased fractionation temperatures did not always yield fractions with increased melting points.
- Refractionation of high-melting milkfat fractions yielded fractions with increased melting points when crystallized from melted milkfat, but yielded fractions with decreased melting points when crystallized from solvent solution; this is consistent with trends observed for very-high-melting milkfat fractions.
- High-melting milkfat fractions obtained using supercritical carbon dioxide extraction had melting points that were similar to those for high-melting milkfat fractions obtained from melted milkfat at fractionation temperatures below 25°C, but were lower than melting points for fractions obtained from melted milkfat at temperatures above 25°C and for all high-melting milkfat fractions obtained from solvent solution.
- The melting point of the residual high-melting milkfat fraction obtained by short-path distillation was slightly greater than the average melting point reported for residual fractions obtained using supercritical carbon dioxide extraction.

Thermal profile Thermal profiles of high-melting milkfat fractions are shown in Figures 5.8, 5.9, 5.10, and 5.11 for fractions obtained from melted milkfat, from solvent solution, by supercritical fluid extraction, and by short-path distillation, respectively.

High-melting milkfat fractions obtained by crystallization from melted milkfat and from solvent solutions exhibited similar trends:

- Small amounts of low-melting and middle-melting triglyceride species were present in these fractions, but there was a sharp dip in the DSC curve around 20°C, indicating a melting of these species prior to the melting of the high-melting triglyceride species.
- Fractions obtained using multiple-step fractionation had smaller peaks for low-melting triglyceride species than did fractions obtained using single-step fractionation.

Kaylegian (1991) produced high-melting milkfat fractions from melted milkfat using single-step and multiple-step fractionation procedures at equivalent temperatures and also produced high-melting milkfat fractions from acetone solution using single-step and multiple-step fractionation procedures, but at different temperatures than the fractions obtained from melted milkfat. The high-melting milkfat fractions obtained from multiple-step fractionation, both from melted milkfat (Figure 5.8) and from solvent solution (Figure 5.9), had narrower peaks for high-melting triglyceride species and smaller peaks for low-melting and middle-melting triglyceride species than did the fractions produced using single-step fractionation. This was probably caused by the presence of a more narrow melting range of triglycerides in the multiple-step fractions. This phenomenon was also observed in the SFC profiles of multiple-step fractions obtained from melted milkfat (Table 5.181), which were more solid than fractions obtained using single-step fractionation at equivalent temperatures (Kaylegian, 1991). The peak for the high-melting triglyceride species shifted to lower temperatures as fractionation temperature decreased for fractions obtained using multiple-step fractions, but this shift was not as apparent for fractions obtained using single-step fractionation, both from melted milkfat and from solvent solutions.

Thermal profiles for high-melting milkfat fractions obtained from melted milkfat did not exhibit perceivable effects between separation techniques. The exceptions were the fractions obtained with the aid of an aqueous detergent solution, but these effects were probably influenced more by the refractionation procedure than by the separation technique employed. High-melting milkfat fractions produced by Sherbon et al. (1972) with the aid of an aqueous detergent solution were both liquid and solid fractions obtained using a refractionation procedure. As expected, the liquid fractions contained larger peaks for the low-melting and middle-melting triglyceride species and smaller peaks for the high-melting triglyceride species than did the solid fractions. The peak for the high-melting triglyceride species became more pronounced at successive steps of the refractionation process, and this correlated with increased melting points reported for these fractions (Table 5.15).

High-melting milkfat fractions obtained from supercritical carbon dioxide extraction exhibited different DSC profiles, depending on whether they were the extracted or residual fractions (Figure 5.10). Thermal profiles for the residual fractions obtained by supercritical carbon dioxide extraction generally had profiles similar to those of the corresponding intact AMF (Figure 5.3) (Chen et al., 1992; Kaufmann et al., 1982; Rizvi et al., 1990), whereas extracted fractions generally had larger peaks for low-melting triglyceride species and smaller peaks for high-melting triglyceride species (Arul et al., 1987; Chen et al., 1992; Rizvi et al., 1990).

The high-melting milkfat fraction obtained using short-path distillation (Figure 5.11) exhibited a thermal profile that had only a large peak for the high-melting triglyceride species and small peaks for the low-and middle-melting triglyceride species. Although this was the residual fraction obtained after the distillation process was complete, the DSC profile appeared more closely related to DSC profiles obtained from solvent solution (Figure 5.9) than to the residual fractions obtained using supercritical carbon dioxide extraction (Figure 5.10).

Solid fat content   The solid fat contents reported for high-melting milkfat fractions have been tabulated and are presented in Tables 5.181, 5.182, 5.183, and 5.184 for fractions obtained by crystallization from melted milkfat, from solvent solution, by supercritical fluid extraction, and by short-path distillation, respectively. Solid fat content profiles for representative fractions from each fractionation method have also been presented graphically in Figure 5.23. High-melting milkfat fractions representing crystallization from melted milkfat ($25S_M$ fraction, Kaylegian, 1991) and from solvent solution ($5S_S$ fraction, Kaylegian, 1991) were chosen for illustration because their SFC profiles appeared to be in the middle range of SFCs profiles reported. The SFC profile of the AMF reported by Kaylegian (1991; Table 5.169) has been included for reference. The solid fat content profiles presented for fractions obtained by supercritical carbon dioxide extraction (Timmen et al., 1984) and by short-path distillation (Arul et al., 1988a) were the only high-melting milkfat fractions for which SFC data were available.

High-melting milkfat fractions obtained from solvent solution exhibited SFC profiles that showed a higher solid fat content than did fractions obtained from melted milkfat (Figure 5.23, Tables 5.182 and 5.181, respectively). The increased SFC of fractions obtained from solvent solution compared to fractions obtained from melted milkfat can be attributed to the increased crystalline purity of these fractions.

High-melting milkfat fractions produced from melted milkfat by Amer et al. (1985) exhibited SFC profiles that were similar to those obtained from solvent solution. This observation could be partly explained by the elevated SFC profile for the AMF employed by Amer et al. compared with the AMFs employed in other studies (Table 5.174). The winter AMF employed by Amer et al. was more solid than the summer AMF studied, but the $29S_S$ fraction produced by Amer et al. using winter AMF was slightly softer than the $29S_S$ produced using summer AMF. The $19S_S$ fractions produced using winter and summer AMF exhibited similar SFC profiles. Thus, it appears that complex parameters govern the influences of parent milkfats on the SFC profiles of milkfat fractions. Still, high-melting milkfat fractions have been produced from a variety of intact AMF sources, but they have exhibited reasonably similar SFC profiles.

High-melting milkfat fractions obtained from multiple-step fractionation from melted milkfat had profiles that showed a greater SFC than did fractions obtained using single-step fractionation at equivalent fractionation temperatures (Table 5.181). Kaylegian (1991) produced high-melting milkfat fractions, using multiple-step fractionation at 30, 25, and 20°C, that had higher SFC profiles than did fractions produced using single-step fractionation at equivalent temperatures (Table 5.179). This was probably caused by the more narrow melting range for the triglycerides in fractions obtained using multiple-step fractionation compared with those obtained using single-step fractionation. For example, the fraction obtained using multiple-step fractionation at 25°C ($25S_M$ fraction) contains only triglycerides that crystallize between 25 and 30°C, whereas the fraction obtained at 25°C using single-step fractionation ($25S_S$ fraction) contains all triglycerides that melt above 25°C. A distinct trend for the SFC profiles with respect to decreasing fractionation temperature was not observed for high-melting milkfat fractions obtained using single-step fractionation.

High-melting milkfat fractions obtained using vacuum filtration and separation with the aid of an aqueous detergent solution generally were more solid than fractions separated by centrifugation and filtration through a milk filter, and this is illustrated by the SFC profiles (Table 5.181). For some fractions obtained using centrifugation and filtration through a milk filter, the solid fat indices (SFI) were reported, rather than the actual solid fat contents reported for fractions obtained using other separation methods, and this might account for some of the differences observed. Interestingly, the AMF used for the fraction obtained by centrifugation (Kankare and Antila, 1988b) was one of the most solid reported, and the AMF used for the fraction obtained using a milk filter (Fouad et al., 1990) was one of the least solid AMF samples reported (Table 5.173). However, both of these milkfat samples produced softer fractions than were produced by other separation methods. Thus, the influence of the AMF source on the resultant fractions may be less important than influences from the separation technologies employed.

High-melting milkfat fractions obtained from melted milkfat exhibited SFC profiles that were similar in shape to those for very-high-melting milkfat fractions, except that the high-melting milkfat fractions melted at lower temperatures. High-melting milkfat fractions obtained from solvent solution exhibited a more gradual melting profile, compared to the sharp melting profiles of very-high-melting milkfat fractions obtained from solvent solution. This can be readily observed in Figures 5.22 and 5.23.

Kaylegian (1991) showed that high-melting milkfat fractions obtained from single-step fractionation from solvent solution exhibited lower SFC profiles as fractionation temperature decreased (Table 5.182), although distinct trends were not observed for fractions obtained from melted milkfat. Only one high-melting milkfat fraction was obtained using multiple-step fractionation from acetone solution (Kaylegian, 1991), and this had a lower SFC profile than did the fractions obtained using single-step fractionation procedures. However, the fractionation temperatures were not equivalent, and a direct comparison of procedures was not possible.

High-melting milkfat fractions obtained using supercritical carbon dioxide extraction and short-path distillation had lower SFC profiles than did high-melting fractions obtained from melted milkfat and from solvent solution. The fraction obtained from supercritical carbon dioxide extraction (Table 5.183) had an increased SFC profile at temperatures below 20°C, compared to the fraction obtained from short-path distillation (Table 5.184). However, the SFC profiles at temperatures above 20°C showed that the fraction obtained from short-path distillation was more solid than the fraction obtained from supercritical carbon dioxide extraction, and the fraction from supercritical carbon dioxide was completely melted at a lower temperature than the fraction obtained from short-path distillation.

Textural characteristics The texture of one high-melting milkfat fraction obtained from melted milkfat was evaluated by cone penetrometry (Table 5.196). This high-melting milkfat fraction was harder than the intact AMF (Table 5.195), as indicated by the decreased penetration readings, 2.2 and 2.7 mm, respectively (Sreebhashyam et al., 1981).

Middle-Melting Milkfat Fractions

A total of 111 middle-melting milkfat fractions (MP between 25 and 35°C) have been reported in the literature (Table 5.6), and this represents approximately 20% of the milkfat fractions that have been characterized by melting behavior. Middle-melting milkfat fractions were produced by:

1. Crystallization from melted milkfat (Table 5.19; 77 fractions), using the following separation techniques:

   - Vacuum filtration (Amer et al., 1985; Barna et al., 1992; Deffense, 1987; deMan and Finoro, 1980; Grall and Hartel, 1992; Kaylegian, 1991; Kaylegian and Lindsay, 1992; Laakso et al., 1992; Makhlouf et al., 1987; Schaap and van Beresteyn, 1970; Sherbon, 1963; Versteeg, 1991)

   - Pressure filtration (deMan, 1968; Stepanenko and Tverdokhleb, 1974)

   - Centrifugation (Antila and Antila, 1970; Branger, 1993; Jordan, 1986; Kankare and Antila, 1986, 1988a; Lakshminarayana and Rama Murthy, 1985; Richardson, 1968)

   - Separation with the aid of an aqueous detergent solution (Dolby, 1970b; Doležálek et al., 1976; Sherbon et al., 1972; Walker, 1972)

   - Filtration through muslin cloth (Khalifa and Mansour, 1988; Sreebhashyam et al., 1981)

   - Filtration through milk filter (Fouad et al., 1990)

- Filtration through a cheese press (Baker, 1970a)
- Filtration through a casein-dewatering press (Black, 1973)
- Separation technique not reported (Deroanne, 1976; Dixon and Black, 1974; Martine, 1982; Riel and Paquet, 1972; Verhagen and Warnaar, 1984)

2. Crystallization from solvent solutions (Table 5.20; 16 fractions) from the following solvents:

- Acetone (Avvakumov, 1974; Kaylegian, 1991; Kaylegian and Lindsay, 1992; Rolland and Riel, 1966; Sherbon and Dolby, 1973; Yi, 1993; Youssef et al., 1977)
- Pentane (Kaylegian, 1991)

3. Supercritical fluid extraction (Table 5.21; 16 fractions)

- Carbon dioxide (Arul et al., 1987; Bhaskar et al., 1993b; Chen et al., 1992; Hamman et al., 1991; Kankare and Antila, 1988a; Kankare et al., 1989; Rizvi et al., 1990)

4. Short-path distillation (Table 5.22; 1 fraction)

- (Arul et al., 1988a)

5. Fractionation method not reported (Table 5.5; 2 fractions) (Walker, 1974).

Fractionation Conditions for Middle-Melting Milkfat Fractions

Crystallization from melted milkfat Middle-melting milkfat fractions obtained by crystallization from melted milkfat were both solid and liquid fractions obtained using single-step and multiple-step fractionation procedures (Table 5.19). The liquid middle-melting fractions had fractionation temperatures greater than 20°C, and solid middle-melting milkfat fractions were obtained between 11 and 25°C. The exceptions were two middle-melting milkfat fractions produced using multiple-step fractionation by Riel and Paquet (1972), which were both solid fractions, obtained at 38 and 4°C, and had cloud points of 32 and 25°C, respectively. In general, the solid fractions obtained at a selected temperature exhibited higher melting points than the fractionation temperature employed. This trend was observed for the solid fraction obtained at 4°C, but not for the solid fraction obtained at 38°C by Riel and Paquet (1972), who did not specify the separation technique used to obtain these fractions; therefore, influences of separation technique on fraction melting point could not be determined.

Vacuum filtration was the most commonly employed separation technique for middle-melting milkfat fractions obtained from melted milkfat. The physical state of the fraction followed a very distinct trend with respect to fractionation temperature for fractions obtained using vacuum filtration: liquid middle-melting milkfat fractions were obtained when fractionation temperatures were greater than 20°C, and solid middle-melting milkfat fractions were obtained at fractionation temperatures below 20°C. The melting points of liquid fractions were within ± 2°C of the fractionation temperature, but solid fractions always exhibited an increase of 2 to 14°C in melting point compared with fractionation temperature. Table 5.19 illustrates the similarity in melting points of the liquid milkfat fractions obtained from melted milkfat at similar fractionation temperature, and the melting points are similar regardless of the separation method employed.

Melting points decreased with decreasing fractionation temperature for middle-melting milkfat fractions when reported by a single investigator, but this trend was not apparent when similar fractions obtained by different investigators were compared. Kaylegian (1991) produced liquid middle-melting milkfat fractions using single-step fractionation at 34 and 30°C and reported melting points of 33.5 and 28.4°C, respectively, whereas deMan and Finoro (1980) produced six liquid middle-melting milkfat fractions using single-step fractionation at 32, 31, 30, 29, 28, and 27°C and reported dropping points of 32.4, 29.5, 28.4, 26.5, 26.8, and

25.1°C, respectively. Versteeg (1991) reported five solid middle-melting milkfat fractions, obtained at 15, 14, 13, 12, and 11.5°C, that also exhibited a decrease in melting behavior, and reported softening points of 26.6, 26.5, 26.3, 26.0 and 25.8°C, respectively.

Melting points of middle-melting milkfat fractions obtained using vacuum filtration ranged from 25.1 to 34°C and did not exhibit any consistent trends with respect to fractionation procedure (single-step or multiple-step) employed. Schaap and van Beresteyn (1970) produced a middle-melting milkfat fraction at 27°C, using multiple-step fractionation, that had a melting point of 29.2°C. Likewise, deMan and Finoro (1980) obtained a middle-melting milkfat fraction at 27°C, but they employed single-step fractionation and reported a dropping point of 25.1°C. The melting behaviors reported for the parent AMF samples were 34.6 and 33.9°C, respectively. It is difficult to determine whether this difference in melting point was caused by influences stemming from the parent AMF or by the fractionation procedure employed. Versteeg (1991) produced middle-melting milkfat fractions at 12°C, using both multiple-step and single-step procedures, and reported softening points of 26°C for both fractions.

Separation with the aid of an aqueous detergent solution was the second most common separation technique that produced middle-melting milkfat fractions (Table 5.19). However, all middle-melting milkfat fractions obtained with the aid of an aqueous detergent solution were liquid fractions obtained between 25 and 30°C, and most of these were obtained using a single-step fractionation procedure. The liquid middle-melting milkfat fractions produced by Sherbon et al. (1972) were obtained by multiple-step fractionation at the first step of a multiple-step procedure or at the second step of a refractionation procedure.

Melting points for middle-melting milkfat fractions separated with the aid of an aqueous detergent solution ranged from 25.0 to 32.2°C, although most middle-melting milkfat fractions had melting points between 25 and 27°C. The fraction with the highest melting point was the liquid fraction obtained by Sherbon and Dolby (1973) from a double fractionation at 30.6°C. Sherbon and Dolby (1973) employed an increasing temperature profile, and this produced middle-melting milkfat fractions with higher melting points at successive steps.

Doležálek et al. (1976) produced liquid middle-melting milkfat fractions using the same AMF sources (1 and 2) at 30 and 25°C, and these exhibited similar (± 1°C) melting points when fractionated at equivalent temperatures (MP = 27.0 and 28.0°C at 30°C, and MP = 25.0 and 25.2°C at 25°C). The melting points of the two AMF samples were 29.8 and 31.9°C, respectively, which correlated to the slight increase in melting points of fractions obtained from AMF (2).

Centrifugation was also a popular separation technique employed to produce middle-melting milkfat fractions (Table 5.19). Most middle-melting milkfat fractions obtained using centrifugation were solid fractions produced between 12 and 25°C. One liquid middle-melting milkfat fraction was obtained using centrifugation at 40°C (Rolland and Riel, 1966).

Middle-melting milkfat fractions produced by *filtration through a muslin cloth* exhibited a decrease in melting point as fractionation temperature decreased (Khalifa and Mansour, 1988; Sreebhashyam et al., 1981). Middle-melting milkfat fractions produced by Riel and Paquet (1972) using an *unspecified fractionation technique* also showed a decrease in melting point as the fractionation temperature decreased for two fractions obtained at 38 and 4°C.

Crystallization from solvent solution Middle-melting milkfat fractions obtained from *acetone* were solid fractions, obtained using multiple-step fractionation between 5 and 10°C, or liquid fractions, obtained using multiple-step or single-step fractionation between 21 and 25°C (Table 5.20). One liquid middle-melting milkfat fraction was obtained at 0°C, but it was at the first step using an increasing temperature profile (Rolland and Riel, 1966).

Solid middle-melting milkfat fractions obtained from acetone solution at equivalent fractionation temperatures exhibited similar melting points. For example, four solid middle-melting milkfat fractions have been obtained at 0°C (Avvakumov, 1974; Kaylegian, 1991; Yi, 1993; Youssef et al., 1977), and all exhibited melting points of 26 to 27°C.

The liquid fraction produced by Sherbon and Dolby (1973) was obtained at the fifth step of a multiple-step fractionation produced from a high-melting glyceride, and it had a melting point of 27°C. Another liquid middle-melting milkfat fraction has been produced at 21°C with a melting point of 27°C, but it was obtained from intact AMF using a single-step fractionation procedure (Kaylegian, 1991).

Both of the liquid middle-melting milkfat fractions obtained from acetone solution exhibited a 2 to 6°C increase in melting point when compared with the fractionation temperature employed, whereas liquid fractions obtained from melted milkfat exhibited a slight increase (± 2°C) or even a decrease in melting point compared with fractionation temperature. This phenomenon was also illustrated by fractions obtained by Kaylegian (1991) from *pentane* solution (Table 5.20), except that the difference between melting point and fractionation temperature was approximately 20°C. Liquid middle-melting milkfat fractions obtained at 5 and 11°C from pentane solution had melting points of 25.3 and 32.2°C, respectively.

Supercritical fluid extraction Middle-melting milkfat fractions obtained using supercritical carbon dioxide extraction were the first through sixth fractions collected at temperatures between 40 and 70°C (Table 5.21). None of the middle-melting milkfat fractions obtained using supercritical carbon dioxide extraction were residual fractions. Although most of the data were reported for a single pressure, many of the investigators reported a range of pressures for the fractions obtained, because they employed an increasing pressure profile during the extraction process. Consequently, precise pressures yielding individual fractions were not available, and comparisons of the fraction properties and fractionation conditions were limited.

Middle-melting milkfat fractions obtained at 50°C showed an increase in melting point as pressure increased. Kankare et al. (1989) reported fractions obtained at 200, 300, and 400 bar with melting points of 27.9, 31.4, and 32.1°C, respectively. Arul et al. (1987) also reported middle-melting milkfat fractions obtained at 50°C, but reported a range of pressures from 100 to 250 bar. The first middle-melting fraction reported by Arul et al. (1987) was obtained at the lower end, and the second fraction at the higher end, of the pressure range. These fractions also showed increased melting points with increases in pressure (Arul et al., 1987).

Kankare and Antila (1988a) also showed an increase in melting point as the pressure profile increased; however, they simultaneously increased the temperature. Increases in both of these variables still resulted in increased melting points. Arul et al. (1987) also reported a fraction obtained at 70°C and 250–250 bar that had a melting point of 34.0°C, a melting point higher than that of any of the fractions obtained by Kankare et al. (1989) at 50°C and 200 and 300 bar (MP = 27.9 and 31.4°C, respectively). Thus, the data show that the melting points of middle-melting milkfat fractions obtained from supercritical carbon dioxide extraction increased as both pressure and temperature increased. The trend was demonstrated by the fraction produced by Hamman et al. (1991) at 40°C and 125 bar (A4 fraction, Table 5.21). This fraction was obtained using the lowest temperature and pressure reported for supercritical fluid extraction of middle-melting fractions, and it also exhibited the lowest melting point (25°C) reported for middle-melting fractions.

Short-path distillation One middle-melting milkfat fraction was obtained using short-path distillation at 265°C and 100 μm Hg (Table 5.22). The fraction was the third, and last, distillate obtained from the fractionation process, and it had a melting point of 26.7°C.

Percentage Yields of Middle-Melting Milkfat Fractions

The percentage yields for middle-melting milkfat fractions obtained by crystallization from melted milkfat (Table 5.19) and from solvent solution (Table 5.20) showed similar trends, and the yields were very dependent upon the physical state of the fraction:

- Liquid fractions had yields that generally ranged from 63 to 97% for fractions obtained both from melted milkfat and from solvent solution, and the yield generally decreased with decreasing fractionation temperature.
- Solid middle-melting milkfat fractions obtained from melted milkfat had yields that ranged from 12 to 58%, and fractions obtained from solvent solution had yields that ranged from 5 to 25%, but these yields did not exhibit trends consistent with fractionation temperature.

Crystallization from melted milkfat The majority of liquid middle-melting milkfat fractions were obtained using single-step fractionation (Table 5.19), and the decrease in yield as fractionation temperatures decreased was consistent with the concurrent increase in yield observed for solid high-melting fractions (Table 5.15) obtained using single-step fractionation. The decrease in yield with decreasing fractionation temperature was observed for liquid fractions reported by individual authors as well as between authors, and was consistent between separation methods. For example, deMan and Finoro (1980) produced liquid middle-melting milkfat fractions using vacuum filtration at 32, 31, 30, 29, 28, and 27°C and reported yields of 90.0, 82.0, 82.7, 81.3, 81.6, and 75.4%. Although some of these yields did not show large decreases as the temperature decreased, the overall trend was evident. Kaylegian (1991) produced three liquid middle-melting milkfat fractions using vacuum filtration at 34, 30, and 16°C that exhibited yields of 97.5, 80.0, and 14.2%, and these data were in agreement with those reported by deMan and Finoro (1980). Also, Doležálek et al. (1976) produced several liquid middle-melting milkfat fractions with the aid of an aqueous detergent solution, which had yields of 87 and 84% at 30°C, and yields of 57 and 50% at 25°C.

Liquid middle-melting milkfat fractions obtained with the aid of an aqueous detergent solution (Table 5.19) also exhibited an overall decrease in yield as fractionation temperature decreased, but there were several exceptions to this. Sherbon et al. (1972) produced a fraction at 30.6°C with a yield of 41%, which was much lower than the 87 to 91% reported for three fractions produced at 30°C by Doležálek et al. (1976). However, the fraction produced by Sherbon et al. (1972) was the liquid fraction obtained using a double fractionation at 30.6°C, and the refractionation process was responsible for the decrease in yield. The solid high-melting milkfat fractions produced by Sherbon et al. (1972; Table 5.15) employed a refractionation process with an increasing temperature profile that resulted in an increase in yield as the fractionation temperature increased. Thus, this approach resulted in a lower yield for the liquid fractions obtained using the refractionation process. Sherbon et al. (1972) also produced two liquid middle-melting milkfat fractions obtained using single-step fractionation at 28 and 25.5°C. The yield for the fraction obtained at 28°C (A1L$_S$ (single fraction at 28°C) [A]) was lower than expected, based on the trends shown in Table 5.19, and was also lower than the yield for the fraction obtained at 25.5°C; therefore, an increase in yield occurred as fractionation temperature decreased. The nature of the separation process with the aid of an aqueous detergent solution is generally to produce relatively pure crystalline fractions containing little liquid milkfat; this would result in a smaller yield for the solid fraction and a greater yield for the liquid fraction. That was not the case for the A1L$_S$ fraction reported by Sherbon et al. (1972), but the reason for this is unclear.

Yields for solid middle-melting milkfat fractions obtained from melted milkfat (Table 5.19) did not exhibit consistent trends as the fractionation temperature decreased, even though yields for solid high-melting milkfat fractions (Table 5.15) increased as the fractionation temperature decreased. However, two solid middle-melting milkfat fractions showed increases in yields as fractionation temperatures decreased for fractions obtained by vacuum filtration; these fractions were produced by Kaylegian (1991) at 16°C (16S$_M$ [W] fraction, 14.2% yield) and by Versteeg (1991) at 12°C (12S$_M$ [Sept] fraction, 31% yield). Sreebhashyam et al. (1981) reported an increase in yields from 34.9 to 38.7% for two fractions obtained by filtration through muslin cloth as the temperature decreased from 25 to 18°C. However, Khalifa and Mansour (1988) also produced a fraction by filtration through muslin cloth at 20°C; they reported a yield of 13.7%, which was lower than both yields reported by Sreebhashyam et al. (1981), even though the fractionation temperature fell between those employed by Sreebhashyam et al. (1981). The solid middle-melting milkfat fractions obtained by centrifugation (Richardson, 1968) exhibited a decrease in yield as fractionation temperature decreased from 25 to 20°C. Lakshminarayana and Rama Murthy (1985) obtained a solid middle-melting milkfat fraction at 23°C and reported a yield of 57.6%. Thus, in conclusion, consistent trends were not reported for the yields of solid middle-melting milkfat fractions obtained from melted milkfat, either between investigators or between separation methods.

Crystallization from solvent solution Solid middle-melting milkfat fractions obtained from solvent solution were all produced using multiple-step fractionation, but the yields did not exhibit consistent trends with respect to fractionation temperature (Table 5.20). Kaylegian (1991) produced two solid middle-melting milkfat fractions, which exhibited a decrease in yield as the fractionation temperature decreased (14.6% at 5°C and 11.1% at 0°C). In contrast, Rolland and Riel (1966) produced two solid middle-melting milkfat fractions that exhibited an increase in yield as the fractionation temperature decreased (5.3% at 10°C and 7.5% at 2°C). Three fractions that have been produced at 0°C have resulted in different yields: 24% (Avvakumov, 1974), 11.1% (Kaylegian, 1991), and 24.6% (Youssef et al., 1977). The fractions produced by Avvakumov (1974) and Youssef et al. (1977) were obtained at the third step, and the fraction obtained by Kaylegian was obtained at the sixth step in a multiple-step fractionation procedure, which would account for the difference in yields. Thus, the number of steps in a multiple-step fractionation procedure influences the yields obtained.

Liquid middle-melting milkfat fractions have been obtained from both acetone and pentane solutions (Table 5.20). Kaylegian (1991) reported one fraction from acetone at 21°C and two fractions from pentane obtained at 11 and 5°C using single-step fractionation, and obtained yields greater than 90%. The liquid fraction obtained by Sherbon and Dolby (1973) from acetone at 25°C provided a much lower yield (12.8%) than did other liquid fractions obtained from solvent solution. However, this fraction was obtained in the fifth step of a multiple-step fractionation procedure, which accounts for the lower yield observed. This is consistent with the trend observed for solid middle-melting milkfat fractions obtained using multiple-step fractionation. The two fractions obtained from pentane solution using single-step fractionation exhibited a decrease in yield as fractionation temperature decreased, which is consistent with trends observed for liquid middle-melting milkfat fractions obtained from melted milkfat (Table 5.19).

Supercritical fluid extraction Middle-melting milkfat fractions obtained from supercritical carbon dioxide extraction (Table 5.21) generally had yields between 10 and 25%, but the reported range was 3.3 to 42.4%. Yields for middle-melting milkfat fractions obtained by supercritical carbon dioxide extraction increased with increasing pressure until a maximum solubility was reached. After that they exhibited a decrease in yield as pressure continued to increase, consistent with the trend observed for high-melting milkfat fractions (Table 5.17). Kankare et al. (1989) reported the yields of 3.3, 21.4 and 17.5% at 200, 300 and 400 bar, respectively, and the fraction obtained at 300 bar gave the greatest yield. In comparison, high-melting milkfat fractions obtained by Chen et al. (1992) provided a maximum yield in the middle of an increasing pressure sequence. Arul et al. (1987) also observed an increase in yield as the pressure sequence increased from 100 to 250 bar (13.1 and 15.7%, respectively).

Yields similarly increased as fractionation temperature increased for fractions obtained using supercritical carbon dioxide extraction. Kankare and Antila (1988a) obtained three fractions over an increasing temperature (50 to 80°C) and pressure profile (100 to 400 bar) and reported an increase in yield (27.9, 31.5, and 32.1%) as the temperature and pressure increased.

Arul et al. (1987) produced a middle-melting milkfat fraction at 70°C over the range of 250–350 bar and reported a yield of 10.7%; this fell between yields reported by Kankare et al. (1989) for fractions obtained at 50°C and at pressures of 200 and 300 bar, which exhibited yields of 3.3 and 21.4%, respectively. Because a range of pressures was used for the fractions obtained by Arul et al. (1987), it was difficult to correlate increases in yields with increasing temperature, but these data appear to be consistent with the data reported by Kankare and Antila (1988a) for increasing temperature and pressures.

The fraction obtained by Rizvi et al. (1990) at 40°C and 241.3 bar exhibited the highest yield (42.4%) reported, even though it was obtained at the lowest temperature employed and at the lower end of the pressure spectrum. The reasons for these deviations from other reports were not apparent, however.

Short-path distillation The middle-melting milkfat fraction obtained at 265°C and 100 μm Hg by short-path distillation (Table 5.22) exhibited a yield of 32.0%. This yield was similar to the yield of solid middle-melting milkfat fractions obtained from melted milkfat, but less than that of liquid middle-melting milkfat fractions. The middle-melting milkfat fraction obtained by short-path distillation was more abundant than middle-melting fractions obtained by supercritical carbon dioxide extraction and crystallization from solvent solution.

Chemical Characteristics of Middle-Melting Milkfat Fractions

Fatty acid composition In general, middle-melting milkfat fractions obtained by crystallization from melted milkfat and from solvent solution exhibited a fatty acid profile that was similar to that of intact AMF. Middle-melting milkfat fractions obtained using supercritical carbon dioxide extraction and short-path distillation contained lower C4–C10 fatty acid contents and higher C12–C15 and C16–C18 saturated fatty acid contents than the parent AMF. Fatty acid composition is presented by individual fatty acids in Tables 5.44 to 5.47 and by selected fatty acid groupings in Tables 5.69 to 5.72 for fractions obtained from melted milkfat, from solvent solution, by supercritical fluid extraction, and by short-path distillation, respectively.

Middle-melting milkfat fractions obtained *from melted milkfat* exhibited a wide range of values for specific fatty acid composition. Variations in fatty acid compositions for middle-melting milkfat fractions obtained from melted milkfat, reported by selected saturated *n*-chain groupings and the C18:1/C18:0 ratios, were similar to the normal variation observed for intact AMF (Table 5.58).

Although there was a slight shift in the C4–C10 and C16–C18 saturated fatty acids for solid and liquid middle-melting milkfat fractions obtained from melted milkfat compared with intact AMF, consistent trends relating to fractionation procedure, separation method, or fractionation temperature were not apparent. Several fractions exhibited atypical values for individual fatty acids compared to the normal ranges, but these were generally related to atypical values observed in their parent AMFs. Causes of atypical values observed for the fatty acid composition of intact milkfat and milkfat fractions were unclear. Interestingly, solid middle-melting milkfat fractions obtained from melted milkfat that did not exhibit a distinct peak for the high-melting glyceride species in the DSC curves (Figure 5.12; Antila and Antila, 1970; Kankare and Antila, 1988a; Kaylegian, 1991) had normal concentrations for all major fatty acid groups, although perhaps a decrease in C16–C18 saturated fatty acids might have been expected. Amer et al. (1985) reported that the majority of the unsaturated fatty acids were present in the *cis* configuration (Table 5.69). Laakso et al. (1992) reported the fatty acid composition of triglyceride classes of the middle-melting milkfat fraction (S2) obtained from melted milkfat (Table 5.44). Trends were as expected for the occurrence of saturated and unsaturated fatty acids in their respective saturated and unsaturated triglycerides, as was illustrated by Laakso et al. (1992) for high-melting milkfat fractions.

Fatty acid composition for middle-melting milkfat fractions obtained *from solvent solution* are presented in Tables 5.45 and 5.70. Middle-melting milkfat fractions obtained from solvent solutions exhibited an increase in C4–C10 fatty acids and a decrease in C16–C18 saturated fatty acids as fractionation temperature decreased (Kaylegian, 1991; Yi, 1993). This was observed for both the liquid and the solid fractions obtained from acetone solution and for the liquid fractions obtained from pentane solution (Kaylegian, 1991).

The liquid middle-melting milkfat fractions obtained from acetone solution (Kaylegian, 1991; Sherbon and Dolby, 1973) exhibited slightly greater C4–C10 fatty acid contents and lower C16–C18 saturated fatty acid contents than did the solid middle-melting fractions, and the C12–C15 fatty acid content stayed relatively stable. These trends were in agreement with trends observed for very-high-melting (Tables 5.38 and 5.63) and high-melting (Tables 5.41 and 5.66) milkfat fractions obtained by crystallization from solvent solution. The liquid middle-melting milkfat fraction produced by Sherbon and Dolby (1973) was obtained at the fifth step of a multiple-step fractionation procedure from a high-melting glyceride and had a greater concentration of C16–C18 saturated fatty acids compared with the liquid fractions obtained from intact AMF (Kaylegian, 1991).

Kaylegian (1991) and Yi (1993) obtained solid middle-melting milkfat fractions at 5°C using multiple-step fractionation from acetone solution; these had relatively low concentrations of C4–C10 fatty acids (8.4 to 8.6%) and relatively high concentrations of C16–C18 saturated fatty acids (53 to 59%) compared to other middle-melting milkfat fractions obtained from acetone, which had C4–C10 fatty acid contents of 10–12% and C16–C18 saturated fatty acid contents of 36–47%. These middle-melting fractions obtained from acetone solution exhibited unusually low C18:1/C18:0 ratios (0.86 and 0.82%) compared with other middle-melting milkfat fractions obtained from acetone solution (1.31–2.36%). The fraction produced by Kaylegian ($5S_M$ [W], 1991) also exhibited a unique DSC profile (Figure 5.13), in that the three main peaks were very distinct and were somewhat more narrow than those observed for other middle-melting milkfat fractions. This fraction exhibited the highest SFC profile for middle-melting milkfat fractions obtained from solvent solution (Kaylegian, 1991; Table 5.186), suggesting that the fraction may exhibit unique functionality.

The causes of the unusual fatty acid composition and melting behaviors observed in these fractions is unclear but may be related to the narrow temperature ranges employed during the multiple-step fractionation procedure, i.e., seven successive fractions were produced from 25 to 0°C at approximately 5°C increments (Kaylegian, 1991; Yi, 1993). However, the solid fraction produced by Kaylegian (1991) at 5°C using single-step fractionation was a high-melting fraction, but it exhibited a thermal profile (Figure 5.9) similar in shape to that of the middle-melting fraction obtained from multiple-step fractionation (Figure 5.13). These data further illustrate the complexity of milkfat fractionation and support the need for increased understanding of the factors affecting milkfat fraction properties.

Middle-melting milkfat fractions obtained by *supercritical carbon dioxide extraction* (Tables 5.46 and 5.71) generally exhibited a small increase in the C4–C10 and C12–C15 fatty acid contents as well as a small decrease in C16–C18 saturated fatty acid, and a slightly greater C18:1/C18:0 ratio, compared with the parent AMFs (Tables 5.35 and 5.60). The C18:1 contents of these middle-melting milkfat fractions were lower than those reported for the parent AMF samples, the very-high-melting milkfat fractions, and the high-melting milkfat fractions. Several fractions exhibited atypical values for individual fatty acids compared to the normal ranges, but these were generally related to atypical values observed in parent AMF samples.

Fatty acid compositions of middle-melting milkfat fractions obtained by supercritical carbon dioxide extraction exhibited decreased C4–C10 and increased C16–C18 saturated fatty acid contents as the fractionation temperature and pressure increased; these were in agreement with the melting points reported for these fractions (Table 5.21).

The fatty acid composition of a middle-melting milkfat fraction obtained by *short-path distillation* (Arul et al., 1988a) is shown in Tables 5.47 and 5.72. This fraction exhibited an increased C4–C10 fatty acid content, a slightly increased C12–C15 fatty acid content, and a decreased C16–C18 saturated fatty acid content compared with the parent AMF (Tables 5.36 and 5.61).

Triglyceride composition   The triglyceride compositions for liquid middle-melting milkfat fractions obtained from *melted milkfat* (Tables 5.93 and 5.117) were similar to intact AMF (Tables 5.83 and 5.106) at fractionation temperatures greater than 30°C. Liquid middle-melting milkfat fractions obtained at fractionation temperatures below 30°C showed a gradual increase in C24–C34 and C36–C40 triglycerides and a gradual decrease in C42–C54 triglycerides as fractionation temperatures decreased. This trend was more evident for fractions obtained by individual investigators than for fractions compared between investigators. For example, Kaylegian (1991) obtained a liquid middle-melting milkfat fraction at 34°C, and deMan and Finoro (1980) obtained a liquid middle-melting milkfat fraction at 32°C, and each of these fractions exhibited triglyceride patterns that were similar to their respective parent AMF samples (Tables 5.83 and 5.106). However, the compositions of the parent AMF samples differed, as did the percent composition for individual triglycerides in the derived fractions.

The solid middle-melting milkfat fraction obtained from melted milkfat ($16S_M$ [W], Kaylegian, 1991) had the lowest concentration of C24–C34 triglycerides and a relatively high concentration of C36–C40 triglycerides compared to liquid middle-melting milkfat fractions, although the C42–C54 triglyceride concentration was similar. Makhlouf et al. (1987) also produced a solid middle-melting milkfat fraction ($15S_M$ (I.1)) that had a relatively high C36–C40 triglyceride composition, but it had a lower C42–C54 concentration. Laakso et al. (1992) reported that the middle-melting milkfat fraction obtained from melted milkfat (S2) was composed of 54% trisaturated (SSS), 36% disaturated (SSU), and 10% di- and triunsaturated (SUU and UUU) triglycerides (Table 5.117).

Triglyceride compositions for fractions obtained from *acetone* and *pentane solutions* exhibited trends were similar to those observed for fatty acid concentration: C24–C34 and C36–C40 triglycerides increased and C42–C54 triglycerides decreased as the fractionation temperature decreased for liquid and solid fractions obtained from acetone and for liquid fractions obtained from pentane solution (Tables 5.94 and 5.118).

The solid middle-melting milkfat fractions obtained from acetone solution had markedly lower concentrations of the C24–C34 triglycerides, offset by smaller increases in both the C36–C40 and C42–C54 triglycerides compared with liquid middle-melting milkfat fractions obtained from acetone solution (Kaylegian, 1991).

Middle-melting milkfat fractions obtained using *supercritical carbon dioxide extraction* showed a decrease in C24–C34 and C36–C40 triglycerides and an increase in C42–C54 triglycerides as fractionation temperature and pressure increased (Tables 5.95 and 5.119). Two fractions obtained by Arul et al. (1987) at 50°C using an increasing pressure profile from 100 to 250 bar and a fraction obtained at 70°C and 250–350 bar illustrated these trends. Kankare et al. (1989) also reported these trends for two fractions obtained at 50°C at 200 and 400 bar. These data further support the concept that lower-melting fractions are the first to be extracted by supercritical carbon dioxide extraction.

Middle-melting milkfat fractions obtained from supercritical carbon dioxide extraction exhibited the highest concentration of C36–C40 triglycerides reported (45.5 to 55.8%), compared with fractions obtained by crystallization from melted milkfat and solvent solutions, which generally had C36–C40 concentrations of 27 to 35%.

The middle-melting milkfat fraction obtained by *short-path distillation* (Tables 5.96 and 5.120) had an increased concentration of C24–C34 triglycerides, a great increase in the C36–C40 triglyceride concentration, and a concurrent decrease in the C42–C54 triglycerides compared with the parent AMF (Table 5.86 and 5.109; Arul et al., 1988a). The concentration of the C36–C40 triglycerides increased from 35.6% in the parent AMF to 61.3% in the middle-melting milkfat fraction, which was greater than the concentration generally observed for middle-melting milkfat fractions obtained by supercritical carbon dioxide extraction (Table 5.95 and 5.119). Consequently, the C42–C54 triglyceride concentration decreased from 52.9% in the AMF to 22.3% in middle-melting fraction. The large decrease in C42–C54 triglycerides was consistent with the lowered SFC profile (Table 5.188) observed for this fraction as well as other chemical and physical characteristics.

Selected chemical characteristics Iodine values, Reichert–Meissl numbers, and Polenske numbers all increased as the fractionation temperature decreased for both liquid and solid middle-melting milkfat fractions from melted milkfat (Table 5.140). These trends were illustrated by liquid middle-melting milkfat fractions obtained using vacuum filtration (deMan and Finoro, 1980; Amer et al., 1985) and liquid fractions separated with the aid of an aqueous detergent solution (Doležálek et al., 1976) as well as by solid middle-melting milkfat fractions filtered through muslin cloth (Sreebhashyam et al. (1981). However, the trends in chemical characteristics did not appear to be substantially influenced by the fraction separation method employed.

Although these chemical characteristics differed for middle-melting fractions between investigators, they also differed for their parent AMF samples (Table 5.131). Thus, the chemical characteristics of the parent AMF greatly influenced the chemical characteristics of milkfat fractions; this has been observed for all categories of milkfat fractions. The chemical characteristics exhibited more narrow ranges for middle-melting milkfat fractions than for high-melting milkfat fractions (Table 5.136). The range of chemical characteristics reported for middle-melting milkfat fractions obtained from melted milkfat were:

CHARACTERISTIC	RANGE	AVERAGE
Iodine value	14 to 38.9	31 to 35
Refractive index	1.4516 to 1.4569	1.4536 to 1.4543
Reichert–Meissl value	23.6 to 31.7	30 to 32
Polenske value	1.5 to 2.0	1.9 to 2.0
Density	0.8935 to 0.9523	0.8935 to 0.8942

Chemical characteristics for fractions obtained from solvent solution are shown in Table 5.141, and for fractions obtained by supercritical fluid extraction in Table 5.142. Iodine values were reported for only a few fractions, and there was not enough information available to determine trends relating to fractionation conditions.

Flavor The potential lactone and methyl ketone concentrations and aldehyde concentrations of middle-melting milkfat fractions obtained from melted milkfat are shown in Tables 5.153, 5.159 and 5.164, respectively. The middle-melting milkfat fractions had slightly increased concentrations of lactones and methyl ketones compared with the intact AMF (Tables 5.150 and 5.156). The middle-melting milkfat fractions also exhibited an increase in aldehyde concentration compared with the intact milkfat (Table 5.161), which was attributed to slight oxidation of the milkfat during processing (Walker, 1972).

Physical Properties of Middle-Melting Milkfat Fractions

Crystal Characteristics The *crystal morphologies* for middle-melting milkfat fractions are shown in Table 5.169. The fractions obtained from melted milkfat (Deroanne, 1976) and from acetone solution (Sherbon and Dolby, 1973) exhibited both the α and β' crystal forms, but the fraction obtained from supercritical carbon dioxide extraction exhibited only the β' form (Hamman et al., 1991). Grall and Hartel (1992) reported that the middle-melting milkfat fraction obtained from melted milkfat with agitation did not exhibit the same needle-like structures as the high-melting milkfat fraction but instead were a conglomerate of smaller, tighter aggregates.

Data for the *crystal sizes* of middle-melting milkfat fractions were not reported.

Melting point The melting points of middle-melting milkfat fractions are presented in Tables 5.19, 5.20, 5.21 and 5.22 for fractions obtained from melted milkfat, from solvent solution, by supercritical fluid extraction, and by short-path distillation, respectively. The relationship between melting point and fractionation conditions was discussed earlier in the section on fractionation conditions for middle-melting milkfat fractions. The general melting point characteristics of melting-melting milkfat fractions were:

- By definition, the melting point is between 25 and 35°C.
- The physical state of the fraction at the time of separation appeared to contribute the most influence on the melting point for fractions obtained from melted milkfat and from solvent solutions.
- Solid middle-melting milkfat fractions obtained from melted milkfat generally exhibited melting points that were 2 to 14°C greater than the fractionation temperature.
- Liquid middle-melting milkfat fractions obtained from melted milkfat generally had melting points that were slightly (± 2°C) above or below fractionation temperature, regardless of separation technique employed.
- Consistent trends were not observed for solid middle-melting milkfat fractions obtained from melted milkfat with respect to fractionation temperature or fractionation method employed, but liquid middle-melting milkfat fractions exhibited a decrease in melting point as fractionation temperature decreased.
- Solid middle-melting milkfat fractions obtained from acetone solution using single-step fractionation exhibited a decrease in melting point as fractionation temperature decreased.
- Melting points of middle-melting milkfat fractions obtained from supercritical fluid extraction increased as both temperature and pressure were increased.

Thermal profile Thermal profiles of middle-melting milkfat fractions are shown in Figures 5.12, 5.13, 5.14, and 5.15 for fractions obtained from melted milkfat, from solvent solution, by supercritical fluid extraction, and by short-path distillation, respectively. Thermal profiles of corresponding parent AMF samples are shown in Figures 5.1, 5.2, 5.3, and 5.4.

In agreement with other characteristics of middle-melting milkfat fractions, DSC profiles of middle-melting milkfat fractions obtained from melted milkfat and from solvent solution showed differences related to the physical state of the fraction at separation (i.e, liquid and solid fractions). Similar trends for thermal profiles were observed within the liquid fraction groups and within the solid fraction groups.

Liquid middle-melting milkfat fractions obtained from melted milkfat and from solvent solution exhibited DSC profiles that were similar in shape and size to their respective parent AMF samples, regardless of separation method or solvent employed. Liquid middle-melting milkfat fractions obtained from melted milkfat using vacuum filtration (Kaylegian, 1991; Schaap and van Beresteyn, 1970), pressure filtration (deMan, 1968), and separation technique not reported (Deroanne, 1976) showed slight increases for the low-melting triglyceride species and slight decreases for the high-melting triglyceride species compared with the parent AMF. These fractions often exhibited more resolution between the low-melting and middle-melting triglyceride peaks compared with the parent AMF, indicating more complete melting of the low-melting triglyceride species prior to the onset of melting for the middle-melting triglyceride species.

Kaylegian (1991) obtained two middle-melting milkfat fractions from melted milkfat at 34 and 30°C (Figure 5.12), and these fractions clearly illustrated the transition from high-melting milkfat fractions, with predominantly high-melting triglyceride species, to the lower-melting fractions, which had greater concentrations of middle- and low-melting triglyceride species. The DSC curve for the fraction obtained 34°C ($34L_S$ [W]) was almost identical to that of the parent AMF (Figure 5.1). As the fractionation temperature decreased to 30°C (30L$_S$ [W]), noticeable decreases in the high-melting triglyceride peak occurred, and the low- and middle-melting triglyceride peaks became less distinguishable. This transition of predominant triglyceride species peaks was also reported for middle-melting milkfat fractions obtained by Kaylegian (1991) both from acetone and from pentane solutions (Figure 5.13).

Middle-melting milkfat fractions obtained with the aid of an aqueous detergent solution have been produced using a single-step procedure and a refractionation procedure that employed an increasing temperature profile (Sherbon et al., 1972). In the Sherbon et al. (1972) study, the liquid fraction produced by the single-step process exhibited trends similar to other liquid middle-melting milkfat fractions. The liquid fraction obtained by the refractionation process showed an increased width for the high-melting glyceride peak compared with the parent AMF; this was probably caused by the large amount of high-melting glycerides in the initial fraction. The liquid middle-melting milkfat fractions obtained from acetone solution by Sherbon and Dolby (1973) was obtained at the fifth step of a multiple-step fractionation procedure that employed a high-melting glyceride as the starting material. This fraction was composed only of low- and middle-melting triglyceride species, because the high-melting species had been removed by previous fractionation steps. Liquid middle-melting milkfat fractions obtained by refractionation using a decreasing temperature profile contain decreased amounts of the high-melting triglyceride species, and fractions obtained by refractionation using an increasing temperature profile contain increased amounts of the high-melting triglyceride species.

The DSC profiles for solid middle-melting milkfat fractions, obtained both from melted milkfat and from solvent solutions, varied in shape and in the size of peaks for the main triglyceride species. In spite of this variation two trends were readily observed:

- Solid middle-melting milkfat fractions exhibited larger peaks for middle-melting triglyceride species compared with the liquid middle-melting milkfat fractions and intact AMFs.
- The peak for the high-melting triglyceride species either was smaller or was not present in the solid middle-melting milkfat fractions, compared with the liquid middle-melting milkfat fractions.

The increased size of the middle-melting glyceride peak is apparent in DSC curves for solid middle-melting milkfat fractions obtained by centrifugation (Antila and Antila, 1970; Kankare and Antila, 1986, 1988a) and vacuum filtration (Kaylegian, 1991; Makhlouf et al., 1987) (Figure 5.11). The peak for the high-melting triglycerides was present in the fractions obtained by Kankare and Antila (1986, 1988a) and absent in DSC curves reported by Antila and Antila (1970), Kaylegian (1991), and Makhlouf et al. (1987). Differences in DSC profiles for fractions obtained using different separation techniques were not apparent.

The predominance of the middle-melting triglyceride species peak was also apparent in the solid middle-melting milkfat fractions obtained from acetone solution (Figure 5.13; Kaylegian, 1991). The fraction obtained at 5°C showed the presence of high-melting triglyceride species, which were not present in the fraction obtained at 0°C. Solid middle-melting milkfat fractions also exhibited a greater resolution between the peaks for the low-melting and middle-melting triglyceride species, as was observed with liquid middle-melting milkfat fractions.

Middle-melting milkfat fractions obtained using supercritical fluid extraction also showed an increase in peak size for the middle-melting triglyceride species for all fractions (Figure 5.14). An increase in the size of the peaks for the high-melting triglyceride species was observed as fractionation temperature and pressure increased, and this was consistent with the increase in melting points of these fractions (Table 5.21). This trend further supports the concept that in supercritical fluid extraction the lower-melting fractions are extracted first, followed by the higher-melting fractions. This can be readily observed in the data for fractions reported by Arul et al. (1987; 50–70°C, 100–350 bar), Kankare and Antila (1988a; 50—80°C, 100–400 bar), and Kankare et al. (1989; 50°C, 200–400 bar).

The DSC profile of the middle-melting milkfat fraction obtained using short-path distillation also showed a predominant peak for the middle-melting triglyceride species, and the peak for the high-melting triglyceride species was absent (Figure 5.15; Arul et al., 1988a).

Solid fat content   Solid fat contents have been tabulated and presented in Tables 5.185, 5.186, 5.187, and 5.188 for middle-melting milkfat fractions obtained by crystallization from melted milkfat, from solvent solution, by supercritical fluid extraction, and by short-path distillation, respectively. Solid fat content profiles for representative fractions from each fractionation method have also been presented graphically in Figure 5.24, and an intact AMF profile has been included for reference (Kaylegian, 1991). Solid and liquid middle-melting milkfat fractions representing crystallization from melted milkfat ($16S_M$ and $34L_S$ fractions, Kaylegian, 1991) and from solvent solution ($25L_S$ and $5S_M$ fraction, Kaylegian, 1991) are presented to illustrate the differences in melting behavior between the solid and liquid middle-melting fractions. The following observations regarding the melting behavior of middle-melting milkfat fractions have been made from Figure 5.24:

- SFC profiles for liquid middle-melting milkfat fractions obtained by melted milkfat and solvent solution were similar to each other and to to that of the parent intact AMF.
- SFC profiles for solid middle-melting fractions obtained by crystallization from melted milkfat and from solvent solution showed that these fractions contained more solid fat at most temperatures compared with the liquid fractions, and then they melted sharply just prior to reaching the point of complete melting (clear point).
- The solid fractions obtained from solvent solution exhibited more elevated SFC profiles than did the solid fractions obtained from melted milkfat.
- The SFC profiles for the middle-melting milkfat fractions obtained by supercritical carbon dioxide extraction showed that the fractions exhibited melting behaviors that were between the solid fractions and the liquid fractions obtained either from melted milkfat or from solvent solutions.
- The SFC profile for the middle-melting milkfat fraction obtained by short-path distillation was similar to profiles for intact AMF and the liquid middle-melting milkfat fractions obtained both from melted milkfat and from solvent solutions.

The SFC profiles of middle-melting milkfat fractions obtained both from melted milkfat and from solvent solutions appear to be more influenced by the physical state of the fraction at separation (i.e., liquid or solid fraction) than other variables. SFC data were reported only for middle-melting milkfat fractions obtained from melted milkfat using vacuum filtration and for one fraction obtained where separation technique was not reported. Thus, inadequate data were available to determine the influence of separation technique on the SFC of these fractions. Middle-melting milkfat fractions obtained using single-step fractionation both from melted milkfat and from solvent solution were all liquid fractions; those obtained using multiple-step fractionation were all solid fractions.

The SFC profiles of liquid middle-melting milkfat fractions obtained from melted milkfat were lowered as fractionation temperatures decreased in studies by individual investigators, but this trend was not consistent when fractions produced by different investigators were compared. This was illustrated by liquid middle-melting milkfat fractions from melted milkfat produced by Kaylegian (1991; 34 and 30°C), and by deMan and Finoro (1980; 32, 31, 30, 29, 28, and 27°C), which generally exhibited lower SFC profiles as the fractionation temperature decreased. However, fractions produced by Amer et al. (1985; 29 and 26°C) and by Versteeg (1991; 28°C) exhibited SFC profiles that were elevated compared with those reported by Kaylegian (1991) and deMan and Finoro (1980) when viewed in sequence as the fractionation temperature decreased from 34 to 26°C.

Liquid middle-melting milkfat fractions obtained both from acetone and from pentane solutions exhibited lower SFC profiles as fractionation temperatures decreased. This was illustrated for middle-melting milkfat fractions produced by Kaylegian (1991) using single-step fractionation from acetone and pentane solutions (Table 5.186). The trend for lowered SFC profiles with decreasing fractionation temperatures for fractions obtained from solvent solutions was in agreement with the trend observed for fractions obtained from melted milkfat in studies by individual authors.

The SFC profiles of solid middle-melting milkfat fractions (Figure 5.24) obtained from solvent solution (Table 5.186) were elevated compared with SFC profiles for fractions produced by melted milkfat (Table 5.185), and this was consistent with the melting behavior seen for very-high-melting (Figure 5.22; Tables 5.178 and 5.179) and high-melting milkfat fractions (Figure 5.23; Tables 5.181 and 5.182). Solid middle-melting milkfat fractions were generally much more solid than the liquid fractions and intact AMF at temperatures above 20 to 25°C, and then appeared to have reasonably similar SFC profiles as the fractions approached the actual melting point. The elevated SFC profiles of the solid fractions compared with the liquid fractions were also consistent with the higher melting points of solid milkfat fractions compared with liquid fractions that were obtained at similar fractionation temperatures.

Solid middle-melting milkfat fractions exhibited lower SFC profiles as the fractionation temperatures decreased. Kaylegian (1991) produced a solid middle-melting milkfat fraction at 15°C from melted milkfat (Table 5.180) that was more solid than the fraction produced at 12°C by Versteeg (1991). Likewise, the solid middle-melting milkfat fractions produced by Kaylegian (1991) from solvent solution (Table 5.186) at 5 and 0°C exhibited lowered SFC profiles as the temperature decreased.

The middle-melting milkfat fraction produced by supercritical fluid extraction exhibited SFC profiles (Table 5.187) that fell between the SFC profiles for the solid and liquid fractions obtained from melted milkfat and from solvent solutions at temperatures below 20°C (Figure 5.24). At temperatures above 20°C the SFC profiles for the middle-melting milkfat fractions obtained from supercritical carbon dioxide extraction were lower compared to most fractions obtained from melted milkfat and solvent solutions (Figure 5.24, Tables 5.185 and 5.186). The three middle-melting milkfat fractions produced by supercritical carbon dioxide extraction were obtained at different temperatures and pressures (Bhaskar et al., 1993b), and so trends relating SFC profiles to fractionation conditions were not discernible.

The middle-melting milkfat fraction produced using short-path distillation by Arul et al. (1988a; Table 5.188) exhibited a SFC profile that was similar in shape to, but slightly lower than, the SFC profile for intact AMF (Figure 5.24). The SFC profile for this fraction was also similar in shape to and lower than the SFC profiles for liquid fractions obtained from melted milkfat and solvent solutions (Figure 5.24).

Textural characteristics Textural characteristics for middle-melting milkfat fractions obtained from melted milkfat are presented in Table 5.197. Two fractions were solid and two fractions were liquid, but all middle-melting fractions were softer than their parent AMFs (Table 5.195), as indicated by cone penetrometer readings (deMan, 1968; Sreebhashyam et al., 1981).

Low-Melting Milkfat Fractions

Approximately one-third of the milkfat fractions characterized by melting behavior in the literature were low-melting milkfat fractions (Table 5.6; 164 fractions). Low-melting milkfat fractions have been defined as fractions that exhibit a melting point below 25°C and above 10°C. Low-melting milkfat fractions have been produced by:

1. Crystallization from melted milkfat (Table 5.23; 125 fractions), using the following separation techniques:

- Vacuum filtration (Amer et al., 1985; Barna et al., 1992; Deffense, 1987, 1989; deMan and Finoro, 1980; Grall and Hartel, 1992; Kaylegian, 1991; Kaylegian and Lindsay, 1992; Laakso et al., 1992; Lund and Danmark, 1987; Makhlouf et al., 1987; Schaap and van Beresteyn, 1970; Sherbon, 1963; Timms, 1978b; Timms and Parekh, 1980; Versteeg, 1991; Vovan and Riel, 1973)
- Pressure filtration (deMan, 1968; Stepanenko and Tverdokhleb, 1974)
- Centrifugation (Antila and Antila, 1970; Baker et al., 1959; Bhat and Rama Murthy, 1983; Black, 1974b; Branger, 1993; Bratland, 1983; Evans, 1986; Jordan, 1986; Kankare and Antila, 1986, 1988a, 1988b; Keogh, 1989; Keogh and Higgins, 1986b; Lakshminarayana and Rama Murthy, 1985; Ramesh and Bindal, 1987b; Richardson, 1968; Riel and Paquet, 1972)
- Separation with the aid of an aqueous detergent solution (Banks et al., 1985; Dolby 1970a, 1970b; Doležálek et al., 1976; Fjaervoll, 1970a, 1970b; Jebson, 1970; Norris et al., 1971; Walker, 1972)
- Filtration through muslin cloth (Khalifa and Mansour, 1988; Sreebhashyam et al., 1981)
- Filtration through milk filter (Fouad et al., 1990)
- Filtration through cheese press (Baker, 1970a)
- Filtration through casein-dewatering press (Black, 1973)
- Separation technique not reported (Deroanne and Guyot, 1974; Dixon and Black, 1974; Riel and Paquet, 1972)

2. Crystallization from solvent solution (Table 5.24; 24 fractions) using the following solvents:
- Acetone (Avvakumov, 1974; Baker et al., 1959; Kaylegian, 1991; Kaylegian and Lindsay, 1992; Lambelet, 1983; Rolland and Riel, 1966; Sherbon, 1963; Timms, 1980a, 1980b; Yi, 1993; Youssef et al., 1977)
- Ethanol (Rolland and Riel, 1966)
- Isopropanol (Norris, 1977)
- Pentane (Kaylegian, 1991)

3. Supercritical fluid extraction (Table 5.25; 10 fractions)
- Carbon dioxide (Arul et al., 1987; Biernoth and Merk, 1985; Chen et al., 1992; Hamman et al., 1991; Kankare and Antila, 1988a; Kankare et al., 1989; Kaufmann et al., 1982; Timmen et al., 1984)

4. Short-path distillation (Table 5.26; 2 fractions)
- (Arul et al., 1988a)

5. Fractionation method not reported (Table 5.5; 3 fractions)
- (Jebson, 1974; Tucker, 1974; Mayhill and Newstead, 1992).

Fractionation Conditions of Low-Melting Milkfat Fractions

Crystallization from melted milkfat Crystallization from melted milkfat was the most popular method employed to produce low-melting milkfat fractions (Table 5.23). Low-melting milkfat fractions were liquid and solid fractions obtained using single-step and multiple-step fractionation procedures. The fractions exhibited differences between separation techniques, but consistent trends within separation techniques were apparent.

Liquid low-melting milkfat fractions were obtained using *vacuum filtration* by single-step and multiple-step fractionation procedures between 7 and 26°C, and solid low-melting milkfat fractions were obtained by multiple-step fractionation between 8 and 13°C. Deffense (1987, 1989, 1993b) reported low-melting milkfat fractions obtained by multiple-step fractionation that were identified as second stearins and as first and second oleins, but fractionation temperatures were not reported.

Liquid low-melting milkfat fractions exhibited lower melting points as fractionation temperatures decreased, and this was consistent with the melting behavior of other liquid fractions obtained from melted milkfat. Generally, liquid fractions exhibit a melting point within a few degrees of their fractionation temperature. Liquid low-melting milkfat fractions obtained at similar temperatures by different investigators exhibited similar melting points (±3°C). Versteeg (1991) produced two liquid low-melting milkfat fractions at 18°C, one with single-step and one with multiple-step fractionation, and obtained similar softening points, 20.4 and 21°C, respectively.

Solid low-melting milkfat fractions had higher melting points than liquid low-melting milkfat fractions obtained at equivalent temperatures. This was observed for solid and liquid fractions reported by Barna et al. (1992) at 20°C (MP = 33.9 and 18.0°C), Grall and Hartel (1992) at 15°C (MP = 19.5 and 12.5°C), Kaylegian (1991) at 13°C (MP = 22.8 and 10.1°C), and Sherbon (1963) at 10°C (MP = 20.0 and 13.2°C). In general, melting points decreased as fractionation temperatures decreased for solid and liquid low-melting milkfat fractions obtained from melted milkfat using vacuum filtration.

Low-melting milkfat fractions obtained from winter milkfats exhibited higher melting points than did fractions obtained from summer milkfats at equivalent temperatures within studies by individual investigators. Amer et al. (1985) reported melting points of 20.6 and 19.7°C for liquid low-melting milkfat fractions obtained at 19°C from winter and summer AMF samples, respectively. Deffense (1989) also observed a higher melting point for the first olein fraction (19.4°C) produced using multiple-step fractionation from winter AMF than for the corresponding fraction obtained from summer AMF (18.8°C). Deffense (1989) also produced second olein fractions from winter and summer AMF samples, but these fractions had melting points of 13.3 and 13.5°C, respectively. Generally, fractions obtained from summer milkfats exhibited slightly lower melting points than fractions obtained from winter milkfats, but the difference was within ±1°C. The significance of this small difference relative to the functionality of milkfat fractions remains to be determined.

Low-melting milkfat fractions obtained from melted milkfat using *pressure filtration* were liquid fractions obtained by single-step fractionation at 24 and 27°C and by multiple-step fractionation at 11°C. The liquid low-melting milkfat fractions obtained by single-step fractionation exhibited a decrease in melting point with a decrease in fractionation temperature (deMan, 1968). The difference between melting point and fractionation temperature was 7 to 9°C, which was greater than the difference generally observed between the melting points of liquid fractions and their fractionation temperature. The liquid low-melting milkfat fraction obtained using multiple-step fractionation by Stepanenko and Tverdokhleb (1974) exhibited a 2°C difference in melting point and fractionation temperature, which was consistent with other liquid milkfat fractions. The effects of cooling rate (i.e., rapid or slow; specific cooling rate in °C/min was not reported) on milkfat fractions were studied by deMan (1968), who reported a minor difference in melting points for liquid low-melting milkfat fractions obtained with rapid cooling compared with slow cooling at 27 and 24°C. The melting point for the fraction obtained by rapid cooling at 27°C was 18.4°C compared with 18.7°C for the fraction obtained by slow cooling. Additionally, melting points were 17.8 and 17.7°C for fractions obtained by rapid and slow cooling at 24°C, respectively.

Low-melting milkfat fractions produced using *centrifugation* were liquid fractions obtained by single-step and multiple-step fractionation procedures between 12 and 32°C, and solid fractions obtained at the second, third, and fourth steps of a multiple-step fractionation procedure at temperatures between 4 and 15°C.

The liquid low-melting milkfat fractions obtained using multiple-step fractionation exhibited differences between melting points and fractionation temperatures of –12 to +6°C, the greatest differences observed within any of the groups of milkfat fractions. This large difference was observed primarily in fractions obtained by Rolland and Riel (1966), who used an increasing temperature profile for the multiple-step fractionation, which began at 0°C and ended at 40°C (Table 5.1). The low-melting milkfat fractions produced by Rolland and Riel were obtained at the second, third, and fourth steps, and they exhibited a progressive decrease in cloud point of 12 to 3°C when compared with fractionation temperature as the fractionation temperature increased. Kankare and Antila (1986, 1988b) obtained two liquid low-melting milkfat fractions at 12°C from the third step of a multiple-step fractionation procedure, and they reported melting points of 17.9 and 18.6°C.

Solid low-melting milkfat fractions exhibited an increase in melting point as fraction temperature decreased, and a difference of 4 to 14°C between melting point and fractionation temperature was observed. These trends were consistent for solid milkfat fractions obtained from melted milkfat. The solid fractions also had increased melting points compared with liquid fractions obtained at the same temperature, and this was consistent with trends identified for solid and liquid milkfat fractions.

Lakshminarayana and Rama Murthy (1985) produced solid and liquid fractions at the third step of a multiple-step procedure at 15°C that had melting points of 19.0 and 14.0°C, respectively. Richardson (1968) also produced solid and liquid fractions at 15°C and reported slip points of 22.5 and 17.0°C, respectively, but these were the fourth step of a multiple-step procedure; these slip points were higher than the melting points reported by Lakshminarayana and Rama Murthy (1985) for fractions obtained at the third fractionation step. The general trend observed for other milkfat fractions obtained using multiple-step processes was for the melting points to decrease with successive fractionation steps, but this was not the case for these studies. The parent AMF employed by Richardson (1968) had a slip point of 33.5°C, whereas the parent AMF employed by Lakshminarayana and Rama Murthy (1985) had a melting point of 34.2°C (Table 5.5). Thus, the melting point of the parent AMF does not provide a readily discernible basis for the melting behavior observed for the derived fractions.

Jordan (1986) produced liquid low-melting milkfat fractions at 24 and 28°C from both winter and summer AMF samples but did not find melting behavior consistent with expected seasonal differences in the AMF. The liquid fraction obtained from summer milkfat at 28°C had a higher slip point (24.2°C) than did the corresponding liquid fraction obtained from winter milkfat (23.9°C). However, the liquid low-melting milkfat fraction obtained from summer milkfat at 24°C had a lower slip point (21.5°C) than did the corresponding liquid fraction obtained from winter milkfat (22.4°C). Consequently, conclusions on the influence of the season of AMF production on fractions obtained from melted milkfat using centrifugation could not be drawn.

All low-melting milkfat fractions obtained by *separation with the aid of an aqueous detergent solution* were liquid fractions obtained using single-step fractionation between 21 and 30°C. This group of fractions exhibited the most narrow range of melting points for low-melting milkfat fractions reported (22 to 24.8°C). These fractions exhibited a slightly greater difference between fractionation temperature and melting point (–6 to +4°C) compared with other milkfat fractions. The melting point of these fractions generally decreased as the fractionation temperature decreased. Norris et al. (1971) produced liquid low-melting milkfat fractions at 26 and 25°C that had slightly lower softening points than other fractions obtained at equivalent temperatures, but this decrease was only 1.3°C, which may not be significant from a functionality standpoint.

Low-melting milkfat fractions were also obtained by filtration through *muslin cloth*, through a *milk filter*, with a *cheese press*, and with a *casein-dewatering press* (Table 5.23). Low-melting milkfat fractions obtained by filtration through muslin cloth exhibited decreased melting points as fractionation temperatures decreased (Khalifa and Mansour, 1988; Sreebhashyam et al., 1981). Data for milkfat fractions obtained using filtration through a milk-fat filter (Fouad et al., 1990), a cheese press (Baker, 1970a), and a casein-dewatering press (Black, 1973) were not adequate to draw conclusions about the relationship between fractionation conditions and melting behavior.

Low-melting milkfat fractions obtained by *separation techniques that were not specified* also showed a decrease in melting points as fractionation temperature decreased.

Crystallization from solvent solution Low-melting milkfat fractions have been obtained from acetone, ethanol, isopropanol, and pentane solutions (Table 5.24).

The low-melting milkfat fractions obtained from *acetone* solution were liquid fractions obtained at fractionation temperatures above 0°C (Baker et al., 1959; Kaylegian, 1991; Rolland and Riel, 1966; Timms, 1980b; Yi, 1993; Youssef et al., 1977) and solid fractions obtained at fractionation temperatures below 0°C (Avvakumov, 1974; Rolland and Riel, 1966). Several other low-melting milkfat fractions were obtained from acetone solution, but their physical states at separation were not readily apparent (S1 fraction, Lambelet, 1983; 2C and 2CJAN [Jan] fractions, Sherbon, 1963; LMF, Timms, 1980a).

Liquid low-melting milkfat fractions have been obtained from acetone solutions using single-step fractionation between 5 and 15°C and by multiple-step fractionations between 0 and 24°C. Rolland and Riel (1966) produced three liquid fractions, which were the second, third, and fourth fractions obtained between 8 and 24°C using increased fractionation temperatures at subsequent steps of the multiple-step fractionation. Youssef et al. (1977) obtained two liquid fractions using multiple-step fractionation from acetone at 18°C that were the first and third fractions obtained, but they both had melting points of 21°C.

Liquid low-melting fractions obtained from acetone solutions exhibited decreased melting points as the fractionation temperatures decreased, and this was consistent with trends observed for other groups of liquid milkfat fractions.

Solid low-melting milkfat fractions have been obtained from acetone solution at the fourth (Avvakumov, 1974; Rolland and Riel, 1966) and fifth (Rolland and Riel, 1966) steps of a multiple-step fractionation procedure at temperatures between –6 and –15°C.

Rolland and Riel (1966) produced three solid low-melting milkfat fractions from *ethanol solution* at the second, third, and fourth step of a multiple-step fractionation procedure between 10 and –6°C, and each exhibited a decrease in melting point as fractionation temperature decreased.

Norris (1977) produced one low-melting milkfat fraction from *isopropanol solution* at –5°C. This fraction was the solid fraction produced at the second step of a multiple-step fractionation procedure.

Kaylegian (1991) obtained one liquid fraction using single-step and one liquid fraction using multiple-step fractionation at 0°C from *pentane solution*, and reported similar melting points for both fractions (21.3 and 21.6°C, respectively).

Supercritical fluid extraction Low-melting milkfat fractions obtained using supercritical fluid extraction (Table 5.25) were either the first or second fractions extracted, and each exhibited increased melting points as pressure increased, which was similar to trends observed for other milkfat fractions obtained by supercritical carbon dioxide extraction. These observations were supported by data reported by Arul et al. (1987) for fractions obtained at 50°C over an increasing pressure profile from 100 to 250 bar. Kaufmann et al. (1982) and Timmen et al. (1984) obtained low-melting milkfat fractions at 80°C and 200 bar, and reported a melting point of 20.4°C for both fractions, similar to the melting point of the fractions obtained at 50°C in the middle of the 100–250 bar profile by Arul et al. (1987). This suggested that increased temperature may not result in an increase in melting point for low-melting milkfat fractions obtained by supercritical carbon dioxide extraction.

Short-path distillation Two low-melting milkfat fractions were obtained using short-path distillation, which were the first and second distillates collected using a decreasing pressure profile from 220 to 100 μm Hg (Arul et al., 1988a; Table 5.26). Arul et al. (1988a) reported an increase in melting point (20.0°C) for the second fraction obtained at 245°C, but at a lower pressure (100 μm Hg) than the first fraction, obtained at 245°C and 220 μm Hg.

Percentage Yields of Low-Melting Milkfat Fractions

Percentage yields of low-melting milkfat fractions obtained from melted milkfat (Table 5.23) and from solvent solution (Table 5.24) were very much influenced by the physical state of the fraction at the time of separation (i.e., solid or liquid) and the fractionation procedure employed:

- Liquid low-melting milkfat fractions obtained from melted milkfat and solvent solution using single-step fractionation decreased in yield as the fractionation temperatures decreased, which correlates to an increase in yields of solid fractions that were observed for corresponding high-melting milkfat fractions (Table 5.15).
- Liquid low-melting milkfat fractions obtained from melted milkfat were generally the last fractions obtained from a multiple-step fractionation procedure, and therefore, yields were lower as fractionation temperatures decreased.
- Solid low-melting milkfat fractions were obtained using multiple-step fractionation from melted milkfat or solvent solution, and they generally gave lower yields than did liquid fractions.

Crystallization from melted milkfat The percentage yields for low-melting milkfat fractions obtained from melted milkfat did not exhibit perceivable differences between separation methods (Table 5.23). Liquid low-melting milkfat fractions provided greater yields than did solid low-melting milkfat fractions, and liquid fractions obtained by single-step fractionation had greater yields than liquid fractions obtained by multiple-step fractionation.

Liquid low-melting milkfat fractions obtained using single-step fractionation gave yields that ranged from 28.8% to 90%. Yields decreased as fractionation temperature decreased within data generated by individual investigators (deMan and Finoro, 1980; Kaylegian, 1991), but trends between investigators were not apparent.

Liquid low-melting milkfat fractions from multiple-step fractionations gave yields that ranged from 10.7% (muslin cloth) to 55.6% (filtration technique not reported); they were generally the last fraction obtained in the sequence (Baker, 1970a; Barna et al., 1992; Branger, 1993; Grall and Hartel, 1992; Kankare and Antila, 1986, 1988a, 1988b; Kaylegian, 1991; Kaylegian and Lindsay, 1992; Khalifa and Mansour, 1988; Makhlouf et al., 1987; Richardson, 1968; Schaap and van Beresteyn, 1970; Sherbon, 1963; Sreebhashyam et al., 1981; Stepanenko and Tverdokhleb, 1974; Versteeg, 1991).

Yields for solid low-melting milkfat fractions generally ranged from 12.4% to 18.5%, and these fractions were obtained in the second through the fourth steps, which were also the last steps, of a multiple-step fractionation procedure (Deffense, 1987, 1989, 1993b; Grall and Hartel, 1992; Kaylegian, 1991; Kaylegian and Lindsay, 1992; Lakshminarayana and Rama Murthy, 1985; Makhlouf et al., 1987; Richardson, 1968; Riel and Paquet, 1972; Versteeg, 1991). Riel and Paquet (1972) produced a solid low-melting milkfat fractions at the second step of a multiple-step fractionation procedure at 4°C and reported a yield of 51%, which was greater than the yields reported for other solid low-melting milkfat fractions obtained from melted milkfat.

Crystallization from solvent solution Low-melting milkfat fractions have been obtained from acetone, ethanol, isopropanol, and pentane solutions (Table 5.24), and have exhibited trends similar to those observed for low-melting milkfat fractions obtained from melted milkfat.

Liquid low-melting milkfat fractions obtained using single-step fractionation from *acetone* solution gave yields that ranged from 10.5 to 88.9% and generally decreased as fractionation temperatures decreased (Baker et al., 1959; Kaylegian, 1991; Rolland and Riel, 1966; Timms, 1980b; Yi, 1993; Youssef et al., 1977). Two liquid low-melting milkfat fractions were obtained at the first and third steps of a multiple-step fractionation procedure at 18°C by Youssef et al. (1977), who reported yields of 88.5 and 10.5%, respectively. Such data further illustrated the decrease in yields for successive steps in a multiple-step fractionation procedure. Rolland and Riel (1966) obtained liquid low-melting milkfat fractions at the second, third, and fourth steps of a multiple-step fractionation procedure that employed an increasing temperature profile and refractionation of the solid phase. These fractions gave yields of 19.0, 14.0, and 21%, respectively (Rolland and Riel, 1966), and trends in yields with respect to fractionation temperature were not apparent.

Solid low-melting milkfat fractions have been obtained at the fourth and fifth steps of multiple-step fractionation procedures, giving yields of 9.8 to 22%, but these fractions did not exhibit trends based on fractionation temperature (Avvakumov, 1974; Rolland and Riel, 1966).

Three low-melting milkfat fractions have been obtained from *ethanol solution*; they were solid fractions obtained at the second, third, and fourth steps of a multiple-step fractionation procedure and gave yields of 10.7, 16.3 and 11.0%, respectively (Rolland and Riel, 1966).

Two low-melting milkfat fractions have been obtained from *pentane solution* at 0°C, one fraction using single-step fractionation and one fraction using multiple-step fractionation (Kaylegian, 1991). These fractions gave similar yields (87.3 and 86.8%, respectively) (Kaylegian, 1991).

Supercritical fluid extraction Percentage yields for low-melting milkfat fractions obtained using supercritical carbon dioxide extraction are shown in Table 5.25, and yields ranged from 0.2 to 20%. Percentage yields increased as pressure and temperature increased, which was consistent with data reported for high-melting (Table 5.17) and middle-melting milkfat fractions (Table 5.21) obtained by supercritical fluid extraction. Data reported by Arul et al. (1987) increased in yield from 4.1 to 9.4% as the pressure increased from 100 to 250 bar. Timmen et al. (1984) reported a 20% yield for the fraction obtained at 80°C and 200 bar, which was a greater yield, but also a higher fractionation temperature, compared with the fractions produced by Arul et al. (1987).

Short-path distillation Percentage yields for low-melting milkfat fractions obtained by short-path distillation are shown in Table 5.26. Two fractions were obtained at 245°C by Arul et al. (1988a), and yields increased as the absolute chamber pressure was lowered. The first distillate fraction was obtained at 220 μm Hg and gave a yield of 2.1%, and the second distillate fraction was obtained at 100 μm Hg and gave a yield of 9.5%. These yields were similar to those for low-melting milkfat fractions obtained from solvent solution (Table 5.24).

Chemical Characteristics of Low-Melting Milkfat Fractions

Fatty acid composition The fatty acid compositions for low-melting milkfat fractions are presented by individual fatty acids in Tables 5.48 to 5.51 and by selected fatty acid groupings in Tables 5.73 to 5.76 for fractions obtained from melted milkfat, solvent solution, supercritical fluid extraction, and short-path distillation, respectively. Overall general trends observed for all low-melting milkfat fractions were an increase in C4–C10 and C12–C15 fatty acids and a decrease in C16–C18 saturated fatty acids compared with:

- Intact AMF (Tables 5.33 to 5.36 and 5.58 to 5.61)
- Very-high-melting milkfat fractions (Tables 5.37 to 5.39 and 5.62 to 5.64)

- High-melting milkfat fractions (Tables 5.40 to 5.43 and 5.65 to 5.68)
- Middle-melting milkfat fractions (Tables 5.44 to 5.47 and 5.69 to 5.72)

The C18:1/C18:0 ratio was also generally greater for low-melting fractions than for other fractions.

Among low-melting milkfat fractions obtained by *crystallization from melted milkfat*, liquid fractions obtained using single-step fractionation exhibited an increase in C4–C10 and C12–C15 fatty acids with a concurrent decrease in the saturated C16–C18 fatty acids (Tables 5.48 and 5.73) compared with intact milkfat (Table 5.33 and 5.58), for fractions obtained by:

- Vacuum filtration (Amer et al., 1985; deMan and Finoro, 1980; Kaylegian, 1991; Lund and Danmark, 1987; Timms and Parekh, 1980)
- Centrifugation (Jordan, 1986; Ramesh and Bindal, 1987b)
- Separation with the aid of an aqueous detergent solution (Banks et al., 1985; Dolby, 1970a; Norris et al., 1971)
- Separation technique not reported (Deroanne, 1976; Deroanne and Guyot, 1974).

The fractions obtained by centrifugation exhibited lower concentrations of C4–C10 and C12–C15 fatty acids, higher concentrations of C16–C18 saturated fatty acids, and lower C18:1/C18:0 ratios than fractions obtained by vacuum filtration. Liquid low-melting milkfat fractions obtained with the aid of an aqueous detergent solution exhibited the lowest concentrations of C4–C10 fatty acids, and they also exhibited the highest concentrations of C16–C18 saturated fatty acids observed for fractions obtained from melted milkfat.

Solid low-melting milkfat fractions obtained using multiple-step fractionation had greater concentrations of C16–C18 saturated fatty acids and lower C18:1/C18:0 ratios than the liquid fractions obtained by multiple-step fractionation at the same temperature (Table 5.73). Solid and liquid low-melting milkfat fractions were obtained at the same temperature by vacuum filtration (Grall and Hartel, 1992; Kaylegian, 1991; Sherbon, 1963), and centrifugation (Lakshminarayana and Rama Murthy, 1985; Richardson, 1968).

Liquid low-melting milkfat fractions have been obtained using single-step and multiple-step fractionation from *acetone solution* between 0 and 15°C (Kaylegian, 1991; Sherbon, 1963; Timms, 1980b; Yi, 1993). These fractions showed an increase in C4-C10 fatty acid contents, a decrease in C16–C18 fatty acid contents, and an increase in the C18:1/C18:0 ratio as fractionation temperature decreased (Tables 5.49 and 5.74). Timms (1980b) and Yi (1993) produced liquid low-melting milkfat fractions using multiple-step fractionation and reported similar fatty acid compositions for selected fatty acid groupings (Table 5.74).

Kaylegian (1991) obtained liquid low-melting milkfat fractions from *pentane solution* at 0°C by multiple-step and single-step fractionation (Tables 5.49 and 5.74). The fraction obtained by multiple-step fractionation had a greater content of C4–C10 fatty acids and a higher C18:1/C18:0 ratio than did the fraction obtained by single-step fractionation, which had greater concentrations of C12–C15 and C16–C18 saturated fatty acids.

Fatty acid compositions for low-melting milkfat fractions obtained by *supercritical carbon dioxide extraction* are presented in Tables 5.50 and 5.75, and these fractions exhibited a wide range of fatty acid compositions. General trends for low-melting milkfat fractions obtained by supercritical carbon dioxide extraction included an increase in C4–C10 and C12–C15 fatty acids and a decrease in C16–C18 fatty acids compared with the parent AMF (Tables 5.35 and 5.60). Arul et al. (1987) reported unusually high quantities of C4–C10 fatty acids, but this was also observed for the parent AMF employed by Arul et al. compared with other parent AMF samples.

The low-melting milkfat fractions exhibited decreased C4–C10 and C12–C15 fatty acid concentrations and increased C16–C18 saturated fatty acid contents as fractionation pressure increased (Arul et al., 1987), in agreement with trends observed for other milkfat fractions obtained by supercritical carbon dioxide extraction.

The C18:1 fatty acid content was unusually high in the very-high-melting milkfat fractions (Tables 5.39 and 5.64) and high-melting milkfat fractions (Tables 5.42 and 5.67) obtained from supercritical carbon dioxide extraction, and the C18:1 fatty acid content was unusually low in the low-melting milkfat fractions (Table 5.50 and 5.75) when each fraction was compared with the parent AMF samples (Tables 5.35 and 5.60). The first fractions collected by supercritical carbon dioxide extraction exhibited low-melting behavior and had increased concentrations of C4–C10 fatty acids. However, the subsequent fractions had increased concentrations of both saturated and unsaturated C16–C18 saturated fatty acids, which implied that separation of fractions by supercritical carbon dioxide extraction also relies on molecular weights, like short-path distillation.

The fatty acid content of low-melting milkfat fractions obtained by *short-path distillation* (Table 5.51 and 5.76) exhibited increased concentrations of C4–C10 and C12–C15 fatty acids, a concurrent decrease in C16 and C18 saturated fatty acids, and decreased C18:1/C18:0 ratios compared with the parent AMF (Table 5.36 and 5.61). The concentrations of C4–C10 and C12–C15 fatty acids and the C18:1/C18:0 ratio decreased, with a concurrent increase in the C16–C18 saturated fatty acid content, as fractionation pressure increased for the second distillate fraction over that for the first distillate fraction obtained by Arul et al. (1988a).

Triglyceride composition The triglyceride compositions (individual triglycerides and selected triglyceride groupings, respectively) are presented in Tables 5.97 to 5.100 and 5.121 to 5.124 for fractions obtained from melted milkfat, solvent solution, supercritical fluid extraction and short-path distillation. Overall general trends observed for all low-melting milkfat fractions were increased C24–C34 and C36–C40 triglyceride concentrations and decreased C42–C54 triglyceride concentrations compared with:

- Intact AMF (Tables 5.83 to 5.86 and 5.106 to 5.109)
- Very-high-melting milkfat fractions (Tables 5.87 to 5.88 and 5.110 to 5.112)
- High-melting milkfat fractions (Tables 5.89 to 5.92 and 5.113 to 5.116)
- Middle-melting milkfat fractions (Tables 5.93 to 5.96 and 5.117 to 5.120)

Low-melting milkfat fractions obtained by *crystallization from melted milkfat* did not appear to exhibit consistent trends when data from all investigators were compared, but the trends were more distinct for data from individual investigators that had reported triglyceride compositions for several fractions (Tables 5.97 and 5.121). Some of the differences in data between investigators arose from the differences among parent AMF samples employed (Tables 5.83 and 5.106). Additional influences of the chemical composition of the parent AMF on the resultant fractions were also observed for the fatty acid composition of low-melting milkfat fractions and for the fatty acid and triglyceride compositions of other milkfat fractions.

Liquid low-melting milkfat fractions obtained by single-step fractionation from melted milkfat using vacuum filtration exhibited an overall increase in the C24–C34 and C36–C40 triglycerides, with a concurrent decrease in the C42–C54 triglycerides (deMan and Finoro, 1980; Kaylegian, 1991; Lund and Danmark, 1987; Makhlouf et al., 1987). Differences in specific triglyceride composition between fractions produced by these investigators were evident in the composition of the corresponding parent AMF samples (Table 5.83 and 5.106). Kaylegian (1991) obtained a solid and a liquid low-melting milkfat fraction using multiple-step fractionation at 13°C and reported an increased concentration of the C42–C54 triglycerides for the solid fraction compared with the liquid fraction, consistent with other trends observed for these fractions.

The low-melting milkfat fractions obtained from melted milkfat by centrifugation (Branger, 1993; Jordan, 1985; Kankare and Antila, 1986; Keogh and Higgins, 1986b) exhibited a triglyceride composition similar to that of the low-melting milkfat fractions obtained using vacuum filtration. The fraction obtained with the aid of an aqueous detergent solution (Banks et al., 1985) exhibited the lowest concentration of C24–C34 and the highest concentration of C36–C40 triglycerides reported for fractions obtained from melted milkfat (Tables 5.97 and 5.121). The DSC curve of the fraction separated with the aid of an aqueous detergent solution (Figure 5.16) showed a small peak for the low-melting triglyceride species and a broad peak for the middle-melting triglycerides species, illustrating that the triglyceride composition of the fraction is well represented by the DSC curve.

Laakso et al. (1992) reported that the two low-melting milkfat fractions obtained from melted milkfat were composed of 35 to 40% SSS, 44 to 47% SSU, 11 to 14% SUU (2:1 ratio of $SM^cM^c:SM^cM^t$) and 5% other triglyceride species (Table 5.121).

Liquid low-melting milkfat fractions obtained by *crystallization from solvent solution* using single-step fractionation exhibited an increase in C24–C34 and C36–C40 triglycerides with a concurrent decrease in C42–C54 triglycerides (Tables 5.98 and 5.122). Triglyceride compositions for liquid fractions obtained by single-step fractionation from acetone at 15, 11, and 5°C (Kaylegian, 1991) were similar to or slightly greater than the compositions observed for C24–C34 and C36–C40 triglyceride compositions for low-melting fractions obtained from melted milkfat (Tables 5.97 and 5.121). The liquid low-melting milkfat fractions obtained by multiple-step fractionation exhibited triglyceride compositions that were similar to the composition of fractions obtained by single-step fractionation.

Kaylegian (1991) obtained two low-melting milkfat fractions at 0°C from pentane solution and reported similar triglyceride compositions for the fractions obtained by single-step and multiple-step fractionations.

A large increase in the C24–C34 triglyceride contents, with concurrent large decrease in the C42–C54 triglyceride contents, has been observed for the low-melting milkfat fractions obtained by *supercritical fluid extraction* (Tables 5.99 and 5.123) compared with the corresponding parent AMF samples (Tables 5.85 and 5.108). Arul et al. (1987) reported an increase in the C24–C34 triglyceride concentration from 12.6% for the parent AMF to 49.9% for the first low-melting milkfat fraction extracted by supercritical carbon dioxide. Similarly, increases in the C24–C34 concentration were reported by Kankare et al. (1989), Kaufmann et al. (1982), and Timmen et al. (1984). Differences in concentrations of the triglycerides for the low-melting milkfat fractions obtained by these investigators appeared to be related to differences in the parent AMF (Tables 5.85 and 5.108).

These fractions were generally the first fractions collected during the extraction process, and the triglyceride profiles were in agreement with the fatty acid compositions reported for these fractions (Tables 5.50 and 5.75). It appears that supercritical carbon dioxide extraction expresses considerable selectivity that is based on the molecular weight of the triglyceride species, and high-molecular-weight species are increasingly extracted as the fractionation progresses. This is illustrated by the increase in C24–C34 triglyceride concentrations and supported by the decrease in C42–C54 triglyceride concentrations, which is in agreement with the decrease in the C16 and C18 saturated and unsaturated fatty acid contents that have been reported for such fractions.

Arul et al. (1987) reported a decrease in C24–C34 triglycerides and an increase in C36–C40 and C42–C54 triglycerides as fractionation pressure increased from 100 to 250 bar at 50°C, and this was consistent with other physical and chemical trends observed for milkfat fractions obtained by supercritical carbon dioxide extraction as pressure increased. Kaufmann et al. (1982) obtained a low-melting milkfat fraction at 200 bar and 80°C, but the differences in triglyceride compositions observed, compared with the fractions reported by Arul et al. (1987), may have been caused more by differences in the parent AMF composition than by differences in fractionation temperature.

The triglyceride compositions of low-melting milkfat fractions obtained by *short-path distillation* are shown in Tables 5.100 and 5.124 and exhibit trends similar to those observed for low-melting milkfat fractions obtained by supercritical carbon dioxide extraction (Tables 5.99 and 5.123). The first fraction obtained by Arul et al. (1988a) had a C24–C34 triglyceride composition of 66.6% and a C42–C54 content of 4.5%, compared with contents of 11.6% and 52.9%, respectively, for the parent AMF samples (Tables 5.87 and 5.110). The second distillate fraction obtained by Arul et al. (1988a) exhibited a decrease in the C24–C34 triglyceride content and an increase in the C36–C40 and C42–C54 triglyceride contents. However, although this second distillate exhibited similar concentrations of C36–C40 triglycerides compared with the parent AMF, the C24–C34 triglyceride content was still enriched, and the C42–C54 triglyceride content was still lower, compared with the parent AMF.

Selected chemical characteristics   Iodine values, Reichert–Meissl numbers, and Polenske numbers generally increased for low-melting milkfat fractions obtained from melted milkfat (Table 5.143) compared with the corresponding parent AMF (Table 5.131). Iodine values generally increased for low-melting milkfat fractions obtained from melted milkfat as fractionation temperature decreased (Amer et al., 1985; Lund and Danmark, 1987). The chemical characteristics of the parent AMF exert a direct influence on the chemical characteristics of the resultant fractions, and this also was observed for other milkfat fractions.

Chemical characteristics for low-melting milkfat fractions obtained from solvent solution are shown in Table 5.144. Youssef et al. (1977) reported an increase in iodine value compared with the parent AMF (Table 5.132) for the first low-melting milkfat fractions obtained by multiple-step fractionation at 18°C from acetone solution, but the iodine value for the fraction obtained at the third step was slightly lower than the parent AMF.

Chemical characteristics for low-melting milkfat fractions obtained by supercritical fluid extraction are shown in Table 5.145. Both Kaufmann et al. (1982) and Timmen et al. (1984) reported a decrease in iodine value from 30.3 in the parent AMF (Table 5.133) to 19.6 in the low-melting milkfat fraction, which further supports the finding that the unsaturated C18:1 fatty acid concentrations decrease in low-melting milkfat fractions obtained by supercritical fluid extraction.

Flavor   The potential lactone and methyl ketone concentrations and the aldehyde concentrations of low-melting milkfat fractions obtained from melted milkfat are shown in Tables 5.154, 5.160 and 5.161, respectively. The potential lactone and methyl ketone concentrations of the low-melting milkfat fractions were slightly greater than their concentrations in the parent AMF (Tables 5.150 and 5.156, respectively). Kankare and Antila (1988a) reported that the free lactones were concentrated in the first extractions using supercritical carbon dioxide extraction, but they did not report specific values for the low-melting milkfat fractions or the intact milkfat. The aldehyde concentration of the low-melting milkfat fraction (Table 5.165) was twice that of the parent AMF (Table 5.161), and some of this increase was caused by oxidation of the milkfat during processing (Walker, 1972).

Jordan (1986) employed sensory panels to evaluate the relative qualities of the low-melting milkfat fractions obtained from melted milkfat. They reported that the flavor of the low-melting milkfat fraction was not significantly different from that of the AMF, but that it was less pleasant-tasting, which was probably caused by an elevated concentration of free fatty acids and oxidized fatty acid products (aldehydes, etc.) in the fraction (Jordan, 1986). The oxidative stability of milkfat fractions is discussed later in this chapter.

Physical Properties of Low-Melting Milkfat Fractions

Crystal Characteristics Crystal morphologies   for two low-melting milkfat fractions obtained by supercritical carbon dioxide extraction are presented in Table 5.170. Both fractions were extracted fractions, and each exhibited only the β' polymorph. Grall and Hartel (1992) described the crystals of the low-melting milkfat fractions obtained by crystallization from melted milkfat as uniform spheres with a single birefringent cross in polarized light.

Data for the *crystal sizes* of low-melting milkfat fractions were not reported.

Melting point The melting points of low-melting milkfat fractions are presented in Tables 5.23, 5.24, 5.25, and 5.26 for fractions obtained by crystallization from melted milkfat, from solvent solution, by supercritical fluid extraction, and by short-path distillation, respectively. The relationship between melting points and fractionation conditions was discussed above in the section on fractionation conditions of low-melting milkfat fractions. The general melting point characteristics of low-melting milkfat fractions are:

- By definition, the melting point is between 10 and 25°C.
- The physical state of the fraction and the fractionation procedure employed had the greatest influences on the melting points for fractions obtained from melted milkfat and from solvent solution.
- Liquid low-melting milkfat fractions obtained from melted milkfat generally exhibited a melting point similar (±3°C) to the fractionation temperature, and therefore, melting points for fractions obtained at similar temperatures had similar melting points, regardless of fractionation method or separation technique employed.
- Solid low-melting milkfat fractions had higher melting points than did liquid low-melting milkfat fractions obtained at equivalent temperatures.
- Low-melting milkfat fractions obtained from melted milkfat and from solvent solution generally exhibited a decrease in melting point as fractionation temperature decreased.
- Low-melting milkfat fractions obtained by supercritical carbon dioxide extraction exhibited increased melting points as pressure increased.
- Low-melting milkfat fractions obtained by short-path distillation had melting points that increased as pressure decreased.

Thermal profile The majority of low-melting milkfat fractions exhibited DSC curves that showed the presence of predominantly low-melting triglyceride species, and some middle-melting triglyceride species; very few of these fractions exhibited a peak for the high-melting triglyceride species. Thermal profiles obtained by DSC are shown in Figures 5.16, 5.17, 5.18, and 5.19 for fractions obtained from melted milkfat, from solvent solution, by supercritical carbon dioxide extraction, and by short-path distillation, respectively. Low-melting milkfat fractions obtained from all fractionation methods exhibited similar DSC profiles. Differences between curves were observed primarily for differences in the physical state of the fraction at the time of separation and in the type of fractionation sequence employed (i.e., single-step or multiple-step).

Liquid low-melting milkfat fractions obtained from melted milkfat and solvent solution generally exhibited a decrease in the high-melting triglyceride species, followed by a decrease in the middle-melting triglyceride species, as fractionation temperature decreased. The fraction obtained by single-step fractionation from melted milkfat at 25°C by Kaylegian (1991) exhibited the presence of high-melting triglycerides in the form of a shoulder on the peak of the middle-melting triglyceride species, which was absent in the fraction at 20°C (Figure 5.16). The shift in triglyceride species in the DSC curves was more apparent for fractions obtained by each individual investigator. Schaap and van Beresteyn (1970) produced two liquid low-melting milkfat fractions, obtained by multiple-step fractionation from melted milkfat, that exhibited a decrease in the peak for the middle-melting triglyceride species as the fractionation temperature decreased from 17 to 7°C.

This trend also was apparent for liquid low-melting milkfat fractions obtained by single-step fractionation from acetone solution at 15, 11, and 5°C (Kaylegian, 1991; Figure 5.17). The fraction obtained at 15°C showed the presence of the high-melting triglycerides species as a shoulder on the middle-melting triglyceride peak, which was not present in the fraction obtained at 11°C. The fraction obtained at 11°C clearly exhibited peaks for the low-melting and middle-melting triglyceride species, and the fraction obtained at 5°C contained only a peak for the low-melting triglyceride species.

DSC curves for liquid low-melting milkfat fractions showed a shift towards lower temperatures in the highest-melting peak observed for each fraction as fractionation temperature decreased, and this agreed with the small change in melting point compared with fractionation temperature that is normally observed for liquid milkfat fractions.

The DSC curves for solid low-melting milkfat fractions obtained by multiple-step fractionation exhibited a broader peak for the middle-melting triglyceride species than did the corresponding liquid fraction. This trend has been reported by Grall and Hartel (1992), Kaylegian (1991), and Sherbon (1963).

Kaylegian (1991) obtained fractions from pentane solution using single-step and multiple-step procedures at 0°C that exhibited DSC profiles similar to those obtained from acetone solution.

The first low-melting milkfat fractions extracted by supercritical carbon dioxide, were composed of only low-melting glyceride species (Figure 5.18). The second and third low-melting milkfat fractions extracted by supercritical carbon dioxide showed an increased presence of middle-melting triglyceride species as the fractionation pressure increased (Arul et al., 1987), and this corresponded to an increase in the melting point (Table 5.25) observed for these fractions.

Low-melting milkfat fractions produced by short-path distillation were the first and second fractions and exhibited thermal profiles (Figure 5.19) that were similar to those observed for low-melting milkfat fractions produced by supercritical carbon dioxide extraction (Figure 5.18). Arul et al. (1988a) reported a DSC profile composed only of the peak for the low-melting triglyceride species for the first distillate fraction obtained at 220 μm Hg, and the peak for the middle-melting triglyceride species became apparent only in the second distillate fraction obtained at 100 μm Hg.

Solid fat content   Solid fat contents of low-melting milkfat fractions have been tabulated and are presented in Tables 5.189, 5.190, 5.191, and 5.192 for fractions obtained from melted milkfat, from solvent solution, by supercritical fluid extraction, and by short-path distillation, respectively. Solid fat contents for very-high-melting, high-melting, and middle-melting milkfat fractions have been portrayed graphically to illustrate representative SFC curves, but the SFC profiles for low-melting milkfat fractions exhibited such a wide range of melting profiles that selection of meaningful representative curves was precluded.

However, the data for fractions obtained from melted milkfat (Table 5.189) and solvent solutions (Table 5.190) exhibited some consistencies in their SFC profiles, and the following observations were drawn:

- SFC profiles for fractions obtained from melted milkfat and solvent solution exhibited trends that depended on the physical state of the fraction at separation and the fractionation method employed.

- Liquid low-melting milkfat fractions obtained using single-step fractionation both from melted milkfat and from solvent solutions exhibited SFC profiles that indicated a gradual melting of the fractions and that were lower in SFC as the fractionation temperature decreased for some individual investigators, but this trend was not as evident when data for several investigators was compared.

- Liquid low-melting milkfat fractions obtained using multiple-step fractionation from melted milkfat and solvent solutions generally had lowered SFC content profiles compared with fractions obtained from single-step fractionation at equivalent temperatures.

- Solid low-melting milkfat fractions obtained by multiple-step fractionation generally exhibited SFC curves that were somewhat similar to the SFC curves for liquid low-melting milkfat fractions obtained by single-step fractionation.

The strong influences of the physical state of the fraction at the time of separation (i.e., solid or liquid) and the fractionation procedure employed (i.e., single-step or multiple-step) on the properties of the low-melting milkfat fractions obtained from melted milkfat and from solvent solutions, compared with other variables, has been evident for all aspects of these fractions. Middle-melting milkfat fractions also exhibited distinct behaviors and trends based on physical state and fractionation procedure. Although very-high-melting milkfat fractions and high-melting milkfat fractions were generally all solid fractions, they often exhibited differences in properties between fractions obtained by single-step fractionation procedures and fractions obtained by multiple-step procedures. Consequently, the physical state of milkfat fractions at the time of separation (liquid or solid) and the fractionation procedure employed (single-step or multiple-step) appear to be critical in predicting the behavior and functionality of milkfat fractions obtained from melted milkfat or from solvent solution. Still, factors including separation method, solvent solution employed, and the source of intact AMF are influential and must also be taken into consideration.

Liquid low-melting milkfat fractions obtained by single-step fractionation from melted milkfat exhibited lowered SFC profiles as fractionation temperatures decreased when the data within studies by individual investigators were compared. This was illustrated by SFC profiles reported by Amer et al. (1985) for fractions obtained at 23 and 19°C and by Kaylegian (1991) for fractions obtained at 25 and 20°C. However, the SFC data reported by Amer et al. (1985) indicated that both fractions were much more solid than the fractions reported by Kaylegian (1991), but the parent AMF employed by Amer et al. was also more solid (Table 5.174). Therefore, the differences in the physical properties of the parent AMF employed by different investigators appears to account for some differences observed in the melting behavior of the corresponding low-melting milkfat fractions. The influence of AMF source on the SFC profile of low-melting milkfat fractions was also observed for fractions reported by deMan and Finoro (1980) and by Lund and Danmark (1987), who each obtained liquid low-melting milkfat fractions using single-step fractionation at 26°C (Table 5.189). However, the fraction produced by deMan and Finoro (1980) was softer than the fraction produced by Lund and Danmark (1987), and this was consistent with the differences observed in the SFC profiles of corresponding parent AMF samples (Table 5.174).

Liquid low-melting milkfat fractions obtained by single-step fractionation from acetone solution also exhibited a decrease in SFC as the fractionation temperature decreased (Table 5.190). Liquid low-melting milkfat fractions obtained from single-step fractionation from acetone solution (Table 5.190) generally exhibited a lower SFC than did liquid fractions obtained by single-step fractionation from melted milkfat (Table 5.189), but this probably resulted from the lower fractionation temperatures that were employed for fractionation from acetone solution compared with fractionation from melted milkfat.

Liquid low-melting milkfat fractions obtained by multiple-step fractionation generally exhibited the lowest SFC profiles observed for fractions obtained from melted milkfat (Table 5.189) and from solvent solution (Table 5.190). Kaylegian (1991) reported an SFC of 9.5% at 0°C and 0.6% at 5°C for the fraction obtained by multiple-step fractionation from melted milkfat at 13°C (Table 5.189), and this was considerably softer than other fractions obtained from melted milkfat. Deffense (1987) described an olein fraction, produced at the second step of a multiple-step procedure from melted milkfat and filtered at 10°C, that had an SFC of 18.5% at 5°C, and the fraction was completely melted by 15°C. In comparison, the olein fraction, produced by Deffense (1987) at the second step of a multiple-step procedure from melted milkfat but filtered at 6°C instead of 10°C, was characterized as a very-low-melting milkfat fraction (Table 5.193).

Several low-melting milkfat fractions have been obtained from melted milkfat (Deffense 1987; Kaylegian, 1991) and from acetone solution (Yi, 1993) that exhibited SFC profiles similar to the SFC profiles of very-low-melting milkfat fractions (Table 5.190), but these fractions had melting points of just over 10°C and therefore were classified as low-melting milkfat fractions.

Two solid low-melting milkfat fractions obtained by multiple-step fractionation from melted milkfat (Deffense, 1987; Kaylegian, 1991) exhibited SFC curves in ranges similar to those observed for liquid low-melting milkfat fractions. However, other trends relative to these two fractions were not apparent.

Liquid low-melting milkfat fractions obtained by multiple-step processes, both from acetone and from pentane solutions (Table 5.190), were the softest fractions reported for each solvent solution (Kaylegian, 1991; Yi, 1993). However, the fractions obtained from pentane were more solid than the low-melting milkfat fractions obtained from acetone solutions.

The SFC profiles of the low-melting milkfat fractions (Table 5.191) obtained by supercritical carbon dioxide extraction at 80°C and 200 bar (Kaufmann et al., 1982; Timmen et al., 1984) exhibited profiles that fell at the lower end of the range observed in the SFC profiles for low-melting fractions obtained from melted milkfat (Table 5.189). Two low-melting fractions have been produced by supercritical carbon dioxide extraction using the same fractionation conditions, and consequently, conclusions on the effect of processing conditions on the SFC of low-melting milkfat fractions produced by supercritical fluid extraction were not possible.

The SFC profiles of low-melting milkfat fractions obtained by short-path distillation are presented in Table 5.192. The first extracted fraction exhibited a much lower SFC profile than the second fraction, which was obtained at a lower pressure. Because the process distills the lower-molecular-weight species first, followed by the higher-molecular-weight species, differences observed in the SFC profiles for these fractions may not be entirely related to the changes in pressure during fractionation but instead may reflect consequences inherent to the distillation process.

Textural characteristics The textural characteristics of low-melting milkfat fractions have been determined by cone penetrometry (Table 5.198), and they were found to be softer than the corresponding parent AMF (Table 5.195). Low-melting milkfat fractions obtained by pressure filtration (deMan, 1968) exhibited marked decreases in penetrometer readings compared with the intact AMF (0.17 and 1.02 kg/cm^2, respectively). Sreebhashyam et al. (1981) produced a low-melting milkfat fraction by filtration through muslin cloth and reported penetration values of 5.6 mm and 2.7 mm for the fraction and the parent AMF, respectively.

Very-Low-Melting Milkfat Fractions

Only 18 very-low-melting milkfat fractions (MP < 10°C) have been reported in the literature, and this represents 3.7% of the milkfat fractions that have been characterized for melting behavior (Table 5.6). Very-low-melting milkfat fractions were produced by:

1. Crystallization from melted milkfat (Table 5.27; 7 fractions), using the following separation methods:

    - Vacuum filtration (Deffense, 1987, 1993b; Guyot, 1982; Jamotte and Guyot, 1980; Schaap and van Beresteyn, 1970; Versteeg, 1991)
    - Centrifugation (Riel and Paquet, 1972)

2. Crystallization from solvent solution (Table 5.28; 11 fractions), using the following solvents:

    - Acetone (Kaylegian, 1991; Kaylegian and Lindsay, 1992; Rolland and Riel, 1966; Sherbon, 1963; Youssef et al., 1977)
    - Ethanol (Rolland and Riel, 1966)
    - Isopropanol (Norris, 1977)

Fractionation Conditions for Very-Low-Melting Milkfat Fractions

Crystallization from melted milkfat Very-low-melting milkfat fractions obtained from melted milkfat were the liquid fractions obtained at the last step (second through fifth) in multiple-step fractionation procedures (Table 5.27). These fractions did not necessarily exhibit a decrease in melting point as fractionation temperature decreased, but the melting points were within ±3°C of the fractionation temperature. For example, Versteeg (1991) obtained a very-low-melting milkfat fraction at 8°C with a softening point of 5°C, whereas Guyot (1982) produced a very-low-melting milkfat fraction at 5°C that had a drop point of 6°C.

Deffense (1987) produced two second olein fractions that were filtered at 10 and 6°C, respectively, that exhibited a decrease in dropping point as fractionation temperature decreased (10.0 and 8.5°C, respectively). The fraction that was filtered at 10°C has been classified as a low-melting milkfat fraction (Table 5.23), whereas the fraction that was filtered at 6°C has been classified as a very-low-melting milkfat fraction. However, the fraction filtered at 10°C exhibited melting characteristics (MP and SFC) that were more similar to the very-low-melting milkfat fractions (Tables 5.27 and 5.193) than to the low-melting milkfat fractions (Tables 5.23 and 5.189). This behavior was also observed for the low-melting milkfat fractions obtained at 10°C by Kaylegian (1991) and at 0°C by Yi (1993).

Riel and Paquet (1972) produced a very-low-melting milkfat fraction at the second step of a multiple-step fractionation at 4°C and reported a cloud point of 9°C. The difference between cloud point and fractionation temperature was 5°C, which was slightly greater than normally observed for liquid milkfat fraction (± 2°C).

Crystallization from solvent solution Very-low-melting milkfat fractions were liquid fractions obtained from *acetone solution* using single-step and multiple-step fractionation, generally between 0 and –15°C (Table 5.28). Two very-low-melting milkfat fractions were produced by Rolland and Riel (1966) at 32 and 40°C using multiple-step fractionation with an increasing temperature profile in which the solid fraction instead of the liquid fraction was refractionated at subsequent steps. Kaylegian (1991) obtained a very-low-melting milkfat fraction by single-step and multiple-step fractionation at 0°C and reported melting points of 6.8 and 9.8°C, respectively. Youssef et al. (1977) also obtained a very-low-melting milkfat fraction using multiple-step fractionation at 0°C with a melting point of 8°C. Youssef et al. (1977) employed an unusual temperature profile of 18, 0, and 18°C for their multiple-step fractionation sequence, and the very-low-melting milkfat fraction was obtained at the lowest temperature in the sequence (0°C), which was not the last step of the fractionation.

Rolland and Riel (1966) obtained two very-low-melting milkfat fractions at –15°C from *ethanol solution* at the fifth step of a multiple-step fractionation. The solid fraction had a cloud point of 8°C and the liquid fraction had a cloud point of 0°C, which is in agreement with the trends observed for solid and liquid milkfat fractions obtained at the same temperature.

Norris (1977) produced a very-low-melting milkfat fraction using multiple-step fractionation at –5°C from *isopropanol solution* and reported a melting point of less than 0°C based on DSC curves.

Percentage Yields of Very-Low-Melting Milkfat Fractions

The percentage yields of very-low-melting milkfat fractions were influenced by the physical state of the fractions at the time of separation (i.e., solid or liquid) and the fractionation procedure (i.e., single-step or multiple-step) employed.

The yields for two very-low-melting milkfat fractions produced by multiple-step fractionation from melted milkfat using vacuum filtration gave yields of 24 and 12%, respectively, for the fractions obtained at the third step at 8°C (Versteeg, 1991) and at the fourth step at 5°C (Guyot, 1982). The fraction obtained from centrifugation at 4°C (Riel and Paquet, 1972) exhibited the highest yield (36%) of the very-low-melting milkfat fractions obtained from melted milkfat.

Liquid very-low-melting milkfat fractions obtained by multiple-step fractionation from solvent solution (Table 5.28) generally gave greater yields than liquid low-melting milkfat fractions obtained by single-step fractionation. A very-low-melting milkfat fraction obtained by single-step fractionation at 0°C from acetone solution (Kaylegian, 1991) gave a lower yield (52.1%) than did fractions obtained by multiple-step fractionation at 0°C, which had yields of 62.5% (Kaylegian, 1991) and 63.9% (Youssef et al., 1977).

The solid very-low-melting milkfat fraction obtained at –15°C from ethanol solution gave a lower yield (9.3%) than the liquid fraction obtained at the same temperature, which had a yield of 40.0% (Rolland and Riel, 1966). This is in agreement with trends observed for solid and liquid low-melting fractions (Table 5.24) obtained by multiple-step fractionation at the same temperature.

The very-low-melting milkfat fraction obtained from isopropanol solution using multiple-step fractionation gave a yield of 34% (Norris, 1977).

Chemical Characteristics of Very-Low-Melting Milkfat Fractions

Fatty acid composition The fatty acid composition (percent fatty acid and selected groupings) of very-low-melting milkfat fractions are presented in Tables 5.52 and 5.77 for fractions obtained from melted milkfat and in Tables 5.53 and 5.78 for fractions obtained from solvent solution. All very-low-melting milkfat fractions showed similar trends compared with their parent AMF, except that the change in percent composition for individual fatty acids was slightly greater for fractions obtained from solvent solution:

- C4–10 fatty acid content increased.
- C12–C15 fatty acid content increased.
- C18:1 fatty acid content and the C18:1/C18:0 ratio increased.
- C16–C18 saturated fatty acid content decreased.

The C4–C10 concentration was approximately 8–9% in the corresponding parent AMF samples (Table 5.58), compared with a concentration of 10–12% in very-low-melting milkfat fractions obtained from melted milkfat (Table 5.77). The C18:1 content increased from 27–28% in parent AMF samples to 31–36% in the very-low-melting milkfat fractions. The C18:1/C18:0 content also increased for very-low-melting milkfat fractions compared with the parent AMF samples, and these were similar to the ratios reported for liquid low-melting milkfat fractions obtained using multiple-step processes (Table 5.73).

The very-low-melting milkfat fraction obtained by single-step fractionation from acetone (Tables 5.53 and 5.78) had a slightly greater C4–C10 concentration (14.4%) and a slightly lower C16–C18 saturated fatty acid concentration (27.6%) than did very-low-melting milkfat fractions obtained from melted milkfat, which had concentrations of 10–12% and 29–34%, respectively.

The very-low-melting milkfat fractions obtained by Sherbon (1963) from acetone solution had an unusually high C4–C10 fatty acid concentration and C18:1/C18:0 ratio. Although Sherbon did not report fatty acid composition for the parent AMF, the unusual compositional behavior was observed for all fractions obtained by Sherbon and therefore must have been caused either by differences in the parent AMF employed by Sherbon or by inaccuracies in fatty acid analyses.

Triglyceride composition The triglyceride composition has been reported for one very-low-melting milkfat fractions obtained from melted milkfat (Table 5.125) and for two fractions obtained from solvent solution (Tables 5.101 and 5.126). The triglyceride composition of the very-low-melting milkfat fraction obtained from melted milkfat was composed of 24% trisaturated (SSS), 52% monounsaturated (SSU), and 34% di- and triunsaturated (SUU and UUU) triglycerides (Deffense, 1993b). The very-low-melting milkfat fractions obtained from acetone solution (Kaylegian, 1991) exhibited an increase in C24–C34 and C36–C40 triglyceride contents and a decrease in the C42–C54 content compared with the parent AMF (Tables 5.84 and 5.107). The very-low-melting milkfat fractions obtained from acetone solution had a slightly greater C24–C34 triglyceride content, with a slight decrease in the C42–C54 triglyceride content, compared with low-melting milkfat fractions obtained from solvent solution (Tables 5.98 and 5.122).

Selected chemical characteristics   Iodine values, refractive index, and Reichert–Meissl numbers of very-low-melting milkfat fractions obtained from melted milkfat are shown in Table 5.146, and iodine values of very-low-melting milkfat fractions obtained from solvent solution are shown in Table 5.147. Overall, iodine values exhibited an increase for the very-low-melting milkfat fractions compared with the parent AMFs (Tables 5.131 and 5.132) and appeared to increase as fractionation temperature decreased.

The solid very-low-melting milkfat fractions obtained from multiple-step fractionation from ethanol solution had a smaller iodine value than did the liquid fraction obtained at the same temperature (Rolland and Riel, 1966). These trends were in agreement compared with other milkfat fractions.

Flavor   Data on the potential lactone and methyl ketone concentrations and aldehyde concentrations in very-low-melting milkfat fractions were not reported.

Physical Properties of Very-Low-Melting Milkfat Fractions

Crystal Characteristics   Data on the *crystal morphologies* and *crystal sizes* of very-low-melting milkfat fractions were not reported.

Melting point   The melting points of very-low-melting milkfat fractions are presented in Table 5.27 for fractions obtained from melted milkfat and in Table 5.28 for fractions obtained from solvent solution. The relationship between melting point and fractionation conditions was discussed earlier in the section on fractionation conditions of very-low-melting milkfat fractions. The general melting point characteristics of very-low-melting milkfat fractions were:

- By definition, the MP is less than 10°C, and melting points ranged from 0 to 9°C.
- Liquid very-low-melting milkfat fractions obtained from multiple-step fractionation from melted milkfat exhibited a difference of ±4°C between fractionation temperature and melting point.
- Liquid very-low-melting milkfat fractions obtained both by single-step and by multiple-step fractionations exhibited an increase of 5 to 8°C for fractions obtained from acetone and isopropanol solutions, and an increase of 15°C for the fraction obtained from ethanol solution, when compared with fractionation temperature.
- A solid very-low-melting milkfat fraction obtained by multiple-step fractionation exhibited a higher melting point than the liquid fraction obtained by multiple-step fractionation at the same temperature.

Thermal profile   Thermal profiles of very-low-melting milkfat fractions are shown in Figure 5.20 for two fractions obtained from melted milkfat and in Figure 5.21 for two fractions obtained from solvent solution. Although these very-low-melting milkfat fractions were obtained using different fractionation methods (i.e., melted milkfat and solvent solution), by different fractionation procedures (i.e., single-step and multiple-step), and from different AMF sources, they exhibited very similar DSC profiles. These fractions exhibited only the peak for the low-melting triglyceride species, with a small leading shoulder. The shoulder that was visible on the leading edge of the peak for the low-melting triglyceride species appears to be present as a similar peak in some parent AMF thermal profiles (Figures 5.1 and 5.2) and on broad peaks for low-melting triglyceride species observed in low-melting milkfat fractions (Figures 5.16 and 5.17) and middle-melting milkfat fractions (Figures 5.12 and 5.13).

Solid fat content The solid fat contents of very-low-melting milkfat fractions are presented in Tables 5.193 and 5.194 for fractions obtained from melted milkfat and from solvent solution, respectively. SFC data were available for only three very-low-melting milkfat fractions and therefore were not illustrated graphically; nor were conclusions formed about the influences of fractionation methods on the properties of very-low-melting milkfat fractions.

The very-low-melting milkfat fractions obtained both from melted milkfat and from solvent solution exhibited similar SFC profiles, and these were considerably softer than other milkfat fractions and the parent AMF samples (Tables 5.174 and 5.175, respectively). Very-low-melting milkfat fractions obtained both from melted milkfat (Table 5.193) and from solvent solution (Table 5.194) had SFC values of 5–10% at 5°C, and these fractions were melted at 10°C (Deffense, 1987; Kaylegian, 1991; Versteeg, 1991). In comparison, SFC values for the parent AMF samples (Tables 5.174 and 5.175) ranged from 50.3 to 62.0% at 5°C, and the AMF samples melted between 35 and 40°C. Low-melting milkfat fractions obtained from melted milkfat (Table 5.189) had SFC values which ranged from 30–60% at 5°C, and low-melting milkfat fractions obtained from solvent solution (Table 5.190) were approximately 30–40% solid at 5°C.

Textural characteristics Data for the textural characteristics of very-low-melting milkfat fractions were not reported.

Unknown-Melting Milkfat Fractions

Approximately one-third of the milkfat fractions reported in the literature (342 fractions, Table 5.6) were characterized as unknown-melting milkfat fractions because melting behavior data were not reported. Unknown-melting milkfat fractions were produced by:

1. Crystallization from melted milkfat (Table 5.29; 164 fractions)

   - Vacuum filtration (Black, 1975b; Bumbalough, 1989; Frede et al., 1980; Guyot, 1982; Hayakawa and deMan, 1982; Kupranycz et al., 1986; Schaap, 1975; Schultz and Timmen, 1966; Versteeg, 1991; Voss et al., 1971)
   - Pressure filtration (Badings et al., 1983b; Deffense, 1993b; deMan, 1968; El-Ghandour et al., 1976; Helal et al., 1977, 1984; Kuwabara et al., 1991)
   - Centrifugation (Baker et al., 1959; Bratland, 1983; Kankare and Antila, 1974; Kankare et al., 1974a; Keshava Prasad and Bhat, 1987a, 1987b; Lakshminarayana and Rama Murthy, 1986)
   - Separation with the aid of an aqueous detergent solution (Banks et al., 1985; Jebson, 1974a; Schaap et al., 1975)
   - Filtration through muslin cloth (Kulkarni and Rama Murthy, 1987)
   - Filtration through milk filter (Fouad et al., 1990)
   - Filtration through a casein-dewatering press (Thomas, 1973a)
   - Separation technique not reported (Lechat et al., 1975; Timmen, 1974; Verhagen and Bodor, 1984; Verhagen and Warnaar, 1984)

2. Crystallization from solvent solution (Table 5.30; 96 fractions)

   - Acetone (Avvakumov, 1974; Baker et al., 1959; Bhalerao et al., 1959; Chen and deMan, 1966; Jensen et al., 1967; Lansbergen and Kemps, 1984; Larsen and Samuelsson, 1979; Lechat et al., 1975; Mattsson et al., 1969; Munro et al., 1978; Norris, 1977; Parodi, 1974a; Schaap et al., 1975; Sherbon, 1963; Walker et al., 1977, 1978; Woodrow and deMan, 1968)
   - Ethanol (Bhalerao et al., 1959)

- Hexane (Muuse and van der Kamp, 1985)
- Isopropanol (Norris, 1977)
3. Supercritical fluid extraction (Table 5.31; 71 fractions)
    - Carbon dioxide (Bhaskar et al., 1993; Bradley, 1989; Chen and Schwartz, 1991; Ensign, 1989; Hamman et al., 1991; Kankare et al., 1989; Rizvi et al., 1993; Shishikura et al., 1986; Shukla et al., 1994)
    - Propane (Biernoth and Merk, 1985)
4. Short-path distillation (Table 5.32; 6 fractions) (McCarthy et al., 1962)

Fractionation Conditions for Unknown-Melting Milkfat Fractions

The fractionation conditions for milkfat fractions that have been classified for melting behavior have been discussed relative to the melting behavior of the fractions. Since melting behavior was not reported for the unknown-melting milkfat fractions, conclusions on the influence of processing conditions on these fractions were not drawn. It is assumed that the influences of processing conditions for the production of unknown-melting milkfat fractions were similar to those observed for milkfat fractions that were classified by melting behavior.

Crystallization from melted milkfat   Unknown-melting milkfat fractions were solid and liquid fractions produced from melted milkfat using single-step and multiple-step fractionation procedures between 4.5 and 32°C (Table 5.29).

Crystallization from solvent solution   Unknown-melting milkfat fractions were solid and liquid fractions produced from acetone solution using single-step and multiple-step fractionation procedures between –70 and 28°C (Table 5.30). Solid and liquid unknown-melting milkfat fractions also were produced from ethanol solution at 20°C and from hexane solution at 12.5°C. One solid unknown-melting milkfat fraction was produced from isopropanol solution at 20°C (Norris, 1977). The melting behavior of this fraction was determined by the original investigator from DSC curves (not reported), but the melting range (> 20°C) was too broad to allow classification of this fraction (Norris, 1977).

Supercritical fluid extraction   Unknown-melting milkfat fractions were extracted and residual fractions obtained from supercritical carbon dioxide extraction at temperatures between 35 and 80°C and from supercritical propane extraction at 125°C (Table 5.31).

Percentage Yields of Unknown-Melting Milkfat Fractions

The percentage yields for unknown-melting milkfat fractions obtained from melted milkfat, solvent solution, and by supercritical fluid extraction are presented in Tables 5.29, 5.30 and 5.31, respectively. The yields reported for unknown-melting milkfat fractions appeared to follow general trends similar to those obtained by corresponding fractionation methods for milkfat fractions that were classified for melting behavior.

Chemical Characteristics of Unknown-Melting Milkfat Fractions

For fractions classified for melting behavior, the fractions were sorted by fractionation temperature in the fatty acid and triglyceride compositional tables in order to visualize trends relating to the fractionation conditions and melting behavior. However, the relationship between fractionation and melting behavior is unclear for the unknown-melting milkfat fractions, and so these fractions have been presented in the tables for fatty acid and triglyceride compositions sorted by author. Thus, this data provides a good illustration of the change in fatty acid and triglyceride composition as the fractionation sequence progresses for individual studies.

Fatty acid compositions The fatty acid compositions (individual fatty acids and selected fatty acid groupings) of unknown-melting milkfat fractions are presented in Tables 5.54 and 5.79, 5.55 and 5.80, and 5.56 and 5.81, and Table 5.57 and 5.82 for fractions obtained from melted milkfat, from solvent solution, by supercritical fluid extraction, and by short-path distillation, respectively.

Unknown-melting milkfat fractions generally exhibited trends in fatty acid composition that were consistent with trends observed for fractions that had been characterized by melting behavior with respect to fractionation conditions and parent AMF data. Fatty acid composition of the corresponding parent AMF samples are presented in Tables 5.33 and 5.58, 5.34 and 5.59, 5.35 and 5.60 and 5.36 and 5.61 for fractions obtained from melted milkfat, from solvent solution, by supercritical fluid extraction, and by short-path distillation, respectively.

Triglyceride compositions The triglyceride compositions (individual triglycerides and selected triglyceride groupings) of unknown-melting milkfat fractions are presented in Tables 5.102 and 5.127, 5.103 and 5.128, and 5.104 and 5.129 and Tables 5.102 and 5.130 for fractions obtained from melted milkfat, from solvent solution, by supercritical fluid extraction, and by short-path distillation respectively.

Unknown-melting milkfat fractions generally exhibited trends in triglyceride composition that were consistent with trends observed for fractions that had been characterized by melting behavior with respect to fractionation conditions and corresponding parent AMF data. Triglyceride composition of the parent AMF samples are presented in Tables 5.83 and 5.106, 5.84 and 5.107, 5.85 and 5.108 and 5.86 and 5.109 for fractions obtained from melted milkfat, solvent solution, and by supercritical fluid extraction, and short-path distillation, respectively.

Selected chemical characteristics Iodine values, refractive indices, and densities of some unknown-melting milkfat fractions are presented in Table 5.148 for fractions obtained from melted milkfat and in Table 5.149 for fractions obtained from solvent solution. These chemical characteristics exhibited trends with respect to the parent AMF samples that were similar to those observed for milkfat fractions that have been classified for melting behavior. For example, iodine values for unknown-melting milkfat fractions (Tables 5.148 and 5.149) were elevated for liquid fractions and depressed for solid fractions when compared with iodine values of the corresponding parent AMF samples (Tables 5.131 and 5.132).

Flavor The free and potential lactone concentrations for unknown-melting milkfat fractions obtained from solvent solution and by supercritical fluid extraction are presented in Table 5.155. Lactone concentration data were not reported for unknown-melting milkfat fractions obtained from melting milkfat. Methyl ketone and aldehyde concentration data for unknown-melting milkfat fractions were also not reported.

Walker et al. (1977) produced eight solid and three liquid unknown-melting milkfat fractions from three parent AMF samples using multiple-step fractionation from acetone solution and reported decreased free and potential lactone concentrations for the solid fractions and increased free and potential lactone concentrations for the liquid fractions (Table 5.155) when compared with the parent AMF (Table 5.150). Free and potential lactone data were reported for two of the solid fractions that were obtained at the first step (14°C) and two of the liquid fractions that were obtained at the third step (–13°C) of multiple-step fractionations from two parent AMF samples. The solid fractions had total free lactone concentrations of 8.8 and 18.4 μg/g and total potential lactone concentrations of 28.5 and 34.1 μg/g, and the liquid fractions had total free lactone concentrations of 69.3 and 188.5 μg/g and total potential lactone concentrations of 134.6 and 254.0 μg/g. In comparison, the parent AMF samples had total free lactone concentrations of 27.4 and 92.9 μg/g and total potential lactone concentrations of 82.6 and 129.2 μg/g, respectively.

The unknown-melting milkfat fractions obtained by supercritical fluid dioxide extraction exhibited marked differences in total free lactone concentrations between the extracted and residual fractions (Table 5.155) and between each fraction and the parent AMF (Table 5.150).

Kankare et al. (1989) produced four extracted fractions and 1 residual fraction at 50°C and over the pressure range of 100 to 400 bar, and reported free lactone concentrations for the first two extracted fractions (the 2E1 and 2E2 fractions) and for the residual fraction (2R fraction) obtained in this study. The total free lactone concentrations of the extracted fractions were 867.1 μg/g and 418.5 μg/g for the first and second fractions, respectively, and the residual fraction only had a total free lactone concentration of 31.8 μg/g. In comparison, the parent AMF had a total free lactone concentration of 59–64 μg/g. Rizvi et al. (1993) produced four extracted fractions and a residual fraction between 40 and 60°C and 3.4 to 24.1 MPa, and collected another volatile fraction using a cold trap at –10°C and 0.14 MPa. The concentration of δ-lactones was greater in the fractions obtained at 40°C than at 60°C, and the highest concentration was found in the trap fraction (620.7 ppm). The fractions obtained at 40°C (S3 and S4) had greater lactone concentrations (Table 5.155) than the parent AMF (Table 5.150), and the fractions obtained at 60°C (S1 and S2) had lower lactone concentrations than the parent AMF. Rizvi et al (1993) measured the solubilities and distribution coefficients of lactones in supercritical carbon dioxide and reported good agreement between the coefficients and experimental data.

The data for unknown-melting milkfat fractions obtained from acetone solution by Walker et al. (1977) and from supercritical carbon dioxide by Kankare et al. (1989) clearly illustrate the separation of flavor compounds by these fractionation methods. Although most of the free and potential lactones were concentrated in the lower-melting fractions, it is important to note that some of the free and potential lactones remained in the higher-melting fractions. Furthermore, since the relative amounts of free and bound lactones were not determined, elevated concentrations of lactones in extracted fractions probably reflect concentrations of free lactones that had been liberated from triglyceride precursors.

Crystallization from solvent solution has been shown to effect more distinctive separations of the solid and liquid fractions than crystallization from melted milkfat, and supercritical fluid extraction has been shown to extract the lowest-melting fractions at the beginning of the extraction, resulting in more discrete fractions. Crystallization from melted milkfat does not produce solid and liquid fractions that are as distinctly separated as fraction obtained by other methods, and consequently the solid fractions generally contain more liquid milkfat than do fractions obtained from solvent solution or by supercritical fluid extraction. Therefore, trends observed for free and potential lactone concentration for fractions obtained from acetone solution and supercritical fluid extraction may be diminished in fractions obtained from melted milkfat.

Physical Properties of Unknown-Melting Milkfat Fractions

Crystal characteristics The *crystal morphologies* of unknown-melting milkfat fractions obtained from melted milkfat and from solvent solution are shown in Table 5.171. Most fractions exhibited the α, β', and β forms. The β forms were more common in the fractions obtained from solvent solution and in fractions that had been stored prior to analysis. El-Ghandour et al. (1976) evaluated the crystal morphology of milkfat fractions using microscopy (photographs not reprinted in this monograph) and observed no differences in crystal morphology between fractions obtained from Egyptian and from Russian milkfats.

Crystal sizes in unknown-melting milkfat fractions obtained from melted milkfat are presented in Table 5.173. Crystal sizes varied considerably with fractionation parameters. For example, Black (1975b) reported that fractions obtained with precooling exhibited larger crystals than fractions that were obtained without precooling and that crystals were also larger when exponential cooling was employed instead of programmed cooling. Cooling time does not appear have an appreciable effect on crystal size. However, crystal sizes were larger when the agitation speed was decreased from 20 to 10 rpm (Black, 1975b). Kankare and Antila (1974a) reported that crystal size decreased as fractionation temperature decreased at successive steps in a multiple-step fractionation procedure.

Melting behavior By definition, melting behavior data (i.e., melting points, thermal profiles, and solid fat contents) were not reported for these fractions.

Textural characteristics Hayakawa and deMan (1982) reported constant speed and cone penetrometer values for solid and liquid unknown-melting milkfat fractions obtained by single-step fractionation from melted milkfat between 32 and 25°C (Table 5.199). As expected, the liquid fractions were softer than the solid fractions obtained at the same temperature, and hardness values decreased as the fractionation temperature decreased.

Oxidative Stability of Experimental Milkfat Fractions

The oxidation of milkfat leads to the deterioration of milkfat quality through the production of off-flavors and the formation of polymers. The importance of fatty acid composition to the oxidative stability of milkfat has been the driving force for several studies on the oxidative stability of milkfat fractions (Tables 5.200 and 5.201). In general, the oxidation rate of milkfat fractions increases as the unsaturated fatty acid content increases. The low-melting milkfat fractions oxidized faster than the intact milkfat, and the high-melting milkfat fractions oxidized more slowly than intact milkfat. Oxidation of intact milkfat was discussed in Chapter 2.

Factors Affecting the Oxidative Stability of Milkfat Fractions

Oxidation studies on milkfat fractions have dealt with the effects of storage temperature and heat; light; presence of heavy metals; use of antioxidants; and availability of oxygen over various time periods ranging from several minutes to 13 weeks (Table 5.200). The data reported for milkfat fractions followed the expected trends for most of these factors when the oxidation states of milkfat fractions were measured by peroxide value (PV) or thiobarbituric acid (TBA) value. The effects of some factors important to the stability of milkfat fractions include the following (Kaylegian, 1991; Kehagias and Radema, 1973: Keshava Prasad and Bhat, 1987b):

1. Fraction type

   - Liquid fractions had elevated PV compared to solid fractions obtained using single-step fractionation.
   - PV gradually increased for fractions obtained at successive steps using multiple-step fractionation, with the lowest PV reported for the most solid fraction and the highest PV reported for the most liquid fraction.

2. Storage temperature and effects of heat

   - PV increased as storage temperature increased (–40 to 60°C).

3. Presence of light

   - PV increased for samples stored under fluorescent light, compared to samples stored in the dark.
   - PV were greater for samples stored in the dark at 60°C compared with samples stored at 25°C under fluorescent light.

4. Presence of heavy metals

   - The presence of copper and iron increased PV compared with control untreated milkfat fractions.
   - Copper-treated samples showed increased PV compared with iron-treated samples.
   - Copper-treated samples showed increased PV compared with oxygen-treated samples.

Kehagias and Radema (1973) evaluated the availability of oxygen by comparing milkfat samples stored in tins with and without carbon dioxide flushing prior to sealing. They reported that the carbon dioxide flushing did not affect the peroxide values of the samples. Storage at 28°C allowed oxidation to progress much more rapidly than at –40°C, and this effect on peroxide value was much greater than the effect of the carbon dioxide flushing treatment. During storage, the peroxide values of the solid fractions were slightly elevated compared with the liquid fractions.

Lakshminarayana and Rama Murthy (1986) noted that the low-melting milkfat fractions contained more naturally occurring tocopherol and carotene than the high-melting milkfat fractions. However, they reported that low-melting milkfat fractions oxidized faster than high-melting milkfat fractions because the influence of the increased unsaturated fatty acid content overpowered the antioxidant effects of the tocopherol and carotene.

Kaylegian (1991) evaluated the effect of antioxidants in retarding the oxidation of milkfat and milkfat fractions. The butylated hydroxyanisole (BHA)–based antioxidant [Tenox 2 (BHA, propyl gallate, and citric acid); 200 ppm] was effective in retarding oxidation in most samples. A tocopherol-based antioxidant [Covi-Ox T70 (mixed tocopherols); 200 ppm] was only slightly effective in the dark and under light at 25°C, but it became a prooxidant in milkfat fraction samples at 60°C. Tocopherols have been shown to have antioxidant properties at low concentrations but prooxidant properties at high concentrations.

Oxidation Products Formed in Milkfat Fractions

The production of off-flavor-producing compounds (carbonyls, ketoglycerides) and other selected compounds have been monitored in milkfat fractions during oxidation (Table 5.201).

The production of carbonyl compounds during oxidation was found to be greater in the low-melting milkfat fractions than in the high-melting milkfat fractions (Bhat and Rama Murthy, 1983; Keshava Prasad and Bhat, 1987b). Bhat and Rama Murthy (1993) suggested that the increased presence of the monocarbonyls in low-melting milkfat fractions may have been caused by their preferential solubility in liquid fat, whereas the presence of monocarbonyls in high-melting milkfat fractions may be due to liquid fat entrapped in the crystals. Keshava Prasad and Bhat (1987b) reported that the production of carbonyl compounds was greatest in copper-treated samples, followed by the oxygen-treated samples, and then the control untreated samples.

Polymer formation in fats is a concern associated with oils employed for deep-fat frying, which is conducted at temperatures above 180°C. Some polymers in fats, formed during thermal oxidation, have been shown to be toxic and are of concern because these compounds are often ingested with food (Kupranycz et al., 1986). Kupranycz et al. (1986) monitored the change in concentrations of polymeric triglycerides formed in milkfat fractions during heating at 185°C for 16 h. They reported that the solid and liquid fractions both had more stability toward intermolecular polymerization than did intact milkfat after 8 hours of oxidation. After 16 hours, the ratio of trimeric and higher oligomeric triglycerides to dimeric triglycerides increased with increasing degree of unsaturation in the milkfat fractions and with increased time of heating. Similar trends were observed with respect to the degree of intramolecular polymerization. Kupranycz et al. (1986) noted that the degree of unsaturation alone did not control the extent or rate of polymerization reactions during thermal oxidation of milkfat.

Nutritional Properties of Experimental Milkfat Fractions

Nutritional Components of Milkfat Fractions

Data for some nutritional components of milkfat fractions obtained by crystallization from melted milkfat, from solvent solution, by supercritical fluid extraction, and by short-path distillation are shown in Table 5.202. The cholesterol, carotene, vitamin A, and vitamin E contents of milkfat fractions obtained from melted milkfat were found in greater concentrations in the liquid fractions than in solid fractions obtained using single-step fractionation. A gradual increase in the concentration of these compounds was observed with successive steps using multiple-step fractionation procedures, and the highest concentration was found in the most liquid fraction, obtained at the last fractionation step. These trends were also observed for fractions obtained by crystallization from solvent solution using single-step and multiple-step fractionation. Chen and deMan (1966) determined the amount of free and esterified cholesterol in the fractions obtained by a seven-step fractionation from acetone solution. They reported that a greater proportion of esterified cholesterol, rather than free cholesterol, was present in the fractions obtained at the first several steps and that fractions obtained at the later steps had a greater proportion of free cholesterol, rather than esterified cholesterol (Chen and deMan, 1966).

The concentration of cholesterol into the more liquid fractions was also evident in fractions obtained by supercritical fluid extraction and short-path distillation (Table 5.202). The cholesterol contents were greatest in the first fractions obtained by both methods, and the content continued to decrease with successive fractions. The fractions from supercritical fluid extraction were often obtained using an increasing pressure profile, and the cholesterol showed preferential solubility in CO_2 at lower pressures (Bradley, 1989; Chen et al., 1992; Chen and Schwartz, 1991; Ensign, 1989). The cholesterol remaining in the residual fractions obtained after extraction has been attributed mainly to esterified cholesterol (Rizvi et al., 1990).

Nutritional Aspects of Milkfat Fractions

The effects of milkfat fractions on different aspects of rat metabolism have been investigated, and topics include the effect of thermal oxidative polymerization on the growth-promoting value of some fractions of butterfat (Bhalerao et al. (1959); the nutritional value of various milkfat fractions (Antila and Antila, 1970); hepatic secretion and plasma clearance of butterfat triacylglycerols (Ney and Lai, 1992); and the effects of defined milkfat fractions on postprandial lipemia in meal-fed rats (Lai and Ney, 1992).

Growth-depressing effects in rats, caused by toxic materials formed during heat polymerization of fats, have been observed (Bhalerao et al., 1959). The toxicity of some thermally oxidized fats, particularly the dimeric triglycerides, has been briefly noted earlier in the section on oxidation products formed in milkfat fractions (Kupranycz et al., 1986). Bhalerao et al. (1959) studied the effects of heat-polymerized (200°C, 24 h) milkfat and milkfat fractions obtained from acetone and ethanol solutions on the growth of rats. They reported that thermally polymerized intact milkfat and polymerized solid fractions ($5S_S$ and $20S_S$) did not cause differences in growth compared with fresh milkfats. A marked depression in growth was noted in rats fed the corresponding polymerized liquid fractions ($5L_S$ and $20L_S$). In comparison, thermally oxidized corn oil or soybean oil showed significant growth-depressing effects compared with rats fed fresh oils. When these polymerized oils were blended with 30% of the polymerized solid milkfat fractions that had initially been obtained from acetone solution, the growth-depressing effects of the oxidized blends were lessened, indicating a positive effect for the polymerized solid milkfat fraction. The data presented by Bhalerao et al. (1959) was in agreement with the increased production of trimeric and dimeric triglyceride species during thermal oxidation in the low-melting milkfat fractions, compared with the thermally oxidized high-melting milkfat fractions, as reported by Kupranycz et al. (1986; Table 5.201).

Antila and Antila (1970) determined the general nutritional value of different milkfat fractions obtained from melted milkfat. They reported that the chemical composition of the milkfat fractions directly affected the iodine value and fatty acid composition of the animal's depot fats. Iodine values and oleic acid content increased for rats fed low-melting milkfat fractions, and decreased for rats fed very-high-melting milkfat fractions, compared with intact milkfat. Histopathological examination of the hearts, kidneys, and livers showed no differences based on diet.

Ney and Lai (1992) and Lai and Ney (1992) have briefly reported on the effects of feeding intact milkfat, milkfat fractions, and various fats and oils on the plasma cholesterol and triglyceride levels in fasting and meal-fed rats. Rats were fed intact milkfat or other dietary fats for 6 weeks and then sampled after a 10-hour fast (Ney and Lai, 1992). Rats fed intact milkfat had significantly lower plasma cholesterol and triglyceride levels than did rats fed beef tallow or palm oil. The high-density lipoproteins of rats fed intact milkfat contained less esterified cholesterol and triglycerides than did those of rats that were fed beef tallow or palm oil (Ney and Lai, 1992). Fasting plasma lipoprotein cholesterol levels were higher in rats fed palm oil, solid milkfat fractions, and intact milkfat than in rats fed corn oil or liquid milkfat fractions. Fasting and postprandial plasma triglyceride levels were higher in rats fed palm oil and all types of milkfat compared with corn oil. Determination of post-heparin plasma lipoprotein lipase activity suggested that delayed plasma triglyceride clearance with the ingestion of palm oil or milkfat fractions may account, in part, for the higher postprandial triglyceride levels observed in rats fed saturated fat compared to corn oil. Meal-fed rats showed a different pattern in the postprandial time course of plasma cholesterol, triglyceride, and insulin levels with ingestion of a liquid milkfat fraction that was enriched in 18:1n9 than with a solid milkfat fraction that was enriched in 16:0 and 18:0.

Blends of Experimental Milkfat Fractions with Other Fats and Oils

Milkfat fractions are blended to alter the functionality of milkfat ingredients and other fats, add flavor to vegetable oils, and decrease costs in specialty products. Blending of milkfat fractions with other milkfat fractions and other fats and oils involves crystallization and melting behavior phenomenon. Mixed crystallization of fats is complex and not well understood for many fats, including milkfat fractions. Consequently, the characteristics of a milkfat blend are difficult to predict accurately using traditional analyses (e.g., solid fat content). Incompatible fat mixtures exhibit a decreased ability to crystallize properly, resulting from either solubility effects of the liquid fats or eutectic formation with solid fats. Fat compatibility is extremely important when blending fats for specific applications, and the final testing of a fat system should always be performance testing.

Blend formulas and general data for mixtures of milkfat fractions with other fats are presented in Table 5.203. The effects of blending milkfat fractions with other milkfat fractions, cocoa butter, and selected fats and oils on melting behavior have been evaluated for solid fat content profiles (Table 5.204), thermal profiles (Figure 5.25), isosolid diagrams (Figure 5.26), and phase diagrams (Figure 5.27).

Milkfat Fraction–Intact Milkfat and Milkfat Fraction–Milkfat Fraction Blends

A blend of a very-high-melting milkfat fraction and intact milkfat (Mixture 1; Kankare and Antila, 1986) and a blend of very-high-melting and middle-melting milkfat fractions (Mixture 2; Kankare and Antila, 1986) exhibited increases in the concentrations of the middle-melting and high-melting triglyceride species compared with the parent AMF samples, as illustrated by the DSC profiles of these blends (Figure 5.25). A blend of 75% low-melting milkfat fraction and 25% very-high-melting milkfat fraction (Mixture 3; Kankare and Antila, 1986) exhibited a DSC profile with prominent peaks for the low-melting and high-melting species and an absence of the peak for the middle-melting peak. Avvakumov (1974) noted that the addition of fractions to intact milkfat changed the solidification properties of the separate groups of glycerides, and the more the concentration of a fraction in a mixture was increased, the more it influenced the physical properties of the mixture.

Milkfat Fraction–Cocoa Butter Blends

Blends of intact milkfat and cocoa butter have been relatively well-studied by the confectionery industry, and constitute the majority of data available on milkfat fraction blends (Table 5.203). Blends of milkfat fractions and cocoa butter have been described by solid fat content profiles (Table 5.204), DSC profiles (Figure 5.25), isosolid diagrams (Figure 5.26), and phase diagrams (Figure 5.27).

The addition of high-melting milkfat fractions to cocoa butter has resulted in increased solid fat content profiles compared to similar mixtures made with intact milkfat, and lower solid fat content profiles have resulted when cocoa butter was blended with low-melting milkfat fractions obtained by crystallization from melted milkfat or from acetone solutions (Table 5.204). These effects were more pronounced as the concentration of the milkfat fraction in the blend increased.

The solid fat content curves of mixtures of milkfat fractions and cocoa butter did not give the results expected based on the data of the individual components (Yi, 1993). The solid fat content curves of mixtures of milkfat and cocoa butter gave almost identical profiles, although the curves of the individual fractions were very different (Yi, 1993).

Thermal profiles obtained by differential scanning calorimetry illustrated the softening effect of low-melting milkfat fractions in cocoa butter, which was shown by the lower melting points observed for these mixtures containing low-melting milkfat fractions than for mixtures made with high-melting milkfat fractions (Figure 5.25).

Isosolid diagrams of milkfat fraction–cocoa butter blends are presented in Figure 5.26. Timms (1980b) reported that the low-melting milkfat fractions have been found to cause softening of the mixtures by dissolving the cocoa butter crystals in the milkfat, whereas the middle-melting milkfat fractions have been shown to cause softening by the formation of eutectics. On the other hand, the high-melting milkfat fractions have been considered not to have a significant contribution to the phase behavior of milkfat fraction–cocoa butter mixtures (Timms, 1980b). Barna et al. (1992) reported similar effects for the low-melting and middle-melting milkfat fractions when mixed with cocoa butter, but they observed the formation of eutectic mixtures with the high-melting milkfat fractions as well. The use of fractionation to remove the liquid triglycerides, which are responsible for the softening effect in cocoa butter, from milkfat results in the concentration of the middle-melting components of milkfat fractions, and as a result, the eutectic softening effect is then increased (Timms and Parekh, 1980).

Phase diagrams have been determined for mixtures of cocoa butter with intact milkfat and with a middle-melting milkfat fraction (Figure 5.27). A stable cocoa butter solid solution (type β-3) was observed in milkfat–cocoa butter blends with the addition of up to 50% milkfat (Timms, 1980b). The triglycerides in the middle-melting milkfat fraction were responsible for the eutectic interaction and for the limited solubility in the solid phase observed in the phase diagrams. The eutectic interaction between cocoa butter and the middle-melting milkfat fraction caused some additional softening, especially in mixtures above 30% milkfat (Timms, 1980b).

Milkfat Fraction–Other Fats and Oils Blends

Milkfat fractions have been blended with rapeseed oil, sunflower oil, soybean oil, and fractionated beef tallow (Table 5.203). These combinations have been investigated primarily from the standpoint of producing butter–margarine spreads and have been described by solid fat content profiles (Table 5.204), DSC curves (Figure 5.25) and isosolid diagrams (Figure 5.26). Solid fat contents and DSC curves of blends of solid milkfat fractions and rapeseed, sunflower, and soybean oils were found to be suitable for the production of spreads (Kankare and Antila, 1988b; Verhagen and Bodor, 1984). The isosolid diagram of a low-melting milkfat fraction combined with a low-melting tallow fraction exhibited some unusual results, in that a 50:50 mixture of the two fats was more liquid than either fraction; such behavior is not normally observed for blends of intact milkfat and intact tallow (Versteeg, 1991; Figure 5.26).

The compatibility of milkfat fractions with lauric fats was evaluated by microscopy and X-ray diffraction (Bartsch et al., 1990). Commercial milkfat fractions obtained by crystallization from melted milkfat and an experimental milkfat fraction obtained by supercritical fluid extraction were blended with a commercial lauric fat. Bartsch et al. (1990) reported that milkfat fractions could be added to the lauric fat at levels of 20 to 40% without disturbing the crystallization behavior. Medium-melting and high-melting milkfat fractions were more compatible with the lauric fat than low-melting milkfat fractions because of the decreased content of oleic acid. Oleic acid interferes with the packing structure in the crystal lattice of the lauric fat and results in improper crystallization. However, a low-melting milkfat fraction obtained by supercritical fluid extraction had a low oleic content and was found to be more compatible with lauric fats than were the middle-melting milkfat fractions obtained by crystallization from melted milkfat.

Fairley et al. (1994) and Simoneau et al. (1994) evaluated mixtures of milkfat and milkfat fractions with tripalmitin. The tripalmitin was added to simulate milkfat fractionation and provide model systems for the study of thermal behavior, phase behavior, and mechanical properties. Fairley et al. (1994) and Simoneau et al. (1994) both reported that the addition of tripalmitin to milkfat yielded model systems that exhibited behavior similar to hard milkfat fractions. They also noted that the functionality of the hard milkfat fractions appeared to be associated with the enrichment of the long-chain saturated triglycerides in these fractions, and therefore, the addition of tripalmitin to milkfat provided an alternative to milkfat fractions in some applications (Fairley et al., 1994; Simoneau et al., 1994).

Commercially Prepared Milkfat Ingredients

Technical data on the chemical and physical characteristics of commercially available milkfat ingredients is not easily accessible, and consequently the scope of this section is limited. Although milkfat fractionation is practiced in a number of countries throughout the world (Tirtiaux, undated), technical data sheets were obtained from only three commercial companies (Alaco in New Zealand, Aveve and Corman in Belgium; Table 5.205). Some characterization and applications data for commercially-prepared fractions was available in the research literature. Data relating to physical and chemical characteristics are discussed in this chapter, and data relating to applications is presented in the appropriate sections of Chapter 8.

Commercially available milkfat fractions are made by crystallization from melted milkfat. Commercially prepared milkfat ingredients differ from the fractions discussed to this point in that they are finished products that may be composed of blends of milkfat fractions and intact milkfat. In addition to fractionation and blending, these products have been recrystallized and texturized where appropriate. The conditions used in the final processing steps also effect the properties of the products (Chapter 8). Information regarding milkfat fractions used for blending, formulations, and production of these products is proprietary.

The ability to manipulate fractionation, as well as blending and texturizing, allows the production of milkfat ingredients tailored for specific uses. Commercially available milkfat ingredients are used primarily in the bakery, confectionery, and dairy industries (Table 5.205). These ingredients represent the low-melting, middle-melting, and high-melting milkfat fraction categories.

Chemical Characteristics of Commercially-Prepared Milkfat Ingredients

The fatty acid compositions, triglyceride compositions, and selected chemical characteristics of commercially-prepared milkfat ingredients are presented in Tables 5.206, 5.207 and 5.208, respectively. The data and trends presented in these tables are in agreement with the data for corresponding experimental low-melting, middle-melting, and high-melting milkfat fractions. Data provided by the commercial companies was presented as ranges instead of specific values.

Physical Characteristics of Commercially Prepared Milkfat Ingredients

The melting behavior of commercially prepared milkfat ingredients has been described by melting points (Table 5.205), thermal profiles (Figure 5.28), and solid fat content profiles (Table 5.209). Melting points are one of the primary identifiers for commercial milkfat ingredients, and a range of low-melting, middle-melting, and high-melting milkfat ingredients is represented by the commercial products (Table 5.205). High-melting ingredients are recommended for pastry use, whereas the middle-melting fractions are recommended for general-purpose baking and dairy applications.

The DSC profiles of middle-melting and high-melting commercial milkfat ingredients showed two distinct peaks: a broad peak for the low-to-middle-melting triglycerides and a narrower peak for the high-melting triglyceride species (Büning-Pfaue, 1989; Figure 5.28). The peak for the high-melting triglyceride species becomes larger and more prominent as the melting points of the milkfat ingredients increase, and this is in agreement with the data for experimental milkfat fractions.

Solid fat content profiles are also important for assessing functionality and are used as identifiers for milkfat ingredients, particularly for pastry applications. The solid fat contents of commercial milkfat ingredients are shown in Table 5.209, and they follow trends reported for experimental low-melting, middle-melting, and high-melting milkfat fractions. Lambelet (1983) evaluated the solid fat content of several commercial fractions (Corman, undated) using nuclear magnetic resonance and differential scanning calorimetry methods. The values obtained with DSC were greater than those obtained by NMR methods. However, the values provided by the NMR method were more in line with those reported by Corman (undated) for ingredients with similar melting points.

Oxidative Stability of Commercially Prepared Milkfat Ingredients

Oxidative stability standards are commonly found in the edible oils industry, and commercially produced milkfat ingredients fall within the standards set for peroxide value and free fatty acid value. The peroxide values for commercial milkfat ingredients range from 0.1 to 0.5 mEq/kg fat, and free fatty acid values range from 0.2 to 0.35% as oleic acid (Aveve, undated; Corman, undated).

Nutritional Composition of Commercially Prepared Milkfat Ingredients

The nutritional components reported for milkfat ingredients include fatty acid compositions, triglyceride compositions, and selected nutrients. The fatty acid compositions and triglyceride compositions were presented in Tables 5.206 and 5.207 and already discussed in the section on chemical characteristics of commercially prepared milkfat fractions. Corman (undated) indicated that selected nutrients in all of its milkfat ingredients fall into the following ranges:

- Vitamin A: 6 to 20 mg/kg
- Vitamin D: traces
- Vitamin E: 5 to 100 mg/kg
- Vitamin K: 1 mg/kg.

With the introduction of more definitive nutritional labeling requirements in the U.S.A., it is likely that expanded information of this type will soon emerge.

Technical Data for Experimental and Commercial Milkfat Fractions 213

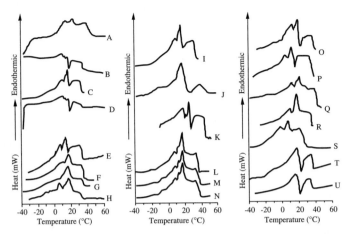

Figure 5.1 Thermal profiles (DSC curves) for intact anhydrous milkfats used to produce milkfat fractions from melted milkfat. *Vacuum filtration:* A, AMF[1] [W] (Amer et al., 1985); B, AMF [W] (Barna et al., 1992); C, AMF (Deffense, 1993b); D, AMF [S] (Grall and Hartel, 1992); E, AMF [W] (Kaylegian, 1991); F, AMF (Makhlouf et al., 1987); G, AMF (Schaap and van Beresteyn, 1970); H, AMF (Sherbon, 1963). Pressure filtration: I, AMF (Deffense, 1993b); J, AMF (deMan, 1968). *Centrifugation:* K, AMF (Antila and Antila, 1970); L, AMF [S] (Kankare and Antila, 1986); M, AMF [S] (Kankare and Antila, 1988a); N, AMF [S] (Kankare and Antila, 1988b). *Separation with the aid of an aqueous detergent solution:* O, AMF [A] (Banks et al., 1985); P, AMF [B] (Banks et al., 1985); Q, AMF [A late S] (Norris et al., 1971); R, AMF [A] (Sherbon et al., 1972). *Filtration through milk filter:* S, AMF (Fouad et al., 1990). *Separation technique not reported:* T, AMF [W] (Deroanne, 1976); U, AMF [S] (Deroanne and Guyot, 1974). [1]AMF = anhydrous milkfat. Designations inside the brackets following the fraction designation refers to the intact anhydrous milkfat source [sample number or season (W = winter, S = summer].

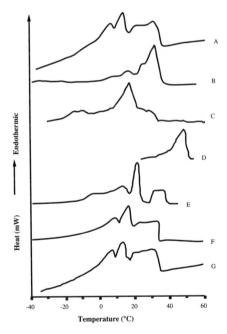

Figure 5.2 Thermal profiles (DSC curves) for intact anhydrous milkfats (AMF) and other starting materials used to produce milkfat fractions from solvent solutions. *Acetone:* A, AMF (Kaylegian, 1991); B, AMF (Lambelet, 1983); C, AMF (Sherbon, 1963); D, HMG (high-melting glyceride) (Sherbon and Dolby, 1973); E, AMF (Timms, 1980a); F, AMF (Yi, 1993). *Pentane:* G, AMF [W] (Kaylegian, 1991).

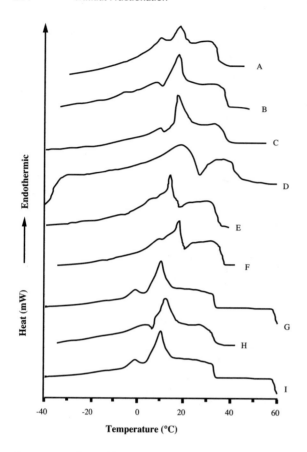

Figure 5.3 Thermal profiles (DSC curves) of intact anhydrous milkfats (AMF) used to produce milkfat fractions by supercritical fluid extraction. *Carbon dioxide:* A, AMF (Arul et al., 1988b); B, AMF (Bhaskar et al., 1993b); C, AMF (Büning-Pfaue et al., 1989); D, AMF (Chen et al., 1992); E, AMF (Kankare and Antila, 1988a); F, AMF (Kankare et al., 1989); G, AMF (Kaufmann et al., 1982); H, AMF (Rizvi et al., 1990); I, AMF (Timmen et al., 1984).

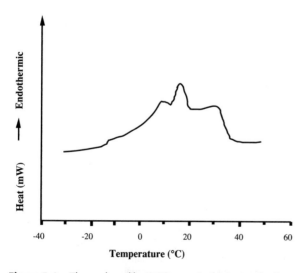

Figure 5.4 Thermal profile (DSC curve) of intact anhydrous milkfat used to produce milkfat fractions by short-path distillation (Arul et al., 1988 a).

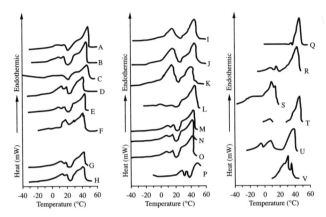

Fig. 5.5 Thermal profiles (DSC curves) of very-high-melting milkfat fractions obtained from melted milkfat. *Vacuum filtration* [1]: A, $34S_M$ [W] (Kaylegian, 1991); B, $34S_S$ [W] (Kaylegian, 1991); C, $27S_M$ (Schaap and van Beresteyn, 1970); D, $26S_M$ (S.1) (Makhlouf et al., 1987); E, $23S_M$ (Sherbon, 1963); F, $23ppt_M$ (refractionated) (Sherbon, 1963); G, $21S_M$ (S.2) (Makhlouf et al., 1987); H, F.S (>21) (Makhlouf et al., 1987). Pressure filtration: I, $30S_S$(A) slow cooling (deMan, 1968); J, $27S_S$ (C) slow cooling (deMan, 1968); K, $24S_S$ (E) slow cooling (deMan, 1968); L, 1st stearin $_M$ (Deffense, 1993b). *Centrifugation:* M, $24S_M$ [S] (Kankare and Antila, 1986); N, $24S_M$ [S] (Kankare and Antila, 1988a); O, $24S_M$ [S] (Kankare and Antila, 1988b); P, D (Antila and Antila, 1970). *Separation with the aid of an aqueous detergent solution:* Q, $28S_S$ (Schaap et al., 1975); R, $21S_S$ [A] High level of detergent (Banks et al., 1985); S, 21 semi-S_S [A] high level of detergent (Banks et al., 1985); T, $16S_S$ [B] high level of detergent (Banks et al., 1985); U, $16S_S$ [B] low level of detergent (Banks et al., 1985). *Separation technique not reported:* V, 1st stearin (Martine, 1982). [1] The number indicates fractionation temperature, the letter indicates physical state (S = solid), and the subscription indicates fractionation procedure used (M = multiple-step, S = single-step). Other descriptions used are those of the original author. Designations in parenthesis are fraction designations given by the original author. The designation inside the brackets refers to the intact anhydrous milkfat source [sample number or season (W = winter, S = summer)].

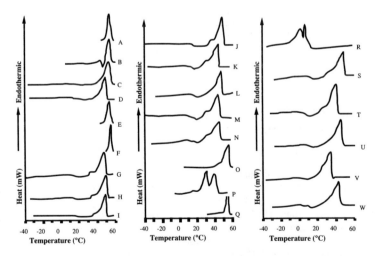

Fig. 5.6 Thermal profiles (DSC curves) of very-high-melting milkfat fractions obtained from solvent solution. *Acetone:* A, $32S_M$ (I) [HMG] (Sherbon and Dolby, 1973); B, $28S_S$ (Schaap et al., 1975); C, $25S_M$ [W] (Kaylegian, 1991); D, $25S_S$ [W] (Kaylegian, 1991); E, $25S_M$ (II) [HMG] (Sherbon and Dolby, 1973); F, $25S_M$ (III refractionated) [HMG] (Sherbon and Dolby, 1973); G, $25S_M$ [S] (Yi, 1993); H, $21S_M$ [W] (Kaylegian, 1991); I, $21S_S$ [W] (Kaylegian, 1991); J, $20S_M$ [S] (Yi, 1993); K, $15S_M$ [W] (Kaylegian, 1991), L, $15S_S$ [W] (Kaylegian, 1991); M, $15S_M$ [S] (Yi, 1993); N, $11S_M$ [W] (Kaylegian, 1991); O, HMF (Timms, 1980a); P, 2B (Sherbon, 1963); Q, 2AJAN [Jan] (Sherbon, 1963). *Pentane:* R, $11S_M$ [W] (Kaylegian, 1991); S, $11S_S$ [W] (Kaylegian, 1991); T, $5S_M$ [W] (Kaylegian, 1991); U, $5S_S$ [W] (Kaylegian, 1991); V, OS_M [W] (Kaylegian, 1991); W, OS_S [W] (Kaylegian, 1991).

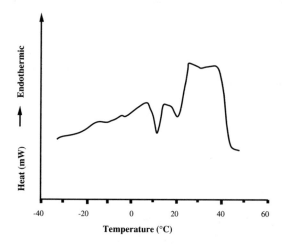

Fig. 5.7 Thermal profile (DSC curve) of very-high-melting milkfat fractions obtained by supercritical fluid extraction. *Carbon dioxide:* S1 (Bhaskar et al., 1993b).

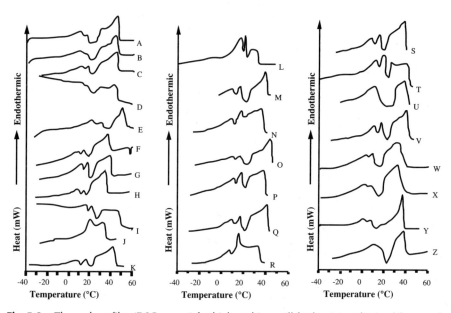

Fig. 5.8 Thermal profiles (DSC curves) for high-melting milkfat fractions obtained from melted milkfat. *Vacuum filtration:* A, $30S_M$ [W] (Barna et al., 1992); B, $30S_M$ [S] (Grall and Hartel, 1992); C, $30S_M$ [W] (Kaylegian, 1991); D, $30S_S$ [W] (Kaylegian, 1991); E, $29S_S$ [W] (Amer et al., 1985); F, $25S_M$ [W] (Kaylegian, 1991); G, $25S_S$ [W] (Kaylegian, 1991); H, $20S_M$ [W] (Kaylegian, 1991); I, $20S_S$ [W] (Kaylegian, 1991); J, $17S_M$ (Schaap and van Beresteyn, 1970); K, 1st stearin$_M$ (Deffense, 1993b). *Centrifugation:* L, $12S_M$ [S] (Kankare and Antila, 1988b). *Separation with the aid of an aqueous detergent solution:* M, $B3S_M$ (triple fractionation at 31.7°C) [B] (Sherbon et al., 1972); N, $B3L_M$ (triple fractionation at 31.7°C) [B] (Sherbon et al., 1972); O, $C2S_M$ (double fractionation at 30.6°C) [C] (Sherbon et al., 1972); P, $A1S_M$ (single fractionation at 28°C) [A] (Sherbon et al., 1972); Q, $B2S_M$ (double fractionation at 28°C) [B] (Sherbon et al., 1972); R, $B2L_M$ (double fractionation at 28°C) [B] (Sherbon et al., 1972); S, $B1S_M$ (single fractionation at 25.5°C) [B] (Sherbon et al., 1972); T, $25S_S$ [A late S] (Norris et al., 1971); U, 16 semi-S_S [B] high level of detergent (Banks et al., 1985); V, $C1S_M$ original fraction [C] (Sherbon et al., 1972). *Filtration through milk filter:* W, $29S_S$ (Fouad et al., 1990); X, $25S_S$ (Fouad et al., 1990). *Separation technique not reported:* Y, $21S_S$ [W] (Deroanne, 1976); Z, $21S_S$ [S] (Deroanne and Guyot, 1974).

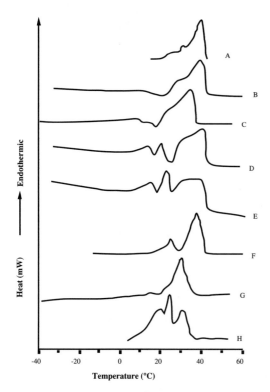

Fig. 5.9 Thermal profiles (DSC curves) of high-melting milkfat fractions obtained from solvent solution. *Acetone:* A, $25S_M$ (IV refractionated) [HMG] (Sherbon and Dolby, 1973); B, $11S_M$ [W] (Kaylegian, 1991); C, $10S_M$ [S] (Yi, 1993); D, $5S_S$ [W] (Kaylegian, 1991); E, OS_S [W] (Kaylegian, 1991); F, MMF (Timms, 1980a); G, $S1_M$ (Lambelet, 1983); H, 2BJAN [Jan] (Sherbon, 1963).

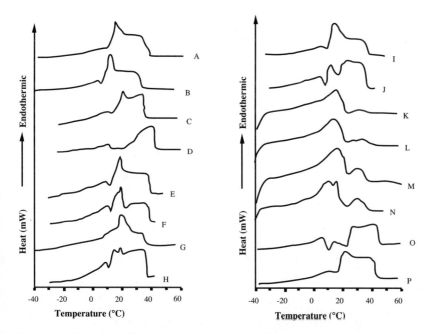

Fig. 5.10 Thermal profiles (DSC curves) of high-melting milkfat fractions obtained by supercritical fluid extraction. *Carbon dioxide:* A, I (residue) (Kaufmann et al., 1982); B, Long-chain fraction (Timmen et al., 1984); C, S2 (Arul et al., 1988b); D, S3 (Arul et al., 1988b); E, S2 (Bhaskar et al., 193b); F, Res (Kankare and Antila, 1988a); G, Extractor fraction (Büning-Pfaue et al., 1989); H, R (Kankare et al., 1989); I, Extract-2 (Rizvi et al., 1990); J, Residue (Rizvi et al., 1990); K, 13.8 (Chen et al., 1992); L, 20.7 (Chen et al., 1992); M, 24.1 (Chen et al., 1992); N, 27.6 (Chen et al., 1992); O, Raffinate (Shukla et al., 1994); P, Column fraction (Büning-Pfaue et al., 1989).

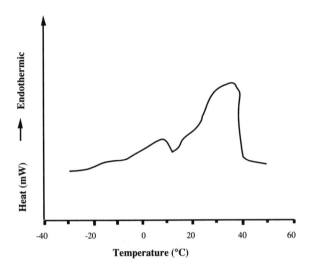

Fig. 5.11 Thermal profile (DSC curve) of high-melting milkfat fractions obtained by short-path distillation. A, SF residue (Arul et al., 1988a).

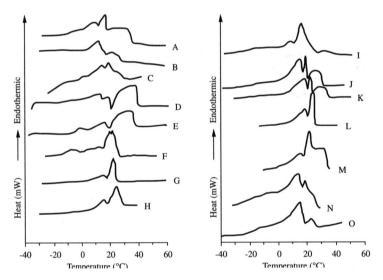

Fig. 5.12 Thermal profiles (DSC curves) of middle-melting milkfat fractions obtained from melted milkfat. *Vacuum filtration*[1]: A, $34L_S$ [W] (Kaylegian, 1991); B, $30L_S$ [W] (Kaylegian, 1991); C, $27L_M$ (Schaap and van Beresteyn, 1970); D, $20S_M$ [W] (Barna et al., 1992); E, $20S_M$ [S] Grall and Hartel, 1992); F, $17S_M$ (Sherbon, 1963); G, $16S_M$ [W] (Kaylegian, 1991); H, I.1 (21-15) (Makhlouf et al., 1987); *Pressure filtration:* I, $30L_S$ (A) slow cooling (deMan, 1968). *Centrifugation:* J, $12S_M$ [S] (Kankare and Antila, 1986); K, $12S_M$ [S] (Kankare and Antila, 1988a); L, C (Antila and Antila, 1970). *Separation with the aid of an aqueous detergent solution:* M, $C2L_M$ (double fractionation at 30.6°C) [C] (Sherbon et al., 1972); N, $A1L_S$ (single fractionation at 28°C) [A] (Sherbon et al., 1972). *Separation technique not reported:* O, $21L_S$ [W] (Deroanne, 1976). [1] The number indicates fractionation temperature, the letter indicates physical state (L = liquid, S = solid), and the subscript indicates fractionation procedure used (M = multiple-step, S = single-step). Other descriptions used are those of the original author. Designations in parenthesis are fraction designations given by the original author. The designation inside the brackets refer to the intact anhydrous milkfat source [sample number or season (W = winter, S = summer)].

Technical Data for Experimental and Commercial Milkfat Fractions   219

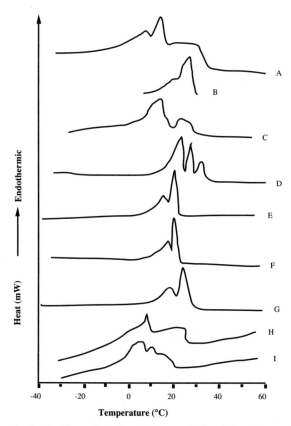

Fig. 5.13 Thermal profiles (DSC curves) of middle-melting milkfat fractions obtained from solvent solution. *Acetone:* A, 25L$_S$ [W] (Kaylegian, 1991); B, 25L$_M$ (V) [HMG] (Sherbon and Dolby, 1973); C, 21L$_S$ [W] (Kaylegian, 1991); D, 5S$_M$ [W] (Kaylegian, 1991); E, 5S$_M$ [S] (Yi, 1993); F, OS$_M$ [W] (Kaylegian, 1991); G, OS$_M$ [S] (Yi, 1993). *Pentane:* H, 11L$_S$ [W] (Kaylegian, 1991); I, 5L$_S$ [W] (Kaylegian, 1991).

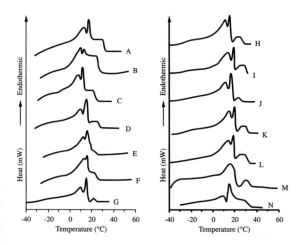

Fig. 5.14 Thermal profiles (DSC curves) of middle-melting milkfat fractions obtained by supercritical fluid extraction. *Carbon dioxide:* A, S3 (Bhaskar et al., 1993b); B, S1 (Arul et al., 1988b); C, S4 (Bhaskar et al., 1993b); D, S5 (Bhaskar et al., 1993b); E, I2 (Arul et al., 1988b); F, I3 (Arul et al., 1988b); G, E2 (Kankare and Antila, 1988a); H, E3 (Kankare and Antila, 1988a); I, E4 (Kankare and Antila, 1988a); J, E2 (Kankare et al., 1989); K, E3 (Kankare et al., 1989); L, E4 (Kankare et al., 1989); M, 17.2 (Chen et al., 1992); N, Extract-1 (Rizvi et al., 1990).

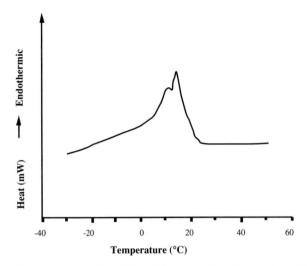

Fig. 5.15 Thermal profile (DSC curve) of middle-melting milkfat fractions obtained by short-path distillation. A, IF Distillate III (Arul et al., 1988a).

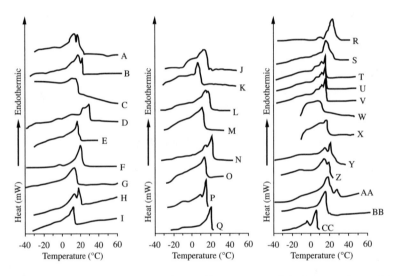

Fig. 5.16 Thermal profiles (DSC curves) of low-melting milkfat fractions obtained from melted milkfat. *Vacuum filtration* [1]: A, $25L_S$ [W] (Kaylegian, 1991); B, $20L_M$ [W] (Barna et al., 1992); C, $2OL_S$ [W] (Kaylegian, 1991); D, $19L_S$ [W] (Amer et al., 1985); E, $17L_M$ (Schaap and van Beresteyn, 1970); F, $15S_M$ [S] (Grall and Hartel, 1992); G, $15L_M$ [S] (Grall and Hartel, 1992); H, $13S_M$ [W] (Kaylegian, 1991); I, $13L_M$ [W] (Kaylegian, 1991); J, $10S_M$ (Sherbon, 1963); K, $10L_M$ (Sherbon, 1963); L, $9S_M$ (I.2) (Makhlouf et al., 1987); M, $9L_M$ (F.L.) (Makhlouf et al., 1987); N, F.1 (21-9) (Makhlouf et al., 1987); O, $7L_M$ (Schaap and van Beresteyn, 1970); P, 1st olein (Deffense, 1993b); Q, 2nd stearin (Deffense, 1993b). *Pressure filtration:* R, $27L_S$ (C) slow cooling (deMan, 1968); S, $24L_S$ (E) slow cooling (deMan, 1968). *Centrifugation:* T, $12L_M$ [S] (Kankare and Antila, 1986); U, $12L_M$ [S] (Kankare and Antila, 1988a); V, $12L_M$ [S] (Kankare and Antila, 1988b); W, A (Antila and Antila, 1970); X, B (Antila and Antila, 1970). *Separation with the aid of an aqueous detergent solution:* Y, $25L_S$ [A late S] (Norris et al., 1971); Z, $21L_S$ A high level of detergent (Banks et al., 1985). *Separation technique not reported:* AA, $21L_S$ (Deroanne, 1976); BB, $21L_S$ [S] (Deroanne and Guyot, 1974); CC, Olein (Martine, 1982).
[1] The number indicates fractionation temperature, the letter indicates physical state (L = liquid, S = solid), and the subscript indicates fractionation procedure used (M = multiple-step, S = single-step). Other descriptions used are those of the original author. Designations in parenthesis are fraction designations given by the original author. The designation inside the brackets refers to the intact anhydrous milkfat source [sample number or season (W = winter, S = summer)].

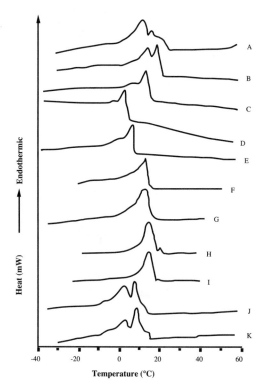

Fig. 5.17 Thermal profiles (DSC curves) of low-melting milkfat fractions obtained from solvent solution. *Acetone:* A, 15L$_S$ [W] (Kaylegian, 1991); B, 11L$_S$ [W] (Kaylegian, 1991); C, 5L$_S$ [W] (Kaylegian, 1991); D, OL$_M$ [W] (Kaylegian, 1991); E, OL$_M$ [S] (Yi, 1993); F, LMF (Timms, 1980a); G, L1$_M$ (Lambelet, 1983); H, 2C (Sherbon, 1963); I, 2CJAN [Jan] (Sherbon, 1963). *Pentane:* J, OL$_M$ [W] (Kaylegian, 1991); K, OL$_S$ [W] (Kaylegian, 1991).

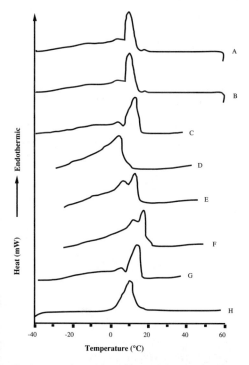

Fig. 5.18 Thermal profiles (DSC curves) of low-melting milkfat fractions obtained by supercritical fluid extraction. *Carbon dioxide:* A, II (extract) (Kaufmann et al., 1982); B, short-chain fraction (Timmen et al., 1984); C, E1 (Kankare and Antila, 1988a); D, L1 (Arul et al., 1987); E, L2 (Arul et al., 1987); F, I1 (Arul et al., 1987); G, E1 (Kankare et al., 1989); H, 10.3 (Chen et al., 1992).

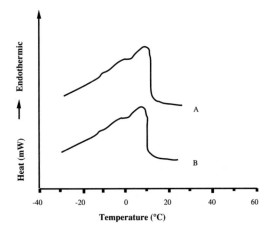

Fig. 5.19 Thermal profiles (DSC curves) of low-melting milkfat fractions obtained by short-path distillation. A, LF1 Distillate I (Arul et al., 1988a); B, LF2 Distillate II (Arul et al., 1988a).

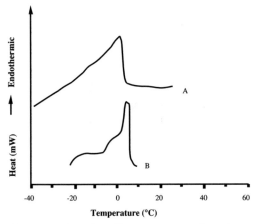

Fig. 5.20 Thermal profiles (DSC curves) of very-low-melting milkfat fractions obtained from melted milkfat. *Vacuum filtration:* A, $2L_M$ (Schaap and van Beresteyn, 1970); B, 2nd olein (Deffense, 1993b).

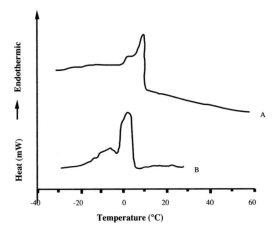

Fig. 5.21 Thermal profiles (DSC curves) for very-low-melting milkfat fractions obtained from solvent solution. *Acetone:* A, OL_S [W] (Kaylegian, 1991); B, 2DJAN [Jan] (Sherbon, 1963).

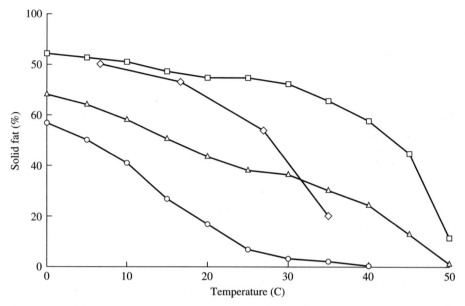

Fig. 5.22 Solid fat content profiles of very-high-melting milkfat fractions. *Crystallization from melted milkfat:* △, 34S$_M$ [W] (Kaylegian, 1991). *Crystallization from solvent solution:* □, 21S$_M$ [W] (Kaylegian, 1991). *Supercritical fluid extraction:* ◇, S1 (Bhaskar et al., 1993b). *Intact milkfat reference:* ○, AMF (Kaylegian, 1991).

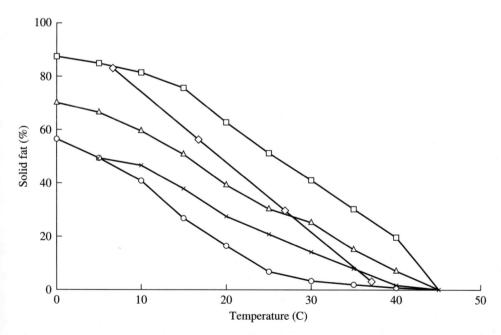

Fig. 5.23 Solid fat content profiles of high-melting milkfat fractions. *Crystallization from melted milkfat:* △, 25S$_M$ [W] (Kaylegian, 1991). *Crystallization from solvent solution:* □, 5S$_M$ [W] (Kaylegian, 1991). *Supercritical fluid extraction:* ◇, S2 (Bhaskar et al., 1993b). *Short-path distillation:* X, SF Residue (Arul et al., 1988a). *Intact milkfat reference:* ○, AMF (Kaylegian, 1991).

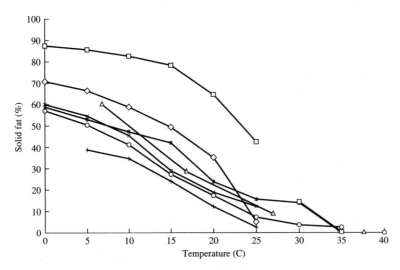

Fig. 5.24 Solid fat content profiles of middle-melting milkfat fractions. *Crystallization from melted milkfat:* X, $34L_M$ [W] (Kaylegian, 1991); ◊, $16S_M$ [W] (Kaylegian, 1991). *Crystallization from solvent solution:* *, $25L_M$ [W] (Kaylegian, 1991); □, $5S_M$ [W] (Kaylegian, 1991). *Supercritical fluid extraction:* △, S3 (Bhaskar et al., 1993b). *Short-path distillation:* +, IF Distillate III (Arul et al., 1988a). *Intact milkfat reference:* ○, AMF (Kaylegian, 1991).

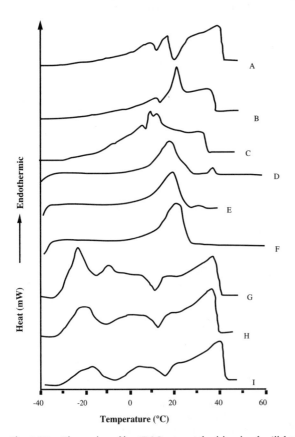

Fig. 5.25 Thermal profiles (DSC curves) for blends of milkfat fractions with other fats and oils. *Milkfat fractions–intact milkfat and milkfat fractions–milkfat fractions:* A, Mixture 1 (Kankare and Antila, 1986); B, Mixture 2 (Kankare and Antila, 1986); C, Mixture 3 (Kankare and Antila, 1986). *Milkfat fractions–cocoa butter:* D, Mixture 1 (Barna et al., 1992); E, Mixture 2 (Barna et al., 1992); F, Mixture 3 (Barna et al., 1992). *Milkfat fractions–other fats and oils:* G, Mixture 1 (Kankare and Antila, 1988b); H, Mixture 2 (Kankare and Antila, 1988b); I, Mixture 3 (Kankare and Antila, 1988b).

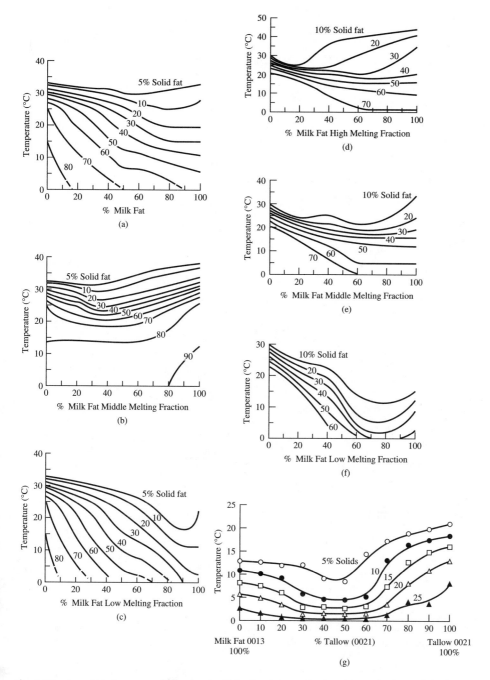

Fig. 5.26 Isosolid diagrams of blends of milkfat fractions and other fats and oils. *Milkfat fractions–cocoa butter:* (a), Mixture 1 (Timms 1980b); (b), Mixture 2 (Timms, 1980b); (c), Mixture 3 (Timms, 1980b); (d), Mixture 4 (Barna et al., 1992), (e), Mixture 5 (Barna et al., 1992); (f), Mixture 6 (Barna et al., 1992). *Milkfat fractions–other fats and oils:* (g), Mixture 1 (Versteeg, 1991). All figures reprinted with permission.

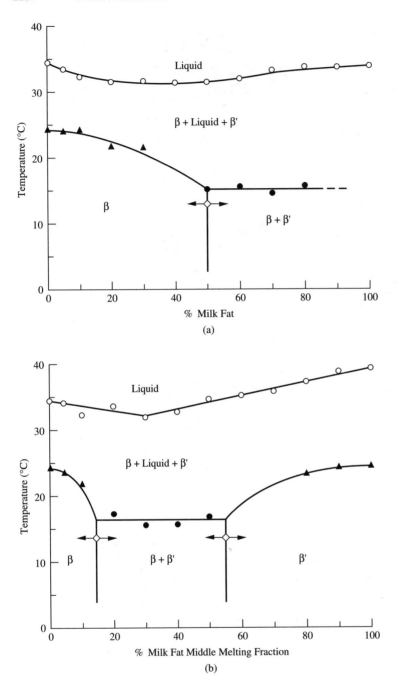

Fig. 5.27 Phase diagrams of milkfat fraction and cocoa butter mixtures. *Milkfat fractions–cocoa butter:* (a), Mixture 1 (Timms, 1980a); (b), Mixture 2 (Timms, 1980a). All figures reprinted with permission.

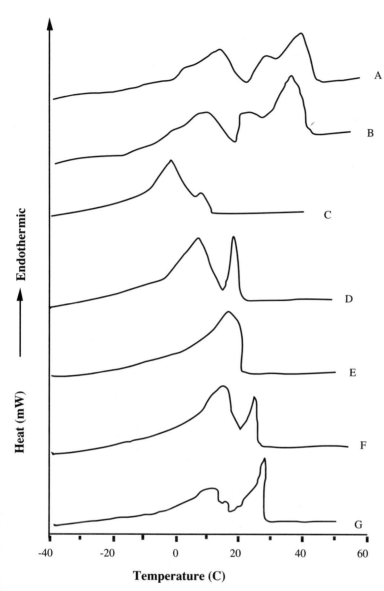

Fig. 5.28 Thermal profiles (DSC curves) of commercially available milkfat fraction products. A, Corman fraction S2 (Lambelet, 1983); B, Corman fraction S3 (Lambelet, 1983); C, Corman fraction MF-35 (Büning-Pfaue et al., 1989); D, Corman fraction MF-38 (Büning-Pfaue et al., 1989); E, Saumweber fraction MF-28 (Büning-Pfaue et al., 1989); F, Saumweber fraction MF-36 (Büning-Pfaue et al., 1989); G, Unspecified fraction MF-42 (Büning-Pfaue et al., 1989).

TABLE 5.1 Experimental Milkfat Fractions Obtained by Crystallization from Melted Milkfat [1]

Fraction separation method Author, year, location detailed data available[2]	Season[3]	Fraction designation[4]	Yield (%)	Melting behavior (°C)[5]	General fraction property[6]
Vacuum filtration					
Sherbon, 1963, USA	—[7]	AMF		34.1 (DSC)	
Technical data:		$23S_M$	—	48.9	VH
Fatty acid composition		$23ppt_M$	—	51.8	VH
Melting behavior		refractionated			
Thermal profile		$17S_M$	—	25.3	M
		$10S_M$	—	20.0	L
		$10L_M$	—	13.2	L
Schultz and Timmen, 1966, Germany	—	AMF		—	
Technical data:		$30S_M$	20	—	U
Yield		$25S_M$	20	—	U
		$20S_M$	20	—	U
		$15S_M$	20	—	U
		$10S_M$	10	—	U
		$7S_M$	4	—	U
		$4.5S_M$	3.5	—	U
		$4.5L_M$	2.5	—	U
Schaap and van Beresteyn, 1970, The Netherlands	—	AMF		34.6 (DSC)	
Technical data:		$27S_M$	—	49.2	VH
Fatty acid composition		$27L_M$	—	29.2	M
Melting behavior		$17S_M$	—	37.9	H
Thermal profile		$17L_M$	—	20.0	L
		$7L_M$	—	17.8	L
		$2L_M$	—	5.7	VL
Voss et al., 1971, Germany	—	AMF		—	
Technical data:		$30S_S$	47.7	—	U
Yield		$30L_S$	52.3	—	U
Selected chemical characteristics					
Vovan and Riel, 1973, Canada	—	AMF		20.0 (CP)[8]	
Technical data:		5% removal of hard fraction (>21°C)	—	18.0	L
Fatty acid composition					
Selected chemical characteristics					
Melting behavior				15.9	L
Solid fat content		10% removal of hard fraction (>21°C)	—		
Applications data:					
Cold-spreadable butter		15% removal of hard fraction (>21°C)		15.4	L
		5% removal of intermediate fraction (10–21°C)	—	20.0	L
		10% removal of intermediate fraction (10–21°C)		21.2	L

Continued

TABLE 5.1 (Continued)

Fraction separation method Author, year, location detailed data available[2]	Season[3]	Fraction designation[4]	Yield (%)	Melting behavior (°C)[5]	General fraction property[6]
		15% removal of intermediate fraction (10–21°C)	—	22.5	L
		33% removal of intermediate fraction (10–21°C)	—	24.0	L
Black, 1975b, Australia	—	AMF		—	
Technical data:		$25S_S$ with precooling	36.4	—	U
Yield		$25S_S$ no precooling	40.4	—	U
Crystal size		$25S_S$ exponential cooling	31.4	—	U
		$25S_S$ programmed cooling	45.4	—	U
		$25S_S$ 8 h cooling time	35.1	—	U
		$25S_S$ 20 h cooling time	41.7	—	U
		$25S_S$ 20 rpm agitator speed	40.4	—	U
		$25S_S$ 10 rpm agitator speed	36.3	—	U
Schaap, 1982, The Netherlands					
Applications data:	—	AMF		—	
Bakery		$24S_S$	—	—	U
		$24L_S$	—	—	U
Timms, 1978a, Australia					
Technical data:	—	AMF		—	
Fatty acid composition					
Melting behavior		Hard	—	39.0 (SP)	H
Thermal profile		Soft	—	24.1	L
deMan and Finoro, 1980, Canada					
Technical data:	—	AMF		33.9 (DP)	
Yield		$32S_S$	10.0	42.8	H
Fatty acid composition		$32L_S$	90.0	32.4	M
Triglyceride composition		$31S_S$	18.0	42.4	H
Selected chemical characteristics		$31L_S$	82.0	29.5	M
Melting behavior		$30S_S$	17.3	44.1	H
Solid fat content		$30L_S$	82.7	28.4	M
		$29S_S$	18.7	44.5	H
		$29L_S$	81.3	26.5	M
		$28S_S$	18.4	44.0	H
		$28L_S$	81.6	26.8	M
		$27S_S$	24.6	42.3	H
		$27L_S$	75.4	25.1	M
		$26S_S$	31.4	41.4	H
		26LS	68.6	24.5	L
		25SS	38.2	41.3	H
		25LS	61.8	23.3	L
Frede et al., 1980, Germany					
Technical data:	S	AMF		—	
Yield					
Selected chemical characteristics		$28S_S$	35	—	U

Continued

TABLE 5.1 (Continued)

Fraction separation method Author, year, location detailed data available[2]	Season[3]	Fraction designation[4]	Yield (%)	Melting behavior (°C)[5]	General fraction property[6]
Applications data: Cold-spreadable butter		$28L_S$	65	—	U
Jamotte and Guyot, 1980, Belgium Technical data: Fatty acid composition Selected chemical characteristics Melting behavior Applications data:	—	Commercial liquid fraction (L_2) doubly fractionated at 15°C		—	
Cold-spreadable butter		L_4 (made from L_2)	—	5.0 (DP)	VL
		Commercial solid fraction A fractionated at 25°C	—	42.3	H
		Commercial solid fraction B fractionated at 25°C	—	43.8	H
Timms and Parekh, 1980, Australia Technical data: Yield Fatty acid composition Melting behavior Solid fat content Milkfat fraction blends Applications data: Chocolate	S	AMF		34.5 (SP)	
		$25S_S$	36	39.2	H
		$25L_S$	64	24.3	L
Guyot, 1982, Belgium Technical data: Yield Melting behavior Applications data: Cold-spreadable butter	—	AMF	—	—	
		$23S_M$	—	—	U
		$15S_M$	—	—	U
		$10S_M$	—	—	U
		$5S_M$	—	—	U
		$5L_M$	12	6 (DP)	VL
Hayakawa and deMan, 1982, Canada Technical data: Textural characteristics	—	AMF	—	—	
		$32S_S$	—	—	U
		$32L_S$	—	—	U
		$31S_S$	—	—	U
		$31L_S$	—	—	U
		$30S_S$	—	—	U
		$30L_S$	—	—	U
		$29S_S$	—	—	U
		$29L_S$	—	—	U
		$28S_S$	—	—	U
		$28L_S$	—	—	U
		$27S_S$	—	—	U
		$27L_S$	—	—	U
		$26S_S$	—	—	U
		$26L_S$	—	—	U
		$25S_S$	—	—	U
		$25L_S$	—	—	U

Continued

TABLE 5.1 (Continued)

Fraction separation method Author, year, location detailed data available[2]	Season[3]	Fraction designation[4]	Yield (%)	Melting behavior (°C)[5]	General fraction property[6]
Amer et al., 1985, Canada					
Technical data:	W	AMF		34.6 (MP)	
Yield		$29S_S$	20.6	43.5	H
Fatty acid composition		$29L_S$	79.4	29.7	M
Triglyceride composition		$26S_S$	36.2	41.5	H
Selected chemical characteristics		$26L_S$	63.8	27.0	M
Melting behavior		$23S_S$	56.0	39.0	H
Thermal profile		$23L_S$	44.0	22.9	L
Solid fat content		$19S_S$	61.4	38.2	H
		$19L_S$	38.6	20.6	L
	S	AMF		33.4	
		$29S_S$	14.9	44.4	H
		$29L_S$	85.1	28.8	M
		$19S_S$	53.7	38.4	H
		$19L_S$	46.3	19.7	L
Kupranycz et al., 1986, Canada					
Technical data:	W	AMF	—	—	
Oxidative stability		$29S_S$	—	—	U
		$29L_S$	—	—	U
		$19S_S$	—	—	U
		$19L_S$	—	—	U
	S	AMF	—	—	
		$29S_S$	—	—	U
		$29L_S$	—	—	U
		$19S_S$	—	—	U
		$19L_S$	—	—	U
Deffense, 1987, Belgium					
Technical data:	—	AMF [1]		33.6 (DP)	
Fatty acid composition		Stearin #1_S	—	43.1	H
Selected chemical characteristics		Olein #1_S	—	27.7	M
Melting behavior		Stearin #2_S	—	42.3	H
Solid fat content		Olein #2_S	—	24.3	L
Applications data:		Stearin #3_S	—	41.1	H
Cold-spreadable butter		Olein #3_S	—	21.1	L
		Stearin #4_S	—	41.0	H
		Olein #4_S	—	20.6	L
	—	AMF [2]		32.0	
		1st stearin$_M$	—	40.0	H
		1st olein$_M$	—	18.0	L
		2nd stearin$_M$	—	24.0	L
		2nd olein$_M$ filtered at 10°C	—	10.0	L
		2nd olein$_M$ filtered at 6°C	—	8.5	VL
Lund and Danmark, 1987, Denmark					
Technical data:	—	AMF [1]		33.0 (DP)	
Yield		$26S_S$	25.5	42.2	H
Fatty acid composition		$26L_S$	74.5	24.1	L
Triglyceride composition	—	AMF [2]		33.0	
Selected chemical characteristics		$18S_S$	42.5	39.8	H
Melting behavior		$18L_S$	57.5	19.1	L
Solid fat content					

Continued

TABLE 5.1 (Continued)

Fraction separation method Author, year, location detailed data available[2]	Season[3]	Fraction designation[4]	Yield (%)	Melting behavior (°C)[5]	General fraction property[6]
Makhlouf et al., 1987, Canada					
Technical data:	—	AMF		37.8 (DSC)	
Yield					
Fatty acid composition		$26S_M$ (S.1)	10.7	45.1	VH
Triglyceride composition		$21S_M$ (S.2)	25.6	48.3	VH
Melting behavior		F.S (>21)	36.3	45.1	VH
Thermal profile		$15S_M$ (I.1)	20.1	29.3	M
Solid fat content		$9S_M$ (I.2)	18.5	20.5	L
Nutritional composition		$9L_M$ (F.L.)	25.1	15.1	L
Applications data:					
Cold-spreadable butter		F.1 (21–9)	38.6	24.4	L
Bumbalough, 1989, USA					
Technical data:	—	AMF [1]		—	
Yield					
Melting behavior		$29.4S_S$ [1]	13.3	43.9 (MP)	H
Solid fat content					
Applications data:		AMF + $29.4L_S$ [2]		—	
Cold-spreadable butter		$18S_M$ [2]	—	—	U
		$12.5S_M$ [2]	—	—	U
		$12.5L_M$ [2]	25.9	—	U
Deffense, 1989, Belgium					
Technical data:	W	AMF		33.9 (MP)	
Selected chemical characteristics		1st stearin$_M$	—	41.2	H
Melting behavior		1st olein$_M$	—	19.4	L
Applications data:					
Cold-spreadable butter		2nd stearin$_M$	—	24.0	L
Bakery		2nd olein$_M$	—	13.3	L
	S	AMF		32.5	
		1st stearin$_M$	—	40.9	H
		1st olein$_M$	—	18.8	L
		2nd stearin$_M$	—	23.1	L
		2nd olein$_M$	—	13.5	L
Kaylegian, 1991, USA					
Technical data:	W	AMF		34.4 (MP)	
Yield		$34S_M$	3.7	49.0	VH
Fatty acid composition		$30S_M$	17.7	44.2	H
Triglyceride composition		$25S_M$	11.2	42.0	H
Melting behavior		$20S_M$	7.0	40.5	H
Thermal profile		$16S_M$	14.2	26.9	M
Solid fat content		$13S_M$	13.3	22.8	L
Oxidative stability		$13L_M$	33.0	10.1	L
Aapplications data:		$34S_S$	2.5	46.6	VH
Cold-spreadable butter		$34L_S$	97.5	33.5	M
		$30S_S$	20.0	44.2	H
		$30L_S$	80.0	28.4	M
		$25S_S$	46.5	40.6	H
		$25L_S$	53.5	21.8	L
		$20S_S$	60.0	39.6	H
		$20L_S$	40.0	18.3	L
Versteeg, 1991, Australia					
Technical data:	Spr	AMF		—	
Yield		$28S_S$	15	—	U
Fatty acid composition		$28L_S$	85	—	U

TABLE 5.1 (Continued)

Fraction separation method Author, year, location detailed data available[2]	Season[3]	Fraction designation[4]	Yield (%)	Melting behavior (°C)[5]	General fraction property[6]
Melting behavior		$24S_S$	23	—	U
Solid fat content		$24L_S$	77	—	U
Milkfat fraction blends		$22S_S$	26	—	U
		$22L_S$	74	—	U
		$18S_S$	29	—	U
		$18L_S$	71	—	U
	S	$28S_S$	23	—	U
		$28L_S$	77	—	U
		$24S_S$	33	—	U
		$24L_S$	67	—	U
		$22S_S$	39	—	U
		$22L_S$	61	—	U
	Sept	AMF		32 (SP)	
		$18S_M$	29 (overall)	43	H
		$18L_M$	(71 for this step)	21	L
		$12S_M$	31 (overall) (44 for this step)	26	M
		$12L_M$	(56 for this step)	14	L
		$8S_M$	16 (overall) (41 for this step)	19	L
		$8L_M$	24 (overall) (59 for this step)	5	VL
		$28S_S$	—	43.2	H
		$28L_S$	—	28.6	M
		$26S_S$	—	42.2	H
		$26L_S$	—	25.4	M
		$24S_S$	—	42.3	H
		$24L_S$	—	24.2	L
		$22S_S$	—	42.0	H
		$22L_S$	—	22.2	L
		$20S_S$	—	41.1	H
		$20L_S$	—	21.9	L
		$18S_S$	—	41.2	H
		$18L_S$	—	20.4	L
		$15S_M$	—	26.6	M
		$15L_M$	—	17.2	L
		$14S_M$	—	26.5	M
		$14L_M$	—	14.4	L
		$13S_M$	—	26.3	M
		$13L_M$	—	14.1	L
		$12S_M$	—	26.0	M
		$12L_M$	—	13.2	L
		$11.5S_M$	—	25.8	M

Continued

TABLE 5.1 (Continued)

Fraction separation method Author, year, location detailed data available[2]	Season[3]	Fraction designation[4]	Yield (%)	Melting behavior (°C)[5]	General fraction property[6]
Barna et al., 1992, USA					
Technical data:	W	AMF		33 (DSC)[9]	
Yield					
Fatty acid composition		30S$_M$	20	43	H
Melting behavior					
Thermal profile		20S$_M$	25	33.9	M
Milkfat fraction blends					
Applications data:					
Chocolate		20L$_M$	55	18	L
Deffense, 1993b, Belgium					
Technical data:		AMF		32 (DP)	
Fatty acid composition		1st stearin$_M$	—	42	H
Triglyceride composition		1st olein$_M$	—	18	L
Melting behavior		2nd stearin$_M$	—	24	L
Thermal profile		2nd olein$_M$	—	9	VL
Grall and Hartel, 1992, USA					
Technical data:	S	AMF		33.1 (DSC)	
Fatty acid composition		30S$_M$	—	43.0	H
Melting behavior		20S$_M$	—	33.9	M
Thermal profile		15S$_M$	—	19.5	L
		15L$_M$	—	12.5	L
Kaylegian and Lindsay, 1992, USA					
Technical data:	W	AMF		34.4 (MP)[10]	
Yield		34S$_M$	3.7[9]	49.0	VH
Melting behavior		30S$_M$	17.7	44.2	H
Solid fat content		25S$_M$	11.2	42.0	H
Applications data:		20S$_M$	7.0	40.5	H
Cold-spreadable butter		16S$_M$	14.2	26.9	M
		13S$_M$	13.3	22.8	L
		13L$_M$	33.0	10.1	L
		34S$_S$	2.5	46.6	VH
		34L$_S$	97.5	33.5	M
		30S$_S$	20.0	44.2	H
		30L$_S$	80.0	28.4	M
		25S$_S$	46.5	40.6	H
		25L$_S$	53.5	21.8	L
		20S$_S$	60.0	39.6	H
		20L$_S$	40.0	18.3	L
Laakso et al., 1992, Finland					
Technical data:	—	AMF		34 (DP)	
Fatty acid composition		S1	—	42	H
Triglyceride composition		L1	—	21	L
Melting behavior		S2	—	27	M
		L2	—	16	L

Pressure filtration					
deMan, 1968, Canada					
Technical data:	—	AMF		32.9 (SP)	
Yield		30S$_S$ (A) slow cooling	11.2	50.0 (DSC)	VH
Fatty acid composition					

Continued

TABLE 5.1 (Continued)

Fraction separation method Author, year, location detailed data available[2]	Season[3]	Fraction designation[4]	Yield (%)	Melting behavior (°C)[5]	General fraction property[6]
Melting behavior Thermal profile Textural characteristics		$30L_S$ (A) slow cooling	88.2	28.3 (SP)	M
		$30S_S$ (B) rapid cooling	27.9	—	U
		$30L_S$ (B) rapid cooling	72.1	27.9 (SP)	M
		$27S_S$ (C) slow cooling	9.9	47.6 (DSC)	VH
		$27L_S$ (C) slow cooling	90.1	18.7 (SP)	L
		$27S_S$ (D) rapid cooling	11.3	—	U
		$27L_S$ (D) rapid cooling	88.7	18.4 (SP)	L
		$24S_S$ (E) slow cooling	30.0	46.7 (DSC)	VH
		$24L_S$ (E) slow cooling	70.0	17.7	L
		$24S_S$ (F) rapid cooling	29.8	—	U
		$24L_S$ (F) rapid cooling	70.2	17.8	L
Stepanenko and Tverdokhleb, 1974, USSR Technical data: Yield Selected chemical characteristics Melting behavior	—	AMF $20S_M$ $11S_M$ $11L_M$	29 30 41	31.2 (MP) 42.3 29.1 13.2	H M L
El-Ghandour et al., 1976, Egypt Technical data: Crystal morphology	Egyptian milkfat	AMF $25S_M$ $15S_M$ $15L_M$	— — — —	— — — —	U U U
	Russian milkfat	AMF $25S_M$ $15S_M$ $15L_M$	— — — —	— — — —	U U U
Helal et al., 1977, Egypt Technical data: Crystal morphology	S	AMF $25S_M$ $15S_M$ $15L_M$	— — —	— — —	U U U
	W	AMF $25S_M$ $15S_M$ $15L_M$	— — —	— — —	U U U
Badings et al., 1983b, The Netherlands Technical data: Yield Fatty acid composition Triglyceride composition	W	AMF $23S_M$ $23L_M$	35.9 (overall) (64.1 this step)	— — —	U U

Continued

TABLE 5.1 (Continued)

Fraction separation method Author, year, location detailed data available[2]	Season[3]	Fraction designation[4]	Yield (%)	Melting behavior (°C)[5]	General fraction property[6]
Selected chemical characteristics		18.5S$_M$	23.3 (overall) (36.3 this step)	—	U
		18.5L$_M$	(63.6 this step)	—	U
		15S$_M$	14.6 (overall) (35.8 this step)	—	U
		15L$_M$	(64.2 this step)	—	U
		11.3S$_M$	9.0 (overall) (34.4 this step)	—	U
		11.3L$_M$	(65.6 this step)	—	U
		5.5S$_M$	10.4 (overall) (60.5 this step)	—	U
		5.5L$_M$	6.8 (overall) (39.5 this step)	—	U
	S	AMF		—	U
		21.5S$_M$	30.3 (overall)	—	U
		21.5L$_M$	(69.7 this step)	—	U
		15.3S$_M$	10.3 (overall) (14.8 this step)	—	U
		15.3L$_M$	(85.2 this step)	—	U
		13S$_M$	26.5 (overall) (44.7 this step)	—	U
		13L$_M$	(55.3 this step)	—	U
		9.3S$_M$	9.6 (overall) (29.2 this step)	—	U
		9.3L$_M$	(70.8 this step)	—	U
		7.1S$_M$	9.0 (overall) (38.6 this step)	—	U

Continued

TABLE 5.1 (Continued)

Fraction separation method Author, year, location detailed data available[2]	Season[3]	Fraction designation[4]	Yield (%)	Melting behavior (°C)[5]	General fraction property[6]
		7.1L$_M$	14.3 (overall) (61.4 this step)	—	U
Helal et al., 1984, Egypt					
Technical data:	S	AMF		—	
Crystal morphology		15S$_M$	—	—	U
	W	AMF	—	—	
		15S$_M$	—	—	U
Kuwabara et al., 1991, Japan					
Technical data:	—	23S$_S$	22.2	—	U
Yield		23L$_S$	77.8	—	U
Selected chemical characteristics		20S$_S$	46.6	—	U
		20L$_S$	53.4	—	U
Deffense, 1993b, Belgium					
Technical data:		AMF		32 (DP)	
Fatty acid composition					
Triglyceride composition		1st stearin	—	46 (DP)	VH
Melting behavior					
Thermal profile		1st olein	—	—	U
Centrifugation					
Baker et al., 1959, Canada					
Technical data:	W	AMF		30 (MP)	
Yield		28S$_S$	20.0	—	U
Melting behavior		28L$_S$	80.0	—	U
		22S$_S$	27.0	—	U
		22L$_S$	73.0	24	L
Richardson, 1968, USA					
Technical data:	W	AMF		33.5 (SLP)	
Yield		30S$_M$	40.3	38.5	H
Fatty acid composition		25S$_M$	14.9	32.0	M
Melting behavior		20S$_M$	11.9	28.0	M
		15S$_M$	15.8	22.5	L
		15L$_M$	17.1	17.0	L
Antila and Antila, 1970, Finland					
Technical data:	—	AMF		45.1 (DSC)	
Fatty acid composition					
Selected chemical characteristics		A	—	17.9	L
Melting behavior		B	—	18.9	L
Thermal profile		C	—	30.2	M
Nutritional composition		D	—	> 50	VH
Riel and Paquet, 1972, Canada					
Technical data:	—	AMF		24 (CP)	
Yield					
Fatty acid composition		38S$_M$	13	35	H
Selected chemical characteristics		4S$_M$	51	23	L
Melting behavior		4L$_M$	36	9	VL
Black, 1974b, Australia					
Technical data:	—	AMF		—	
Yield		23S$_S$	22	42.1 (SP)	H
Melting behavior		23L$_S$	78	24.7	L

Continued

TABLE 5.1 (Continued)

Fraction separation method Author, year, location detailed data available[2]	Season[3]	Fraction designation[4]	Yield (%)	Melting behavior (°C)[5]	General fraction property[6]
Kankare and Antila, 1974a, Finland					
Technical data:	S	AMF		—	
Yield		$24S_M$	15.7	—	U
Crystal size		$18S_M$	6.3	—	U
Applications data:		$12S_M$	29.0	—	U
Cold-spreadable butter		$12L_M$	49.0	—	U
	W	AMF		—	
		$24S_M$	30.2	—	U
		$18S_M$	9.1	—	U
		$12S_M$	27.9	—	U
		$12L_M$	32.8	—	U
Evans, 1976, New Zealand					
Technical data:	S	AMF [1]		34.4 (SP)	
Yield		$25S_S$	44.8	40.2	H
Melting behavior		$25L_S$	55.2	23.1	L
	S	AMF [2]		34.3	
		$25S_S$	—	40.2	H
		$25L_S$	—	23.0	L
	S	AMF [3]		34.3	
		$25S_S$	44.4	40.5	H
		$25L_S$	55.6	22.8	L
	S	AMF [4]		34.3	
		$25S_S$	42.4	40.7	H
		$25L_S$	57.6	22.9	L
Bhat and Rama Murthy, 1983, India					
Technical data:	—	AMF		—	
Yield					
Melting behavior		$28S_S$	≈50	41 (SP)	H
Oxidative stability		$28L_S$	≈50	23	L
Bratland, 1983, England					
Applications data:	—	AMF		—	
Cold-spreadable butter		$28S_S$	—	—	U
Dairy-based spreads		$28L_S$	—	—	U
Dairy products		$25S_S$	—	—	U
		$25L_S$	—	—	U
		$24S_S$	—	—	U
		$24L_S$	—	—	U
		$20S_S$	—	—	U
		$20L_S$	—	—	U
Lakshminarayana and Rama Murthy, 1985, India					
Technical data:	—	AMF		34.2 (MP)	
Yield		$31S_M$	10.6	36.5	H
Fatty acid composition		$23S_M$	57.6	30.5	M
Selected chemical characteristics		$15S_M$	12.4	19.0	L
Melting behavior		$15L_M$	19.4	14.0	L
Jordan, 1986, Britain					
Technical data:	S	AMF		32.4 (SLP)	
Fatty acid composition		$28S_S$	—	37.0	H
Triglyceride composition		$28L_S$	—	24.2	L
Melting behavior		$24S_S$	—	37.9	H
Applications data:		$24L_S$	—	21.5	L
Bakery		$15S_S$	—	26.6	M
Chocolate					

Continued

TABLE 5.1 (Continued)

Fraction separation method Author, year, location detailed data available[2]	Season[3]	Fraction designation[4]	Yield (%)	Melting behavior (°C)[5]	General fraction property[6]
		$15L_S$	—	19.9	L
	W	AMF		33.0	
		$28S_S$	—	39.0	H
		$28L_S$	—	23.9	L
		$24S_S$	—	36.6	H
		$24L_S$	—	22.4	L
Kankare and Antila, 1986, Finland					
Technical data:	S	AMF		37.6 (DSC)	
Fatty acid composition					
Triglyceride composition		$24S_M$	—	46.3	VH
Selected chemical characteristics					
Melting behavior		$12S_M$	—	34.1	M
Thermal profile					
Milkfat fraction blends		$12L_M$	—	17.9	L
Keogh and Higgins, 1986b, Ireland					
Technical data:	—	AMF		31.4 (SLP)	
Fatty acid composition		Fraction 1	—	44.5	H
Trglyceride composition		Fraction 2	—	42.3	H
Melting behavior		Fraction 3	—	22.8	L
		Fraction 4	—	18.5	L
Lakshminarayana and Rama Murthy, 1986, India					
Technical data:	—	AMF		—	
Oxidative stability		$31S_M$	—	—	U
		$23S_M$	—	—	U
		$15S_M$	—	—	U
		$15L_M$	—	—	U
Keshava Prasad and Bhat, 1987a, India					
Technical data:	—	AMF		—	
Oxidative stability		$28S_S$	—	—	U
		$28L_S$	—	—	U
		$23S_S$	—	—	U
		$23L_S$	—	—	U
Keshava Prasad and Bhat, 1987b, India					
Technical data:	—	AMF		—	
Oxidative stability		$28S_S$	—	—	U
		$28L_S$	—	—	U
		$23S_S$	—	—	U
		$23L_S$	—	—	U
Ramesh and Bindal, 1987b, India					
Technical data:	—	Ghee		35.6 (MP)	
Yield				33.7 (SP)	
Fatty acid composition		$28S_S$	69.5	37.3 (MP)	H
Selected chemical characteristics				35.9 (SP)	
Melting behavior		$28L_S$	28.8	23.3 (MP)	L
				22.2 (SP)	
Kankare and Antila, 1988a, Finland					
Technical data:	S	AMF		—	
Fatty acid composition					
Selected chemical characteristics		$24S_M$		47 (MP)	VH
Melting behavior		$12S_M$	—	34	M
Thermal profile		$12L_M$	—	14.4	L

Continued

TABLE 5.1 (Continued)

Fraction separation method Author, year, location detailed data available[2]	Season[3]	Fraction designation[4]	Yield (%)	Melting behavior (°C)[5]	General fraction property[6]
Kankare and Antila, 1988b, Finland					
Technical data:	S	AMF		38.0 (MP)	
Fatty acid composition					
Triglyceride composition		$24S_M$	—	48.0	VH
Selected chemical characteristics					
Melting behavior		$12S_M$	—	35.2	H
Thermal profile					
Solid fat content		$12L_M$	—	18.6	L
Milkfat fraction blends					
Keogh, 1989, Ireland					
Applications data:	—	Soft fraction$_S$	—	20 (SLP)	L
Deep-fat frying					
Keogh and Morrissey, 1990, Ireland					
Applications data:	S	AMF			
Bakery		Fraction 1	—	41 (SLP)	H
		Fraction 2	—	38	H
		Fraction 3	—	35	H
Branger, 1993, USA					
Technical data:	—	AMF [1]		40 (SFC)	
Triglyceride composition		bench-top			
Solid fat content		$32S_M$ [1]	—	>40	H
Nutritional composition		bench-top			
		$22S_M$ [1]	—	40	H
		bench-top			
		$22L_M$ [1]	—	30-35	M
		bench-top			
		AMF [2]		40	
		bench-top			
		$32S_M$ [2]	—	>40	H
		bench-top			
		$20S_M$ [2]	—	35	H
		bench-top			
		$20L_M$ [2]	—	15	L
		bench-top			
		AMF [3]		40	
		$32S_M$ [3]	—	>40	H
		bench-top			
		$22S_M$ [3]	—	40	H
		bench-top			
		$12S_M$ [3]	—	25	M
		bench-top			
		$12L_M$ [3]	—	20	L
		bench-top			
		AMF		35	
		pilot-scale			
		$32S_S$	—	>40	H
		pilot-scale			
		$32L_S$	—	30	M
		pilot-scale			
		$25S_S$	—	40	H
		pilot-scale			
		$25L_S$	—	25	M
		pilot-scale			

Continued

TABLE 5.1 (Continued)

Fraction separation method Author, year, location detailed data available[2]	Season[3]	Fraction designation[4]	Yield (%)	Melting behavior (°C)[5]	General fraction property[6]
		$20S_S$ pilot-scale	—	30	M
		$20L_S$ pilot-scale	—	20	L
		$17S_S$ pilot-scale	—	30	M
		$17L_S$ pilot-scale	—	20	L
Separation with the aid of an aqueous detergent solution					
Dolby, 1970a, Australia					
Technical data:	Midseason	AMF		34.1 (SP)	
Fatty acid composition					
Selected chemical characteristics		Solid	—	38.6	H
Melting behavior					
Nutritional composition		Liquid	—	24.8	L
Dolby, 1970b, Australia					
Technical data:	—	AMF [1]		—	
Melting behavior		#1	—	43.5 (SP)	H
Applications data:		#2	—	37.5	H
Cold-spreadable butter		#3	—	26.6	M
		#4	—	23.9	L
	Midseason	AMF [2]		33.8	
		#1	—	38.2	H
		#2	—	29.3	M
Fjaervoll, 1970a, Sweden					
Technical data:	—	AMF		28.2 (MP)	
Fatty acid composition		HMF	—	38.0	H
Selected chemical characteristics		LMF	—	22.1	L
Melting behavior					
Fjaervoll, 1970b, Sweden					
Technical data:	—	AMF		28.2 (MP)	
Selected chemical characteristics		HMF	—	38.0	H
Melting behavior		LMF	—	22.1	L
Jebson, 1970, New Zealand					
Technical data:	—	AMF		—	
Melting behavior		$27S_S$	—	40 (SP)	H
		$27L_S$	—	24	L
		$24S_S$	—	38	H
		$24L_S$	—	22	L
Norris et al., 1971, New Zealand					
Technical data:	Late S	AMF [A]		33.8 (SP)	
Fatty acid composition		$25S_S$	—	36.2	H
Selected chemical characteristics		$25L_S$	—	22.6	L
Melting behavior	Late S	AMF [B]		33.8	
Solid fat content		$26S_S$	—	37.5	H
Nutritional composition		$26L_S$	—	22.7	L
	Early F	AMF [C]		33.4	
		$26S_S$	—	37.6	H
		$26L_S$	—	22.9	L

Continued

TABLE 5.1 (Continued)

Fraction separation method Author, year, location detailed data available[2]	Season[3]	Fraction designation[4]	Yield (%)	Melting behavior (°C)[5]	General fraction property[6]
Sherbon et al., 1972, New Zealand Technical data: Yield Selected chemical characteristics Melting behavior Thermal profile	—	AMF [A]		34.2 (SP)	
		$A1S_S$ (single fractionation at 28°C)	47	38.0	H
		$A1L_S$ (single fractionation at 28°C)	53	25.0	M
	—	AMF [B]		35.0	
		$B1S_M$ (single fractionation at 25.5°C)	31	41.0	H
		$B1L_M$ (single fractionation at 25.5°C)	63.5	25.2	M
		$B2S_M$ (double fractionation at 28°C)	92	40.9	H
		$B2L_M$ (double fractionation at 28°C)	8	35.0	H
		$B3S_M$ (triple fractionation at 31.7°C)	94	40.6	H
		$B3L_M$ (triple fractionation at 31.7°C)	6	35.4	H
		$C1S_M$ original fraction	—	41.2	H
		$C2S_M$ (double fractionation at 30.6°C)	59	43.9	H
		$C2L_M$ (double fractionation at 30.6°C)	41	32.2	M
Walker, 1972, New Zealand Technical data: Yield Flavor Melting behavior	—	AMF [F1]		32.1 (SP)	
	—	AMF [F2]		32.9	
		$26.5S_S$ [F2S]	17	44.2	H
		$26.5L_S$ [F2L]	83	25.2	M
		$26.5S_S$ [F2S´]	18	44.5	H
		$26.5L_S$ [F2L´]	82	25.2	M

Continued

Technical Data for Experimental and Commercial Milkfat Fractions 243

TABLE 5.1 (Continued)

Fraction separation method Author, year, location detailed data available[2]	Season[3]	Fraction designation[4]	Yield (%)	Melting behavior (°C)[5]	General fraction property[6]
		AMF [F3]		33.1	
		26.5S_S [F3S]	—	45.4	VH
		26.5L_S [F3L]	—	24.3	L
Jebson, 1974a, New Zealand					
		25S_S	—	—	U
		25L_S	—	—	U
Schaap et al., 1975, The Netherlands					
Technical data:	—	AMF		—	
Yield					
Fatty acid composition		28S_S	10	54.0 (DSC)	VH
Crystal morphology					
Melting behavior		28L_S	90	—	U
Doležálek et al., 1976, Czechoslovakia					
Technical data:	—	AMF [1]		—	
Yield		30S_S	13	41.5 (MP)	H
Fatty acid composition		30L_S	87	27.0	M
Selected chemical characteristics		27S_S	24	42.0	H
Melting behavior		27L_S	76	25.4	M
Crystal size		25S_S	43	37.5	H
		25L_S	57	25.0	M
	—	AMF [2]		—	
		30S_S	16	41.5	H
		30L_S	84	28.0	M
		27S_S	26	40.0	H
		27L_S	74	24.3	L
		25S_S	50	35.5	H
		25L_S	50	25.2	M
	—	AMF [3]		—	
		30S_S	10	40.5	H
		30L_S	90	24.3	L
		27S_S	26	40.0	H
		27L_S	74	24.8	L
		25S_S	47	35.1	H
		25L_S	53	24.0	L
	—	AMF [4]		—	
		30S_S	9	43.5	H
		30L_S	91	26.5	M
		27S_S	20	41.8	H
		27L_S	80	24.5	L
		25S_S	29	40.4	H
		25L_S	71	23.0	L
Banks et al., 1985, UK					
Technical data:	—	AMF [A]		—	
Yield		Low level			
Fatty acid composition		of detergent			
Triglyceride composition		21S_S	—	—	U
Melting behavior		21L_S	—	—	U
Thermal profile		AMF [B]		—	
		16S_S	—	45.4 (DSC)	VH
		16L_S	—	—	U
		High level	8	51.6	VH
		of detergent			
		21S_S [A]			

Continued

TABLE 5.1 (Continued)

Fraction separation method Author, year, location detailed data available[2]	Season[3]	Fraction designation[4]	Yield (%)	Melting behavior (°C)[5]	General fraction property[6]
		21 semi-S_S [A]	27	53.0	VH
		21L_S [A]	65	22.3	L
		16S_S [B]	—	52.3	VH
		16 semi-S_S [B]	—	44.7	H
		16L_S [B]	—	—	U
Filtration through muslin cloth					
Sreebhashyam et al., 1981, India					
Technical data:	—	AMF		31.3 (MP)	
Yield		30S_M (IV)	25.8	37.3	H
Selected chemical characteristics		25S_M (III)	34.9	29.6	M
Melting behavior		18S_M (II)	38.7	25.8	M
Textural characteristics		18L_M (I)	10.7	14.3	L
Kulkarni and Rama Murthy, 1987, India					
Technical data:	—	AMF		—	
Selected chemical characteristics		23S_S	—	—	U
		23L_S	—	—	U
Khalifa and Mansour, 1988, Egypt					
Technical data:	—	AMF		36.0 (MP)	
Yield		30S_M	29.32	41.0	H
Selected chemical characteristics		25S_M	31.78	37.5	H
Melting behavior		20S_M	13.74	32.0	M
		20L_M	25.16	15.0	L
Filtration through milk filter					
Fouad et al., 1990, Canada					
Technical data:	—	AMF		—	
Yield		29S_S	10	40.7 (DSC)	H
Fatty acid composition		29L_S	90	25–30 (SFC)	M
Triglyceride composition		25S_S	12	39.8	H
Melting behavior		25L_S	88	—	U
Thermal profile		21S_S	43.5	—	U
Solid fat content		21L_S	57.5	—	U
		17S_S	52.3	30–35	M
		17L_S	47.7	20–25	L
Filtration through cheese press					
Baker, 1970a, Australia					
Technical data:	—	AMF		—	
Yield		13L_M	35	<13 (SP)	L
Selected chemical characteristics		13S_M	40	28	M
Melting behavior		30S_M	25	40	H
Filtration through casein-dewatering press					
Black, 1973, Australia					
Technical data:	—	AMF		—	
Yield		Hard #1	—	38.7 (SP)	H
Fatty acid composition		Hard #2	—	39.8	H
Selected chemical characteristics		Hard #3	—	42.8	H
Crystal size		Soft #1	—	21.5	L
Melting behavior		Soft #2	—	21.1	L
		Soft #3	—	21.0	L

Continued

TABLE 5.1 (Continued)

Fraction separation method Author, year, location detailed data available[2]	Season[3]	Fraction designation[4]	Yield (%)	Melting behavior (°C)[5]	General fraction property[6]
		IT = 33.5°C S_S	41.7	37.5	H
		IT = 33.5°C L_S	58.3	27.6	M
		IT = 29.5°C S_S	42.9	36.8	H
		IT = 29.5°C L_S	57.1	27.1	M
		FT = 21.0°C S_S	40	37.4	H
		FT = 21.0°C L_S	60	29.1	M
		FT = 18.5°C S_S	44.6	36.9	H
		FT = 18.5°C L_S	55.4	25.6	M
		t = 21 h S_S	39.6	37.2	H
		t = 21 h L_S	60.4	27.7	M
		t = 16 h S_S	45.0	37.0	H
		t = 16 h L_S	55.0	27.0	M
Separation technique not reported					
Riel and Paquet, 1972, Canada					
Technical data:	—	AMF		24 (CP)	
Yield					
Fatty acid composition		$38S_M$	24	32	M
Selected chemical characteristics		$4S_M$	21	25	M
Melting behavior		$4L_M$	55	11	L
Thomas, 1973a, Australia					
Applications data:	—	AMF		—	
Dairy products		$20S_S$	—	—	U
		$20L_S$	—	—	U
Deroanne and Guyot, 1974, Belgium					
Technical data:	W	AMF		33.3 (SP)	
Fatty acid composition		$21S_S$	—	36.1	H
Selected chemical characteristics		$21L_S$	—	21.4	L
Melting behavior	S	AMF		32.5	
		$21S_S$	—	38.6	H
		$21L_S$	—	18.1	L
Dixon and Black, 1974, Australia					
Technical data:	—	AMF		—	
Yield		HMF	22.0	40 (SP)	H
Melting behavior		MMF	42.0	32	M
Applications data:		LMF	36.0	20	L
Cold-spreadable butter					
Dairy-based spreads					
Timmen, 1974, Germany					
Technical data:	W	AMF		—	
Yield		$28S_S$	30–40	—	U
Fatty acid composition		$28L_S$	60–70	—	U
	S	AMF		—	
		$22S_S$	51	—	U
		$22L_S$	49	—	U
Lechat et al., 1975, France					
Technical data:	Sept	AMF		—	
Fatty acid composition		Most liquid fraction	—	—	U
Triglyceride composition					
Selected chemical characteristics	Feb	AMF		—	
		Most solid fraction	—	—	U
		Most liquid fraction	—	—	U

Continued

TABLE 5.1 (Continued)

Fraction separation method Author, year, location detailed data available[2]	Season[3]	Fraction designation[4]	Yield (%)	Melting behavior (°C)[5]	General fraction property[6]
Deroanne, 1976, Belgium					
Technical data:	W	AMF		34.3 (MP)	
Fatty acid composition				32.1 (DP)	
Selected chemical characteristics		$21S_S$	—	41.0 (MP)	H
Crystal morphology				40.8 (DP)	
Melting behavior		$21L_S$	—	27.0 (MP)	M
Thermal profile				24.8 (DP)	
El-Ghandour and El-Nimr, 1982, Egypt					
Technical data:	—	AMF		—	
Melting behavior		$25S_M$	—	>50 (SFC)	VH
Solid fat content		$15S_M$	—	>40	H
		$15L_M$	—	>20	L
Martine, 1982, France					
Technical data:	—	AMF		30.5 (MP)	
Fatty acid composition					
Melting behavior		1st stearin	—	42	H
Thermal profile		2nd stearin	—	37.6	H
Solid fat content		Olein	—	28	M
Verhagen and Bodor, 1984, The Netherlands					
Technical data:	—	$25S_S$	—	>35 (SFC)	H
Melting behavior		$25L_S$	—	—	U
Solid fat content					
Milkfat fraction blends		$15S_S$	—	>35	H
Applications data:					
Dairy-based spreads		$15L_S$	—	—	U
Verhagen and Warnaar, 1984, The Netherlands					
Technical data:	—	$25S_S$	—	—	U
Melting behavior		$25L_S$	—	<30 (SFC)	M
Solid fat content					
Applications data:					
Dairy-based spreads					

[1]Sorted by fraction separation method employed and listed in chronological order.

[2]Technical data are presented in Chapter 5, and applications data are presented in Chapter 8.

[3]F = fall, S = summer, Spr = Spring, W = winter, months are indicated using conventional abbreviations.

[4]AMF = anydrous milkfat. The number indicates fractionation temperature, the letter indicates physical state (L = liquid, S = solid), and the subscript indicates fractionation procedure used (M = multiple-step, S = single-step). Other designations are those of the original author. Designations in brackets refer to the anhydrous milkfat source or sample number. Designations in parentheses are fraction designations given by the original author.

[5]CP = cloud point, DP = dropping point, DSC = melting behavior estimated using differential scanning calorimetry curves provided by the original investigator, MP = melting point, SFC = melting behavior estimated from solid fat content data provided by the original investigators, SLP = slip point, SP = softening point.

[6]VH = very-high-melting milkfat fractions (MP > 45°C), H = high-melting milkfat fractions (MP between 35 and 45°C), M = middle-melting milkfat fractions (MP between 25 and 35°C), L = low-melting milkfat fractions (MP between 10 and 25°C), VL = very-low-melting milkfat fractions (MP < 10°C), U = unknown-melting milkfat fractions (MP not available).

[7]Data not available.

[8]Data estimated from graphs provided by the original investigators.

[9]Melting behavior taken at DSC peak temperature as reported by Barna et al. (1992).

[10]Yield and melting behavior data not reported in Kaylegian and Lindsay (1992) but were the same as in Kaylegian (1991).

TABLE 5.2 Experimental Milkfat Fractions Obtained by Crystallization from Solvent Solution

Type of solvent solution Author, year, location detailed data available[2]	Season[3]	Fraction designation[4]	Yield (%)	Melting behavior (°C)[5]	General fraction property[6]
Acetone					
Baker et al., 1959, Canada					
Technical data:	S	AMF		—[7]	
Yield					
Melting behavior		$5S_S$	21	—	U
Applications data:					
Dairy products		$5L_S$	79	20 (MP)	L
Bhalerao et al., 1959, USA					
Technical data:	—	AMF		—	
Selected chemical characteristics		$0S_S$	—	—	U
		$0L_S$	—	—	U
Sherbon, 1963, USA					
Technical data:	—	AMF		—	
Fatty acid composition		2A	—	—	U
Melting behavior		2B	—	45.0 (DSC)	VH
Thermal profile		2C	—	22.0	L
		2D	—	—	U
		2D'	—	—	U
	Jan	AMFJAN		—	
		2AJAN	—	56.0	VH
		2BJAN	—	38.9	H
		2CJAN	—	19.8	L
		2DJAN	—	6.5	VL
		2EJAN	—	—	U
		2FJAN	—	—	U
Chen and deMan, 1966, Canada					
Technical data:	—	AMF		—	
Yield		$15S_M$	9	—	U
Fatty acid composition		$5S_M$	10	—	U
Triglyceride composition		$-5S_M$	14	—	U
Nutritional composition		$-15S_M$	27	—	U
		$-25S_M$	13	—	U
		$-35S_M$	9	—	U
		$-45S_M$	5	—	U
		$-45L_M$	13	—	U
Rolland and Riel, 1966, Canada					
Technical data:	—	AMF (1)		—	
Yield		$18S_M$ (1)	3.0	42 (CP)	H
Melting behavior		$10S_M$ (1)	5.3	34	M
Selected chemical characteristics		$2S_M$ (1)	7.5	26	M
		$-6S_M$ (1)	9.8	20	L
		$-15S_M$ (1)	15.8	15	L
		$-15L_M$ (1)	58.8	5	VL
	—	AMF (2)		—	
		$0L_M$ (2)	7.0	33	M
		$8L_M$ (2)	19.0	24	L
		$16L_M$ (2)	14.0	17	L
		$24L_M$ (2)	21.0	13	L
		$32L_M$ (2)	8.0	5	VL
		$40L_M$ (2)	31.0	3	VL

Continued

TABLE 5.2 (Continued)

Type of solvent solution Author, year, location detailed data available[2]	Season[3]	Fraction designation[4]	Yield (%)	Melting behavior (°C)[5]	General fraction property[6]
Jensen et al., 1967, USA					
Technical data:	—	AMF		—	
Fatty acid composition		$-25S_M$	—	—	U
		$-70S_M$	—	—	U
		$-70L_M$	—	—	U
Woodrow and deMan, 1968, Canada					
Technical data:	—	AMF		—	
Crystal morphology		$15S_S$	—	—	U
		$15L_S$	—	—	U
Mattsson et al., 1969, Sweden					
Technical data:	Aug	AMF		—	
Fatty acid composition		$15S_S$	—	—	U
		$15L_S$	—	—	U
	Dec	AMF		—	
		$15S_S$	—	—	U
		$15L_S$	—	—	U
Sherbon and Dolby, 1973, New Zealand					
Technical data:	—	HMG		46 (MP)	
Yield		(starting			
Fatty acid composition		material)			
Melting behavior		$32S_M$ (I)	16.1	56	VH
Thermal profile		$25S_M$ (II)	33.1	54	VH
		$25S_M$ (III refractionated)	2.7	48	VH
		$25S_M$ (IV refractionated)	35.2	41	H
		$25L_M$ (V)	12.8	27	M
Avvakumov, 1974, USSR					
Technical data:	—	AMF		—	
Yield		$20S_M$	—	—	U
Selected chemical characteristics		$10S_M$	15	42.2 (MP)	H
Melting behavior		$0S_M$	24	26	M
		$-10S_M$	22	17	L
		$-20S_M$	—	—	U
		$-20L_M$	—	—	U
Parodi, 1974a, Australia					
Technical data:	May	AMF		33.7 (SP)	
Yield					
Fatty acid composition		$20S_S$	4.7	—	U
Triglyceride composition		$20L_S$	95.3	—	U
Lechat et al., 1975, France					
Technical data:	Sept	AMF		—	
Yield		$-9L_M$	57	—	U
Fatty acid composition		$-9S_M$	43	—	U
Triglyceride composition		$0L_M$	16	—	U
		$0S_M$	27	—	U
		$10L_M$	6.5	—	U
		$10S_M$	20.5	—	U
	April	AMF		—	
		$-9L_M$	43	—	U
		$-9S_M$	57	—	U

Continued

TABLE 5.2 (Continued)

Type of solvent solution Author, year, location detailed data available[2]	Season[3]	Fraction designation[4]	Yield (%)	Melting behavior (°C)[5]	General fraction property[6]
		$0L_M$	26	—	U
		$0S_M$	31	—	U
		$10L_M$	13	—	U
		$10S_M$	18	—	U
Schaap et al., 1975, The Netherlands					
Technical data:	—	AMF		—	
Fatty acid composition					
Melting behavior		28SS	—	58.6 (DSC)	VH
Crystal morphology		28LS	—	—	U
Norris, 1977, New Zealand					
Technical data:	Spr	AMF		32.0 (SP)	
Yield		$22S_M$	1.4	—	U
Applications data:		$12S_M$	8.3	—	U
Cold-spreadable butter		$-14S_M$	46.3	—	U
		$-14L_M$	43.9	—	U
	S	AMF		—	
		$22S_M$	6.8	—	U
		$11.5S_M$	10.9	—	U
		$-15S_M$	42.9	—	U
		$-15L_M$	39.4	—	U
Walker et al., 1977, New Zealand					
Technical data:	—	AMF [A]		—	
Yield		$14S_M$ (HMG)	10	—	U
Selected chemical characteristics		$2S_M$ (IMG$_1$)	26	—	U
Flavor		$-13S_M$ (IMG$_2$)	30	—	U
Applications data:		$-13L_M$ (LMG)	34	—	U
Cold-spreadable butter		AMF [B]		—|	
		$14S_M$ (HMG)	9	—	U
		$2S_M$ (IMG$_1$)	24	—	U
		$-13S_M$ (IMG$_2$)	30	—	U
		$-13L_M$ (LMG)	37	—	U
		AMF [C] (1)		—	
		$14S_M$ (HMG)	11	—	U
		$2S_M +$ $-13S_M$ (IMG$_{1+2}$)	62	—	U
		$-13L_M$ (LMG)	27	—	U
		AMF [C] (3)			
		$14S_M$ [HMG]	10	—	U
		$2S_M +$ $-13S_M$ (IMG$_{1+2}$)	60	—	
		$-13L_M$ (LMG)	30	—	U
Youssef et al., 1977, Egypt					
Technical data:	Nov	AMF		36 (MP)	
Yield		$18S_M$ (A)	11.5	51	VH
Selected chemical characteristics		$18L_M$ (B)	88.5	21	L
Melting behavior		$0S_M$ (C)	24.6	27	M
		$0L_M$ (D)	63.9	8	VL
		$18S_M$ (E)	14.2	46	VH
		$18L_M$ (F)	10.5	21	L
Munro et al., 1978, New Zealand					
Technical data:	—	AMF		—	
Yield		$10S_M$	12	—	U

Continued

TABLE 5.2 (Continued)

Type of solvent solution Author, year, location detailed data available[2]	Season[3]	Fraction designation[4]	Yield (%)	Melting behavior (°C)[5]	General fraction property[6]
Applications data:		-15S$_M$	52	—	U
Cold-spreadable butter		-15L$_M$	36	—	U
Walker et al., 1978, New Zealand					
Applications data:	—	AMF		—	
Cold-spreadable butter		14S$_M$	—	—	U
		2S$_M$	—	—	U
		-13S$_M$	—	—	U
		-13L$_M$			U
Larsen and Samuelsson, 1979, Denmark					
Technical data:	—	AMF		—	
Fatty acid composition		21.4S$_S$	—	—	U
		15.0S$_S$	—	—	U
		9.6S$_S$	—	—	U
		5.0S$_S$	—	—	U
		-5.0L$_S$	—	—	U
		-8.0L$_S$	—	—	U
		-16.4L$_S$	—	—	U
		-21.0L$_S$	—	—	U
Timms, 1980a, Australia					
Technical data:	—	AMF		40.0 (DSC)	
Yield					
Crystal morphology		HMF	6	60.3	VH
Melting behavior		MMF	24	40.8	H
Thermal profile		LMF	70	14.4	L
Timms, 1980b, Australia					
Technical data:	S	AMF		—	
Yield					
Fatty acid composition		20S$_M$	6	>50 (SFC)	VH
Triglyceride composition					
Melting behavior		3S$_M$	22	35–40	H
Solid fat content					
Milkfat fraction blends		3L$_M$	72	>15	L
Lambelet, 1983, Switzerland					
Technical data:	—	AMF		—	
Triglyceride composition		S1$_M$	—	40 (SFC)	H
Solid fat content		L1$_M$	—	>15	L
Lansbergen and Kemps, 1984, The Netherlands					
Technical data:	Hardened	AMF		—	
Fatty acid composition	butterfat				
Triglyceride composition		12S$_M$ [1]	—	—	U
Applications data:		12L$_M$ [1]	—	—	U
Dairy-based spreads		6S$_M$ [1]	—	—	U
		6L$_M$ [1]	—	—	U
		12S$_M$ [2]	—	—	U
		12L$_M$ [2]	—	—	U
		0S$_M$ [2]	—	—	U
		0L$_M$ [2]	—	—	U
	Natural butterfat	AMF		—	
		Top fraction	—	—	U
		Mid fraction	—	—	U
		Olein fraction	—	—	U

Continued

TABLE 5.2 (Continued)

Type of solvent solution Author, year, location detailed data available[2]	Season[3]	Fraction designation[4]	Yield (%)	Melting behavior (°C)[5]	General fraction property[6]
Kaylegian, 1991, USA					
Technical data:	W	AMF		34.4 (MP)	
Yield		$25S_M$	1.7	53.7	VH
Fatty acid composition		$21S_M$	3.0	50.7	VH
Triglyceride composition		$15S_M$	4.5	46.2	VH
Melting behavior		$11S_M$	2.7	41.5	H
Thermal profile		$5S_M$	14.6	31.9	M
Solid fat content		$0S_M$	11.1	26.1	M
		$0L_M$	62.5	9.8	VL
		$25S_S$	2.2	53.0	VH
		$25L_S$	97.8	31.9	M
		$21S_S$	4.9	51.1	VH
		$21L_S$	95.1	27.6	M
		$15S_S$	11.1	48.5	VH
		$15L_S$	88.9	23.1	L
		$11S_S$	19.3	47.0	VH
		$11L_S$	80.7	19.8	L
		$5S_S$	30.7	43.7	H
		$5L_S$	69.3	13.3	L
		$0S_S$	47.9	40.6	H
		$0L_S$	52.1	6.8	VL
Kaylegian and Lindsay, 1992, USA					
Technical data:	W	AMF		34.4 (MP)[8]	
Yield		$25S_M$	1.7[9]	53.7	VH
Melting behavior		$21S_M$	3.0	50.7	VH
Solid fat content		$15S_M$	4.5	46.2	VH
Applications data:		$11S_M$	2.7	41.5	H
Cold-spreadable butter		$5S_M$	14.6	31.9	M
		$0S_M$	11.1	26.1	M
		$0L_M$	62.5	9.8	VL
Yi, 1993, USA					
Technical data:	S	AMF		33.7 (MP)	
Fatty acid composition		$25S_M$	—	51.5	VH
Triglyceride composition		$20S_M$	—	50.4	VH
Melting behavior		$15S_M$	—	45.4	VH
Thermal profile		$10S_M$	—	41.0	H
Solid fat content		$5S_M$	—	30.2	M
Milkfat fraction blends		$0S_M$	—	26.7	M
Applications data: Chocolate		$0L_M$	—	11.0	L
Ethanol					
Bhalerao et al., 1959, USA					
Technical data:	—	AMF	—	—	
Selected chemical characteristics		$20S_S$	—	—	U
		$20L_S$	—	—	U
Rolland and Riel, 1966, Canada					
Technical data:	—	AMF			
Yield		$18S_M$	12.7	35	H
Selected chemical characteristics		$10S_M$	10.7	20	L
Melting behavior		$2S_M$	16.3	16	L
		$-6S_M$	11.0	12	L
		$-15S_M$	9.3	8	VL
		$-15L_M$	40.0	0	VL

Continued

TABLE 5.2 (Continued)

Type of solvent solution Author, year, location detailed data available[2]	Season[3]	Fraction designation[4]	Yield (%)	Melting behavior (°C)[5]	General fraction property[6]
Hexane					
Muuse and van der Kamp, 1985, The Netherlands					
Technical data:	—	AMF		—	
Fatty acid composition					
Trigylceride composition		$12.5S_S$	—	—	U
Nutritional composition		$12.5L_S$	—	—	U
Isopropanol					
Norris, 1977, New Zealand					
Technical data:	—	AMF		—	
Yield		$20S_M$	19.0	>20 (DSC)[9]	U
Selected chemical characteristics					
Melting behavior		$-5S_M$	46.6	0-20	L
Applications data:					
Cold-spreadable butter		$-5L_M$	34.4	<0	VL
Pentane					
Kaylegian, 1991, USA					
Technical data:	W	AMF		34.4 (MP)	
Yield		$11S_M$	0.9	52.2	VH
Fatty acid composition		$5S_M$	8.8	49.0	VH
Triglyceride composition		$0S_M$	3.5	45.3	VH
Melting behavior		$0L_M$	86.8	21.6	L
Thermal profile		$11S_S$	2.6	50.6	VH
Solid fat content		$11L_S$	97.4	32.2	M
		$5S_S$	9.8	48.9	VH
		$5L_S$	90.2	25.3	M
		$0S_S$	12.7	47.8	VH
		$0L_S$	87.3	21.3	L

[1] Sorted by type of solvent solution employed and listed in chronological order.

[2]Technical data are presented in Chapter 5; applications data are presented in Chapter 8.

[3]W = winter, S = summer.

[4]AMF = anydrous milkfat. The sample number indicates fractionation temperature; the letter indicates physical state (L = liquid, S = solid); and the subscript indicates fractionation procedure used (M = multiple-step, S = single-step). Other primary designations and designations in parentheses are those of the original author.

[5]CP = cloud point, DSC = melting behavior estimated using differential scanning calorimetry curves provided by the original investigators, MP = melting point, SFC = melting behavior estimated using solid fat content data provided by the original investigators, SP = softening point.

[6]VH = very-high-melting milkfat fractions (MP > 45°C), H = high-melting milkfat fractions (MP between 35 and 45°C), M = middle-melting milkfat fractions (MP between 25 and 35°C), L = low-melting milkfat fractions (MP between 10 and 25°C), VL = very-low-melting milkfat fractions (MP < 10°C), U = unknown-melting milkfat fractions (MP not available).

[7]Data not available.

[8]Yield and melting behavior data not reported in Kaylegian and Lindsay (1992) but were the same as in Kaylegian (1991).

[9]Melting behavior data given as ranges determined by original author based on DSC curves (not reported). The ranges for the $20S_M$ and $-5S_M$ fractions were too broad to assign a definitive general fraction category.

TABLE 5.3 Experimental Milkfat Fractions Obtained by Supercritical Fluid Extraction[1]

Type of solvent solution Author, year, location detailed data available[2]	Fraction designation[3]	Fractionation conditions		Yield (%)	Melting behavior (°C)[4]	General fraction property[5]
		Temperature (°C)	Pressure			
Carbon dioxide						
Kaufmann et al., 1982, Germany						
Technical data:	AMF				34.3 (DSC)	
Fatty acid composition						
Triglyceride composition	I (residue)	80	200 bar	—[6]	36.0	H
Selected chemical characteristics						
Melting behavior						
Thermal profile						
Solid fat content	II (extract)	80	200	—	20.4	L
Oxidative stability						
Nutritional composition						
Timmen et al., 1984, Germany						
Technical data:	AMF				35.6 (DSC)	
Yield						
Fatty acid composition						
Triglyceride composition	Long-chain	80	200 bar	80	36.2	H
Selected chemical characteristics	fraction					
Melting behavior						
Thermal profile						
Solid fat content	Short-chain	80	200	20	20.4	L
Oxidative stability	fraction					
Nutritional composition						
Biernoth and Merk, 1985, Germany						
Technical data:	AMF				—	
Yield						
Fatty acid composition	Extract	80	200 bar	19.4	15 (SFC)	L
Triglyceride composition						
Selected chemical characteristics						
Melting behavior	Residual	80	200	80.6	>35	H
Solid fat content						
Shishikura et al., 1986, Japan						
Technical data:	AMF				—	
Fatty acid composition	F'1	40	300 kg/cm²	—	—	U
	F'2	40	300	—	—	U
	F'3	40	300	—	—	U
	F'4	40	300	—	—	U
	F'5	40	300	—	—	U
	F'6	40	300	—	—	U
	Residue	40	300	—	—	U
Arul et al., 1987, Canada						
Technical data:	AMF				35.0 (DSC)	
Yield	L1	50	100–250 bar	4.1	15.0	L
Fatty acid composition	L2	50	100–250	8.1	20.5	L
Triglyceride composition	I1	50	100–250	9.4	21.5	L
Melting behavior	I2	50	100–250	13.1	26.0	M
Thermal profile	I3	50	100–250	15.7	30.0	M
	S1	70	250–350	10.7	34.0	M
	S2	70	250–350	17.3	39.3	H
	S3	70	250–350	21.5	43.3	H

Continued

TABLE 5.3 (Continued)

Type of solvent solution Author, year, location detailed data available[2]	Fraction designation[3]	Fractionation conditions		Yield (%)	Melting behavior (°C)[4]	General fraction property[5]
		Temperature (°C)	Pressure			
Kankare and Antila, 1988a, Finland					38.3 (DSC)	
Technical data:	AMF					
Fatty acid composition	E1	50–80	100–400 bar	—	17.6	L
Melting behavior	E2	50–80	100–400	—	27.9	M
Thermal profile	E3	50–80	100–400	—	31.5	M
	E4	50–80	100–400	—	32.1	M
	Res	50–80	100–400	—	39.0	H
Bradley, 1989, USA						
Technical data:	AMF				—	
Yield	WM-14					
Fatty acid composition	WM-14, 1	80	2300 psig	14.6	—	U
Nutritional composition	WM-14, 2	80	2500	2.2	—	U
	WM-14, 3	80	2800	5.9	—	U
	WM-14, 4	80	3050	8.6	—	U
	WM-14, 5	80	3350	17.1	—	U
	WM-14, 6	80	3750	16.8	—	U
	WM-14, 7	80	6000	34.7	—	U
	AMF				—	
	WM-21					
	WM-21, 1	80	2300	11.1	—	U
	WM-21, 2	80	2500	8.7	—	U
	WM-21, 3	80	2800	14.5	—	U
	WM-21, 4	80	3050	11.6	—	U
	WM-21, 5	80	3400	9.2	—	U
	WM-21, 6	80	6000	45.1	—	U
	AMF				—	
	WM-38					
	WM-38, 1	60	2150	—	—	U
	WM-38, 2	60	2300	—	—	U
	WM-38, 3	60	2400	—	—	U
	WM-38, 4	60	2400	—	—	U
	WM-38, 5	60	2500	—	—	U
	WM-38, 6	60	2500	—	—	U
	WM-38, 7	60	2650	—	—	U
	WM-38, 8	60	2850	—	—	U
	WM-38, 9	60	2900	—	—	U
	WM-38, 10	60	2900	—	—	U
	WM-38, 11	60	2900	—	—	U
	WM-38, 12	60	2900	—	—	U
	WM-38, 13	60	2900	—	—	U
	WM-38, 14	60	2900	—	—	U
	WM-38, 15	60	—	—	—	U
Büning-Pfaue et al., 1989, Germany						
Technical data:	Parent					
Fatty acid composition	milkfat				42.3 (DSC)	
Triglyceride composition						
Melting behavior	Extract	50	25,000 kPa	—	38.5	H
Thermal profile	Residue	30	20,000	—	44.9	H
Ensign, 1989, USA						
Technical data:	AMF				—	
Fatty acid composition	WM-38					
Triglyceride composition	WM-38, 1	60	2150 psi	—	—	U
Nutritional composition	WM-38, 2	60	2300	—	—	U

Continued

TABLE 5.3 (Continued)

Type of solvent solution Author, year, location detailed data available[2]	Fraction designation[3]	Fractionation conditions		Yield (%)	Melting behavior (°C)[4]	General fraction property[5]
		Temperature (°C)	Pressure			
	WM-38, 3	60	2400	—	—	U
	WM-38, 4	60	2400	—	—	U
	WM-38, 5	60	2500	—	—	U
	WM-38, 6	60	2500	—	—	U
	WM-38, 7	60	2650	—	—	U
	WM-38, 8	60	2850	—	—	U
	WM-38, 9	60	2900	—	—	U
	WM-38, 10	60	2900	—	—	U
	WM-38, 11	60	2900	—	—	U
	WM-38, 12	60	2900	—	—	U
	WM-38, 13	60	2900	—	—	U
	WM-38, 14	60	2900	—	—	U
	WM-38, 15	60	—	—	—	U
	AMF WM-39				—	
	WM-39, 1	60	2000	—	—	U
	WM-39, 2	60	6000	—	—	U
	AMF WM-43				—	
	WM-43, 1	40	1700	—	—	U
	WM-43, 2	40	6000	—	—	U
Kankare et al., 1989, Finland						
Technical data:	AMF				39.2 (DSC)	
Yield	E1	50	100 bar	0.2	17.7	L
Fatty acid composition	E2	50	200	3.3	27.9	M
Triglyceride composition	E3	50	300	21.4	31.4	M
Flavor	E4	50	400	17.5	32.1	M
Melting behavior	R	50	400	57.7	39.8	H
Thermal profile	AMF2				—	
Nutritional composition	2E1	48	120 bar (3 h 12 min) + 150 bar (4 h)	4.2	—	U
	2E2	48	200	9.5	—	U
	2R	48	200	86.2	—	U
Rizvi et al., 1990, USA						
Technical data: Yield	AMF				36.1 (MP)	
Triglyceride composition	Extract-1	40	241.3 bar	42.4	32.8	M
Selected chemical characteristics						
Melting behavior	Extract-2	48	275.8	28.7	37.1	H
Thermal profile						
Nutritional composition	Residue	—	—	28.9	40.0	H
Chen and Schwartz, 1991, USA						
Technical data:	AMF				—	
Fatty acid composition						
Nutritional composition	unknown	35, 40, 50, 60	1500, 2000, 2500, 3000, 3500, 4000 psi	—	—	U
Hamman et al., 1991, Sweden						
Technical data:	AMF [A]				35 (DSC)	
Fatty acid composition	A1	40	125 bar	—	17	L
Crystal morphology	A4	40	125	—	25	M
Melting behavior	A_{res}	40	125	—	—	U

Continued

TABLE 5.3 (Continued)

Type of solvent solution Author, year, location detailed data available[2]	Fraction designation[3]	Fractionation conditions		Yield (%)	Melting behavior (°C)[4]	General fraction property[5]
		Temperature (°C)	Pressure			
	AMF [B]				—	
	B1	40	350	—	—	U
	B$_{res}$	40	350	—	48	VH
Chen et al., 1992, USA						
Technical data:	AMF				42.8 (DSC)	
Yield	10.3	40	10.3 MPa	7.3	18.2	L
Fatty acid composition	13.8	40	13.8	21.2	39.8	H
Melting behavior	17.2	40	17.2	24.8	34.0	M
Thermal profile	20.7	40	20.7	32.8	39.8	H
Nutritional composition	24.1	40	24.1	21.4	37.9	H
	27.6	40	27.6	1.5	39.8	H
Bhaskar et al., 1993a, USA						
Technical data:	AMF				—	
Triglyceride composition	Extract	40	24.1 MPa	—	—	U
	Raffinate	40	24.1 MPa	—	—	U
Bhaskar et al., 1993b, USA						
Technical data:	AMF				40 (DSC)[8]	
Yield						
Fatty acid composition						
Triglyceride composition	S1 (raffinate)	40	24.1 MPa	21.0	45	VH
Selected chemical characteristics	S2	60	24.1	15.0	38	H
Melting behavior	S3	75	17.2	48.0	32	M
Thermal profile	S4	60	6.9	4.0	31	M
Solid fat content	S5	60	3.5	11.0	30	M
Nutritional composition						
Rizvi et al., 1993, USA						
Technical data:	AMF				—	
Yield	S1	60	17.2 MPa	14	—	U
Flavor	S2	60	13.8	47	—	U
	S3	40	6.9	4	—	U
	S4	40	3.4	2	—	U
	Raffinate	40	24.1	20	—	U
	Trap	−10	0.14	1	—	U
Shukla et al., 1994, USA						
Technical data:	AMF				—	
Yield						
Fatty acid composition						
Triglyceride composition	Extract	40	24.1 MPa	78	—	U
Melting behavior						
Thermal profile						
Solid fat content						
Nutritional composition	Raffinate	40	24.1 MPa	21	37 (DSC)	H
Applications data:						
Dairy-based spreads						

Propane

Biernoth and Merk, 1985, Germany						
Technical data:	AMF				—	
Yield	Extract 1	125	85 bar	13.3	—	U
Triglyceride composition	Extract 2	125	85	20.0	—	U
	Residual	125	85	66.7	—	U

Continued

[1] Sorted by type of solvent employed and listed in chronological order.

[2] Technical data are presented in Chapter 5, and applications data are presented in Chapter 8.

[3] AMF = anhydrous milkfat. Other designations used are those given by the original author.

[4] DSC = melting behavior estimated using differential scanning calorimetry curves provided by the original investigators, MP = melting point, SFC = melting behavior estimated using solid fat content data provided by the original investigators.

[5] VH = very-high-melting milkfat fractions (MP > 45°C), H = high-melting milkfat fractions (MP between 35 and 45°C), M = middle-melting milkfat fractions (MP between 25 and 35°C), L = low-melting milkfat fractions (MP between 10 and 25°C), VL = very-low-melting milkfat fractions (MP < 10°C), U = unknown-melting milkfat fractions (MP not available).

[6] Data not available.

[7] Melting behavior estimated from solid fat content data provided by the original authors.

[8] Melting behavior determined by the original investigators using DSC melting thermograms. Melting point was taken at the temperature which all the fat was in liquid state.

TABLE 5.4 Experimental Milkfat Fractions Obtained by Short-Path Distillation

Author, year, location detailed data available[1]	Fraction designation[2]	Fractionation conditions		Yield (%)	Melting behavior (°C)[3]	General fraction property[4]
		Temperature (°C)	Pressure			
McCarthy et al., 1962, Canada					—[5]	
Technical data:	AMF					
Yield	R-1	150	15μm	2.5	—	U
Fatty acid composition	R-2	160	15μm	2.5	—	U
	R-3	180	12μm	2.5	—	U
Triglyceride composition	R-4	185	12μm	2.5	—	U
	D-2	177–182	5–7μm	39	—	U
	D-3	—	—	57	—	U
Arul et al., 1988a, Canada						
Technical data:	AMF				35.7 (DSC)	
Yield	LF1	245	220	2.1	16.8	L
Fatty acid composition	Distillate I		μm Hg			
Triglyceride composition						
Melting behavior						
Thermal profile	LF2	245	100	9.5	20.0	L
Solid fat content	Distillate II		μm Hg			
Nutrition composition						
	IF	265	100	32.0	26.7	M
	Distillate III		μm Hg			
	SF			56.4	42.9	H
	Residue					

[1]Technical data are presented in Chapter 5, and applications data are presented in Chapter 8.

[2]AMF = anhydrous milkfat. Other designations used are those given by the original author.

[3]DSC = melting behavior estimated using differential scanning calorimetry curves provided by the original investigator.

[4]H = high-melting milkfat fractions (MP between 35 and 45°C), M = middle-melting milkfat fractions (MP between 25 and 35°C), L = low-melting milkfat fractions (MP bewteen 10 and 25°C).

[5]Data not reported.

TABLE 5.5 Experimental Milkfat Fractions Obtained by Unspecified Fractionation Methods[1]

Author, year, location detailed data available[2]	Fraction designation[3]	Yield (%)	Melting behavior (°C)[4]	General fraction property[5]
Baker, 1970b, Australia				
Applications data:	30S$_M$	—[6]	—	U
Dairy products	13S$_M$	—	—	U
Deep-fat frying	13L$_M$	—	—	U
Jebson, 1974b, New Zealand				
Applications data:	Hard fraction	—	45.5 (SP)	VH
Chocolate	Soft fraction	—	19	L
Kehagias and Radema, 1973, The Netherlands				
Technical data:	22S$_S$	—	—	U
Oxidative stability	22L$_S$	—	—	U
Tucker, 1974, Australia				
Applications data:				
Bakery	Hard fraction	—	36–39 (MP)	H
Dairy products	Soft fraction	—	18–19	L
Pedersen, 1988, Denmark				
Applications data:				
Bakery	Stearin fraction	—	36 (SP)	H
Mayhill and Newstead, 1992, New Zealand				
Applications data:	Hard fraction	—	46.1 (DP)	VH
Dairy products	Soft fraction	—	22.8	L
Walker, 1974, New Zealand				
Technical data:	High-melting fraction	20	33 (SP)	M
Flavor	Low-melting fraction	80	24–27	M

[1] Listed in chronological order.

[2] Technical data presented in Chapter 5 and applications data presented in Chapter 8.

[3] The number indicates fractionation temperature, the letter indicates physical state (L = liquid, S = solid), and the subscript indicates fractionation procedure used (M = multiple-step, S = single-step). Other designations are that of the original author.

[4] DP = dropping point, MP = melting point, SP = softening point.

[5] VH = very-high-melting milkfat fractions (MP > 45°C), H = high-melting milkfat fractions (MP between 35 and 45°C), M = middle-melting milkfat fractions (MP between 25 and 35°C), L = low-melting milkfat fractions (MP between 10 and 25°C), U = unknown-melting milkfat fractions (MP not reported).

[6] Data not available.

TABLE 5.6 Overall Summary of Experimental Milkfat Fractions Produced by Various Methods[1]

General method of fraction production and separation	Number of milkfat fractions produced							Total Fractions
	Very-high-melting[2]	High-melting	Middle-melting	Low-melting	Very-low-melting	Unknown-melting	
Melted milkfat							
Vacuum filtration	10	52	30	61	6	67	226
Pressure filtration	4	1	3	5	0	42	55
Centrifugation	4	28	13	29	1	31	107
Separation with the aid of an aqueous detergent solution	6	34	13	17	0	7	77
Filtration through muslin cloth	0	3	3	2	0	2	10
Filtration through milk filter	0	2	2	1	0	3	8
Filtration through cheese press	0	1	1	1	0	0	3
Filtration through casein-dewatering press	0	9	6	3	0	2	20
Filtration technique not reported	1	9	6	6	0	10	31
Total melted milkfat	25	139	77	125	7	164	537
Solvent solution							
Acetone	23	12	14	18	8	91	166
Ethanol	0	1	0	3	2	2	8
Hexane	0	0	0	0	0	2	2
Isopropanol	0	0	0	1	1	1	3
Pentane	6	0	2	2	0	0	10
Total solvent solution	29	13	16	24	11	96	189
Supercritical fluid extraction							
Carbon dioxide	2	17	15	10	0	68	112
Propane	0	0	0	0	0	3	3
Total supercritical fluid extraction	2	17	15	10	0	71	115

Short-path distillation	0	1	1	2	0	6	10
Fractionation method not reported	2	2	2	3	0	5	14
Total Fractions	58	172	111	164	18	342	865

[1]See Tables 5.1 through 5.5 for summary of original literature reports.

[2]Very-high-melting = MP > 45°C, high-melting = MP between 35 and 45°C, middle-melting = MP between 25 and 35°C, low-melting = MP between 10 and 25°C, very-low-melting = MP < 10°C, unknown melting = melting behavior not reported.

TABLE 5.7 Selected Melting Behavior of Anhydrous Milkfats Used to Produce Milkfat Fractions

Location	Season of production	Melting behavior[1]	Author
Canada	Winter	34.6 (MP)	Amer et al., 1985
		30 (MP)	Baker et al., 1959
	Summer	33.4 (MP)	Amer et al., 1985
	Unspecified	37.8 (DSC)	Makhlouf et al., 1987
		35 (DSC)	Arul et al., 1987
		33.9 (DP)	deMan and Finoro, 1980
		32.9 (SP)	deMan, 1968
		24 (CP)	Riel and Paquet, 1972
Belgium	Winter	34.3 (MP)	Deroanne, 1976
		33.9 (MP)	Deffense, 1989
		33.3 (SP)	Deroanne and Guyot, 1974
	Summer	32.5 (MP)	Deffense, 1989
		32.5 (SP)	Deroanne and Guyot, 1974
	Unspecified	33.6 (DP)	Deffense, 1987
		32.0 (SP)	Deffense, 1987
Australia	September	32 (SP)	Versteeg, 1991
	May	33.7 (SP)	Parodi, 1974a
	Summer	34.5 (SP)	Timms and Parekh, 1980
	Mid-season	34.1 (SP)	Dolby, 1970a
		33.8 (SP)	Dolby, 1970b
	Unspecified	40.0 (DSC)	Timms, 1980a
		33.8 (SP)	Dolby, 1970b

[1]CP = cloud point, DP = drop point, DSC = melting point determined by the authors using DSC curves provided by the original investigators, MP = melting point, SP = softening point.

TABLE 5.8 Melting Behavior of Intact Anhydrous Milkfats Used to Produce Milkfat Fractions from Melted Milkfat

Fraction separation method Author	Fraction designation[1]	Melting behavior (°C)[2]
Vacuum filtration		
Amer et al., 1985	AMF [W]	34.6 (MP)
	AMF [S]	33.4
Barna et al., 1992	AMF [W]	33 (DSC)[3]
Deffense, 1987	AMF [1]	33.6 (DP)
	AMF [2]	32.0
Deffense, 1989	AMF [W]	33.9 (MP)
	AMF [S]	32.5
Deffense, 1993b	AMF	32 (DP)
deMan and Finoro, 1980	AMF	33.9 (DP)
Grall and Hartel, 1992	AMF [S]	33.1 (MP)
Kaylegian, 1991	AMF [W]	34.4 (MP)
Kaylegian and Lindsay, 1992	AMF [W]	34.4 (MP)[4]
Laakso et al., 1992	AMF	34 (DP)
Lund and Danmark, 1987	AMF [1]	33.0 (DP)
	AMF [2]	33.0
Makhlouf et al., 1987	AMF	37.8 (DSC)
Schaap and van Beresteyn, 1970	AMF	34.6 (DSC)
Sherbon, 1963	AMF	34.1 (DSC)
Timms and Parekh, 1980	AMF [S]	34.5 (SP)
Versteeg, 1991	AMF [Sept]	32 (SP)
Vovan and Riel, 1973	AMF	20.0 (CP)
Pressure filtration		
Deffense, 1993b	AMF	32 (DP)
deMan, 1968	AMF	32.9 (SP)
Stepanenko and Tverdokhleb, 1974	AMF	31.2 (MP)
Centrifugation		
Antila and Antila, 1970	AMF	45.1 (DSC)
Baker et al., 1959	AMF [W]	30 (MP)
Evans, 1976	AMF [S 1]	34.4 (SP)
	AMF [S 2]	34.3
	AMF [S 3]	34.3
	AMF [S 4]	34.3
Jordan, 1986	AMF [S]	32.4 (SLP)
	AMF [W]	33.0
Kankare and Antila, 1986	AMF [S]	37.6 (DSC)
Kankare and Antila, 1988b	AMF [S]	38.0 (MP)
Keogh and Higgins, 1986b	AMF	31.4 (SLP)
Lakshminarayana and Rama Murthy, 1985	AMF	34.2 (MP)
Ramesh and Bindal, 1987b	Ghee	35.6 (MP) 33.7 (SP)

Continued

TABLE 5.8 (Continued)

Fraction separation method Author	Fraction designation[1]	Melting behavior (°C)[2]
Richardson, 1968	AMF [W]	33.5 (SLP)
Riel and Paquet, 1972	AMF	24 (CP)
Separation with the aid of an aqueous detergent solution		
Dolby, 1970a	AMF	34.1 (SP)
Dolby, 1970b	AMF [2]	33.8
Fjaervoll, 1970a	AMF	28.2 (MP)
Fjaervoll, 1970b	AMF	28.2 (MP)
Norris et al., 1971	AMF [A]	33.8 (SP)
	AMF [B]	33.8
	AMF [C]	33.4
Sherbon et al., 1972	AMF [A]	34.2 (SP)
	AMF [B]	35.0
	C1S$_M$ original fraction	41.2
Walker, 1972	AMF [F1]	32.1 (SP)
	AMF [F2]	32.9
	AMF [F3]	33.1
Filtration through muslin cloth		
Khalifa and Mansour, 1988	AMF	36.0 (MP)
Sreebhashyam et al., 1981	AMF	31.3 (MP)
Separation technique not reported		
Martine, 1982	AMF	30.5 (MP)
Deroanne, 1976	AMF	34.3 (MP) 32.1 (DP)
Deroanne and Guyot, 1974	AMF [W]	33.3 (SP)
	AMF [S]	32.5
Riel and Paquet, 1972	AMF	24 (CP)

[1] AMF = anydrous milkfat. Designations in brackets refer to the intact milkfat source [sample number or season (S = summer, W = winter)].

[2] CP = cloud point, DP = drop point, DSC = melting point determined by the authors of the monograph using differential scanning calorimetry curves provided by the original investigator, MP = melting point, SLP = slip point, SP = softening point.

[3] Melting behavior taken at DSC peak temperature as reported by Barna et al. (1992).

[4] Yield and melting behavior data not reported in Kaylegian and Lindsay (1992) but was the same as Kaylegian (1991).

TABLE 5.9 Melting Behavior of the Intact Anhydrous Milkfats Used to Produce Milkfat Fractions from Solvent Solution

Type of solvent solution Author	Fraction designation[1]	Melting behavior (°C)[2]
Acetone		
Kaylegian, 1991	AMF [W]	34.4 (MP)
Kaylegian and Lindsay, 1992	AMF [W]	34.4 (MP)
Norris, 1977	AMF	32.0 (SP)
Parodi, 1974	AMF [May]	33.7 (SP)
Sherbon and Dolby, 1973	AMF	46 (DP)
Timms, 1980a	AMF	40.0 (DSC)
Yi, 1993	AMF [S]	33.7 (MP)
Youssef et al., 1977	AMF [Nov]	36 (MP)
Pentane		
Kaylegian, 1991	AMF [W]	34.4 (MP)

[1]AMF = anhydrous milkfat. The designation inside the brackets following the fraction designation refers to the milkfat source [month or season (W = winter, S = summer)].

[2]DP = dropping point, DSC = melting point determined by the authors of the monograph using differential scanning calorimetry curves provided by the original investigator, MP = melting point, SP = softening point.

TABLE 5.10 Melting Behavior of the Intact Anhydrous Milkfats Used to Produce Milkfat Fractions by Supercritical Fluid Extraction

Type of solvent Author	Fraction designation[1]	Melting behavior (°C)[2]
Carbon dioxide		
Arul et al., 1987	AMF	35.0 (DSC)
Bhaskar et al., 1993b	AMF	40 (DSC)[3]
Büning-Pfaue et al., 1989	AMF	42.3 (DSC)
Chen et al., 1992	AMF	42.8 (DSC)
Hamman et al., 1991	AMF [A]	35 (DSC)
Kankare and Antila, 1988a	AMF	38.3 (DSC)
Kankare et al., 1989	AMF	39.2 (DSC)
Kaufmann et al., 1982	AMF	34.3 (DSC)
Rizvi et al., 1990	AMF	36.1 (MP)
Timmen et al., 1984	AMF	35.6 (DSC)

[1]AMF = anydrous milkfat. Designations in brackets are those given by the original author.

[2]DSC = melting point has been determined by the authors of the monograph using differential scanning calorimetry curves provided by the original investigators, MP = melting point.

[3]Melting behavior determined by the original investigators using DSC melting thermograms. Melting point taken at the temperature which all the fat is in the liquid state.

TABLE 5.11 Melting Point of the Intact Anhydrous Milkfat Used to Produce Milkfat Fractions by Short-Path Distillation

Author	Fraction designation[1]	Melting behavior (°C)[2]
Arul et al., 1988a	AMF	35.7 (DSC)

[1] AMF = anhydrous milkfat.

[2] DSC = Melting point has been determined by the author of the monograph using differential scanning calorimetry curves provided by the original investigator.

TABLE 5.12 Yield and Melting Behavior of Very-High-Melting Milkfat Fractions Obtained from Melted Milkfat

Fraction separation method Author	Fraction designation[1]	Yield (%)	Melting behavior (°C)[2]
Vacuum filtration			
Kaylegian, 1991	$34S_M$ [W]	3.7	49.0 (MP)
	$34S_S$ [W]	2.5	46.6
Kaylegian and Lindsay, 1992	$34S_M$ [W]	3.7	49.0 (MP)
	$34S_S$ [W]	2.5	46.6
Schaap and van Beresteyn, 1970	$27S_M$	—[3]	49.2 (DSC)
Makhlouf et al., 1987	$26S_M$ (S.1)	10.7	45.1 (DSC)
Sherbon, 1963	$23S_M$	—	48.9 (DSC)
	$23Sppt_M$ refractionated		51.8
Makhlouf et al., 1987	$21S_M$ (S.2)	25.6	48.3 (DSC)
	F.S (>21)	36.3	45.1
Pressure filtration			
deMan, 1968	$30S_S$ (A) slow cooling	11.2	50.0 (DSC)
	$27S_S$ (C) slow cooling	9.9	47.6
	$24S_S$ (E) slow cooling	30.0	46.7
Deffense, 1993b	1st stearin$_M$	—	46 (DP)
Centrifugation			
Kankare and Antila, 1988a	$24S_M$ [S]	—	47 (MP)
Kankare and Antila, 1988b	$24S_M$ [S]	—	48.0 (MP)
Kankare and Antila, 1986	$24S_M$ [S]	—	46.3 (DSC)
Antila and Antila, 1970	D	—	> 50 (DSC)
Separation with the aid of an aqueous detergent solution			
Schaap et al., 1975	$28S_S$	10	54.0 (DSC)
Walker, 1972	$26.5S_S$ [F3S]	—	45.4 (SP)
Banks et al., 1985	$21S_S$ [A] high level of detergent	8	51.6 (DSC)
	21 semi-S_S [A] high level of detergent	27	53.0
	$16S_S$ [B] high level of detergent	—	52.3
	$16S_S$ [B] low level of detergent	—	45.4

Continued

TABLE 5.12 (Continued)

Fraction separation method Author	Fraction designation[1]	Yield (%)	Melting behavior (°C)[2]
Separation technique not reported			
El-Ghandour and El-Nimr, 1982	$25S_M$	—	>50 (SFC)

[1]The number indicates fractionation temperature, the letter indicates physical state (S = solid), and the subscript indicates fractionation procedure used (M = multiple-step, S = single-step). Other primary designations and designations in parentheses are those of the original author. The designation inside the brackets following the fraction designation refers to the intact anhydrous milkfat source [sample letter or season (S = summer. W = winter)].

[2]DP = dropping point, DSC = melting behavior estimated using differential scanning calorimetry curves provided by the original investigator; MP = melting point, SFC = melting behavior estimated using solid fat content data provided by the original investigator.

[3]Data not available.

TABLE 5.13 Yield and Melting Behavior of Very-High-Melting Milkfat Fractions Obtained from Solvent Solution

Type of solvent solution Author	Fraction designation[1]	Yield (%)	Melting behavior (°C)[2]
Acetone			
Sherbon and Dolby, 1973	$32S_M$ (I) [HMG]	16.1	56 (MP)
Schaap et al., 1975	$28S_S$	—[3]	58.6 (DSC)
Kaylegian, 1991	$25S_M$ [W]	1.7	53.7 (MP)
	$25S_S$ [W]	2.2	53.0
Kaylegian and Lindsay, 1992	$25S_M$ [W]	1.7	53.7 (MP)
Sherbon and Dolby, 1973	$25S_M$ (II) [HMG]	33.1	54 (MP)
	$25S_M$ (III refractionated) [HMG]	2.7	48
Yi, 1993	$25S_M$ [S]	—	51.5 (MP)
Kaylegian, 1991	$21S_M$ [W]	3.0	50.7 (MP)
	$21S_S$ [W]	4.9	51.1
Kaylegian and Lindsay, 1992	$21S_M$ [W]	3.0	50.7 (MP)
Timms, 1980b	$20S_M$	6	>50 (SFC)
Yi, 1993	$20S_M$ [S]	—	50.4 (MP)
Youssef et al., 1977	$18S_M$ (A) [Nov]	11.5	51 (MP)
	$18S_M$ (E) [Nov]	14.2	46
Kaylegian, 1991	$15S_M$ [W]	4.5	46.2 (MP)
	$15S_S$ [W]	11.1	48.5
Kaylegian and Lindsay, 1992	$15S_M$ [W]	4.5	46.2 (MP)
Yi, 1993	$15S_M$ [S]	—	45.4 (MP)
Kaylegian, 1991	$11S_S$ [W]	19.3	47.0 (MP)
Timms, 1980a	HMF	6	60.3 (DSC)
Sherbon, 1963	2B	—	45.0 (DSC)
	2AJAN [Jan]	—	56.0
Pentane			
Kaylegian, 1991	$11S_M$ [W]	0.9	52.2 (MP)
	$11S_S$ [W]	2.6	50.6
	$5S_M$ [W]	8.8	49.0
	$5S_S$ [W]	9.8	48.9
	$0S_M$ [W]	3.5	45.3
	$0S_S$ [W]	12.7	47.8

[1]The number indicates fractionation temperature, the letter indicates physical state (S = solid), and the subscript indicates fractionation procedure used (M = multiple-step, S = single-step). Other primary designations and designations in parentheses are those of the original author. The designation inside the brackets following the fraction designation refers to the intact anhydrous milkfat source (HMG = high-melting glyceride, S = summer, W = winter).

[2]DSC = melting behavior estimated using differential scanning calorimetry curves provided by the original investigators, MP = melting point, SFC = melting behavior estimated using solid fat content data provided by the original investigators.

[3]Data not available.

TABLE 5.14 Yield and Melting Behavior of Very-High-Melting Milkfat Fractions Obtained by Supercritical Fluid Extraction

Type of solvent Author	Fraction designation[1]	Fractionation conditions		Yield (%)	Melting behavior (°C)[2]
		Temperature (°C)	Pressure		
Carbon dioxide					
Bhaskar et al., 1993b	S1 (raffinate)	40	24.1 MPa	21.0	45 (DSC)[3]
Hamman et al., 1991	B_{res}	40	350 bar	—[4]	48 (DSC)

[1]Fraction designations used are those given by the original author.

[2]DSC = melting point has been determined by the author of the monograph using differential scanning calorimetry curves provided by the original investigators.

[3]Melting behavior determined by original investigators using DSC melting thermograms. Melting point taken at the temperature which all the fat is in the liquid state.

[4]Data not available.

TABLE 5.15 Yield and Melting Behavior of High-Melting Milkfat Fractions Obtained from Melted Milkfat

Fraction separation method Author	Fraction designation[1]	Yield (%)	Melting behavior (°C)[2]
Vacuum filtration			
deMan and Finoro, 1980	$32S_S$	10.0	42.8 (DP)
	$31S_S$	18.0	42.4
	$30S_S$	17.3	44.1
Barna et al., 1992	$30S_M$ [W]	20	43 (DSC)[3]
Grall and Hartel, 1992	$30S_M$ [S]	—[4]	43.0 (MP)
Kaylegian, 1991	$30S_M$ [W]	17.7	44.2 (MP)
	$30S_S$ [W]	20.0	44.2
Kaylegian and Lindsay, 1992	$30S_M$ [W]	17.7	44.2 (MP)
	$30S_S$ [W]	20.0	44.2
Bumbalough, 1989	$29.4S_S$ [1]	13.3	43.9 (MP)
Amer et al., 1985	$29S_S$ [W]	20.6	43.5 (MP)
	$29S_S$ [S]	14.9	44.4
deMan and Finoro, 1980	$29S_S$	18.7	44.5 (DP)
	$28S_S$	18.4	44.0
Versteeg, 1991	$28S_S$ [Sept]	—	43.2 (SP)
deMan and Finoro, 1980	$27S_S$	24.6	42.3 (DP)
Amer et al., 1985	$26S_S$ [W]	36.2	41.5 (MP)
deMan and Finoro, 1980	$26S_S$	31.4	41.4 (DP)
Lund and Danmark, 1987	$26S_S$ [1]	25.5	42.2 (DP)
Versteeg, 1991	$26S_S$ [Sept]	—	42.2 (SP)
deMan and Finoro, 1980	$25S_S$	38.2	41.3 (DP)
Jamotte and Guyot, 1980	Commercial solid fraction A fractionated at 25°C	—	43.8 (DP)
	Commercial solid fraction B fractionated at 25°C	—	42.3
Kaylegian, 1991	$25S_M$ [W]	11.2	42.0 (MP)
	$25S_S$ [W]	46.5	40.6
Kaylegian and Lindsay, 1992	$25S_M$ [W]	11.2	42.0 (MP)
	$25S_S$ [W]	46.5	40.6
Timms and Parekh, 1980	$25S_S$ [S]	36	39.2 (SP)
Versteeg, 1991	$24S_S$ [Sept]	—	42.3 (SP)
Amer et al., 1985	$23S_S$ [W]	56.0	39.0 (MP)
Versteeg, 1991	$22S_S$ [Sept]	—	42.0 (SP)
Kaylegian, 1991	$20S_M$ [W]	7.0	40.5 (MP)
	$20S_S$ [W]	60.0	39.6
Kaylegian and Lindsay, 1992	$20S_M$ [W]	7.0	40.5 (MP)
	$20S_S$ [W]	60.0	39.6
Versteeg, 1991	$20S_S$ [Sept]	—	41.1 (SP)
Amer et al., 1985	$19S_S$ [W]	61.4	38.2 (MP)
	$19S_S$ [S]	53.7	38.4

Continued

TABLE 5.15 (Continued)

Fraction separation method Author	Fraction designation[1]	Yield (%)	Melting behavior (°C)[2]
Lund and Danmark, 1987	$18S_S$ [2]	42.5	39.8 (DP)
Versteeg, 1991	$18S_M$ [Sept]	29	43 (SP)
	$18S_S$ [Sept]	—	41.2
Schaap and van Beresteyn, 1970	$17S_M$	—	37.9 (DSC)
Deffense, 1989	1st stearin$_M$ [W]	—	41.2 (MP)
	1st stearin$_M$ [S]	—	40.9
Deffense, 1987	1st stearin$_M$ [2]	—	40.0 (DP)
	Stearin #1$_S$ [1]	—	43.1
	Stearin #2$_S$ [1]	—	42.3
	Stearin #3$_S$ [1]	—	41.1
	Stearin #4$_S$ [1]	—	41.0
Deffense, 1993b	1st stearin$_M$	—	42 (DP)
Laakso et al., 1992	S1	—	42 (DP)
Timms, 1978a	Hard	—	39.0 (SP)
Pressure filtration			
Stepanenko and Tverdokhleb, 1974	$20S_M$	29	42.3 (MP)
Centrifugation			
Riel and Paquet, 1972	$38S_M$	13	35 (CP)
Branger, 1993	$32S_M$ (1) bench-top	—	>40 (SFC)
	$32S_M$ (2) bench-top	—	>40
	$32S_M$ (3) bench-top	—	>40
	$32S_S$ pilot-scale	—	>40
Lakshminarayana and Rama Murthy, 1985	$31S_M$	10.6	36.5 (MP)
Richardson, 1968	$30S_M$ [W]	40.3	38.5 (SLP)
Bhat and Rama Murthy, 1983	$28S_S$	≈50	41 (SP)
Jordan, 1986	$28S_S$ [S]	—	37.0 (SLP)
	$28S_S$ [W]	—	39.0
Ramesh and Bindal, 1987b	$28S_S$ [Ghee]	69.5	37.3 (MP)
			35.9 (SP)
Evans, 1976	$25S_S$ [1 S]	44.8	40.2 (SP)
	$25S_S$ [2 S]	—	40.2
	$25S_S$ [3 S]	44.4	40.5
	$25S_S$ [4 S]	42.4	40.7
Jordan, 1986	$24S_S$ [S]	—	37.9 (SLP)
	$24S_S$ [W]	—	36.6

Continued

TABLE 5.15 (Continued)

Fraction separation method Author	Fraction designation[1]	Yield (%)	Melting behavior (°C)[2]
Branger, 1993	$25S_S$ pilot-scale	—	40 (SFC)
Black, 1974b	$23S_S$	22	42.1 (SP)
Branger, 1993	$22S_M$ (1) bench-top	—	40 (SFC)
	$22S_M$ (3) bench-top	—	40
	$20S_M$ (2) bench-top	—	35
Kankare and Antila, 1988b	$12S_M$ [S]	—	35.2 (MP)
Keogh and Higgins, 1986	Fraction 1	—	44.5 (SLP)
	Fraction 2	—	42.3
Keogh and Morrissey, 1990	Fraction 1	—	41 (SLP)
	Fraction 2	—	38
	Fraction 3	—	35
Separation with the aid of an aqueous detergent solution			
Sherbon et al., 1972	$B3S_M$ (triple fractionation at 31.7°C) [B]	94	40.6 (SP)
	$B3L_M$ (triple fractionation at 31.7°C) [B]	6	35.4
	$C2S_M$ (double fractionation at 30.6°C) [C]	59	43.9
Doležálek et al., 1976	$30S_S$ [1]	13	41.5 (MP)
	$30S_S$ [2]	16	41.5
	$30S_S$ [3]	10	40.5
	$30S_S$ [4]	9	43.5
Sherbon et al., 1972	$A1S_S$ (single fractionation at 28°C) [A]	47	38.0 (SP)
	$B2S_M$ (double fractionation at 28°C) [B]	92	40.9
	$B2L_M$ (double fractionation at 28°C) [B]	8	35.0
Doležálek et al., 1976	$27S_S$ [1]	24	42.0 (MP)
	$27S_S$ [2]	26	40.0
	$27S_S$ [3]	26	40.0
	$27S_S$ [4]	20	41.8
Jebson, 1970	$27S_S$	—	40 (SP)
Walker, 1972	$26.5S_S$ [F2S]	17	44.2 (SP)
	$26.5S_S$ [F2S']	18	44.5
Norris et al., 1971	$26S_S$ [B late S]	—	37.5 (SP)
	$26S_S$ [C early F]	—	37.6

Continued

TABLE 5.15 (Continued)

Fraction separation method Author	Fraction designation[1]	Yield (%)	Melting behavior (°C)[2]
Sherbon et al., 1972	$B1S_M$ (single fractionation at 25.5°C) [B]	31	41.0 (SP)
Doležálek et al., 1976	$25S_S$ [1]	43	37.5 (MP)
	$25S_S$ [2]	50	35.5
	$25S_S$ [3]	47	35.1
	$25S_S$ [4]	29	40.4
Norris et al., 1971	$25S_S$ [A late S]	—	36.2 (SP)
Jebson, 1970	$24S_S$	—	38 (SP)
Banks et al., 1985	16 semi-S_S [B] high level of detergent	—	44.7 (DSC)
Fjaervoll, 1970a	HMF	—	38.0 (MP)
Fjaervoll, 1970b	HMF	—	38.0 (MP)
Dolby, 1970a	Solid [mid-season]	—	38.6 (SP)
Dolby, 1970b	#1 [1]	—	43.5 (SP)
	#1 [mid-season]	—	38.2
	#2 [1]	—	37.5
Sherbon et al., 1972	$C1S_M$ original fraction	—	41.2 (SP)
Filtration through muslin cloth			
Khalifa and Mansour, 1988	$30S_M$	29.32	41.0 (MP)
Sreebhashyam et al., 1981	$30S_M$	25.8	37.3 (MP)
Khalifa and Mansour, 1988	$25S_M$	31.78	37.5 (MP)
Filtration through milk filter			
Fouad et al., 1990	$29S_S$	10	40.7 (DSC)
	$25S_S$	12	39.8
Filtration through cheese press			
Baker, 1970a	$30S_M$	25	40 (SP)
Filtration through casein-dewatering press			
Black, 1973	Hard #1	—	38.7 (SP)
	Hard #2	—	39.8
	Hard #3	—	42.8
	IT = 33.5°C S_S	41.7	37.5
	IT = 29.5°C S_S	42.9	36.8
	FT = 21.0°C S_S	40	37.4
	FT = 18.5°C S_S	44.6	36.9
	t = 21 h S_S	39.6	37.2
	t = 16 h S_S	45.0	37.0

Continued

TABLE 5.15 (Continued)

Fraction separation method Author	Fraction designation[1]	Yield (%)	Melting behavior (°C)[2]
Separation technique not reported			
Verhagen and Bodor, 1984	$25S_S$	—	>35 (SFC)
Deroanne, 1976	$21S_S$ [W]	—	41.0 (MP) 40.8 (DP)
Deroanne and Guyot, 1974	$21S_S$ [W] $21S_S$ [S]	— —	36.1 (SP) 38.6
El-Ghandour and El-Nimr, 1982	$15S_M$	—	>40 (SFC)
Verhagen and Bodor, 1984	$15S_S$	—	>35 (SFC)
Dixon and Black, 1974	HMF	22.0	40 (SP)
Martine, 1982	1st stearin 2nd stearin	— —	42 (MP) 37.6

[1] The number indicates fractionation temperature, the letter indicates physical state (L = liquid, S = solid), and the subscript indicates fractionation procedure used (M = multiple-step, S = single-step). Other primary designations and designations in parentheses are those of the original author. The designation inside the brackets following the fraction designation refers to the intact anhydrous milkfat source [sample number, letter, or season (S = summer, W = winter)].

[2] CP = cloud point, DP = dropping point, DSC = melting behavior estimated using differential scanning calorimetry curves provided by the original investigator, MP = melting point, SFC = melting behavior estimated using solid fat content data provided by the original investigators, SLP = slip point, SP = softening point.

[3] Melting behavior taken at DSC peak temperature as reported by Barna et al. (1992).

[4] Data not available.

TABLE 5.16 Yield and Melting Behavior of High-Melting Milkfat Fractions Obtained from Solvent Solution

Type of solvent solution Author	Fraction designation[1]	Yield (%)	Melting behavior (°C)[2]
Acetone			
Sherbon and Dolby, 1973	25S_M (IV refractionated) [HMG]	35.2	41 (MP)
Rolland and Riel, 1966	18S_M (1)	3.0	42 (CP)
Kaylegian, 1991	11S_M [W]	2.7	41.5 (MP)
Kaylegian and Lindsay, 1992	11S_M [W]	2.7	41.5 (MP)
Avvakumov, 1974	10S_M	15	42.2 (MP)
Yi, 1993	10S_M [S]	—[3]	41.0 (MP)
Timms, 1980b	3S_M	22	35–40 (SFC)
Kaylegian, 1991	5S_S [W] 0S_S [W]	30.7 47.9	43.7 (MP) 40.6
Timms, 1980a	MMF	24	40.8 (DSC)
Lambelet, 1983	S1$_M$	—	40 (SFC)
Sherbon, 1963	2BJAN [Jan]	—	38.9 (DSC)
Ethanol			
Rolland and Riel, 1966	18S_M	12.7	35 (CP)

[1] The number indicates fractionation temperature, the letter indicates physical state (S = solid), and the subscript indicates fractionation procedure used (M = multiple-step, S = single-step). Other primary designations and designations in parentheses are those of the original author. The designation inside the brackets following the fraction designation refers to the intact anhydrous milkfat source (HMG = high-melting glyceride, S = summer, W = winter).

[2] CP = cloud point, DSC = melting behavior estimated differential scanning calorimetry curves provided by the original investigator, MP = melting point, SFC = melting behavior estimated from solid fat content data provided by the original investigator.

[3] Data not available.

TABLE 5.17 Yield and Melting Behavior of High-Melting Milkfat Fractions Obtained by Supercritical Fluid Extraction

Type of solvent Author	Fraction designation[1]	Fractionation conditions		Yield (%)	Melting behavior (°C)[2]
		Temperature (°C)	Pressure		
Carbon dioxide					
Biernoth and Merk, 1985	Residual	80	200 bar	19.4	>35 (SFC)
Kaufmann et al., 1982	I (residue)	80	200 bar	—[3]	36.0 (DSC)
Timmen et al., 1984	Long-chain fraction	80	200 bar	80	36.2 (DSC)
Arul et al., 1987	S2	70	250–350 bar	17.3	39.3 (DSC)
	S3	70	250–350	21.5	43.3
Bhaskar et al., 1993b	S2	60	24.1 MPa	15.0	38 (DSC)[4]
Kankare and Antila, 1988a	Res	50–80	100–400 bar	—	39.0 (DSC)
Büning-Pfaue et al., 1989	Extractor fraction	50	25,000 kPa	—	38.5 (DSC)
Kankare et al., 1989	R	50	400 bar	57.7	39.8 (DSC)
Rizvi et al., 1990	Extract-2	48.0	275.8 bar	28.7	37.1 (MP)
	Residue	48	275.8	28.9	40.0
Chen et al., 1992	13.8	40	13.8 MPa	21.2	39.8 (DSC)
	20.7	40	20.7	32.8	39.8
	24.1	40	24.1	21.4	37.9
	27.6	40	27.6	1.5	39.8
Shukla et al., 1994	Raffinate	40	24.1 MPa	21	37 (DSC)
Büning-Pfaue et al., 1989	Column fraction	30	20,000 kPa	—	44.9 (DSC)

[1]Fraction designations used are those given by the original author.

[2]DSC = melting behavior estimated from differential scanning calorimetry curves provided by the original investigator; MP = melting point, SFC = melting behavior estimated from solid fat content data provided by the original investigators.

[3]Data not available.

[4]Melting behavior determined by original investigator using DSC melting thermograms. Melting point taken at the temperature which all the fat is in the liquid state.

TABLE 5.18 Yield and Melting Behavior of High-Melting Milkfat Fraction Obtained by Short-Path Distillation

Author	Fraction designation[1]	Yield (%)	Melting behavior (°C)[2]
Arul et al., 1988a	SF Residue	56.4	42.9 (DSC)

[1]Fraction designation are those given by the original author.

[2]DSC = melting behavior estimated using differential scanning calorimetry curves provided by the original investigator.

278   *Milkfat Fractionation*

TABLE 5.19 Yield and Melting Behavior of Middle-Melting Milkfat Fractions Obtained from Melted Milkfat

Fraction separation method Author	Fraction designation[1]	Yield (%)	Melting behavior (°C)[2]
Vacuum filtration			
Kaylegian, 1991	34L$_S$ [W]	97.5	33.5 (MP)
Kaylegian and Lindsay, 1992	34L$_S$ [W][3]	97.5	33.5 (MP)
deMan and Finoro, 1980	32L$_S$	90.0	32.4 (DP)
	31L$_S$	82.0	29.5
	30L$_S$	82.7	28.4
Kaylegian, 1991	30L$_S$ [W]	80.0	28.4 (MP)
Kaylegian and Lindsay, 1992	30L$_S$ [W][3]	80.0	28.4 (MP)
Amer et al., 1985	29L$_S$ [W]	79.4	29.7 (MP)
	29L$_S$ [S]	85.1	28.8
deMan and Finoro, 1980	29L$_S$	81.3	26.5 (DP)
	28L$_S$	81.6	26.8
Versteeg, 1991	28L$_S$ [Sept]	—[4]	28.6 (SP)
deMan and Finoro, 1980	27L$_S$	75.4	25.1 (DP)
Schaap and van Beresteyn, 1970	27L$_M$	—	29.2 (DSC)
Amer et al., 1985	26L$_S$ [W]	63.8	27.0 (MP)
Versteeg, 1991	26L$_S$ [Sept]	—	25.4 (SP)
Barna et al., 1992	20S$_M$ [W]	25	33.9 (DSC)[5]
Grall and Hartel, 1992	20S$_M$ [S]	—	33.9 (MP)
Sherbon, 1963	17S$_M$	—	25.3 (DSC)
Kaylegian, 1991	16S$_M$ [W]	14.2	26.9 (MP)
Kaylegian and Lindsay, 1992	16S$_M$ [W]3	14.2	26.9 (MP)
Makhlouf et al., 1987	15S$_M$ (I.1)	20.1	29.3 (DSC)
Versteeg, 1991	15S$_S$ [Sept]	—	26.6 (SP)
	14S$_S$ [Sept]	—	26.5
	13S$_S$ [Sept]	—	26.3
	12S$_M$ [Sept]	31	26 (SP)
	12S$_S$ [Sept]	—	26.0
	11.5S$_S$ [Sept]	—	25.8 (SP)
Deffense, 1987	Olein #1$_S$ [1]	—	27.7 (DP)
Laakso et al., 1992	S2	—	27 (DP)
Pressure filtration			
deMan, 1968	30L$_S$ (A) slow cooling	88.2	28.3 (SP)
	30L$_S$ (B) rapid cooling	72.1	27.9
Stepanenko and Tverdokhleb, 1974	11S$_M$	30	29.1 (MP)
Centrifugation			
Branger, 1993	32L$_S$ pilot-scale	—	30 (SFC)
	25L$_S$ pilot-scale	—	25

Continued

TABLE 5.19 (Continued)

Fraction separation method Author	Fraction designation[1]	Yield (%)	Melting behavior (°C)[2]
Richardson, 1968	25S$_M$ [W]	14.9	32.0 (SLP)
Lakshminarayana and Rama Murthy, 1985	23S$_M$	57.6	30.5 (MP)
Branger, 1993	22L$_M$ [1] bench-top	—	30–35 (SFC)
	20S$_S$ pilot-scale	—	30
Richardson, 1968	20S$_M$ [W]	11.9	28.0 (SLP)
Branger, 1993	17S$_S$ pilot-scale	—	30 (SFC)
Jordan, 1986	15S$_S$ [S]	—	26.6 (SLP)
Branger, 1993	12S$_M$ [3] bench-top	—	20 (SFC)
Kankare and Antila, 1986	12S$_M$ [S]	—	34.1 (DSC)
Kankare and Antila, 1988a	12S$_M$ [S]	—	34 (MP)
Antila and Antila, 1970	C	—	30.2 (DSC)
Separation with the aid of an aqueous detergent solution			
Sherbon et al., 1972	C2L$_M$ (double fractionation at 30.6°C) [C]	41	32.2 (SP)
Doležálek et al., 1976	30L$_S$ [1]	87	27.0 (MP)
	30L$_S$ [2]	84	28.0
	30L$_S$ [4]	91	26.5
Sherbon et al., 1972	A1L$_S$ (single fractionation at 28°C) [A]	53	25.0 (SP)
Doležálek et al., 1976	27L$_S$ [1]	76	25.4 (MP)
Walker, 1972	26.5L$_S$ [F2L]	83	25.2 (SP)
	26.5L$_S$ [F2L']	82	25.2
Sherbon et al., 1972	B1L$_M$ (single fractionation at 25.5°C) [B]	63.5	25.2 (SP)
Doležálek et al., 1976	25L$_S$ [1]	57	25.0 (MP)
	25L$_S$ [2]	50	25.2
Dolby, 1970b	#3 [1]	—	26.6 (SP)
	#2 [mid-season]	—	29.3
Filtration through muslin cloth			
Sreebhashyam et al., 1981	25S$_M$	34.9	29.6 (MP)
Khalifa and Mansour, 1988	20S$_M$	13.7	32.0 (MP)
Sreebhashyam et al., 1981	18S$_M$	38.7	25.8 (MP)
Filtration through milk filter			
Fouad et al., 1990	29L$_S$	90	25–30 (SFC)
	17S$_S$	52.3	30–35
Filtration through cheese press			
Baker, 1970a	13S$_M$	40	28 (SP)

Continued

TABLE 5.19 (Continued)

Fraction separation method Author	Fraction designation[1]	Yield (%)	Melting behavior (°C)[2]
Filtration through casein-dewatering press			
Black, 1973	IT = 33.5°C L_S	58.3	27.6 (SP)
	IT = 29.5°C L_S	57.1	27.1
	FT = 21.0°C L_S	60	29.1
	FT = 18.5°C L_S	55.4	25.6
	t = 21 h L_S	60.4	27.7
	t = 16 h L_S	55.0	27.0
Separation technique not reported			
Riel and Paquet, 1972	38S_M	24	32 (CP)
Verhagen and Warnaar, 1984	25L_S	—	<30 (SFC)
Deroanne, 1976	21L_S [W]	—	27.0 (MP)
			24.8 (DP)
Riel and Paquet, 1972	4S_M	21	25
Dixon and Black, 1974	MMF	42.0	32 (SP)
Martine, 1982	Olein	—	28 (MP)

[1]The number indicates fractionation temperature, the letter indicates physical state (L = liquid, S = solid), and the subscript indicates fractionation procedure used (M = multiple-step, S = single-step). Other primary designations and designations in parentheses are those of the original author. The designation inside the brackets following the fraction designation refers to the intact anhydrous milkfat source [sample number or letter, month or season (W = winter, S = summer).

[2]CP = cloud point, DP = dropping point, DSC = melting behavior estimated using differential scanning calorimetry curves provided by the original investigators, MP = melting point, SFC = melting behavior estimated from solid fat content data provided by the original investigators, SLP = slip point, SP = softening point.

[3]Yield and melting point data not reported in Kaylegian and Lindsay (1992) but is the same as Kaylegian (1991).

[4]Data not available.

[5]Melting behavior taken at DSC peak temperature as reported by Barna et al. (1992).

TABLE 5.20 Yield and Melting Behavior of Middle-Melting Milkfat Fractions Obtained from Solvent Solution

Type of solvent solution Author	Fraction designation[1]	Yield (%)	Melting behavior (°C)[2]
Acetone			
Rolland and Riel, 1966	$40L_M$ (2)	7.0	33 (CP)
Kaylegian, 1991	$25L_S$ [W]	97.8	31.9 (MP)
Sherbon and Dolby, 1973	$25L_M$ (V) [HMG]	12.8	27 (MP)
Kaylegian, 1991	$21L_S$ [W]	95.1	27.6 (MP)
Rolland and Riel, 1966	$10S_M$ (1)	5.3	34 (CP)
Kaylegian, 1991	$5S_M$ [W]	14.6	31.9 (MP)
Kaylegian and Lindsay, 1992	$5S_M$ [W][3]	14.6	31.9 (MP)
Yi, 1993	$5S_M$ [S]	—[4]	30.2 (MP)
Rolland and Riel, 1966	$2S_M$ (1)	7.5	26 (CP)
Avvakumov, 1974	$0S_M$	24	26 (MP)
Kaylegian, 1991	$0S_M$ [W]	11.1	26.1 (MP)
Kaylegian and Lindsay, 1992	$0S_M$ [W][3]	11.1	26.1 (MP)
Yi, 1993	$0S_M$ [S]	—	26.7 (MP)
Youssef et al., 1977	$0S_M$ (C) [Nov]	24.6	27 (MP)
Pentane			
Kaylegian, 1991	$11L_S$ [W]	97.4	32.2 (MP)
	$5L_S$ [W]	90.2	25.3

[1]The number indicates fractionation temperature, the letter indicates physical state (L = liquid, S = solid), and the subscript indicates fractionation procedure used (M = multiple-step, S = single-step). Other primary designations and designations in parentheses are those of the original author. The designation inside the brackets following the fraction designation refers to the intact anhydrous milkfat source (HMG = high-melting glyceride, S = summer, W = winter).

[2]CP = cloud point, MP = melting point.

[3]Yield and melting behavior data not reported in Kaylegian and Lindsay (1992) but same as Kaylegian (1991).

[4]Data not available.

TABLE 5.21 Yield and Melting Behavior of Middle-Melting Milkfat Fractions Obtained by Supercritical Fluid Extraction

Type of solvent Author	Fraction designation[1]	Fractionation conditions		Yield (%)	Melting behavior (°C)[2]
		Temperature (°C)	Pressure		
Carbon dioxide					
Bhaskar et al., 1993b	S3	75	17.2 MPa	48.0	32 (DSC)[3]
Arul et al., 1987	S1	70	250–350 bar	10.7	34.0 (DSC)
Bhaskar et al., 1993b	S4	60	6.9 MPa	4.0	31 (DSC)[3]
	S5	60	3.5	11.0	30
Kankare and Antila, 1988a	E2	50–80	100–400 bar	—[4]	27.9 (DSC)
	E3	50–80	100–400	—	31.5
	E4	50–80	100–400	—	32.1
Arul et al., 1987	I2	50	100–250 bar	13.1	26.0 (DSC)
	I3	50	100–250	15.7	30.0
Kankare et al., 1989	E2	50	200 bar	3.3	27.9 (DSC)
	E3	50	300	21.4	31.4
	E4	50	400	17.5	32.1
Chen et al., 1992	17.2	40	17.2 MPa	24.8	34.0 (DSC)
Hamman et al., 1991	A4	40	125 bar	—	25 (DSC)
Rizvi et al., 1990	Extract-1	40	241.3 bar	42.4	32.8 (MP)

[1]Fraction designations are those given by the original author.

[2]DSC = melting behavior estimated using differential scanning calorimetry curves provided by the original investigators, MP = melting point.

[3]Melting behavior determined by original investigators using DSC melting thermograms at temperature which all the fat is in liquid state.

[4]Data not available.

TABLE 5.22 Yield and Melting Behavior of Middle-Melting Milkfat Fractions Obtained by Short-Path Distillation

Author	Fraction designation[1]	Fractionation conditions		Yield (%)	Melting behavior (°C)[2]
		Temperature (°C)	Pressure		
Arul et al., 1988a	IF Distillate III	265	100 μm Hg	32.0	26.7 (DSC)

[1]Fraction designation is that given by the original author.

[2]DSC = melting point determined by the author of the monograph using differential scanning calorimetry curves provided by the original investigator.

TABLE 5.23 Yield and Melting Behavior of Low-Melting Milkfat Fractions Obtained from Melted Milkfat

Fraction separation method Author	Fraction designation[1]	Yield (%)	Melting behavior (°C)[2]
Vacuum filtration			
deMan and Finoro, 1980	26L$_S$	68.6	24.5 (DP)
Lund and Danmark, 1987	26L$_S$ [1]	74.5	24.1 (DP)
deMan and Finoro, 1980	25L$_S$	61.8	23.3 (DP)
Kaylegian, 1991	25L$_S$ [W]	53.5	21.8 (MP)
Kaylegian and Lindsay, 1992	25L$_S$ [W]	53.5[3]	21.8 (MP)
Timms and Parekh, 1980	25L$_S$ [S]	64	24.3 (SP)
Versteeg, 1991	24L$_S$ [Sept]	—[4]	24.2 (SP)
Amer et al., 1985	23L$_S$ [W]	44.0	22.9 (MP)
Versteeg, 1991	22L$_S$ [Sept]	—	22.2 (SP)
Barna et al., 1992	20L$_M$ [W]	55	19 (DSC)[5]
Kaylegian, 1991	20L$_S$ [W]	40.0	18.3 (MP)
Kaylegian and Lindsay, 1992	20L$_S$ [W]	40.0[3]	18.3 (MP)
Versteeg, 1991	20L$_S$ [Sept]	—	21.9 (SP)
Amer et al., 1985	19L$_S$ [W]	38.6	20.6 (MP)
	19L$_S$ [S]	46.3	19.7
Lund and Danmark, 1987	18L$_S$ [2]	57.5	19.1 (DP)
Versteeg, 1991	18L$_M$ [Sept]	71 (for this step)	21 (SP)
	18L$_S$ [Sept]	—	20.4
Schaap and van Beresteyn, 1970	17L$_M$	—	20.0 (DSC)
Grall and Hartel, 1992	15S$_M$ [S]	—	19.5 (MP)
	15L$_M$ [S]	—	12.5
Versteeg, 1991	15L$_M$ [Sept]	—	17.2 (SP)
	14L$_M$ [Sept]	—	14.4
Kaylegian, 1991	13S$_M$ [W]	13.3	22.8 (MP)
	13L$_M$ [W]	33.0	10.1
Kaylegian and Lindsay, 1992	13S$_M$ [W]	13.3[3]	22.8 (MP)
	13L$_M$ [W]	33.0	10.1
Versteeg, 1991	13L$_M$ [Sept]	—	14.1 (SP)
Versteeg, 1991	12L$_M$ [Sept]	56 (for this step)	14 (SP)
	12L$_M$ [Sept]	—	13.2
	11.5L$_M$ [Sept]	—	12.0
Deffense, 1987	2nd olein$_M$ filtered at 10°C [2]	—	10.0 (DP)
Sherbon, 1963	10S$_M$	—	20.0 (DSC)
	10L$_M$	—	13.2
Makhlouf et al., 1987	9S$_M$ (I.2)	18.5	20.5 (DSC)
	9L$_M$ (F.L.)	25.1	15.1
	F.1 (21-9)	38.6	24.4

Continued

TABLE 5.23 (Continued)

Fraction separation method Author	Fraction designation[1]	Yield (%)	Melting behavior (°C)[2]
Versteeg, 1991	8S$_M$ [Sept]	16 (overall) 41 (for this step)	19
Schaap and van Beresteyn, 1970	7L$_M$	—	17.8 (DSC)
Deffense, 1989	2nd stearin$_M$ [W]	—	24.0 (MP)
	2nd stearin$_M$ [S]	—	23.1
Deffense, 1987	2nd stearin$_M$ [2]	—	24.0 (DP)
Deffense, 1993b	2nd stearin$_M$	—	24 (DP)
Deffense, 1989	1st olein$_M$ [W]	—	19.4 (MP)
	1st olein$_M$ [S]	—	18.8
Deffense, 1987	1st olein$_M$ [2]	—	18.0 (DP)
Deffense, 1993b	1st olein$_M$	—	18 (DP)
Deffense, 1989	2nd olein$_M$ [W]	—	13.3 (MP)
	2nd olein$_M$ [S]	—	13.5
Deffense, 1987	Olein #2$_S$ [1]	—	24.3
	Olein #3$_S$ [1]	—	21.1
	Olein #4$_S$ [1]	—	20.6
Timms, 1978b	Soft	—	24.1 (SP)
Vovan and Riel, 1973	5% removal of hard fraction (>21°C)	—	18.0 (CP)
	10% removal of hard fraction (>21°C)	—	15.9
	15% removal of hard fraction (>21°C)	—	15.4
	5% removal of intermed. fraction (10–21°C)	—	20.0
	10% removal of intermed. fraction (10–21°C)	—	21.2
	15% removal of intermed. fraction (10–21°C)	—	22.5
	33% removal of intermed. fraction (10–21°C)	—	24.0
Laakso et al., 1992	L1	—	21 (DP)
	L2	—	16 (MP)
Pressure filtration			
deMan, 1968	27L$_S$ (C) slow cooling	90.1	18.7 (SP)
	27L$_S$ (D) rapid cooling	88.7	18.4
	24L$_S$ (E) slow cooling	70.0	17.7
	24L$_S$ (F) rapid cooling	70.2	17.8
Stepanenko and Tverdokhleb, 1974	11L$_M$	41	13.2 (MP)

Continued

TABLE 5.23 (Continued)

Fraction separation method / Author	Fraction designation[1]	Yield (%)	Melting behavior (°C)[2]
Centrifugation			
Bhat and Rama Murthy, 1983	28L$_S$	≈50	21 (SP)
Jordan, 1986	28L$_S$ [W]	—	23.9 (SLP)
	28L$_S$ [S]	—	24.2
Ramesh and Bindal, 1987b	28L$_S$	28.8	23.3 (MP)
			22.2 (SP)
Evans, 1976	25L$_S$ [1 S]	55.2	23.1 (SP)
	25L$_S$ [2 S]	—	23.0
	25L$_S$ [3 S]	55.6	22.8
	25L$_S$ [4 S]	57.6	22.9
Jordan, 1986	24L$_S$ [S]	—	21.5 (SLP)
	24L$_S$ [W]	—	22.4
Black, 1974b	23L$_S$	78	24.7 (SP)
Baker et al., 1959	22L$_S$ [W]	73.0	24 (MP)
Branger, 1993	20L$_M$ [2] bench-top	—	15 (SFC)
	20L$_S$ pilot-scale	—	20
	17L$_S$ pilot-scale	—	20
Jordan, 1986	15L$_S$ [S]	—	19.9 (SLP)
Lakshminarayana and Rama Murthy, 1985	15S$_M$	12.4	19.0 (MP)
	15L$_M$	19.4	14.0
Richardson, 1968	15S$_M$ [W]	15.8	22.5 (SLP)
	15L$_M$ [W]	17.1	17.0
Branger, 1993	12L$_M$ [3] bench-top	—	20 (SFC)
Kankare and Antila, 1986	12L$_M$ [S]	—	17.9 (DSC)
Kankare and Antila, 1988a	12L$_M$ [S]	—	14.4 (MP)
Kankare and Antila, 1988b	12L$_M$ [S]	—	18.6 (MP)
Antila and Antila, 1970	A	—	12.6 (DSC)
	B	—	18.9
Keogh, 1989	Soft fraction$_S$	—	20 (SLP)
Keogh and Higgins, 1986b	Fraction 3	—	22.8 (SLP)
	Fraction 4	—	18.5
Separation with the aid of an aqueous detergent solution			
Doležálek et al., 1976	30L$_S$ [3]	90	24.3 (MP)
	27L$_S$ [2]	74	24.3
	27L$_S$ [3]	74	24.8
	27L$_S$ [4]	80	24.5
Jebson, 1970	27L$_S$	—	24 (SP)
Walker, 1972	26.5L$_S$ [F3L]	—	24.3 (SP)
Norris et al., 1971	26L$_S$ [B late S]	—	22.7 (SP)
	26L$_S$ [C early F]	—	22.9

Continued

TABLE 5.23 (Continued)

Fraction separation method / Author	Fraction designation[1]	Yield (%)	Melting behavior (°C)[2]
Doležálek et al., 1976	25L$_S$ [3]	53	24.0 (MP)
	25L$_S$ [4]	71	23.0
Norris et al., 1971	25L$_S$ [A late S]	—	22.6 (SP)
Jebson, 1970	24L$_S$	—	22 (SP)
Banks et al., 1985	21L$_S$ [A] high level of detergent	65	22.3 (DSC)
Dolby, 1970a	Liquid [midseason]	—	24.8 (SP)
Fjaervoll, 1970a	LMF	—	22.1 (MP)
Fjaervoll, 1970b	LMF	—	22.1 (MP)
Dolby, 1970b	#4 [1]	—	23.9 (SP)
Filtration through muslin cloth			
Khalifa and Mansour, 1988	20L$_M$	25.16	15.0 (MP)
Sreebhashyam et al., 1981	18L$_M$	10.7	14.3 (MP)
Filtration through milk filter			
Fouad et al., 1990	17L$_S$	47.7	20–25 (SFC)
Filtration through cheese press			
Baker, 1970a	13L$_M$	35	<13 (SP)
Filtration through casein-dewatering press			
Black, 1973	Soft #1	—	21.5 (SP)
	Soft #2	—	21.1
	Soft #3	—	21.0
Separation technique not reported			
Deroanne and Guyot, 1974	21L$_S$ [W]	—	21.4 (SP)
	21L$_S$ [S]	—	18.1
El-Ghandour and El-Nimr, 1982	15L$_M$	—	>20 (SFC)
Riel and Paquet, 1972	4S$_M$	51	23 (CP)
	4L$_M$	55	11
Dixon and Black, 1974	LMF	36.0	20 (SP)

[1]The number indicates fractionation temperature, the letter indicates physical state (L = liquid, S = solid), and the subscript indicates fractionation procedure used (M = multiple-step, S = single-step). Other primary designations and designations in parentheses are those of the original authors. The designation inside the brackets following the fraction designation refers to the intact anhydrous milkfat source [sample number or letter, month or season (S = summer, W = winter)].

[2]CP = cloud point, DP = dropping point, DSC = melting behavior estimated using differential scanning calorimetry curves provided by the original investigator, MP = melting point, SFC = melting behavior estimated using solid fat content data provided by the original investigator, SLP = slip point, SP = softening point.

[3]Yield and melting behavior not reported in Kaylegian and Lindsay (1992) but same as Kaylegian (1991).

[4]Data not available.

[5]Melting behavior taken at DSC peak temperature as reported by Barna et al. (1992).

TABLE 5.24 Yield and Melting Behavior of Low-Melting Milkfat Fractions Obtained from Solvent Solution

Type of solvent solution Author	Fraction designation[1]	Yield (%)	Melting behavior (°C)[2]
Acetone			
Rolland and Riel, 1966	$24L_M$ (2)	21.0	13 (CP)
Youssef et al., 1977	$18L_M$ (B) [Nov]	88.5	21 (MP)
	$18L_M$ (F) [Nov]	10.5	21
Rolland and Riel, 1966	$16L_M$ (2)	14.0	17 (CP)
Kaylegian, 1991	$15L_S$ [W]	88.9	23.1 (MP)
	$11L_S$ [W]	80.7	19.8 (MP)
Rolland and Riel, 1966	$8L_M$ (2)	19.0	24 (CP)
Baker et al., 1959	$5L_S$ [S]	79	20 (MP)
Kaylegian, 1991	$5L_S$ [W]	69.3	13.3 (MP)
Timms, 1980b	$3L_M$	72	>15 (SFC)
Yi, 1993	$0L_M$ [S]	—[3]	11.3 (MP)
Rolland and Riel, 1966	$-6S_M$ (1)	9.8	20 (CP)
Avvakumov, 1974	$-10S_M$	22	17 (MP)
Rolland and Riel, 1966	$-15S_M$ (1)	15.8	15 (CP)
Timms, 1980a	LMF	70	14.4 (DSC)
Lambelet, 1983	S1	—	>15 (SFC)
Sherbon, 1963	2C	—	22.0 (DSC)
	2CJAN [Jan]	—	19.8
Ethanol			
Rolland and Riel, 1966	$10S_M$	10.7	20 (CP)
	$2S_M$	16.3	16
	$-6S_M$	11.0	12
Isopropanol			
Norris, 1977	-5LM	34.4	<0 (DSC)[4]
Pentane			
Kaylegian, 1991	$0L_M$ [W]	86.8	21.6 (MP)
	$0L_S$ [W]	87.3	21.3

[1]The number indicates fractionation temperature, the letter indicates physical state (L = liquid, S = solid), and the subscript indicates fractionation procedure used (M = multiple-step, S = single-step). Other primary descriptions and designations in parentheses are those of the original author. The designation inside the brackets following the fraction designation refers to the intact anhydrous milkfat source (S = summer, W = winter).

[2]CP = cloud point, DSC = melting behavior estimated using differential scanning calorimetry curves provided by original investigator, MP = melting point, SFC = melting behavior estimated using solid fat content data provided by the original investigator.

[3]Data not available.

[4]Melting behavior data given as ranges determined by original author based on DSC curves (not reported; Norris, 1977).

TABLE 5.25 Yield and Melting Behavior of Low-Melting Milkfat Fractions Obtained by Supercritical Fluid Extraction

Type of solvent Author	Fraction designation[1]	Fractionation conditions		Yield (%)	Melting behavior (°C)[2]
		Temperature (°C)	Pressure		
Carbon dioxide					
Biernoth and Merk, 1985	Extract	80	200 bar	19.4	>15 (SFC)
Kaufmann et al., 1982	II (extract)	80	200 bar	—[3]	20.4 (DSC)
Timmen et al., 1984	Short chain fraction	80	200 bar	20	20.4 (DSC)
Kankare and Antila, 1988a	E1	50–80	100–400 bar	—	17.6 (DSC)
Arul et al., 1987	L1	50	100–250 bar	4.1	15.0 (DSC)
	L2	50	100–250	8.1	20.5
	I1	50	100–250	9.4	21.5
Kankare et al., 1989	E1	50	100 bar	0.2	17.7 (DSC)
Chen et al., 1992	10.3	40	10.3 MPa	7.3	18.2 (DSC)
Hamman et al., 1991	A1	40	125 bar	—	17 (DSC)

[1]Fraction designations used are those given by the original author.

[2]DSC = melting behavior estimated using differential scanning calorimetry curves provided by the original investigator; SFC = melting behavior estimated from solid fat content data provided by the original investigators.

[3]Data not available.

TABLE 5.26 Yield and Melting Behavior of Low-Melting Milkfat Fractions Obtained by Short-Path Distillation

Author	Fraction designation[1]	Fractionation conditions		Yield (%)	Melting behavior (°C)[2]
		Temperature (°C)	Pressure		
Arul et al., 1988a	LF1 Distillate I	245	220 µm Hg	2.1	16.8 (DSC)
	LF2 Distillate II	245	100	9.5	20.0

[1]Fraction designations are those given by the original author.

[2]DSC = melting behavior estimated using differential scanning calorimetry curves provided by the original investigator.

TABLE 5.27 Yield and Melting Behavior of Very-Low-Melting Milkfat Fractions Obtained from Melted Milkfat

Fraction separation method Author	Fraction designation[1]	Yield (%)	Melting behavior (°C)[2]
Vacuum filtration			
Versteeg, 1991	$8L_M$ [Sept]	24 (overall) (59 for this step)	5 (SP)
Deffense, 1987	2nd olein$_M$ filtered at 6°C [2]	—[3]	8.5 (DP)
Guyot, 1982	$5L_M$	12	6 (DP)
Schaap and van Beresteyn, 1970	$2L_M$	—	5.7 (DSC)
Jamotte and Guyot, 1980	L4	—	5.0 (DP)
Deffense, 1993b	2nd olein$_M$	—	9 (DP)
Centrifugation			
Riel and Paquet, 1972	$4L_M$	36	9 (CP)

[1]The number indicates fractionation temperature, the letter indicates physical state (L = liquid), and the subscript indicates fractionation procedure used (M = multiple-step). Other descriptions used are those of the original author. The designation inside the brackets following the fraction designation refers to the intact anhydrous milkfat source.

[2]CP = cloud point, DP = dropping point, DSC = melting behavior estimated using DSC curves provided by the original author, SP = softening point.

[3]Data not available.

TABLE 5.28 Yield and Melting Behavior of Very-Low-Melting Milkfat Fractions Obtained from Solvent Solution

Type of solvent solution Author	Fraction designation[1]	Yield (%)	Melting behavior (°C)[2]
Acetone			
Rolland and Riel, 1966	$40L_M$ (2)	31.0	3 (CP)
	$32L_M$ (2)	8.0	5
Kaylegian, 1991	$0L_M$ [W]	62.5	9.8 (MP)
	$0L_S$ [W]	52.1	6.8
Kaylegian and Lindsay, 1992	$0L_M$ [W]	62.5[3]	9.8 (MP)
Youssef et al., 1977	$0L_M$ (D) [Nov]	63.9	8 (MP)
Rolland and Riel, 1966	$-15L_M$ (1)	58.8	5 (CP)
Sherbon, 1963	2DJAN [Jan]	—[4]	6.5 (DSC)
Ethanol			
Rolland and Riel, 1966	$-15S_M$	9.3	8 (CP)
	$-15L_M$	40.0	0
Isopropanol			
Norris, 1977	$-5L_M$	34.4	<0[5]

[1]The number indicates fractionation temperature, the letter indicates physical state (L = liquid, S = solid), and the subscript indicates fractionation procedure used (M = multiple-step). Other primary designations and designations in parentheses are those of the original author. The designation inside the brackets following the fraction designation refers to the intact anhydrous milkfat source (W = winter).

[2]CP = cloud point, DSC = melting behavior estimated using DSC curves provided by the original author, MP = melting point.

[3]Yield and melting behavior data not reported in Kaylegian and Lindsay (1992), but same as Kaylegian (1991).

[4]Data not available.

[5]Melting behavior data reported as range determined by original investigator based on DSC curves (not reported)

TABLE 5.29 Yield and Melting Behavior of Unknown-Melting Milkfat Fractions Obtained from Melted Milkfat

Fraction separation method Author	Fraction designation[1]	Yield (%)	Melting behavior (°C)
Vacuum filtration			
Hayakawa and deMan, 1982	$32S_S$	—	—[2]
	$32L_S$	—	—
	$31S_S$	—	—
	$31L_S$	—	—
	$30S_S$	—	—
	$30L_S$	—	—
Voss et al., 1971	$30S_S$	47.7	—
	$30L_S$	52.3	—
Schultz and Timmen, 1966	$30S_M$	20	—
Hayakawa and deMan, 1982	$29S_S$	—	—
	$29L_S$	—	—
Kupranycz et al., 1986	$29S_S$ [W]	—	—
	$29L_S$ [W]	—	—
	$29S_S$ [S]	—	—
	$29L_S$ [S]	—	—
Frede et al., 1980	$28S_S$ [S]	35	—
	$28L_S$ [S]	65	—
Hayakawa and deMan, 1982	$28S_S$	—	—
	$28L_S$	—	—
Versteeg, 1991	$28S_S$ [Spr]	15	—
	$28L_S$ [Spr]	85	—
	$28S_S$ [S]	23	—
	$28L_S$ [S]	77	—
Hayakawa and deMan, 1982	$27S_S$	—	—
	$27L_S$	—	—
	$26S_S$	—	—
	$26L_S$	—	—
	$25S_S$	—	—
	$25L_S$	—	—
Schultz and Timmen, 1966	$25S_M$	20	—
Black, 1975b	$25S_S$ with precooling	36.4	—
	$25S_S$ no precooling	40.4	—
	$25S_S$ exponential cooling	31.4	—
	$25S_S$ programmed cooling	45.4	—
	$25S_S$ 8 h cooling time	35.1	—
	$25S_S$ 20 h cooling time	41.7	—
	$25S_S$ 20 rpm agitator speed	40.4	—
	$25S_S$ 10 rpm agitator speed	36.3	—

Continued

TABLE 5.29 (Continued)

Fraction separation method Author	Fraction designation[1]	Yield (%)	Melting behavior (°C)
Schaap, 1975	$24S_S$	—	—
	$24L_S$	—	—
Versteeg, 1991	$24S_S$ [Spr]	23	—
	$24L_S$ [Spr]	77	—
	$24S_S$ [S]	33	—
	$24L_S$ [S]	67	—
Guyot, 1982	$23S_M$	—	—
Versteeg, 1991	$22S_S$ [Spr]	26	—
	$22L_S$ [Spr]	74	—
	$22S_S$ [S]	39	—
	$22L_S$ [S]	61	—
Schultz and Timmen, 1966	$20S_M$	20	—
Kupranycz et al., 1986	$19S_S$ [W]	—	—
	$19L_S$ [W]	—	—
	$19S_S$ [S]	—	—
	$19L_S$ [S]	—	—
Bumbalough, 1989	$18S_M$ [2]	—	—
Versteeg, 1991	$18S_S$ [Spr]	29	—
	$18L_S$ [Spr]	71	—
Guyot, 1982	$15S_S$	—	—
Schultz and Timmen, 1966	$15S_M$	20	—
Bumbalough, 1989	$12.5S_M$ [2]	—	—
	$12.5L_M$ [2]	25.9	—
Guyot, 1982	$10S_M$	—	—
Schultz and Timmen, 1966	$10S_M$	10	—
	$7S_M$	4	—
Guyot, 1982	$5S_M$	—	—
Schultz and Timmen, 1966	$4.5S_M$	3.5	—
	$4.5L_M$	2.5	—
Pressure filtration			
deMan, 1968	$30S_S$ (B) rapid cooling	27.9	—
	$27S_S$ (D) rapid cooling	11.3	—
El-Ghandour et al., 1976	$25S_M$ [Egyptian milkfat]	—	—
	$25S_M$ [Russian milkfat]	—	—
Helal et al., 1977	$25S_M$ [S]	—	—
	$25S_M$ [W]	—	—
deMan, 1968	$24S_S$ (F) rapid cooling	29.8	—
Badings et al., 1983b	$23S_M$ [W]	35.9 overall	—
	$23L_M$ [W]	64.1 this step	—
Kuwabara et al., 1991	$23S_S$	22.2	—
	$23L_S$	77.8	—

Continued

TABLE 5.29 (Continued)

Fraction separation method Author	Fraction designation[1]	Yield (%)	Melting behavior (°C)
Badings et al., 1983b	21.5S$_M$ [S]	30.3 overall	—
	21.5L$_M$ [S]	69.7 this step	—
Kuwabara et al., 1991	20S$_S$	46.6	—
	20L$_S$	53.4	—
Badings et al., 1983b	18.5S$_M$ [W]	23.3 overall 36.3 this step	—
	18.5L$_M$ [W]	63.6 this step	—
	15.3S$_M$ [S]	10.3 overall 14.8 this step	—
	15.3L$_M$ [S]	85.2 this step	—
	15S$_M$ [W]	14.6 overall 35.8 this step	—
	15L$_M$ [W]	64.2 this step	—
El-Ghandour et al., 1976	15S$_M$ [Egyptian milkfat]	—	—
	15L$_M$ [Egyptian milkfat]	—	—
	15S$_M$ [Russian milkfat]	—	—
	15L$_S$ [Russian milkfat]	—	—
Helal et al., 1977	15S$_M$ [S]	—	—
	15L$_M$ [S]	—	—
	15S$_M$ [W]	—	—
	15L$_M$ [W]	—	—
Helal et al., 1984	15S$_M$ [S]	—	—
	15S$_M$ [W]	—	—
Badings et al., 1983b	13S$_M$ [S]	26.5 overall 44.7 this step	—
	13L$_M$ [S]	55.3 this step	—
	11.3S$_M$ [W]	9.0 overall 34.4 this step	—
	11.3L$_M$ [W]	65.6 this step	—

Continued

TABLE 5.29 (Continued)

Fraction separation method Author	Fraction designation[1]	Yield (%)	Melting behavior (°C)
	9.3S_M [S]	9.6 overall 29.2 this step	—
	9.3L_M [S]	70.8 this step	—
	7.1S_M [S]	9.0 overall 38.6 this step	—
	7.1L_M [S]	14.3 overall 61.4 this step	—
	5.5S_M [W]	10.4 overall 60.5 this step	—
	5.5L_M [W]	6.8 overall 39.5 this step	—
Deffense, 1993b	1st olein	—	—
Centrifugation			
Lakshminarayana and Rama Murthy, 1986	31S_M	—	—
Baker et al., 1959	28S_S 28L_S	20.0 80.0	— —
Bratland, 1983	28S_S 28L_S	— —	— —
Keshava Prasad and Bhat, 1987a	28S_S 28L_S	— —	— —
Keshava Prasad and Bhat, 1987b	28S_S 28L_S	— —	— —
Bratland, 1983	25S_S 25L_S 24S_S 24L_S	— — — —	— — — —
Kankare and Antila, 1974a	24S_M [S] 24S_M [W]	15.7 30.2	— —
Keshava Prasad and Bhat, 1987a	23S_S 23L_S	— —	— —
Keshava Prasad and Bhat, 1987b	23S_S 23L_S	— —	— —
Lakshminarayana and Rama Murthy, 1986	23S_M	—	—
Baker et al., 1959	22S_S	27.0	—
Bratland, 1983	20S_S 20L_S	— —	— —

Continued

TABLE 5.29 (Continued)

Fraction separation method Author	Fraction designation[1]	Yield (%)	Melting behavior (°C)
Kankare and Antila, 1974	18S$_M$ [S]	6.3	—
	18S$_M$ [W]	9.1	—
Lakshminarayana and Rama Murthy, 1986	15S$_M$ 15L$_M$	— —	— —
Kankare and Antila, 1974a	12S$_M$ [S]	29.0	—
	12L$_M$ [S]	49.0	—
	12S$_M$ [W]	27.9	—
	12L$_S$ [W]	32.8	—
Separation with the aid of an aqueous detergent solution			
Schaap et al., 1975	28L$_S$	90	—
Jebson, 1974a	25S$_S$	—	—
	25L$_S$	—	—
Banks et al., 1985	21S$_S$ [A] low level of detergent	—	—
	21L$_S$ [A] low level of detergent	—	—
	16L$_S$ [B] low level of detergent	—	—
	16L$_S$ [B] high level of detergent	—	—
Filtration through muslin cloth			
Kulkarni and Rama Murthy, 1987	23S$_S$ 23L$_S$	— —	— —
Filtration through milk filter			
Fouad et al., 1990	25L$_S$	88	—
	21S$_S$	43.5	—
	21S$_S$	56.5	—
Separation technique not reported			
Timmen, 1974	28S$_S$ [W]	30–40	—
	28L$_S$ [W]	60–70	—
Thomas, 1973a	20S$_S$	—	—
	20L$_S$	—	—
Verhagen and Bodor, 1984	25L$_S$	—	—
Verhagen and Warnaar, 1984	25S$_S$	—	—
Timmen, 1974	22S$_S$ [S]	51	—
	22L$_S$ [S]	49	—
Verhagen and Bodor, 1984	15L$_S$	—	—

Continued

TABLE 5.29 (Continued)

Fraction separation method Author	Fraction designation[1]	Yield (%)	Melting behavior (°C)[2]
Lechat et al., 1975	Most liquid fraction [Sept]	—	—
	Most solid fraction [Feb]	—	—
	Most liquid fraction [Feb]	—	—

[1]The number indicates fractionation temperature, the letter indicates physical state (L = liquid, S = solid), and the subscript indicates fractionation procedure used (M = multiple-step, S = single-step). The designation inside the brackets following the fraction designation refers to the intact anhydrous milkfat source (S = summer, W = winter).

[2]Data not available.

TABLE 5.30 Yield and Melting Behavior of Unknown-Melting Milkfat Fractions Obtained from Solvent Solution

Type of solvent solution Author	Fraction designation[1]	Yield (%)	Melting behavior (°C)
Acetone			
Schaap et al., 1975	28L_S	—	—[2]
Norris, 1977	22S_M [Spr]	1.4	—
	22S_M [S]	6.8	—
Larsen and Samuelsson, 1979	21.4S_S	—	—
Avvakumov, 1974	20S_M	—	—
Parodi, 1974a	20S_S	4.7	—
	20L_S	95.3	—
Chen and deMan, 1966	15S_M	9	—
Larsen and Samuelsson, 1979	15S_S	—	—
Mattsson et al., 1969	15S_S [Aug]	—	—
	15L_S [Aug]	—	—
	15S_S [Dec]	—	—
	15L_S [Dec]	—	—
Woodrow and deMan, 1968	15S_S	—	—
	15L_S	—	—
Walker et al., 1977	14S_M (HMG) [A]	10	—
	14S_M (HMG) [B]	9	—
	14S_M (HMG) [C1]	11	—
	14S_M (HMG) [C2]	10	—
Walker et al., 1978	14S_M	—	—
Norris, 1977	12S_M [Spr]	8.3	—
Lansbergen and Kemps, 1984	12S_M [1, hardened butterfat]	—	—
	12L_M [1, hardened butterfat]	—	—
	12S_M [2, hardened butterfat]	—	—
	12L_M [2, hardened butterfat]	—	—
Norris, 1977	11.5S_M [S]	10.9	—
Lechat et al., 1975	10L_M [Sept]	6.5	—
	10S_M [Sept]	20.5	—
	10L_M [April]	13	—
	10S_M [April]	18	—
Munro et al., 1978	10S_M	12	—
Larsen and Samuelsson, 1979	9.6S_S	—	—
Lansbergen and Kemps, 1984	6S_M [1, hardened butterfat]	—	—
	6LM [1, hardened butterfat]	—	—

Continued

TABLE 5.30 (Continued)

Type of solvent solution Author	Fraction designation[1]	Yield (%)	Melting behavior (°C)
Baker et al., 1959	$5S_S$	21	—
Chen and deMan, 1966	$5S_M$	10	—
Larsen and Samuelsson, 1979	$5S_S$	—	—
Walker et al., 1977	$2S_M$ (IMG_1) [A]	26	—
	$2S_M$ (IMG_1) [B]	24	—
	$2S_M + -13S_M$ (IMG_{1+2}) [C1]	62	—
	$2S_M + -13S_M$ (IMG_{1+2}) [C3]	60	—
Walker et al., 1978	$2S_M$	—	—
Bhalerao et al., 1959	$0S_S$	—	—
	$0L_S$	—	—
Lansbergen and Kemps, 1984	$0S_M$ [2, hardened butterfat]	—	—
	$0L_M$ [2, hardened butterfat]	—	—
Lechat et al., 1975	$0L_M$ [Sept]	16	—
	$0S_M$ [Sept]	27	—
	$0L_M$ [April]	26	—
	$0S_M$ [April]	31	—
Chen and deMan, 1966	$-5S_M$	14	—
Larsen and Samuelsson, 1979	$-5.0L_S$	—	—
	$-8.0L_S$	—	—
Lechat et al., 1975	$-9L_M$ [Sept]	57	—
	$-9S_M$ [Sept]	43	—
	$-9L_M$ [April]	43	—
	$-9S_M$ [April]	57	—
Walker et al., 1977	$-13S_M$ (IMG_2) [A]	30	—
	$-13L_M$ (LMG) [A]	37	—
	$-13S_M$ (IMG_2) [B]	30	—
	$-13L_M$ (LMG) [B]	37	—
	$-13L_M$ (LMG) [C1]	27	—
	$-13L_M$ (LMG) [C3]	30	—
Walker et al., 1978	$-13S_M$	—	—
	$-13L_M$	—	—
Norris, 1977	$-14S_M$ [Spr]	46.3	—
	$-14L_M$ [Spr]	43.9	—
Chen and deMan, 1966	$-15S_M$	27	—
Norris, 1977	$-15S_M$ [S]	42.9	—
	$-15L_M$ [S]	39.4	—
Munro et al., 1978	$-15S_M$	52	—
	$-15L_M$	36	—

Continued

TABLE 5.30 (Continued)

Type of solvent solution Author	Fraction designation[1]	Yield (%)	Melting behavior (°C)
Larsen and Samuelsson, 1979	-16.4L_S	—	—
Avvakumov, 1974	-20S_M	—	—
	-20L_M	—	—
Larsen and Samuelsson, 1979	-21.0L_S	—	—
Chen and deMan, 1966	-25S_M	13	—
Jensen et al., 1967	-25S_M	—	—
Chen and deMan, 1966	-35S_M	9	—
	-45S_M	5	—
	-45L_M	13	—
Jensen et al., 1967	-70S_M	—	—
	-70L_M	—	—
Lansbergen and Kemps, 1984	Top fraction [natural butterfat]	—	—
	Mid fraction [natural butterfat]	—	—
	Olein fraction [natural butterfat]	—	—
Sherbon, 1963	2A	—	—
	2D	—	—
	2D'	—	—
	2EJAN [Jan]	—	—
	2FJAN [Jan]	—	—
Ethanol			
Bhalerao, 1959	20S_S	—	—
	20L_S	—	—
Hexane			
Muuse and van der Kamp, 1985	12.5S_S	—	—
	12.5L_S	—	—
Isopropanol			
Norris, 1977	20S_M	19.0	>20 (DSC)[3]

[1]The number indicates fractionation temperature, the letter indicates physical state (L = liquid, S = solid), and the subscript indicates fractionation procedure used (M = multiple-step, S = single-step). The designation inside the brackets following the fraction designation refers to the intact anhydrous milkfat source.

[2]Data not available.

[3]Melting behavior data were given as ranges determined by original author based on DSC curves (not reported; Norris, 1977). The range for the 20S_M fractions was too broad to assign a definitive general fraction category.

TABLE 5.31 Yield and Melting Behavior of Unknown-Melting Milkfat Fractions Obtained by Supercritical Fluid Extraction

Type of solvent Author	Fraction designation[1]	Fractionation conditions		Yield (%)	Melting behavior (°C)
		Temperature (°C)	Pressure		
Carbon dioxide					
Bradley, 1989	WM-14, 1	80	2300 psig	14.6	—[2]
	WM-14, 2	80	2500	2.2	—
	WM-14, 3	80	2800	5.9	—
	WM-14, 4	80	3050	8.6	—
	WM-14, 5	80	3350	17.1	—
	WM-14, 6	80	3750	16.8	—
	WM-14, 7	80	6000	34.7	—
	WM-21, 1	80	2300	11.1	—
	WM-21, 2	80	2500	8.7	—
	WM-21, 3	80	2800	14.5	—
	WM-21, 4	80	3050	11.6	—
	WM-21, 5	80	3400	9.2	—
	WM-21, 6	80	6000	45.1	—
	WM-38, 1	60	2150	—	—
	WM-38, 2	60	2300	—	—
	WM-38, 3	60	2400	—	—
	WM-38, 4	60	2400	—	—
	WM-38, 5	60	2500	—	—
	WM-38, 6	60	2500	—	—
	WM-38, 7	60	2650	—	—
	WM-38, 8	60	2850	—	—
	WM-38, 9	60	2900	—	—
	WM-38, 10	60	2900	—	—
	WM-38, 11	60	2900	—	—
	WM-38, 12	60	2900	—	—
	WM-38, 13	60	2900	—	—
	WM-38, 14	60	2900	—	—
	WM-38, 15	60	—	—	—
Ensign, 1989	WM-38, 1	60	2150 psig	—	—
	WM-38, 2	60	2300	—	—
	WM-38, 3	60	2400	—	—
	WM-38, 4	60	2400	—	—
	WM-38, 5	60	2500	—	—
	WM-38, 6	60	2500	—	—
	WM-38, 7	60	2650	—	—
	WM-38, 8	60	2850	—	—
	WM-38, 9	60	2900	—	—
	WM-38, 10	60	2900	—	—
	WM-38, 11	60	2900	—	—
	WM-38, 12	60	2900	—	—
	WM-38, 13	60	2900	—	—
	WM-38, 14	60	2900	—	—
	WM-38, 15	60	—	—	—
	WM-39, 1	60	2000	—	—
	WM-39, 2	60	6000	—	—
Rizvi et al., 1993	S1	60	17.2 MPa	14	—
	S2	60	13.8	6.9	—
Kankare et al., 1989	2E1	48	120 bar (3 h 12 min) + 150 bar (4 h)	4.2	—
	2E2	48	200	9.5	—
	2R	48	200	86.2	—

Continued

TABLE 5.31 (Continued)

Type of solvent Author	Fraction designation[1]	Fractionation conditions Temperature (°C)	Pressure	Yield (%)	Melting behavior (°C)
Bhaskar et al., 1993a	Extract	40	24.1 MPa	—	—
	Raffinate	40	24.1	—	—
Ensign, 1989	WM-43, 1	40	1700 psig	—	—
	WM-43, 2	40	6000	—	—
Hamman et al., 1991	A_{res}	40	125 bar	—	—
	B1	40	350	—	—
Rizvi et al., 1993	S3	40	6.9 MPa	4	—
	S4	40	3.4	12	—
	Raffinate	40	24.1	20	—
Shishikura et al., 1986	F'1	40	300 kg/cm^2	—	—
	F'2	40	300	—	—
	F'3	40	300	—	—
	F'4	40	300	—	—
	F'5	40	300	—	—
	F'6	40	300	—	—
	Residue	40	300	—	—
Shukla et al., 1994	Extract	40	24.1 MPa	—	—
Chen and Schwartz, 1991	Unknown	35, 40, 50, 60	1500, 2000, 2500, 3000 3500, 4000 psi	—	—
Rizvi et al., 1993	Trap	–10	0.14	1	—
Propane					
Biernoth and Merk, 1985	Extract 1	125	85 bar	13.3	—
	Extract 2	125	85	20.0	—
	Residual	125	85	66.7	—

[1]Designations used are those given by the original author.

[2]Data not available.

TABLE 5.32 Yield and Melting Behavior of Unknown Melting Milkfat Fractions Obtained by Short-Path Distillation

Author	Fraction designation[1]	Fractionation conditions Temperature (°C)	Pressure	Yield (%)	Melting behavior (°C)
McCarthy et al., 1962	R-1	150	15μ	2.5	—[2]
	R-2	160	15	2.5	—
	R-3	185	12	2.5	—
	R-4	185	12	2.5	—
	D-2	177–182	5–7	39	—
	D-3	—	—	51	—

[1]Designations used are those given by the original author.

[2]Data not reported.

TABLE 5.33 Individual Fatty Acid Composition of Intact Anhydrous Milkfats Used to Produce Milkfat Fractions from Melted Milkfat

Fraction separation method Author	Fraction designation[1]	Fatty acid (%)[2]													
		4:0	6:0	8:0	10:0	12:0	14:0	15:0	16:0	16:1	17:0	18:0	18:1	18:2	18:3
Vacuum filtration																
Barna et al., 1992	AMF [W]	4.4[n]	2.5	1.4	3.0	3.4	11.2	—[3]	30.1	1.4	—	12.0	23.0	3.0	—
Deffense, 1987	AMF [2]	3.5[w]	1.6	1.0	2.2	2.8	9.6	1.2	24.8	2.5	1.0	11.8	27.6	3.0	0.7
Deffense, 1993b	AMF	3.5[w]	1.6	1.0	2.2	2.8	9.6	1.3	24.8	2.5	1.0	11.8	27.6	3.0	0.7
deMan and Finoro, 1980	AMF	4.1[w]	2.4	1.4	2.9	3.5	11.4	—	23.2	—	—	12.4	25.2	2.6	0.9
Grall and Hartel, 1992	AMF [S]	4.4[a]	2.5	1.4	3.0	3.4	11.2	—	30.1	1.4	—	12.0	23.0	3.0	—
Jamotte and Guyot, 1980	Commercial liquid fraction (L2) doubly fractionated at 15°C	3.6[n]	2.2	1.3	3.0	3.5	10.6	1.5	27.5	3.0	0.8	10.0	26.0	2.2	0.9
Kaylegian, 1991	AMF [W]	4.2[a]	2.3	1.3	2.8	3.3	10.5	—	28.5	1.6	—	10.3	22.1	2.9	0.3
Laakso et al., 1992	AMF	8.5[m]	4.0	1.8	3.2	3.2	10.4	1.0	26.4	1.2	0.9	11.3	21.1	2.8 incl C19:0	0.4
	AMF SSS[4]	11.0[m]	5.3	2.4	4.5	4.5	14.8	1.8	37.0	—	1.6	15.7	0.5	0.3 incl C19:0	—
	AMF SSM[t]	3.8[m]	3.9	1.7	3.2	3.0	10.0	1.5	24.5	2.4 incl C17:0	—	12.1	27.9	4.4 incl C19:0, C19:1	—
	AMF SSM[c]	7.9[m]	3.6	1.6	2.8	2.6	9.1	2.7 incl C14:1	23.5	2.4	0.9	10.1	29.9	1.1 incl C19:0m	—
	AMF SM[c]M[t]	3.9[m]	1.9	0.8	1.3	1.3	5.7	2.2 incl C14:1	12.2	2.2	1.5	5.8	49.0	7.4 incl C19:1	0.6

Reference	Fraction														
	AMF SM^{c}M^c	4.3^m	1.8	0.8	1.3	1.2	4.2	3.0 incl C14:1	12.3	3.6	1.4	5.0	53.6	2.5 incl C19:0, C19:1	0.2
	AMF UUU	4.1^m	2.0	0.9	1.6	1.5	4.7	1.9 incl C14:1	12.2	1.9	1.1	5.5	34.8	21.1 incl C19:0	2.6
Lund and Danmark, 1987	AMF [1]	4.1^w	2.5	1.5	2.9	3.5	11.1	—	30.0	—	—	11.0	24.0	2.3	1.0
Schaap and van Beresteyn, 1970	AMF	3.2^n	2.7	1.6	2.7	3.9	12.2	—	28.1	—	—	13.3	27.8	2.4	2.4
Timms, 1978a	AMF	4.0^w	2.8	1.3	2.6	3.0	10.4	1.5	26.0	2.5 incl C17br	1.2	13.3	24.4	1.7	1.8
Timms and Parekh, 1980	AMF [S]	4.0^w	2.8	1.3	2.6	3.0	10.4	1.5	26.1	2.5 incl C17br	1.2	13.3	24.4	1.7	1.8 incl C18:2 conj
Versteeg, 1991	AMF [Sept]	3.3^n	2.0	1.3	2.6	3.2	9.5	1.4	21.9	2.6	0.8	12.2	27.9	1.7	1.9
Vovan and Riel, 1973	AMF	3.8^n	2.5	1.6	3.6	3.9	11.6	0.9	27.4	2.2	0.3	12.6	25.2	1.1	1.3
Pressure filtration															
Badings et al., 1983b	AMF [W]	4.6^w	2.7	1.5	3.4	4.2	11.6	1.1	29.1	0.7	0.9	9.5	21.3	1.6	1.4
	AMF [S]	4.2^w	2.3	1.3	2.6	3.0	9.6	1.1	23.3	1.0	0.8	11.9	28.1	1.3	1.1
Deffense, 1993	AMF	3.5^w	1.6	1.0	2.2	2.8	9.6	1.3	24.8	2.5	1.0	11.8	27.6	3.0	0.7
deMan, 1968	AMF	3.3^n	1.6	0.9	1.9	2.1	9.3	—	27.7	—	—	13.4	30.7	3.1	0.5
Centrifugation															
Antila and Antila, 1970	AMF	5.2^n	2.7	1.5	3.3	4.2	12.6	1.7	29.3	2.2	2.5	6.0	20.9	3.2	0.8
Jordan, 1986	AMF [W]	3.3^w	2.4	1.2	2.4	2.7	9.9	1.0	29.7	1.9	0.6	11.3	23.4	1.7	0.4
Kankare and Antila, 1986	AMF [S]	3.3^n	2.2	1.4	2.9	3.2	10.7	—	25.3	1.6	—	12.4	26.6	3.0	0.7
Kankare and Antila, 1988a	AMF [S]	3.5^n	—	—	—	—	—	—	26.0	—	—	12.5	26.5	—	—
Kankare and Antila, 1988b	AMF [S]	3.5^n	2.4	—	—	—	—	—	26.0	—	—	12.5	26.5	—	—
Keogh and Higgins, 1986b	AMF	4.1^w	2.4	1.4	3.0	3.4	11.0	—	26.0	2.4	0.9	11.6	24.3	2.0	0.9
Lakshminarayena and Rama Murthy, 1985	AMF	3.8^w	2.0	1.1	2.1	2.9	13.3	—	32.2	—	—	11.3	30.3	—	—
Richardson, 1968	AMF [W]	4.0^w	2.4	1.4	2.8	3.1	8.4	1.3	30.1	2.8	1.0	11.7	24.9	1.8	1.2
Riel and Paquet, 1972	AMF	1.2^n	2.1	1.4	3.7	4.6	13.5	—	32.0	—	—	14.2	28.5	—	—

Continued

TABLE 5.33 (Continued)

Fraction separation method Author	Fraction designation[1]	4:0	6:0	8:0	10:0	12:0	14:0	15:0	16:0	16:1	17:0	18:0	18:1	18:2	18:3
Separation with the aid of an aqueous detergent solution															
Banks et al., 1985	AMF [A]	3.6[w]	2.1	0.8	1.5	1.7	7.8	—	38.2	2.2	—	10.9	27.7 cis = 25.0 trans = 2.7	2.3 incl C18:3	—
	AMF [B]	2.9[w]	1.8	1.0	1.4	2.4	10.1	—	26.2	0.9	—	18.0	32.6 cis = 24.6 trans = 8.0	0.9 incl C18:3	—
Dolby, 1970a	AMF [mid-season]	—	—	3.2[n] incl C4,C6	3.3	4.4	15.5	—	36.2	0.7	—	12.7	21.3	—	—
Doležálek et al., 1976	AMF [1]	1.7[w]	2.0	1.2	3.6	4.8	13.2	1.8	32.4	2.6	1.1	7.0	20.0	2.3	0.4
Fjaervoll, 1970a	AMF	3.0[n]	2.1	1.4	3.4	4.4	12.5	—	26.7	—	—	15.5	28.4	1.2	1.4
Norris et al., 1971	AMF [A late S]	2.0[a]	1.8	1.1	2.6	2.8	10.3	1.0	27.5	1.0	—	14.9	30.6	0.9	1.6
	AMF [B late S]	1.7[a]	1.6	1.1	2.6	2.6	9.5	1.0	27.9	1.3	—	15.1	30.6	1.0	1.8
	AMF [C early F]	1.9[a]	1.9	1.3	3.4	4.2	11.9	1.1	28.0	1.2	—	13.2	26.0	1.5	2.1
Schaap et al., 1975	AMF	3.4[w]	2.2	1.4	4.0	3.5	11.0	1.1	23.8	1.9 incl C17br	0.4	12.6	27.1	1.5	2.1
Filtration through muslin cloth															
Kulkarni and Rama Murthy, 1987	AMF	3.8[n]	2.3	1.3	2.4	3.1	13.3	—	32.4	—	—	11.2	29.2	1.0	—
Filtration through milk filter															
Fouad et al., 1990	AMF	5.1[w]	3.1	1.4	3.2	3.5	11.1	—	27.5	1.1	—	10.8	22.5	1.7	0.5
Separation technique not reported															
Deroanne, 1976	AMF [W]	2.3[n]	1.6	1.1	2.4	2.9	10.5	1.2	28.3	2.8	0.7	10.6	28.1	2.5	1.4 incl C20

Technical Data for Experimental and Commercial Milkfat Fractions 305

Source	Sample														
Deroanne and Cuyot, 1974	AMF [W]	3.8[a]	2.3	1.2	3.2	4.1	12.5	1.5	30.4	3.1 incl C17:0	—	8.4	20.8	1.9	1.1 incl C20:1
	AMF [S]	3.8[a]	2.3	1.2	3.0	3.4	11.0	1.4	23.8	2.9 incl C17:0	—	11.5	27.0	2.0	1.5 incl C20:1
Lechat et al., 1975	AMF [Sept]	3.3[n]	2.0	1.4	2.6	3.0	10.8	1.5	23.8	2.5	1.1	12.3	26.8	2.7	0.6
	AMF [Feb]	3.1[n]	1.6	1.3	2.3	2.8	9.5	1.6	22.7	1.6	1.0	13.6	28.3	2.9	1.1
Martine, 1982	AMF	0.9[n]	0.6	1.0	2.1	2.9	10.6	1.2	30.3	2.2	0.9	11.0	27.5	1.8	0.5
Riel and Paquet, 1972	AMF	1.2[n]	2.1	1.4	3.7	4.6	13.5	—	32.0	—	—	14.2	28.5	—	—

[1] AMF = anhydrous milkfat. The designation inside the brackets refers to the intact anhydrous milkfat source [sample number or letter, season (F = fall, S = summer, W = winter)].

[2] br = branched, conj = conjugated, incl = includes. Units of chemical composition are designated in the first data column using these abbreviations: a = area %, m = mole %, n = % not specified, w = wt %.

[3] Data not available.

[4] S = saturated, M = monounsaturated, U = polyunsaturated, c = cis, t = trans.

TABLE 5.34 Individual Fatty Acid Composition of Intact Anhydrous Milkfats Used to Produce Milkfat Fractions from Solvent Solution

Type of solvent solution Author	Fraction designation[1]	Fatty acid (%)[2]														
		4:0	6:0	8:0	10:0	12:0	14:0	15:0	16:0	16:1	17:0	18:0	18:1	18:2	18:3
Acetone																
Chen and deMan, 1966	AMF	2.5[w]	1.6	1.0	2.4	2.9	10.1	—[3]	23.4	—	—	14.3	29.5	3.9	2.8
	SSS fraction of AMF[4]	4.4[w]	2.3	1.5	3.6	4.7	16.0	—	36.6	—	—	23.7	0.0	0.0	0.0
Jensen et al., 1967	AMF	11.3[m]	5.1	2.4	4.0	3.8	10.5	—	22.8	2.4	—	9.8	20.2	2.7	0.8
Kaylegian, 1991	AMF [W]	4.2[a]	2.3	1.3	2.8	3.3	10.5	—	28.5	1.6	—	10.3	22.1	2.9	0.3
Lansbergen and Kemps, 1984	Hardened butterfat	3.3[w]	2.2	1.4	3.2 incl C10:1	3.7	11.3	2.1 incl C15br	26.4 incl C16:1	—	1.9 incl C17br	43.2	—	—	—
	Natural butterfat	3.8[w]	2.5	1.5	3.7	5.2	12.5	1.2	30.9	2.4 incl C17br	0.7	8.9	20.2	2.1	0.7
Larsen and Samuelsson, 1979	AMF	6.4[n]	3.4	2.0	3.4	6.2	9.7	—	25.0	5.8	—	11.3	23.9	—	—
Lechat et al., 1975	AMF [Sept]	3.3[n]	2.0	1.4	2.6	3.0	10.8	1.5	23.8	2.5	1.1	12.3	26.8	2.7	0.6
	AMF [Apr]	3.4[n]	2.0	1.0	2.6	2.9	9.7	1.1	25.3	1.9	1.4	13.7	26.5	3.0	0.7
Mattsson et al., 1969	AMF [Aug]	9.5[m]	3.3	1.9	3.7	3.8	10.9	—	22.7	—	—	9.2	21.5	1.5	2.1
	AMF [Dec]	9.4[m]	3.7	2.2	4.4	4.7	12.4	—	28.1	—	—	6.8	16.0	1.3	1.1
Parodi, 1974a	AMF [May]	10.3[m]	4.4	2.0	3.5	3.4	10.8	1.4	24.2	2.6 incl C17br	1.0	10.5	18.9	1.3	1.5
Schaap et al., 1975	AMF	3.4[w]	2.2	1.4	4.0	3.5	11.0	1.1	23.8	1.9 incl C17br	0.4	12.6	27.1	1.5	2.1
Timms, 1980b	AMF [S]	4.0[w]	2.8	1.3	2.6	3.0	10.4	1.5	26.1	2.5 incl C17br	1.2	13.3	24.4	1.7	1.8 incl C18:2 conj
Yi, 1993	AMF [S]	3.3[a]	1.2	1.3	2.9	3.4	11.6	—	32.6	2.1	—	12.8	23.7	3.0	0.9

Hexane														
Muuse and van der Kamp, 1985	AMF	4.0[a]	2.2	1.3	2.7	3.3	10.0	—	24.5	—	12.1	27.5	—	
Pentane														
Kaylegian, 1991	AMF [W]	4.2[a]	2.3	1.3	2.8	3.3	10.5	—	28.5	1.6	10.3	22.1	2.9	0.3

[1] AMF = anhydrous milkfat. The designation inside the brackets following the fraction designation refers to the intact anhydrous milkfat source (W = winter, S = summer).

[2] br = branched, corj = conjugated, incl = includes. Units of chemical composition are designated in the first data column using these abbreviations: a = area %, m = mole %, n = % not specified, w = wt %.

[3] Data not available

[4] SSS = trisaturated triglyceride.

TABLE 5.35 Individual Fatty Acid Composition of Intact Anhydrous Milkfats Used to Produce Milkfat Fractions by Supercritical Fluid Extraction

Type of solvent Author	Fraction designation[1]	Fatty acid (%)[2]													
		4:0	6:0	8:0	10:0	12:0	14:0	15:0	16:0	16:1	17:0	18:0	18:1	18:2	18:3
Carbon dioxide															
Arul et al., 1987	AMF	11.8[m]	6.8	2.3	5.1	4.4	12.6	1.0	26.0	1.2	—[3]	8.5	17.8	1.0	—
Bradley, 1989	AMF [WM-38]	2.9[w]	1.6	0.9	2.2	2.7	7.3	1.0	27.0	1.6	0.5	17.4	29.3 incl C18:2, C18:3	—	—
Biernoth and Merk, 1985	AMF	2.8[a]	2.2	1.5	3.3	4.1	11.8	—	30.0	—	—	8.9	24.4	2.7	1.3
Büning-Pfaue et al., 1989	AMF	3.1[w]	1.7	0.9	2.6	3.4	12.4	—	36.9	—	—	8.7	24.4	2.3	—
Ensign, 1989	AMF [WM-38]	2.9[w]	1.6	0.9	2.2	2.7	7.3	1.0	27.0	1.6	0.5	17.4	29.3 incl C18:2, C18:3	—	—
Hamman et al., 1991	AMF [A]	1.9[w]	1.9	1.3	2.7	3.1	10.8	—	28.1	1.4	—	15.1	29.8	1.3	—
	AMF [B]	1.9[w]	1.8	1.3	3.3	3.8	12.4	—	32.1	1.7	—	11.9	23.3	1.4	—
Kankare and Antila, 1988a	AMF	3.1[n]	—	—	—	—	—	—	25.5	—	—	12.4	26.5	—	—
Kankare et al., 1989	AMF	3.2[n]	2.2	1.4	3.0	3.4	12.1	—	29.3	1.4	—	11.2	23.8	2.4	0.4
Kaufmann et al., 1982	AMF	10.2[w]	4.8	2.2	4.2	4.6	11.6	1.0	27.0	1.5	0.5	7.0	17.7	2.1	0.9
Shishikura et al., 1986	AMF	3.5[n]	2.8	1.6	3.9	4.2	14.1	—	31.3	—	—	12.4	25.8	0.5	—
Timmen et al., 1984	AMF	3.8[w]	2.2	1.3	3.0	4.1	11.5	—	30.3	—	—	8.8	22.0	2.6	1.2

[1] AMF = anhydrous milkfat. Designations in brackets refer those used by the original author.
[2] incl = includes. Units of chemical composition are designated in the first data column using these abbreviations: a = area %, m = mole %, n = % not specified, w = wt %.
[3] Data not available.

TABLE 5.36 Individual Fatty Acid Composition of Intact Anhydrous Milkfat Used to Produce Milkfat Fractions by Short-Path Distillation

Author	Fraction designation[1]	Fatty acid (%)[2]												
		4:0	6:0	8:0	10:0	12:0	14:0	15:0	16:0	16:1	17:0	18:0	18:1	18:2	18:3
Arul et al., 1988a	AMF	8.3m	4.0	1.8	3.1	3.8	11.4	—[3]	26.2	1.2	—	10.6	20.1	2.0	0.8
McCarthy et al., 1962	AMF	2.0w	2.2	1.4	3.3	3.5	12.9	1.1	32.5	1.7	—	11.6	25.0	0.5	—

[1] AMF = anhydrous milkfat.

[2] m = mole %, w = wt %.

[3] Data not available.

TABLE 5.37 Individual Fatty Acid Composition of Very-High-Melting Milkfat Fractions Obtained from Melted Milkfat

Fraction separation method Author	Fraction designation[1]	4:0	6:0	8:0	10:0	12:0	14:0	15:0	16:0	16:1	17:0	18:0	18:1	18:2	18:3
Vacuum filtration															
Kaylegian, 1991	34S$_M$ [W]	3.1[a]	1.7	0.9	2.2	2.8	10.5	—[3]	33.3	1.1	—	16.4	15.9	2.0	0.3
	34S$_S$ [W]	2.7[a]	1.6	0.9	2.2	3.0	10.6	—	32.6	1.1	—	16.0	16.8	2.3	0.2
Schaap and van Beresteyn, 1970	27S$_M$	3.0[n]	1.9	1.0	2.6	3.6	13.0	—	32.1	—	—	18.3	21.7	1.3	1.5
Sherbon, 1963	23S$_M$	9.2[m]	1.0	0.7	2.7	2.6	11.0	0.9	35.4	1.2	1.0	13.2	16.7	1.4	0.5
	23ppt$_M$ refractionated	4.2[m]	1.0	1.2	3.0	2.7	10.8	1.0	33.5	0.8	0.9	10.5	17.2	1.2	0.6
Pressure filtration															
deMan, 1968	30S$_S$ (A) slow cooling	2.0[n]	0.8	0.4	1.3	1.8	9.7	—	35.6	—	—	21.4	21.6	1.6	—
	27S$_S$ (C) slow cooling	2.3[n]	1.2	0.6	1.7	2.1	9.6	—	32.9	—	—	18.0	24.5	1.1	—
	24S$_S$ (E) slow cooling	2.9[n]	1.3	0.7	1.8	2.2	10.0	—	30.7	—	—	15.5	26.9	3.6	—
Deffense, 1993b	1st stearin$_M$	1.4[w]	1.0	0.7	1.9	2.7	10.9	1.5	33.0	1.1	1.1	18.6	19.0	2.0	0.4
Centrifugation															
Kankare and Antila, 1986	24S$_M$ [S]	2.0[n]	1.5	1.0	2.4	3.1	12.0	—	30.5	1.3	—	16.6	19.8	2.5	0.5
Kankare and Antila, 1988a	24S$_M$ [S]	2.5[n]	—	—	—	—	—	—	30.5	—	—	16.0	20.0	—	—
Kankare and Antila, 1988b	24S$_M$ [S]	2.0[n]	—	—	—	—	—	—	31.0	—	—	16.5	19.5	—	—
Antila and Antila, 1970	D	2.3[n]	1.2	0.8	2.2	3.1	11.8	1.5	42.9	1.4	0.4	13.9	11.0	1.9	0.4
Separation with the aid of an aqueous detergent solution															
Schaap et al., 1975	28S$_S$	1.2[w]	tr[4]	0.1	1.0	3.0	14.2	1.8	39.7	0.9 incl C:17br	0.5	27.6	6.9	0.5	tr
Banks et al., 1985	21S$_S$ [A] high level of detergent	1.9[w]	1.7	0.6	1.3	1.9	9.4	—	45.6	1.5	—	15.5	18.8 cis = 16.2 trans = 2.6	0.7 incl C18:3	—

Fraction											
21 semi-S$_S$ [A] high level of detergent	2.9^w	2.1	0.8	1.5	1.8	8.4	—	40.8	2.0	—	12.5	24.7 cis = 22.0 trans = 2.7	1.3 incl C18:3	—
16S$_S$ [B] low level of detergent	2.1^w	1.2	0.9	1.5	2.4	10.4	—	29.9	0.7	—	21.9	26.5 cis = 18.6 trans = 7.9	0.6 incl C18:3	—
16S$_S$ [B] high level of detergent	1.5^w	1.2	1.1	1.4	2.1	11.9	—	34.0	0.6	—	24.1	18.7 cis = 11.2 trans = 7.5	0.6 incl C18:3	—

[1]The number indicates fractionation temperature, the letter indicates physical state (S = solid), and the subscript indicates fractionation procedure used (M = multiple-step, S = single-step). Other primary designations and designations in parentheses are those of the original author. The designation inside the brackets following the fraction designation refers to the intact anhydrous milkfat source [sample letter or season (S = summer, W = winter)].

[2]br = branched, incl = includes, a = area %, m = mole %, n = % not specified, w = wt %.

[3]Data not available.

[4]tr = trace.

TABLE 5.38 Individual Fatty Acid Composition of Very-High-Melting Milkfat Fractions Obtained from Solvent Solution

Type of solvent solution Author	Fraction designation[1]	Fatty acid (%)[2]														
		4:0	6:0	8:0	10:0	12:0	14:0	15:0	16:0	16:1	17:0	18:0	18:1	18:2	18:3
Acetone																
Sherbon and Dolby, 1973	32S$_M$ (I) [HMG]	0.2[m]	—[3]	—	—	0.7	7.3	1.9	48.0	—	1.5	36.4	1.0	0.2	0.5
Schaap et al., 1975	28S$_S$	0.5[w]	—	0.1	0.1	0.7	10.1	1.4	37.5	0.5	0.2	41.4	4.4	0.7	tr[4]
Kaylegian, 1991	25S$_M$ [W]	1.2[a]	0.6	0.4	0.9	1.4	9.1	—	38.7	0.5	—	28.6	7.4	1.2	0.1
	25S$_S$ [W]	1.2[a]	0.7	0.4	1.0	1.6	9.6	—	38.4	0.5	—	27.7	7.5	1.3	0.1
Sherbon and Dolby, 1973	25S$_M$ (II) [HMG]	0.5[m]	0.5	tr	0.5	2.0	14.1	2.2	47.3	—	1.4	23.2	4.3	0.2	0.3
	25S$_M$ (III) refractionated [HMG]	—	—	tr	0.7	3.0	18.6	2.1	40.3	—	1.3	18.7	8.7	0.4	0.1
Yi, 1993	25S$_M$ [S]	0.9[a]	0.3	0.4	0.9	1.9	9.7	—	40.1	0.6	—	25.8	11.1	4.0	2.1
Kaylegian, 1991	21S$_M$ [W]	1.3[a]	0.7	0.4	0.9	1.9	11.6	—	39.9	0.4	—	22.7	8.1	1.9	0.1
	21S$_S$ [W]	1.2[a]	0.6	0.4	1.0	2.0	11.2	—	39.2	0.5	—	23.7	8.9	1.2	tr
Timms, 1980b	20S$_M$ [S]	—	0.1	0.1	0.6	1.6	10.5	1.6	40.2	—	1.2	31.6	8.9	0.6	0.2 incl C18:2 conj
Yi, 1993	20S$_M$ [S]	0.8[a]	0.2	0.4	1.2	2.1	12.9	—	44.4	0.6	—	22.7	12.6	1.4	0.2
Kaylegian, 1991	15S$_M$ [W]	1.3[a]	0.8	0.5	1.5	3.2	13.8	—	37.5	0.7	—	17.0	11.9	1.6	0.1
	15S$_S$ [W]	1.7[a]	0.9	0.5	1.5	2.7	12.3	—	37.2	0.5	—	19.4	11.5	1.7	0.1
Yi, 1993	15S$_M$ [S]	0.9[a]	0.2	0.5	1.8	3.4	15.0	—	42.0	0.8	—	18.0	15.2	1.4	0.2
Kaylegian, 1991	11S$_S$ [W]	2.1[a]	1.1	0.7	1.9	3.0	12.0	—	35.0	1.0	—	18.0	13.4	1.7	0.3
Sherbon, 1963	2B	8.2[m]	1.5	1.1	4.0	3.4	12.0	1.2	29.0	—	1.0	25.3	12.2	—	—
	2AJAN [Jan]	8.7[m]	—	—	—	0.5	8.2	1.0	44.8	—	1.0	33.4	—	0.7	—
Pentane																
Kaylegian, 1991	11S$_M$ [W]	1.2[a]	0.6	0.3	0.9	1.7	10.6	—	40.1	0.6	—	24.9	7.7	1.3	tr
	11S$_S$ [W]	1.7[a]	1.0	0.6	1.4	2.2	11.2	—	36.7	0.9	—	21.1	11.4	1.7	0.2
	5S$_M$ [W]	1.5[a]	0.8	0.5	1.3	2.7	13.1	—	38.7	0.8	—	18.8	10.3	1.4	0.1
	5S$_S$ [W]	1.5[a]	0.8	0.5	1.4	2.6	13.0	—	36.7	0.9	—	18.7	11.7	1.9	0.1

			4:0	6:0	8:0	10:0	12:0	14:0	15:0	16:0	16:1	17:0	18:0	18:1	18:2	18:3
OS_M [W]			1.5^a	0.9	0.5	2.0	3.7	13.8	—	35.7	0.8	—	16.9	12.3	1.6	0.2
OS_S [W]			1.3^a	0.8	0.5	1.6	3.1	13.5	—	36.4	0.9	—	18.2	11.5	1.9	0.3

[1]Fraction designations; the number indicates fractionation temperature, the letter indicates physical state (S = solid), and the subscript indicates fractionation procedure used (S = single-step, M = multiple-step). Other descriptions used are those of the original author. The designation inside the brackets refers to the intact anhydrous milkfat source (W = winter, HMG = high-melting glyceride).

[2]br = branched, conj = conjugated, incl = includes, a = area %, m = mole %, w = wt %.

[3]Data not available.

[4]tr = trace.

TABLE 5.39 Individual Fatty Acid Composition of Very-High-Melting Milkfat Fractions Obtained by Supercritical Fluid Extraction

Type of solvent Author	Fraction designation[1]	Fractionation conditions		Fatty acid (%)[2]													
		Temperature (°C)	Pressure	4:0	6:0	8:0	10:0	12:0	14:0	15:0	16:0	16:1	17:0	18:0	18:1	18:2	18:3
Carbon dioxide																	
Hamman et al., 1991	B_{res}	40	350 bar	0.0^w	0.0	0.0	1.3	2.1	9.4	—[3]	31.6	1.3	—	16.6	30.0	1.5	—

[1]Fraction designations are those given by the original author.

[2]w = wt %.

[3]Data not available.

TABLE 5.40 Individual Fatty Acid Composition of High-Melting Milkfat Fractions Obtained from Melted Milkfat

Fraction separation method Author	Fraction designation[1]	Fatty acid (%)[2]													
		4:0	6:0	8:0	10:0	12:0	14:0	15:0	16:0	16:1	17:0	18:0	18:1	18:2	18:3
Vacuum filtration															
deMan and Finoro, 1980	32S$_S$	3.9^w	2.3	1.5	3.1	3.8	11.2	—[3]	21.8	—	—	15.4	21.9	2.8	1.1
	31S$_S$	3.4^w	1.8	1.0	2.4	3.2	11.9	—	27.0	—	—	15.3	22.8	2.0	0.8
	30S$_S$	3.6^w	1.9	1.1	2.5	3.2	12.0	—	26.2	—	—	15.8	21.9	2.0	0.7
Barna et al., 1992	30S$_M$ [W]	3.3^n	1.9	1.2	2.6	3.2	11.9	—	35.3	1.1	—	16.2	16.7	2.4	—
Grall and Hartel, 1992	30S$_M$ [S]	3.3^a	1.9	1.2	2.6	3.2	11.9	—	35.3	1.1	—	16.2	16.7	2.4	—
Kaylegian, 1991	30S$_M$ [W]	3.3^a	1.9	1.1	2.4	3.1	10.9	—	32.5	1.3	—	14.0	17.1	2.2	0.3
	30S$_S$ [W]	3.8^a	2.0	1.0	2.4	3.0	10.5	—	31.5	1.4	—	13.9	18.1	2.2	0.3
deMan and Finoro, 1980	29S$_S$	2.9^w	1.7	1.1	2.5	3.3	12.2	—	25.4	—	—	16.2	22.0	2.2	0.7
	28S$_S$	2.6^w	1.6	0.9	2.5	3.3	12.3	—	27.3	—	—	16.5	21.7	2.0	0.5
	27S$_S$	2.8^w	2.2	1.4	3.3	3.9	13.0	—	23.5	—	—	15.0	21.0	2.3	0.9
deMan and Finoro, 1980	26S$_S$	3.0^w	1.9	1.1	2.7	3.7	12.5	—	24.7	—	—	15.3	22.9	2.2	0.8
Lund and Danmark, 1987	26S$_S$ [1]	3.2^w	2.0	1.0	2.5	3.5	11.5	—	34.0	—	—	13.5	20.0	2.0	0.8
deMan and Finoro, 1980	25S$_S$	4.1^w	2.4	1.5	3.3	4.2	11.5	—	20.7	—	—	14.9	22.1	2.9	1.0
Jamotte and Guyot, 1980	Commercial solid fraction A fractionated at 25°C	2.3^n	1.6	1.1	2.6	4.0	13.3	1.1	34.6	2.0	0.8	12.6	18.8	1.6	1.1
	Commercial solid fraction B fractionated at 25°C	2.5^n	1.4	1.0	2.3	3.3	11.5	1.4	30.8	2.1	1.0	15.8	21.5	1.1	0.9
Kaylegian, 1991	25S$_M$ [W]	3.4^a	1.9	1.1	2.7	3.5	11.7	—	31.8	1.3	—	12.4	17.7	2.1	0.4
	25S$_S$ [W]	4.2^a	2.1	1.1	2.5	3.1	10.3	—	29.7	1.5	—	12.3	20.6	2.5	—
Timms and Parekh, 1980	25S$_S$ [S]	3.4^w	2.3	1.3	3.2	4.0	12.3	1.3	29.1	2.0 incl C17:0br	1.0	13.9	19.4	1.6	1.6 incl C18:2 conj
Kaylegian, 1991	20S$_M$ [W]	2.5^a	1.5	1.0	2.7	4.1	13.2	—	32.0	1.2	—	12.5	16.7	2.5	0.3
	20S$_S$ [W]	4.1^a	2.2	1.2	2.6	3.2	10.3	—	29.0	1.6	—	11.7	21.5	2.3	0.3
Vacuum filtration															

Versteeg, 1991	18S$_M$ [Sept]	3.2^n	1.9	1.1	2.8	3.1	10.5	1.2	25.3	2.3	1.1	15.7	22.0	1.6	1.7
Schaap and van Beresteyn, 1970	17S$_M$	3.4^n	2.2	1.4	3.0	3.5	11.9	—	31.0	—	—	14.9	24.0	2.2	2.5
Deffense, 1987	1st stearin$_M$ [2]	2.7^w	1.2	0.8	2.0	2.8	10.1	1.3	28.6	2.2	1.0	15.1	22.8	2.5	0.7
Timms, 1978b	Hard	3.4^w	2.3	1.2	2.7	3.4	11.7	1.7	31.0	2.7 incl C17:0br	1.4	12.0	19.7	1.3	1.2
Laakso et al., 1992	S1	7.4^m	3.4	1.6	3.0	3.3	11.4	2.1 incl 14:1	30.3	0.9	1.4	13.8	17.1	2.0 incl C19:0	0.3
	S1 SSS^4	8.1^m	3.9	1.9	3.9	4.4	15.3	1.8	40.0	—	1.5	18.5	0.1	0.1 incl C19:0	—
	S1 SSM^t	1.7^m	2.4	1.2	2.5	2.5	10.0	1.4	27.8	2.4 incl C17:0	—	14.1	28.4	4.0 incl C19:0, C19:1	—
	S1 SSM^c	6.8^m	3.0	1.4	2.4	2.4	8.9	2.6 incl C14:1	24.9	2.3	1.0	11.2	29.6	1.3 incl C19:0	—
	S1 SM^{c}M^t	3.4^m	1.6	0.7	1.9	1.2	5.9	2.5 incl C14:1	12.1	2.8	1.2	7.0	50.6	6.9 incl C19:1	—
	S1 SM^{c}M^c	4.3^m	1.8	0.8	1.4	1.2	4.3	3.3 incl C14:1	12.1	3.2	1.1	5.8	53.7	3.5 incl C19:0, C19:1	0.2
	S1 UUU	3.9^m	1.9	0.9	1.9	1.4	4.9	1.9 incl C14:1	12.5	1.7	0.8	6.6	36.2	20.8 incl C19:0	2.2
Centrifugation															
Riel and Paquet, 1972	38S$_M$	0.5^n	1.6	1.2	3.0	4.0	14.1	—	34.8	—	—	19.8	20.6	—	—
Lakshminarayana and Rama Murthy, 1985	31S$_M$	3.2^w	1.7	0.8	2.2	2.7	11.9	—	34.7	—	—	12.5	29.1	—	—

Continued

TABLE 5.40 (Continued)

Fraction separation method Author	Fraction designation[1]	Fatty acid (%)[2]												
		4:0	6:0	8:0	10:0	12:0	14:0	15:0	16:0	16:1	17:0	18:0	18:1	18:2	18:3
Richardson, 1968	30S$_M$ [W]	3.2^w	2.0	1.2	2.7	2.9	9.7	1.2	32.1	2.8	1.1	12.9	22.4	1.6	1.2
Jordan, 1986	28S$_S$ [W]	2.8^w	2.1	1.0	2.0	2.6	10.2	1.1	31.9	1.8	0.7	12.9	22.0	1.6	0.4
Ramesh and Bindal, 1987b	28S$_S$ [Ghee]	2.3^w	2.0	1.6	2.5	3.7	12.7	1.2	30.3	0.6	—	13.0	23.4	0.8	0.2
Kankare and Antila, 1988b	12S$_M$ [S]	3.0^n	—	—	—	—	—	—	28.0	—	—	14.5	21.8	—	—
Keogh and Higgins, 1986b	Fraction 1	2.2^w	1.3	0.8	2.1	3.0	12.9	—	37.4	1.9	1.1	13.9	15.0	1.3	0.3
	Fraction 2	2.7^w	1.7	1.0	2.5	3.3	12.1	—	31.0	1.9	0.9	14.8	19.0	1.3	0.5
Separation with the aid of an aqueous detergent solution															
Doležálek et al., 1976	27S$_S$ [1]	1.2^w	1.4	0.8	2.7	4.1	12.3	1.8	35.7	2.1	1.0	9.0	18.3	1.8	0.3
Norris et al., 1971	26S$_S$ [B late S]	1.8^a	1.5	1.0	2.6	3.1	10.9	1.2	30.0	1.0	—	15.3	27.0	0.8	1.5
	26S$_S$ [C early F]	1.3^a	1.5	1.1	2.9	3.4	10.9	1.1	29.0	1.0	—	16.2	25.5	1.6	2.4
	25S$_S$ [A late S]	1.4^a	1.3	0.9	2.2	1.5	10.8	1.2	29.7	1.0	—	17.0	28.6	0.8	1.7
Banks et al., 1985	16 semi-S$_S$ [B] high level of detergent	2.6^w	1.2	1.3	2.1	2.7	11.4	—	28.5	0.7	—	18.0	29.0 cis = 20.9 trans = 8.1 C18:3	0.7 incl	—
Fjaervoll, 1970a	HMF	2.5^n	1.9	1.3	3.0	4.3	12.8	—	28.8	—	—	17.3	25.3	1.3	1.5
Dolby, 1970a	Solid [mid-season]	—	—	2.7^n incl C:4 + C:6	3.3	5.3	17.1	—	35.5	tr^5	—	16.1	16.9	—	—
Filtration through milk filter															
Fouad et al., 1990	29S$_S$	2.1^w	0.9	0.7	2.3	3.4	12.8	—	36.6	0.9	—	16.5	16.1	1.2	0.4
	25S$_S$	3.8^w	2.1	1.2	2.6	3.4	12.5	—	34.6	0.9	—	14.9	16.5	1.1	0.4
Filtration through casein-dewatering press															

Technical Data for Experimental and Commercial Milkfat Fractions

Black, 1973	Hard #1	2.6^w	1.5	1.2	2.8	3.7	12.1	—	28.3	3.2	—	16.2	19.5	3.1	1.4
	Hard #2	2.5^w	1.7	1.2	2.9	3.9	12.0	—	26.5	3.5	—	16.7	19.8	3.2	1.6
	Hard #3	2.0^w	1.2	0.8	2.3	3.1	11.7	—	29.4	2.6	—	19.5	20.1	2.7	1.4
Separation technique not reported															
Deroanne, 1976	21S$_S$ [W]	1.8^n	1.3	0.9	2.1	2.8	10.8	1.3	31.5	2.5	0.9	13.1	24.1	2.2	1.3 incl C20
Deroanne and Guyot, 1974	21S$_S$ [W]	3.5^a	2.2	1.2	3.0	4.2	12.7	1.5	32.3	2.7 incl C17:0	—	9.4	20.0	1.5	0.8 incl C20:1
	21S$_S$ [S]	3.5^a	2.1	1.2	3.2	3.5	12.0	1.4	27.0	2.7 incl C17:0	—	13.0	23.3	1.7	1.1 incl C20:1
Martine, 1982	1st stearin	1.1^n	0.7	0.8	2.1	2.9	12.3	0.1	29.4	0.7	0.3	39.7	1.6	1.3	1.1
	2nd stearin	1.0^n	0.7	0.8	1.8	2.6	11.4	0.1	27.4	0.7	0.3	42.5	1.7	1.4	1.5

[1]The number indicates fractionation temperature, the letter indicates physical state (S = solid), and the subscript indicates fractionation procedure used (M = multiple-step, S = single-step). Other descriptions used are those of the original author. The designation inside the brackets refers to the intact anhydrous milkfat source (S = summer, W = winter)].

[2]br = branched, con = conjugated, incl = includes, a = area %, m = mole %, n = % not specified, w = wt %.

[3]Data not available.

[4]S = saturated, M = monounsaturated, U = polyunsaturated, c = cis, t = trans.

[5]tr = trace.

TABLE 5.41 Individual Fatty Acid Composition of High-Melting Milkfat Fractions Obtained from Solvent Solution

Type of solvent solution Author	Fraction designation[1]	Fatty acid (%)[2]													
		4:0	6:0	8:0	10:0	12:0	14:0	15:0	16:0	16:1	17:0	18:0	18:1	18:2	18:3
Acetone																
Sherbon and Dolby, 1973	25S$_M$ (IV) refractionated [HMG]	0.7^m	0.7	0.5	2.2	4.2	15.1	1.9	37.7	0.2	1.0	17.8	12.8	0.5	0.6
Kaylegian, 1991	11S$_M$ [W]	1.8^a	1.1	0.7	2.5	4.4	13.7	—[3]	33.9	0.9	—	13.6	15.4	1.9	0.2
Yi, 1993	10S$_M$ [S]	1.5^a	0.4	0.8	2.8	4.2	14.0	—	37.3	1.3	—	15.0	19.8	1.8	0.4
Kaylegian, 1991	5S$_S$ [W]	2.9^a	1.7	0.9	2.2	2.9	10.5	—	34.1	1.1	—	16.8	14.9	1.8	0.2
Timms, 1980b	3S$_M$ [S]	2.1^w	1.5	0.8	2.0	2.5	10.2	1.5	35.1	1.6 incl C17br	1.2	22.0	14.9	0.9	0.7 incl C18:2 conj
Kaylegian, 1991	0S$_S$ [W]	3.9^a	2.1	1.1	2.3	2.7	10.1	—	33.4	1.2	—	14.4	16.7	1.9	0.1
Sherbon, 1963	2BJAN [Jan]	5.3^m	10.3	1.4	3.0	2.4	10.3	1.1	37.6	0.7	1.2	14.2	9.1	1.1	0.4

[1]The number indicates fractionation temperature, the letter indicates physical state (S = solid), and the subscript indicates fractionation procedure used (M = multiple-step, S = single-step). Designations in parentheses are those of the original author. The designation inside the brackets following the fraction designation refers to the intact anhydrous milkfat source (HMG = high-melting glyceride, S = summer, W = winter).

[2]br = branched, conj = conjugated, incl = includes, a = area %, m = mole %, w = wt %.

[3]Data not available.

TABLE 5.42 Individual Fatty Acid Composition of High-Melting Milkfat Fractions Obtained by Supercritical Fluid Extraction

Type of solvent Author	Fraction designation[1]	Fractionation conditions		Fatty acid (%)[2]													
		Temperature (°C)	Pressure	4:0	6:0	8:0	10:0	12:0	14:0	15:0	16:0	16:1	17:0	18:0	18:1	18:2	18:3
Carbon dioxide																	
Biernoth and Merk, 1985	Residual	80	200	2.0[a]	1.8	1.1	2.8	13.5	10.7	—[3]	29.9	—	—	9.8	27.0	2.9	1.5
Kaufmann et al., 1982	I (residue)	80	200 bar	7.6[w]	4.1	2.0	4.0	4.4	12.0	1.1	30.5	1.7	0.6	8.9	13.4	2.8	1.2
Timmen et al., 1984	Long chain fraction	80	200 bar	2.7[w]	1.8	1.1	2.7	3.3	10.4	—	30.1	—	—	9.6	24.6	3.0	1.4
Arul et al., 1987	S2	70	250–350 bar	6.0[m]	7.0	3.3	6.3	4.8	13.4	tr[4]	30.4	0.0	—	9.0	18.4	tr	—
	S3	70	250–350 bar	9.3[m]	1.8	0.0	2.7	2.9	10.6	tr	29.3	1.4	—	12.2	27.2	1.1	—
Kankare and Antila, 1988a	Res	50–80	100–400 bar	2.4[n]	—	—	3.1	—	—	—	29.5	—	—	11.3	24.9	—	—
Büning-Pfaue et al., 1989	Extractor fraction	50	25,000 kPa	3.5[w]	2.2	1.2	3.1	4.0	13.3	—	36.1	—	—	6.8	20.6	1.8	—
Kankare et al., 1989	R	50	400 bar	2.4[n]	1.9	1.2	3.0	3.3	11.5	—	29.0	1.8	—	11.7	25.1	2.9	0.4
Büning-Pfaue et al., 1989	Column fraction	30	20,000 kPa	1.6[w]	1.0	0.7	2.0	2.8	11.1	—	35.8	—	—	9.3	26.1	2.2	—

[1]Fraction designations used are those given by the original author.

[2]a = area %, m = mole %, n = % not specified, w = wt %.

[3]Data not available.

[4]tr = trace.

TABLE 5.43 Individual Fatty Acid Composition of High-Melting Milkfat Fractions Obtained by Short-Path Distillation

Author	Fraction designation[1]	Fatty acid (%)[2]												
		4:0	6:0	8:0	10:0	12:0	14:0	15:0	16:0	16:1	17:0	18:0	18:1	18:2	18:3
Arul et al., 1988a	SF Residue	2.0[m]	1.5	1.0	2.4	2.8	9.8	—[3]	28.0	1.5	—	13.2	27.0	2.4	1.1

[1]Fraction designations are those given by the original author.

[2]m = mole %.

[3]Data not available.

TABLE 5.44 Individual Fatty Acid Composition of Middle-Melting Milkfat Fractions Obtained from Melted Milkfat

Fraction separation method Author	Fraction designation[1]	Fatty acid (%)[2]												
		4:0	6:0	8:0	10:0	12:0	14:0	15:0	16:0	16:1	17:0	18:0	18:1	18:2	18:3
Vacuum filtration															
Kaylegian, 1991	34L$_S$ [W]	5.2[a]	2.7	1.3	2.8	3.2	10.0	—[3]	27.6	1.5	—	10.4	22.9	2.2	—
deMan and Finoro, 1980	32L$_S$	4.2[w]	2.6	1.5	3.3	3.9	11.0	—	20.7	—	—	12.8	24.4	2.8	1.1
	31L$_S$	4.6[w]	2.8	1.6	3.3	3.8	11.7	—	21.2	—	—	11.5	25.3	2.9	1.0
	30L$_S$	3.7[w]	2.2	1.4	3.2	3.6	11.6	—	24.6	—	—	10.5	26.0	2.4	1.0
Kaylegian, 1991	30L$_S$ [W]	5.4[a]	2.8	1.4	3.0	3.4	10.0	—	26.6	1.5	—	9.7	23.1	2.8	0.4
deMan and Finoro, 1980	29L$_S$	4.9[w]	2.5	1.5	3.1	3.6	11.3	—	21.3	—	—	12.0	25.8	2.8	1.1
	28L$_S$	4.4[w]	2.5	1.4	3.0	3.5	11.1	—	22.5	—	—	11.2	26.6	2.8	1.0
	27L$_S$	4.9[w]	3.0	1.5	3.2	3.7	11.1	—	20.6	—	—	11.7	25.2	3.0	1.0
Barna et al., 1992	20S$_M$ [W]	3.5[n]	2.0	1.2	3.0	3.8	12.7	—	33.2	1.2	—	13.8	18.6	2.5	—
Grall and Hartel, 1992	20S$_M$ [S]	3.5[a]	2.0	1.2	3.0	3.8	12.7	—	33.2	1.2	—	13.8	18.6	2.5	—
Sherbon, 1963	17S$_M$	13.4[m]	2.4	2.3	3.8	2.3	8.9	0.8	30.6	0.5	1.0	7.0	22.0	1.5	0.9
Kaylegian, 1991	16S$_M$ [W]	4.5[a]	2.5	1.4	2.7	2.8	9.4	—	33.1	1.3	—	12.3	17.1	2.5	0.4
Laakso et al., 1992	S2	8.5[m]	4.0	1.7	2.9	2.7	11.9	2.5 incl C14:1	26.1	1.0	1.5	12.5	19.9	3.7 incl C19:0, C19:1	0.4

S2 SSS[4]	12.5[m]	5.8	2.4	4.1	3.7	13.0	2.0	35.3	—	1.7	18.3	0.6	0.1 incl C19:0		
S2 SSM[t]	4.8[m]	4.4	1.9	3.4	3.0	9.8	1.7	22.0	2.7 incl 17:0	—	12.1	27.9	4.0 incl C19:0, C19:1	—	
S2 SSM[c]	6.9[m]	2.8	1.2	2.2	2.3	8.8	3.0 incl C14:1	23.6	2.4	1.0	12.2	30.1	1.6 incl C19:0	—	
S2 SM[c]M[t]	3.6[m]	1.7	0.7	1.2	1.2	4.7	1.9 incl C14:1	12.4	1.4	1.5	7.0	51.6	7.4 incl C19:1	—	
S2 SM[c]M[c]	4.3[m]	1.8	0.8	1.4	1.3	4.5	3.2 incl C14:1	11.2	2.8	1.3	5.6	54.5	4.0 incl C19:0, C19:1	0.1	
S2 UUU	3.9[m]	1.8	0.8	1.2	1.2	4.0	1.9 incl C14:1	10.9	1.8	0.7	6.4	38.8	22.2 incl C19:0, C19:1	2.1	

Pressure filtration

deMan, 1968	30L$_S$ (B) rapid cooling	3.8[n]	1.9	1.2	2.1	2.3	9.4	—	27.4	—	—	11.7	31.5	1.9	0.5

Centrifugation

Richardson, 1968	25S$_M$ [W]	4.3[w]	2.5	1.3	2.8	3.3	9.4	1.4	29.6	2.8	1.0	12.0	23.9	1.7	1.1
Lakshminarayara and Rama Murthy, 1985	23S$_M$	3.8[w]	1.7	1.1	2.4	2.8	14.8	—	31.7	—	—	10.5	31.3	—	—
Richardson, 1968	20S$_M$ [W]	5.3[w]	2.9	1.5	3.1	3.3	9.0	1.2	30.6	2.5	0.8	11.8	23.3	1.4	0.9
Kankare and Artila, 1986	12S$_M$ [S]	3.2[n]	2.3	1.4	3.1	3.2	11.0	—	28.2	1.6	—	13.6	22.0	2.6	0.5
Kankare and Artila, 1988a	12S$_M$ [S]	3.5[n]	—	—	—	—	—	—	28.0	—	—	13.0	22.0	—	—
Antila and Antila, 1970	C	2.3[n]	1.7	1.1	2.7	3.3	12.3	1.5	35.9	1.4	0.5	9.6	18.3	2.7	0.3

Continued

TABLE 5.44 (Continued)

Fraction separation method Author	Fraction designation[1]	Fatty acid (%)[2]													
		4:0	6:0	8:0	10:0	12:0	14:0	15:0	16:0	16:1	17:0	18:0	18:1	18:2	18:3
Separation with the aid of an aqueous detergent solution															
Doležálek et al., 1976	27L$_S$ [1]	1.7^w	1.9	1.3	3.6	4.6	12.4	1.7	30.5	2.8	1.0	7.0	22.0	2.4	0.4
Filtration through milk filter															
Fouad et al., 1990	29L$_S$	5.3^w	2.3	1.3	3.1	3.6	11.7	—	29.4	1.3	—	10.9	22.9	1.7	0.5
	17S$_S$	3.9^w	2.2	1.2	3.0	3.5	12.2	—	31.8	1.2	—	12.7	20.2	1.5	0.5
Separation technique not reported															
Riel and Paquet, 1972	38S$_M$	1.0^n	1.9	1.2	2.7	3.6	13.2	—	33.1	—	—	18.8	24.3	—	—
Deroanne, 1976	21L$_S$ [W]	2.5^n	1.8	1.1	2.4	2.9	10.3	1.1	27.3	3.0	0.8	9.5	28.9	3.4	1.5 incl C20
Riel and Paquet, 1972	4S$_M$	1.8^n	2.4	1.6	3.7	4.5	14.5	—	31.1	—	—	14.9	25.3	—	—
Martine, 1982	Olein	—	—	1.3^n	3.0	3.7	12.3	1.4	33.2	2.0	0.8	9.4	23.4	1.5	0.1

[1]The number indicates fractionation temperature, the letter indicates physical state (L = liquid, S = solid), and the subscript indicates fractionation procedure used (M = multiple-step, S = single-step). Other primary designations and designations in parentheses are those of the original author. The designation inside the brackets refers to the intact anhydrous milkfat source [sample number or letter or season (S = summer, W = winter)].

[2]incl = includes, a = area %, m = mole %, n = % not specified, w = wt %.

[3]Data not available.

[4]S = saturated, M = monounsaturated, U = polyunsaturated, c = cis, t = trans.

TABLE 5.45 Individual Fatty Acid Composition of Middle-Melting Milkfat Fractions Obtained from Solvent Solution

Type of solvent solution Author	Fraction designation[1]	\								Fatty acid (%)[2]						
		4:0	6:0	8:0	10:0	12:0	14:0	15:0	16:0	16:1	17:0	18:0	18:1	18:2	18:3
Acetone															
Kaylegian, 1991	25L$_S$ [W]	4.6[a]	2.6	1.4	3.0	3.5	10.8	—[3]	26.9	1.7	—	9.7	21.9	3.2	0.8
Sherbon and Dolby, 1973	25L$_M$ (V) [HMG]	3.4[m]	2.4	1.4	3.3	3.3	12.4	1.7	34.0	0.8	1.0	12.9	17.7	0.6	1.2
Kaylegian, 1991	21L$_S$ [W]	5.1[a]	2.8	1.5	3.2	3.4	10.2	—	26.4	1.7	—	9.6	22.7	3.1	0.3
	5S$_M$ [W]	3.1[a]	1.9	1.0	2.4	2.7	9.1	—	36.9	0.8	—	16.1	13.8	2.0	0.2
Yi, 1993	5S$_M$ [S]	2.9[a]	1.8	1.1	2.8	2.9	9.4	—	40.0	1.2	—	19.0	15.5	2.1	0.4
Kaylegian, 1991	0S$_M$ [W]	4.3[a]	2.5	1.2	2.4	2.7	10.8	—	36.5	1.0	—	11.3	14.8	2.3	0.2
Yi, 1993	0S$_M$ [S]	3.6[a]	1.9	1.1	2.6	2.5	11.3	—	42.9	1.2	—	13.8	15.9	1.9	0.4
Pentane															
Kaylegian, 1991	11L$_S$ [W]	3.8[a]	2.2	1.3	3.0	3.2	10.2	—	26.8	2.0	—	10.2	23.3	3.5	0.6
	5L$_S$ [W]	4.1[a]	2.5	1.4	3.1	3.4	10.2	—	26.2	2.0	—	9.4	23.6	3.5	0.7

[1]The number indicates fractionation temperature, the letter indicates physical state (L = liquid, S = solid), and the subscript indicates fractionation procedure used (M = multiple-step, S = single-step). Any designations in parentheses are fraction designations given by the original author. The designation inside the brackets refers to the intact anhydrous milkfat source (HMG = high-melting glyceride, S = summer, W = winter).

[2]a = area %, m = mole %.

[3]Data not available.

TABLE 5.46 Individual Fatty Acid Composition of Middle-Melting Milkfat Fractions Obtained by Supercritical Fluid Extraction

Type of solvent Author	Fraction designation[1]	Fractionation conditions		Fatty acid (%)[2]												
		Temperature (°C)	Pressure	4:0	6:0	8:0	10:0	12:0	14:0	15:0	16:0	16:1	17:0	18:0	18:1	18:2	18:3

Type of solvent Author	Fraction designation[1]	Temperature (°C)	Pressure	4:0	6:0	8:0	10:0	12:0	14:0	15:0	16:0	16:1	17:0	18:0	18:1	18:2	18:3
Carbon dioxide																	
Arul et al., 1987	S1	70	250–350 bar	12.3[m]	9.1	3.1	4.8	3.7	12.1	tr[3]	31.1	0.0	—[4]	7.4	16.5	tr	—
Kankare and Antila, 1988a	E2	50–80	100–400 bar	5.8[n]	—	—	4.3	—	—	—	29.2	—	—	7.9	17.2	—	—
	E3	50–80	100–400	4.9[n]	—	—	3.9	—	—	—	29.7	—	—	9.1	19.2	—	—
	E4	50–80	100–400	4.8[n]	—	—	3.6	—	—	—	29.5	—	—	9.2	20.3	—	—
Arul et al., 1987	I2	50	100–250 bar	17.6[m]	10.2	3.6	5.5	4.6	13.6	0.8	26.9	tr	—	5.2	10.6	tr	—
	I3	50	100–250	12.5[m]	8.3	3.2	5.3	4.1	13.0	1.0	27.2	1.2	—	7.2	14.6	0.8	—
Kankare et al., 1989	E2	50	200 bar	5.9[n]	3.6	2.2	4.6	4.8	14.3	—	29.7	1.2	—	8.2	17.5	1.6	0.4
	E3	50	300	5.0[n]	3.2	1.9	4.0	4.2	13.5	—	30.4	1.2	—	9.2	19.5	1.9	0.4
	E4	50	400	4.6[n]	3.1	1.8	3.8	4.1	13.1	—	30.0	1.6	—	9.4	20.4	2.3	0.4
Hamman et al., 1991	A4	40	125 bar	1.7[w]	1.9	1.3	3.1	3.1	9.9	—	27.2	1.8	—	14.3	30.9	1.4	—

[1]Fraction descriptions used are those given by the original author.

[2]Units of chemical composition are designated using these abbreviations: m = mole %, n = % not specified, w = wt %.

[3]tr = trace.

[4]Data not available.

TABLE 5.47 Individual Fatty Acid Composition of Middle-Melting Milkfat Fraction Obtained by Short-Path Distillation

Author	Fraction designation[1]	Fractionation conditions		Fatty acid (%)[2]											
		Temperature (°C)	Pressure	4:0	6:0	8:0	10:0	12:0	14:0	15:0	16:0	16:1	18:0	18:1	18:2	18:3
Arul et al., 1988a	IF Distillate III	265	100 μm Hg	9.8[m]	4.9	2.2	4.2	4.2	11.8	—[3]	29.2	1.2	8.6	15.0	1.5	0.7

[1]Fraction descriptions used are those given by the original author.

[2]m = mole %.

TABLE 5.48 Individual Fatty Acid Composition of Low-Melting Milkfat Fractions Obtained from Melted Milkfat

Fraction separation method Author	Fraction designation[1]	4:0	6:0	8:0	10:0	12:0	14:0	15:0	16:0	16:1	17:0	18:0	18:1	18:2	18:3
Vacuum filtration															
deMan and Finoro, 1980	26L$_S$	4.4[w]	2.4	1.4	3.0	3.4	11.2	—[3]	22.6	—	—	11.1	27.3	2.8	1.0
Lund and Danmark, 1987	26L$_S$ [1]	4.6[w]	2.8	1.5	3.0	3.8	11.0	—	29.0	—	—	10.0	25.5	2.5	1.0
deMan and Finoro, 1980	25L$_S$	5.3[w]	2.7	1.5	3.2	3.7	11.9	—	20.5	—	—	11.8	25.6	3.1	1.0
Kaylegian, 1991	25L$_S$ [W]	4.5[a]	2.6	1.4	3.1	3.6	10.3	—	26.0	1.8	—	8.5	24.4	3.5	0.4
Timms and Parekh, 1980	25L$_S$ [S]	4.3[w]	3.0	1.7	3.7	4.0	11.4	1.5	25.4	2.3 incl C17:0br	1.0	10.7	22.4	2.1	2.0 incl C18:2 conj
Barna et al., 1992	20L$_M$ [W]	5.2[n]	2.9	1.6	3.3	3.5	10.9	—	28.5	1.6	—	10.6	26.6	3.2	—
Kaylegian, 1991	20L$_S$ [W]	6.3[a]	3.2	1.6	3.2	3.4	9.5	—	24.5	1.9	—	8.9	24.7	2.5	0.4
Schaap and van Beresteyn, 1970	17L$_M$	4.0[n]	2.7	1.7	3.3	4.0	11.5	—	24.2	—	—	10.9	32.9	2.6	2.2
Grall and Hartel, 1992	15S$_M$ [S]	5.2[a]	3.1	1.6	3.2	3.3	10.7	—	29.4	1.6	—	11.0	25.6	3.0	—
	15L$_M$ [S]	5.1[a]	2.9	1.7	3.5	3.8	11.2	—	25.6	1.8	—	9.4	29.2	3.6	—
Kaylegian, 1991	13S$_M$ [W]	4.8[a]	2.5	1.2	2.8	3.2	10.5	—	29.4	1.7	—	9.8	21.6	2.5	0.2
	13L$_M$ [W]	5.4[a]	2.9	1.6	3.4	3.7	10.3	—	22.1	2.0	—	6.4	28.4	3.3	0.4
Sherbon, 1963	10S$_M$	7.9[m]	4.5	2.0	4.5	3.2	9.9	0.8	27.7	0.6	0.8	5.8	27.0	1.7	1.0
	10L$_M$	9.9[m]	3.9	1.7	4.4	3.1	9.3	0.7	22.5	0.7	0.8	4.3	32.5	2.0	1.2
Deffense, 1987	2nd olein$_M$ filtered at 10°C [2]	3.9[w]	1.8	1.2	2.7	3.2	9.2	1.1	18.5	3.0	0.9	8.0	34.8	4.4	1.1
Timms, 1978b	1st Olein$_M$ [2]	4.0[w]	1.8	1.2	2.6	3.1	9.8	1.2	22.3	2.8	0.9	9.9	32.0	3.7	1.0
	Soft	4.7[w]	3.0	1.5	2.9	3.2	10.4	1.5	26.8	3.0 incl C17:0br	1.2	10.4	24.4	1.6	1.5
Laakso et al., 1992	L1	9.8[m]	4.5	2.0	3.4	3.2	10.1	2.2 incl C14:1	25.0	1.3	1.3	10.0	22.1	2.5 incl C19:0	0.4

Continued

TABLE 5.48 (Continued)

Fraction separation method Author	Fraction designation[1]	Fatty acid (%)[2]													
		4:0	6:0	8:0	10:0	12:0	14:0	15:0	16:0	16:1	17:0	18:0	18:1	18:2	18:3
	L1 SSS[4]	14.4[m]	6.4	2.8	4.8	4.6	14.4	1.8	35.2	—	1.4	12.8	0.3	0.2 incl C19:0	—
	L1 SSM[t]	6.9[m]	5.5	2.4	4.2	3.4	9.8	1.5	20.7	2.4 incl C17:0	—	9.2	28.2	4.1 incl C19:0, C19:1	—
	L1 SSM[c]	8.0[m]	3.6	1.6	2.7	2.7	9.0	2.7 incl C14:1	23.0	2.1	0.9	9.8	30.5	1.3 incl C19:0	—
	L1 SM[c]M[t]	3.9[m]	2.0	0.8	1.5	1.3	4.7	2.0 incl C14:1	12.3	2.5	2.1	6.4	49.6	6.9 incl C19:1	0.4
	L1 SM[c]M[c]	4.2[m]	1.8	0.8	1.3	1.2	4.3	3.1 incl C14:1	11.8	3.3	1.1	5.2	55.8	2.7 incl C19:0, C19:1	0.1
	L1 UUU	4.1[m]	2.6	1.1	1.5	1.4	4.7	2.2 incl C14:1	11.7	1.9	1.0	5.5	35.1	21.0 incl C19:0, C19:1	2.6
	L2	9.7[m]	4.5	2.0	3.6	3.3	9.8	2.7 incl C14:1	20.2	1.5	1.4	8.2	27.2	3.1 incl C19:0, C19:1	0.5
	L2 SSS	14.4[m]	7.2	3.3	6.0	5.7	16.0	2.6	30.4	—	1.9	10.7	0.7	0.3 incl C19:0	—
	L2 SSM[t]	8.0[m]	6.6	2.7	4.6	3.7	9.7	1.8	17.4	2.8 incl C17:0	—	8.3	30.0	3.0 incl C19:0, C19:1	—

L2 SSM^c	8.8^m	3.8	1.7	3.0	2.9	9.4	2.9 incl C14:1	20.4	2.4	1.5	8.9	30.9	1.4 incl C19:0	—
L2 SM^{c}M^t	4.1^m	2.0	0.8	1.3	1.4	4.6	3.1 incl C14:1	10.8	2.2	1.3	6.4	51.2	7.5 incl C19:1	0.3
L2 SM^{c}M^c	4.4^m	1.9	0.8	1.9	1.3	4.7	3.5 incl C14:1	10.9	3.0	0.5	5.2	54.7	3.2 incl C19:0, C19:1	0.2
L2 UUU	4.3^m	2.3	0.8	1.7	1.2	3.8	2.2 incl C14:1	9.3	2.4	1.3	4.3	40.1	20.2 incl C19:0	1.6
Vovan and Riel, 1973														
5% removal of hard fraction (>21°C)	3.6^n	2.6	1.5	4.0	4.2	11.5	0.9	25.6	2.5	0.7	12.8	24.9	1.8	1.6
10% removal of hard fraction (>21°C)	3.2^n	2.4	1.7	4.1	4.5	11.8	1.0	24.0	2.4	0.7	12.1	26.9	1.4	1.8
15% removal of hard fraction (>21°C)	3.2^n	2.1	1.3	3.5	3.5	11.1	1.1	25.1	3.0	0.9	13.5	25.7	1.9	2.0
5% removal of intermed. fraction (10–21°C)	2.7^n	2.1	1.3	3.5	4.1	10.9	1.2	25.9	2.9	0.9	12.8	25.9	1.9	1.6
10% removal of intermed. fraction (10–21°C)	4.6^n	2.7	1.6	3.7	3.6	12.7	1.1	24.6	2.6	0.8	12.3	25.4	1.4	1.7
15% removal of intermed. fraction (10–21°C)	3.8^n	1.7	1.0	3.1	3.9	11.1	1.1	25.9	2.9	0.5	13.2	26.6	1.4	1.4
33% removal of intermed. fraction (10–21°C)	3.8^n	1.9	1.3	3.3	4.1	12.3	1.0	26.6	2.3	0.7	12.0	25.4	1.8	1.6

Continued

TABLE 5.48 (Continued)

Fraction separation method Author	Fraction designation[1]	Fatty acid (%)[2]													
		4:0	6:0	8:0	10:0	12:0	14:0	15:0	16:0	16:1	17:0	18:0	18:1	18:2	18:3
Pressure filtration															
deMan, 1968	27L$_S$ (C) slow cooling	3.8[n]	1.8	1.0	2.7	2.3	9.2	—	26.7	—	—	11.3	33.4	2.9	0.5
	27L$_S$ (D) rapid cooling	4.3[n]	2.0	1.1	2.2	2.0	9.0	—	26.9	—	—	10.9	33.4	2.3	0.5
	24L$_S$ (F) rapid cooling	4.3[n]	2.0	1.1	2.3	2.4	9.2	—	26.1	—	—	10.6	32.3	3.0	0.4
Centrifugation															
Jordan, 1986	28L$_S$ [W]	4.2[w]	3.0	1.4	2.5	2.7	10.1	1.0	29.7	1.9	0.5	9.4	24.4	1.8	0.5
Ramesh and Bindal, 1987b	28L$_S$ [Ghee]	4.1[w]	3.9	2.7	3.9	4.5	12.1	0.5	20.6	0.7	—	10.0	31.4	1.3	1.1
Lakshminarayana and Rama Murthy, 1985	15S$_M$	4.3[w]	2.5	1.4	2.6	3.0	12.8	—	30.5	—	—	10.9	31.7	—	—
	15L$_M$	4.5[w]	2.6	1.5	3.0	3.2	12.9	—	29.2	—	—	9.9	32.4	—	—
Richardson, 1968	15S$_M$ [W]	5.3[w]	2.7	1.4	2.7	3.0	9.6	1.3	29.3	2.7	1.0	10.6	24.3	2.0	1.2
	15L$_M$ [W]	5.3[w]	2.8	1.7	3.2	3.5	9.4	1.0	24.3	3.1	0.8	8.7	29.2	2.2	1.7
Kankare and Antila, 1986	12L$_M$ [S]	3.5[n]	2.4	1.5	3.1	3.3	10.7	—	22.3	2.0	—	9.3	30.7	3.3	0.9
Kankare and Antila, 1988a	12L$_M$ [S]	4.0[n]	—	—	—	—	—	—	22.5	—	—	9.0	31.0	—	—
Kankare and Antila, 1988b	12L$_M$ [S]	3.8[n]	—	—	—	—	—	—	22.0	—	—	9.0	30.0	—	—
Riel and Paquet, 1972	4S$_M$	2.0[n]	2.8	1.8	4.2	4.5	13.4	—	29.0	—	—	14.2	28.0	—	—
Antila and Antila, 1970	A	4.9[n]	2.1	1.9	3.4	4.6	10.7	1.8	19.7	2.9	3.0	4.5	24.5	4.7	1.5
	B	7.1[n]	5.8	3.2	5.2	4.7	11.5	1.7	20.3	2.0	1.4	4.4	19.2	3.0	1.4
Keogh and Higgins, 1986b	Fraction 3	4.5[w]	2.6	1.4	3.2	3.5	11.0	1.2	25.0	2.5	0.8	9.7	25.0	2.1	0.8
	Fraction 4	4.5[w]	2.7	1.6	2.4	3.7	10.9	1.2	21.1	2.3	0.8	10.7	26.3	2.2	1.1
Separation with the aid of an aqueous detergent solution															
Norris et al., 1971	26L$_S$ [B late S]	1.9[a]	1.8	1.2	2.8	3.2	10.6	1.2	26.4	1.7	—	11.9	30.9	1.4	2.6

Technical Data for Experimental and Commercial Milkfat Fractions   329

	26L$_S$ [C early F]	2.4^a	2.3	1.6	3.7	3.8	10.8	1.1	25.1	1.3	—	12.4	28.9	1.6	2.8
	25L$_S$ [A late S]	2.5^a	2.0	1.3	2.9	3.0	10.1	1.1	26.2	1.7	—	12.5	31.0	1.3	2.1
Banks et al., 1985	21L$_S$ [A] high level of detergent	3.9^w	2.4	1.0	1.7	1.9	7.8	—	36.0	2.4	—	8.9	30.5 cis = 27.8 trans = 2.7	2.4 incl C18:3	—
Dolby, 1970a	Liquid [mid-season]	—	—	3.3^w incl C:4, C:6	2.7	3.7	13.9	—	32.4	0.7	—	14.3	25.3	—	—
Fjaervoll, 1970a	LMF	4.0^n	2.8	1.6	4.0	4.7	12.6	—	24.3	—	—	12.1	30.8	1.4	1.7
Filtration through casein-dewatering press																
Black, 1973	Soft #1	3.7^w	2.1	1.3	3.2	3.9	12.2	—	24.6	2.7	—	14.0	21.1	3.9	2.5
	Soft #2	3.4^w	2.3	1.5	3.2	4.5	12.3	—	23.5	2.8	—	12.7	23.4	3.5	2.7
	Soft #3	3.0^w	2.0	1.5	3.1	4.1	10.9	—	23.1	2.7	—	13.4	25.1	4.1	2.9
Filtration through milk filter																
Fouad et al., 1990	17L$_S$	7.0^w	3.9	1.6	3.4	3.4	10.8	—	26.7	1.4	—	9.4	24.2	1.9	0.6
Separation technique not reported																
Deroanne, 1976	21L$_S$ [W]	2.5^n	1.8	1.1	2.4	2.9	10.4	1.1	27.3	3.0	0.8	10.0	28.9	28.9	1.5 incl C20:0
Deroanne and Guyot, 1974	21L$_S$ [W]	4.2^a	2.3	1.3	3.2	4.2	12.3	1.3	29.2	3.6 incl C17:0	—	7.1	22.7	2.0	1.1 incl C20:1

Continued

TABLE 5.48 (Continued)

Type of solvent solution Author	Fraction designation[1]	Fatty acid (%)[2]												
		4:0	6:0	8:0	10:0	12:0	14:0	15:0	16:0	16:1	17:0	18:0	18:1	18:2	18:3
	21L$_S$ [S]	4.0[a]	2.4	1.3	3.0	3.3	10.2	1.3	23.0	3.4 incl C17:0	—	10.8	29.0	2.0	1.6 incl C20:1

[1]The number indicates fractionation temperature, the letter indicates physical state (L = liquid, S = solid), and the subscript indicates fractionation procedure used (M = multiple-step, S = single-step). Other primary descriptions and descriptions in parentheses are those of the original author. The designation inside the brackets refers to the intact anhydrous milkfat source [sample number or letter or season (S = summer, W = winter)].

[2]br = branched, conj = conjugated, incl = includes. Units of chemical composition are designated using these abbreviations: a = area %, m = mole %, n = % not specified, w = wt %.

[3]Data not available.

[4]S = saturated, M = monounsaturated, U = polyunsaturated, c = cis, t = trans.

TABLE 5.49 Individual Fatty Acid Composition of Low-Melting Milkfat Fractions Obtained from Solvent Solution

Type of solvent solution Author	Fraction designation[1]	Fatty acid (%)[2]														
		4:0	6:0	8:0	10:0	12:0	14:0	15:0	16:0	16:1	17:0	18:0	18:1	18:2	18:3
Acetone																
Kaylegian, 1991	15L$_S$ [W]	4.8[a]	2.6	1.4	3.1	3.4	10.5	—[3]	27.1	1.7	—	8.7	23.3	2.8	0.5
	11L$_S$ [W]	4.8[a]	2.7	1.5	3.2	3.7	10.5	—	26.5	1.8	—	7.8	24.1	3.1	0.3
	5L$_S$ [W]	5.5[a]	3.1	1.6	3.2	3.3	9.9	—	23.9	1.9	—	7.2	26.1	3.7	0.7
Timms, 1980b	3L$_M$ [S]	4.9[w]	3.0	1.5	3.1	3.4	10.5	1.4	22.7	2.6 incl C17br	1.1	9.2	28.0	2.1	2.1 incl C18:2 conj
Yi, 1993	0L$_M$ [S]	4.7[a]	2.7	1.5	3.9	4.2	11.3	—	25.5	2.7	—	7.4	29.9	3.7	1.0
Sherbon, 1963	2C	4.9[m]	3.4	2.3	4.7	4.3	12.6	0.9	22.2	0.8	0.2	11.0	29.2	1.5	—
	2CJAN [Jan]	6.8[m]	3.5	1.8	4.6	3.8	10.3	1.0	27.2	2.1	0.3	8.3	26.4	1.7	—
Pentane																
Kaylegian, 1991	0L$_M$ [W]	5.4[a]	3.0	1.6	3.2	3.1	9.5	—	25.8	1.6	—	8.8	24.3	3.0	0.6
	0LS [W]	4.3[a]	2.5	1.4	3.0	3.4	10.0	—	26.1	2.0	—	9.3	23.7	3.6	0.7

[1]Fraction description: the fraction number indicates fractionation temperature, the letter indicates physical state (L = liquid), and the subscript indicates fractionation procedure used (S = single-step, M = multiple-step). The designation inside the brackets refers to the intact anhydrous milkfat source (S = summer, W = winter). Other descriptions used are those of the original author. Units of chemical composition are designated using these abbreviations: a = area %, m = mole %, w = wt %.

[2]br = branched, conj = conjugated, incl = includes.

[3]Data not available.

TABLE 5.50 Individual Fatty Acid Composition of Low-Melting Milkfat Fractions Obtained by Supercritical Fluid

Type of solvent Author	Fraction designation[1]	Fractionation conditions		Fatty acid (%)[2]													
		Temperature (°C)	Pressure	4:0	6:0	8:0	10:0	12:0	14:0	15:0	16:0	16:1	17:0	18:0	18:1	18:2	18:3
Carbon dioxide																	
Biernoth and Merk, 1985	Extract	80	200 bar	6.4[a]	4.3	3.0	5.9	6.9	16.0	—[3]	29.2	—	—	4.9	13.5	1.7	0.6
Kaufmann et al., 1982	II (extract)	80	200 bar	19.8[w]	7.7	3.5	5.9	6.2	13.3	1.0	23.3	1.1	0.3	3.4	8.6	1.1	0.4
Timmen et al., 1984	Short-chain fraction	80	200 bar	8.7[w]	4.3	2.5	5.1	6.2	15.4	—	30.3	—	—	5.0	12.5	1.6	0.6
Kankare and Antila, 1988a	E1	50–80	100–400 bar	6.6[n]	—	—	5.1	—	—	—	29.4	—	—	7.7	14.0	—	—
Arul et al., 1987	L1	50	100–250 bar	26.4[m]	12.7	5.6	7.6	5.4	12.6	tr[4]	19.2	tr	—	1.4	5.6	3.4	—
	L2	50	100–250	27.4[m]	13.4	4.9	6.7	4.9	12.3	tr	21.5	tr	—	1.9	7.1	tr	—
	I1	50	100–250	20.7[m]	11.4	3.7	5.4	5.0	13.8	tr	26.8	tr	—	4.5	8.9	tr	—
Kankare et al., 1989	E1	50	100 bar	6.4[n]	3.9	2.6	5.1	5.4	15.8	—	29.5	1.3	—	7.8	14.0	1.8	0.5
Hamman et al., 1991	A1	40	125 bar	3.4[w]	2.8	1.4	3.9	4.6	13.9	—	30.1	1.6	—	11.1	21.3	1.2	—

[1] Fraction designations used are those given by the original author.

[2] Units of chemical composition are designated using these abbreviations: a = area %, m = mole %, n = % not specified, w = wt %.

[3] Data not available.

[4] tr = trace.

TABLE 5.51 Individual Fatty Acid Composition of Low-Melting Milkfat Fractions Obtained by Short-Path Distillation

Author	Fraction designation[1]	Fractionation conditions			Fatty acid (%)[2]												
		Temperature (°C)	Pressure	4:0	6:0	8:0	10:0	12:0	14:0	15:0	16:0	16:1	17:0	18:0	18:1	18:2	18:3
Arul et al., 1988a	LF1 Distillate I	245	220 μm Hg	18.6m	7.3	4.0	6.8	7.3	15.8	—[3]	21.6	0.9	—	3.9	7.1	0.7	0.3
	LF2 Distillate II	245	100	15.6m	6.3	3.2	5.6	6.1	14.9	—	24.2	1.0	—	5.8	9.6	0.9	0.3

[1]Designations used are those given by the original author.
[2]m = mole %.
[3]Data not available.

TABLE 5.52 Individual Fatty Acid Composition of Very-Low-Melting Milkfat Fractions Obtained from Melted Milkfat

Fraction separation method Author	Fraction designation[1]	Fatty acid (%)[2]													
		4:0	6:0	8:0	10:0	12:0	14:0	15:0	16:0	16:1	17:0	18:0	18:1	18:2	18:3
Vacuum filtration															
Versteeg, 1991	8L_M [Sept]	4.3n	2.5	1.2	4.1	3.5	9.1	0.9	19.6	2.7	0.8	8.8	31.6	2.0	2.1
Deffense, 1987	2nd olein_M filtered at 6°C [2]	4.0w	1.9	1.3	2.8	3.2	8.9	1.0	17.6	3.1	0.9	7.4	36.3	4.6	1.2
Deffense, 1993b	2nd olein_M	3.1w	2.1	1.3	2.9	3.4	9.8	1.1	19.1	2.4	0.8	7.6	37.4	1.8	0.7
Centrifugation															
Riel and Paquet, 1972	4I_M	2.5n	3.2	2.4	4.8	5.2	13.3	—[3]	24.6	—	—	10.2	33.6	—	—

[1]The number indicates fractionation temperature, the letter indicates physical state (L = liquid), and the subscript indicates fractionation procedure used (M = multiple-step). Other descriptions used are those of the original author. The designation inside the brackets refers to the intact anhydrous milkfat source (sample number or month).
[2]n = % not specified, w = wt %.
[3]Data not available.

TABLE 5.53 Individual Fatty Acid Composition of Very-Low-Melting Milkfat Fractions Obtained from Solvent Solution

Type of solvent solution Author	Fraction designation[1]	\multicolumn{12}{c}{Fatty acid (%)[2]}
		4:0	6:0	8:0	10:0	12:0	14:0	15:0	16:0	16:1	17:0	18:0	18:1	18:2	18:3

Type of solvent solution Author	Fraction designation[1]	4:0	6:0	8:0	10:0	12:0	14:0	15:0	16:0	16:1	17:0	18:0	18:1	18:2	18:3
Acetone															
Kaylegian, 1991	0L$_M$ [W]	4.7[a]	2.7	1.6	3.4	3.9	10.2	—[3]	20.6	2.4	—	6.0	28.4	5.5	0.8
	0L$_S$ [W]	6.1[a]	3.3	1.7	3.3	3.5	9.7	—	21.0	2.1	—	6.6	26.2	3.9	0.7
Sherbon, 1963	2DJAN [Jan]	11.9^m	5.9	2.5	5.1	3.3	8.5	0.9	20.8	1.7	2.2	3.3	28.5	1.2	0.9

[1]The number indicates fractionation temperature, the letter indicates physical state (L = liquid), and the subscript indicates fractionation procedure used (M = multiple-step, S = single-step). Other descriptions used are those of the original author. The designation inside the brackets refers to the intact anhydrous milkfat source (W = winter).

[2]a = area %, m = mole %.

[3]Data not available.

TABLE 5.54 Individual Fatty Acid Composition of Unknown-Melting Milkfat Fractions Obtained from Melted Milkfat

Fraction separation method Author	Fraction designation[1]	4:0	6:0	8:0	10:0	12:0	14:0	15:0	16:0	16:1	17:0	18:0	18:1	18:2	18:3
Pressure filtration															
deMan, 1968	30L$_S$ (A) slow cooling	3.6^n	1.7	1.0	2.0	2.1	9.2	—[3]	26.8	—	—	12.2	31.6	3.2	0.5
	30S$_S$ (B) rapid cooling	2.6^n	1.1	0.5	1.5	1.9	9.8	—	34.4	—	—	19.5	24.0	0.3	—
	27S$_S$ (D) rapid cooling	2.9^n	1.4	0.8	1.7	2.2	10.0	—	31.1	—	—	16.1	26.4	2.1	—
	24L$_S$ (E) slow cooling	4.5^n	2.2	1.0	2.3	2.2	9.0	—	25.8	—	—	11.0	32.0	4.2	0.5
	24S$_S$ (F) rapid cooling	2.9^n	1.3	0.6	1.9	2.2	10.0	—	30.6	—	—	15.6	26.8	2.5	—
Badings et al., 1983b	23S$_M$ [W]	3.6^w	2.3	1.3	3.2	4.3	12.4	1.3	31.8	0.7	0.7	11.2	18.3	1.4	1.3
	23L$_M$ [W]	5.1^w	3.0	1.7	3.6	4.2	11.2	1.2	27.3	0.7	0.7	8.5	22.6	1.7	1.5
	21.5S$_M$ [S]	3.2^w	1.8	1.1	2.3	3.0	10.5	1.3	26.1	0.9	0.8	14.6	25.0	1.0	1.0

Technical Data for Experimental and Commercial Milkfat Fractions

Sample														
21.5L$_M$ [S]	4.5^w	2.4	1.4	2.8	3.1	9.3	1.1	22.1	1.0	0.7	10.7	29.6	1.4	1.2
18.5S$_M$ [W]	4.8^w	3.0	1.8	3.3	3.8	10.7	1.1	31.1	0.7	0.8	10.0	19.7	1.6	1.4
18.5L$_M$ [W]	5.0^w	3.0	1.7	3.7	4.4	11.5	1.2	25.0	0.7	0.7	7.9	24.5	2.0	1.6
15.3S$_M$ [S]	4.2^w	2.3	1.2	2.4	2.6	8.9	1.1	28.3	0.9	0.8	15.2	23.0	1.0	1.0
15.3L$_M$ [S]	4.5^w	2.4	1.4	2.8	3.2	9.5	1.1	21.2	1.0	0.7	9.9	30.7	1.5	1.2
15S$_M$ [W]	4.7^w	2.7	1.5	3.3	3.9	11.6	1.2	30.7	0.6	0.8	9.5	20.3	1.6	1.3
15L$_M$ [W]	5.0^w	3.0	1.8	3.8	4.6	11.5	1.1	22.7	0.8	0.8	7.0	26.2	2.1	1.6
13S$_M$ [S]	4.5^w	2.4	1.3	2.7	3.1	9.7	1.1	23.1	0.9	0.6	11.3	28.2	1.4	1.0
13L$_M$ [S]	4.6^w	2.5	1.4	2.9	3.3	9.3	1.1	19.0	1.0	0.7	8.6	32.8	1.6	1.3
11.3S$_M$ [W]	4.6^w	2.7	1.6	3.5	4.4	12.3	1.2	26.2	0.8	0.8	8.4	23.2	1.8	1.5
11.3L$_M$ [W]	5.3^w	3.1	1.8	3.9	4.7	11.2	1.1	20.8	0.7	0.6	6.7	27.9	2.3	1.8
9.3S$_M$ [S]	4.5^w	2.5	1.4	2.8	3.3	9.9	1.2	21.1	1.0	0.6	9.7	30.3	1.5	1.1
9.3L$_M$ [S]	4.7^w	2.5	1.5	2.9	3.4	9.2	1.0	18.4	1.0	0.6	8.2	33.7	1.7	1.3
7.1S$_M$ [S]	4.6^w	2.4	1.4	2.8	3.4	9.6	1.1	19.5	1.0	0.7	9.1	32.0	1.6	1.2
7.1L$_M$ [S]	4.7^w	2.6	1.5	3.1	3.5	9.1	1.0	17.6	1.2	0.5	7.6	34.4	1.7	1.4
5.5S$_M$ [W]	4.9^w	2.8	1.7	3.7	4.8	11.9	1.2	22.8	0.7	0.8	7.2	26.3	2.0	0.3
5.5L$_M$ [W]	5.1^w	3.1	2.0	4.0	4.8	10.1	1.0	18.7	0.8	0.7	5.6	30.6	2.7	0.8

Separation with the aid of an aqueous detergent solution

Banks et al., 1985														
16L$_S$ [B] low level of detergent	2.3^w	1.5	1.2	2.2	2.8	10.9	—	25.0	0.9	—	14.4	35.9 cis = 27.5 trans = 8.4	1.2 incl C18:3	—

Filtration through milk filter

Fouad et al., 1990														
25L$_S$	6.3^w	3.5	1.4	3.3	3.5	11.4	—	28.3	1.3	—	10.5	22.0	1.9	0.6
21S$_S$	3.8^w	2.1	1.0	2.8	3.5	12.2	—	32.2	1.0	—	13.2	20.2	1.5	0.4
21L$_S$	7.9^w	3.9	2.1	3.6	3.5	10.7	—	26.2	1.3	—	9.4	22.7	1.9	0.6

Separation technique not reported

Lechat et al., 1975														
Most solid fraction [Feb]	1.8^n	0.8	0.9	1.9	2.5	10.0	1.7	28.1	1.6	1.5	18.8	21.5	2.2	0.7

Continued

TABLE 5.54 (Continued)

Fraction separation method Author	Fraction designation[1]	Fatty acid (wt %)[2]													
		4:0	6:0	8:0	10:0	12:0	14:0	15:0	16:0	16:1	17:0	18:0	18:1	18:2	18:3
	Most liquid fraction [Sept]	4.0[n]	2.6	1.3	2.8	3.3	9.7	1.4	20.0	2.7	0.8	8.7	32.6	3.3	0.6
	Most liquid fraction [Feb]	4.1[n]	1.7	1.6	3.0	3.2	8.5	1.2	15.7	2.8	0.9	8.1	35.5	3.9	1.6

[1]The number indicates fractionation temperature, the letter indicates physical state (L = liquid, S = solid), and the subscript indicates fractionation procedure used (M = multiple-step, S = single-step). Designations in parentheses are those of the original author. The designation inside the brackets refers to the intact anhydrous milkfat source (S = summer, W = winter).

[2]incl = includes. Units of chemical composition are designated using these abbreviations: n = % not specified, w = wt %.

[3]Data not available.

TABLE 5.55 Individual Fatty Acid Composition of Unknown-Melting Milkfat Fractions Obtained from Solvent Solution

Type of solvent solution Author	Fraction designation[1]	Fatty acid (%)[2]													
		4:0	6:0	8:0	10:0	12:0	14:0	15:0	16:0	16:1	17:0	18:0	18:1	18:2	18:3
Acetone																
Larsen and Samuelsson, 1979	21.4S$_S$	3.9[n]	3.3	1.9	2.6	3.3	9.9	—[3]	29.3	4.1	—	15.9	23.3	—	—
Parodi, 1974a	20S$_S$ [May]	—	—	0.1[m]	0.8	1.7	11.4	2.0	41.0	1.8 incl C17br	1.9	29.4	6.2	—	—
	SSS fraction of 20S$_S$ [May][4]	—	—	0.2[m]	0.9	2.1	12.7	2.1	43.0	1.7 incl C17br	1.8	31.7	—	—	—
	SSM[f] fraction of 20S$_S$ [May]	—	—	0.1[m]	0.4	0.9	6.4	1.6	31.5	1.8 incl C17br	1.7	22.5	27.8	—	—
	SSM[c] fraction of 20S$_S$ [May]	—	—	0.2[m]	0.4	1.0	6.1	1.5	32.4	2.1 incl C17br	2.8	21.3	27.5	—	—

		10.6^m	4.6	2.1	3.7	3.5	10.8	1.4	23.5	2.6 incl C17br	0.9	9.7	19.3	1.3	1.5
	20L$_S$ [May]														
Chen and deMan, 1966	15S$_M$	0.0^w	0.0	0.0	0.9	2.5	11.2	—	33.8	—	—	28.6	14.4	3.3	0.3
	SSS fraction of 15S$_M$	0.0^w	0.0	0.0	0.6	2.4	12.2	—	41.0	—	—	39.9	0.0	0.0	0.0
Larsen and Samuelsson, 1979	15.0S$_S$	2.8^n	2.0	1.4	2.4	3.5	10.7	—	32.5	3.8	—	17.5	19.8	—	—
Mattsson et al., 1969	15S$_S$ [Aug]	1.1^m	0.3	0.4	1.5	3.0	13.3	—	35.4	—	—	23.0	10.8	0.5	0.6
	15S$_S$ [Dec]	1.8^m	0.6	0.6	2.4	4.5	15.8	—	41.4	—	—	15.0	7.3	0.3	0.4
Lansbergen and Kemps, 1984	12S$_M$ [1]	0.7^w	0.7	0.4	1.3 incl C10:1	2.0	8.7	1.9 incl C15br	24.7 incl C16:1	—	2.3 incl C17br	54.7	—	—	—
	12L$_M$ [1]	6.0^w	3.6	2.2	4.6 incl C10:1	2.8	10.6	1.9 incl C15br	27.3 incl C16:1	—	2.3 incl C17br	31.0	—	0.1	0.1
	12S$_M$ [2]	1.2^w	0.9	0.6	1.7 incl C10:1	2.2	8.8	1.5 incl C15br	24.6 incl C16:1	—	1.9 incl 17br	54.7	—	—	—
	12L$_M$ [2]	5.8^w	3.6	2.3	4.8 incl C10:1	5.4	13.9	2.6 incl C15br	28.4 incl C16:1	—	1.9 incl C17br	30.3	0.1	—	—
Lechat et al., 1975	10S$_M$ [Sept]	1.3^n	0.9	0.5	1.4	2.0	10.4	1.5	37.4	0.7	0.8	27.2	12.7	0.8	tr^5
	10L$_M$ [Sept]	2.7^n	1.8	0.9	1.9	2.0	10.3	1.7	34.8	1.5	1.0	20.1	16.2	1.5	0.1
	10S$_M$ [Apr]	0.7^n	0.5	0.2	0.9	1.8	9.8	1.6	36.4	0.7	1.2	28.6	12.6	1.8	0.1
	10L$_M$ [Apr]	3.2^n	2.1	1.1	2.0	2.3	9.6	2.0	30.4	1.0	1.9	15.2	22.4	2.0	0.2
Larsen and Samuelsson, 1979	9.6S$_S$	3.8^n	2.8	1.8	3.2	3.8	10.9	—	34.1	3.6	—	16.7	16.9	—	—
Lansbergen and Kemps, 1984	6S$_M$ [1]	5.3^w	3.1	1.6	3.5 incl C10:1	2.8	10.6	1.9 incl C15br	31.1 incl C16:1	—	1.9 incl C17br	36.9	—	—	—
	6L$_M$ [1]	6.9^w	4.5	3.5	7.2 incl C10:1	9.3	18.1	3.1 incl C15br	23.3 incl C16:1	—	2.0 incl C17br	20.9	—	—	—
Chen and deMan, 1966	5S$_M$	1.3^w	0.9	0.5	1.5	2.6	9.3	—	32.9	—	—	26.9	16.2	2.0	0.6
	SSS fraction of 5S$_M$	2.3^w	1.4	0.8	2.4	3.5	11.9	—	39.0	—	—	32.9	0.0	0.0	0.0
Larsen and Samuelsson, 1979	5.0S$_S$	4.6^n	2.6	1.8	3.4	4.6	7.4	—	32.2	4.0	—	17.2	19.2	—	—

Continued

TABLE 5.55 (Continued)

Type of solvent solution Author	Fraction designation[1]	4:0	6:0	8:0	10:0	12:0	14:0	15:0	16:0	16:1	17:0	18:0	18:1	18:2	18:3
								Fatty acid (%)[2]							
Lansbergen and Kemps, 1984	$0S_M$ [2]	4.9^w	3.0	1.7	3.8 incl C10:1	3.5	9.8	1.8 incl C15br	30.9 incl C16:1	—	1.9 incl C17br	37.2	0.1	—	—
	$0L_M$ [2]	6.3^n	4.0	2.9	6.0	7.4	17.6	3.0 incl C15br	25.8 incl C16:1	—	2.0 incl C17br	23.9	0.1	—	—
Lechat et al., 1975	$0S_M$ [Sept]	1.6^n	1.1	0.6	1.5	2.0	10.1	1.9	34.4	1.5	1.3	23.3	15.1	1.3	0.2
	$0L_M$ [Sept]	3.5^n	2.1	1.2	2.3	3.1	11.1	1.7	22.8	2.3	1.0	10.3	27.6	3.4	1.0
	$0S_M$ [Apr]	2.1^n	1.4	0.5	1.6	2.3	10.9	1.9	34.0	0.8	1.4	19.6	17.0	1.8	0.2
	$0L_M$ [Apr]	3.9^n	2.0	1.9	3.3	3.8	11.2	2.0	20.8	2.2	1.6	10.9	21.4	3.6	1.7
Chen and deMan, 1966	$-5S_M$	3.2^w	1.8	0.9	1.9	2.1	9.5	—	29.8	—	—	20.1	20.9	3.5	1.2
	SSS fraction of $-5S_M$	4.9^w	2.9	1.4	3.0	3.0	10.4	—	41.9	—	—	26.8	0.0	0.0	0.0
Larsen and Samuelsson, 1979	$-5.0L_S$	5.9^n	4.0	2.5	3.6	5.7	10.3	—	21.8	5.3	—	8.2	29.2	—	—
	$-8.0L_S$	5.3^n	3.4	2.5	3.6	6.4	10.1	—	20.3	5.7	—	7.3	32.1	—	—
Lechat et al., 1975	$-9S_M$ [Sept]	2.1^n	1.4	0.9	1.7	2.4	10.3	1.6	29.5	1.8	1.3	18.1	19.2	3.0	0.5
	$-9L_M$ [Sept]	3.7^n	2.4	1.3	2.2	2.6	9.1	1.3	17.6	2.6	0.9	7.1	36.3	4.4	1.1
	$-9L_M$ [Apr]	4.7^n	2.0	1.3	2.4	3.0	8.4	1.6	16.5	2.8	1.4	8.7	32.4	5.0	1.6
Chen and deMan, 1966	$-15S_M$	3.9^w	2.1	1.2	2.6	3.3	12.7	—	24.0	—	—	12.0	26.7	4.5	1.4
	SSS fraction of $-15S_M$	5.2^w	2.9	1.5	3.7	4.8	18.2	—	39.8	—	—	17.9	0.0	0.0	0.0
Larsen and Samuelsson, 1979	$-16.4L_S$	6.0^n	3.8	2.6	4.5	7.1	9.0	—	17.3	6.3	—	6.4	33.4	—	—
	$-21.0L_S$	6.0^n	3.9	2.6	4.7	7.7	8.6	—	16.9	6.6	—	5.8	33.6	—	—
Chen and deMan, 1966	$-25S_M$	4.4^w	2.5	1.6	3.5	4.1	10.0	—	17.0	—	—	8.7	34.9	4.4	2.6
	SSS fraction of $-25S_M$	5.9^w	3.6	2.2	5.7	8.1	21.4	—	32.5	—	—	13.8	0.0	0.0	0.0
Jensen et al., 1967	$-25S_M$	6.8^m	3.3	1.4	3.0	3.5	13.1	—	33.0	1.4	—	16.2	12.7	1.0	0.4
Chen and deMan, 1966	$-35S_M$	4.3^w	2.6	1.7	3.5	3.6	7.9	—	17.5	—	—	6.7	37.0	5.2	3.0
	SSS fraction of $-35S_M$	6.0^w	4.4	3.0	7.9	9.8	17.5	—	31.0	—	—	12.6	0.0	0.0	0.0
	$-45S_M$	5.0^w	2.9	1.8	3.6	3.6	8.4	—	14.7	—	—	6.0	38.4	5.1	2.8
	SSS fraction of $-45S_M$	6.7^w	4.7	3.6	8.0	9.2	17.4	—	29.5	—	—	13.0	0.0	0.0	0.0

Technical Data for Experimental and Commercial Milkfat Fractions

	Fraction													
		8.7^w	6.5	5.6	11.2	10.2	19.8	—	23.5	—	7.0	0.0	0.0	0.0
	SSS fraction of -45L$_M$													
Jensen et al., 1967	-70S$_M$	12.5^m	5.6	2.7	4.6	4.1	10.1	—	19.2	2.4	7.1	22.8	2.8	1.1
	-70L$_M$	21.6^m	9.0	4.2	5.3	3.8	6.8	—	5.5	3.6	0.6	27.3	5.6	1.6
Sherbon, 1963	2A	—	—	—	0.2^m incl C10:1	1.0	13.1	1.7	44.5	0.8	29.9	5.1	0.5	1.1
	2D	4.2^m	3.7	3.4	5.0	4.2	8.5	0.5	22.3	2.8	6.2	34.1	2.1	—
	2D'	7.8^m	3.2	2.1	4.0	2.6	10.5	0.9	33.1	—	13.3	20.4	0.8	—
	2E[Jan]	5.1^m	13.8	3.7	5.4	2.9	6.6	0.7	15.4	2.0	4.5	32.4	2.4	1.4
	2F[Jan]	17.4^m	6.8	3.7	4.5	2.8	6.2	0.6	9.5	2.8	0.5	35.0	2.6	1.5
Lansbergen and Kemps, 1984	Top fraction	0.1^w	—	0.2	0.9 incl C10:1	3.3	14.1	1.6	49.2	0.7 incl C17br	20.5	5.4	0.8	0.3
	Mid fraction	2.9^w	2.1	1.1	2.8 incl C10:1	3.9	12.9	1.3	42.0	1.6 incl C17br	12.6	12.3	1.4	0.2
	Olein fraction	4.6^w	2.9	1.9	4.5 incl C10:1	5.9	12.3	1.0	21.7	3.0 incl C17br	5.4	27.8	2.9	0.9
Hexane														
Muuse and van der Kamp, 1985	12.5S$_S$	3.3^a	1.8	1.1	2.4	3.3	10.4	—	27.0	—	14.5	24.5	—	—
	12.5L$_S$	4.2^a	2.3	1.3	2.8	3.4	9.9	—	23.9	—	11.2	28.0	—	—

[1]The number indicates fractionation temperature, the letter indicates physical state (L = liquid, S = solid), and the subscript indicates fractionation procedure used (M = multiple-step, S = single-step). Other descriptions used are those of the original author. The designation inside the brackets following the fraction designation refers to the intact anhydrous milkfat source (sample or month).

[2]br = branched, conj = conjugated, incl = includes. Units of chemical composition are designated using these abbreviations: a = area %, m = mole %, n = % not specified, w = wt %.

[3]Data not available

[4]SSS = trisaturated triglyceride.

[5]tr = trace.

TABLE 5.56 Individual Fatty Acid Composition of Unknown-Melting Milkfat Fractions Obtained by Supercritical Fluid Extraction

Type of solvent Author	Fraction designation[1]	Fractionation conditions		Fatty acid (%)[2]														
		Temperature (°C)	Pressure	4:0	6:0	8:0	10:0	12:0	14:0	15:0	16:0	16:1	17:0	18:0	18:1	18:2	18:3
Carbon dioxide																		
Bradley, 1989	WM-38, 1	60	2150 psi	6.5[w]	3.3	2.3	4.7	5.1	10.5	1.2	27.6	1.6	0.4	9.9	18.5 incl C18:2, C18:3	—[3]	—
	WM-38, 2	60	2300	6.1[w]	2.9	1.8	3.7	4.2	9.8	1.2	28.7	1.7	0.4	11.1	20.4 incl C18:2, C18:3	—	—
	WM-38, 3	60	2400	6.6[w]	2.0	1.7	3.5	4.0	9.6	1.2	29.7	1.7	0.4	11.9	21.8 incl C18:2, C18:3	—	—
	WM-38, 4	60	2400	6.2[w]	2.9	1.5	3.1	3.7	9.1	1.2	29.6	1.7	0.4	12.5	22.5 incl C18:2, C18:3	—	—
	WM-38, 5	60	2500	4.8[w]	2.5	1.3	2.8	3.1	7.6	1.0	24.9	1.4	0.3	13.1	19.8 incl C18:2, C18:3	—	—
	WM-38, 6	60	2500	5.6[w]	2.6	1.3	2.8	3.1	7.9	1.0	26.8	1.5	0.4	12.3	21.9 incl C18:2, C18:3	—	—
	WM-38, 7	60	2650	6.2[w]	2.7	1.3	2.7	3.1	8.1	1.0	26.9	1.6	0.4	13.0	21.3 incl C18:2, C18:3	—	—

Sample															
WM-38, 8	60	2850	4.4[w]	2.4	1.2	2.5	2.7	7.4	1.0	27.3	1.6	0.4	13.4	23.8 incl C18:2, C18:3	—
WM-38, 9	60	2900	3.8[w]	2.2	1.2	2.5	2.6	6.9	0.9	26.5	1.6	0.4	14.4	24.8 incl C18:2, C18:3	—
WM-38, 10	60	2900	2.6[w]	1.9	1.2	2.5	2.6	6.6	0.9	25.6	1.5	0.4	15.1	25.8 incl C18:2, C18:3	—
WM-38, 11	60	2900	2.1[w]	1.6	1.0	2.5	2.6	6.6	0.9	25.1	1.6	0.4	16.5	27.9 incl C18:2, C18:3	—
WM-38, 12	60	2900	1.1[w]	1.1	0.8	2.3	2.6	6.7	0.9	24.7	1.6	0.5	17.4	29.7 incl C18:2, C18:3	—
WM-38, 13	60	2900	0.4[w]	0.6	0.6	1.9	2.6	6.9	0.9	25.0	1.8	0.5	18.1	32.2 incl C18:2, C18:3	—
WM-38, 14	60	2900	0.0[w]	0.2	0.3	1.3	2.0	6.3	0.9	24.4	1.8	0.5	18.4	34.5 incl C18:2, C18:3	—
WM-38, 15	60	—	0.0[w]	tr[4]	tr	0.3	0.8	3.7	0.6	20.5	1.4	0.5	23.2	38.2 incl C18:2, C18:3	—
Chen and Schwartz, 1991	Unknown	35, 40, 50, 60	1500, 2000, 2500, 3000, 3500, 4000 psi	Fractions obtained at lower extraction pressure, such as 1500 psi, contained significantly larger amounts of C4–C8 and C10–C12 fatty acids and smaller amounts of C14–C18 fatty acids.											

Continued

TABLE 5.56 (Continued)

Type of solvent Author	Fraction designation[1]	Fractionation conditions			Fatty acid (%)[2]													
		Temperature (°C)	Pressure	4:0	6:0	8:0	10:0	12:0	14:0	15:0	16:0	16:1	17:0	18:0	18:1	18:2	18:3
Ensign, 1989	WM-38, 1	60	2150 psi	6.5w	3.3	2.3	4.7	5.1	10.5	1.2	27.6	1.6	0.4	9.9	18.5 incl C18:2, C18:3	—	—
	WM-38, 2	60	2300	6.1w	2.9	1.8	3.7	4.2	9.8	1.2	28.7	1.7	0.4	11.1	20.4 incl C18:2, C18:3	—	—
	WM-38, 3	60	2400	6.6w	3.0	1.7	3.5	4.0	9.6	1.2	29.7	1.7	0.4	11.9	21.8 incl C18:2, C18:3	—	—
	WM-38, 4	60	2400	6.2w	2.9	1.5	3.1	3.7	9.1	1.2	29.6	1.7	0.4	12.5	22.5 incl C18:2, C18:3	—	—
	WM-38, 5	60	2500	4.8w	2.5	1.3	2.8	3.1	7.6	1.0	24.9	1.4	0.3	13.0	19.8 incl C18:2, C18:3	—	—
	WM-38, 6	60	2500	5.6w	2.6	1.3	2.8	3.1	7.9	1.0	26.8	1.5	0.4	12.3	21.9 incl C18:2, C18:3	—	—
	WM-38, 7	60	2650	6.2w	2.7	1.3	2.8	3.1	8.1	1.0	26.9	1.6	0.4	13.0	21.3 incl C18:2, C18:3	—	—
	WM-38, 8	60	2850	4.4w	2.4	1.2	2.7	2.7	7.4	1.0	27.3	1.6	0.4	13.4	23.8 incl C18:2, C18:3	—	—
	WM-38, 9	60	2900	3.8w	2.2	1.2	2.5	2.6	6.9	0.9	26.5	1.6	0.4	14.4	24.8 incl C18:2, C18:3	—	—
	WM-38, 10	60	2900	2.6w	1.9	1.2	2.5	2.6	6.6	0.9	25.6	1.5	0.4	15.1	25.8 incl C18:2, C18:3	—	—
	WM-38, 11	60	2900	2.1w	1.6	1.0	2.5	2.6	6.6	0.9	25.1	1.6	0.4	16.5	27.8 incl C18:2, C18:3	—	—

Technical Data for Experimental and Commercial Milkfat Fractions

Sample																
WM-38, 12	60	2900	1.1w	1.1	0.8	2.3	2.6	6.7	0.9	24.7	1.6	0.5	17.4	29.7 incl C18:2, C18:3	—
WM-38, 13	60	2900	0.4w	0.6	0.6	1.9	2.6	6.9	0.9	25.0	1.8	0.5	18.1	32.2 incl C18:2, C18:3	—
WM-38, 14	60	2900	0.0w	0.2	0.3	1.2	2.0	6.3	0.9	24.4	1.8	0.5	18.4	34.5 incl C18:2, C18:3	—
WM-38, 15	60	—	0.0w	tr	tr	0.3	0.8	3.7	0.6	20.5	1.4	0.5	23.2	38.2 incl C18:2, C18:3	—
Hamman et al., 1991 A$_{res}$	40	125 bar	0.0w	0.0	0.0	0.0	0.0	6.2	—	27.6	0.0	—	28.3	34.4	2.1
B1	40	350	2.8w	2.4	1.6	3.9	4.4	14.2	—	32.6	1.7	—	9.4	20.3	1.3
Shishikura et al. 1986 F1	40	300 kg/cm²	8.6n	6.1	3.3	6.2	6.0	15.1	—	29.6	—	—	5.9	17.9	1.4
F2	40	300	6.9n	5.2	2.7	5.3	5.4	14.8	—	31.4	—	—	6.9	20.1	1.4
F3	40	300	5.3n	4.1	2.2	4.6	4.9	14.0	—	32.1	—	—	8.2	23.1	1.6
F4	40	300	4.3n	3.3	1.9	4.1	4.5	13.6	—	32.4	—	—	9.0	25.1	1.7
F5	40	300	4.4n	3.2	1.8	4.0	4.4	13.8	—	33.8	—	—	9.9	27.4	1.8
F6	40	300	3.3n	2.4	1.4	3.4	3.9	12.9	—	32.9	—	—	10.4	27.6	1.8
Residue	40	300	1.2n	0.8	0.5	1.6	2.5	10.8	—	34.5	—	—	14.4	31.7	1.9

[1]Designations used are those given by the original author.

[2]incl = includes. Units of chemical composition are designated using these abbreviations: n = % not specified, w = wt %.

[3]Data not available.

[4]tr = trace.

TABLE 5.57 Individual Fatty Acid Composition of Unknown-Melting Milkfat Fractions Obtained by Short-Path Distillation

Author	Fraction designation[1]	Fractionation conditions		Fatty acid (%)[2]												
		Temperature (°C)	Pressure	4:0	6:0	8:0	10:0	12:0	14:0	15:0	16:0	16:1	17:0	18:0	18:1	18:2	18:3
McCarthy et al., 1962	R-1	150	15 μ	7.3w	6.5	6.1	9.5	9.1	18.5	1.5	24.9	1.3	—[3]	4.0	7.7	tr[4]	—
	R-2	160	15	8.3w	5.2	3.4	7.2	7.3	22.3	tr	31.3	1.4	—	3.8	8.1	tr	—
	R-3	180	12	3.7w	3.5	2.3	4.1	5.5	18.4	1.6	36.2	1.6	—	6.8	13.5	0.9	—
	R-4	185	12	2.9w	3.8	1.8	3.3	3.7	7.6	1.5	42.2	1.5	—	10.1	19.8	tr	—
	D-2	177–182	5–7	2.7w	3.3	1.7	3.8	3.9	14.2	1.3	35.4	1.6	—	10.2	19.4	tr	—
	D-3	—	—	0.7w	0.9	0.8	2.3	2.7	11.0	1.0	30.0	1.8	—	13.8	32.0	0.9	—

[1]Designations used are those given by the original author.

[2]Units of chemical composition are designated using these abbreviations: w = wt %.

[3]Data not reported.

[4]tr = trace.

TABLE 5.58 Fatty Acid Composition (Selected Fatty Acid Groupings) of Intact Anhydrous Milkfats Used to Produce Milkfat Fractions from Melted Milkfat

Fraction separation method Author	Fraction designation[1]	Selected saturated n–chain groupings (%)[2]				Selected ratio (%)	Selected unsaturated groupings (%)	
		C4–C10	C12–C15	C12,C14,C16	C16–C18	C18:1/ C18:0	cis	trans
Vacuum filtration								
Amer et al., 1985	AMF [W]	13.0[w]	16.9	—[3]	40.7	—	24.9 C10:1– C18:3	1.6 C18:1
	AMF [S]	11.8[w]	15.4	—	39.5	—	26.4 C10:1– C18:3	3.4 C18:1
Barna et al., 1992	AMF [W]	11.30[n]	14.60	44.70	42.10	1.92	—	—
Deffense, 1987	AMF [2]	8.30[w]	13.60	37.20	37.60	2.34	—	—
Deffense, 1993b	AMF	8.30[w]	13.70	37.20	37.60	2.34	—	—
deMan and Finoro, 1980	AMF	10.80[w]	14.90	38.10	35.60	2.03	—	—
Grall and Hartel, 1992	AMF [S]	11.30[a]	14.60	44.70	42.10	1.92	—	—
Jamotte and Guyot, 1980	Commercial liquid fraction (L$_2$) doubly fractionated at 15°C	10.10[n]	15.60	41.60	38.30	2.60	—	—
Kaylegian, 1991	AMF [W]	10.60[a]	13.80	42.30	38.80	2.15	—	—
Laakso et al., 1992	AMF	17.50[m]	14.60	40.00	38.60	1.87	—	—
	AMF SSS[3]	23.20[m]	21.10	56.30	54.30	0.03	—	—
	AMF SSM[t]	12.60[m]	14.50	37.50	36.60	2.31	—	—
	AMF SSM[c]	15.90[m]	14.4	35.20	34.50	2.96	—	—
	AMF SM[c]M[t]	7.90[m]	9.2	19.20	19.50	8.45	—	—
	AMF SM[c]M[c]	8.20[m]	8.4	17.70	18.70	10.72	—	—
	AMF UUU	8.60[m]	8.1	18.40	18.80	6.33	—	—
Lund and Danmark, 1987	AMF [1]	11.00[w]	14.60	44.60	41.00	2.18	—	—
Makhlouf et al., 1987	AMF	24.0[m] C4–C12	—	—	75.0 C14–C18	—	—	—
Schaap and van Beresteyn, 1970	AMF	10.20[n]	16.10	44.20	41.40	2.09	—	—
Timms, 1978b	AMF	10.70[w]	14.90	39.40	40.50	1.83	—	—
Timms and Parekh, 1980	AMF [S]	10.70[w]	14.90	39.50	40.60	1.83	—	—
Versteeg, 1991	AMF [Sept]	9.20[n]	14.10	34.60	34.90	2.29	—	—
Vovan and Riel, 1973	AMF	11.50[n]	16.40	42.90	40.30	2.00	—	—
Pressure filtration								
Badings et al., 1983b	AMF [W]	12.20[w]	16.90	44.90	39.50	2.24	—	—
	AMF [S]	10.40[w]	13.70	35.90	36.00	2.36	—	—
Deffense, 1993b	AMF	8.30[w]	13.70	37.20	37.60	2.34	—	—
deMan, 1968	AMF	7.70[n]	11.40	39.10	41.10	2.29	—	—

Continued

TABLE 5.58 (Continued)

Fraction separation method Author	Fraction designation[1]	Selected saturated n–chain groupings (%)[2]				Selected ratio (%)	Selected unsaturated groupings (%)	
		C4–C10	C12–C15	C12,C14,C16	C16–C18	C18:1/C18:0	cis	trans
Centrifugation								
Antila and Antila, 1970	AMF	12.70[n]	18.50	46.10	37.80	3.48	—	—
Jordan, 1986	AMF [W]	9.30[w]	13.60	42.30	41.60	2.07	—	—
	AMF [S]	—	—	—	—	2.21[w]	—	—
Kankare and Antila, 1986	AMF [S]	9.80[n]	13.90	39.20	37.70	2.15	—	—
Kankare and Antila, 1988a	AMF [S]	—	—	—	38.50[n]	2.12	—	—
Kankare and Antila, 1988b	AMF [S]	—	—	—	38.50[n]	2.12	—	—
Keogh and Higgins, 1986b	AMF	10.90[w]	14.40	40.40	38.50	2.09	—	—
Lakshminarayana and Rama Murthy, 1985	AMF	9.00[w]	16.20	48.40	43.50	2.68	—	—
Richardson, 1968	AMF [W]	10.60[w]	12.80	41.60	42.80	2.13	—	—
Riel and Paquet, 1972	AMF	8.40[n]	18.10	50.10	46.20	2.01	—	—
Separation with the aid of an aqueous detergent solution								
Banks et al., 1985	AMF [A]	8.00[w]	9.50	47.70	49.10	2.54	25.0 C18:1	2.7 C18:1
	AMF [B]	7.10[w]	12.50	38.70	44.20	1.81	24.6 C18:1	8.0 C18:1
Dolby, 1970a	AMF [mid-season]	9.30[n]	19.90	56.10	48.90	1.68	—	—
Doležálek et al., 1976	AMF [1]	8.50[w]	19.80	50.40	40.50	2.86	—	—
Fjaervoll, 1970a	AMF	9.90[n]	16.90	43.60	42.20	1.83	—	—
Norris et al., 1971	AMF [A late S]	7.50[a]	14.10	40.60	42.40	2.05	—	—
	AMF [B late S]	7.00[a]	13.10	40.00	43.00	2.03	—	—
	AMF [C early F]	8.50[a]	17.20	44.10	41.20	1.97	—	—
Schaap et al., 1975	AMF	11.00[w]	15.60	38.30	36.80	2.15	—	—
Filtration through muslin cloth								
Kulkarni and Rama Murthy, 1987	AMF	9.80[n]	16.40	48.80	43.60	2.61	—	—
Filtration through milk filter								
Fouad et al., 1990	AMF	12.80[w]	14.60	42.10	38.30	2.08	—	—
Separation technique not reported								
Deroanne, 1976	AMF [W]	7.40[n]	14.60	41.70	39.60	2.65	—	—
Deroanne and Guyot, 1974	AMF [W]	10.50[a]	18.10	47.00	38.80	2.48	—	—
	AMF [S]	10.30[a]	15.80	38.20	35.30	2.35	—	—

Continued

TABLE 5.58 (Continued)

Fraction separation method Author	Fraction designation[1]	Selected saturated n–chain groupings (%)[2]				Selected ratio (%)	Selected unsaturated groupings (%)	
		C4–C10	C12–C15	C12,C14,C16	C16–C18	C18:1/C18:0	cis	trans
Lechat et al., 1975	AMF [Sept]	9.30n	15.30	37.60	37.20	2.18	—	—
	AMF [Feb]	8.30n	13.90	35.00	37.30	2.08	—	—
Martine, 1982	AMF	4.60n	14.70	43.80	42.20	2.50	—	—
Riel and Paquet, 1972	AMF	8.40n	18.10	50.10	46.20	2.01	—	—
Timmen, 1974	AMF [W]	—	—	—	—	2.38n	—	—
	AMF [S]	—	—	—	—	2.38n	—	—

[1]AMF = anhydrous milkfat. The designation inside the brackets following the fraction designation refers to the intact anhydrous milkfat source [sample number or letter or season (F = fall, S = summer, W = winter)].

[2]Units of chemical composition are designated using these abbreviations: a = area %, m = mole %, n = % not specified, w = wt %.

[3]Data not available.

[4]S = saturated, M = monounsaturated, U = polyunsatured, c = cis, t = trans.

TABLE 5.59 Fatty Acid Composition (Selected Fatty Acid Groupings) of Intact Anhydrous Milkfats Used to Produce Milkfat Fractions from Solvent Solution

Type of solvent solution Author	Fraction designation[1]	Selected saturated n–chain groupings (%)				Selected ratio (%)
		C4–C10	C12–C15	C12,C14,C16	C16–C18	C18:1/C18:0
Acetone						
Chen and deMan, 1966	AMF	7.50w	13.00	36.40	37.70	2.06
Jensen et al., 1967	AMF	22.80m	14.30	37.10	32.60	2.06
Kaylegian, 1991	AMF [W]	10.60a	13.80	42.30	38.80	2.15
Lansbergen and Kemps, 1984	Hardened butterfat	10.1w	17.1	41.4	71.5	—[3]
	Natural butterfat	11.50w	18.90	48.60	40.50	2.27
Larsen and Samuelsson, 1979	AMF	15.20n	15.90	40.90	36.30	2.12
Lechat et al., 1975	AMF [Sept]	9.30n	15.30	37.60	37.20	2.18
	AMF [Apr]	9.00n	13.70	37.90	40.40	1.93
Mattsson et al., 1969	AMF [Aug]	18.40m	14.70	37.40	31.90	2.34
	AMF [Dec]	19.70m	17.10	45.20	34.90	2.35
Parodi, 1974a	AMF [May]	20.20m	15.60	38.40	35.70	1.80
Schaap et al., 1975	AMF	11.00w	15.60	38.30	36.80	2.15
Timms, 1980b	AMF [S]	10.70w	14.90	39.50	40.60	1.83
Yi, 1993	AMF [S]	8.70a	15.00	47.60	45.40	1.85
Pentane						
Kaylegian, 1991	AMF [W]	10.60a	13.80	42.30	38.80	2.15
Hexane						
Muuse and van der Kamp, 1985	AMF	10.20a	13.30	37.80	36.60	2.27

[1]AMF = anhydrous milkfat. The designation inside the brackets following the fraction designation refers to the intact anhydrous milkfat source (W = winter, S = summer).

[2]Units of chemical composition are designated using these abbreviations: a = area %, m = mole %, n = % not specified, w = wt %.

[3]Data not available.

TABLE 5.60 Fatty Acid Composition (Selected Fatty Acid Groupings) of Intact Anhydrous Milkfats Used to Produce Milkfat Fractions by Supercritical Fluid Extraction

Type of solvent Author	Fraction designation[1]	Selected saturated n–chain groupings (%)[2]				Selected ratio (wt %)
		C4–C10	C12–C15	C12,C14,C16	C16–C18	C18:1/18:0
Carbon dioxide						
Arul et al., 1987	AMF	26.00[m]	18.00	43.00	34.50	2.09
Bhaskar et al., 1993b	AMF	8.55[w] C4–C8	4.60 C10–C12	—	86.85 C14:0–C18:3	—
Biernoth and Merk, 1985	AMF	9.80[a]	15.90	45.90	38.90	2.74
Bradley, 1989	AMF [WM-38]	7.60[w]	11.00	37.00	44.90	10.52
Büning-Pfaue et al., 1989	AMF	8.30[w]	15.80	52.70	45.60	2.80
Chen et al., 1992	AMF	21.9[n] C4–C8	6.5 C10–C12	—[3]	71.9 C14–C18	—
Ensign, 1989	AMF [WM-38]	7.60[w]	11.00	37.00	44.90	10.52
Hamman et al., 1991	AMF [A]	7.80[w]	13.90	42.00	43.20	1.97
	AMF [B]	8.30[w]	16.20	48.30	44.00	1.96
Kankare and Antila, 1988a	AMF	3.10[n]	-0.00	25.50	37.90	2.14
Kankare et al., 1989	AMF	9.80[n]	15.50	44.80	40.50	2.13
Kaufmann et al., 1982	AMF	21.40[w]	17.20	43.20	34.50	2.53
Shishikura et al., 1986	AMF	11.80[n]	18.30	49.60	43.70	2.08
Timmen et al., 1984	AMF	10.30[w]	15.60	45.90	39.10	2.50

[1] AMF = anhydrous milkfat. Designations in brackets are those used by the original author.

[2] Units of chemical composition are designated in the first data column using these abbreviations: a = area %, m = mole %, n = % not specified, w = wt %.

[3] Data not available.

TABLE 5.61 Fatty Acid Composition (Selected Fatty Acid Groupings) of Intact Anhydrous Milkfat Used to Produce Milkfat Fractions by Short-Path Distillation

Author	Fraction designation[1]	Selected saturated n–chain groupings (%)				Selected ratio (wt %)
		C4–C10	C12–C15	C12,C14,C16	C16–C18	C18:1/18:0
Arul et al., 1988a	AMF	17.20[m]	15.20	41.40	36.80	2.01
McCarthy et al., 1962	AMF	8.9[w]	17.5	48.9	44.1	2.16

[1] AMF = anhydrous milkfat.

[2] Units of chemical composition are designated in the first data column using these abbreviations: m = mole %, w = wt %.

TABLE 5.62 Fatty Acid Composition (Selected Fatty Acid Groupings) of Very-High-Melting Milkfat Fractions Obtained from Melted Milkfat

Fraction separation method Author	Fraction designation[1]	Selected saturated n–chain groupings (%)[2]				Selected ratio (%)	Selected unsaturated groupings (%)	
		C4–C10	C12–C15	C12,C14,C16	C16–C18	C18:1/C18:0	cis	trans
Vacuum filtration								
Kaylegian, 1991	$34S_M$ [W]	7.90[a]	13.30	46.60	49.70	0.97	—[3]	—
	$34S_S$ [W]	7.40[a]	13.60	46.20	48.60	1.05	—	—
Schaap and van Beresteyn, 1970	$27S_M$	8.50[n]	16.60	48.70	50.40	1.19	—	—
Makhlouf et al., 1987	$26S_M$ (S.1)	24.0[m] C4–C12	—	—	75.0 C14–C18	—	—	—
Sherbon, 1963	$23S_M$	13.60[m]	14.50	49.00	49.60	1.27	—	—
	$23ppt_M$ refractionated	9.40[m]	14.50	47.00	44.90	1.64	—	—
Mahklouf et al., 1987	$21S_M$ (S.2)	19.0[m]	—	—	81.0	—	—	—
	F.S (>21)	20.0[m]	—	—	80.0	—	—	—
Pressure filtration								
deMan, 1968	$30S_S$ (A) slow cooling	4.50[n]	11.50	47.10	57.00	1.01	—	—
	$27S_S$ (C) slow cooling	5.80[n]	11.70	44.60	50.90	1.36	—	—
	$24S_S$ (E) slow cooling	6.70[n]	12.20	42.90	46.20	1.74	—	—
Deffense, 1993b	1st stearin$_M$	5.00[w]	15.10	46.60	52.70	1.02	—	—
Centrifugation								
Kankare and Antila, 1986	$24S_M$ [S]	6.90[n]	15.10	45.60	47.10	1.19	—	—
Kankare and Antila, 1988a	$24S_M$ [S]	—	—	—	46.50[n]	1.25	—	—
Kankare and Antila, 1988b	$24S_M$ [S]	—	—	—	47.50	1.18	—	—
Antila and Antila, 1970	D	6.50[n]	16.40	57.80	57.20	0.79	—	—
Separation with the aid of an aqueous detergent solution								
Schaap et al., 1975	$28S_S$	4.30[w]	19.00	56.90	67.80	0.25	—	—
Banks et al., 1985	$21S_S$ [A] high level of detergent	5.50[w]	11.30	56.90	61.10	1.21	16.2 C18:1	2.6 C18:1
	21 semi-S_S [A] high level of detergent	7.30[w]	10.20	51.00	53.30	1.98	22.0	2.7
	$16S_S$ [B] low level of detergent	5.70[w]	12.80	42.70	51.80	1.21	18.6	7.9

Continued

TABLE 5.62 (Continued)

Fraction separation method Author	Fraction designation[1]	Selected saturated n–chain groupings (%)[2]				Selected ratio (%)	Selected unsaturated groupings (%)	
		C4–C10	C12–C15	C12,C14,C16	C16–C18	C18:1/C18:0	cis	trans
	16S$_S$ [B] high level of detergent	5.20^w	14.00	48.00	58.10	0.78	11.2	7.5

[1]The number indicates fractionation temperature, the letter indicates physical state (S = solid), and the subscript indicates fractionation procedure used (M = multiple-step, S = single-step). Other primary designations and designations in parentheses are those of the original author. The designation inside the brackets refers to the intact anhydrous milkfat source [sample letter or season (S = summer, W = winter)].

[2]Units of chemical composition are designated using these abbreviations: a = area %, m = mole %, n = % not specified, w = wt %.

[3]Data not available.

TABLE 5.63 Fatty Acid Composition (Selected Fatty Acid Groupings) of Very-High-Melting Milkfat Fractions Obtained from Solvent Solution

Type of solvent solution Author	Fraction designation[1]	Selected saturated n–chain groupings (%)[2]				Selected ratio (%)
		C4–C10	C12–C15	C12,C14,C16	C16–C18	C18:1/C18:0
Acetone						
Sherbon and Dolby, 1973	$32S_M$ (I) [HMG]	0.20^m	9.90	56.00	85.90	0.03
Schaap et al., 1975	$28S_S$	0.70^w	12.20	48.30	79.10	0.11
Kaylegian, 1991	$25S_M$ [W]	3.10^a	10.50	49.20	67.30	0.26
	$25S_S$ [W]	3.30^a	11.20	49.60	66.10	0.27
Sherbon and Dolby, 1973	$25S_M$ (II) [HMG]	3.50^m	18.30	63.40	71.90	0.19
	$25S_M$ (III) refractionated [HMG]	0.70^m	23.70	61.90	60.30	0.47
Yi, 1993	$25S_M$ [S]	2.50^a	11.60	51.70	65.90	0.43
Kaylegian, 1991	$21S_M$ [W]	3.30^a	13.50	53.40	62.60	0.36
	$21S_S$ [W]	3.20^a	13.20	52.40	62.90	0.38
Timms, 1980b	$20S_M$ [S]	0.80^w	13.70	52.30	73.00	0.28
Yi, 1993	$20S_M$ [S]	2.60^a	15.00	59.40	67.10	0.56
Kaylegian, 1991	$15S_M$ [W]	4.10^a	17.00	54.50	54.50	0.70
	$15S_S$ [W]	4.60^a	15.00	52.20	56.60	0.59
Yi, 1993	$15S_M$ [S]	3.40^a	18.40	60.40	60.00	0.84
Kaylegian, 1991	$11S_S$ [W]	5.80^a	15.00	50.00	53.00	0.74
Sherbon, 1963	2B	14.80^m	16.60	44.40	55.30	0.48
	2AJAN [Jan]	8.70^m	9.70	53.50	79.20	—[3]
Pentane						
Kaylegian, 1991	$11S_M$ [W]	3.00^a	12.30	52.40	65.00	0.31
	$11S_S$ [W]	4.70^a	13.40	50.10	57.80	0.54
	$5S_M$ [W]	4.10^a	15.80	54.50	57.50	0.55
	$5S_S$ [W]	4.20^a	15.60	52.30	55.40	0.63
	$0S_M$ [W]	4.90^a	17.50	53.20	52.60	0.73
	$0S_S$ [W]	4.20^a	16.60	53.00	54.60	0.63

[1]The number indicates fractionation temperature, the letter indicates physical state (S = solid), and the subscript indicates fractionation procedure used (M = multiple-step, S = single-step). Other descriptions used are those of the original author. The designation inside the brackets refers to the intact anhydrous milkfat source (HMG = high-melting glyceride, W = winter).

[2]Units of chemical composition are designated in the first data column using these abbreviations: a = area %, m = mole %, w = wt %.

[3]Data not available.

TABLE 5.64 Fatty Acid Composition (Selected Fatty Acid Groupings) of Very-High-Melting Milkfat Fractions Obtained by Supercritical Fluid Extraction

Type of solvent Author	Fraction designation[1]	Fractionation conditions		Selected saturated n–chain groupings (%)[2]				Selected ratio (%)
		Temperature (°C)	Pressure	C4–C10	C12–C15	C12,C14,C16	C16–C18	C18:1/18:0
Carbon dioxide								
Bhaskar et al., 1993b	S1	40	24.1 MPa	1.22[w] C4–C8	1.95 C10–C12	—	96.83 C14:0–C18:3	—
Hamman et al., 1991	B$_{res}$	40	350 bar	1.30[w]	11.50	43.10	48.20	1.81

[1]Fraction designations are those used by the original author.

[2]w = wt %.

TABLE 5.65 Fatty Acid Composition (Selected Fatty Acid Groupings) of High-Melting Milkfat Fractions Obtained from Melted Milkfat

Fraction separation method Author	Fraction designation[1]	Selected saturated n–chain groupings (%)[2]				Selected ratio (%)	Selected unsaturated groupings (%)	
		C4–C10	C12–C15	C12,C14,C16	C16–C18	C18:1/C18:0	cis	trans
Vacuum filtration								
deMan and Finoro, 1980	32S$_S$	10.80^w	15.00	36.80	37.20	1.42	—[3]	—
	31S$_S$	8.60^w	15.10	42.10	42.30	1.49	—	—
	30SS$_S$	9.10^w	15.20	41.40	42.00	1.39	—	—
Barna et al., 1992	30S$_M$ [W]	9.00^n	15.10	50.40	51.50	1.03	—	—
Grall and Hartel, 1992	30S$_M$ [S]	9.00^a	15.10	50.40	51.50	1.03	—	—
Kaylegian, 1991	30S$_M$ [W]	8.70^a	14.00	46.50	46.50	1.22	—	—
	30S$_S$ [W]	9.20^a	13.50	45.00	45.40	1.30	—	—
Amer et al., 1985	29S$_S$ [W]	13.0^w	16.9	—	40.7	—	20.4 C10:1–C18:3	1.5 C18:1
	29S$_S$ [S]	9.4^w	15.9	—	47.0	—	21.1	3.1
deMan and Finoro, 1980	29S$_S$	8.20^w	15.50	40.90	41.60	1.36	—	—
	28S$_S$	7.60^w	15.60	42.90	43.80	1.32	—	—
	27S$_S$	9.70^w	16.90	40.40	38.50	1.40	—	—
Amer et al., 1985	26S$_S$ [W]	11.5^w	17.3	—	44.8	—	22.0 C10:1–C18:3	1.4 C18:1
deMan and Finoro, 1980	26S$_S$	8.70^w	16.20	40.90	40.00	1.50	—	—
Lund and Danmark, 1987	26S$_S$ [1]	8.70^w	15.00	49.00	47.50	1.48	—	—
deMan and Finoro, 1980	25S$_S$	11.30^w	15.70	36.40	35.60	1.48	—	—
Jamotte and Guyot, 1980	Commercial solid fraction A fractionated at 25°C	7.60^n	18.40	51.90	48.00	1.49	—	—
	Commercial solid fraction B fractionated at 25°C	7.20^n	16.20	45.60	47.60	1.36	—	—
Kaylegian, 1991	25S$_M$ [W]	9.10^a	15.20	47.00	44.20	1.43	—	—
	25S$_S$ [W]	9.90^a	13.40	43.10	42.00	1.67	—	—
Timms and Parekh, 1980	25S$_S$ [S]	10.20^w	17.60	45.40	44.00	1.40	—	—
Amer et al., 1985	23S$_S$ [W]	12.1^w	17.2	—	43.5	—	22.9 C10:1–C18:3	1.5 C18:1
Kaylegian, 1991	20S$_M$ [W]	7.70^a	17.30	49.30	44.50	1.34	—	—
	20S$_S$ [W]	10.10^a	13.50	42.50	40.70	1.84	—	—
Amer et al., 1985	19S$_S$ [W]	12.2^w	17.1	—	42.9	—	23.1 C10:1–C18:3	1.6 C18:1
	19S$_S$ [S]	11.1^w	15.8	—	41.8	—	24.7	3.1
Versteeg, 1991	18S$_M$ [Sept]	9.00^n	14.80	38.90	42.10	1.40	—	—
Schaap and van Beresteyn, 1970	17S$_M$	10.00^n	15.40	46.40	45.90	1.61	—	—

Continued

TABLE 5.65 (Continued)

Fraction separation method Author	Fraction designation[1]	Selected saturated n-chain groupings (%)[2]				Selected ratio (%)	Selected unsaturated groupings (%)	
		C4–C10	C12–C15	C12,C14,C16	C16–C18	C18:1/C18:0	cis	trans
Deffense, 1987	1st stearin$_M$ [2]	6.70^w	14.20	41.50	44.70	1.51	—	—
Timms, 1978b	Hard	9.60^w	16.80	46.10	44.40	1.64	—	—
Laakso et al., 1992	S1	15.40^m	16.8	45.00	45.50	1.24	—	—
	S1 SSS[4]	17.80^m	21.8	59.70	60.00	0.01	—	—
	S1 SSM^t	7.80^m	13.9	40.3	41.90	2.01	—	—
	S1 SSM^c	13.60^m	13.9	36.20	37.10	2.64	—	—
	S1 SM^{c}M^t	7.60^m	9.6	19.20	20.30	7.23	—	—
	S1 SM^{c}M^c	8.30^m	8.8	17.60	19.00	9.26	—	—
	S1 UUU	8.60^m	8.2	18.80	19.90	5.48	—	—
Centrifugation								
Riel and Paquet, 1972	38S$_M$	6.30^n	18.10	52.90	54.60	1.04	—	—
Lakshminarayana and Rama Murthy, 1985	31S$_M$	7.90^w	14.60	49.30	47.20	2.33	—	—
Richardson, 1968	30S$_M$ [W]	9.10^w	13.80	44.70	46.10	1.74	—	—
Jordan, 1986	28S$_S$ [S]	—	—	—	—	1.76^w	—	—
	28S$_S$ [W]	7.90^w	13.90	44.70	45.50	1.71	—	—
Ramesh and Bindal, 1987b	28S$_S$ [Ghee]	8.40^w	17.60	46.70	43.30	1.80	—	—
Jordan, 1986	24S$_S$ [S]	—	—	—	—	1.98^w	—	—
	24S$_S$ [W]	—	—	—	—	1.89^w	—	—
Kankare and Antila, 1988b	12S$_M$ [S]	3.00^n	—	28.00	42.50	1.50	—	—
Keogh and Higgins, 1986b	Fraction 1	6.40^w	15.90	53.30	52.40	1.08	—	—
	Fraction 2	7.90^w	15.40	46.40	46.70	1.28	—	—
Separation with the aid of an aqueous detergent solution								
Doležálek et al., 1976	27S$_S$ [1]	6.10^w	18.20	52.10	45.70	2.03	—	—
Norris et al., 1971	26S$_S$ [B late S]	6.90^a	15.20	44.00	45.30	1.76	—	—
	26S$_S$ [C early F]	6.80^a	15.40	43.30	45.20	1.57	—	—
	25S$_S$ [A late S]	5.80^a	13.50	42.00	46.70	1.68	—	—
Banks et al., 1985	16 semi-S$_S$ [B] high level of detergent	7.20^w	14.10	42.60	46.50	1.61	20.9 C18:1	8.1 C18:1
Fjaervoll, 1970a	HMF	8.70^n	17.10	45.90	46.10	1.46	—	—
Dolby, 1970a	Solid [midseason]	9.30^n	22.40	57.90	51.60	1.05	—	—
Filtration through milk filter								
Fouad et al., 1990	29S$_S$	6.00^w	16.20	52.80	53.10	0.98	—	—
	25S$_S$	9.70^w	15.90	50.50	49.50	1.11	—	—
Filtration through casein-dewatering press								
Black, 1973	Hard #1	8.10^w	15.80	44.10	44.50	1.20	—	—

Continued

TABLE 5.65 (Continued)

Fraction separation method Author	Fraction designation[1]	Selected saturated n-chain groupings (%)[2]				Selected ratio (%)	Selected unsaturated groupings (%)	
		C4–C10	C12–C15	C12,C14,C16	C16–C18	C18:1/C18:0	cis	trans
	Hard #2	8.30w	15.90	42.40	43.20	1.19	—	—
	Hard #3	6.30w	14.80	44.20	48.90	1.03	—	—
Separation technique not reported								
Deroanne, 1976	21S$_S$ [W]	6.10n	14.90	45.10	45.50	1.84	—	—
Deroanne and Guyot, 1974	21S$_S$ [W]	9.90a	18.40	49.20	41.70	2.13	—	—
	21S$_S$ [S]	10.00a	16.90	42.50	40.00	1.79	—	—
Martine, 1982	1st stearin	4.70n	15.30	44.60	69.40	0.04	—	—
	2nd stearin	4.30n	14.1	41.4	70.2	0.04	—	—

[1]The number indicates fractionation temperature, the letter indicates physical state (S = solid), and the subscript indicates fractionation procedure used (M = multiple-step, S = single-step). Other descriptions used are those of the original investigator. The designation inside the brackets following the fraction designation refers to the intact anhydrous milkfat source [sample number, letter or season (F = fall, S = summer, W = winter)].

[2]Units of chemical composition are designated using these abbreviations: a = area %, m = mole %, n = % not specified, w = wt %.

[3]Data not available.

[4]S = saturated, M = monounsaturated, U = polyunsaturated, c = cis, t = trans.

TABLE 5.66 Fatty Acid Composition (Selected Fatty Acid Groupings) of High-Melting Milkfat Fractions Obtained from Solvent Solution

Type of solvent solution Author	Fraction designation[1]	Selected saturated n-chain groupings (%)[2]				Selected ratio (%)
		C4–C10	C12–C15	C12,C14,C16	C16–C18	C18:1/C18:0
Acetone						
Sherbon and Dolby, 1973	25S$_M$ (IV) refractionated [HMG]	4.10m	21.20	57.00	56.50	0.72
Kaylegian, 1991	11S$_M$ [W]	6.10a	18.10	52.00	47.50	1.13
Yi, 1993	10S$_M$ [S]	5.50a	18.20	55.50	52.30	1.32
Kaylegian, 1991	5S$_S$ [W]	7.70a	13.40	47.50	50.90	0.89
Timms, 1980b	3S$_M$ [S]	6.40w	14.20	47.80	58.30	0.68
Kaylegian, 1991	0S$_S$ [W]	9.40a	12.80	46.20	47.80	1.16
Sherbon, 1963	2BJAN [Jan]	20.00m	13.80	50.30	53.00	0.64

[1]The number indicates fractionation temperature, the letter indicates physical state (S = solid), and the subscript indicates fractionation procedure used (M = multiple-step, S = single-step). Other primary designations and designations in parentheses are those of the original author. The designation inside the brackets refers to the intact anhydrous milkfat source (HMG = high-melting glyceride, W = winter).

[2]Units of chemical composition are designated using these abbreviations: a = area %, m = mole %, w = wt %.

TABLE 5.67 Fatty Acid Composition (Selected Fatty Acid Groupings) of High-Melting Milkfat Fractions Obtained by Supercritical Fluid Extraction

Type of solvent Author	Fraction designation[1]	Fractionation conditions		Selected saturated n–chain groupings (%)[2]				Selected ratio (%)
		Temperature (°C)	Pressure	C4–C10	C12–C15	C12,C14,C16	C16–C18	C18:1/18:0
Carbon dioxide								
Biernoth and Merk, 1985	Residual	80	200 bar	7.70[a]	24.20	54.10	39.70	2.76
Kaufmann et al., 1982	I (residue)	80	200 bar	17.70[w]	17.50	46.90	40.00	1.51
Timmen et al., 1984	long chain fraction	80	200 bar	8.30[w]	13.70	43.80	39.70	2.56
Arul et al., 1987	S2	70	250–350 bar	22.60[m]	18.20	48.60	39.40	2.04
	S3	70	250–350	13.80[m]	13.50	42.80	41.50	2.23
Bhaskar et al., 1993b	S2	60	24.1 MPa	5.97[w] C4–C8	4.17 C10–C12	—[3]	89.86 C14:0–C18:3	—
Kankare and Antila, 1988a	Res	50–80	100–400 bar	5.50[n]	—	29.50	40.80	2.20
Büning-Pfaue et al., 1989	Extractor fraction	50	25,000 kPa	10.00[w]	17.30	53.40	42.90	3.03
Kankare et al., 1989	R	50	400 bar	8.50[n]	14.80	43.80	40.70	2.15
Chen et al., 1992	13.8	40	13.8 MPa	24.9[n] C4–C8	10.1 C10–C12	—	65.1 C14–C18	—
	20.7	40	20.7	18.1[n]	6.6	—	75.3	—
	24.1	40	24.1	17.0[n]	5.8	—	76.9	—
	27.6	40	27.6	17.5[n]	6.5	—	76.4	—
Shukla et al., 1994	Raffinate	40	24.1 MPa	1.22[w] C4–C8	1.95 C10–C12	—	96.83 C14:0–C18:3	—
Büning-Pfaue et al., 1989	Column fraction	30	20,000 kPa	5.30[w]	13.90	49.70	45.10	2.81

[1]Fraction designations used are those given by the original author.

[2]Units of chemical composition are designated using these abbreviations: a = area %, m = mole %, n = % not specified, w = wt %.

[3]Data not available.

TABLE 5.68 Fatty Acid Composition (Selected Fatty Acid Groupings) of High-Melting Milkfat Fractions Obtained by Short-Path Distillation

Author	Fraction designation[1]	Selected saturated n–chain groupings (%)[2]				Selected ratio (%)
		C4–C10	C12–C15	C12,C14,C16	C16–C18	C18:1/18:0
Arul et al., 1988a	SF Residue	6.90[m]	12.60	40.60	41.20	2.05

[1]Fraction designations are those given by the original author.

[2]m = mole %.

TABLE 5.69 Fatty Acid Composition (Selected Fatty Acid Groupings) of Middle-Melting Milkfat Fractions Obtained from Melted Milkfat

Fraction separation method Author	Fraction designation[1]	Selected saturated n-chain groupings (%)[2]				Selected ratio (%)	Selected unsaturated groupings (%)	
		C4–C10	C12–C15	C12,C14,C16	C16–C18	C18:1/C18:0	cis	trans
Vacuum filtration								
Kaylegian, 1991	34L$_S$ [W]	12.00[a]	13.20	40.80	38.00	2.20	—[3]	—
deMan and Finoro, 1980	32L$_S$	11.60[w]	14.90	35.60	33.50	1.91	—	—
	31L$_S$	12.30[w]	15.50	36.70	32.70	2.20	—	—
	30L$_S$	10.50[w]	15.20	39.80	35.10	2.48	—	—
Kaylegian, 1991	30L$_S$ [W]	12.60[a]	13.40	40.00	36.30	2.38	—	—
Amer et al., 1985	29L$_S$ [W]	13.3[w]	16.5	—	38.9	—	26.5 C10:1–C18:3	1.6 C18:1
	29L$_S$ [S]	13.0[w]	15.3	—	37.0	—	27.8	3.2
deMan and Finoro, 1980	29L$_S$	12.00[w]	14.90	36.20	33.30	2.15	—	—
	28L$_S$	11.30[w]	14.60	37.10	33.70	2.38	—	—
	27L$_S$	12.60[w]	14.80	35.40	32.30	2.15	—	—
Amer et al., 1985	26L$_S$ [W]	13.8[w]	16.5	—	38.5	—	26.9 C10:1–C18:3	1.3 C18:1
Deroanne, 1976	21L$_S$ [W]	7.80[n]	14.30	40.50	37.60	3.04	—	—
Barna et al., 1992	20S$_M$ [W]	9.70[n]	16.50	49.70	47.00	1.35	—	—
Grall, 1992	20S$_M$ [S]	9.70[a]	16.50	49.70	47.00	1.35	—	—
Amer et al., 1985	19L$_S$ [W]	12.2[w]	17.1	—	42.9	—	29.2 C10:1–C18:3	2.6 C18:1
Sherbon, 1963	17S$_M$	21.90[m]	12.00	41.80	38.60	3.14	—	—
Kaylegian, 1991	16S$_M$ [W]	11.10[a]	12.20	45.30	45.40	1.39	—	—
Makhlouf et al., 1987	15S$_M$ (I.1)	16.0[m] C4–C12	—	—	82.0 C14–C18	—	—	—
Laakso et al., 1992	S2	17.10[m]	17.1	40.70	40.10	1.59	—	—
	S2 SSS[4]	24.80[m]	18.7	52.00	55.30	0.03	—	—
	S2 SSM[t]	14.50[m]	14.5	34.80	34.10	2.31	—	—
	S2 SSM[c]	13.10[m]	14.1	34.70	36.80	2.47	—	—
	S2 SM[c]M[t]	7.20[m]	7.8	18.30	20.90	7.37	—	—
	S2 SM[c]M[c]	8.30[m]	9.0	17.00	18.10	9.73	—	—
	S2 UUU	7.70[m]	7.1	16.10	18.00	6.06	—	—
Pressure filtration								
deMan, 1968	30L$_S$ (B) rapid cooling	9.00[n]	11.70	39.10	39.10	2.69	—	—
Centrifugation								
Richardson, 1968	25S$_M$ [W]	10.90[w]	14.10	42.30	42.60	1.99	—	—
Lakshminarayana and Rama Murthy, 1985	23S$_M$	9.00[w]	17.60	49.30	42.20	2.98	—	—
Richardson, 1968	20S$_M$ [W]	12.80[w]	13.50	42.90	43.20	1.97	—	—
Jordan, 1986	15S$_S$ [S]	—	—	—	—	2.48[w]	—	—
Kankare and Antila, 1986	12S$_M$ [S]	10.00[n]	14.20	42.40	41.80	1.62	—	—

Continued

TABLE 5.69 (Continued)

Fraction separation method Author	Fraction designation[1]	Selected saturated n–chain groupings (%)[2]				Selected ratio (%)	Selected unsaturated groupings (%)	
		C4–C10	C12–C15	C12,C14,C16	C16–C18	C18:1/C18:0	cis	trans
Kankare and Antila, 1988a	12S$_M$ [S]	—	—	—	41.00^n	1.69	—	—
Antila and Antila, 1970	C	7.80^n	17.10	51.50	46.00	1.91	—	—
Separation with the aid of an aqueous detergent solution								
Doležálek et al., 1976	27L$_S$ [1]	8.50^w	18.70	47.50	38.50	3.14	—	—
Filtration through milk filter								
Fouad et al., 1990	29L$_S$	12.00^w	15.30	44.70	40.30	2.10	—	—
	17S$_S$	10.30^w	15.70	47.50	44.50	1.59	—	—
Separation technique not reported								
Riel and Paquet, 1972	38S$_M$	6.80^n	16.80	49.90	51.90	1.29	—	—
	4S$_M$	9.50^n	19.00	50.10	46.00	1.70	—	—
Martine, 1982	Olein	4.30^n	17.40	49.20	43.40	2.49	—	—

[1]The number indicates fractionation temperature, the letter indicates physical state (L = liquid, S = solid), and the subscript indicates fractionation procedure used (M = multiple-step, S = single-step). Other designations are those of the original author. The designation inside the brackets refers to the intact anhydrous milkfat source [sample number, letter or season (S = summer, W = winter)].

[2]Units of chemical composition are designated using these abbreviations: a = area %, m = mole %, n = % not specified, w = wt %.

[3]Data not available.

[4]S = saturated, M = monounsaturated, U = polyunsaturated, c = cis, t = trans.

TABLE 5.70 Fatty Acid Composition (Selected Fatty Acid Groupings) of Middle-Melting Milkfat Fractions Obtained from Solvent Solution

Type of solvent solution Author	Fraction designation[1]	Selected saturated n–chain groupings (wt %)[2]				Selected ratio (wt %)
		C4–C10	C12–C15	C12,C14,C16	C16–C18	C18:1/C18:0
Acetone						
Kaylegian, 1991	25L$_S$ [W]	11.60^a	14.30	41.20	36.60	2.26
Sherbon and Dolby, 1973	25L$_M$ (V) [HMG]	10.50^m	17.40	49.70	47.90	1.37
Kaylegian, 1991	21L$_S$ [W]	12.60^a	13.60	40.00	36.00	2.36
	5S$_M$ [W]	8.40^a	11.80	48.70	53.00	0.86
Yi, 1993	5S$_M$ [S]	8.60^a	12.30	52.30	59.00	0.82
Kaylegian, 1991	0S$_M$ [W]	10.40^a	13.50	50.00	47.80	1.31
Yi, 1993	0S$_M$ [S]	9.20^a	13.80	56.70	56.70	1.15
Pentane						
Kaylegian, 1991	11L$_S$ [W]	10.30^a	13.40	40.20	37.00	2.28
	5L$_S$ [W]	11.10^a	13.60	39.80	35.60	2.51

[1]The number indicates fractionation temperature, the letter indicates physical state (L = liquid, S = solid), and the subscript indicates fractionation procedure used (M = multiple-step, S = single-step). Any designations in parentheses are fraction designations given by the original author. The designation inside the brackets refers to the intact anhydrous milkfat source (HMG = high-melting glyceride, S = summer, W = winter).

[2]Units of chemical composition are designated using these abbreviations: a = area %, m = mole %.

TABLE 5.71 Fatty Acid Composition (Selected Fatty Acid Groupings) of Middle-Melting Milkfat Fractions Obtained by Supercritical Fluid Extraction

Type of solvent Author	Fraction designation[1]	Fractionation conditions		Selected saturated n–chain groupings (wt %)[2]				Selected ratio (wt %)
		Temperature (°C)	Pressure	C4–C10	C12–C15	C12,C14,C16	C16–C18	C18:1/18:0
Carbon dioxide								
Bhaskar et al., 1993b	S3	75	17.2 MPa	10.14^w C4–C8	5.34 C10–C12	—[3]	84.52 C14:0–C18:3	—
Arul et al., 1987	S1	70	250–350 bar	29.30^m	17.80	46.90	38.50	2.23
Bhaskar et al., 1993b	S4	60	6.9 MPa	10.67^w C4–C8	5.91 C10–C12	—	83.42 C14:0–C18:3	—
	S5	60	3.5	12.42^w	5.88	—	81.69	—
Kankare and Antila, 1988a	E2	50-80	100–400 bar	10.10^n	—[3]	29.20	37.10	2.18
	E3	50–80	100–400	8.80^n	—	29.70	38.80	2.11
	E4	50–80	100–400	8.40^n	—	29.50	38.70	2.21
Arul et al., 1987	I2	50	100–250 bar	36.90^m	19.00	45.10	32.10	2.04
	I3	50	100–250	29.30^m	18.10	44.30	34.40	2.03
Kankare et al., 1989	E2	50	200 bar	16.30^n	19.10	48.80	37.90	2.13
	E3	50	300	14.10^n	17.70	48.10	39.60	2.12
	E4	50	400	13.30^n	17.20	47.20	39.40	2.17
Chen et al., 1992	17.2	40	17.2 MPa	17.1^n C4–C8	6.3 C10–C12	—	77.1 C14–C18	—
Hamman et al., 1991	A4	40	125 bar	8.00^w	13.00	40.20	41.50	2.16

[1]Fraction descriptions used are those used by the original author.

[2]Units of chemical composition are designated using these abbreviations: m = mole %, n = % not specified, w = wt %.

[3]Data not available.

TABLE 5.72 Fatty Acid Composition (Selected Fatty Acid Groupings) of Middle-Melting Milkfat Fraction Obtained by Short-Path Distillation

Author	Fraction designation[1]	Fractionation conditions		Selected saturated n–chain groupings (%)[2]				Selected ratio (%)
		Temperature (°C)	Pressure	C4–C10	C12–C15	C12,C14,C16	C16–C18	C18:1/18:0
Arul et al., 1988a	IF Distillate III	265	100 μm Hg	21.10^m	16.00	45.20	37.80	1.74

[1]Fraction descriptions used are those used by the original author.

[2]m = mole %.

TABLE 5.73 Fatty Acid Composition (Selected Fatty Acid Groupings) of Low-Melting Milkfat Fractions Obtained from Melted Milkfat

Fraction separation method Author	Fraction designation[1]	Selected saturated n-chain groupings (%)[2]				Selected ratio (%)	Selected unsaturated groupings (%)	
		C4–C10	C12–C15	C12,C14,C16	C16–C18	C18:1/C18:0	cis	trans
Vacuum filtration								
deMan and Finoro, 1980	26L$_S$	11.20[w]	14.60	37.20	33.70	2.46	—[3]	—
Lund and Danmark, 1987	26L$_S$ [1]	11.90[w]	14.80	43.80	39.00	2.55	—	—
deMan and Finoro, 1980	25L$_S$	12.70[w]	15.60	36.10	32.30	2.17	—	—
Kaylegian, 1991	25L$_S$ [W]	11.60[a]	13.90	39.90	34.50	2.87	—	—
Timms and Parekh, 1980	25L$_S$ [S]	12.70[w]	16.90	40.80	37.10	2.09	—	—
Amer et al., 1985	23L$_S$ [W]	14.3[w]	16.2	—	37.5	—	27.6 C10:1–C18:3	1.4 C18:1
Barna et al., 1992	20L$_M$ [W]	13.00[n]	14.40	42.90	39.10	2.51	—	—
Kaylegian, 1991	20L$_S$ [W]	14.30[a]	12.90	37.40	33.40	2.78	—	—
Amer et al., 1985	19L$_S$ [W]	14.8[w]	16.1	—	36.7	—	27.7 C10:1–C18:3	1.4 C18:1
	19L$_S$ [S]	14.0[w]	15.0	—	35.7	—	29.2	2.6
Schaap and van Beresteyn, 1970	17L$_M$	11.70[n]	15.50	39.70	35.10	3.02	—	—
Grall and Hartel, 1992	15S$_M$ [S]	13.10[a]	14.00	43.40	40.40	2.33	—	—
	15L$_M$ [S]	13.20[a]	15.00	40.60	35.00	3.11	—	—
Kaylegian, 1991	13S$_M$ [W]	11.30[a]	13.70	43.10	39.20	2.20	—	—
	13L$_M$ [W]	13.30[a]	14.00	36.10	28.50	4.44	—	—
Sherbon, 1963	10S$_M$	18.90[m]	13.90	40.80	34.30	4.66	—	—
	10L$_M$	19.90[m]	13.10	34.90	27.60	7.56	—	—
Deffense, 1987	2nd olein$_M$ filtered at 10·C [2]	9.60[w]	13.50	30.90	27.40	4.35	—	—
Makhlouf et al., 1987	9S$_M$ (I.2)	26.0[m]	—	—	73.0	—	—	—
	9L$_M$ (F.L.)	28.0[m]	—	—	70.0	—	—	—
	F.1 (21-9)	24.0[m]	—	—	75.0	—	—	—
Deffense, 1987	1st Olein$_M$ [2]	9.60[w]	14.10	35.20	33.10	3.23	—	—
Timms, 1978b	Soft	12.10[w]	15.10	40.40	38.40	2.35	—	—
Laakso et al., 1992	L1	19.70[m]	15.5	38.30	36.30	2.21	—	—
	L1 SSS[4]	28.40[m]	20.8	54.20	49.40	0.02	—	—
	L1 SSM[t]	19.00[m]	14.7	33.90	29.90	3.07	—	—
	L1 SSM[c]	15.90[m]	14.4	34.70	33.70	3.11	—	—
	L1 SM[c]M[t]	8.20[m]	8.0	18.30	20.80	7.75	—	—
	L1 SM[c]M[c]	8.10[m]	8.6	17.30	18.10	10.73	—	—
	L1 UUU	9.30[m]	8.3	17.80	18.20	6.38	—	—

Continued

TABLE 5.73 (Continued)

Fraction separation method Author	Fraction designation[1]	Selected saturated n–chain groupings (%)[2]				Selected ratio (%)	Selected unsaturated groupings (%)	
		C4–C10	C12–C15	C12,C14,C16	C16–C18	C18:1/C18:0	cis	trans
	L2	19.80[m]	15.8	33.30	29.80	3.32	—	—
	L2 SSS	30.90[m]	24.3	52.10	43.00	0.07	—	—
	L2 SSM[t]	21.90[m]	15.2	30.80	25.70	3.61	—	—
	L2 SSM[c]	17.30[m]	15.2	32.70	30.80	3.47	—	—
	L2 SM[c]M[t]	8.20[m]	9.1	16.80	18.50	8.00	—	—
	L2 SM[c]M[c]	9.00[m]	9.5	16.90	16.60	10.52	—	—
	L2 UUU	9.10[m]	7.2	14.30	14.90	9.33	—	—
Vovan and Riel, 1973	5% removal of hard fraction (>21°C)	11.70[n]	16.60	41.30	39.10	1.95	—	—
	10% removal of hard fraction (>21°C)	11.40[n]	17.30	40.30	36.80	2.22	—	—
	15% removal of hard fraction (>21°C)	10.10[n]	15.70	39.70	39.50	1.90	—	—
	5% removal of intermed. fraction (10–21°C)	9.60[n]	16.20	40.90	39.60	2.02	—	—
	10% removal of intermed. fraction (10–21°C)	12.60[n]	17.40	40.90	37.70	2.07	—	—
	15% removal of intermed. fraction (10–21°C)	9.60[n]	16.10	40.90	39.60	2.02	—	—
	33% removal of intermed. fraction (10–21°C)	10.30[n]	17.40	43.00	39.30	2.12	—	—
Pressure filtration								
deMan, 1968	27L$_S$ (C) slow cooling	9.30[n]	11.50	38.20	38.00	2.96	—	—
	27L$_S$ (D) rapid cooling	9.60[n]	11.00	37.90	37.80	3.06	—	—
	24L$_S$ (F) rapid cooling	9.70[n]	11.60	37.70	36.70	3.05	—	—
Centrifugation								
Jordan, 1986	28L$_S$ [S]	—	—	—	—	2.66[w]	—	—
	28L$_S$ [W]	11.10[w]	13.80	42.50	39.60	2.60	—	—
Ramesh and Bindal, 1987b	28L$_S$ [Ghee]	14.60[w]	17.10	37.20	30.60	3.14	—	—
Jordan, 1986	24L$_S$ [S]	—	—	—	—	2.78[w]	—	—
	24L$_S$ [W]	—	—	—	—	2.47[w]	—	—
	15L$_S$ [S]	—	—	—	—	2.79[w]	—	—
Lakshminarayana and Rama Murthy, 1985	15S$_M$	10.80[w]	15.80	46.30	41.40	2.91	—	—
	15L$_M$	11.60[w]	16.10	45.30	39.10	3.27	—	—
Richardson, 1968	15S$_M$ [W]	12.10[w]	13.90	41.90	40.90	2.29	—	—
	15L$_M$ [W]	13.00[w]	13.90	37.20	33.80	3.36	—	—

Continued

TABLE 5.73 (Continued)

Fraction separation method Author	Fraction designation[1]	Selected saturated n–chain groupings (%)[2]				Selected ratio (%)	Selected unsaturated groupings (%)	
		C4–C10	C12–C15	C12,C14,C16	C16–C18	C18:1/C18:0	cis	trans
Kankare and Antila, 1988a	12L$_M$ [S]	—	—	—	31.50[n]	3.44	—	—
Kankare and Antila, 1988b	12L$_M$ [S]	—	—	—	31.00[n]	3.33	—	—
Kankare and Antila, 1986	12L$_M$ [S]	10.50[n]	14.00	36.30	31.60	3.30	—	—
Riel and Paquet, 1972	4S$_M$	10.80[n]	17.90	46.90	43.20	1.97	—	—
Antila and Antila, 1970	A	12.30[n]	17.10	35.00	27.20	5.44	—	—
	B	21.30[n]	17.90	36.50	26.10	4.36	—	—
Keogh and Higgins, 1986b	Fraction 3	11.70[w]	15.70	39.50	35.50	2.58	—	—
	Fraction 4	11.20[w]	15.80	35.70	32.60	2.46	—	—
Separation with the aid of an aqueous detergent solution								
Norris et al., 1971	26L$_S$ [B late S]	7.70[a]	15.00	40.20	38.30	2.60	—	—
	26L$_S$ [C early F]	10.00[a]	15.70	39.70	37.50	2.33	—	—
	25L$_S$ [A late S]	8.70[a]	14.20	39.30	38.70	2.48	—	—
Banks et al., 1985	21L$_S$ [A] high level of detergent	9.00[w]	9.70	45.70	44.90	3.43	27.8 C18:1	2.7 C18:1
Dolby, 1970a	Liquid [midseason]	8.70[n]	17.60	50.00	46.70	1.77	—	—
Fjaervoll, 1970a	LMF	12.40[n]	17.30	41.60	36.40	2.55	—	—
Filtration through casein-dewatering press								
Black, 1973	Soft #1	10.30[w]	16.10	40.70	38.60	1.51	—	—
	Soft #2	10.40[w]	16.80	40.30	36.20	1.84	—	—
	Soft #3	9.60[w]	15.00	38.10	36.50	1.87	—	—
Filtration through milk filter								
Fouad et al., 1990	17L$_S$	15.90[w]	14.20	40.90	36.10	2.57	—	—
Separation technique not reported								
Deroanne, 1976	21L$_S$ [W]	7.80[n]	14.40	40.60	38.10	2.89	—	—
Deroanne and Guyot, 1974	21L$_S$ [W]	11.00[a]	17.80	45.70	36.30	3.20	—	—
	21L$_S$ [S]	10.70[a]	14.80	36.50	33.80	2.69	—	—
Riel and Paquet, 1972	4L$_M$	11.30[n]	18.10	43.60	37.70	2.70	—	—

[1]The number indicates fractionation temperature, the letter indicates physical state (L = liquid, S = solid), and the subscript indicates fractionation procedure used (M = multiple-step, S = single-step). Other primary descriptions and descriptions in parentheses are those of the original author. The designation inside the brackets refers to the intact anhydrous milkfat source [sample number, letter or season (F = fall, S = summer, W = winter)].

[2]Units of chemical composition are designated using these abbreviations: a = area %, m = mole %, n = % not specified, w = wt %.

[3]Data not available.

[4]S = saturated, M = monounsaturated, U = polyunsaturated, c = cis, t = trans.

TABLE 5.74 Fatty Acid Composition (Selected Fatty Acid Groupings) of Low-Melting Milkfat Fractions Obtained from Solvent Solution

Type of solvent solution Author	Fraction designation[1]	Selected saturated n-chain groupings (%)[2]				Selected ratio (%)
		C4–C10	C12–C15	C12,C14,C16	C16–C18	C18:1/C18:0
Acetone						
Kaylegian, 1991	15L$_S$ [W]	11.90^a	13.90	41.00	35.80	2.68
	11L$_S$ [W]	12.20^a	14.20	40.70	34.30	3.09
	5L$_S$ [W]	13.40^a	13.20	37.10	31.10	3.63
Timms, 1980b	3L$_M$ [S]	12.50^w	15.30	36.60	33.00	3.04
Yi, 1993	0L$_M$ [S]	12.80^a	15.50	41.00	32.90	4.04
Sherbon, 1963	2C	15.30^m	17.80	39.10	33.40	2.65
	2CJAN [Jan]	16.70^m	15.10	41.30	35.80	3.18
Pentane						
Kaylegian, 1991	0L$_M$ [W]	13.20a	12.60	38.40	34.60	2.76
	0L$_S$ [W]	11.20a	13.40	39.50	35.40	2.55

[1]The number indicates fractionation temperature, the letter indicates physical state (L = liquid), and the subscript indicates fractionation procedure used (M = multiple-step, S = single-step). Other descriptions used are those of the original author. The designation inside the brackets refers to the intact anhydrous milkfat source (S = summer, W = winter).

[2]Units of chemical composition are designated using these abbreviations: a = area %, m = mole %, w = wt %.

TABLE 5.75 Fatty Acid Composition (Selected Fatty Acid Groupings) of Low-Melting Milkfat Fractions Obtained by Supercritical Fluid Extraction

Type of solvent Author	Fraction designation[1]	Fractionation conditions		Selected saturated n-chain groupings (%)[2]				Selected ratio (%)
		Temperature (°C)	Pressure	C4–C10	C12–C15	C12,C14,C16	C16–C18	C18:1/18:0
Carbon dioxide								
Biernoth and Merk, 1985	Extract	80	200 bar	19.60^a	22.90	52.10	34.10	2.76
Kaufmann et al., 1982	II (extract)	80	200 bar	36.90^w	20.50	42.80	27.00	2.53
Timmen et al., 1984	Short chain fraction	80	200 bar	20.60^w	21.60	51.90	35.30	2.50
Kankare and Antila, 1988a	E1	50–80	100–400 bar	11.70^n	—[3]	29.40	37.10	1.82
Arul et al., 1987	L1	50	100–250 bar	52.30^m	20.00	37.20	20.60	4.00
	L2	50	100–250	52.40^m	17.20	38.70	23.40	3.74
	I1	50	100–250	41.20^m	18.80	45.60	31.30	1.98
Kankare et al., 1989	E1	50	100 bar	18.00^n	21.20	50.70	37.30	1.79
Chen et al., 1992	10.3	40	10.3 MPa	38.9^n C4–C8	10.1 C10–C12	—	51.9 C14–C18	—
Hamman et al., 1991	A1	40	125 bar	11.50^w	18.50	48.60	41.20	1.92

[1]Fraction descriptions used are those given by the original author.

[2]Units of chemical composition are designated using these abbreviations: a = area %, m = mole %, n = % not specified, w = wt %.

[3]Data not available.

TABLE 5.76 Fatty Acid Composition (Selected Fatty Acid Groupings) of Low-Melting Milkfat Fractions Obtained by Short-Path Distillation

Author	Fraction designation[1]	Fractionation conditions		Selected saturated n–chain groupings (%)[2]				Selected ratio (%)
		Temperature (°C)	Pressure	C4–C10	C12–C15	C12,C14,C16	C16–C18	C18:1/18:0
Arul et al., 1988a	LF1 Distillate I	245	220 μm Hg	36.70[m]	23.10	44.70	25.50	1.82
	LF2 Distillate II	245	100	30.70[m]	21.00	45.20	30.00	1.66

[1]Designations used are those given by the original author.

[2]m = mole %.

TABLE 5.77 Fatty Acid Composition (Selected Fatty Acid Groupings) of Very-Low-Melting Milkfat Fractions Obtained from Melted Milkfat

Fraction separation method Author	Fraction designation[1]	Selected saturated n–chain groupings (%)[2]				Selected ratio (%)
		C4–C10	C12–C15	C12,C14,C16	C16–C18	C18:1/C18:0
Vacuum filtration						
Versteeg, 1991	8L$_M$ [Sept]	12.10[n]	13.50	32.20	29.20	3.59
Deffense, 1987	2nd olein$_M$ filtered at 6°C [2]	10.00[w]	13.10	29.70	25.90	4.91
Deffense, 1993b	2nd olein$_M$	9.40[w]	14.30	32.30	27.50	4.92
Centrifugation						
Riel and Paquet, 1972	4L$_M$	12.90[n]	18.50	43.10	34.80	3.29

[1]The number indicates fractionation temperature, the letter indicates physical state (L = liquid), and the subscript indicates fractionation procedure used (M = multiple-step). Other descriptions used are those of the original author. The designation inside the brackets refers to the intact anhydrous milkfat source (sample number or month).

[2]Units of chemical composition are designated using these abbreviations: n = % not specified, w = wt %.

TABLE 5.78 Fatty Acid Composition (Selected Fatty Acid Groupings) of Very-Low-Melting Milkfat Fractions Obtained from Solvent Solution

Type of solvent solution Author	Fraction designation[1]	Selected saturated n–chain groupings (%)[2]				Selected ratio (%)
		C4–C10	C12–C15	C12, C14, C16	C16–C18	C18:1/C18:0
Acetone						
Kaylegian, 1991	0L$_M$ [W]	12.40[a]	14.10	34.70	26.60	4.73
	0L$_S$ [W]	14.40[a]	13.20	34.20	27.60	3.97
Sherbon, 1963	2DJAN [Jan]	25.40[m]	12.70	32.60	26.30	8.64

[1]The number indicates fractionation temperature, the letter indicates physical state (L = liquid), and the subscript indicates fractionation procedure used (M = multiple-step, S = single-step). Other descriptions used are those of the original author. The designation inside the brackets refers to the intact anhydrous milkfat source (W = winter).

[2]Units of chemical composition are designated using these abbreviations: a = area %, m = mole %.

TABLE 5.79 Fatty Acid Composition (Selected Fatty Acid Groupings) of Unknown-Melting Milkfat Fractions Obtained from Melted Milkfat

Fraction separation method Author	Fraction designation[1]	Selected saturated n-chain groupings (%)[2]				Selected ratio (%)	Selected unsaturated groupings (%)	
		C4–C10	C12–C15	C12, C14, C16	C16–C18	C18:1/C18:0	cis	trans
Pressure filtration								
deMan, 1968	30L$_S$ (A) slow cooling	8.30^n	11.30	38.10	39.00	2.59	—[3]	—
	30S$_S$ (B) rapid cooling	5.70^n	11.70	46.10	53.90	1.23	—	—
	27S$_S$ (D) rapid cooling	6.80^n	12.20	43.30	47.20	1.64	—	—
	24L$_S$ (E) slow cooling	10.00^n	11.20	37.00	36.80	2.91	—	—
	24S$_S$ (F) rapid cooling	6.70^n	12.20	42.80	46.20	1.72	—	—
Badings et al., 1983b	23S$_M$ [W]	10.40^w	18.00	48.50	43.70	1.63	—	—
	23L$_M$ [W]	13.40^w	16.60	42.70	36.50	2.66	—	—
	21.5S$_M$ [S]	8.40^w	14.80	39.60	41.50	1.71	—	—
	21.5L$_M$ [S]	11.10^w	13.50	34.50	33.50	2.77	—	—
	18.5S$_M$ [W]	12.90^w	15.60	45.60	41.90	1.97	—	—
	18.5L$_M$ [W]	13.40^w	17.10	40.90	33.60	3.10	—	—
	15.3S$_M$ [S]	10.10^w	12.60	39.80	44.30	1.51	—	—
	15.3L$_M$ [S]	11.10^w	13.80	33.90	31.80	3.10	—	—
	15S$_M$ [W]	12.20^w	16.70	46.20	41.00	2.14	—	—
	15L$_M$ [W]	13.60^w	17.20	38.80	30.50	3.74	—	—
	13S$_M$ [S]	10.90^w	13.90	35.90	35.00	2.50	—	—
	13L$_M$ [S]	11.40^w	13.70	31.60	28.30	3.81	—	—
	11.3S$_M$ [W]	12.40^w	17.90	42.90	35.40	2.76	—	—
	11.3L$_M$ [W]	14.10^w	17.00	36.70	28.10	4.16	—	—
	9.3S$_M$ [S]	11.20^w	14.40	34.30	31.40	3.12	—	—
	9.3L$_M$ [S]	11.60^w	13.60	31.00	27.20	4.11	—	—
	7.1S$_M$ [S]	11.20^w	14.10	32.50	29.30	3.52	—	—
	7.1L$_M$ [S]	11.90^w	13.60	30.20	25.70	4.53	—	—
	5.5S$_M$ [W]	13.10^w	17.90	39.50	30.80	3.65	—	—
	5.5L$_M$ [W]	14.20^w	15.90	33.60	25.00	5.46	—	—
Separation with the aid of an aqueous detergent solution								
Banks et al., 1985	16L$_S$ [B] low level of detergent	7.20^w	13.70	38.70	39.40	2.49	24.6 C18:1	8.0 C18:1
Filtration through milk filter								
Fouad et al., 1990	25L$_S$	14.50^w	14.90	43.20	38.80	2.10	—	—
	21S$_S$	9.70^w	15.70	47.90	45.40	1.53	—	—
	21L$_S$	17.50^w	14.20	40.40	35.60	2.41	—	—
Separation technique not reported								
Timmen, 1974	28S$_S$ [W]	—	—	—	—	1.82^n	—	—
	28L$_S$ [W]	—	—	—	—	2.70^n	—	—
	24S$_S$ [S]	—	—	—	—	2.50^n	—	—
	24L$_S$ [S]	—	—	—	—	2.94^n	—	—

Continued

TABLE 5.79 (Continued)

Fraction separation method Author	Fraction designation[1]	Selected saturated n–chain groupings (%)[2]				Selected ratio (%)	Selected unsaturated groupings (%)	
		C4–C10	C12–C15	C12, C14, C16	C16–C18	C18:1/C18:0	cis	trans
Lechat et al., 1975	Most solid fraction [Feb]	5.40^n	14.20	40.60	48.40	1.14	—	—
	Most liquid fraction [Sept]	10.70^n	14.40	33.00	29.50	3.75	—	—
	Most liquid fraction [Feb]	10.4^n	12.90	27.40	24.70	4.38	—	—

[1]The number indicates fractionation temperature, the letter indicates physical state (L = liquid, S = solid), and the subscript indicates fractionation procedure used (M = multiple-step, S = single-step). Other designations are those of the original author. The designation inside the brackets refers to the intact anhydrous milkfat source (S = summer, W = winter).

[2]Units of chemical composition are designated using these abbreviations: n = % not specified, w = wt %.

[3]Data not available.

TABLE 5.80 Fatty Acid Composition (Selected Fatty Acid Groupings) of Unknown-Melting Milkfat Fractions Obtained by Short-Path Distillation

Fraction separation method Author	Fraction designation[1]	Selected saturated n–chain groupings (%)[2]				Selected ratio
		C4–C10	C12–C15	C12,C14,C16	C16–C18	C18:1/C18:0
McCarthy et al., 1962	R-1	29.4^w	29.1	52.5	28.9	1.93
	R-2	24.1^w	29.6	60.9	35.1	2.13
	R-3	13.6^w	25.5	60.1	43.0	1.99
	R-4	11.8^w	12.8	53.5	52.3	1.96
	D-2	11.5^w	19.4	53.5	45.6	1.90
	D-3	4.7^w	14.7	43.7	43.8	2.32

[1]Designations used are those given by the original author.

[2]Units of chemical composition are designated using this abbreviation: w = wt %.

TABLE 5.81 Fatty Acid Composition (Selected Fatty Acid Groupings) of Unknown-Melting Milkfat Fractions Obtained from Solvent Solution

Type of solvent solution Author	Fraction designation[1]	Selected saturated n-chain groupings (%)[2]				Selected ratio (%)
		C4–C10	C12–C15	C12, C14, C16	C16–C18	C18:1/C18:0
Acetone						
Larsen and Samuelsson, 1979	21.4S_S	11.70[n]	13.20	42.50	45.20	1.47
Parodi, 1974a	20S_S [May]	0.90[m]	15.10	54.10	72.30	0.21
	SSS fraction of 20S_S [May][3]	1.1[m]	16.9	57.8	76.5	—[4]
	SSM^t fraction of 20S_S [May]	0.5[m]	8.9	38.8	55.7	1.24
	SSM^c fraction of 20S_S [May]	0.6[m]	8.6	39.5	56.5	1.29
	20L_S [May]	21.00[m]	15.70	37.80	34.10	1.99
Chen and deMan, 1966	15S_M	0.90[w]	13.70	47.50	62.40	0.50
	SSS fraction of 15S_M	0.6[w]	14.6	55.6	80.9	—
Larsen and Samuelsson, 1979	15.0S_S	8.60[n]	14.20	46.70	50.00	1.13
Mattsson et al., 1969	15S_S [Aug]	3.30[m]	16.30	51.70	58.40	0.47
	15S_S [Dec]	5.40[m]	20.30	61.70	56.40	0.49
Lansbergen and Kemps, 1984	12S_M [1]	3.1[w]	12.6	35.4	81.7	—
	12L_M [1]	16.4[w]	15.3	40.7	60.6	—
	12S_M [2]	4.4[w]	12.5	35.6	81.2	—
	12L_M [2]	16.5[w]	21.9	47.7	60.6	—
Lechat et al., 1975	10S_M [Sept]	4.10[n]	13.90	49.80	65.40	0.47
	10L_M [Sept]	7.30[n]	14.00	47.10	55.90	0.81
	10S_M [Apr]	2.30[n]	13.20	48.00	66.20	0.44
	10L_M [Apr]	8.40[n]	13.90	42.30	47.50	1.47
Larsen and Samuelsson, 1979	9.6S_S	11.60[n]	14.70	48.80	50.80	1.01
Lansbergen and Kemps, 1984	6S_M [1]	13.5[w]	15.3	44.5	69.9	—
	6L_M [1]	22.1[w]	30.5	50.7	46.2	—
Chen and deMan, 1966	5S_M	4.20[w]	11.90	44.80	59.80	0.60
	SSS fraction of 5S_M	6.9[w]	15.4	54.4	71.9	—
Larsen and Samuelsson, 1979	5.0S_S	12.40[n]	12.00	44.20	49.40	1.12
Lansbergen and Kemps, 1984	0S_M [2]	13.4[w]	15.1	44.2	70.0	—
	0L_M [2]	19.2[w]	28.0	50.8	51.7	—
Lechat et al., 1975	0S_M [Sept]	4.80[n]	14.00	46.50	59.00	0.65
	0L_M [Sept]	9.10[n]	15.90	37.00	34.10	2.68
	0S_M [Apr]	5.60[n]	15.10	47.20	55.00	0.87
	0L_M [Apr]	11.10[n]	17.00	35.80	33.30	1.96
Chen and deMan, 1966	-5S_M	7.80[w]	11.60	41.40	49.90	1.04
	SSS fraction of -5S_M	12.2[w]	13.4	55.3	68.7	—
Larsen and Samuelsson, 1979	-5.0L_S	16.00[n]	16.00	37.80	30.00	3.56
	-8.0L_S	14.80[n]	16.50	36.80	27.60	4.40

Continued

TABLE 5.81 (Continued)

Type of solvent solution Author	Fraction designation[1]	Selected saturated n-chain groupings (%)[2]				Selected ratio (%)
		C4–C10	C12–C15	C12, C14, C16	C16–C18	C18:1/C18:0
Lechat et al., 1975	-9S$_M$ [Sept]	6.10[n]	14.30	42.20	48.90	1.06
	-9L$_M$ [Sept]	9.60[n]	13.00	29.30	25.60	5.11
	-9L$_M$ [Apr]	10.40[n]	13.00	27.90	26.60	3.72
Chen and deMan, 1966	-15S$_M$	9.80[w]	16.00	40.00	36.00	2.23
	SSS fraction of -15S$_M$	13.3[w]	23.0	62.8	57.7	—
Larsen and Samuelsson, 1979	-16.4L$_S$	16.90[n]	16.10	33.40	23.70	5.22
	-21.0L$_S$	17.20[n]	16.30	33.20	22.70	5.79
Chen and deMan, 1966	-25S$_M$	12.00[w]	14.10	31.10	25.70	4.01
	SSS fraction of -25S$_M$	17.4[w]	29.5	62.0	46.3	—
Jensen et al., 1967	-25S$_M$	14.50[m]	16.60	49.60	49.20	0.78
Chen and deMan, 1966	-35S$_M$	12.10[w]	11.50	29.00	24.20	5.52
	SSS fraction of -35S$_M$	21.3[w]	27.3	58.3	43.6	—
	-45S$_M$	13.30[w]	12.00	26.70	20.70	6.40
	SSS fraction of -45S$_M$	23.0[w]	26.6	56.1	42.5	—
	-45L$_M$	16.90[w]	10.60	20.00	11.40	22.50
	SSS fraction of -45L$_M$	32.0[w]	30.0	53.5	30.5	—
Jensen et al., 1967	-70S$_M$	25.40[m]	14.20	33.40	26.30	3.21
	-70L$_M$	40.10[m]	10.60	16.10	6.10	45.50
Sherbon, 1963	2A	0.20[m]	15.80	58.60	76.10	0.17
	2D	16.30[m]	13.20	35.00	28.50	5.50
	2D'	17.10[m]	14.00	46.20	46.40	1.53
	2EJAN [Jan]	28.00[m]	10.20	24.90	21.00	7.20
	2FJAN [Jan]	32.40[m]	9.60	18.50	11.40	70.00
Lansbergen and Kemps, 1984	Top fraction	1.2[w]	19.00	66.60	70.70	0.26
	Mid fraction	8.8[w]	18.10	58.80	55.40	0.98
	Olein fraction	13.9[w]	19.20	39.90	27.70	5.15
Hexane						
Muuse and van der Kamp, 1985	12.5S$_S$	8.60[a]	13.70	40.70	41.50	1.69
	12.5L$_S$	10.60[a]	13.30	37.20	35.10	2.50

[1]The number indicates fractionation temperature, the letter indicates physical state (L = liquid, S = solid), and the subscript indicates fractionation procedure used (M = multiple-step, S = single-step). Other descriptions used are those of the original author. The designation inside the brackets following the fraction designation refers to the intact anhydrous milkfat source.

[2]Units of chemical composition are designated using these abbreviations: a = area %, m = mole %, n = % not specified, w = wt %.

[3]S = saturated, M = monounsaturated, c = cis, t = $trans$.

[4]Data not available.

TABLE 5.82 Fatty Acid Composition (Selected Fatty Acid Groupings) of Unknown-Melting Milkfat Fractions Obtained by Supercritical Fluid Extraction

Type of solvent Author	Fraction designation[1]	Fractionation conditions		Selected saturated n–chain groupings (%)[2]				Selected ratio (%)
		Temperature (°C)	Pressure	C4–C10	C12–C15	C12, C14, C16	C16–C18	C18:1/18:0
Carbon dioxide								
Bradley, 1989	WM-38, 1	60	2150 psi	16.80[w]	16.80	43.20	37.90	1.82
	WM-38, 2	60	2300	14.50[w]	15.20	42.70	40.20	1.84
	WM-38, 3	60	2400	13.80[w]	14.80	43.30	42.00	1.83
	WM-38, 4	60	2400	13.70[w]	14.00	42.40	42.50	1.80
	WM-38, 5	60	2500	11.40[w]	11.70	35.60	38.30	1.51
	WM-38, 6	60	2500	12.30[w]	12.00	37.80	39.50	1.83
	WM-38, 7	60	2650	12.90[w]	12.20	38.10	40.30	1.64
	WM-38, 8	60	2850	10.50[w]	11.10	37.40	41.10	1.78
	WM-38, 9	60	2900	9.70[w]	10.40	36.00	41.30	1.72
	WM-38, 10	60	2900	8.20[w]	10.10	34.80	41.10	1.71
	WM-38, 11	60	2900	7.20[w]	10.10	34.30	42.00	1.69
	WM-38, 12	60	2900	5.30[w]	10.20	34.00	42.60	1.71
	WM-38, 13	60	2900	3.50[w]	10.40	34.50	43.60	1.78
	WM-38, 14	60	2900	1.80[w]	9.20	32.70	43.30	1.87
	WM-38, 15	60	—[3]	0.30[w]	5.10	25.00	44.20	1.65
Ensign, 1989	WM-38, 1	60	2150 psi	16.80[w]	16.80	43.20	37.90	1.87
	WM-38, 2	60	2300	14.50[w]	15.20	42.70	40.20	1.84
	WM-38, 3	60	2400	14.80[w]	14.80	43.30	42.00	1.83
	WM-38, 4	60	2400	13.70[w]	14.00	42.40	42.50	1.80
	WM-38, 5	60	2500	11.40[w]	11.70	35.60	38.20	1.52
	WM-38, 6	60	2500	12.30[w]	12.00	37.80	39.50	1.78
	WM-38, 7	60	2650	13.00[w]	12.20	38.10	40.30	1.64
	WM-38, 8	60	2850	10.70[w]	11.10	37.40	41.10	1.78
	WM-38, 9	60	2900	9.70[w]	10.40	36.00	41.30	1.72
	WM-38, 10	60	2900	8.20[w]	10.10	34.80	41.10	1.71
	WM-38, 11	60	2900	7.20[w]	10.10	34.30	42.00	1.68
	WM-38, 12	60	2900	5.30[w]	10.20	34.00	42.60	1.71
	WM-38, 13	60	2900	3.50[w]	10.40	34.50	43.60	1.78
	WM-38, 14	60	2900	1.70[w]	9.20	32.70	43.30	1.88
	WM-38, 15	60	—	0.30[w]	5.1	25.0	44.2	1.65
Hamman et al., 1991	Ares	40	125 bar	0.00[w]	6.20	33.80	55.90	1.22
	B1	40	350	10.70[w]	18.60	51.20	42.00	2.16
Shishikura et al., 1986	F'1	40	300 kg/cm^2	24.20[n]	21.10	50.70	35.50	3.03
	F'2	40	300	20.10[n]	20.20	51.60	38.30	2.91
	F'3	40	300	16.20[n]	18.90	51.00	40.30	2.82
	F'4	40	300	13.60[n]	18.10	50.50	41.40	2.79
	F'5	40	300	13.40[n]	18.20	52.00	43.70	2.77
	F'6	40	300	10.50[n]	16.80	49.70	43.30	2.65
	Residue	40	300	4.10[n]	13.30	47.80	48.90	2.20

[1] Designations used are those given by the original author.

[2] Units of chemical composition are designated using these abbreviations: n = % not specified, w = wt %.

[3] Data not available.

TABLE 5.83 Triglyceride Composition (Individual Triglyceride) of Intact Anhydrous Milkfats Used to Produce Milkfat Fractions from Melted Milkfat

Fraction separation method Author	Fraction designation[1]	Triglyceride (%)[2]															
		C24	C26	C28	C30	C32	C34	C36	C38	C40	C42	C44	C46	C48	C50	C52	C54
Vacuum filtration																		
Amer et al., 1985	AMF [W]	tr[3]	0.2[a]	0.5	1.2	2.3	6.6	11.4	13.7	10.2	7.3	6.6	6.6	8.1	8.4	6.1	1.6
deMan and Finoro, 1980	AMF	0.2	0.1[w]	0.4	0.7	1.7	4.5	10.2	12.6	10.6	6.5	6.0	6.8	9.1	12.8	13.1	4.6
Kaylegian, 1991	AMF [W]	—[4]	0.7[a]	0.8	1.4	2.4	5.3	9.3	10.8	9.1	7.1	7.6	9.6	11.9	12.1	8.7	3.3
Lund and Danmark, 1987	AMF [1]	—	—	0.4[w]	1.1	2.4	5.3	10.0	12.7	10.0	6.4	5.9	6.5	7.6	9.9	9.7	4.7
	AMF [2]	—	—	0.4[w]	0.9	2.2	4.3	9.0	12.0	10.0	6.7	5.6	6.3	7.5	10.2	10.7	5.7
Pressure filtration																		
Badings et al., 1983b	AMF [W]	0.4[w]	0.3	0.7	1.2	2.9	6.6	11.7	13.2	10.3	7.6	7.0	7.3	8.4	9.8	8.6	3.9
	AMF [S]	0.4[w]	0.2	0.5	1.1	2.4	5.2	10.0	12.7	11.0	6.5	5.6	6.3	7.8	11.0	11.5	7.8
Centrifugation																		
Jordan, 1986	AMF [W]	—	0.6[w]	1.3	1.4	2.5	4.7	9.8	11.5	9.1	6.6	6.2	6.6	8.5	10.2	9.6	5.7
Kankare and Antila, 1986	AMF [S]	—	0.2[n]	0.5	1.1	3.0	4.6	10.7	13.3	10.1	6.2	4.1	6.5	7.7	10.9	12.5	6.3
Keogh and Higgins, 1986b	AMF	—	0.3[w]	0.4	1.1	2.1	4.8	9.5	12.9	11.0	6.9	6.0	6.6	8.0	10.8	9.5	4.6
Separation with the aid of an aqueous detergent solution																		
Banks et al., 1985	AMF [A]	—	—	—	0.3[w]	1.1	4.4	12.8	18.4	9.8	5.1	4.6	6.0	9.9	14.8	11.0	1.9
	AMF [B]	—	—	—	0.1[w]	0.4	2.0	4.5	8.8	9.1	4.4	3.6	5.6	8.7	17.8	18.3	15.7
Filtration through milk filter																		
Fouad et al., 1990	AMF	—	0.2[w]	0.6	1.1	2.3	5.6	10.3	11.5	9.4	6.7	6.2	6.5	7.6	9.2	8.5	3.6
Separation technique not reported																		
Lechat et al., 1975	AMF [Sept]	—	0.5[n]	0.5	1.0	1.8	5.0	10.8	13.3	13.6	9.0	6.8	7.1	7.5	10.1	9.0	4.1
	AMF [Feb]	—	0.5[n]	1.0	1.7	3.6	8.0	14.8	17.0	13.4	8.0	7.1	6.9	6.5	6.3	3.7	0.9

[1] AMF = anhydrous milkfat. The designation inside the brackets following the fraction designation refers to the intact anhydrous milkfat source [sample number, letter or season (W = winter, S = summer)].

[2] Units of chemical composition are designated using these abbreviations: a = area %, n = % not specified, w = wt %.

[3] tr = trace.

[4] Data not reported.

TABLE 5.84 Triglyceride Composition (Individual Triglyceride) of Intact Anhydrous Milkfats Used to Produce Milkfat Fractions from Solvent Solution

Type of solvent solution Author	Fraction designation[1]	Triglyceride (%)[2]														
		C24	C26	C28	C30	C32	C34	C36	C38	C40	C42	C44	C46	C48	C50	C52	C54
Acetone																	
Kaylegian, 1991	AMF	—[3]	0.7[a]	0.8	1.4	2.4	5.3	9.3	10.8	9.1	7.1	7.6	9.6	11.9	12.1	8.7	3.3
Lambelet, 1983	AMF	—	0.3[n]	0.6	1.1	2.2	4.8	9.1	—	—	—	—	—	—	—	—	—
Lansbergen and Kemps, 1984	Hardened butterfat	0.4[n]	0.2	0.6	1.2	3.1	7.2	13.0	12.7	9.6	7.5	7.3	7.9	9.0	10.1	7.9	3.0
	Natural butterfat	tr[4,n]	0.2	0.2	1.2	3.1	7.2	12.0	12.8	9.8	7.6	7.4	8.1	9.0	9.7	7.5	3.0
Lechat et al., 1975	AMF [Apr]	—	0.2[n]	0.3	1.1	2.2	5.9	12.6	14.5	13.2	7.5	6.1	6.8	8.5	9.7	7.8	3.5
Parodi, 1974a	AMF [May]	—	0.3[m]	0.7	1.2	2.7	6.5	12.9	14.7	10.6	6.7	6.0	6.4	7.9	10.2	9.5	4.0
Timms, 1980b	AMF [S]	0.3[w]	0.3	0.5	0.9	2.2	5.4	10.6	12.9	10.0	6.2	5.7	6.6	9.0	12.3	11.3	5.5
Yi, 1993	AMF [S]	—	—	0.3[a]	1.0	2.6	6.5	12.8	15.0	11.5	6.7	7.2	8.3	7.0	9.8	8.7	—
Pentane																	
Kaylegian, 1991	AMF [W]	—	0.7[a]	0.8	1.4	2.4	5.3	9.3	10.8	9.1	7.1	7.6	9.6	11.9	12.1	8.7	3.3
Hexane																	
Muuse and van der Kamp, 1985	AMF	—	0.3[a]	0.7	1.5	3.2	6.9	12.3	13.7	10.0	7.3	6.5	6.6	6.7	7.5	5.2	1.8

[1] AMF = anhydrous milkfat. The designation inside the brackets following the fraction designation refers to the intact anhydrous milkfat source (S = summer, W = winter).
[2] Units of chemical composition are designated using these abbreviations: a = area %, m = mole %, n = % not specified, w = wt %.
[3] Data not available.
[4] tr = trace.

TABLE 5.85 Triglyceride Composition (Individual Triglyceride) of Intact Anhydrous Milkfats Used to Produce Milkfat Fractions by Supercritical Fluid Extraction

Type of solvent Author	Fraction designation[1]	Triglyceride (%)[2]															
		C24	C26	C28	C30	C32	C34	C36	C38	C40	C42	C44	C46	C48	C50	C52	C54
Carbon dioxide																		
Arul et al., 1988b	AMF	0.5[w]	0.3	0.8	1.5	3.0	6.5	10.9	14.0	10.5	7.0	6.8	7.0	8.3	10.3	9.0	3.6
Bhaskar et al., 1993a	AMF	0.5[w]	0.7	1.0	1.9	4.0	8.7	17.3	20.3	13.1	7.7	5.0	5.2	5.4	5.6	3.9	1.1
Biernoth and Merk, 1985	AMF	—[3]	0.5[n]	0.8	1.3	2.9	5.9	10.9	12.7	10.0	7.2	6.7	7.7	8.6	10.0	8.6	4.2
Büning-Pfaue et al., 1989	Parent milkfat	—	0.34[w]	0.7	1.3	2.5	5.7	10.6	11.4	8.3	6.3	6.0	6.7	8.1	10.8	8.4	3.2
		—	—	—	—	6.75[5a] <C34	4.4	12.6	18.0	14.1	7.8	8.4	10.2	11.6	5.9	0.3	—
Ensign, 1989	AMF WM-38	—	0.3[a]	0.6	1.2	2.1	5.1	10.7	14.3	11.2	6.8	5.8	6.8	7.6	10.6	10.5	4.2
	AMF WM-39	—	0.2[a]	0.6	1.1	1.8	4.8	9.4	13.5	10.4	6.5	5.5	6.1	7.2	10.0	10.6	3.8
	AMF WM-43	—	0.2[a]	0.6	1.0	1.9	4.2	9.5	13.4	10.5	6.3	5.6	5.8	7.2	10.7	10.4	4.6
Kankare et al., 1989	AMF	—	0.1[w]	0.7	1.5	4.8	8.1	14.0	14.2	11.3	6.1	5.1	4.9	6.7	9.3	7.1	5.0
Kaufmann et al., 1982	AMF	0.1	0.4[m]	1.0	1.8	3.8	8.0	13.1	13.7	10.3	7.7	7.0	7.2	7.7	8.2	7.2	2.3
Rizvi et al., 1990	AMF	—	—	0.6[w]	1.2	2.5	6.1	13.8	16.7	13.2	6.6	5.8	6.2	7.7	7.2	5.6	1.8
Timmen et al., 1984	AMF	0.2[m]	0.6	1.0	1.7	3.7	7.9	12.8	13.6	10.3	7.7	6.9	7.1	7.6	8.0	7.1	2.4
Propane																		
Biernoth and Merk, 1985	AMF	—	—	0.4[n]	1.1	2.5	5.9	11.7	12.1	9.1	6.9	7.1	8.1	9.2	10.0	7.3	2.7

[1]Sample description: AMF = anhydrous milkfat.

[2]Units of chemical composition are designated using these abbreviations: a = area %, m = mole %, n = % not specified, w = wt %.

[3]Data not available.

[4]Triglyceride determination by GC analysis (weight %).

[5]Triglyceride determination by HPLC analysis (area %).

TABLE 5.86 Triglyceride Composition (Individual Triglyceride) of Intact Anhydrous Milkfat Used to Produce Milkfat Fractions by Short-Path Distillation

Author	Fraction designation[1]	Triglyceride (%)[2]														
		C24	C26	C28	C30	C32	C34	C36	C38	C40	C42	C44	C46	C48	C50	C52	C54
Arul et al., 1988a	AMF	0.4w	0.2	0.8	1.3	2.6	6.3	12.0	13.4	10.2	7.8	6.6	7.2	8.5	9.9	8.7	4.2
McCarthy et al., 1962	AMF	0.1w	0.3	0.6	1.0	2.1	5.6	13.3	15.8	10.6	6.9	6.0	6.8	8.5	11.2	8.0	3.2

[1] AMF = anhydrous milkfat.
[2] Units of chemical composition are designated using this abbreviation: w = wt %.

TABLE 5.87 Triglyceride Composition (Individual Triglyceride) of Very-High-Melting Milkfat Fractions Obtained from Melted Milkfat

Fraction separation method Author	Fraction designation[1]	Triglyceride (%)[2]														
		C24	C26	C28	C30	C32	C34	C36	C38	C40	C42	C44	C46	C48	C50	C52	C54
Vacuum filtration																	
Kaylegian, 1991	34S$_M$ [W]	—[3]	0.3a	0.4	0.7	1.4	3.2	6.0	7.2	6.5	6.2	8.3	12.4	15.9	16.6	11.2	4.0
	34S$_S$ [W]	—	0.3a	0.4	0.8	1.7	3.7	6.8	8.1	7.2	6.6	8.4	12.1	15.4	15.4	10.0	3.1
Centrifugation																	
Kankare and Antila, 1986	24S$_M$ [S]	—	0.1n	0.2	0.6	1.5	3.1	6.5	8.7	6.8	6.1	7.1	9.4	11.8	15.6	14.3	6.9
Separation with the aid of an aqueous detergent solution																	
Banks et al., 1985	21S$_S$ [A] high level of detergent	—	—	—	0.2w	0.6	2.0	5.3	8.4	4.8	4.6	5.9	11.6	15.5	24.0	14.3	2.8
	21 semi-S$_S$ [A] high level of detergent	—	—	—	0.7w	1.7	5.5	15.6	19.1	10.0	5.5	3.9	4.1	7.1	12.3	11.4	3.1
	16S$_S$ [B] high level of detergent	—	—	—	0.0w	0.0	0.2	1.0	1.9	2.0	2.7	4.5	9.4	15.4	23.7	25.8	13.4

[1] The number indicates fractionation temperature, the letter indicates physical state (S = solid), and the subscript indicates fractionation procedure used (M = multiple-step, S = single-step). Other descriptions are that of the original investigator. The designation inside the brackets refers to the intact anhydrous milkfat source [sample letter or season (S = summer, W = winter)].
[2] Units of chemical composition are designated using these abbreviations: a = area %, n = % not specified, w = wt %.
[3] Data not available

TABLE 5.88 Triglyceride Composition (Individual Triglyceride) of Very-High-Melting Milkfat Fractions Obtained from Solvent Solution

Type of solvent solution Author	Fraction designation[1]	Triglyceride (%)[2]															
		C24	C26	C28	C30	C32	C34	C36	C38	C40	C42	C44	C46	C48	C50	C52	C54
Acetone																		
Kaylegian, 1991	25S$_M$ [W]	—[3]	0.1[a]	0.2	0.3	0.6	1.6	3.0	3.5	3.2	4.1	7.2	13.4	21.2	23.4	14.5	3.7
	25S$_S$ [W]	—	0.1[a]	0.2	0.3	0.7	1.7	3.1	3.7	3.4	4.5	6.6	12.6	20.9	22.4	14.6	5.1
Yi, 1993	25S$_M$ [S]	—	—	0.5[a]	0.8	1.4	2.9	4.2	5.2	4.1	3.5	5.5	11.0	19.0	23.4	15.5	3.1
Kaylegian, 1991	21S$_M$ [W]	—	0.2[a]	0.2	0.5	0.9	2.1	3.6	4.2	3.9	5.0	9.1	16.4	21.7	19.0	10.0	3.3
	21S$_S$ [W]	—	0.1[a]	0.2	0.3	0.7	1.6	3.0	3.6	3.6	5.1	8.6	16.5	21.7	20.5	11.5	2.9
Timms, 1980b	20S$_M$ [S]	—	—	—	—	—	0.1[w]	0.2	0.3	0.6	1.9	5.2	12.4	20.7	28.3	22.4	7.3
Yi, 1993	20S$_M$ [S]	—	—	0.0[a]	0.0	0.6	1.6	2.9	3.6	3.1	3.5	7.4	16.5	23.2	22.8	12.3	2.5
Kaylegian, 1991	15S$_M$ [W]	—	0.1[a]	0.2	0.4	0.8	1.9	3.5	4.4	4.7	6.8	12.1	18.3	19.5	16.0	9.0	2.4
	15S$_S$ [W]	—	0.2[a]	0.2	0.4	0.9	2.1	3.9	4.7	4.8	6.1	10.3	16.1	19.4	17.8	10.4	2.8
Yi, 1993	15S$_M$ [S]	—	—	0.0[a]	0.0	0.5	1.7	3.6	4.0	4.2	6.5	13.1	20.7	17.5	16.2	9.7	2.3
Kaylegian, 1991	11S$_S$ [W]	—	0.2[a]	0.2	0.5	0.9	2.2	4.2	5.1	5.1	7.1	10.8	15.6	18.2	16.8	10.2	2.9
Pentane																		
Kaylegian, 1991	11S$_M$ [W]	—	0.1[a]	0.1	0.2	0.6	1.5	2.7	3.3	3.6	4.7	8.8	16.1	22.2	21.2	11.8	3.1
	11S$_S$ [W]	—	0.2[a]	0.3	0.5	1.1	2.5	4.4	5.2	5.0	5.4	8.8	14.8	19.7	18.6	10.7	2.8
	5S$_M$ [W]	—	0.1[a]	0.2	0.3	0.8	1.9	3.4	4.2	4.5	6.0	11.0	17.1	20.3	17.4	10.0	2.7
	5S$_S$ [W]	—	0.1[a]	0.2	0.3	0.9	2.1	3.8	4.5	5.0	6.2	10.7	16.6	19.9	17.5	9.8	2.4
	0S$_M$ [W]	—	0.1[a]	0.2	0.3	1.0	2.4	3.7	4.4	5.4	8.2	13.2	17.1	17.6	14.4	9.4	2.6
	0S$_S$ [W]	—	0.1[a]	0.2	0.3	0.9	2.1	3.5	4.4	4.9	6.9	11.6	16.7	19.3	16.7	9.9	2.4

[1]The number indicates fractionation temperature, the letter indicates physical state (S = solid), and the subscript indicates fractionation procedure used (M = multiple-step, S = single-step). The designation inside the brackets refers to the intact anhydrous milkfat source (S = summer, W = winter).
[2]Units of chemical composition are designated using these abbreviations: a = area %, w = wt %.
[3]Data not available.

TABLE 5.89 Triglyceride Composition (Individual Triglyceride) of High-Melting Milkfat Fractions Obtained from Melted Milkfat

Fraction separation method Author	Fraction designation[1]	C24	C26	C28	C30	C32	C34	C36	C38	C40	C42	C44	C46	C48	C50	C52	C54
Vacuum filtration																	
deMan and Finoro, 1980	32S$_S$	0.2[w]	0.2	0.4	0.8	1.6	3.7	7.3	9.5	7.9	5.6	6.1	8.6	10.8	16.5	15.5	5.3
	31S$_S$	0.3[w]	0.2	0.5	0.7	1.4	3.4	7.0	8.7	7.2	5.2	5.8	8.1	11.2	16.9	16.6	7.3
	30S$_S$	0.2[w]	0.1	0.3	0.7	1.3	3.1	6.3	8.0	6.8	5.2	6.0	9.0	12.1	17.7	16.5	6.9
Kaylegian, 1991	30S$_M$ [W]	—[3]	0.3[a]	0.4	0.8	1.7	3.9	6.9	8.6	7.6	7.1	9.1	12.4	14.8	14.2	9.6	2.6
	30S$_S$ [W]	—	0.3[a]	0.5	0.9	1.8	4.0	7.3	8.8	7.8	7.2	8.9	11.9	14.5	14.2	9.1	2.9
Amer et al., 1985	29S$_S$ [W]	tr[4a]	0.1	0.4	0.8	1.7	4.8	9.1	10.1	7.8	6.5	7.3	9.1	11.0	11.7	7.0	2.1
deMan and Finoro, 1980	29S$_S$	0.2[w]	0.1	0.4	0.6	1.3	3.1	6.1	8.6	7.3	5.5	6.3	8.9	12.5	17.9	15.8	5.5
	28S$_S$	0.2[w]	0.2	0.3	0.6	1.3	2.9	5.8	8.1	7.1	5.5	6.5	9.3	12.7	18.1	16.0	5.5
	27S$_S$	0.2[w]	0.2	0.4	0.7	1.4	3.6	7.6	9.8	8.2	5.8	6.3	8.3	11.3	16.4	14.8	5.1
	26S$_S$	0.2[w]	0.2	0.4	0.7	1.4	3.5	7.4	9.6	8.2	5.9	6.5	8.2	11.3	16.0	14.8	5.7
Lund and Danmark, 1987	26S$_S$ [1]	—	—	0.3[w]	0.7	1.6	3.7	6.6	9.5	8.0	6.0	5.8	8.4	10.0	13.0	11.1	4.5
deMan and Finoro, 1980	25S$_S$	0.3[w]	0.2	0.4	0.8	1.7	3.9	8.0	10.6	8.7	5.9	6.0	7.5	10.0	14.7	14.9	6.4
Kaylegian, 1991	25S$_M$ [W]	—	0.3[a]	0.3	0.7	1.4	3.4	6.3	7.9	8.0	8.7	10.5	12.8	14.4	13.3	9.3	2.7
	25S$_S$ [W]	—	0.4[a]	0.5	0.9	2.1	4.9	8.9	10.9	9.3	7.6	8.3	10.2	12.2	12.3	8.5	2.9
	20S$_M$ [W]	—	0.3[a]	0.3	0.7	1.4	3.4	6.3	7.9	8.0	8.7	10.5	12.8	14.4	13.3	9.3	2.7
	20S$_S$ [W]	—	0.5[a]	0.6	1.1	2.2	4.9	8.7	10.4	8.9	7.4	8.4	10.5	12.6	12.6	8.7	2.8
Amer et al., 1985	19S$_S$ [W]	0.1[a]	0.1	0.6	1.0	2.2	5.8	10.9	12.1	9.3	7.1	7.0	7.6	8.9	9.4	6.4	2.1
Lund and Danmark, 1987	18S$_S$ [2]	—	—	0.4[w]	0.9	1.7	3.6	7.5	10.5	8.8	6.0	6.2	8.2	9.1	12.0	11.4	5.0
Centrifugation																	
Jordan, 1986	28S$_S$ [W]	—	0.6[w]	0.7	1.1	2.0	4.3	9.2	10.8	8.7	6.7	6.9	8.3	10.1	12.7	10.5	5.6
Keogh and Higgins, 1986b	Fraction 2	—	0.1[w]	0.3	0.8	1.4	3.0	6.2	8.3	7.5	6.0	6.4	8.4	12.0	15.0	12.1	5.0
Separation with the aid of an aqueous detergent solution																	
Banks et al., 1985	16 semi-S$_S$ [B] high level of detergent	—	—	—	0.1[w]	0.3	1.3	3.8	6.7	7.4	3.8	4.0	6.5	10.7	18.7	21.3	15.4

Continued

TABLE 5.89 (Continued)

Fraction separation method Author	Fraction designation[1]	Triglyceride (%)[2]														
		C24	C26	C28	C30	C32	C34	C36	C38	C40	C42	C44	C46	C48	C50	C52	C54
Filtration through milk filter																	
Fouad et al., 1990	$29S_s$	—	0.1^w	0.4	0.4	1.1	3.2	6.2	7.5	6.6	5.9	7.1	9.6	11.8	13.9	10.6	4.0
	$25S_s$	—	0.1^w	0.4	0.7	1.0	3.8	6.8	8.3	7.3	6.6	7.6	9.6	11.4	13.3	9.3	3.5

[1]The number indicates fractionation temperature, the letter indicates physical state (S = solid), and the subscript indicates fractionation procedure used (M = multiple-step, S = single-step). The designation inside the brackets refers to the intact anhydrous milkfat source [sample number, letter or season (W = winter)].

[2]Units of chemical composition are designated using these abbreviations: a = area %, w = wt %.

[3]Data not available.

[4]tr = trace.

TABLE 5.90 Triglyceride Composition (Individual Triglyceride) of High-Melting Milkfat Fractions Obtained from Solvent Solution

Type of solvent solution Author	Fraction designation[1]	Triglyceride (%)[2]														
		C24	C26	C28	C30	C32	C34	C36	C38	C40	C42	C44	C46	C48	C50	C52	C54
Acetone																	
Kaylegian, 1991	11S$_M$ [W]	—[3]	0.2^a	0.3	0.6	1.1	2.4	4.5	5.7	6.1	8.9	13.1	15.5	15.5	14.3	9.3	2.6
Yi, 1993	10S$_M$ [S]	—	—	0.4^a	0.8	1.6	4.2	6.6	7.3	6.7	8.3	12.6	13.9	12.2	13.2	9.6	2.7
Kaylegian, 1991	5S$_S$ [W]	—	0.2^a	0.3	0.6	1.1	2.8	6.4	8.1	7.0	7.7	9.9	12.7	15.1	15.4	9.6	3.1
Lambelet, 1983	S1$_M$	—	—	0.1^n	0.4	1.8	8.0	—	—	—	—	—	—	—	—	—	—
Timms, 1980b	3S$_M$ [S]	—	—	—	—	0.1^w	1.1	6.2	10.3	7.6	7.0	8.5	9.9	12.2	17.5	14.9	4.6
Kaylegian, 1991	0S$_S$ [W]	—	0.3^a	0.4	0.7	1.4	3.9	9.1	10.3	8.2	7.7	8.8	10.9	13.5	13.7	8.7	2.5

[1]The number indicates fractionation temperature, the letter indicates physical state (S = solid), and the subscript indicates fractionation procedure used (M = multiple-step, S = single-step). The designation inside the brackets refers to the intact anhydrous milkfat source (S = summer, W = winter).

[2]Units of chemical composition are designated using these abbreviations: a = area %, n = % not specified, w = wt %.

[3]Data not available.

TABLE 5.91 Triglyceride Composition (Individual Triglyceride) of High-Melting Milkfat Fractions Obtained by Supercritical Fluid Extraction

Type of solvent Author	Fraction designation[1]	Fractionation conditions		Triglyceride (%)[2]															
		Temperature (°C)	Pressure	C24	C26	C28	C30	C32	C34	C36	C38	C40	C42	C44	C46	C48	C50	C52	C54
Carbon dioxide																			
Biernoth and Merk, 1985	Residual	80	200 bar	—[3]	—	—	0.2[n]	0.9	3.1	8.1	11.3	10.5	8.3	8.1	9.0	10.3	11.7	10.0	5.6
Kaufmann et al., 1982	I (residue)	80	200 bar	—	—	0.1[m]	0.4	1.5	4.6	9.9	13.0	11.1	9.0	8.6	9.0	9.6	10.5	9.4	3.1
Timmen et al., 1984	Long-chain fraction	80	200 bar	0.0	0.0	0.1[m]	0.5	1.5	4.5	9.7	12.9	11.0	9.0	8.5	8.9	9.4	9.8	9.3	3.1
Arul et al., 1987	S2	70	250–350 bar	0.1[w]	tr[4]	tr	0.1	0.4	2.0	6.5	12.4	12.7	10.6	10.4	11.2	11.1	12.0	8.7	1.8
	S3	70	250–350 bar	tr[w]	tr	tr	tr	0.2	0.5	1.4	3.5	4.9	5.2	7.7	10.9	15.9	22.1	18.0	9.7
Büning-Pfaue et al., 1989	Extractor fraction	50	25,000 kPa	—	0.3[5w]	1.2	2.0	3.9	8.3	14.2	14.2	9.7	6.7	5.8	5.7	6.0	6.9	5.1	—
				—	—	—	—	6.4[6a] <C34	3.5	19.0	24.0	14.5	8.0	5.7	7.5	7.6	2.9	0.8	—
Kankare et al., 1989	R	50	400 bar	—	0.0[w]	0.0	1.4	2.6	6.2	9.8	15.1	14.4	6.5	4.6	5.6	8.0	11.1	10.1	4.1
Rizvi et al., 1990	Extract-2	48.0	275.8 bar	—	—	0.5[w]	0.9	1.9	6.1	12.1	18.0	14.0	7.4	6.7	6.7	8.4	6.7	4.4	0.9
	Residue	48.0	275.8 bar	—	—	0.6[w]	0.4	0.9	2.6	6.7	8.8	8.4	5.7	5.8	8.5	13.7	15.7	12.9	5.1
Büning-Pfaue et al., 1989	Column fraction	30	20,000 kPa	—	—	0.3[5w]	0.5	1.4	3.7	7.3	8.4	7.0	6.0	6.4	7.8	10.2	14.3	11.2	4.6
				—	—	—	—	2.9[6a] <C34	2.8	10.0	16.8	12.5	8.6	8.7	13.6	16.7	6.5	1.0	—

[1]Designations used are those given by the original author.
[2]Units of chemical composition are designated using these abbreviations: a = area %, m = mole %, n = % not specified, w = wt %.
[3]Data not available.
[4]tr = trace.
[5]Triglyceride determination by GC analysis (weight %).
[6]Triglyceride determination by HPLC analysis (area %).

TABLE 5.92 Triglyceride Composition (Individual Triglyceride) of High-Melting Milkfat Fraction Obtained by Short-Path Distillation

Author	Fraction designation[1]	Fractionation conditions		Triglyceride (%)[2]													
		Temperature (°C)	Pressure	C24	C26	C28	C30	C32	C34	C36	C38	C40	C42	C44	C46	C48	C50	C52	C54
Arul et al., 1988a	SF Residue	—[3]	—	0.1w	0.0	0.1	0.1	0.3	0.8	3.3	7.1	8.9	8.0	8.9	10.5	13.1	16.7	14.8	7.4

[1]Designations used are those given by the original author.
[2]Units of chemical composition are designated using this abbreviation: w = wt %.
[3]Data not available.

TABLE 5.93 Triglyceride Composition (Individual Triglyceride) of Middle-Melting Milkfat Fractions Obtained from Melted Milkfat

Fraction separation method Author	Fraction designation[1]	Triglyceride (%)[2]															
		C24	C26	C28	C30	C32	C34	C36	C38	C40	C42	C44	C46	C48	C50	C52	C54
Vacuum filtration																		
Kaylegian, 1991	34L$_S$ [W]	—[3]	0.5a	0.6	1.2	2.4	5.1	9.4	10.9	9.2	7.1	7.6	9.7	11.8	12.2	9.1	3.2
deMan and Finor, 1980	32L$_S$	0.3w	0.2	0.6	1.0	1.9	4.8	9.6	12.6	10.1	6.2	5.6	6.4	8.3	12.1	13.7	6.5
	31L$_S$	0.4w	0.3	0.6	1.1	2.1	5.1	10.0	13.1	10.5	6.4	5.6	6.1	7.5	11.9	13.6	5.9
	30L$_S$	0.4w	0.2	0.5	1.1	2.1	5.1	10.3	13.3	10.6	6.4	5.6	6.1	7.9	11.7	13.1	5.7
Kaylegian, 1991	30L$_S$ [W]	—	0.7a	0.8	1.4	2.8	5.9	10.2	11.6	9.3	7.0	7.2	8.8	10.6	11.7	8.7	3.2
Amer et al., 1985	29L$_S$ [W]	0.1a	0.2	0.7	1.3	2.6	6.7	12.5	13.7	10.3	7.1	5.9	5.9	7.2	8.1	5.8	2.0
deMan and Finor, 1980	29L$_S$	0.4w	0.2	0.6	1.0	2.1	5.0	10.3	13.3	10.7	6.4	5.7	6.0	7.7	11.7	12.9	6.0
	28L$_S$	0.3w	0.2	0.7	1.0	2.1	5.0	10.4	13.5	10.8	6.5	5.4	5.8	7.5	11.5	13.1	6.2
	27L$_S$	0.4w	0.2	0.5	1.1	2.1	5.1	10.4	13.6	10.9	6.4	5.4	5.7	7.5	11.4	13.0	6.5
Kaylegian, 1991	16S$_M$ [W]	—	0.3a	0.4	0.7	1.7	5.2	12.6	14.3	10.2	7.5	6.9	8.1	10.2	11.5	7.9	2.3
Centrifugation																		
Kankare and Antila, 1986	12S$_M$ [S]	—	0.1n	0.4	0.8	1.9	4.9	10.4	14.5	9.2	6.7	5.0	7.0	8.7	10.8	11.6	6.3
Filtration through milk filter																		
Fouad et al., 1990	29L$_S$	—	0.3w	0.8	1.4	2.3	6.4	11.7	13.6	11.2	7.5	6.3	6.0	7.2	8.7	7.9	3.6
	17S$_S$	—	0.2w	0.6	1.1	1.8	5.3	9.2	10.1	9.1	6.9	7.2	7.7	9.3	10.0	8.3	3.5

[1]The number indicates fractionation temperature, the letter indicates physical state (L = liquid, S = solid), and the subscript indicates fractionation procedure used (M = multiple-step, S = single-step). The designation inside the brackets refers to the intact anhydrous milkfat source (S = summer, W = winter).
[2]Units of chemical composition are designated using these abbreviations: a = area %, n = % not specified, w = wt %.
[3]Data not available.

TABLE 5.94 Triglyceride Composition (Individual Triglyceride) of Middle-Melting Milkfat Fractions Obtained from Solvent Solution

Type of solvent solution Author	Fraction designation[1]	Triglyceride (%)[2]															
		C24	C26	C28	C30	C32	C34	C36	C38	C40	C42	C44	C46	C48	C50	C52	C54
Acetone																	
Kaylegian, 1991	25L$_S$ [W]	—[3]	0.6[a]	0.7	1.3	2.6	5.7	10.2	11.8	9.6	7.4	7.5	9.2	10.8	11.1	8.3	3.1
	21L$_S$ [W]	—	0.6[a]	0.8	1.4	2.7	5.9	10.4	12.1	9.8	7.5	7.5	8.8	10.5	11.0	8.4	2.6
	5S$_M$ [W]	—	0.1[a]	0.2	0.3	0.7	2.3	7.9	10.7	8.8	8.6	9.2	9.7	12.3	15.1	10.5	3.5
Yi, 1993	5S$_M$ [S]	—	—	0.1[a]	0.4	1.0	3.6	11.7	14.5	11.1	9.8	9.0	6.9	8.0	13.3	9.0	1.7
Kaylegian, 1991	0S$_M$ [W]	—	0.2[a]	0.2	0.4	1.2	5.4	14.4	13.5	9.4	7.7	6.6	8.0	11.6	12.0	7.0	2.3
Yi, 1993	0S$_M$ [S]	—	—	0.1[a]	0.3	0.9	5.2	16.9	16.5	10.9	8.0	6.6	5.1	9.0	12.0	6.9	1.5
Pentane																	
Kaylegian, 1991	11L$_S$ [W]	—	0.5[a]	0.6	1.1	2.4	5.4	9.6	11.5	9.6	7.4	7.8	9.4	11.4	11.6	8.6	2.9
	5L$_S$ [W]	—	0.6[a]	0.7	1.3	2.6	5.8	10.3	12.2	10.0	7.5	7.4	8.8	10.8	11.0	8.5	2.6

[1]The number indicates fractionation temperature, the letter indicates physical state (L = liquid, S = solid), and the subscript indicates fractionation procedure used (M = multiple-step, S = single-step). The designation inside the brackets refers to the intact anhydrous milkfat source (S = summer, W = winter).

[2]Units of chemical composition are designated using this abbreviation: a = area.

[3]Data not available.

TABLE 5.95 Triglyceride Composition (Individual Triglyceride) of Middle-Melting Milkfat Fractions Obtained by Supercritical Fluid Extraction

Type of solvent Author	Fraction designation[1]	Fractionation conditions		Triglyceride (%)[2]														
		Temperature (°C)	Pressure	C24	C26	C28	C30	C32	C34	C36	C38	C40	C42	C44	C46	C48	C50	C52	C54
Carbon dioxide																			
Arul et al., 1987	S1	70	250–350 bar	0.3[w]	0.3	tr[3]	0.2	1.1	4.4	10.9	18.0	16.6	11.0	9.1	8.1	7.9	7.3	3.3	1.5
	I2	50	100–250	0.8[w]	0.2	1.2	2.3	4.8	12.0	20.3	20.0	14.2	7.7	5.1	4.5	3.1	2.2	1.1	0.5
	I3	50	100–250	0.6[w]	0.3	0.5	1.1	3.0	8.0	15.1	20.3	16.9	9.2	6.9	5.7	4.8	4.0	2.6	1.0
Kankare et al., 1939	E2	50	200 bar	—[4]	2.2[w]	5.0	5.8	8.8	12.8	19.1	18.8	11.9	5.9	3.1	2.1	2.2	1.6	1.0	0.0
	E4	50	400	—	0.0[w]	0.8	2.0	6.2	10.3	17.5	18.0	14.3	7.5	6.0	4.0	4.3	4.0	2.3	2.6
Rizvi et al., 1990	Extract-1	40.0	241.3 bar	—	—	2.6[w]	2.5	5.4	10.9	20.4	21.4	14.0	5.2	3.5	3.6	3.4	1.9	0.8	0.0

[1]Designations used are those given by the original author.

[2]Units of chemical composition are designated using this abbreviation: w = wt %.

[3]tr = trace.

[4]Data not available

TABLE 5.96 Triglyceride Composition (Individual Triglyceride) of Middle-Melting Milkfat Fraction Obtained by Short-Path Distillation

Author	Fraction designation[1]	Fractionation conditions		Triglyceride (%)[2]														
		Temperature (°C)	Pressure	C24	C26	C28	C30	C32	C34	C36	C38	C40	C42	C44	C46	C48	C50	C52	C54
Arul et al., 1988a	IF Distillate III	265	100 µm Hg	0.3[w]	0.1	0.5	1.5	3.6	10.5	22.8	24.1	14.4	8.2	4.8	3.4	2.7	2.2	1.0	0.0

[1]Designations used are those given by the original author.

[2]Units of chemical composition are designated using this abbreviation: w = wt %.

TABLE 5.97 Triglyceride Composition (Individual Triglyceride) of Low-Melting Milkfat Fractions Obtained from Melted Milkfat

Fraction separation method Author	Fraction designation[1]	Triglyceride (%)[2]															
		C24	C26	C28	C30	C32	C34	C36	C38	C40	C42	C44	C46	C48	C50	C52	C54
Vacuum filtration																	
deMan and Finoro, 1980	26L$_S$	0.3^w	0.2	0.5	1.0	2.0	5.2	10.7	13.8	11.1	6.6	5.6	5.8	7.5	11.6	12.4	5.7
Lund and Danmark, 1987	26L$_S$ [1]	—[3]	—	0.4^w	1.2	2.5	5.5	11.0	13.7	10.8	6.4	5.6	5.7	6.7	8.9	9.7	5.3
deMan and Finoro, 1980	25L$_S$	0.4^w	0.3	0.6	1.1	2.2	5.2	10.9	14.2	11.2	6.5	5.4	5.6	6.7	11.1	12.9	5.9
Kaylegian, 1991	25L$_S$ [W]	—	0.6^a	0.7	1.4	2.7	5.9	10.6	12.5	10.1	7.3	7.1	8.5	10.6	11.0	8.4	2.6
	20L$_S$ [W]	—	0.6^a	0.7	1.4	2.8	6.4	11.4	13.3	10.5	7.1	6.7	7.9	9.9	10.7	7.8	2.7
Amer et al., 1985	19L$_S$ [W]	0.1	0.2^a	0.8	1.5	2.8	7.5	14.0	15.4	11.3	6.8	5.4	5.0	6.2	6.9	5.7	2.2
Lund and Danmark, 1987	18L$_S$ [2]	—	—	0.5^w	1.1	2.5	5.0	10.2	14.0	11.5	6.3	5.0	5.0	5.9	8.3	9.7	5.6
Kaylegian, 1991	13S$_M$ [W]	—	0.5^a	0.6	1.0	2.3	5.9	11.2	12.1	9.5	7.2	6.9	9.0	10.9	11.9	8.0	3.0
	13L$_M$ [W]	—	0.7^a	0.8	1.5	2.9	5.8	9.0	11.2	9.7	6.7	6.7	8.7	10.5	11.6	9.6	4.7
Centrifugation																	
Jordan, 1986	28L$_S$ [W]	—	0.7^w	1.4	1.6	2.6	5.1	10.7	12.4	9.7	6.6	6.0	6.0	7.4	9.3	9.2	5.7
Kankare and Antila, 1986	12L$_M$ [S]	—	0.4^n	0.7	1.6	2.1	7.5	10.6	16.9	8.4	7.4	5.2	3.4	6.3	9.0	10.8	7.0
Keogh and Higgins, 1986b	Fraction 3	—	0.3^w	0.4	1.1	2.1	3.0	9.7	13.4	11.0	6.9	5.5	5.6	6.9	9.6	9.5	6.2
Separation with the aid of an aqueous detergent solution																	
Banks et al., 1985	21L$_S$ A high level of detergent	—	—	—	0.5^w	1.5	5.0	14.9	18.7	10.5	5.8	4.0	4.4	7.7	11.9	11.9	3.2
Filtration through milk filter																	
Fouad et al., 1990	17L$_S$	—	0.3^w	0.9	1.4	2.8	7.2	12.9	14.1	11.7	6.4	—	4.9	5.6	7.6	7.2	3.5

[1]The number indicates fractionation temperature, the letter indicates physical state (L = liquid, S = solid), and the subscript indicates fractionation procedure used (M = multiple-step, S = single-step). Other descriptions and descriptions in parentheses are those of the original author. The designation inside the brackets refers to the intact anhydrous milkfat source [sample number or season (S = summer, W = winter)].

[2]Units of chemical composition are designated using these abbreviations a = area %, n = % not specified, w = wt %.

[3]Data not available.

TABLE 5.98 Triglyceride Composition (Individual Triglyceride) of Low-Melting Milkfat Fractions Obtained from Solvent Solution

Type of solvent solution Author	Fraction designation[1]	Triglyceride (%)[2]														
		C24	C26	C28	C30	C32	C34	C36	C38	C40	C42	C44	C46	C48	C50	C52	C54
Acetone																	
Kaylegian, 1991	15L$_S$ [W]	—[3]	0.6[a]	0.7	1.3	2.6	5.9	10.6	12.6	10.3	7.7	7.4	8.6	10.3	10.8	8.1	2.5
	11L$_S$ [W]	—	0.7[a]	0.8	1.6	3.0	6.4	11.2	12.9	10.3	7.3	6.8	8.0	9.9	10.5	8.1	2.7
	5L$_S$ [W]	—	0.8[a]	0.9	1.8	3.4	7.2	11.8	13.2	10.4	7.0	6.4	7.7	9.5	9.8	7.7	2.6
Timms, 1980b	3L$_M$ [S]	0.4[w]	0.4	0.6	1.4	2.9	7.0	12.9	15.0	11.7	6.2	4.8	5.3	6.7	9.2	9.4	6.1
Yi, 1993	0L$_M$ [S]	—	—	0.0[a]	1.5	4.1	9.1	13.6	16.4	13.1	7.0	5.8	5.9	6.7	7.1	7.5	2.4
Lambelet, 1983	L1$_M$	—	0.4[n]	0.7	1.3	2.6	5.7	9.3	—	—	—	—	—	—	—	—	—
Pentane																	
Kaylegian, 1991	0L$_M$ [W]	—	0.6[a]	0.7	1.3	2.6	5.8	10.3	12.1	10.0	7.3	7.0	8.8	10.8	11.6	8.2	3.0
	0L$_S$ [W]	—	0.6[a]	0.7	1.2	2.6	5.9	10.5	12.5	10.2	7.4	7.2	8.8	10.5	11.3	8.0	2.7

[1]The number indicates fractionation temperature, the letter indicates physical state (L = liquid), and the subscript indicates fractionation procedure used (M = multiple-step, S = single-step). The designation inside the brackets refers to the intact anhydrous milkfat source (S = summer, W = winter).
[2]Units of chemical composition are designated using these abbreviations: a = area %, n = % not specified, w = wt %.
[3]Data not available.

TABLE 5.99 Triglyceride Composition (Individual Triglyceride) of Low-Melting Milkfat Fractions Obtained by Supercritical Fluid Extraction

Type of solvent Author	Fraction designation[1]	Fractionation conditions		Triglyceride (%)[2]														
		Temperature (°C)	Pressure	C24	C26	C28	C30	C32	C34	C36	C38	C40	C42	C44	C46	C48	C50	C52	C54
Carbon dioxide																			
Biemoth and Merk, 1985	Extract	0	200 bar	—[3]	1.3[n]	2.9	5.7	11.0	19.9	23.7	17.1	8.8	3.8	1.9	1.0	0.8	0.5	0.4	0.3
Kaufmann et al., 1932	II (extract)	80	200 bar	0.7[m]	2.0	4.4	7.2	12.3	20.3	23.7	16.3	7.3	2.9	1.2	0.6	0.3	0.1	—	—
Timmen et al., 1984	Short-chain fraction	80	200 bar	0.8[m]	2.1	4.3	7.1	12.2	20.2	23.5	16.2	7.7	2.8	1.2	0.6	0.2	0.0	0.0	0.0
Arul et al., 1987	L1	50	100–250 bar	2.9[w]	3.7	6.8	8.9	11.3	16.3	19.0	13.7	6.8	2.8	1.4	0.9	0.4	0.8	1.7	2.6
	L2	50	100–250	1.6[w]	2.0	3.8	6.1	10.1	11.1	22.7	19.2	10.5	4.6	2.8	1.7	1.3	0.8	0.8	0.9
	I1	50	100–250	1.1[w]	0.7	2.0	4.2	7.9	14.9	22.4	20.3	11.7	5.5	3.2	2.2	1.6	1.1	0.8	0.4
Kankare et al., 1989	E1	50	100 bar	—	3.6[w]	4.0	6.4	11.5	15.4	20.0	15.3	8.8	3.8	2.5	1.2	2.7	1.9	0.0	0.0

[1]Designations used are those given by the original author.
[2]Units of chemical composition are designated using these abbreviations: m = mole %, n = % not specified, w = wt %.
[3]Data not available.

TABLE 5.100 Triglyceride Composition (Individual Triglyceride) of Low-Melting Milkfat Fractions Obtained by Short-Path Distillation

Author	Fraction designation[1]	Fractionation conditions		Triglyceride (%)[2]														
		Temperature (°C)	Pressure	C24	C26	C28	C30	C32	C34	C36	C38	C40	C42	C44	C46	C48	C50	C52	C54
Arul et al., 1988a	LF1 Distillate 1	245	220µm Hg	8.2w	6.1	10.8	12.1	13.5	15.9	14.9	10.1	3.7	1.6	0.9	0.7	0.6	0.7	0.0	0.0
	LF2 Distillate 2	245	100µm Hg	1.8w	1.3	3.9	6.7	10.3	15.2	22.4	15.7	7.2	5.9	1.9	1.2	1.7	2.2	2.9	0.8

[1]Designations used are those given by the original author.

[2]Units of chemical composition are designated using this abbreviation: w = wt %.

TABLE 5.101 Triglyceride Composition (Individual Triglyceride) of Very-Low-Melting Milkfat Fractions Obtained from Solvent Solution

Type of solvent solution Author	Fraction designation[1]	Triglyceride (%)[2]														
		C24	C26	C28	C30	C32	C34	C36	C38	C40	C42	C44	C46	C48	C50	C52	C54
Acetone																	
Kaylegian, 1991	0L$_M$ [W]	—[3]	0.8a	0.9	1.8	3.6	7.2	10.6	13.1	11.1	7.3	6.8	7.9	9.3	8.9	7.7	2.9
	0L$_S$ [W]	—	0.9a	1.1	2.0	3.8	7.5	10.9	13.1	10.9	6.9	6.6	7.7	9.1	9.3	7.6	2.7

[1]The number indicates fractionation temperature, the letter indicates physical state (L = liquid), and the subscript indicates fractionation procedure used (M = multiple-step, S = single-step). The designation inside the brackets refers to the intact anhydrous milkfat source (W = winter).

[2]Units of chemical composition are designated using this abbreviation: a = area %.

[3]Data not available.

Technical Data for Experimental and Commercial Milkfat Fractions 385

TABLE 5.102 Triglyceride Composition (Individual Triglyceride) of Unknown-Melting Milkfat Fractions Obtained by Short-Path Distillation

Author	Fraction designation[1]	Fractionation conditions		Triglyceride (%)[2]															
		Temperature (°C)	Pressure	C24	C26	C28	C30	C32	C34	C36	C38	C40	C42	C44	C46	C48	C50	C52	C54
McCarthy et al., 1962	R-1	150	15μ	2.8w	8.8	15.6	18.0	19.4	17.7	11.8	4.5	1.4	—[3]	—	—	—	—	—	—
	R-2	160	15	1.0w	1.6	5.4	11.5	20.2	26.2	21.6	9.8	2.6	—	—	—	—	—	—	—
	R-3	180	12	—	—	—	3.0	10.4	25.0	32.4	20.1	6.1	2.4	—	—	—	—	—	—
	R-4	185	12	—	—	0.6w	—	—	8.5w	30.2	33.4	17.1	6.2	3.1	—	—	—	—	—
	D-2	177-182	5-7	—	—	—	0.4w	2.2	9.3	24.8	28.8	16.4	7.1	3.5	1.5	1.8	1.8	1.7	—
	D-3	—	—	—	—	—	—	—	—	2.2w	5.1	6.8	7.6	9.0	11.7	15.6	21.0	14.6	6.4

[1] Designations used are those given by original author.
[2] w = wt %.
[3] Data not reported.

TABLE 5.103 Triglyceride Composition (Individual Triglyceride) of Unknown-Melting Milkfat Fractions Obtained from Melted Milkfat

Fraction separation method Author	Fraction designation[1]	Triglyceride (%)[2]															
		C24	C26	C28	C30	C32	C34	C36	C38	C40	C42	C44	C46	C48	C50	C52	C54
Pressure filtration																		
Badings et al., 1983b	23S$_M$ [W]	0.3w	0.2	0.5	1.0	2.3	5.2	9.3	10.6	8.8	7.7	8.4	9.6	10.8	12.0	9.2	4.0
	23L$_M$ [W]	0.4w	0.3	0.7	1.4	3.3	7.4	13.1	14.5	11.1	7.5	6.2	6.0	7.1	8.6	8.2	4.1
	21.5S$_M$ [S]	0.3w	0.2	0.5	0.9	2.0	4.3	8.0	10.3	9.1	6.2	6.3	7.9	10.3	13.6	12.8	7.3
	21.5L$_M$ [S]	0.4w	0.2	0.6	1.1	2.5	5.7	10.8	13.9	11.9	6.7	5.3	5.5	6.8	9.9	10.9	7.8
	18.5S$_M$ [W]	0.4w	0.2	0.6	1.2	2.8	6.8	13.9	15.0	11.0	7.8	6.4	6.0	6.9	9.1	8.1	3.7
	18.5L$_M$ [W]	0.5w	0.3	0.8	1.6	3.7	7.8	12.5	14.1	11.1	7.3	6.1	5.6	7.0	8.4	8.3	4.5
	15.3S$_M$ [S]	0.3w	0.2	0.5	0.8	1.6	4.6	11.2	15.0	11.6	7.0	5.6	5.4	7.0	11.1	11.5	6.4
	15.3L$_M$ [S]	0.5w	0.3	0.7	1.3	2.7	5.9	11.0	13.8	11.9	6.6	5.1	5.9	6.7	9.6	10.7	7.8
	15S$_M$ [W]	0.3w	0.2	0.6	1.2	2.9	7.4	14.2	14.8	10.8	7.6	6.1	5.4	7.1	9.1	8.0	3.7
	15L$_M$ [W]	0.5w	0.4	0.9	1.7	4.0	7.9	11.8	14.0	11.2	7.2	6.0	5.9	7.0	7.9	8.5	4.8
	13S$_M$ [S]	0.4w	0.2	0.6	1.2	2.4	5.8	11.3	14.0	11.6	6.7	5.2	6.0	7.0	10.1	10.7	7.4
	13L$_M$ [S]	0.5w	0.4	0.7	1.4	2.9	6.2	10.5	13.6	12.2	6.6	5.0	5.4	6.5	9.1	10.6	9.4
	11.3S$_M$ [W]	0.4w	0.3	0.7	1.4	3.3	7.7	12.6	13.9	11.0	7.4	6.1	6.1	7.5	8.7	8.4	4.3

Continued

TABLE 5.103 (Continued)

Fraction separation method Author	Fraction designation[1]	Triglyceride (%)[2]															
		C24	C26	C28	C30	C32	C34	C36	C38	C40	C42	C44	C46	C48	C50	C52	C54
	11.3L_M [W]	0.6^w	0.4	1.0	2.1	4.2	8.0	11.5	14.0	11.5	7.0	5.9	5.9	6.7	7.5	8.6	4.9
	9.3S_M [S]	0.4^w	0.3	0.7	1.3	2.8	6.2	11.1	13.6	11.8	6.6	5.0	5.4	6.8	9.6	10.6	7.8
	9.3L_M [S]	0.5^w	0.3	0.8	1.5	3.1	6.1	10.1	13.6	12.4	6.6	5.1	5.4	6.4	8.9	10.7	8.5
	7.1S_M [S]	0.5^w	0.3	0.7	1.4	2.9	6.2	10.7	13.4	11.9	6.5	5.0	5.5	6.8	9.4	10.6	8.2
	7.1L_M [S]	0.6^w	0.4	0.8	1.5	3.2	6.1	9.8	13.6	12.6	6.7	5.1	5.4	6.2	8.5	10.6	8.9
	5.5S_M [W]	0.5^w	0.4	0.9	1.7	4.0	8.3	12.0	13.9	11.2	6.9	5.8	6.0	7.2	7.9	8.5	4.7
	5.5L_M [W]	0.7^w	0.5	1.2	2.3	4.6	7.6	10.8	14.5	11.9	7.0	5.9	5.7	5.8	6.9	8.7	5.6
Separation with the aid of an aqueous detergent solution																		
Banks et al., 1985	16L_S [B] high level of detergent	—[3]	—	—	0.3^w	0.6	2.1	6.1	10.8	11.4	4.6	3.7	4.5	7.2	13.0	18.1	17.6
Filtration through milk filter																		
Fouad et al., 1990	25L_S	—	0.2^w	0.8	1.3	2.7	6.4	10.9	13.3	10.3	7.1	4.7	5.8	6.9	8.9	7.5	3.8
	21S_S	—	0.2^w	0.5	0.9	1.9	4.6	9.0	10.7	8.4	6.5	6.6	7.8	10.8	11.1	8.6	3.5
	21L_S	—	0.2^w	0.6	1.1	2.6	5.7	10.8	12.6	9.8	6.0	5.2	4.6	5.5	7.1	7.5	3.5
Separation technique not reported																		
Lechat et al., 1975	Most solid fraction [Feb]	—	0.3^n	0.4	0.6	0.7	1.2	5.5	9.5	7.9	7.1	8.7	12.0	15.0	17.2	11.7	2.4
	Most liquid fraction [Sept]	—	0.5^n	0.9	1.5	3.1	5.8	10.0	14.5	14.1	6.8	4.6	4.6	6.4	8.9	10.4	7.5
	Most liquid fraction [Feb]	—	1.1^n	1.8	2.7	5.1	9.0	15.1	19.5	17.0	10.0	6.1	4.1	3.2	3.0	2.0	0.2

[1]The number indicates fractionation temperature, the letter indicates physical state (L = liquid, S = solid), and the subscript indicates fractionation procedure used (M = multiple-step, S = single-step). Other designations are those of the original author. The designation inside the brackets refers to the intact anhydrous milkfat source (S = summer, W = winter).

[2]Units of chemical composition are designated using these abbreviations: n = % not specified, w = wt %.

[3]Data not available.

TABLE 5.104 Triglyceride Composition (Individual Triglyceride) of Unknown-Melting Milkfat Fractions Obtained from Solvent Solution

Type of solvent solution Author	Fraction designation[1]	C24	C26	C28	C30	C32	C34	C36	C38	C40	C42	C44	C46	C48	C50	C52	C54
Acetone																	
Parodi, 1974a	20S$_S$ [May]	—[3]	—	—	—	—	—	—	—	0.6^m	1.9	5.0	12.0	20.1	28.0	23.0	7.6
	SSS fraction of 20S$_S$ [May][4]	—	—	—	—	—	—	—	—	0.6^m	2.1	6.0	14.5	22.9	27.7	19.1	5.5
	SSM^t fraction of 20S$_S$ [May]	—	—	—	—	—	—	—	—	0.5^m	0.4	0.9	2.8	10.8	30.7	38.9	14.3
	SSM^c fraction of 20S$_S$ [May]	—	—	—	—	—	—	—	—	0.1^m	0.3	1.1	3.7	11.5	29.7	38.1	14.9
	20L$_S$ [May]	—	0.3^m	0.7	1.3	2.9	6.8	13.5	15.2	11.0	6.9	6.0	6.1	7.3	9.3	8.9	3.8
Lansbergen and Kemps, 1984	12S$_M$ [1]	—	—	—	—	0.1^n	0.5	1.5	4.1	6.3	4.0	5.2	8.9	14.2	20.7	21.1	12.4
	12L$_M$ [1]	0.8^n	0.3	1.0	2.0	4.6	10.4	18.5	20.9	14.7	9.2	6.4	4.4	2.8	2.1	1.1	0.7
	12S$_M$ [2]	—	—	0.1^n	0.1	0.3	1.0	3.0	5.7	7.5	4.5	5.4	8.9	14.0	19.4	18.7	10.7
	12L$_M$ [2]	0.8^n	0.4	1.1	2.1	5.1	10.8	19.2	20.9	14.3	9.4	6.6	4.1	2.1	1.1	1.3	0.6
Lechat et al., 1975	10S$_M$ [Apr]	—	—	—	—	0.5^n	0.2^n	1.2	3.9	4.1	5.0	8.1	11.6	17.3	23.0	18.5	6.1
Lansbergen and Kemps, 1984	6S$_M$ [1]	2.4^n	1.2	3.1	5.6	11.6	20.5	21.6	25.0	18.2	11.4	8.6	6.3	3.9	2.8	1.4	0.7
	6L$_M$ [1]	0.1^n	—	0.3	0.4	1.0	3.6	12.4	13.8	8.7	5.4	2.6	0.7	0.8	0.7	0.5	0.4
	0S$_M$ [2]	1.4^n	0.8	2.0	3.8	8.1	16.7	23.5	22.7	17.2	10.9	8.9	6.9	4.6	4.6	4.1	2.2
	0L$_M$ [2]	—	0.3^n	0.9	2.0	4.1	7.8	13.0	17.2	10.6	7.3	4.2	2.1	0.5	0.8	0.4	0.2
Lechat et al., 1975	-9L$_M$ [Apr]	—	—	—	—	0.1^n	0.5	1.1	17.1	15.9	8.8	5.8	5.1	5.3	5.7	5.4	2.3
Lansbergen and Kemps, 1984	Top fraction	tr^{5n}	tr	0.1	0.2	0.7	4.3	12.7	12.1	1.5	4.7	11.3	19.8	22.9	22.0	11.4	2.6
	Mid fraction	—	—	—	—	—	—	—	1.0	9.0	9.2	8.9	8.7	10.8	12.8	7.7	2.1
	Olein fraction	0.1^n	0.4	1.0	2.2	4.9	9.5	12.6	14.4	11.1	6.8	6.0	6.1	6.3	6.4	7.1	4.0
Hexane																	
Muuse and van der Kamp, 1985	12.5S$_S$	—	0.3^a	0.6	1.2	2.5	5.6	9.9	11.3	8.5	7.4	7.8	8.7	9.1	9.3	5.0	1.7
	12.5L$_S$	—	0.4^a	0.9	1.6	3.5	7.5	13.3	14.9	10.7	7.2	5.9	5.5	5.8	6.3	5.1	1.8

[1]The number indicates fractionation temperature, the letter indicates physical state (L = liquid, S = solid), and the subscript indicates fractionation procedure used (M = multiple-step, S = single-step). Other designations are those given by the original author. The designation inside the brackets refers to the intact anhydrous milkfat source (S = summer).
[2]Units of chemical composition are designated using these abbreviations: a = area %, m = mole %, n = % not specified.
[3]Data not available.
[4]S = saturated, M = monounsaturated, c = cis, t = trans.
[5]tr = trace.

TABLE 5.105 Triglyceride Composition (Individual Triglyceride) of Unknown-Melting Milkfat Fractions Obtained by Supercritical Fluid Extraction

Type of solvent Author	Fraction designation[1]	Fractionation conditions			Triglyceride (%)[2]														
		Temperature (°C)	Pressure	C24	C26	C28	C30	C32	C34	C36	C38	C40	C42	C44	C46	C48	C50	C52	C54
Carbon dioxide																			
Bhaskar et al., 1993a	Extract	40	24.1 MPa	0.6[w]	0.9	1.1	2.0	4.1	10.0	18.1	21.4	14.8	7.7	5.0	4.2	4.0	3.5	2.2	0.5
	Raffinate	40	24.1	0.0[w]	0.0	0.0	0.0	0.0	0.0	5.5	6.0	5.7	5.1	6.9	13.8	14.2	19.0	18.3	5.8
Ensign, 1989	WM-38, 1	60	2150 psi	—[3]	3.4[a]	5.8	6.8	8.8	12.6	15.9	15.2	8.7	3.6	1.8	1.3	1.1	1.0	0.6	0.2
	WM-38, 2	60	2300	—	1.2[a]	2.9	4.4	7.3	12.3	18.0	18.2	10.1	4.5	2.6	1.9	1.7	1.9	1.4	0.5
	WM-38, 3	60	2400	—	0.7[a]	2.0	3.6	6.1	11.8	18.8	19.8	12.0	5.2	2.9	2.2	1.9	2.1	1.4	0.4
	WM-38, 4	60	2400	—	0.4[a]	1.3	2.8	5.2	10.2	18.4	20.6	12.9	5.7	3.3	2.5	2.3	2.5	1.9	0.4
	WM-38, 5	60	2500	—	0.3[a]	0.9	2.1	4.2	9.5	17.4	20.8	13.1	6.5	4.1	3.0	2.7	2.9	2.3	0.9
	WM-38, 6	60	2500	—	0.1[a]	0.6	1.6	3.6	8.7	11.6	20.4	13.5	7.0	4.6	3.5	3.2	3.4	2.8	1.0
	WM-38, 7	60	2650	—	tr[4a]	0.3	1.3	3.5	9.3	18.2	21.7	14.4	6.4	4.2	3.2	2.8	2.6	2.1	0.7
	WM-38, 8	60	2850	—	0.0[a]	0.1	0.4	1.5	5.8	14.8	21.3	16.3	8.0	5.4	4.2	3.7	4.1	3.3	1.1
	WM-38, 9	60	2900	—	0.0[a]	0.0	0.1	0.6	3.7	11.5	19.4	17.0	9.3	6.8	5.8	5.1	5.0	4.3	1.5
	WM-38, 10	60	2900	—	0.0[a]	0.0	0.1	0.3	2.2	8.3	16.2	16.6	10.2	8.0	6.9	6.5	7.2	5.3	2.0
	WM-38, 11	60	2900	—	0.0[a]	0.0	0.0	0.1	1.0	4.6	11.3	12.8	10.2	9.3	9.6	10.3	11.1	7.9	2.2
	WM-38, 12	60	2900	—	0.0[a]	0.0	0.0	0.0	0.4	1.6	6.2	9.2	8.9	9.7	11.2	12.8	15.0	10.9	3.9
	WM-38, 13	60	2900	—	0.0[a]	0.0	0.0	0.0	0.2	0.9	2.6	4.8	6.0	8.4	11.5	15.1	18.7	15.0	4.9
	WM-38, 14	60	2900	—	0.0[a]	0.0	0.0	0.0	tr	0.3	0.7	1.6	2.8	5.5	9.8	15.9	22.4	20.3	7.7
	WM-38, 15	60	—	—	0.0[a]	0.0	0.0	0.0	0.0	0.1	0.2	0.2	0.4	1.3	3.7	9.0	22.6	32.9	19.2
	WM-39, 1	60	2000	—	4.4[a]	8.5	10.2	11.8	9.2	17.0	12.5	5.7	2.1	1.1	0.7	0.6	0.5	0.4	0.0
	WM-39, 2	60	6000	—	tr[a]	0.3	0.7	1.5	4.2	9.5	12.4	10.5	6.6	5.8	6.0	7.7	10.5	5.7	5.0
	WM-43, 1	40	1700	—	3.4[a]	6.7	8.3	10.3	13.7	17.0	15.5	8.6	2.6	1.8	1.1	0.9	0.5	0.0	0.0
	WM-43, 2	40	6000	—	0.0[a]	0.2	0.6	1.6	4.0	8.0	11.6	10.8	6.6	4.8	6.2	8.4	12.5	12.2	5.3
Propane																			
Biernoth and Merk, 1985	Extract-1	125	85 bar	—	—	1.5[n]	2.3	5.0	9.4	16.0	15.0	10.1	7.1	6.0	5.9	5.8	5.5	4.1	1.3
	Extract-2	125	85	—	—	1.5[n]	2.7	4.4	10.0	16.8	15.6	10.5	6.8	6.0	5.7	5.6	5.0	3.3	1.0
	Residual	125	85	—	—	0.4[n]	0.7	1.6	4.6	10.3	11.3	9.1	7.4	7.4	8.6	10.1	11.6	8.7	2.2

[1]Designations used are those given by the original author.

[2]Units of chemical composition are designated using these abbreviations: a = area %, n = % not specified, w = wt %.

[3]Data not available.

[4]tr = trace.

TABLE 5.106 Triglyceride Composition (Selected Triglyceride Groupings) of Intact Anhydrous Milkfats Used to Produce Milkfat Fractions from Melted Milkfat

Fraction separation method Author	Fraction designation[1]	Selected triglyceride groupings (%)[2]					Composition of triglyceride classes (%)[3]			
		C24–C34	C36–C40	≤C40	C42–C54	SSS	SSU	SUU	UUU
Vacuum filtration										
Amer et al., 1985	AMF [W]	12.80[a]	35.30	48.10	44.70	—[3]	—	—	—
Deffense, 1993b	AMF	—	—	—	—	33.2[w] C24–C42 = 24.8 C44–C56 = 8.4	49.2 C24–C42 = 24.8 C44–C56 = 8.4	SUU +UUU = 17.6 C24–C42 = 7.6 C44–C56 = 10.0	
deMan and Finoro, 1980	AMF	7.60[w]	33.40	41.00	58.90	—	—	—	—
Kaylegian, 1991	AMF [W]	10.60[a]	29.20	39.80	60.30	—	—	—	—
Laakso et al., 1992	AMF	—	—	—	—	44.9[w]	41.6 SSM[t] = 3.4 SSM[c] = 38.2	8.6 SM[c]M[t] = 2.5 SM[c]M[c] = 6.1	4.9 others
Lund and Danmark, 1987	AMF [1]	9.20[w]	32.70	41.90	50.70	—	—	—	—
	AMF [2]	7.80[w]	31.00	38.80	52.70	—	—	—	—
Makhlouf et al., 1987	AMF	12.0[w]	36.0	48.0	52.0	—	—	—	—
Pressure filtration										
Badings et al., 1983b	AMF [W]	12.10[w]	35.20	47.30	52.60	—	—	—	—
	AMF [S]	9.80[w]	33.70	43.50	56.50	—	—	—	—
Deffense, 1993b	AMF	—	—	—	—	33.2[w] C24–C42 = 24.8 C44–C56 = 8.4	49.2 C24–C42 = 24.8 C44–C56 = 8.4	SUU +UUU = 17.6 C24–C42 = 7.6 C44–C56 = 10.0	
Centrifugation										
Branger, 1993	AMF [1] bench-top	7.90[a] C26–C34	42.79 C36–C44	—	35.61 C46–C54	—	—	—	—
	AMF [2] bench-top	7.47[a]	41.19	—	39.32	—	—	—	—
	AMF [3] bench-top	7.58[a]	43.49	—	36.24	—	—	—	—
	AMF pilot-scale	8.83[a]	39.56	—	37.4	—	—	—	—

Continued

TABLE 5.106 (Continued)

Fraction separation method Author	Fraction designation[1]	Selected triglyceride groupings (%)[2]				Composition of triglyceride classes (%)[3]			
		C24–C34	C36–C40	≤C40	C42–C54	SSS	SSU	SUU	UUU
Jordan, 1986	AMF [W]	10.50^w	30.40	40.90	53.40	—	—	—	—
Kankare and Antila, 1986	AMF [S]	9.40^n	34.10	43.50	54.20	—	—	—	—
Kankare and Antila, 1988b	AMF	—	—	—	—	35^n	49	12.5	2
Keogh and Higgins, 1986b	AMF	8.70^n	33.40	42.10	52.40	—	—	—	—
Separation with the aid of an aqueous detergent solution									
Banks et al., 1985	AMF [A]	5.80^w	41.00	46.80	53.30	—	—	—	—
	AMF [B]	2.50^w	22.40	24.90	74.10	—	—	—	—
Filtration through milk filter									
Fouad et al., 1990	AMF	9.80^w	31.20	41.00	48.30	—	—	—	—
Separation technique not reported									
Lechat et al., 1975	AMF [Sept]	8.80^n	37.70	46.50	53.60	—	—	—	—
	AMF [Feb]	14.80^n	45.20	60.00	39.40	—	—	—	—

[1] AMF = anhydrous milkfat. The designation inside the brackets refers to the intact anhydrous milkfat source [sample number, letter or season (S = summer, W = winter)].

[2] Units of chemical composition are designated using these abbreviations: a = area %, n = % not specified, w = wt %.

[3] S = saturated, U = unsaturated, M = monoenoic, c = cis, t = trans.

[4] Data not available.

TABLE 5.107 Triglyceride Composition (Selected Triglyceride Groupings) of Intact Anhydrous Milkfats Used to Produce Milkfat Fractions from Solvent Solution

Type of solvent solution Author	Fraction designation[1]	Selected triglyceride groupings (%)[2]				Composition of trigylceride classes (%)[3]			
		C24–C34	C36–C40	≤C40	C42–C54	SSS	SSU	SUU	UUU
Acetone									
Chen and deMan, 1966	AMF	—[4]	—	—	—	38.4[w]	—	—	—
Kaylegian, 1991	AMF	10.60[a]	29.20	39.80	60.30	—	—	—	—
Lambelet, 1983	AMF	9.00[n]	9.10	18.10	0.00	—	—	—	—
Lansbergen and Kemps, 1984	Hardened butterfat	12.70[n]	35.30	48.00	52.70	—	—	—	—
	Natural butterfat	13.90[n]	34.60	48.50	52.30	—	—	—	—
Lechat et al., 1975	AMF [Apr]	9.70[n]	40.30	50.00	49.90	—	—	—	—
Parodi, 1974a	AMF [May]	11.40[m]	38.20	49.60	50.70	—	—	—	—
Timms, 1980b	AMF [S]	9.60[w]	33.50	43.10	56.60	—	—	—	—
Yi, 1993	AMF [S]	10.40[a]	39.30	49.70	47.70	—	—	—	—
Pentane									
Kaylegian, 1991	AMF [W]	10.60[a]	29.20	39.80	60.30	—	—	—	—
Hexane									
Muuse and van der Kamp, 1985	AMF	12.60[a]	36.00	48.60	41.60	—	—	—	—

[1]AMF = anhydrous milkfat. The designation inside the brackets following the fraction designation refers to the intact anhydrous milkfat source (S = summer, W = winter).

[2]Units of chemical composition are designated using these abbreviations: a = area %, m = mole %, n = % not specified, w = wt %.

[3]S = saturated, U = unsaturated.

[4]Data not available.

Milkfat Fractionation

TABLE 5.108 Triglyceride Composition (Selected Triglyceride Groupings) of Intact Anhydrous Milkfats Used to Produce Milkfat Fractions by Supercritical Fluid Extraction

Type of solvent Author	Fraction designation[1]	Selected triglyceride groupings (%)[2]			
		C24–C34	C36–C40	≤C40	C42–C54
Carbon dioxide					
Arul et al., 1987	AMF	12.60[w]	35.40	48.00	52.00
Bhaskar et al., 1993a	AMF	16.80[w]	50.70	67.50	33.90
Bhaskar et al., 1993b	AMF	16.72[w]	50.85	67.57	32.93
Biernoth and Merk, 1985	AMF	11.40[n]	33.60	45.00	53.00
Büning-Pfaue et al., 1989	Parent milkfat	10.53[3w]	30.30	40.83	49.50
		11.1[4a]	44.70	55.80	44.20
Chen et al., 1992	AMF	28.4[n] ≤C36	—[5]	—	71.7 >C36
Ensign, 1989	AMF WM-38	9.30[a]	36.20	45.50	52.30
	AMF WM-39	8.50[a]	33.30	41.80	49.70
	AMF WM-43	7.90[a]	33.40	41.30	50.60
Kankare et al., 1989	AMF	15.20[w]	39.50	54.70	44.20
Kaufmann et al., 1982	AMF	15.10[m]	37.10	52.20	47.30
Rizvi et al., 1990	AMF	10.40[w]	43.70	54.10	40.90
Timmen et al., 1984	AMF	15.10[m]	36.70	51.80	46.80
Propane					
Biernoth and Merk, 1985	AMF	9.90[n]	32.90	42.80	51.30

[1]Sample description: AMF = anhydrous milkfat.

[2]Units of chemical composition are designated using these abbreviations: a = area %, m = mole %, n = % not specified, w = wt %.

[3]Triglyceride determination by GC analysis (weight %).

[4]Triglyceride determination by HPLC analysis (area %).

[5]Data not available.

TABLE 5.109 Triglyceride Composition (Selected Triglyceride Groupings) of Intact Anhydrous Milkfat Used to Produce Milkfat Fractions by Short-Path Distillation

Author	Fraction designation[1]	Selected triglyceride groupings (%)[2]			
		C24–C34	C36–C40	≤C40	C42–C54
Arul et al., 1988a	AMF	11.60[w]	35.60	47.20	52.90
McCarthy et al., 1962	AMF	9.7[w]	39.7	49.4	50.6

[1]AMF = anhydrous milkfat.

[2]Units of chemical composition are designated using this abbreviations: w = wt %.

TABLE 5.110 Triglyceride Composition (Selected Triglyceride Groupings) of Very-High-Melting Milkfat Fractions Obtained from Melted Milkfat

Fraction separation method Author	Fraction designation[1]	Selected triglyceride groupings (%)[2]				Composition of triglyceride classes (%)[3]			
		C24–C34	C36–C40	≤C40	C42–C54	SSS	SSU	SUU	UUU
Vacuum filtration									
Kaylegian, 1991	34S$_M$ [W]	6.00[a]	19.70	25.70	74.60	—[4]	—	—	—
	34S$_S$ [W]	6.90[a]	22.10	29.00	71.00	—	—	—	—
Makhlouf et al. 1987	26S$_M$ (S.1)	3.0^w	25.0	28.0	72.0	—	—	—	—
	21S$_M$ (S.2)	5.0^w	28.0	33.0	61.0	—	—	—	—
	F.S (>21)	12.0^w	28.0	40.0	60.0	—	—	—	—
Pressure filtration									
Deffense, 1993	1st stearin	—	—	—	—	53.9^w C24–C42 = 11.6 C44–C56 = 42.3	37.3 C24–C42 = 8.8 C44–C56 = 28.5	SUU +UUU = 8.8 C24–C42 = 2.0 C44–C56 = 6.8	
Centrifugation									
Kankare and Antila, 1986	24S$_M$ [S]	5.50[n]	22.00	27.50	71.20	—	—	—	—
Kankare and Antila, 1988b	24S$_M$	—	—	—	—	51[n]	38	7	1
Filtration with the aid of an aqueous detergent solution									
Banks et al., 1985	21S$_S$ [A] high level of detergent	2.80^w	18.50	21.30	73.10	—	—	—	—
	21 semi-S$_S$ [A] high level of detergent	7.90^w	44.70	52.60	47.40	—	—	—	—
	16S$_S$ [B] high level of detergent	0.20^w	4.90	5.10	94.90	—	—	—	—

[1]The number indicates fractionation temperature, the letter indicates physical state (S = solid), and the subscript indicates fractionation procedure used (M = multiple-step, S = single-step). Other designations are those of the original author. The designation inside the brackets refers to the intact anhydrous milkfat source [sample letter or season (W = winter)].

[2]a = area %, n = % not specified, w = wt %.

[3]S = saturated, U = unsaturated.

[4]Data not available.

TABLE 5.111 Triglyceride Composition (Selected Triglyceride Groupings) of Very-High-Melting Milkfat Fractions Obtained from Solvent Solution

Type of solvent solution Author	Fraction designation[1]	Selected triglyceride groupings (%)[2]			
		C24–C34	C36–C40	≤C40	C42–C54
Acetone					
Kaylegian, 1991	25S_M [W]	2.80[a]	9.70	12.50	87.50
	25S_S [W]	3.00[a]	10.20	13.20	86.70
Yi, 1993	25S_M [S]	5.60[a]	13.50	19.10	81.00
Kaylegian, 1991	21S_M [W]	3.90[a]	11.70	15.60	84.50
	21S_S [W]	2.90[a]	10.20	13.10	86.80
Timms, 1980b	20S_M [S]	0.10[w]	1.10	1.20	98.20
Yi, 1993	20SM [S]	2.20[a]	9.60	11.80	88.20
Kaylegian, 1991	15S_M [W]	3.40[a]	12.60	16.00	84.10
	15S_S [W]	3.80[a]	13.40	17.20	82.90
Yi, 1993	15S_M [S]	2.20[a]	11.80	14.00	86.00
Kaylegian, 1991	11S_S [W]	4.00[a]	14.40	18.40	81.60
Pentane					
Kaylegian, 1991	11S_M [W]	2.50[a]	9.60	12.10	87.90
	11S_S [W]	4.60[a]	14.60	19.20	80.80
	5S_M [W]	3.30[a]	12.10	15.40	84.50
	5S_S [W]	3.60[a]	13.30	16.90	83.10
	0S_M [W]	4.00[a]	13.50	17.50	82.50
	0S_S [W]	3.60[a]	12.80	16.40	83.50

[1]The number indicates fractionation temperature, the letter indicates physical state (S = solid), and the subscript indicates fractionation procedure used (M = multiple-step, S = single-step). The designation inside the brackets refers to the intact anhydrous milkfat source (S = summer, W = winter).

[2]Units of chemical composition are designated using these abbreviations: a = area %, w = wt %.

TABLE 5.112 Triglyceride Composition (Selected Triglyceride Groupings) of Very-High-Melting Milkfat Fractions Obtained by Supercritical Fluid Extraction

Type of solvent Author	Fraction designation[1]	Fractionation conditions		Selected triglyceride groupings (%)[2]			
		Temperature (°C)	Pressure	C24–C34	C36–C40	≤C40	C42–C54
Carbon dioxide							
Bhaskar et al., 1993b	S1	40	24.1 MPa	tr[3w]	17.07	17.07	82.93

[1]Designations used are those given by the original author.

[2]Units of chemical composition are designated using this abbreviation: w = wt %.

[3]tr = trace.

TABLE 5.113 Triglyceride Composition (Selected Triglyceride Groupings) of High-Melting Milkfat Fractions Obtained from Melted Milkfat

Fraction separation method Author	Fraction designation[1]	Selected triglyceride groupings (%)[2]					Composition of triglyceride classes (%)[2]			
		C24–C34	C36–C40	≤C40	C42–C54	SSS	SSU	SUU	UUU
Vacuum filtration										
deMan and Finoro, 1980	32S$_S$	6.90^w	24.70	31.60	68.40	—[4]	—	—	—
	31S$_S$	6.50^w	22.90	29.40	71.10	—	—	—	—
	30S$_S$	5.70^w	21.10	26.80	73.40	—	—	—	—
Kaylegian, 1991	30S$_M$ [W]	7.10^a	23.10	30.20	69.80	—	—	—	—
	30S$_S$ [W]	7.50^a	23.90	31.40	68.70	—	—	—	—
Amer et al., 1985	29S$_S$ [W]	9.80^a	27.00	36.80	54.70	—	—	—	—
deMan and Finoro, 1980	29S$_S$	5.70^w	22.00	27.70	72.40	—	—	—	—
	28S$_S$	5.50^w	21.00	26.50	73.60	—	—	—	—
	27S$_S$	6.50^w	25.60	32.10	68.00	—	—	—	—
	26S$_S$	6.40^w	25.20	31.60	68.40	—	—	—	—
Lund and Danmark, 1987	26S$_S$ [1]	6.30^w	24.10	30.40	58.80	—	—	—	—
	25S$_S$	7.30^w	27.30	34.60	65.40	—	—	—	—
deMan and Finoro, 1980	25S$_M$ [W]	6.10^a	22.20	28.30	71.70	—	—	—	—
Kaylegian, 1991	25S$_S$ [W]	8.80^a	29.10	37.90	62.00	—	—	—	—
	20S$_M$ [W]	6.10^a	22.20	28.30	71.70	—	—	—	—
	20S$_S$ [W]	9.30^a	28.00	37.30	63.00	—	—	—	—
Amer et al., 1985	19S$_S$ [W]	9.80^a	32.30	42.10	48.50	—	—	—	—
Lund and Danmark, 1987	18S$_S$ [2]	6.60^w	26.80	33.40	57.90	—	—	—	—
Deffense, 1993b	1st stearin	—	—	—	—	48.6^w C24–C42 = 15.3 C44–C56 = 33.3	40.9 C24–C42 = 16.4 C44–C56 = 24.5	—	SUU +UUU = 10.5 C24–C42 = 2.7 C44–C56 = 7.8
Laakso et al., 1992	S1	—	—	—	—	58.9^w	30.7 SSM^t = 2.8 SSM^c = 27.9	7.5 SM^{c}M^t = 2.1 SM^{c}M^c = 5.4	2.8 others
Centrifugation										
Branger, 1993	32S$_M$ [1] bench-scale	7.21^a C26–C34	38.79 C36–C44	—	39.45 C46–C54	—	—	—	—

Continued

TABLE 5.113 (Continued)

Fraction separation method Author	Fraction designation[1]	Selected triglyceride groupings (%)[2]					Composition of triglyceride classes (%)[3]				
		C24-C34	C36-C40	≤C40	C42-C54	SSS	SSU	SUU	SUU	UUU
	$32S_M$ [2] bench-scale	7.00[a]	37.93	—	42.65	—	—	—	—	—
	$32S_M$ [3] bench-scale	6.37[a]	37.41	—	42.85	—	—	—	—	—
	$32S_S$ pilot-scale	6.80[a]	34.03	—	45.29	—	—	—	—	—
Jordan, 1986	$28S_S$ [W]	8.70[w]	28.70	37.40	60.80	—	—	—	—	—
Branger, 1993	$25S_S$ pilot-scale	7.92[a]	38.60	—	39.82	—	—	—	—	—
	$22S_M$ [1] bench-scale	7.75[a]	42.5	—	36.68	—	—	—	—	—
	$22S_M$ [3] bench-scale	6.89[a]	40.32	—	38.92	—	—	—	—	—
	$20S_M$ [2] bench-scale	7.03[a]	42.29	—	38.29	—	—	—	—	—
Kankare and Antila, 1988b	$12S_M$	—	—	—	—	41.5[n]	45	10	1	—
Keogh and Higgins, 1986b	Fraction 2	5.60[w]	22.00	27.60	64.90	—	—	—	—	—

Separation with the aid of an aqueous detergent solution

Banks et al., 1985	16 semi-S_S [B] high level of detergent	1.70[w]	17.90	19.60	80.40	—	—	—	—	—

Filtration through milk filter

Fouad et al., 1990	$29S_S$	5.20[w]	20.30	25.50	62.90	—	—	—	—	—
	$25S_S$	6.00[w]	22.40	28.40	61.30	—	—	—	—	—

[1] The number indicates fractionation temperature, the letter indicates physical state (S = solid), and the subscript indicates fractionation procedure used (M = multiple-step, S = single-step). The designation inside the brackets refers to the intact anhydrous milkfat source [sample number, letter or season (W = winter)].

[2] a = area %, w = wt %.

[3] S = saturated, M = monounsaturated, U = polyunsaturated, c = cis, t = trans.

TABLE 5.114 Triglyceride Composition (Selected Triglyceride Groupings) of High-Melting Milkfat Fractions Obtained from Solvent Solution

Type of solvent solution Author	Fraction designation[1]	Selected triglyceride groupings (%)[2]			
		C24–C34	C36–C40	≤C40	C42–C54
Acetone					
Kaylegian, 1991	11S_M [W]	4.60[a]	16.30	20.90	79.20
Yi, 1993	10S_M [S]	7.00[a]	20.60	27.60	72.50
Kaylegian, 1991	5S_S [W]	5.00[a]	21.50	26.50	73.50
Lambelet, 1983	$S1_M$	10.30[n]	0.00	10.30	0.00
Timms, 1908b	3S_M [S]	1.20[w]	24.10	25.30	74.60
Kaylegian, 1991	0S_S [W]	6.70[a]	27.60	34.30	65.80

[1]The number indicates fractionation temperature, the letter indicates physical state (S = solid), and the subscript indicates fractionation procedure used (M = multiple-step, S = single-step). The designation inside the brackets refers to the intact anhydrous milkfat source (S = summer, W = winter).

[2]Units of chemical composition are designated using these abbreviations: a = area %, n = % not specified, w = wt %.

TABLE 5.115 Triglyceride Composition (Selected Triglyceride Groupings) of High-Melting Milkfat Fractions Obtained by Supercritical Fluid Extraction

Type of solvent Author	Fraction designation[1]	Fractionation conditions		Selected triglyceride groupings (%)[2]			
		Temperature (°C)	Pressure	C24–C34	C36–C40	≤C40	C42–C54
Carbon dioxide							
Biernoth and Merk, 1985	Residual	80	200	4.20[n]	29.90	34.10	63.00
Kaufmann et al., 1982	I (residue)	80	200 bar	6.60[m]	34.00	40.60	59.20
Timmen et al., 1984	Long-chain fraction	80	200 bar	6.60[m]	33.60	40.20	58.00
Arul et al., 1987	S2	70	250–350 bar	4.60[w]	31.60	36.20	65.80
	S3	70	250–350	0.70[w]	9.80	10.50	89.50
Bhaskar et al., 1993b	S2	60	24.1 MPa	10.18[w]	49.94	60.12	39.88
Büning-Pfaue et al., 1989	Extractor fraction	50	25,000 kPa	15.73[3w]	38.10	53.83	36.20
				9.9[4a]	57.50	67.40	32.50
Kankare et al., 1989	R	50	400 bar	10.20[w]	39.30	49.50	50.00
Rizvi et al., 1990	Extract-2	48.0	275.8 bar	9.40[w]	44.10	53.50	41.20
	Residue	48.0	275.8	4.50[w]	23.90	28.40	67.40
Chen et al., 1992	13.8	40	13.8 MPa	39.6[n] ≤C36	—[5]	—	60.4 >C36
	20.7	40	20.7	38.5[n] ≤C36[n]	—	—	61.6 >C36
	24.1	40	24.1	29.5[n] ≤C36	—	—	71.5 >C36
	27.6	40	27.6	31.2[n] ≤C36	—	—	69.6 >C36
Shukla et al., 1994	Raffinate	40	24.1 MPa	tr[6w]	17.07	17.07	82.93

Continued

TABLE 5.115 (Continued)

Type of solvent Author	Fraction designation[1]	Fractionation conditions		Selected triglyceride groupings (wt %)			
		temperature (°C)	pressure	C24–C34	C36–C40	≤C40	C42–C54
Büning-Pfaue et al., 1989	Column fraction	30	20,000 kPa	5.93[3w] 5.7[4a]	22.70 39.30	28.63 45.00	60.50 55.10

[1]Designations used are those given by the original author.

[2]Units of chemical composition are designated using these abbreviations: a = area %, m = mole %, n = % not specified, w = wt %.

[3]Triglyceride determination by GC analysis (weight %).

[4]Triglyceride determination by HPLC analysis (area %).

[5]Data not available.

[6]tr = trace.

TABLE 5.116 Triglyceride Composition (Selected Triglyceride Groupings) of High-Melting Milkfat Fractions Obtained by Short-Path Distillation

Author	Fraction designation[1]	Fractionation conditions		Selected triglyceride groupings (%)[2]			
		Temperature (°C)	Pressure	C24–C34	C36–C40	≤C40	C42–C54
Arul et al., 1988a	SF Residue	—[3]	—	1.40[w]	19.30	20.70	50.80

[1]Designations used are those given by the original author.

[2]Units of chemical composition are designated using this abbreviation: w = wt %.

[3]Data not available.

TABLE 5.117 Triglyceride Composition (Selected Triglyceride Groupings) of Middle-Melting Milkfat Fractions Obtained from Melted Milkfat

Fraction separation method Author	Fraction designation[1]	Selected triglyceride groupings (%)[2]				Composition of triglyceride classes (%)[3]			
		C24–C34	C36–C40	≤C40	C42–C54	SSS	SSU	SUU	UUU
Vacuum filtration									
Kaylegian, 1991	34L_S [W]	9.80[a]	29.50	39.30	60.70	—[4]	—	—	—
deMan and Finoro, 1980	32L_S	8.80[w]	32.30	41.10	58.80	—	—	—	—
	31L_S	9.60[w]	33.60	43.20	57.00	—	—	—	—
	30L_S	9.40[w]	34.20	43.60	56.50	—	—	—	—
Kaylegian, 1991	30L_S [W]	11.60[a]	31.10	42.70	57.20	—	—	—	—
Amer et al., 1985	29L_S [W]	11.60[a]	36.50	48.10	42.00	—	—	—	—
deMan and Finoro, 1980	29L_S	9.30[w]	34.30	43.60	56.40	—	—	—	—
	28L_S	9.30[w]	34.70	44.00	56.00	—	—	—	—
	27L_S	9.40[w]	34.90	44.30	55.90	—	—	—	—
Kaylegian, 1991	16S_M [W]	8.30[a]	37.10	45.40	54.40	—	—	—	—
Makhlouf et al., 1987	15S_M (L1)	11.0[w]	41.0	52.0	48.0	—	—	—	—
Laakso et al., 1992	S2	—	—	—	—	53.8[w]	35.8 SSM^t = 2.5 SSM^c = 33.3	7.0 SM^cM^t = 1.8 SM^cM^c = 5.2	3.4 others
Centrifugation									
Branger, 1993	32L_S pilot-scale	9.38[a]	41.97	—	35.64	—	—	—	—
	25L_S pilot-scale	10.19[a]	43.46	—	34.61	—	—	—	—
	22L_M [1] bench-top	9.39[a]	47.85	—	30.79	—	—	—	—
	20S_S pilot-scale	9.00[a]	38.82	—	37.22	—	—	—	—
	17S_S pilot-scale	8.63[a]	41.23	—	37.25	—	—	—	—
	12S_M [3] bench-top	7.85[a]	47.01	—	33.19	—	—	—	—
Kankare and Antila, 1986	12S_M [S]	8.10[n]	34.10	42.20	56.10	—	—	—	—

Continued

TABLE 5.117 (Continued)

Fraction separation method Author	Fraction designation[1]	Selected triglyceride groupings (%)				Composition of triglyceride classes (%)[3]				
		C24–C34	C36–C40	≤C40	C42–C54	SSS	SSU	SUU	SUU	UUU

Filtration through milk filter

Fouad et al., 1990	29L$_S$	11.20^w	36.50	47.70	47.20	—	—	—	—	—
	17S$_S$	9.00^w	28.40	37.40	52.90	—	—	—	—	—

[1]The number indicates fractionation temperature, the letter indicates physical state (L = liquid, S = solid), and the subscript indicates fractionation procedure used (M = multiple-step, S = single-step). Any designations in parentheses are fraction designations given by the original author. The designation inside the brackets refers to the intact anhydrous milkfat source (S = summer, W = winter).

[2]a = area %, n = % not specified, w = wt %.

[3]S = saturated, U = unsaturated, M = monoenoic, c = *cis*, t = *trans*.

[4]Data not available.

TABLE 5.118 Triglyceride Composition (Selected Triglyceride Groupings) of Middle-Melting Milkfat Fractions Obtained from Solvent Solution

Type of solvent solution Author	Fraction designation[1]	Selected triglyceride groupings (%)[2]			
		C24–C34	C36–C40	≤C40	C42–C54
Acetone					
Kaylegian, 1991	25L$_S$ [W]	10.90^a	31.60	42.50	57.40
	21L$_S$ [W]	11.40^a	32.30	43.70	56.30
	5S$_M$ [W]	3.60^a	27.40	31.00	68.90
Yi, 1993	5S$_M$ [S]	5.10^a	37.30	42.40	57.70
Kaylegian, 1991	0S$_M$ [W]	7.40^a	37.30	44.70	55.20
Yi, 1993	0S$_M$ [S]	6.50^a	44.30	50.80	49.10
Pentane					
Kaylegian, 1991	11L$_S$ [W]	10.00^a	30.70	40.70	59.10
	5L$_S$ [W]	11.00^a	32.50	43.50	56.60

[1]The number indicates fractionation temperature, the letter indicates physical state (L = liquid, S = solid), and the subscript indicates fractionation procedure used (M = multiple-step, S = single-step). The designation inside the brackets refers to the intact anhydrous milkfat source (S = summer, W = winter).

[2]Units of chemical composition are designated using this abbreviations: a = area %.

TABLE 5.119 Triglyceride Composition (Selected Triglyceride Groupings) of Middle-Melting Milkfat Fractions Obtained by Supercritical Fluid Extraction

Type of solvent Author	Fraction designation[1]	Fractionation conditions		Selected triglyceride groupings (%)[2]			
		Temperature (°C)	Pressure	C24–C34	C36–C40	≤C40	C42–C54
Carbon dioxide							
Bhaskar et al., 1993b	S3	75	17.2 MPa	18.82^w	56.19	75.01	24.99
Arul et al., 1987	S1	70	250–350 bar	8.30^w	45.50	53.80	48.20
Bhaskar et al., 1993b	S4	60	6.9 MPa	24.30^w	53.62	77.92	22.08
	S5	60	3.5	26.39^w	54.22	80.61	19.39
Arul et al., 1987	I2	50	100–250	21.30^w	54.50	75.80	24.20
	I3	50	100–250	13.50^w	52.30	65.80	34.20
Kankare et al., 1989	E2	50	200 bar	34.60^w	49.80	84.40	15.90
	E4	50	400	19.30^w	49.80	69.10	30.70
Chen et al., 1992	17.2	40	17.2 MPa	33.3^n ≤C36	—[3]	—	66.8 >C36
Rizvi et al., 1990	Extract-1	40.0	241.3 bar	21.40^w	55.80	77.20	18.40

[1]Designations used are those given by the original author.

[2]Units of chemical composition are designated using these abbreviations: n = % not specified, w = wt %.

[3]Data not available.

TABLE 5.120 Triglyceride Composition (Selected Triglyceride Groupings) of Middle-Melting Milkfat Fraction Obtained by Short-Path Distillation

Author	Fraction designation[1]	Fractionation conditions		Selected triglyceride groupings (%)[2]			
		Temperature (°C)	Pressure	C24–C34	C36–C40	≤C40	C42–C54
Arul et al., 1988a	IF Distillate III	265	100 μm Hg	16.50^w	61.30	77.80	22.30

[1]Designations used are those given by the original author.

[2]Units of chemical composition are designated using this abbreviation: w = wt %.

TABLE 5.121 Triglyceride Composition (Selected Triglyceride Groupings) of Low-Melting Milkfat Fractions Obtained from Melted Milkfat

Fraction separation method Author	Fraction designation[1]	Selected triglyceride groupings (%)[2]				Composition of trigylceride classes (%)[3]			
		C24–C34	C36–C40	≤C40	C42–C54	SSS	SSU	SUU	UUU
Vacuum filtration									
deMan and Finoro, 1980	26L$_S$	9.20^w	35.60	44.80	55.20	—[4]	—	—	—
Lund and Danmark, 1987	26L$_S$ [1]	9.60^w	35.50	45.10	48.30	—	—	—	—
deMan and Finoro, 1980	25L$_S$	9.80^w	36.30	46.10	54.10	—	—	—	—
Kaylegian, 1991	25L$_S$ [W]	11.30^a	33.20	44.50	55.50	—	—	—	—
	20L$_S$ [W]	11.90^a	35.20	47.10	52.80	—	—	—	—
Amer et al., 1985	19L$_S$ [W]	12.90^a	40.70	53.60	38.20	—	—	—	—
Lund and Danmark, 1987	18L$_S$ [2]	9.10^w	35.70	44.80	45.80	—	—	—	—
Kaylegian, 1991	13S$_M$ [W]	10.30^a	32.80	43.10	56.90	—	—	—	—
	13L$_M$ [W]	11.70^a	29.90	41.60	58.50	—	—	—	—
Makhlouf et al., 1987	9S$_M$ (I.2)	15.0^w	38.0	53.0	47.0	—	—	—	—
	9L$_M$ (F.L.)	18.0^w	37.0	55.0	45.0	—	—	—	—
	F.1 (21-9)	14.0^w	38.0	53.0	48.0	—	—	—	—
Laakso et al., 1992	L1	—	—	—	—	39.7^w SSM^t = 2.7 SSM^c = 41.4	44.1 SM^{c}M^t = 2.7 SM^{c}M^c = 8.7	11.4	4.8 others
	L2	—	—	—	—	35.0^w SSM^t = 2.5 SSM^c = 44.1	46.6 SM^{c}M^t = 4.4 SM^{c}M^c = 9.1	13.5	5.0 others
Centrifugation									
Jordan, 1986	28L$_S$ [W]	11.40^w	32.80	44.20	50.20	—	—	—	—
Branger, 1993	20L$_M$ [2] bench-top	9.49^a C26–C34	46.04 C36–C44	—	34.31 C46–C54	—	—	—	—
	20L$_S$ pilot-scale	10.28^a	44.10	—	34.10	—	—	—	—
	17L$_S$ pilot-scale	10.92^a	45.27	—	32.63	—	—	—	—
	12L$_M$ [3] bench-top	9.66^a	45.57	—	32.96	—	—	—	—
Kankare and Antila, 1986	12L$_M$ [S]	12.30^n	35.90	48.20	49.10	—	—	—	—
Kankare and Antila, 1988b	12LM	—	—	—	—	26^n	50	20	2.5
Keogh and Higgins, 1986b	Fraction 3	6.90^w	34.10	41.00	50.20	—	—	—	—
Separation with the aid of an aqueous detergent solution									
Banks et al., 1985	21L$_S$ A high level of detergent	7.00^w	44.10	51.10	48.90	—	—	—	—

Continued

TABLE 5.121 (Continued)

Fraction separation method Author	Fraction designation[1]	Selected triglyceride groupings (%)[2]				Composition of triglyceride classes (%)[3]			
		C24–C34	C36–C40	≤C40	C42–C54	SSS	SSU	SUU	UUU
Filtration through milk filter									
Fouad et al., 1990	17L$_S$	12.60^w	38.70	51.30	35.20	—[4]	—	—	—

[1]The number indicates fractionation temperature, the letter indicates physical state (L = liquid, S = solid), and the subscript indicates fractionation procedure used (M = multiple-step, S = single-step). Other primary descriptions and descriptions in parentheses are those of the original author. The designation inside the brackets following the fraction designation refers to the intact anhydrous milkfat source [sample number or season (S = summer, W = winter)].

[2]a = area %, n = % not specified, w = wt %.

[3]S = saturated, M = monounsaturated, U = unsaturated, c = cis, t = trans.

[4]Data not available.

TABLE 5.122 Triglyceride Composition (Selected Triglyceride Groupings) of Low-Melting Milkfat Fractions Obtained from Solvent Solution

Type of solvent solution Author	Fraction designation[1]	Selected triglyceride groupings (%)[2]			
		C24–C34	C36–C40	≤C40	C42–C54
Acetone					
Kaylegian, 1991	15L$_S$ [W]	11.10^a	33.50	44.60	55.40
	11L$_S$ [W]	12.50^a	34.40	46.90	53.30
	5L$_S$ [W]	14.10^a	35.40	49.50	50.70
Timms, 1980b	3L$_M$ [S]	12.70^w	39.60	52.30	47.70
Yi, 1993	0L$_M$ [S]	14.70^a	43.10	57.80	42.40
Lambelet, 1983	L1$_M$	10.70^n	9.30	20.00	0.00
Pentane					
Kaylegian, 1991	0L$_M$ [W]	11.00^a	32.40	43.40	56.70
	0L$_S$ [W]	11.00^a	33.20	44.20	55.90

[1]The number indicates fractionation temperature, the letter indicates physical state (L = liquid), and the subscript indicates fractionation procedure used (M = multiple-step, S = single-step). The designation inside the brackets refers to the intact anhydrous milkfat source (S = summer, W = winter).

[2]Units of chemical composition are designated using these abbreviations: a = area %, n = % not specified, w = wt %.

TABLE 5.123 Triglyceride Composition (Selected Triglyceride Groupings) of Low-Melting Milkfat Fractions Obtained by Supercritical Fluid Extraction

Type of solvent Author	Fraction designation[1]	Fractionation conditions		Selected triglyceride groupings (%)[2]			
		Temperature (°C)	Pressure	C24–C34	C36–C40	≤C40	C42–C54
Carbon dioxide							
Biernoth and Merk, 1985	Extract	80	200 bar	40.80[n]	49.60	90.40	8.70
Kaufmann et al., 1982	II (extract)	80	200 bar	46.90[m]	47.30	94.20	5.10
Timmen et al., 1984	Short-chain fraction	80	200 bar	46.70[m]	47.40	94.10	4.80
Arul et al., 1987	L1	50	100–250 bar	49.90[w]	39.50	89.40	10.60
	L2	50	100–250	34.70[w]	52.40	87.10	12.90
	I1	50	100–250	30.80[w]	54.40	85.20	14.80
Kankare et al., 1989	E1	50	100 bar	40.90[w]	44.10	85.00	12.10
Chen et al., 1992	10.3	40	10.3 MPa	43.3[n] ≤C36	—[3]	—	56.4 >C36

[1]Designations used are those given by the original author.

[2]Units of chemical composition are designated using these abbreviations: m = mole %, n = % not specified, w = wt %.

[3]Data not reported.

TABLE 5.124 Triglyceride Composition (Selected Triglyceride Groupings) of Low-Melting Milkfat Fractions Obtained by Short-Path Distillation

Author	Fraction designation[1]	Fractionation conditions		Selected triglyceride groupings (%)[2]			
		Temperature (°C)	Pressure	C24–C34	C36–C40	≤C40	C42–C54
Arul et al., 1988a	LF1 Distillate 1	245	220 μm Hg	66.60[w]	28.70	95.30	4.50
	LF2 Distillate 2	245	100	39.20[w]	45.30	84.50	16.60

[1]Designations used are those given by the original author.

[2]Units of chemical composition are designated using this abbreviation: w = wt %.

TABLE 5.125 Triglyceride Composition (Selected Triglyceride Groupings) of Very-Low-Melting Milkfat Fractions Obtained from Melted Milkfat

Fraction separation method Author	Fraction designation[1]	Composition of triglyceride classes (%)[2]			
		SSS	SSU	SUU	UUU
Vacuum filtration					
Deffense, 1993b	2nd olein	24.4[w] C24–C42 = 22.9 C44–C56 = 1.5	51.8 C24–C42 = 35.1 C44–C56 = 16.7	SUU +UUU = 23.8 C24–C42 = 6.9 C44–C56 = 16.9	

[1]Designations are that of original author.

[2]S = saturated, U = unsaturated, w = wt %.

TABLE 5.126 Triglyceride Composition (Selected Triglyceride Groupings) of Very-Low-Melting Milkfat Fractions Obtained from Solvent Solution

Type of solvent solution Author	Fraction designation[1]	Selected triglyceride groupings (%)[2]			
		C24–C34	C36–C40	≤C40	C42–C54
Acetone					
Kaylegian, 1991	0L$_M$ [W]	14.30[a]	34.80	49.10	50.80
	0L$_S$ [W]	15.30[a]	34.90	50.20	49.90

[1]The number indicates fractionation temperature, the letter indicates physical state (L = liquid), and the subscript indicates fractionation procedure used (M = multiple-step, S = single-step). The designation inside the brackets refers to the intact anhydrous milkfat source (W = winter).

[2]Units of chemical composition are designated using this abbreviation: a = area %.

TABLE 5.127 Triglyceride Composition (Selected Triglyceride Groupings) of Unknown-Melting Milkfat Fractions Obtained from Melted Milkfat

Fraction separation method Author	Fraction designation[1]	Selected triglyceride groupings (%)[2]			
		C24–C34	C36–C40	≤C40	C42–C54
Pressure filtration					
Badings et al., 1983b	23S$_M$ [W]	9.50[w]	28.70	38.20	61.70
	23L$_M$ [W]	13.50[w]	38.70	52.20	47.70
	21.5S$_M$ [S]	8.20[w]	27.40	35.60	64.40
	21.5L$_M$ [S]	10.50[w]	36.60	47.10	52.90
	18.5S$_M$ [W]	12.00[w]	39.90	51.90	48.00
	18.5L$_M$ [W]	14.70[w]	37.70	52.40	47.60
	15.3S$_M$ [S]	8.00[w]	37.80	45.80	54.20
	15.3L$_M$ [S]	11.40[w]	36.70	48.10	51.90
	15S$_M$ [W]	12.60[w]	39.80	52.40	47.50
	15L$_M$ [W]	15.40[w]	37.00	52.40	47.40
	13S$_M$ [S]	10.60[w]	36.90	47.50	52.50
	13L$_M$ [S]	12.10[w]	36.30	48.40	52.60
	11.3S$_M$ [W]	13.80[w]	37.50	51.30	48.50
	11.3L$_M$ [W]	16.30[w]	37.00	53.30	46.50
	9.3S$_M$ [S]	11.70[w]	36.50	48.20	51.80
	9.3L$_M$ [S]	12.30[w]	36.10	48.40	51.60
	7.1S$_M$ [S]	12.00[w]	36.00	48.00	52.00
	7.1L$_M$ [S]	12.60[w]	36.00	48.60	51.40
	5.5S$_M$ [W]	15.80[w]	37.10	52.90	47.00
	5.5L$_M$ [W]	16.90[w]	37.20	54.10	45.60
Separation with the aid of an aqueous detergent solution					
Banks et al., 1985	16L$_S$ [B] high level of detergent	3.00[w]	28.30	31.30	68.70
Filtration through milk filter					
Fouad et al., 1990	25L$_S$	11.40[w]	34.50	45.90	44.70
	21S$_S$	8.10[w]	28.10	36.20	54.90
	21L$_S$	10.20[w]	33.20	43.40	39.40

Continued

TABLE 5.127 (Continued)

Fraction separation method Author	Fraction designation[1]	Selected triglyceride groupings (%)[2]			
		C24–C34	C36–C40	≤C40	C42–C54
Separation technique not reported					
Lechat et al., 1975	Most solid fraction [Feb]	3.20^n	22.90	26.10	62.10
	Most liquid fraction [Sept]	11.80^n	38.60	50.40	49.20
	Most liquid fraction [Feb]	19.70^n	51.60	71.30	28.60

[1]The number indicates fractionation temperature, the letter indicates physical state (L = liquid, S = solid), and the subscript indicates fractionation procedure used (M = multiple-step, S = single-step). Other designations are those of the original author. The designation inside the brackets refers to the intact anhydrous milkfat source (S = summer, W = winter).

[2]Units of chemical composition are designated using these abbreviations: n = % not specified, w = wt %.

TABLE 5.128 Triglyceride Composition (Selected Triglyceride Groupings) of Unknown-Melting Milkfat Fractions Obtained from Solvent Solution

Type of solvent solution Author	Fraction designation[1]	Selected triglyceride groupings (%)[2]				Composition of triglyceride classes (%)[3]			
		C24–C34	C36–C40	≤C40	C42–C54	SSS	SSU	SUU	UUU
Acetone									
Parodi, 1974a	20S$_S$ [May]	0.00^m	0.60	0.60	97.60	79.2	19.6^3 SSM^t = 13.3 SSM^c = 6.3	—[4]	1.2 Residue
	SSS fraction of 20S$_S$ [May]	0.0^m	0.6	0.6	97.8	—	—	—	—
	SSM^t fraction of 20S$_S$ [May]	0.0^m	0.5	0.5	99.3	—	—	—	—
	SSM^c fraction of 20S$_S$ [May]	0.0^m	0.1	0.1	99.3	—	—	—	—
	20L$_S$ [May]	12.00^m	39.70	51.70	48.30	—	—	—	—
Chen and deMan, 1966	15S$_M$	—	—	—	—	67.2^w	—	—	—
Lansbergen and Kemps, 1984	12S$_M$ [1]	0.60^n	11.90	12.50	86.50	—	—	—	—
	12L$_M$ [1]	19.10^n	54.10	73.20	26.70	—	—	—	—
	12S$_M$ [2]	1.50^n	16.20	17.70	81.60	—	—	—	—
	12L$_M$ [2]	20.30^n	54.40	74.70	25.20	—	—	—	—
Lechat et al., 1975	10S$_M$ [Apr]	0.20^n	9.20	9.40	89.60	—	—	—	—
Lansbergen and Kemps, 1984	6S$_M$ [1]	4.80^n	59.80	64.60	35.10	—	—	—	—
	6L$_M$ [1]	44.40^n	44.10	88.50	11.10	—	—	—	—
Chen and deMan, 1966	5S$_M$	—	—	—	—	68.6^w	—	—	—
	0S$_M$ [2]	5.40^w	52.30	57.70	42.20	—	—	—	—
	0L$_M$ [2]	32.80^w	51.30	84.10	15.50	—	—	—	—
Lechat et al., 1975	-9L$_M$ [Apr]	15.10^n	46.00	61.10	38.40	—	—	—	—
Chen and deMan, 1966	-15S$_M$	—	—	—	—	43.7^w	—	—	—
	-25S$_M$	—	—	—	—	32.2^w	—	—	—
	-35S$_M$	—	—	—	—	30.6^w	—	—	—
	-45S$_M$	—	—	—	—	27.5^w	—	—	—
	-45L$_M$	—	—	—	—	27.2^w	—	—	—
Lansbergen and	Top fraction	0.60^n	3.60	4.20	94.70	—	—	—	—

Continued

TABLE 5.128 (Continued)

Type of solvent solution Author	Fraction designation[1]	Selected triglyceride groupings (%)[2]				Composition of triglyceride classes (%)[3]			
		C24–C34	C36–C40	≤C40	C42–C54	SSS	SSU	SUU	UUU
Kemps, 1984	Mid fraction	5.30[n]	33.80	39.10	60.20	—	—	—	—
	Olein fraction	18.10[n]	38.10	56.20	42.70	—	—	—	—
Hexane									
Muuse and van der Kamp, 1985	12.5S$_S$	10.20[a]	29.70	39.90	49.00	—	—	—	—
	12.5L$_S$	13.90[a]	38.90	52.80	37.60	—	—	—	—

[1]Sample description: the sample number indicates fractionation temperature, the letter indicates physical state (S = solid, L = liquid), and the subscript indicates fractionation procedure used (S = single-step, M = multiple-step). Other designations are those given by the original author. The designation inside the brackets refers to the intact anhydrous milkfat source.

[2]a = area %, m = mole %, n = % not specified, w = wt %.

[3]S = saturated, M = monounsaturated, U = unsaturated, c = *cis*, t = *trans*.

[4]Data not reported.

TABLE 5.129 Triglyceride Composition (Selected Triglyceride Groupings) of Unknown-Melting Milkfat Fractions Obtained by Supercritical Fluid Extraction

Type of solvent Author	Fraction designation[1]	Fractionation conditions		Selected triglyceride groupings (%)[2]			
		Temperature (°C)	Pressure	C24–C34	C36–C40	≤C40	C42–C54
Carbon dioxide							
Bhaskar et al., 1993a	Extract	40	24.1 MPa	18.70[a]	54.30	73.00	27.10
	Raffinate	40	24.1	0.00[a]	17.20	17.20	83.10
Ensign, 1989	WM-38, 1	60	2150 psi	37.40[a]	39.80	77.20	9.60
	WM-38, 2	60	2300	28.10[a]	46.30	74.40	14.50
	WM-38, 3	60	2400	24.20[a]	50.60	74.80	16.10
	WM-38, 4	60	2400	19.90[a]	51.90	71.80	18.60
	WM-38, 5	60	2500	17.00[a]	51.30	68.30	22.40
	WM-38, 6	60	2500	14.60[a]	45.50	60.10	25.50
	WM-38, 7	60	2650	17.40[a]	54.30	71.70	22.00
	WM-38, 8	60	2850	7.80[a]	52.40	60.20	29.80
	WM-38, 9	60	2900	4.40[a]	47.90	52.30	37.80
	WM-38, 10	60	2900	2.60[a]	41.10	43.70	46.10
	WM-38, 11	60	2900	1.10[a]	28.70	29.80	60.60
	WM-38, 12	60	2900	0.40[a]	17.00	17.40	72.40
	WM-38, 13	60	2900	0.20[a]	8.30	8.50	79.60
	WM-38, 14	60	2900	0.00[a]	2.60	2.60	84.40
	WM-38, 15	60	—[3]	0.00[a]	0.50	0.50	89.10
	WM-39, 1	60	2000	44.10[a]	35.20	79.30	5.40
	WM-39, 2	60	6000	6.70[a]	32.40	39.10	47.30
	WM-43, 1	40	1700	42.40[a]	41.10	83.50	6.90
	WM-43, 2	40	6000	6.40[a]	30.40	36.80	56.00
Propane							
Biernoth and Merk, 1985	Extract-1	125	85 bar	18.20[n]	41.10	59.30	35.70
	Extract-2	125	85	18.60[n]	42.90	61.50	33.40
	Residual	125	85	7.30[n]	30.70	38.00	56.00

[1]Designations used are those given by the original author.

[2]Units of chemical composition are designated using these abbreviations: a = area %, n = % not specified.

[3]Data not available.

TABLE 5.130 Triglyceride Composition (Selected Triglyceride Groupings) of Unknown-Melting Milkfat Fractions Obtained from Short-Path Distillation

Author	Fraction Designation[1]	Fractionation conditions		Selected triglyceride groupings (%)[2]			
		Temperature (°C)	Pressure	C24–C34	C36–C40	≤C40	C42–C54
McCarthy et al., 1962	R-1	150	15μ	82.3w	17.7	100	—[3]
	R-2	160	15	65.9w	34.0	99.9	—
	R-3	180	12	39.0w	58.6	97.6	2.4
	R-4	185	12	8.5w	80.7	89.2	10.8
	D-2	177–182	5–7	2.6w	70.0	72.6	18.1
	D-3	—	—	—	14.1w	14.1	85.9

[1]Designations used are those given by the original investigator.
[2]Units of chemical composition are designated using this abbreviation: w = wt %.
[3]Data not reported.

TABLE 5.131 Selected Chemical Characteristics of the Intact Anhydrous Milkfats Used to Produce Milkfat Fractions from Melted Milkfat

Fraction separation method Author	Fraction designation[1]	Iodine value	Refractive index	Reichert–Meissl number	Polenske number	Density (g/mL)
Vacuum filtration						
Amer et al., 1985	AMF [W]	30.26	—[2]	—	—	—
	AMF [S]	35.78	—	—	—	—
Deffense, 1987	AMF [2]	37.8	—	—	—	—
Deffense, 1989	AMF [W]	30	—	—	—	—
	AMF [S]	35	—	—	—	—
deMan and Finoro, 1980	AMF	33.0	1.4539	29.96	1.88	0.8936 (25°C)
Frede et al., 1980	AMF [S]	37	—	—	—	—
Jamotte and Guyot, 1980	Commercial liquid fraction (L₂) doubly fractionated at 15°C	40.5	—	—	—	—
Lund and Danmark, 1987	AMF [1]	34.0	—	—	—	—
	AMF [2]	36.1	—	—	—	—
Voss et al., 1971	AMF	30.05	1.4518	—	—	—
Vovan and Riel, 1973	AMF	35.6	1.4495 (40°C)	—	—	—
Pressure filtration						
Badings et al., 1983b	AMF [W]	31.1	1.4534	—	—	0.9078 (35°C)
	AMF [S]	39.0	1.4546	—	—	0.9074
Stepanenko and Tverdokhleb, 1974	AMF	34.42	1.4555 (40°C) 1.4503 (60°C)	20.02	—	—

Continued

TABLE 5.131 (Continued)

Fraction separation method Author	Fraction designation[1]	Iodine value	Refractive index	Reichert–Meissl number	Polenske number	Density (g/mL)
Centrifugation						
Antila and Antila, 1970	AMF	29.0	—	—	—	—
Kankare and Antila, 1986	AMF [S]	36.1	—	—	—	—
Kankare and Antila, 1988a	AMF [S]	36.1	—	—	—	—
Kankare and Antila, 1988b	AMF [S]	36.1	—	—	—	—
Lakshminarayana and Rama Murthy, 1985	AMF	32.2	1.4543 (40°C)	24.6	1.50	0.9624 (30°C)
Ramesh and Bindal, 1987b	Ghee	35.1	—	—	—	—
Riel and Paquet, 1972	AMF	38.6	—	—	—	—
Separation with the aid of an aqueous detergent solution						
Dolby, 1970a	AMF [midseason]	32.5	1.4534	30.5	—	—
Doležálek et al., 1976	AMF [1]	29.8	—	—	—	—
	AMF [2]	31.9	—	—	—	—
	AMF [3]	33.7	—	—	—	—
	AMF [4]	37.2	—	—	—	—
Fjaervoll, 1970a	AMF	37.5	1.4542	—	—	—
Fjaervoll, 1970b	AMF	37.5	1.4520	—	—	—
Norris et al., 1971	AMF [A late S]	39.1	1.4550 (40°C)	26.24	—	—
	AMF [B late S]	38.8	1.4550	26.07	—	—
	AMF [C early F]	38.7	1.4548	27.61	—	—
Sherbon et al., 1972	AMF [A]	31.3	—	—	—	—
	AMF [B]	34.0	—	—	—	—
Filtration through muslin cloth						
Khalifa and Mansour, 1988	AMF	35.07	1.4575 (30°C)	—	—	—
Kulkarni and Rama Murthy, 1987	AMF	32.9	—	—	—	—
Sreebhashyam et al., 1981	AMF	35.9	—	—	—	—
Separation technique not reported						
Deroanne, 1976	AMF [W]	35.8	—	—	—	—
Deroanne and Guyot, 1974	AMF [W]	31.1	41.8 (Zeiss number)	—	—	—
	AMF [S]	39.2	43.8	—	—	—
Lechat et al., 1975	AMF [Feb]	39.5	—	—	—	—
Riel and Paquet, 1972	AMF	38.6	—	—	—	—

[1]AMF = anhydrous milkfat. The designation inside the brackets following the fraction designation refers to the milkfat source [sample number, letter, or season (S = summer, W = winter).

[2]Data not available.

TABLE 5.132 Selected Chemical Charactersitics of the Intact Anhydrous Milkfats Used to Produce Milkfat Fractions from Solvent Solution

Type of solvent solution Author	Fraction designation[1]	Iodine value	Refractive index
Acetone			
Bhalerao et al., 1959	AMF	38	1.4542 (40°C)
Larsen and Samuelsson, 1979	AMF	32.2	—[2]
Walker et al., 1977	AMF [A]	32.1	—
	AMF [B]	34.8	—
	AMF [C]	36.0	—
Youssef et al., 1977	AMF [Nov]	31.7	—
Ethanol			
Bhalerao et al., 1959	AMF	38	1.4542

[1]AMF = anhydrous milkfat. The designation inside the brackets following the fraction designation refers to the milkfat source (sample letter or month).

[2]Data not available.

TABLE 5.133 Selected Chemical Characteristics of the Intact Anhydrous Milkfats Used to Produce Milkfat Fractions by Supercritical Fluid Extraction

Type of solvent Author	Fraction designation[1]	Iodine value
Carbon dioxide		
Bhaskar et al., 1993b	AMF	35.20
Biernoth and Merk, 1985	AMF	30.3
Kaufmann et al., 1982	AMF	30.3
Rizvi et al., 1990	AMF	33
Timmen et al., 1984	AMF	30.3

[1]AMF = anhydrous milkfat.

TABLE 5.134 Selected Chemical Characteristics of Very-High-Melting Milkfat Fractions Obtained from Melted Milkfat

Fraction separation method Author	Fraction designation[1]	Iodine value
Centrifugation		
Antila and Antila, 1970	D	10.5

[1]Designations are those of the original author.

TABLE 5.135 Selected Chemical Characteristics of Very-High-Melting Milkfat Fractions Obtained from Solvent Solution

Type of solvent solution Author	Fraction designation[1]	Iodine value
Acetone		
Youssef et al., 1977	$18S_M$ (A) [Nov]	22.5
	$18S_M$ (E) [Nov]	25.6

[1]The number indicates fractionation temperature, the letter indicates physical state (S = solid), and the subscript indicates fractionation procedure used (M = multiple-step). Designations in parentheses are those of the original author. The designation inside the brackets refers to the intact anhydrous milkfat source.

TABLE 5.136 Selected Chemical Characteristics of Very-High-Melting Milkfat Fractions Obtained by Supercritical Fluid Extraction

Type of solvent Author	Fraction designation[1]	Fractionation conditions		Iodine value
		Temperature (°C)	Pressure	
Carbon dioxide				
Bhaskar et al., 1993a	S1	40	24.1 MPa	42.9

[1]Designations used are those given by the original author.

TABLE 5.137 Selected Chemical Characteristics of High-Melting Milkfat Fractions Obtained from Melted Milkfat

Fraction separation method Author	Fraction designation[1]	Iodine value	Refractive index	Reichert–Meissl number	Polenske number	Density (g/mL)
Vacuum filtration						
deMan and Finoro, 1980	$32S_S$	28.9	1.4536	25.0	1.7	0.8910 (25°C)
	$31S_S$	28.3	1.4537	24.7	1.3	0.8916
	$30S_S$	26.6	1.4538	23.6	1.7	0.8906
Amer et al., 1985	$29S_S$ [W]	24.70	—[2]	—	—	—
	$29S_S$ [S]	29.18	—	—	—	—
deMan and Finoro, 1980	$29S_S$	25.9	1.4537	22.4	1.5	0.8910 (25°C)
	$28S_S$	26.5	1.4537	22.4	1.7	0.8914
deMan and Finoro, 1980	$27S_S$	27.3	1.4538	23.9	1.8	0.8919 (25°C)
Jamotte and Guyot, 1980	Commercial solid fraction A fractionated at 27°C	30.4	—	17.3	—	—
	Commercial solid fracion B fractionated at 27°C	29.6	—	18.0	—	—
Amer et al., 1985	$26S_S$ [W]	26.70	—	—	—	—
deMan and Finoro, 1980	$26S_S$	27.8	1.4538	24.2	1.8	0.8916 (25°C)
Lund and Danmark, 1987	$26S_S$ [1]	27.7	—	—	—	—
deMan and Finoro, 1980	$25S_S$	28.7	1.4539	25.7	1.8	0.8918 (25°C)
Amer et al., 1985	$23S_S$ [W]	27.78	—	—	—	—
	$19S_S$ [W]	27.94	—	—	—	—
	$19S_S$ [S]	32.80	—	—	—	—
Lund and Danmark, 1987	$18S_S$ [2]	31.3	—	—	—	—
Deffense, 1987	1st stearin$_M$ [2]	31.3	—	—	—	—
Pressure filtration						
Stepanenko and Tverdokhleb, 1974	$20S_M$	31.14	1.4550 (40°C) 1.4496 (60°C)	25.19	—	—

Continued

TABLE 5.137 (Continued)

Fraction separation method Author	Fraction designation[1]	Iodine value	Refractive index	Reichert–Meissl number	Polenske number	Density (g/mL)
Centrifugation						
Riel and Paquet, 1972	$38S_M$	27.9	—	—	—	—
Lakshminarayana and Rama Murthy, 1985	$31S_M$	30.1	1.4535 (40°C)	22.5	1.4	0.9390 (30°C)
Ramesh and Bindal, 1987b	$28S_S$ [Ghee]	31.3	—	—	—	—
Separation with the aid of an aqueous detergent solution						
Sherbon et al., 1972	$B3S_M$ (triple fractionation at 31.7°C) [B]	30.1	—	—	—	—
	$B3L_M$ (triple fractionation at 31.7°C) [B]	32.7	—	—	—	—
	$C2S_M$ (double fractionation at 30.6°C) [C]	26.8	—	—	—	—
Doležálek et al., 1976	$30S_S$ [1]	23.7	1.4534 (40°C)	—	—	—
	$30S_S$ [2]	26.4	1.4540	—	—	—
	$30S_S$ [3]	27.9	1.4540	—	—	—
	$30S_S$ [4]	30.5	1.4546	—	—	—
Sherbon et al., 1972	$A1S_S$ (single fractionation at 28°C) [A]	29.6	—	—	—	—
	$B2S_M$ (double fractionation at 28°C) [B]	30.3	—	—	—	—
	$B2L_M$ (double fractionation at 28°C) [B]	33.4	—	—	—	—
Doležálek et al., 1976	$27S_S$ [1]	23.6	1.4533 (40°C)	—	—	—
	$27S_S$ [2]	26.9	1.4540	—	—	—
	$27S_S$ [3]	28.5	1.4540	—	—	—
	$27S_S$ [4]	29.6	1.4547	—	—	—
Norris et al., 1971	$26S_S$ [B late S]	36.9	1.4548 (40°C)	24.37	—	—
	$26S_S$ [C early F]	36.2	1.4548	25.28	—	—
Sherbon et al., 1972	$B1S_M$ (single fractionation at 25.5°C) [B]	30.1	—	—	—	—
Doležálek et al., 1976	$25S_S$ [1]	29.0	1.4536 (40°C)	—	—	—
	$25S_S$ [2]	30.2	1.4540	—	—	—
	$25S_S$ [3]	31.0	1.4540	—	—	—
	$25S_S$ [4]	32.4	1.4548	—	—	—
Norris et al., 1971	$25S_S$ [A late S]	36.3	1.4548 (40°C)	25.19	—	—
Fjaervoll, 1970a	HMF	33.7	1.4544	—	—	—

Continued

TABLE 5.137 (Continued)

Fraction separation method Author	Fraction designation[1]	Iodine value	Refractive index	Reichert–Meissl number	Polenske number	Density (g/mL)
Fjaervoll, 1970b	HMF	33.7	1.4544	—	—	—
Dolby, 1970a	Solid [midseason]	28.9	1.4533	28.1	—	—
Filtration through muslin cloth						
Khalifa and Mansour, 1988	$30S_M$	32.35	1.4572 (30°C)	—	—	—
	$25S_M$	34.43	1.4574 (30°C)	—	—	—
Filtration through cheese press						
Baker, 1970a	$30S_M$	30	—	—	—	—
Filtration through casein-dewatering press						
Black, 1973	Hard #1	—	—	26.3	1.5	—
	Hard #2	—	—	25.4	1.9	—
	Hard #3	—	—	21.7	2.1	—
Separation technique not reported						
Deroanne, 1976	$21S_S$ [W]	31.7	—	—	—	—
Deroanne and Guyot, 1974	$21S_S$ [W]	30.2	41.5 (Zeiss number)	—	—	—
	$21S_S$ [S]	35.8	43.5	—	—	—

[1]The number indicates fractionation temperature, the letter indicates physical state (S = solid), and the subscript indicates fractionation procedure used (M = multiple-step, S = single-step). Other primary designations and designations in parentheses are those of the original author. The designation inside the brackets refers to the intact anhydrous milkfat source [sample number or letter, month or season (F = fall, S = summer, W = winter)].

[2]Data not available.

TABLE 5.138 Selected Chemical Characteristics of High-Melting Milkfat Fractions Obtained from Solvent Solution

Type of solvent solution Author	Fraction designation[1]	Iodine value
Acetone		
Rolland and Riel, 1966	$18S_M$ (1)	14
Avvakumov, 1974	$10S_M$	26.5
Ethanol		
Rolland and Riel, 1966	$18S_M$	28

[1]The number indicates fractionation temperature, the letter indicates physical state (S = solid), and the subscript indicates fractionation procedure used (M = multiple-step). Designations in parentheses are those of the original author.

TABLE 5.139 Selected Chemical Characteristics of High-Melting Milkfat Fractions Obtained by Supercritical Fluid Extraction

Type of solvent Author	Fraction designation[1]	Fractionation conditions		Iodine value
		Temperature (°C)	Pressure	
Carbon dioxide				
Biernoth and Merk, 1985	Residual	80	200	33.6
Kaufmann et al., 1982	I (residue)	80	200 bar	33.6
Timmen et al., 1984	Long-chain fraction	80	200 bar	33.6
Bhaskar et al., 1993a	S2	60	24.1 MPa	39.7
Rizvi et al., 1990	Extract-2	48.0	275.8 bar	33
	Residue	48	275.8	37

[1]Designations used are those given by the original author.

TABLE 5.140 Selected Chemical Characteristics of Middle-Melting Milkfat Fractions Obtained from Melted Milkfat

Fraction separation method Author	Fraction designation[1]	Iodine value	Refractive index	Reichert–Meissl number	Polenske number	Density (g/mL)
Vacuum filtration						
deMan and Finoro, 1980	$32L_S$	33.5	1.4540	30.3	1.9	0.8936 (25°C)
	$31L_S$	33.8	1.4539	31.1	1.9	0.8941
	$30L_S$	34.0	1.4540	31.5	1.9	0.8940
Amer et al., 1985	$29L_S$ [W]	31.44	—[2]	—	—	—
	$29L_S$ [S]	36.13	—	—	—	—
deMan and Finoro, 1980	$29L_S$	34.4	1.4540	31.4	1.9	0.8935 (25°C)
	$28L_S$	34.5	1.4541	31.7	2.0	0.8936
	$27L_S$	34.7	1.4541	31.7	2.0	0.8942 (25°C)
Amer et al., 1985	$26L_S$ [W]	31.95	—	—	—	—
Pressure filtration						
Stepanenko and Tverdokhleb, 1974	$11S_M$	34.07	1.4560 (40°C) 1.4516 (60°C)	28.27	—	—
Centrifugation						
Lakshminarayana and Rama Murthy, 1985	$23S_M$	31.2	1.4540 (40°C)	23.6	1.45	0.9523 (30°C)
Antila and Antila, 1970	C	18.7	—	—	—	—
Separation with the aid of an aqueous detergent solution						
Sherbon et al., 1972	$C2L_M$ (double fractionation at 30.6°C) [C]	33.9	—	—	—	—

Continued

TABLE 5.140 (Continued)

Fraction separation method Author	Fraction designation[1]	Iodine value	Refractive index	Reichert–Meissl number	Polenske number	Density (g/mL)
Doležálek et al., 1976	30L$_S$ [1]	31.8	1.4538 (40°C)	—	—	—
	30L$_S$ [2]	33.1	1.4543	—	—	—
	30L$_S$ [4]	38.5	1.4550	—	—	—
Sherbon et al., 1972	A1L$_S$ (single fractionation at 28°C) [A]	33.3	—	—	—	—
Doležálek et al., 1976	27L$_S$ [1]	31.7	1.4536 (40°C)	—	—	—
Sherbon et al., 1972	B1L$_M$ (single fractionation at 25.5°C) [B]	36.6	—	—	—	—
Doležálek et al., 1976	25L$_S$ [1]	33.5	1.4540 (40°C)	—	—	—
	25L$_S$ [2]	33.5	1.4543	—	—	—
Filtration through muslin cloth						
Sreebhashyam et al., 1981	25S$_M$	34.6	—	—	—	—
Khalifa and Mansour, 1988	20S$_M$	36.12	1.4569 (30°C)	—	—	—
Sreebhashyam et al., 1981	18S$_M$	35.9	—	—	—	—
Filtration through cheese press						
Baker, 1970a	13S$_M$	34	—	—	—	—
Separation technique not reported						
Riel and Paquet, 1972	38S$_M$	32.7	—	—	—	—
Deroanne, 1976	21L$_S$ [W]	38.9	—	—	—	—
Riel and Paquet, 1972	4S$_M$	37.0	—	—	—	—

[1]The number indicates fractionation temperature, the letter indicates physical state (L = liquid, S = solid), and the subscript indicates fractionation procedure used (M = multiple-step, S = single-step). Other primary designations and designations in parentheses are those of the original author. The designation inside the brackets refers to the intact anhydrous milkfat source [sample number, letter, or season (S = summer, W = winter).

[2]Data not available.

TABLE 5.141 Selected Chemical Characteristics of Middle-Melting Milkfat Fractions Obtained from Solvent Solution

Type of solvent solution Author	Fraction designation[1]	Iodine value
Acetone		
Rolland and Riel, 1966	40L$_M$ (2)	14
	10S$_M$ (1)	23
	2S$_M$ (1)	32
Avvakumov, 1974	0S$_M$	34.4
Youssef et al., 1977	0S$_M$ (C) [Nov]	30.8

[1]The number indicates fractionation temperature, the letter indicates physical state (L = liquid, S = solid), and the subscript indicates fractionation procedure used (M = multiple-step). Designations in parentheses are those of the original author. The designation inside the brackets refers to the intact anhydrous milkfat source.

TABLE 5.142 Selected Chemical Characteristics of Middle-Melting Milkfat Fractions Obtained by Supercritical Fluid Extraction

Type of solvent Author	Fraction designation[1]	Fractionation conditions		Iodine value
		Temperature (°C)	Pressure	
Carbon dioxide				
Bhaskar et al., 1993a	S3	75	17.2 MPa	32.05
	S4	60	6.9	28.72
	S5	60	3.5	28.36
Rizvi et al., 1990	Extract-1	40.0	241.3 bar	29

[1]Designations are those given by the original author.

TABLE 5.143 Selected Chemical Characteristics of Low-Melting Milkfat Fractions Obtained from Melted Milkfat

Fraction separation method Author	Fraction designation[1]	Iodine value	Refractive index	Reichert–Meissl number	Polenske number	Density (g/mL)
Vacuum filtration						
deMan and Finoro, 1980	26L$_S$	34.9	1.4541	32.8	2.1	0.8941 (25°C)
Lund and Danmark, 1987	26L$_S$ [1]	35.7	—[2]	—	—	—
deMan and Finoro, 1980	25L$_S$	34.9	1.4541	33.5	1.9	0.8937 (25°C)
Amer et al., 1985	23L$_S$ [W]	32.70	—	—	—	—
	19L$_S$ [W]	33.07	—	—	—	—
	19L$_S$ [S]	37.67	—	—	—	—
Lund and Danmark, 1987	18L$_S$ [2]	39.5	—	—	—	—
Deffense, 1987	2nd olein$_M$ filtered at 10°C [2]	47.8	—	—	—	—
	1st olein$_M$ [2]	42.5	—	—	—	—
Vovan and Riel, 1973	5% removal of hard fraction (>21°C)	36.3	1.4496 (40°C)	—	—	—
	10% removal of hard fraction (>21°C)	36.0	1.4497	—	—	—
	15% removal of hard fraction (>21°C)	36.5	1.4498	—	—	—
	5% removal of intermediate fraction (10–21°C)	36.3	1.4495	—	—	—
	10% removal of intermediate fraction (10–21°C)	36.8	1.4497	—	—	—
	15% removal of intermed. fraction (10–21°C)	36.5	1.4498	—	—	—

Continued

TABLE 5.143 (Continued)

Fraction separation method Author	Fraction designation[1]	Iodine value	Refractive index	Reichert–Meissl number	Polenske number	Density (g/mL)
	33% removal of intermed. fraction (10–21°C)	37.2	1.450	—	—	—
Pressure filtration						
Stepanenko and Tverdokhleb, 1974	11L$_M$	44.31	1.4555 (40°C) 1.4503 (60°C)	25.85	—	—
Centrifugation						
Ramesh and Bindal, 1987b	28L$_S$	38.4	—	—	—	—
Lakshminarayana and Rama Murthy, 1985	15S$_M$	33.4	1.4546 (40°C)	25.0	1.60	0.9638 (30°C)
	15L$_M$	35.2	1.4550	26.4	1.65	0.9746
Riel and Paquet, 1972	4S$_M$	39.0	—	—	—	—
Antila and Antila, 1970	A	42.0	—	—	—	—
	B	32.2	—	—	—	—
Separation with the aid of an aqueous detergent solution						
Doležálek et al., 1976	30L$_S$ [3]	36.2	1.4545 (40°C)	—	—	—
	27L$_S$ [2]	34.1	1.4542	—	—	—
	27L$_S$ [3]	36.0	1.4545	—	—	—
	27L$_S$ [4]	38.8	1.4551	—	—	—
Norris et al., 1971	26L$_S$ [B late S]	41.9	1.4550 (40°C)	28.66	—	—
	26L$_S$ [C early F]	40.7	1.4548	30.03	—	—
Doležálek et al., 1976	25L$_S$ [3]	35.4	1.4545 (40°C)	—	—	—
	25L$_S$ [4]	39.2	1.4550	—	—	—
Norris et al., 1971	25L$_S$ [A late S]	41.6	1.4550 (40°C)	28.93	—	—
Dolby, 1970a	liquid [midseason]	34.3	1.4534	33.3	—	—
Fjaervoll, 1970a	LMF	39.4	1.4540	—	—	—
Fjaervoll, 1970b	LMF	39.4	1.4540	—	—	—
Filtration through muslin cloth						
Khalifa and Mansour, 1988	20L$_M$	38.92	1.4568 (30°C)	—	—	—
Sreebhashyam et al., 1981	18L$_M$	38.6	—	—	—	—
Filtration through cheese press						
Baker, 1970a	13L$_M$	46	—	—	—	—

Continued

TABLE 5.143 (Continued)

Fraction separation method Author	Fraction designation[1]	Iodine value	Refractive index	Reichert–Meissl number	Polenske number	Density (g/mL)
Filtration through casein-dewatering press						
Black, 1973	Soft #1	—	—	34.8	2.6	—
	Soft #2	—	—	32.7	2.0	—
	Soft #3	—	—	34.7	2.4	—
Separation technique not reported						
Deroanne and Guyot, 1974	$21L_S$ [W]	33.6	42.1 (Zeiss number)	—	—	—
	$21L_S$ [S]	42.5	44.1	—	—	—
Riel and Paquet, 1972	$4L_M$	42.9	—	—	—	—

[1]The number indicates fractionation temperature, the letter indicates physical state (L = liquid, S = solid), and the subscript indicates fractionation procedure used (M = multiple-step, S = single-step). Other designations are those of the original authors. The designation inside the brackets following the fraction designation refers to the intact anhydrous milkfat source [sample number, letter, or season (F = fall, S = summer, W = winter)].

[2]Data not available.

TABLE 5.144 Selected Chemical Characteristics of Low-Melting Milkfat Fractions Obtained from Solvent Solution

Type of solvent solution Author	Fraction designation[1]	Iodine value
Acetone		
Rolland and Riel, 1966	$32L_M$ (2)	22
	$24L_M$ (2)	30
Youssef et al., 1977	$18L_M$ (B) [Nov]	33.2
	$18L_M$ (F) [Nov]	31.2
Rolland and Riel, 1966	$16L_M$ (2)	32
	$-6S_M$ (1)	27
Avvakumov, 1974	$-10S_M$	35.2
Rolland and Riel, 1966	$-15S_M$ (1)	34
Ethanol		
Rolland and Riel, 1966	$10S_M$	30
	$2S_M$	32
	$-6S_M$	33

[1]The number indicates fractionation temperature, the letter indicates physical state (L = liquid, S = solid), and the subscript indicates fractionation procedure used (M = multiple-step). Designations in parentheses are those of the original author. The designation inside the brackets refers to the intact anhydrous milkfat source.

TABLE 5.145 Selected Chemical Characteristics of Low-Melting Milkfat Fractions Obtained by Supercritical Fluid Extraction

Type of solvent Author	Fraction designation[1]	Fractionation conditions		Iodine value
		Temperature (°C)	Pressure	
Carbon dioxide				
Biernoth and Merk, 1985	Extract	80	200 bar	18.6
Kaufmann et al., 1982	II (extract)	80	200 bar	19.6
Timmen et al., 1984	Short-chain fraction	80	200 bar	19.6

[1]Designations used are those given by the original author.

TABLE 5.146 Selected Chemical Characteristics of Very-Low-Melting Milkfat Fractions Obtained from Melted Milkfat

Fraction separation method Author	Fraction designation[1]	Iodine value	Refractive index	Reichert–Meissl number
Vacuum filtration				
Deffense, 1987	2nd olein$_M$ filtered at 6°C [2]	49.3	—[2]	—
Jamotte and Guyot, 1980	L4	—	40.0	33.3
Centrifugation				
Riel and Paquet, 1972	4L$_M$	43.0	—	—

[1]The number indicates fractionation temperature, the letter indicates physical state (L = liquid), and the subscript indicates fractionation procedure used (M = multiple-step). Other descriptions used are those of the original author. The designation inside the brackets refers to the intact anhydrous milkfat source.

[2]Data not available.

TABLE 5.147 Selected Chemical Characteristics of Very-Low-Melting Milkfat Fractions Obtained from Solvent Solution

Type of solvent solution Author	Fraction designation[1]	Iodine value
Acetone		
Rolland and Riel, 1966	8L$_M$ (2)	41
	0L$_M$ (2)	43
Youssef et al., 1977	0L$_M$ (D) [Nov]	36.0
Rolland and Riel, 1966	-15L$_M$ (1)	48
Ethanol		
Rolland and Riel, 1966	-15S$_M$	39
	-15L$_M$	53

[1]The number indicates fractionation temperature, the letter indicates physical state (L = liquid, S = solid), and the subscript indicates fractionation procedure used (M = multiple-step). Designations in parentheses are those of the original author. The designation inside the brackets refers to the intact anhydrous milkfat source.

TABLE 5.148 Selected Chemical Characteristics of Unknown-Melting Milkfat Fractions Obtained from Melted Milkfat

Fraction separation method Author	Fraction designation[1]	Iodine value	Refractive index	Density (g/mL)
Vacuum filtration				
Voss et al., 1971	$30S_S$	29.0	—[2]	—
	$30L_S$	30.4	—	—
Frede et al., 1980	$28S_S$ [S]	34.0	—	—
	$28L_S$ [S]	39.8	—	—
Pressure filtration				
Badings et al., 1983b	$23S_M$ [W]	26.7	1.4531	0.9060 (35°C)
	$23L_M$ [W]	33.9	1.4533	0.9089
Kuwabara et al., 1991	$23S_S$	26.9	—	—
	$23L_S$	34.8	—	—
Badings et al., 1983b	$21.5S_M$ [S]	34.0	1.4544	0.9052
	$21.5L_M$ [S]	41.5	1.4546	0.9084
Kuwabara et al., 1991	$20S_S$	28.7	—	—
	$20L_S$	35.9	—	—
Badings et al., 1983b	$18.5S_M$ [W]	29.3	1.4529	0.9082
	$18.5L_M$ [W]	36.3	1.4536	0.9094
	$15.3S_M$ [S]	32.4	1.4539	0.9067
	$15.3L_M$ [S]	42.9	1.4550	0.9087
	$15S_M$ [W]	30.0	1.4530	0.9084
	$15L_M$ [W]	38.5	1.4538	0.9098
	$13S_M$ [S]	39.6	1.4545	0.9082
	$13L_M$ [S]	46.1	1.4553	0.9092
	$11.3S_M$ [W]	33.8	1.4535	0.9089
	$11.3L_M$ [W]	41.5	1.4543	0.9103
	$9.3S_M$ [S]	42.6	1.4549	0.9089
	$9.3L_M$ [S]	47.6	1.4557	0.9094
	$7.1S_M$ [S]	45.2	1.4552	0.9089
	$7.1L_M$ [S]	49.1	1.4558	0.9092
	$5.5S_M$ [W]	39.4	1.4538	0.9097
	$5.5L_M$ [W]	45.9	1.4546	0.9111
Separation technique not reported				
Lechat et al., 1975	Most solid fraction [Feb]	30.5	—	—
	Most liquid fraction [Feb]	52	—	—

[1]The number indicates fractionation temperature, the letter indicates physical state (L = liquid, S = solid), and the subscript indicates fractionation procedure used (M = multiple-step, S = single-step). The designation inside the brackets refers to the intact anhydrous milkfat source (S = summer, W = winter).

[2]Data not available.

TABLE 5.149 Selected Chemical Characteristics of Unknown-Melting Milkfat Fractions Obtained from Solvent Solution

Type of solvent solution Author	Fraction designation[1]	Iodine value	Refractive index
Acetone			
Walker et al., 1977	$14S_M$ [A]	15.9	—[2]
	$14S_M$ [B]	17.6	—
	$14S_M$ [C1]	19.2	—
	$14S_M$ [C2]	17.8	—
	$2S_M$ [A]	24.4	—
Bhalerao et al., 1959	$0S_S$	25	1.4539 (40°C)
	$0L_S$	53	1.4551
Walker et al., 1977	$-13S_M$ [A]	25.4	—
	$-13L_M$ [A]	48.2	—
Ethanol			
Bhalerao et al., 1959	$20S_S$	30	1.4544
	$20L_S$	40	1.4538

[1] The number indicates fractionation temperature, the letter indicates physical state (L = liquid, S = solid), and the subscript indicates fractionation procedure used (M = multiple-step, S = single-step). The designation inside the brackets refers to the intact anhydrous milkfat source.

[2] Data not available.

TABLE 5.150 Lactone Concentration of Anhydrous Milkfats Used to Produce Milkfat Fractions

Fractionation method Author	Fraction designation[1]	Lactone concentration									
		δ-C10		γ-C12		δ-C12		δ-C14		δ-C16		Total	
		Free	Potential	Free	Potential	Free	Potential	Free	Potential	Free	Potential	Free	Potential

Crystallization from melted milkfat

Separation with the aid of an aqueous detergent solution

Walker, 1972	Fresh unheated milkfat (F1)	—[2]	15.8 ppm	—	2.8	—	24.8	—	23.7	—	—	—	67.1 ppm
	Commercial AMF from cream (F2)	—	14.5	—	2.7	—	23.7	—	23.1	—	—	—	64.0
	AMF from butter (F3)	—	13.6	—	2.4	—	22.8	—	21.0	—	—	—	59.8

Crystallization from solvent solution

Acetone

Walker et al., 1977	AMF [A]	4.3 µg/g	11.5	—	0.6	8.7	24.6	9.6	29.1	4.8	16.8	27.4 µg/g	82.6
	AMF [B]	—	12.1	—	1.2	—	26.9	—	33.9	—	17.3	—	91.4
	AMF [C1]	—	14.1	—	3.3	—	32.6	—	48.4	—	26.6	—	125.1
	AMF [C3]	12.1	14.9	2.1	3.6	26.6	34.5	33.7	47.4	18.4	28.8	92.9	129.2

Supercritical fluid extraction

Carbon dioxide

Kankare et al., 1989	AMF	22.1 µg/g	—	—	—	35.6	—	42.3	—	—	—	100.0 µg/g	—
Rizvi et al., 1993	AMF	6.7ppm	—	—	—	10.1	—	18.7	—	18.9	—	54.4ppm	—

Fractionation method not reported

Walker, 1974	AMF	—	—	—	—	—	—	—	—	—	—	—	59–64 mg/kg

[1]AMF = anhydrous milkfat. Designations in parentheses are those of the original author. The designation inside the brackets refers to the intact anhydrous milkfat source.

[2]Data not available.

Technical Data for Experimental and Commercial Milkfat Fractions 423

Lactone Concentration of Very-High-Melting Milkfat Fractions

Fractionation method	Fraction designation[1]	Lactone concentration												
		δ-C10		γ-C12		δ-C12		δ-C14		δ-C16		Total	
Author		Free	Potential	Free	Potential	Free	Potential	Free	Potential	Free	Potential	Free	Potential
Crystallization from melted milkfat													
Separation with the aid of an aqueous detergent solution													
Walker, 1972	26.5S$_S$ (F3S)	—[2]	8.6 ppm	—	1.3	—	14.9	—	11.1	—	—	—	35.9 ppm

[1]The number indicates fractionation temperature, the letter indicates physical state (S = solid), and the subscript indicates fractionation procedure used (S = single-step). Designations in parentheses are those of the original author.

[2]Data not available.

TABLE 5.152 Lactone Concentration of High-Melting Milkfat Fractions

Fractionation method	Fraction designation[1]	Lactone concentration												
		δ-C10		γ-C12		δ-C12		δ-C14		δ-C16		Total	
Author		Free	Potential	Free	Potential	Free	Potential	Free	Potential	Free	Potential	Free	Potential
Crystallization from melted milkfat													
Separation with the aid of an aqueous detergent solution													
Walker, 1972	26.5S$_S$ (F2S)	—[2]	9.2 ppm	—	1.5	—	14.6	—	12.7	—	—	—	38.0 ppm
	26.5S$_S$ (F2S[1])	—	8.5 ppm	—	1.5	—	13.8	—	13.4	—	—	—	37.2
Fractionation method not reported													
Walker, 1974	HMF	—	—	—	—	—	—	—	—	—	—	—	36–38 mg/kg

[1]The number indicates fractionation temperature, the letter indicates physical state (S = solid), and the subscript indicates fractionation procedure used (S = single-step). Designations in parentheses are those of the original author. The designation inside the brackets refers to the intact anhydrous milkfat source.

[2]Data not available.

TABLE 5.153 Lactone Concentration of Middle-Melting Milkfat Fractions

Fractionation method Author	Fraction designation[1]	Lactone concentration											
		δ-C10		γ-C12		δ-C12		δ-C14		δ-C16		Total	
		Free	Potential	Free	Potential	Free	Potential	Free	Potential	Free	Potential	Free	Potential
Crystallization from melted milkfat													
Separation with the aid of an aqueous detergent solution													
Walker, 1972	26.5L$_S$ (F2L)	—[2]	16.3 ppm	—	3.1	—	26.6	—	21.8	—	—	—	67.8 ppm
	26.5L$_S$ (F2L[1])	—	17.0 ppm	—	2.9	—	27.1	—	25.6	—	—	—	72.6
Fractionation method not reported													
Walker, 1974	LMF	—	—	—	—	—	—	—	—	—	—	—	67–72 mg/kg

[1]The number indicates fractionation temperature, the letter indicates physical state (L = liquid), and the subscript indicates fractionation procedure used (S = single-step). Other primary designations and designations in parentheses are those of the original author.

[2]Data not available.

TABLE 5.154 Lactone Concentration of Low-Melting Milkfat Fractions

Fractionation method Author	Fraction designation[1]	Lactone concentration											
		δ-C10		γ-C12		δ-C12		δ-C14		δ-C16		Total	
		Free	Potential	Free	Potential	Free	Potential	Free	Potential	Free	Potential	Free	Potential
Crystallization from melted milkfat													
Separation with the aid of an aqueous detergent solution													
Walker, 1972	26.5L$_S$ (F3L)	—[2]	15.1 ppm	—	3.3	—	25.0	—	23.6	—	—	—	37.0 ppm

[1]The number indicates fractionation temperature, the letter indicates physical state (L = liquid), and the subscript indicates fractionation procedure used (S = single-step). Designations in parentheses are those of the original author.

[2]Data not available.

TABLE 5.155 Lactone Concentrations of Unknown-Melting Milkfat Fractions

Fractionation method Author	Fraction designation[1]	δ-C10 Free	δ-C10 Potential	γ-C12 Free	γ-C12 Potential	δ-C12 Free	δ-C12 Potential	δ-C14 Free	δ-C14 Potential	δ-C16 Free	δ-C16 Potential	Total Free	Total Potential
Crystallization from solvent solution													
Acetone													
Walker et al., 1977	14S$_M$ (HMG) [A]	0.9 μg/g	2.2	—[2]	—	2.6	7.5	3.5	10.9	1.8	7.9	8.8 μg/g	28.5
	2S$_M$ (IMG^1) [A]	—	6.5	—	—	—	13.4	—	16.8	—	10.8	—	47.5
	-13S$_M$ (IMG^2) [A]	—	7.1	—	—	—	17.5	—	20.1	—	12.6	—	57.3
	-13L$_M$ (LMG) [A]	8.9	19.0	0.5	1.4	21.3	40.8	26.5	48.8	12.1	24.6	69.3	134.6
	14S$_M$ (HMG) [B]	—	3.8	—	—	—	9.7	—	13.7	—	8.5	—	35.7
	2S$_M$ (IMG^1) [B]	—	5.5	—	0.5	—	13.8	—	21.2	—	10.2	—	51.2
	-13S$_M$ (IMG^2) [B]	—	6.6	—	0.6	—	16.1	—	22.9	—	11.5	—	57.7
	-13L$_M$ (LMG) [B]	—	16.7	—	1.9	—	44.4	—	51.5	—	27.9	—	142.4
	14S$_M$ (HMG) [C1]	—	4.1	—	0.3	—	9.2	—	22.1	—	11.4	—	48.9
	-13L$_M$ (LMG) [C1]	—	26.0	—	4.6	—	61.6	—	77.5	—	42.3	—	212.0
	14S$_M$ (HMG) [C3]	1.8	1.9	—	—	4.1	5.7	9.2	16.5	3.3	10.0	18.4	34.1
	-13L$_M$ (LMG) [C3]	23.3	32.6	4.0	6.1	55.9	72.7	67.7	91.5	37.6	51.1	188.5	254.0
Supercritical fluid extraction													
Carbon dioxide													
Kankare et al., 1989	2E1	155.6 μg/g	—	—	—	335.3	—	376.2	—	—	—	867.1 μg/g	—
	2E2	78.6	—	—	—	151.1	—	188.8	—	—	—	418.5	—

Continued

TABLE 5.155 (Continued)

Fractionation method Author	Fraction Designation[1]	Lactone concentration												
		δ-C10		γ-C12		δ-C12		δ-C14		δ-C16		Total	
		Free	Potential	Free	Potential	Free	Potential	Free	Potential	Free	Potential	Free	Potential
Rizvi et al., 1993	2R	5.6	—	—	—	10.3	—	15.9	—	—	—	31.8	—
	S1	0.0 ppm	—	—	—	4.0	—	6.0	—	6.5	—	16.5	—
	S2	2.5	—	—	—	4.2	—	7.4	—	7.2	—	21.3	—
	S3	29.1	—	—	—	35.6	—	50.6	—	34.9	—	150.2	—
	S4	55.7	—	—	—	53.1	—	73.4	—	56.5	—	238.7	—
	Raffinate	0.0	—	—	—	1.6	—	1.6	—	0.0	—	3.2	—
	Trap	205.9	—	—	—	154.3	—	15	—	107.6	—	620.7	—

[1]The number indicates fractionation temperature, the letter indicates physical state (L = liquid, S = solid), and the subscript indicates fractionation procedure used (M = multiple-step). Other primary designations and designations in parentheses are those of the original author. The designation inside the brackets refers to the intact anhydrous milkfat source.

[2]Data not available.

TABLE 5.156 Methyl Ketone Concentration of Anhydrous Milkfats Used to Produce Milkfat Fractions

Fractionation method Author	Fraction designation[1]	\multicolumn{8}{c}{Methyl ketone concentration}								
		C3	C4	C5	C7	C9	C11	C13	C15	Total
Crystallization from melted milkfat										
Separation with the aid of an aqueous detergent solution										
Walker, 1972	Fresh unheated milkfat (F1)	1.2 (μmoles/ 10 g fat)	0.2	1.2	2.4	0.8	0.7	1.2	1.9	9.6 (μmoles/ 10 g fat)
	Commercial AMF from cream (F2)	0.8	0.1	1.0	2.3	0.8	0.7	1.1	1.8	8.6
	AMF from butter (F3)	—[2]	—	—	—	—	—	—	—	8.0
Fractionation method not reported										
Walker, 1974	AMF	—	—	—	—	—	—	—	—	115–125 mg/kg

[1] AMF = anhydrous milkfat. Designations in parentheses are those of the original author.

[2] Data not available.

TABLE 5.157 Methyl Ketone Concentration of Very-High-Melting Milkfat Fractions

Fractionation method Author	Fraction designation[1]	Methyl ketone concentration								
		C3	C4	C5	C7	C9	C11	C13	C15	Total
Crystallization from melted milkfat										
Separation with the aid of an aqueous detergent solution										
Walker, 1972	26.5S$_S$ (F3S)	—[2]	—	—	—	—	—	—	—	5.2 (μmoles/ 10 g fat)

[1] The number indicates fractionation temperature, the letter indicates physical state (S = solid), and the subscript indicates fractionation procedure used (S = single-step). Designations in parentheses are those of the original author.

[2] Data not available.

TABLE 5.158 Methyl Ketone Concentration of High-Melting Milkfat Fractions

Fractionation method Author	Fraction designation[1]	Methyl ketone concentration							
		C3	C4	C5	C7	C9	C11	C13	C15	Total
Crystallization from melted milkfat										
Separation with the aid of an aqueous detergent solution										
Walker, 1972	26.5S$_S$ (F2S)	0.2 (μmoles/ 10 g fat)	—	0.5	1.6	0.6	0.5	0.8	2.6	5.8 (μmoles/ 10 g fat)
	26.5S$_S$ (F2S^1)	—	—	—	—	—	—	—	—	6.3
Fractionation method not specified										
Walker, 1974	HMF	—	—	—	—	—	—	—	—	83–92 mg/kg

[1]Fraction designation: the number indicates fractionation temperature, the letter indicates physical state (S = solid), and the subscript indicates fractionation procedure used (S = single-step). Other designations are those of the original author.

TABLE 5.159 Methyl Ketone Concentration of Middle-Melting Milkfat Fractions

Fractionation method Author	Fraction designation[1]	Methyl ketone concentration							
		C3	C4	C5	C7	C9	C11	C13	C15	Total
Crystallization from melted milkfat										
Separation with the aid of an aqueous detergent solution										
Walker, 1972	26.5L$_S$ (F2L)	0.8 (μmoles/ 10 g fat)	0.1	0.9	2.2	0.9	0.7	1.2	2.2	9.0 (μmoles/ 10 g fat)
	26.5L$_S$ (F2L^1)	—[2]	—	—	—	—	—	—	—	9.2
Fractionation method not reported										
Walker, 1974	LMF	—	—	—	—	—	—	—	—	126–138 mg/kg

[1]The number indicates fractionation temperature, the letter indicates physical state (L = liquid), and the subscript indicates fractionation procedure used (S = single-step). Other primary designations and designations in parentheses are those of the original author.

[2]Data not available.

TABLE 5.160 Methyl Ketone Concentration of Low-Melting Milkfat Fractions

Fractionation method Author	Fraction designation[1]	Total methyl ketone concentration
Crystallization from melted milkfat		
Separation with the aid of an aqueous detergent solution		
Walker, 1972	26.5L$_S$ (F3L)	8.4 (μmoles/10 g fat)

[1]The number indicates fractionation temperature, the letter indicates physical state (L = liquid), and the subscript indicates fractionation procedure used (S = single-step). Designations in parentheses are those of the original author.

TABLE 5.161 Aldehyde Concentration of Anhydrous Milkfats Used to Produce Milkfat Fractions

Fractionation method Author	Fraction designation[1]	Total aldehyde concentration
Crystallization from melted milkfat		
Separation with the aid of an aqueous detergent solution		
Walker, 1972	Fresh unheated milkfat (F1)	3.7 (μ moles/10 g fat)
	Commercial AMF from cream (F2)	4.4
	AMF from butter (F3)	6.3

[1] AMF = anhydrous milkfat. Designations in parentheses are those of the original author.

TABLE 5.162 Aldehyde Concentration of Very-High-Melting Milkfat Fractions

Fractionation method Author	Fraction designation[1]	Total aldehyde concentration
Crystallization from melted milkfat		
Separation with the aid of an aqueous detergent solution		
Walker, 1972	$26.5S_S$ (F3S)	7.4 (μmoles/10 g fat)

[1] The number indicates fractionation temperature, the letter indicates physical state (S = solid), and the subscript indicates fractionation procedure used (S = single-step). The designation in parentheses is that of the original author.

TABLE 5.163 Aldehyde Concentration of High-Melting Milkfat Fractions

Fractionation method Author	Fraction designation[1]	Total aldehyde concentration
Crystallization from melted milkfat		
Separation with the aid of an aqueous detergent solution		
Walker, 1972	$26.5S_S$ (F2S) $26.5S_S$ (F2S[1])	5.7 (μmoles/10 g fat) 5.2

[1] The number indicates fractionation temperature, the letter indicates physical state (S = solid), and the subscript indicates fractionation procedure used (S = single-step). Designations in parentheses are those of the original author.

TABLE 5.164 Aldehyde Concentration of Middle-Melting Milkfat Fractions

Fractionation method Author	Fraction designation[1]	Total aldehyde concentration
Crystallization from melted milkfat		
Separation with the aid of an aqueous detergent solution		
Walker, 1972	$26.5L_S$ (F2L) $26.5L_S$ (F2L[1])	7.9 (μ moles/10 g fat) 7.3

[1] The number indicates fractionation temperature, the letter indicates physical state (L = liquid), and the subscript indicates fractionation procedure used (S = single-step). Designations in parentheses are those of the original author.

TABLE 5.165 Aldehyde Concentration of Low-Melting Milkfat Fractions

Fractionation method Author	Fraction designation[1]	Total aldehyde concentration
Crystallization from melted milkfat		
Separation with the aid of an aqueous detergent solution		
Walker, 1972	26.5L$_S$ (F3L)	9.2 (μmoles/10 g fat)

[1]The number indicates fractionation temperature, the letter indicates physical state (L = liquid), and the subscript indicates fractionation procedure used (S = single-step). The designation in parentheses is that of the original author.

TABLE 5.166 Crystal Morphology for Anhydrous Milkfats Used to Produce Milkfat Fractions

Fractionation method Author	Fraction designation[1]	Sample treatment Short spacings (Å)	Polymorph	Sample treatment Short spacings (Å)	Polymorph	Sample treatment Short spacings (Å)	Polymorph
Crystallization from melted milkfat							
Pressure filtration							
Helal et al., 1977		*Rapid cooling at 3–4°C*		*Rapid cooling at 3–4°C and holding for 1 night*		*Rapid cooling at 3–4°C and holding for 10 nights*	
	AMF [S]	4.12	α	4.25, 3.8	α, β'	3.8, 4.2, 4.6	α, β', β
	AMF [W]	4.12	α	4.25, 3.8	α, β'	3.85, 4.25	α, β'
Crystallization from solvent solution							
Acetone							
Timms, 1980a		*16–24 h at 3–5°C*					
	AMF	3.81, 4.19, 4.6	β', β (-2, -3)				
Woodrow and deMan, 1968		*X-ray diffraction*		*Infrared absorption peaks (cm^{-1})*			
	AMF slowly cooled	3.78, 4.17, 4.64	α, β', β	718, 728			
	AMF fast cooled	3.83, 4.11, 4.57	α, β', β	718, 727			
Supercritical fluid extraction							
Carbon dioxide							
Hamman et al., 1991		*Long spacing values at 10°C[2]*		*Long spacing values at 15°C*			
	AMF [A]	42	β'	42	β'	42	β'
		Long spacing values at 20°C		*Long spacing values at 25°C*		*Long spacing values at 30°C*	

Continued

TABLE 5.166 (Continued)

Fractionation method Author	Fraction designation[1]	Sample treatment		Sample treatment		Sample treatment	
		Short spacings (Å)	Polymorph	Short spacings (Å)	Polymorph	Short spacings (Å)	Polymorph
	AMF [A]	42	β'	42	β'	42	—[3]
		After storage at 7°C 3 wk		*After storage at 7°C 6 wk*			
	AMF [A]	—	β'	—	β' + (β)		

[1] AMF = anhydrous milkfat. Designations in brackets refer to the intact milkfat source (S = Summer, W = winter).

[2] Short spacings not available.

[3] Data not available.

TABLE 5.167 Crystal Morphology of Very-High-Melting Milkfat Fractions

Fractionation method Author	Fraction designation[1]	Sample treatment		Sample treatment		Sample treatment	
		Short spacings (Å)	Polymorph	Short spacings (Å)	Polymorph	Short spacings (Å)	Polymorph
Crystallization from melted milkfat							
Separation with the aid of an aqueous detergent solution							
Schaap et al., 1975	28S$_S$	3.8, 4.12, 4.29, 4.43	α, β'				
Crystallization from solvent solution							
Acetone							
Schaap et al., 1975	28S$_M$	3.8, 4.14, 4.28, 4.60	α, β', β				
Timms, 1980a		*16–24 h at 3–5°C*		*16–24 h at 3–5°C + 5 d at 25°C + 7 d at 40°C*			
	HMF	4.10	α-2	3.79, 4.10, 4.25	β'-2		
Supercritical fluid extraction							
Carbon dioxide							
Hamman et al., 1991		*Long spacing values at 10°C*[2]		*Long spacing values at 15°C*			
	B$_{res}$	42	β'	42	β'	42	β'
		Long spacing values at 20°C		*Long spacing values at 25°C*		*Long spacing values at 30°C*	
	B$_{res}$	42	β'	42	β'	42	β'
		Long spacing values at 35°C					
	B$_{res}$	42	—[3]				
		After storage at 7°C 3 wk		*After storage at 7°C 6 wk*			
	B$_{res}$	—	β' + (β)	—	β' + (β)		

[1] The number indicates fractionation temperature, the letter indicates physical state (L = liquid, S = solid), and the subscript indicates fractionation procedure used (M = multiple-step, S = single-step). Other designations are those of the original author.

[2] Short spacings not available.

[3] Data not available.

Milkfat Fractionation

TABLE 5.168 Crystal Morphology of High-Melting Milkfat Fractions

Fractionation method Author	Fraction designation[1]	Sample treatment Short spacings (Å)	Polymorph	Sample treatment Short spacings (Å)	Polymorph
Crystallization from melted milkfat					
Separation technique not reported					
Deroanne, 1976	21S$_S$ [W]	3.8, 4.23	α, β'		
Crystallization from solvent solution					
Acetone					
Timms, 1980a		*16–24 h at 3–5°C*		*16–24 h at 3–5°C + 5 d at 25°C + 7 d at 30°C*	
	MMF	3.80, 4.12	α-2, β'-2	3.81, 4.12, 4.2	β'-2, β'-3

[1]The number indicates fractionation temperature, the letter indicates physical state (L = liquid, S = solid), and the subscript indicates fractionation procedure used (M = multiple-step, S = single-step). Other designations are those of the original author. Designations in brackets refer to the intact milkfat source (W = winter).

TABLE 5.169 Crystal Morphology of Middle-Melting Milkfat Fractions

Fractionation method Author	Fraction designation[1]	Sample treatment Short spacings (Å)	Polymorph	Sample treatment Short spacings (Å)	Polymorph
Crystallization from melted milkfat					
Separation technique not reported					
Deroanne, 1976	21L$_S$ [W]	3.83, 4.13	α, β'		
Supercritical fluid extraction					
Carbon dioxide					
Hamman et al., 1991		*Long spacing values at 10°C*[2]		*Long spacing values at 15°C*	
	A4 [A]	41	β'	—[3]	—
		After storage at 7°C 3 wk		*After storage at 7°C 6 wk*	
	A4	—	β'	—	β'

[1]The number indicates fractionation temperature, the letter indicates physical state (L = liquid), and the subscript indicates fractionation procedure used (S = single-step). Other designations are those of the original author. Designations in brackets refer to the intact milkfat source (W = winter).

[2]Short spacings not available.

[3]Data not available.

TABLE 5.170 Crystal Morphology of Low-Melting Milkfat Fractions

Fractionation method Author	Fraction designation[1]	Sample treatment Short spacings (Å)	Polymorph	Sample treatment Short spacings (Å)	Polymorph
Supercritical fluid extraction					
Carbon dioxide					
Hamman et al., 1991		Long spacing values at 10°C[2]		Long spacing values at 15°C	
	A1 [A]	40	β'	—[3]	—
		After storage at 7°C 3 wk		After storage at 7°C 6 wk	
	A1 [A]	—[3]	β'	—	β'

[1]Fraction designations are those of the original author. Designations in brackets refer to the intact milkfat source.

[2]Short spacings not available.

[3]Data not available.

TABLE 5.171 Crystal Morphology of Unknown-Melting Milkfat Fractions

Fractionation method Author	Fraction designation[1]	Sample treatment Short spacings (Å)	Polymorph	Sample treatment Short spacings (Å)	Polymorph	Sample treatment Short spacings (Å)	Polymorph
Crystallization from melted milkfat							
Pressure filtration							
Helal et al., 1977		Rapid cooling at 3–4°C		Rapid cooling at 3–4°C and holding for 1 night		Rapid cooling at 3–4°C and holding for 10 nights	
	25S$_M$ [S]	4.12	α	4.25, 3.8, 4.6	α, β', β	3.8, 4.2, 4.6	α, β', β
	15S$_M$ [S]	4.12	α	4.25, 3.8	α, β'	3.8, 4.2, 4.6	α, β', β
	15L$_M$ [S]	4.12	α	4.2, 3.8	α, β'	3.8, 4.2, 4.6	α, β', β
	25S$_M$ [W]	4.12	α	3.8, 4.25, 4.6	α, β', β	3.8, 4.2, 4.6	α, β', β
	15S$_M$ [W]	4.12	α	3.8, 4.25	α, β'	3.85, 4.25	α, β'
	15L$_M$ [W]	4.12	α	3.8, 4.2	α, β'	3.8, 4.25	α, β'
Helal et al., 1984		Sudden cooling to 3–4°C and then holding 1 night		Sudden cooling to 3–4°C and holding 10 nights		Sudden cooling to 3–4°C and holding 30 nights	
	15S$_M$ [S]	3.8, 4.2, 4.6	α, β', β	3.8, 4.2, 4.6	α, β', β	3.8, 4.25, 4.6	α, β', β
	15S$_M$ [W]	3.8, 4.2	α, β'	3.8, 4.25	α, β'	3.8, 4.25, 4.6	α, β', β
		Sudden cooling to 3–4°C for 3 h and then raising to 20–22°C and holding 1 night		Sudden cooling to 3–4°C for 3 h and then raising to 20–22°C and holding 10 nights		Sudden cooling to 3–4°C for 3 h and then raising to 20–22°C and holding 30 nights	
	15S$_M$ [S]	4.3, 4.6	β', β	3.8, 4.3, 4.6	α, β', β	3.8, 4.6	α, β

Continued

TABLE 5.171 (Continued)

Fractionation method Author	Fraction designation[1]	Sample treatment		Sample treatment		Sample treatment	
		Short spacings (Å)	Polymorph	Short spacings (Å)	Polymorph	Short spacings (Å)	Polymorph
Crystallization from solvent solution							
Acetone							
Woodrow and deMan, 1968		X-ray diffraction		Infrared absorption peaks (cm^{-1})			
	$15S_S$ slowly cooled	3.84, 4.13, 4.23	α, β'	718, 727			
	$15S_S$ fast cooled	3.85, 4.11	α, β'	720, 727			
	$15S_S$ crystallized from acetone	3.60, 3.75, 3.88, 4.61	β', β	717			

[1]The number indicates fractionation temperature, the letter indicates physical state (L = liquid, S = solid), and the subscript indicates fractionation procedure used (M = multiple-step, S = single-step). Designations in brackets refer to the intact milkfat source (S = Summer, W = winter).

TABLE 5.172 Crystal Size of High-Melting Milkfat Fractions

Fractionation method Author	Fraction designation[1]	Crystal size
Crystallization from melted milkfat		
Separation with the aid of an aqueous detergent solution		
Doležálek et al., 1976	$30S_S$ [1]	64–320 μm
	$27S_S$ [1]	64–128
	$25S_S$ [1]	32–112
	$30S_S$ [2]	64–356
	$27S_S$ [2]	112–208
	$25S_S$ [2]	96–208
	$30S_S$ [3]	54–360
	$27S_S$ [3]	80–200
	$25S_S$ [3]	80–128
	$30S_S$ [4]	80–480
	$27S_S$ [4]	80–192
	$25S_S$ [4]	80—128
Filtration through casein-dewatering press		
Black, 1973	IT = 33.5°C[2]	179 μm
	IT = 29.5°C	119
	FT = 21.0°C	150
	FT = 18.5°C	146
	t = 21 h	162
	t = 16 h	134

[1]The number indicates fractionation temperature, the letter indicates physical state (S = solid), and the subscript indicates fractionation procedure used (S = single-step). Other designations are those of the original author. Designations in brackets refer to the intact milkfat source.

[2]IT = initial temperature, FT = final temperature, t = time.

TABLE 5.173 Crystal Size of Unknown-Melting Milkfat Fractions

Fractionation method Author	Fraction designation[1]	Crystal size
Crystallization from melted milkfat		
Vacuum filtration		
Black, 1975b	$25S_S$ with precooling	571 mm
	$25S_S$ no precooling	546
	$25S_S$ exponential cooling	812
	$25S_S$ programmed cooling	305
	$25S_S$ 8 h cooling time	553
	$25S_S$ 20 h cooling time	564
	$25S_S$ 20 rpm agitator speed	456
	$25S_S$ 10 rpm agitator speed	661
Centrifugation		
Kankare and Antila, 1974a	$24S_M$	0.1–1.0 mm
	$18S_M$	0.05–0.2
	$12S_M$	0.02–0.05

[1]The number indicates fractionation temperature, the letter indicates physical state (S = solid), and the subscript indicates fractionation procedure used (M = multiple-step, S = single-step).

TABLE 5.174 Solid Fat Content of Intact Anhydrous Milkfats Used to Produce Milkfat Fractions from Melted Milkfat

Fraction separation method Author	Fraction designation[1]	Solid fat content (%) at temperature (°C)[2]								
		0	5	10	15	20	25	30	35	40
Vacuum filtration										
Amer et al., 1985	AMF [W]	84.4	75.2	60.7	48.1	35.4	24.2	12.1	2.7	1.9
	AMF [S]	81.8	71.1	57.2	43.4	32.6	21.4	10.8	2.0	0.1
Deffense, 1987	AMF [1]	—[3]	69.7	51.3	38.0	22.1	12.2	5.7	0.8	0.0
	AMF [2]	—	51.2	42.0	38.3	16.5	8.1	4.0	0.0	—
deMan and Finoro, 1980	AMF	—	—	45.4	33.7	18.2	11.1	—	—	1.3
Kaylegian, 1991	AMF [W]	57.1	50.3	41.1	26.8	16.7	6.5	2.8	1.7	0.0
Kaylegian and Lindsay, 1992	AMF [W]	57.1	50.3	41.1	26.8	16.7	6.5	2.8	1.7	0.0
Lund and Danmark, 1987	AMF [1]	—	56.8	49.5	33.8	18.0	9.5	4.9	0.8	0.0
	AMF [2]	—	56.8	49.5	33.8	18.0	9.5	4.9	0.8	0.0
Makhlouf et al., 1987	AMF	—	44.0	40.6	29.9	15.9	8.9	4.8	1.6	0.0
Timms and Parekh, 1980	AMF [S]	—	—	44.9	31.1	17.5	12.0	7.5	2.5	0.0
Versteeg, 1991	AMF [Sept]	65.0	62.0	55.0	41.5	24.5	13.5	7.0	3.0	0.0
Vovan and Riel, 1973	AMF	—	47.0[4] SFI	44.3	35.9	21.6	13.0	7.7	3.2	1.1
Centrifugation										
Branger, 1993	AMF [1] bench-top	57.4	52.8	48.4	33.3	20.0	12.3	12.3	4.6	0.0
	AMF [2] bench-top	56.5	45.6	40.0	30.0	16.5	11.0	9.0	3.0	0.0
	AMF [3] bench-top	61.9	60.0	50.9	41.4	21.4	14.2	10.0	2.8	0.0
	AMF pilot-scale	57.2	50.0	42.5	30.0	15.0	8.8	3.1	0.0	
Kankare and Antila, 1988b	AMF [S]	84.1 SFI	—	64.8	—	33.0	—	11.8	—	0.0
Separation with the aid of an aqueous detergent solution										
Norris et al., 1971	AMF [A late S]	79.8	69.3	58.6	43.8	30.4	19.2	9.0	0.0	—
	AMF [B late S]	80.2	70.1	58.6	43.2	30.5	19.6	9.3	0.0	—
	AMF [C early F]	80.3	70.8	59.7	45.1	31.1	19.7	9.0	0.0	—
Filtration through milk filter										
Fouad et al., 1990	AMF	44.0 SFI	35.5	25.9	17.7	11.9	3.8	0.1	0.0	—
Separation technique not reported										
El-Ghandour and El-Nimr, 1982	AMF	—	—	34.9 SFI	—	19.4	—	4.6	—	0.0
Martine, 1982	AMF	55.3	51.9	45.9	32.6	15.2	9.9	5.9	0.0	—

[1] AMF = anhydrous milkfat. The designation inside the brackets following the fraction designation refers to the intact anhydrous milkfat source [sample number or season (W = winter, S = summer, F = fall)].

[2] SFI = solid fat index.

[3] Data not available.

[4] Data estimated from graphs provided by the original investigator.

TABLE 5.175 Solid Fat Content of Intact Anhydrous Milkfats Used to Produce Milkfat Fractions from Solvent Solution

Type of solvent solution Author	Fraction designation[1]	Solid fat content (%) at temperature (°C)[2]								
		0	5	10	15	20	25	30	35	40
Acetone										
Kaylegian, 1991	AMF [W]	57.1	50.3	41.1	26.8	16.7	6.5	2.8	1.7	0.0
Kaylegian and Lindsay, 1992	AMF [W]	57.1	50.3	41.1	26.8	16.7	6.5	2.8	1.7	0.0
Lambelet, 1983	AMF	47.9[2]	43.7	37.8	27.7	15.2	10.7	7.2	0.0	—[3]
		89.7[4]	78.1	60.1	34.5	22.5	18.1	10.1	0.5	—
		88.4[5]	75.4	55.3	26.5	13.1	11.5	6.4	0.3	—
Timms, 1980b	AMF [S]	56.8	42.0	30.2	19.5	13.1	8.2	5.6	3.5	0.0
Yi, 1993	AMF [S]	55.1	47.8	38.2	29.5	16.7	7.0	3.7	0.7	0.0
Pentane										
Kaylegian, 1991	AMF [W]	57.1	50.3	41.1	26.8	16.7	6.5	2.8	1.7	0.0

[1] AMF = anhydrous milkfat. The designation inside the brackets following the fraction designation refers to the intact anhydrous milkfat source (S = summer, W = winter).

[2] SFC determined by NMR method.

[3] Data not available.

[4] SFC determined by DSC method.

[5] SFC determined by DSC method, values corrected.

TABLE 5.176 Solid Fat Content of Intact Anhydrous Milkfats Used to Produce Milkfat Fractions by Supercritical Fluid Extraction

Type of solvent Author	Fraction designation[1]	Solid fat content (%) at temperature (°C)								
		0	5	10	15	20	25	30	35	40
Carbon dioxide										
Bhaskar et al., 1993b	AMF	—[2]	64.8 280K	—	47.0 290K	—	20.0 300K	—	0 310K	—
Biernoth and Merk, 1985	AMF	—	—	52.4	37.1	18.1	10.5	4.4	0.7	—
Kaufmann et al., 1982	AMF	—	58.4	49.1	34.1	16.8	7.9	3.0	0.0	—
Timmen et al., 1984	AMF	—	58.4	49.1	34.1	16.8	7.9	3.0	0.0	—

[1] AMF = anhydrous milkfat.

[2] Data not available.

TABLE 5.177 Solid Fat Content Profiles of Intact Anhydrous Milkfat Used to Produce Milkfat Fractions by Short-Path Distillation

Author	Fraction designation[1]	Solid fat content (%) at temperature (°C)							
		0	5	10	15	20	25	30	35	40
Arul et al., 1988a	AMF	—[2]	42.1	38.0	28.0	14.6	9.5	5.1	1.2	0.0

[1]AMF = anhydrous milkfat.

[2]Data not available.

TABLE 5.178 Solid Fat Content of Very-High-Melting Milkfat Fractions Obtained from Melted Milkfat

Fraction separation method Author	Fraction designation[1]	Solid fat content (%) at temperature (°C)[2]										
		0	5	10	15	20	25	30	35	40	45	50	55
Vacuum Filtration													
Kaylegian, 1991	$34S_M$ [W]	68.7	64.6	58.5	51.0	43.2	37.8	36.1	29.8	24.1	12.5	1.2	0.0
	$34S_S$ [W]	69.3	65.1	60.8	55.9	45.9	40.2	39.3	33.8	30.6	20.7	0.0	0.0
Kaylegian and Lindsay, 1992	$34S_M$ [W]	68.7	64.6	58.5	51.0	43.2	37.8	36.1	29.8	24.1	12.5	1.2	0.0
Makhlouf et al., 1987	$26S_M$ (S.1)	—[3]	58.1	57.0	51.3	41.6	35.2	29.0	21.0	10.8	0.0	—	—
	$21S_M$ (S.2)	—	50.2	48.3	41.7	30.0	23.1	16.8	8.5	1.5	0.0	—	—
	F.S (>21)	—	51.7	50.8	42.9	32.6	24.5	18.8	12.2	3.1	0.0	—	—
Centrifugation													
Kankare and Antila, 1988b	$24S_M$ [S]	93.5 SFI	—	83.4	—	71.9	—	65.0	—	31.4	—	—	—
Separation technique not reported													
El Ghandour and El-Nimr, 1982	$25S_M$	—	—	79.9 SFI	—	65.8	—	44.8	—	22.7	—	4.3	—

[1]The number indicates fractionation temperature, the letter indicates physical state (S = solid), and the subscript indicates fractionation procedure used (M = multiple-step, S = single-step). Designations in parentheses are those of the original authors. The designation inside the brackets following the fraction designation refers to the intact anhydrous milkfat source (S = summer, W = winter).

[2]SFI = solid fat index.

[3]Data not available.

TABLE 5.179 Solid Fat Content of Very-High-Melting Milkfat Fractions Obtained from Solvent Solution

Type of solvent solution Author	Fraction designation[1]	Solid fat content (%) at temperature (°C)											
		0	5	10	15	20	25	30	35	40	45	50	55
Acetone													
Kaylegian, 1991	25S_M [W]	86.2	85.1	84.0	82.8	81.5	81.0	79.0	76.0	70.8	64.9	51.3	14.5
	25S_S [W]	88.3	85.6	85.3	82.9	81.5	80.9	79.2	75.7	70.4	63.3	47.1	9.5
Yi, 1993	25S_M [S]	82.4	78.3	77.2	76.2	74.6	74.7	71.5	64.4	59.7	43.0	24.5	0.0
Kaylegian, 1991	21S_M [W]	84.7	82.7	80.7	77.4	74.5	74.6	72.0	65.3	57.2	44.2	11.0	0.0
	21S_S [W]	88.4	87.6	86.2	84.4	82.5	81.4	80.0	76.0	69.3	56.9	28.4	0.0
Yi, 1993	20S_M [S]	72.9	67.4	64.3	61.1	57.8	60.9	54.2	48.9	46.4	30.4	13.9	0.0
Timms, 1980b	20S_M [S]	99.7	99.5	—[2]	99.0	—	98.6	—	—	93.0	78.9	38.1	0.0
Kaylegian, 1991	15S_M [W]	82.8	82.2	79.8	76.7	70.6	67.3	65.5	57.8	47.3	21.7	0.0	—
	15S_S [W]	81.8	81.5	78.5	73.7	68.2	64.4	64.0	56.1	45.4	29.8	0.0	—
Kaylegian and Lindsay, 1992	15S_M [W]	82.8	82.8	79.8	76.7	70.6	67.3	65.5	57.8	47.3	21.7	0.0	—
Yi, 1993	15S_M [S]	77.0	77.4	73.8	70.1	64.3	61.3	59.6	49.8	36.7	14.0	0.0	—
Kaylegian, 1991	11S_S [W]	85.7	83.2	80.1	75.9	68.4	65.4	61.4	56.1	44.4	25.7	0.0	—
Kaylegian, 1991	11S_S [W]	85.7	83.2	80.1	75.9	68.4	65.4	61.4	56.1	44.4	25.7	0.0	—
Pentane													
Kaylegian, 1991	11S_M [W]	88.9	87.5	86.0	84.0	81.8	80.8	79.1	75.6	67.4	56.7	29.7	0.0
	11S_S [W]	82.1	80.0	77.1	74.0	71.0	69.6	67.3	62.6	53.2	44.2	21.9	0.0
	5S_M [W]	88.6	87.9	86.0	82.7	77.9	75.6	73.6	67.6	53.8	36.4	8.7	0.0
	5S_S [W]	83.7	82.7	79.5	75.3	72.0	70.3	68.1	61.9	51.5	35.8	4.2	0.0
	0S_M [W]	65.1	65.0	59.4	51.0	37.0	35.1	33.6	29.7	19.3	8.9	0.0	—
	0S_S [W]	84.8	84.2	81.4	76.5	72.6	70.1	67.9	61.4	45.8	26.7	0.0	—

[1]The number indicates fractionation temperature, the letter indicates physical state (S = solid), and the subscript indicates fractionation procedure used (M = multiple-step, S = single-step). The designation inside the brackets following the fraction designation refers to the intact anhydrous milkfat source (S = summer, W = winter).

[2]Data not available.

TABLE 5.180 Solid Fat Content of Very-High-Melting Milkfat Fractions Obtained by Supercritical Fluid Extraction

Type of solvent Author	Fraction designation[1]	Fractionation conditions		Solid fat content (%) at temperature (°C)								
		Temperature (°C)	Pressure	0	5	10	15	20	25	30	35	40	45
Carbon dioxide													
Bhaskar et al., 1993b	S1	40	24.1 MPa	—[2]	80.7 280K	—	73.0 290K	—	54.1 300K	—	20.0 310K	—	—

[1]Designations are those used by the original authors.

[2]Data not available.

TABLE 5.181 Solid Fat Content of High-Melting Milkfat Fractions Obtained from Melted Milkfat

Fraction separation method Author	Fraction designation[1]	Solid fat content (%) at temperature (°C)[2]									
		0	5	10	15	20	25	30	35	40	45
Vacuum filtration											
deMan and Finoro, 1980	32S$_S$	—[3]	—	60.2	48.5	36.2	29.1	—	—	7.4	—
	31S$_S$	—	—	61.9	51.3	38.2	31.7	—	—	9.5	—
	30S$_S$	—	—	65.9	55.0	43.6	36.0	—	—	10.5	—
Kaylegian, 1991	30S$_M$ [W]	67.0	62.4	55.2	46.6	36.7	28.2	25.7	17.3	11.0	1.2
	30S$_S$ [W]	67.7	64.9	59.9	54.6	42.0	34.8	32.4	26.2	20.5	0.0
Kaylegian and Lindsay, 1992	30S$_M$ [W]	67.0	62.4	55.2	46.6	36.7	28.2	25.7	17.3	11.0	1.2
	30S$_S$ [W]	67.7	64.9	59.9	54.6	42.0	34.8	32.4	26.2	20.5	0.0
Bumbalough, 1989	29.4S$_S$ [1]	—	—	47.4 SFI	—	36.5 (21.1°C)	34.9 (26.7°C)	25.6 (33.3°C)	—	12.0	—
Amer et al., 1985	29S$_S$ [W]	91.0	84.8	76.2	68.6	63.4	62.9	53.7	38.4	16.4	—
	29S$_S$ [S]	90.9	84.7	76.5	70.0	67.9	67.4	37.8	43.0	21.3	—
deMan and Finoro, 1980	29S$_S$	—	—	67.3	57.5	47.5	39.5	—	—	13.9	—
	28S$_S$	—	—	67.6	56.9	46.8	39.4	—	—	13.4	—
Versteeg, 1991	28S$_S$ [Sept]	78.5	77.0	73.5	67.0	61.0	51.0	37.0	25.5	8.0	3.0
deMan and Finoro, 1980	27S$_S$	—	—	62.5	50.1	39.4	31.4	—	—	5.3	—
Amer et al., 1985	26S$_S$ [W]	89.3	82.5	72.3	63.8	55.8	52.6	41.1	25.4	7.5	—
deMan and Finoro, 1980	26S$_S$	—	—	61.8	50.8	38.6	30.5	—	—	6.5	—
Lund and Danmark, 1987	26S$_S$ [1]	—	73.4	68.5	57.2	45.8	36.9	26.5	12.5	4.8	—
deMan and Finoro, 1980	25S$_S$	—	—	59.8	47.9	35.2	27.4	—	—	5.0	—
Kaylegian, 1991	25S$_M$ [W]	71.0	66.8	60.0	51.0	39.4	30.2	24.6	14.4	6.4	0.0
	25S$_S$ [W]	64.3	60.0	53.7	48.3	33.2	24.9	20.9	14.2	7.7	0.0
Kaylegian and Lindsay, 1992	25S$_M$ [W]	71.0	66.8	60.0	51.0	39.4	30.2	24.6	14.4	6.4	0.0
Timms and Parekh, 1980	25S$_S$ [S]	—	—	52.8	41.8	31.4	26.9	20.0	11.0	2.0	—
Verhagen and Bodor, 1984	25S$_S$	—	—	65	—	40	—	—	12	—	—
Amer et al., 1985	23S$_S$ [W]	87.9	80.7	70.1	60.7	51.7	45.2	32.6	16.3	4.0	—
Kaylegian, 1991	20S$_M$ [W]	76.9	75.0	69.4	60.1	46.0	29.8	23.4	10.6	0.2	0.0
	20S$_S$ [W]	64.2	60.7	54.3	46.5	30.8	21.3	17.6	10.8	0.8	0.0
Kaylegian and Lindsay, 1992	20S$_M$ [W]	76.9	75.0	69.4	30.1	46.0	29.8	23.4	10.6	0.2	0.0
Amer et al., 1985	19S$_S$ [W]	87.5	80.3	69.5	59.7	51.0	42.7	29.1	12.8	2.8	—
	19S$_S$ [S]	85.8	77.8	67.0	57.0	51.0	43.1	29.9	14.3	3.2	—
Lund and Danmark, 1987	18S$_S$ [2]	—	69.0	64.5	52.8	38.4	28.1	18.3	8.5	1.3	—
Deffense, 1987	1st Stearin$_M$ [2]	—	67.0	63.0	54.1	41.3	30.1	20.0	11.0	3.8	—
	Stearin #1$_S$ [1]	—	74.0	69.2	62.0	52.0	42.3	31.0	20.0	8.7	—
	Stearin #2$_S$ [1]	—	72.3	68.1	60.5	50.0	39.2	28.1	18.0	6.0	—
	Stearin #3$_S$ [1]	—	72.7	67.7	59.5	48.3	36.9	25.0	14.5	3.1	—
	Stearin #4$_S$ [1]	—	72.1	67.0	58.5	46.5	36.0	24.2	13.8	2.9	—
Centrifugation											
Branger, 1993	32S$_M$ [1] bench-top	65.6	60.9	54.6	44.6	33.7	23.7	24.6	16.5	12.0	—

Continued

TABLE 5.181 (Continued)

Fraction separation method Author	Fraction designation[1]	Solid fat content (%) at temperature (°C)[2]								
		0	5	10	15	20	25	30	35	40	45
	$32S_M$ [2] bench-top	56.5	52.0	44.2	35.6	24.6	19.0	16.5	9.0	6.5	—
	$32S_M$ [3] bench-top	67.4	64.6	57.4	50.0	35.1	28.4	22.8	16.0	10.0	—
	$32S_S$ pilot-scale	67.8	61.9	57.0	47.0	36.0	31.6	27.0	20.0	12.5	—
	$25S_S$ pilot-scale	64.4	57.5	50.0	39.4	23.8	19.4	12.8	6.6	0.0	—
	$22S_M$ [1] bench-top	62.8	60.0	54.6	41.0	26.0	17.0	17.0	4.6	0.0	—
	$22S_M$ [3] bench-top	67.0	63.7	55.1	44.6	26.0	17.4	10.9	4.6	0.0	—
	$20S_M$ [2] bench-top	61.4	55.6	40.0	37.4	21.9	5.6	4.6	0.0	0.0	
Kankare and Antila, 1988b	$12S_M$ [S] SFI	82.4	—	68.7	—	33.4	—	8.1	—	0.0	—
Separation with the aid of an aqueous detergent solution											
Norris et al., 1971	$26S_S$ [B late S]	84.5	75.9	65.8	54.1	46.0	36.7	24.2	10.3	—	—
	$26S_S$ [C early F]	84.0	75.8	65.7	55.3	48.2	39.3	26.6	12.6	—	—
	$25S_S$ [A late S]	84.7	76.1	64.5	54.9	46.5	36.1	23.3	9.4	—	—
Filtration through milk filter											
Fouad et al., 1990	$29S_S$ SFI	60.2	57.0	58.7	57.8	49.0	37.8	25.0	10.0	1.0	—
	$25S_S$	57.2	52.5	52.8	48.5	39.2	27.5	13.8	2.5	0.0	—
Separation technique not reported											
El-Ghandour and El-Nimr, 1982	$15S_M$	—	—	57.5	—	44.8	—	23.5	—	5.6	—
Martine, 1982	1st stearin	70.3	69.0	65.2	54.5	42.0	35.5	27.0	18.8	9.3	0.0
	2nd stearin	59.5	57.1	52.9	39.6	26.7	19.9	14.5	8.9	2.8	0.0

[1]The number indicates fractionation temperature, the letter indicates physical state (S = solid), and the subscript indicates fractionation procedure used (M = multiple-step, S = single-step). Other descriptions used are those of the original author. The designation inside the brackets refers to the intact anhydrous milkfat source [sample number or season (F= fall, S = summer, W = winter)].

[2]SFI = solid fat index.

[3]Data not available.

TABLE 5.182 Solid Fat Content of High-Melting Milkfat Fractions Obtained from Solvent Solution

Type of solvent solution Author	Fraction designation[1]	Solid fat content (%) at temperature (°C)									
		0	5	10	15	20	25	30	35	40	45
Acetone											
Kaylegian, 1991	$11S_M$ [W]	81.3	80.9	77.8	72.6	59.1	48.3	43.6	31.4	17.2	0.0
Yi, 1993	$10S_M$ [S]	73.0	69.1	64.4	59.1	48.9	36.9	32.0	16.6	0.0	—[2]
Kaylegian, 1991	$5S_S$ [W]	87.8	85.1	81.5	75.9	62.8	51.3	40.4	29.8	18.8	0.0
Timms, 1980b	$3S_S$ [S]	92.1	90.5	88.6	86.2	80.2	51.2	26.4	16.4	0.0	—
Kaylegian, 1991	$0S_S$ [W]	84.0	81.1	76.2	68.7	50.9	33.4	23.3	10.0	4.1	0.0
Lambelet, 1983	$S1_M$	88.2^3	—	85.1	—	77.2	67.9	26.6	6.7	0.0	—
		98.9^4	—	94.9	—	87.8	76.6	38.6	6.6	1.6	—

[1]The number indicates fractionation temperature, the letter indicates physical state (S = solid), and the subscript indicates fractionation procedure used (M = multiple-step, S = single-step). Other designations used are those of the original author. The designation inside the brackets refers to the intact anhydrous milkfat source (S = summer, W = winter).

[2]Data not available.

[3]SFC determined using NMR method.

[4]SFC determined using DSC method.

TABLE 5.183 Solid Fat Content of High-Melting Milkfat Fractions Obtained by Supercritical Fluid Extraction

Type of solvent Author	Fraction designation[1]	Fractionation conditions		Solid fat content (%) at temperature (°C)									
		Temperature (°C)	Pressure	0	5	10	15	20	25	30	35	40	45
Carbon dioxide													
Biernoth and Merk, 1985	Residue	80	200 bar	—[2]	—	61.1	47.5	26.3	16.5	8.0	1.7	—	—
Kaufmann et al., 1982	I (residue)	80	200 bar	—	64.4	57.1	43.4	23.8	14.0	5.9	0.0	—	—
Timmen et al., 1984	Long chain fraction	80	200 bar	—	64.4	57.1	43.4	23.8	14.0	5.9	0.0	—	—
Bhaskar et al., 1993b	S2	60	24.1 MPa	—	83.0 280K	—	56.0 290K	—	29.0 300K	—	2.6 310K	—	—
Shukla et al., 1994	Raffinate	40	24.1 MPa	—	76.8 280K	—	71.6 290K	—	55.6 300K	—	18.0 310K	—	—

[1]Designations are those used by the original authors.

[2]Data not available.

TABLE 5.184 Solid Fat Content of High-Melting Milkfat Fractions Obtained by Short-Path Distillation

Author	Fraction designation[1]	Fractionation conditions		Solid fat content (%) at temperature (°C)									
		Temperature (°C)	Pressure	0	5	10	15	20	25	30	35	40	45
Arul et al., 1988a	SF Residue	—[2]	—	—	49.5	46.9	37.8	27.6	20.7	14.2	7.3	1.5	0.0

[1]Designations are those given by the original author.

[2]Data not available.

TABLE 5.185 Solid Fat Content of Middle-Melting Milkfat Fractions Obtained from Melted Milkfat

Fraction separation method Author	Fraction designation[1]	Solid fat content (%) at temperature (°C)								
		0	5	10	15	20	25	30	35	40
Vacuum filtration										
Kaylegian, 1991	34L_S [W]	59.0	53.7	47.0	41.6	23.5	14.9	12.6	0.0	—[2]
deMan and Finoro, 1980	32L_S	—	—	44.7	31.4	15.1	8.3	—	—	1.4
	31L_S	—	—	41.5	27.9	11.4	4.3	—	—	1.7
	30L_S	—	—	39.6	27.1	10.9	4.4	—	—	1.2
Kaylegian, 1991	30L_S [W]	48.6	47.8	40.8	33.2	15.4	4.7	0.0	—	—
Kaylegian and Lindsay, 1992	30L_S [W]	48.6	47.8	40.8	33.2	15.4	4.7	0.0	—	—
Amer et al., 1985	29L_S [W]	76.7	67.7	52.9	33.0	19.3	10.7	2.3	1.3	1.2
	29L_S [S]	75.3	65.7	52.4	32.9	20.0	10.9	2.6	0.4	0.1
deMan and Finoro, 1980	29L_S	—	—	36.5	25.0	10.3	3.7	—	—	0.5
	28L_S	—	—	38.3	24.2	11.2	3.4	—	—	0.6
Versteeg, 1991	28L_S [Sept]	69.5	54.5	47.0	30.0	14.0	6.5	1.0	0.0	—
deMan and Finoro, 1980	27L_S	—	—	35.6	22.9	8.9	2.5	—	—	0.7
Amer et al., 1985	26L_S [W]	75.4	65.8	50.7	29.0	14.7	5.3	1.6	0.9	0.6
Verhagen and Warnaar, 1984	25L_S	—	—	36	23	6	1.8	0.0	—	—
Kaylegian, 1991	16S_M [W]	70.8	66.6	59.0	49.6	35.1	4.2	0.0	—	—
Kaylegian and Lindsay, 1992	16S_M [W]	70.8	66.6	59.0	49.6	35.1	4.2	0.0	—	—
Makhlouf et al., 1987	15S_M (I.1)	—	48.5	47.1	39.5	26.0	6.0	0.5	0.0	0.0
Verhagen and Bodor, 1984	15S_S	—	—	30	—	14	—	—	0.7	—
Versteeg, 1991	12S_M [Sept]	66.0	62.0	56.0	43.0	23.0	6.5	0.0	—	—
Deffense, 1987	Olein #1$_S$ [1]	—	65.2	45.7	28.3	11.9	3.8	0.0	—	—
Centrifugation										
Branger, 1993	32L_S pilot-scale	57.5	50.0	42.5	28.0	12.0	7.0	1.0	0.0	—
	25L_S pilot-scale	50.0	42.0	34.1	21.0	4.5	0.0	—	—	—
	22L_M [1] bench-top	44.6	40.0	31.0	12.8	10.0	3.7	1.0	0.0	—
	20S_S pilot-scale	60.0	52.0	45.0	32.0	14.5	7.5	0.0	—	—
	17S_S pilot-scale	61.3	53.8	45.0	32.0	13.0	5.0	0.0	—	—
	12S_M [3] bench-top	55.6	53.7	42.8	31.4	11.4	0.0	—	—	—
Filtration through milk filter										
Fouad et al., 1990	29L_S	44.0	35.0	23.7	14.7	7.7	2.0	0.0	0.0	0.0
	17S_S	48.5	40.2	35.0	28.2	20.0	10.8	2.5	0.0	0.0
Separation technique not reported										
Martine, 1982	Olein	51.7	48.2	41.9	27.2	15.2	9.5	5.0	0.0	—

[1] The number indicates fractionation temperature, the letter indicates physical state (L = liquid, S = solid), and the subscript indicates fractionation procedure used (M = multiple-step, S = single-step). Other primary designations and designations in parentheses are those of the original author. The designation inside the brackets refers to the intact anhydrous milkfat source [sample number or season (S = summer, W = winter)].

[2] Data not available.

TABLE 5.186 Solid Fat Content of Middle-Melting Milkfat Fractions Obtained from Solvent Solution

Type of solvent solution Author	Fraction designation[1]	Solid fat content (%) at temperature (°C)							
		0	5	10	15	20	25	30	35
Acetone									
Kaylegian, 1991	25L$_S$ [W]	59.8	55.1	45.5	28.2	17.8	11.3	10.0	0.0
	21L$_S$ [W]	48.8	48.0	38.0	24.1	11.3	5.4	0.0	—[2]
	5S$_M$ [W]	88.1	86.2	82.9	78.5	64.9	42.0	13.0	0.0
Yi, 1993	5S$_M$ [S]	88.8	85.9	82.0	77.8	70.2	50.3	21.1	0.0
Kaylegian, 1991	0S$_M$ [W]	83.3	83.8	79.6	70.5	48.8	14.8	0.0	—
Yi, 1993	0S$_M$ [S]	89.1	88.9	83.6	76.3	61.1	21.1	0.0	—
Pentane									
Kaylegian, 1991	11L$_S$ [W]	59.9	55.0	47.0	31.1	15.8	8.2	5.1	0.0
	5L$_S$ [W]	48.2	47.1	37.4	22.2	6.2	0.0	—	—

[1]The number indicates fractionation temperature, the letter indicates physical state (L = liquid, S = solid), and the subscript indicates fractionation procedure used (M = multiple-step, S = single-step). The designation inside the brackets refers to the intact anhydrous milkfat source (S = summer, W = winter).

[2]Data not available.

TABLE 5.187 Solid Fat Content of Middle-Melting Milkfat Fractions Obtained by Supercritical Fluid Extraction

Type of solvent Author	Fraction designation[1]	Fractionation conditions		Solid fat content (%) at temperature (°C)									
		Temperature (°C)	Pressure	0	5	10	15	20	25	30	35	40	45
Carbon dioxide													
Bhaskar et al., 1993b	S3	75	17.2 MPa	—[2]	60.0 280K	—	28.0 290K	—	8.5 300K	—	0.0 310K	—	
	S4	60	6.9	—	63.0	—	26.0	—	6.0	—	0.0	—	
	S5	60	3.5	—	58.0	—	18.0	—	3.0	—	0.0	—	

[1]Designations are those used by the original authors.

[2]Data not available.

TABLE 5.188 Solid Fat Content of Middle-Melting Milkfat Fractions Obtained by Short-Path Distillation

Author	Fraction designation[1]	Fractionation conditions		Solid fat content (%) at temperature (°C)						
		Temperature (°C)	Pressure	0	5	10	15	20	25	30	35
Arul et al., 1988a	IF Distillate III	265	100 μm Hg	—[2]	38.9	34.8	23.7	11.6	1.8	0.0	—

[1]Fraction designations are those given by the original author.

[2]Data not available.

TABLE 5.189 Solid Fat Content of Low-Melting Milkfat Fractions Obtained from Melted Milkfat

Fraction separation method Author	Fraction designation[1]	Solid fat content (%) at temperature (°C)[2]								
		0	5	10	15	20	25	30	35	40
Vacuum filtration										
deMan and Finoro, 1980	26L$_S$	—[3]	—	31.3	20.7	6.8	3.8	—	—	0.3
Lund and Danmark, 1987	26L$_S$ [1]	—	49.1	38.8	24.6	6.9	0.0	—	—	—
deMan and Finoro, 1980	25L$_S$	—	—	32.4	20.7	7.2	0.9	—	—	0.3
Kaylegian, 1991	25L$_S$ [W]	37.7	39.1	30.8	23.7	7.3	0.0	—	—	—
Kaylegian and Lindsay, 1992	25L$_S$ [W]	37.7	39.1	30.8	23.7	7.3	0.0	—	—	—
Timms and Parekh, 1980	25L$_S$ [S]	—	—	41.9	24.6	7.6	2.2	0.9	0.0	—
Amer et al., 1985	23L$_S$ [W]	73.1	62.4	46.1	23.2	6.4	1.8	0.5	0.0	—
Kaylegian, 1991	20L$_S$ [W]	33.7	34.3	22.0	15.2	0.0	—	—	—	—
Amer et al., 1985	19L$_S$ [W]	71.9	60.7	42.2	20.7	4.6	1.7	0.5	0.5	0.4
	19L$_S$ [S]	70.7	59.1	43.3	21.2	3.9	1.0	0.0	—	—
Lund and Danmark, 1987	18L$_S$ [2]	—	39.0	27.0	13.0	0.0	—	—	—	—
Versteeg, 1991	18L$_M$ [Sept]	49.5	42.0	31.0	18.0	5.0	1.5	0.0	—	—
Kaylegian, 1991	13S$_M$ [W]	53.7	53.0	43.6	30.5	11.5	0.0	—	—	—
	13L$_M$ [W]	9.5	0.6	0.0	—	—	—	—	—	—
Kaylegian and Lindsay, 1992	13S$_M$ [W]	53.7	53.0	43.6	30.5	11.5	0.0	—	—	—
	13L$_M$ [W]	9.5	0.6	0.0	—	—	—	—	—	—
Versteeg, 1991	12L$_M$ [Sept]	32.5	25.0	16.0	4.0	0.5	0.0	—	—	—
Makhlouf et al., 1987	9S$_M$ (I.2)	—	19.5	14.2	0.0	0.0	0.0	0.0	0.0	0.0
	9L$_M$ (F.L.)	—	17.9	10.1	0.0	0.0	0.0	0.0	0.0	0.0
	F.1 (21-9)	—	31.1	28.3	18.3	0.3	0.0	0.0	0.0	0.0
Deffense, 1987	2nd stearin$_M$ [2]	—	56.5	46.0	30.0	10.0	1.0	0.0	—	—
	1st Olein$_M$ [2]	—	39.2	23.0	7.4	0.0	—	—	—	—
	2nd olein$_M$ filtered at 10°C [2]	—	18.5	2.2	0.0	—	—	—	—	—
	Olein #2$_S$ [1]	—	53.4	42.0	23.5	8.0	1.1	0.0	—	—
	Olein #3$_S$ [1]	—	51.4	37.7	19.0	2.5	0.0	—	—	—
	Olein #4$_S$ [1]	—	50.6	35.3	17.7	0.3	0.0	—	—	—
Vovan and Riel, 1973	5% removal of hard fraction (>21°C)	—	37.9[4] SFI	32.9	23.2	14.1	8.6	3.9	1.3	0.0
	10% removal of hard fraction (>21°C)	—	37.1	31.4	22.3	14.1	8.5	3.0	0.7	0.4
	15% removal of hard fraction (>21°C)	—	37.5	32.5	—	15.1	9.3	4.1	1.1	0.0
	5% removal of intermed. fraction (10–21°C)	—	42.0	37.9	26.8	15.4	10.2	5.0	1.8	—
	10% removal of intermed. fraction (10–21°C)	—	41.6	35.9	23.9	12.7	9.1	3.8	1.3	—
	15% removal of intermed. fraction (10–21°C)	—	40.9	35.9	23.4	12.1	7.9	3.8	—	—

Continued

TABLE 5.189 (Continued)

Fraction separation method Author	Fraction designation[1]	Solid fat content (%) at temperature (°C)[2]								
		0	5	10	15	20	25	30	35	40
Centrifugation										
Branger, 1993	20L$_M$ [2] bench-top	21.4	21.4	10.5	0.0	—	—	—	—	—
	20L$_S$ pilot-scale	47.8	38.5	30.0	16.3	0.0	—	—	—	—
	17L$_S$ pilot-scale	42.8	32.8	22.8	10.0	0.0	—	—	—	—
	12L$_M$ [3] bench-top	23.7	21.4	8.4	4.6	0.0	—	—	—	—
Kankare and Antila, 1988b	12L$_M$ [S]	61.0 SFI	—	31.5	—	0.0	—	—	—	—
Separation with the aid of an aqueous detergent solution										
Norris et al., 1971	26L$_S$ [B late S]	64.4	54.6	39.4	22.8	4.9	0.0	—	—	—
	26L$_S$ [C early F]	64.1	53.3	36.3	19.6	4.0	0.0	—	—	—
	25L$_S$ [A late S]	63.1	52.0	38.7	21.4	5.2	0.0	—	—	—
Filtration through milk filter										
Fouad et al., 1990	17L$_S$	40.0	29.7	17.2	6.5	2.5	0.0	—	—	—
Separation technique not reported										
El Ghandour and El-Nimr, 1982	15L$_M$	—	—	27.5	—	3.6	—	—	—	—

[1]The number indicates fractionation temperature, the letter indicates physical state (L = liquid), and the subscript indicates fractionation procedure used (M = multiple-step, S = single-step). Other primary designations and designations in parentheses are those of the original authors. The designation inside the brackets refers to the intact anhydrous milkfat source [sample number or season (F = fall, S = summer, W = winter)].

[2]SFI = solid fat index.

[3]Data not available.

[4]Data estimated from graphs provided by the original investigator.

TABLE 5.190 Solid Fat Content of Low-Melting Milkfat Fractions Obtained from Solvent Solution

Type of solvent solution Author	Fraction designation[1]	Solid fat content (%) at temperature (°C)					
		0	5	10	15	20	25
Acetone							
Kaylegian, 1991	$15L_S$ [W]	39.8	41.7	31.1	18.8	4.0	0.0
	$11L_S$ [W]	43.2	36.7	25.8	13.8	1.3	0.0
	$5L_S$ [W]	35.7	29.2	18.1	6.4	0.0	—[2]
Yi, 1993	$0L_M$ [S]	1.2	0.0	—	—	—	—
Timms, 1980b	$3L_M$ [S]	26.1	11.6	8.4	6.1	0.0	—
Lambelet, 1983	$L1_M$	25.7[3]	21.4	9.9	3.3	—	—
		66.1[4]	48.0	13.2	0.2	—	—
Pentane							
Kaylegian, 1991	$0L_M$ [W]	19.9	21.2	17.0	10.7	0.4	0.0
	$0L_S$ [W]	43.2	43.7	31.0	19.2	4.9	0.0

[1]The number indicates fractionation temperature, the letter indicates physical state (L = liquid), and the subscript indicates fractionation procedure used (M = multiple-step, S = single-step). Other designations used are those of the original author. The designation inside the brackets refers to the intact anhydrous milkfat source (S = summer, W = winter).

[2]Data not available.

[3]SFI determined by NMR method.

[4]SFI determined by DSC method, noncorrected values.

TABLE 5.191 Solid Fat Content of Low-Melting Milkfat Fractions Obtained by Supercritical Fluid Extraction

Type of solvent Author	Fraction designation[1]	Fractionation conditions		Solid fat content (%) at temperature (°C)					
		Temperature (°C)	Pressure	0	5	10	15	20	25
Carbon dioxide									
Biernoth and Merk, 1985	Residue	80	200 bar	—[2]	—	25.1	10.7	0.4	0.5
Kaufmann et al., 1982	II (extract)	80	200 bar	—	34.3	18.7	10.4	0.0	—
Timmen et al., 1984	Short-chain fraction	80	200 bar	—	34.3	18.7	10.4	0.0	—

[1]Designations are those given by the original author.

[2]Data not available.

TABLE 5.192 Solid Fat Content of Low-Melting Milkfat Fractions Obtained by Short-Path Distillation

Author	Fraction designation[1]	Fractionation conditions		Solid fat content (%) at temperature (°C)					
		Temperature (°C)	Pressure	0	5	10	15	20	25
Arul et al., 1988a	LF1 Distillate I	245	220 μm Hg	—[2]	16.5	7.4	0.0	—	—
	LF2 Distillate II	245	100	—	31.4	24.0	12.7	3.4	0.0

[1]Designations used are those given by the original author.

[2]Data not available.

TABLE 5.193 Solid Fat Content of Very-Low-Melting Milkfat Fractions Obtained from Melted Milkfat

Fraction separation method Author	Fraction designation[1]	Solid fat content (%) at temperature (°C)			
		0	5	10	15
Vacuum filtration					
Versteeg, 1991	$8L_M$ [Sept]	10.5	4.0	0.3	0.0
Deffense, 1987	2nd $olein_M$ filtered at 6°C [2]	—[2]	10.7	0.0	—

[1]The number indicates fractionation temperature, the letter indicates physical state (L = liquid), and the subscript indicates fractionation procedure used (M = multiple-step). Other descriptions used are those of the original author. The designation inside the brackets following the fraction designation refers to the intact anhydrous milkfat source.

[2]Data not availablle.

TABLE 5.194 Solid Fat Content of Very-Low-Melting Milkfat Fractions Obtained from Solvent Solution

Type of solvent solution Author	Fraction designation[1]	Solid fat content (%) at temperature (°C)			
		0	5	10	15
Acetone					
Kaylegian, 1991	$0L_M$ [W]	1.1	0.0	—[2]	—
	$0L_S$ [W]	12.9	8.0	0.0	—

[1]The number indicates fractionation temperature, the letter indicates physical state (L = liquid), and the subscript indicates fractionation procedure used (S = single-step, M = multiple-step). The designation inside the brackets following the fraction designation refers to the intact anhydrous milkfat source (W = winter).

[2]Data not available.

TABLE 5.195 Textural Characteristics of Anhydrous Milkfats Used to Produce Milkfat Fractions

Fractionation method Author	Fraction designation[1]	Type of texture measurement	Temperature (°C)				
			5	10	15	20	25
Crystallization from melted milkfat							
Vacuum filtration							
Hayakawa and deMan, 1982	AMF	Constant-speed penetrometer	14.83 kg/cm^2	6.73	2.58	0.210	0.009
	AMF	Cone penetrometer	21.3[2] (.1 mm)	34.3	52.7	203	—[3]
Pressure filtration							
deMan, 1968	AMF	Cone penetrometer	—	—	1.02 kg/cm^2	—	—
Filtration through muslin cloth							
Sreebhashyam et al., 1981	AMF	Cone penetrometry	—	2.7 mm	—	—	—

[1] AMF = anhydrous milkfat.

[2] Mass of cone assembly = 242.5 g for 5°C reading, 192.5 g for 10°C reading, 52.7 g for 15°C reading, 92.5 g for 20°C reading.

[3] Data not available.

TABLE 5.196 Textural Characteristics of High-Melting Milkfat Fractions

Fractionation method Author	Fraction designation[1]	Type of texture measurement	Temperature (°C)				
			5	10	15	20	25
Crystallization from melted milkfat							
Filtration through muslin cloth							
Sreebhashyam et al., 1981	30S$_M$ (IV)	Cone penetrometry	—[2]	2.2 mm	—	—	—

[1] The number indicates fractionation temperature, the letter indicates physical state (S = solid), and the subscript indicates fractionation procedure used (M = multiple-step). Designations in parentheses are those of the original author.

[2] Data not available.

TABLE 5.197 Textural Characteristics of Middle-Melting Milkfat Fractions

Fractionation method Author	Fraction designation[1]	Type of texture measurement	Temperature (°C)				
			5	10	15	20	25
Crystallization from melted milkfat							
Pressure filtration							
deMan, 1968	$30L_S$ (A) slow cooling	Cone penetrometer	—[2]	—	0.26 kg/cm^2	—	—
	$30L_S$ (B) rapid cooling		—	—	0.24	—	—
Filtration through muslin cloth							
Sreebhashyam et al., 1981	$25S_M$ (III) $18S_M$ (II)	Cone penetrometry	— —	3.0 mm 3.5	— —	— —	— —

[1] The number indicates fractionation temperature, the letter indicates physical state (L = liquid, S = solid), and the subscript indicates fractionation procedure used (M = multiple-step, S = single-step). Designations in parentheses are those of the original author.

[2] Data not available.

TABLE 5.198 Textural Characteristics of Low-Melting Milkfat Fractions

Fractionation method Author	Fraction designation[1]	Type of texture measurement	Temperature (°C)				
			5	10	15	20	25
Crystallization from melted milkfat							
Pressure filtration							
deMan, 1968	$27L_S$ (C) slow cooling	Cone penetrometer	—[2]	—	0.17 kg/cm^2	—	—
	$27L_S$ (D) rapid cooling		—	—	0.17	—	—
	$24L_S$ (E) slow cooling		—	—	0.15	—	—
	$24L_S$ (F) rapid cooling		—	—	0.15	—	—
Filtration through muslin cloth							
Sreebhashyam et al., 1981	$18L_M$ (I)	Cone penetrometry	—	5.6 mm	—	—	—

[1] The number indicates fractionation temperature, the letter indicates physical state (L = liquid), and the subscript indicates fractionation procedure used (M = multiple-step, S = single-step). Designations in parentheses are those of the original author.

[2] Data not available.

TABLE 5.199 Textural Characteristics of Unknown-Melting Milkfat Fractions

Fractionation method Author	Fraction designation[1]	Type of texture measurement	Temperature (°C)				
			5	10	15	20	25
Crystallization from melted milkfat							
Vacuum filtration							
Hayakawa and deMan, 1982	$32S_S$	Constant-speed penetrometer	20.50 kg/cm²	11.21	4.50	1.58	0.67
	$32L_S$		9.62	6.32	1.58	0.151	0.009
	$31S_S$		25.35	13.46	5.83	2.39	1.01
	$31L_S$		10.66	5.66	1.28	0.100	0.005
	$30S_S$		29.78	19.61	7.42	6.59	1.71
	$30L_S$		9.36	5.21	1.37	0.074	0.002
	$29S_S$		30.07	21.34	8.03	3.35	1.79
	$29L_S$		8.67	4.89	1.50	0.062	—[2]
	$28S_S$		29.78	19.72	8.76	4.09	2.17
	$28L_S$		7.95	4.40	1.32	0.032	—
	$27S_S$		29.89	19.53	6.83	2.58	1.22
	$27L_S$		7.75	3.99	1.27	0.022	—
	$26S_S$		27.47	16.40	5.45	2.02	0.73
	$26L_S$		7.08	3.53	1.11	0.012	—
	$25S_S$		25.67	11.82	4.29	1.10	0.34
	$25L_S$		6.55	3.11	0.92	0.007	—
	$32S_S$	Cone penetrometer	11.2[3] (.1 mm)	24.2	29.8	73.2	95.3
	$32L_S$		10.3[4]	27.7	52.2	—	—
	$31S_S$		10.0[3]	21.7	28.8	59.2	67.1
	$31L_S$		12.0[4]	30.0	59.0	—	—
	$30S_S$		8.8[3]	21.5	22.0	43.2	53.7
	$30L_S$		15.3[4]	36.2	65.7	—	—
	$29S_S$		7.5[3]	17.5	20.3	46.8	47.5
	$29L_S$		19.7[4]	42.5	58.8	—	—
	$28S_S$		8.0[3]	16.2	23.8	42.3	44.3
	$28L_S$		18.8[4]	44.0	68.5	—	—
	$27S_S$		9.2[3]	23.0	24.7	52.3	70.8
	$27L_S$		22.5[4]	46.3	71.7	—	—
	$26S_S$		9.7[3]	22.7	27.3	64.2	94.8
	$26L_S$		26.3[4]	51.5	75.2	—	—
	$25S_S$		14.8[3]	34.7	35.0	102.0	144.3
	$25L_S$		31.7[4]	52.8	86.8	—	—

[1]The number indicates fractionation temperature, the letter indicates physical state (L = liquid, S = solid), and the subscript indicates fractionation procedure used (M = multiple-step, S = single-step).

[2]Data not available.

[3]Mass of cone assembly = 242.5 for 5 and 10°C readings, 142.5 g for 15 and 20°C readings, 92.5g for 25°C readings.

[4]Mass of cone assembly = 142.5 g for 5 and 10°C readings, 92.5g for 15 and 20°C readings.

TABLE 5.200 Oxidative and Lipolytic Stability Characterization of Intact Milkfat and Milkfat Fractions

Fractionation method Author	Fraction designation[1]	General fraction property[2]	Peroxide concentration (mm)			Other analysis	
			Treatment #1	Treatment #2	Treatment #3	Treatment #1	Treatment #2
Crystallization from melted milkfat							
Vacuum filtration							
Kaylegian, 1991			Storage in the dark 25°C (μ moles hydroperoxide/ mg fat)	Storage under fluorescent light (450 Lux) at 25°C (μ moles hydroperoxide/ mg fat)	Storage in the dark 60°C (μ moles hydroperoxide/ mg fat)		
	UW AMF control		0 d = 2.5 3 d = 2.4 7 d = 4.6 14 d = 7.2 21 d = 5.6	0 d = 2.5 3 d = 34.1 7 d = 130.4 14 d = 231.8 21 d = —	0 d = 4.7 0.5 d = —[3] 1 d = 5.5 1.5 d = — 2 d = 7.9 3 d = 7.6		
	UW AMF + Tenox 2 (200 ppm)		0 d = 2.5 3 d = 1.9 7 d = 4.8 14 d = 4.0 21 d = 5.9	0 d = 2.5 3 d = 28.0 7 d = 136.4 14 d = 198.8 21 d = —	0 d = 4.7 0.5 d = 6.0 1 d = 6.5 1.5 d = — 2 d = 6.1 3 d = 6.4		
	UW AMF + Covi-Ox T70 (200 ppm)		0 d = 2.5 3 d = 2.5 7 d = 4.9 14 d = 4.2 21 d = 5.1	0 d = 2.5 3 d = 29.6 7 d = 108.2 14 d = 186.7 21 d = —	0 d = 4.7 0.5 d = 5.9 1 d = 6.2 1.5 d = — 2 d = 6.0 3 d = 8.2		
	LV AMF control		0 d = 2.3 3 d = 3.7 7 d = 3.9 14 d = 5.6 21 d = 5.5	0 d = 2.3 3 d = 11.7 7 d = 21.8 14 d = 43.1 21 d = 65.7	0 d = 2.8 0.5 d = 7.3 1 d = 8.8 1.5 d = 10.4 2 d = 11.0 3 d = 8.3		

Sample				
LV AMF + Tenox 2		0 d = 2.3 3 d = 3.7 7 d = 3.9 14 d = 3.5 21 d = 4.0	0 d = 2.3 3 d = 10.7 7 d = 18.9 14 d = 26.7 21 d = 46.6	0 d = 2.8 0.5 d = 4.0 1 d = 6.2 1.5 d = — 2 d = 6.6 3 d = 3.4
LV AMF + Covi-Ox T70		0 d = 2.3 3 d = 3.3 7 d = 3.5 14 d = 5.0 21 d = 5.4	0 d = 2.3 3 d = 10.4 7 d = 21.5 14 d = 36.7 21 d = 66.8	0 d = 2.8 0.5 d = 9.7 1 d = 15.2 1.5 d = 16.6 2 d = 16.6 3 d = 24.0
Blended fractions [LV] control[4]	VH+H+L	0 d = 3.7 3 d = 4.1 7 d = 4.8 14 d = 7.8 21 d = 9.1	0 d = 3.7 3 d = 11.1 7 d = 21.0 14 d = 49.6 21 d = 81.9	0 d = 4.5 0.5 d = 11.7 1 d = 12.8 1.5 d = 15.2 2 d = 15.4 3 d = 20.5
Blended fractions [LV] + Tenox 2	VH+H+L	0 d = 3.7 3 d = 4.0 7 d = 4.3 14 d = 6.0 21 d = 7.3	0 d = 3.7 3 d = 14.8 7 d = 22.8 14 d = 43.0 21 d = 100.4	0 d = 4.5 0.5 d = 5.8 1 d = 6.5 1.5 d = 6.3 2 d = 7.3 3 d = 7.1
Blended fractions [LV] + Covi-Ox T70	VH+H+L	0 d = 3.7 3 d = 4.3 7 d = 4.7 14 d = 6.8 21 d = 10.1	0 d = 3.7 3 d = 11.8 7 d = 18.7 14 d = 39.2 21 d = 81.4	0 d = 4.5 0.5 d = 12.5 1 d = 14.2 1.5 d = 15.8 2 d = 17.9 3 d = 21.5
20L$_S$ [LV] control	L	0 d = 4.8 3 d = 5.4 7 d = 6.6 14 d = 10.1 21 d = 12.2	0 d = 4.8 3 d = 14.9 7 d = 23.7 14 d = 52.4 21 d = 84.8	0 d = 4.7 0.5 d = 13.1 1 d = 15.6 1.5 d = 16.9 2 d = 19.7

Continued

TABLE 5.200 (Continued)

Fractionation method Author	Fraction designation[1]	General fraction property[2]	Peroxide concentration (mm)			Other analysis		
			Treatment #1	Treatment #2	Treatment #3	Treatment #1	Treatment #2	
	$20L_S$ [LV] + Tenox 2	L	0 d = 4.8 3 d = 7.8 7 d = 6.1 14 d = 5.5 21 d = 6.0	0 d = 4.8 3 d = 12.3 7 d = 21.2 14 d = 40.4 21 d = 50.6	3 d = 19.5 0 d = 4.7 0.5 d = 7.2 1 d = 7.0 1.5 d = 8.9 2 d = 14.9 3 d = 6.0			
	$20L_S$ [LV] + Covi-Ox T70	L	0 d = 4.8 3 d = 4.7 7 d = 6.4 14 d = 11.7 21 d = 10.1	0 d = 4.8 3 d = 12.5 7 d = 21.3 14 d = 43.8 21 d = 85.6	0 d = 4.7 0.5 d = 16.0 1 d = 22.5 1.5 d = 26.8 2 d = 25.4 3 d = 28.4			

Centrifugation

Fractionation method Author	Fraction designation[1]	General fraction property[2]	Storage at 60°C (mM/kg fat)
Bhat and Rama Murthy, 1983	AMF		0 d = 0.0 1 d = 0.0 2 d = 0.2 3 d = — 4 d = 0.9 5 d = 1.5 6 d = 2.0 7 d = 2.9 8 d = — 9 d = 3.6
	$28S_S$	H	0 d = 0.0 1 d = 0.0 2 d = 0.0 3 d = 0.1 4 d = 0.5

Technical Data for Experimental and Commercial Milkfat Fractions

	28L$_S$	L	5 d = 1.0
			6 d = 1.2
			7 d = 2.0
			8 d = —
			9 d = 1.8
			0 d = 0.0
			1 d = 0.1
			2 d = 0.5
			3 d = 0.6
			4 d = 1.2
			5 d = 1.8
			6 d = 2.5
			7 d = 5.1
			8 d = 7.1
			9 d = 13.0

Storage at 37°C (mL of 0.1 N alkali)

Lakshminarayana and Rama Murthy, 1986	AMF		0 min = 0.0
			4 min = 4.3
			8 min = 7.9
			12 min = 9.8
			16 min = 13.5
			20 min = 15.4
	31S$_M$	U	0 min = 0.0
			4 min = 3.0
			8 min = 6.0
			12 min = 8.1
			16 min = 10.0
			20 min = 11.1
	23S$_M$	U	0 min = 0.0
			4 min = 3.6
			8 min = 6.8
			12 min = 9.4

Continued

TABLE 5.200 (Continued)

Fractionation method Author	Fraction designation[1]	General fraction property[2]	Peroxide concentration Treatment #1	Peroxide concentration Treatment #2	Peroxide concentration Treatment #3	Other analysis Treatment #1	Other analysis Treatment #2
	$15S_M$	U	16 min = 12.0 20 min = 13.8				
			0 min = 0.0 4 min = 5.0 8 min = 9.4 12 min = 13.5 16 min = 16.9 20 min = 19.3				
	$15L_M$	U	0 min = 0.0 4 min = 5.6 8 min = 10.8 12 min = 15.0 16 min = 18.4 20 min = 21.6				
Keshava Prasad and Bhat, 1987b			*Storage at 50°C* *(μm peroxide/g fat)*				
	$23S_S$ control	U	0 d = 0 15 d = 1.0 30 d = 1.8 45 d = 2.6				
	$23S_S$ oxygen-treated	U	0 d = 0 15 d = 1.3 30 d = 2.4 45 d = 3.4				
	$23S_S$ copper-treated	U	0 d = 0 15 d = 1.1 30 d = 2.1 45 d = 3.7				
	$23L_S$ control	U	0 d = 0 15 d = 1.4 30 d = 2.3 45 d = 2.8				

23L$_S$ oxygen-treated	U	0 d = 0 15 d = 1.8 30 d = 2.8 45 d = 3.6
23L$_S$ copper-treated	U	0 d = 0 15 d = 1.4 30 d = 2.4 45 d = 4.3
28S$_S$ control	U	0 d = 0 15 d = 0.7 30 d = 1.4 45 d = 2.1
28S$_S$ oxygen-treated	U	0 d = 0 15 d = 1.2 30 d = 1.6 45 d = 2.1
28S$_S$ copper-treated	U	0 d = 0 15 d = 1.9 30 d = 1.4 45 d = 4.6
28L$_S$ control	U	0 d = 0 15 d = 1.1 30 d = 1.8 45 d = 2.6
28L$_S$ oxygen-treated	U	0 d = 0 15 d = 1.9 30 d = 2.4 45 d = 2.8
28L$_S$ copper-treated	U	0 d = 0 15 d = 1.1 30 d = 1.6 45 d = 4.7

Continued

TABLE 5.200 (Continued)

Fractionation method Author	Fraction designation[1]	General fraction property[2]	Peroxide concentration			Other analysis	
			Treatment #1	Treatment #2	Treatment #3	Treatment #1	Treatment #2
Supercritical fluid extraction							
Carbon dioxide							
Kaufmann et al., 1982			*Peroxide value*			*Free fatty acid (% oleic)*	
	AMF		0.13			0.32	
	I (residue) 80°C 200 bar	H	1.53			0.46	
	II (extract) 80°C 200 bar	L	1.46			1.51	
Timmen et al., 1984			*Peroxide value*			*Free fatty acid (% oleic)*	
	AMF		0.13			0.32	
	Long-chain fraction 80°C 200 bar	H	1.53			0.46	
	Short-chain fraction 80°C 200 bar	L	1.46			1.51	
Fractionation method not reported							
Kehagias and Radema, 1973	AMF		*Storage at 28°C (mEq oxygen/kg fat)*	*Storage at -40°C (mEq oxygen/kg fat)*		*Thiobarbituric acid value storage at 28°C (extinction value)*	*Thiorbarbituric acid value storage at -40°C (extinction value)*
			0 wk = 0.226	0 wk = 0.226		0 wk = 0.017	0 wk = 0.017
			1 wk = 0.274			1 wk = 0.044	
			2 wk = 0.304	2 wk = 0.309		2 wk = 0.083	2 wk = 0.021
			3 wk = 0.398			3 wk = 0.060	
			4 wk = 0.499			4 wk = 0.129	
			5 wk = 0.604	5 wk = 0.263		5 wk = 0.136	5 wk = 0.016
			7 wk = 0.821			7 wk = 0.171	

Technical Data for Experimental and Commercial Milkfat Fractions

Sample		Column 1	Column 2	Column 3	Column 4
$22S_S$ + air		9 wk = 1.166 10 wk = 1.307 13 wk = 1.637	10 wk = 0.236	9 wk = 0.252 10 wk = 0.208 13 wk = 0.255	10 wk = 0.009
	U	0 wk = 0.368 1 wk = 0.304 2 wk = 0.315 3 wk = 0.423 4 wk = 0.486 5 wk = 0.535 7 wk = 0.657 9 wk = 0.763 10 wk = 0.931 13 wk = 1.435	0 wk = 0.368 2 wk = 0.400 5 wk = 0.385 10 wk = 0.342	0 wk = 0.025 1 wk = 0.022 2 wk = 0.043 3 wk = 0.034 4 wk = 0.060 5 wk = 0.058 7 wk = 0.069 9 wk = 0.099 10 wk = 0.088 13 wk = 0.112	0 wk = 0.025 2 wk = 0.026 5 wk = 0.019 10 wk = 0.016
$22S_S$ + CO_2		0 wk = 0.317 2 wk = 0.279 4 wk = 0.496 5 wk = 0.555 7 wk = 0.680 9 wk = 0.845 10 wk = 0.877 13 wk = 1.161	0 wk = 0.317 2 wk = 0.387 5 wk = 0.372 10 wk = 0.318	0 wk = 0.018 2 wk = 0.050 4 wk = 0.083 5 wk = 0.092 7 wk = 0.102 9 wk = 0.131 10 wk = 0.124 13 wk = 0.108	0 wk = 0.018 2 wk = 0.024 5 wk = 0.019 10 wk = 0.016
	U				
$22L_S$ + air		0 wk = 0.501 1 wk = 0.499 2 wk = 0.468 3 wk = 0.572 4 wk = 0.652 5 wk = 0.706 7 wk = 0.885 9 wk = 1.286 10 wk = 1.409 13 wk = 1.931	0 wk = 0.501 2 wk = 0.486 5 wk = 0.485 10 wk = 0.476	0 wk = 0.018 1 wk = 0.027 2 wk = 0.039 3 wk = 0.039 4 wk = 0.076 5 wk = 0.074 7 wk = 0.078 9 wk = 0.168 10 wk = 0.137 13 wk = 0.214	0 wk = 0.018 2 wk = 0.019 5 wk = 0.013 10 wk = 0.015
	U				
$22L_S$ = $CO2$		0 wk = 0.379 2 wk = 0.450 4 wk = 0.711	0 wk = 0.379 2 wk = 0.391	0 wk = 0.016 2 wk = 0.057 4 wk = 0.096	0 wk = 0.016 2 wk = 0.020
	U				

Continued

TABLE 5.200 (Continued)

Fractionation method Author	Fraction designation[1]	General fraction property[2]	Peroxide concentration			Other analysis	
			Treatment #1	Treatment #2	Treatment #3	Treatment #1	Treatment #2
			5 wk = 0.834	5 wk = 0.94	5 wk = 0.115	5 wk = 0.012	
			7 wk = 1.072		7 wk = 0.131		
			9 wk = 1.228		9 wk = 0.187		
			10 wk = 1.757	10 wk = 0.356	10 wk = 0.231	10 wk = 0.013	
			13 wk = 1.813		13 wk = 0.197		

[1] AMF = anhydrous milkfat. The number indicates the fractionation temperature, the letter indicates physical state (L = liquid, S = solid), and the subscript indicates fractionation procedure used (M = multiple step, S = single-step). Other designations are those of the original author. Designations in brackets refer to the intact milkfat.

[2] VH = very-high-melting milkfat fraction (MP > 45°C), H = high-melting milkfat fraction (MP between 35 and 45°C), L = low-melting milkfat fraction (MP between 10 and 25°C), U = unknown-melting milkfat fractions (MP not available).

[3] Data not available.

[4] Blended fraction sample = 70% 20L_S [W] fraction and 10% 30S_S [W] fraction obtained from melted milkfat + 20% 15S_M [W] fraction obtained from acetone solution.

TABLE 5.201 Oxidation Products Formed in Milkfat and Milkfat Fractions

Fractionation method / Author	Fraction designation[1]	General fraction property[2]	Selected chemical species formed during oxidation[3]				
			Trimeric and higher oligomeric content after heating at 185°C (% of intact sample)	Dimeric content after heating at 185°C (% of intact sample)	Monomeric (TG) content after heating at 185°C (% of intact sample)	Monomeric (TG, DG) content after heating at 185°C (% of intact sample)	Monomeric (FFA) content after heating at 185°C (% of intact sample)
Crystallization from melted milkfat							
Vacuum filtration							
Kupranycz et al., 1986	AMF [W]		0 h = nd[4] 8 h = 4.93 16 h = 9.12	0 h = nd 8 h = 9.74 16 h = 12.81	0 h = 50.01 8 h = 39.63 16 h = 34.23	0 h = 49.54 8 h = 45.70 16 h = 43.84	0 h = 0.45 8 h = tr[5] 16 h = tr
	29S$_S$ [W]	U	0 h = nd 8 h = 4.38 16 h = 7.95	0 h = nd 8 h = 8.86 16 h = 12.48	0 h = 65.86 8 h = 55.38 16 h = 50.39	0 h = 33.74 8 h = 31.30 16 h = 29.18	0 h = 0.39 8 h = 0.11 16 h = tr
	29L$_S$ [W]	U	0 h = nd 8 h = 5.24 16 h = 13.33	0 h = nd 8 h = 8.45 16 h = 11.74	0 h = 44.83 8 h = 35.16 16 h = 27.14	0 h = 54.55 8 h = 50.86 16 h = 47.78	0 h = 0.62 8 h = 0.29 16 h = tr
	19S$_S$ [W]	U	0 h = nd 8 h = 4.62 16 h = 9.04	0 h = nd 8 h = 9.36 16 h = 12.24	0 h = 56.05 8 h = 45.93 16 h = 40.42	0 h = 43.54 8 h = 40.10 16 h = 38.29	0 h = 0.41 8 h = tr 16 h = tr
	19L$_S$ [W]	U	0 h = nd 8 h = 5.00 16 h = 12.22	0 h = nd 8 h = 7.82 16 h = 11.71	0 h = 40.55 8 h = 30.73 16 h = 22.76	0 h = 58.82 8 h = 56.25 16 h = 53.32	0 h = 0.64 8 h = 0.20 16 h = tr
	AMF [S]		0 h = nd 8 h = 5.51 16 h = 11.48	0 h = nd 8 h = 9.34 16 h = 13.16	0 h = 52.26 8 h = 41.63 16 h = 34.27	0 h = 47.24 8 h = 43.23 16 h = 41.03	0 h = 0.50 8 h = 0.28 16 h = 0.05
	29S$_S$ [S]	U	0 h = nd 8 h = 4.75 16 h = 9.90	0 h = nd 8 h = 8.83 16 h = 12.48	0 h = 68.69 8 h = 59.46 16 h = 52.15	0 h = 31.08 8 h = 26.84 16 h = 25.26	0 h = 0.23 8 h = 0.12 16 h = 0.20
	29L$_S$ [S]	U	0 h = nd 8 h = 6.33 16 h = 13.75	0 h = nd 8 h = 8.48 16 h = 11.75	0 h = 49.38 8 h = 39.01 16 h = 31.54	0 h = 50.22 8 h = 46.01 16 h = 42.83	0 h = 0.40 8 h = 0.18 16 h = 0.13

Continued

TABLE 5.201 (Continued)

Fractionation method Author	Fraction designation[1]	General fraction property[2]	Selected chemical species formed during oxidation[3]			
	19S$_S$ [S]	U	Trimeric and higher oligomeric content after heating 16 h at 185°C (% FAME) 0 h = nd 8 h = 5.48 16 h = 10.77	Dimeric content after heating 16 h at 185°C (% FAME) 0 h = nd 8 h = 9.40 16 h = 12.99	Monomeric (fatty acids) content after heating 16 h at 185°C (% FAME) 0 h = 59.22 8 h = 48.34 16 h = 40.90	0 h = 40.45 8 h = 36.57 16 h = 35.09	0 h = 0.30 8 h = 0.21 16 h = 0.26
	19L$_S$ [S]	U	0 h = nd 8 h = 6.62 16 h = 15.34	0 h = nd 8 h = 8.51 16 h = 11.39	0 h = 45.12 8 h = 34.52 16 h = 26.71	0 h = 54.40 8 h = 50.08 16 h = 46.32	0 h = 0.48 8 h = 0.27 16 h = 0.23
	AMF [W]		4.21	9.18	86.61		
	29S$_S$ [W]	U	3.97	9.24	86.74		
	29L$_S$ [W]	U	8.18	7.20	84.62		
	19S$_S$ [W]	U	4.31	8.94	86.75		
	19L$_S$ [W]	U	7.07	7.82	85.11		
	AMF [S]		4.96	8.97	86.07		
	29S$_S$ [S]	U	3.59	7.70	88.71		
	29L$_S$ [S]	U	6.44	8.51	85.03		
	19S$_S$ [S]	U	4.50	8.63	86.88		
	19L$_S$ [S]	U	7.92	8.06	84.02		

Centrifugation

			Total carbonyls	Monocarbonyls	Ketoglycerides
Bhat and Rama Murthy, 1983	28S$_S$	H	4.35	0.386	1.99
	28L$_S$	L	4.37	0.486	1.87
			Alkanals in autoxidized AMF (PV > 40)	Methyl ketones in autoxidized AMF (PV > 40)	
	AMF		C2 = 9.3 C3 = tr C4 = 6.9 C5 = 12.8	C3 = 21.5 C4 = 5.6 C5 = 8.2 C6 = 8.5	

Technical Data for Experimental and Commercial Milkfat Fractions

28S$_S$		C6 = 17.8 C7 = 14.1 C8 = 24.2 C9 = 14.0 ≥C10 = 11.9	C7 = 8.3 C8 = 9.6 C9 = 1.2 C11 = 10.5 C13 = 10.0 C15 = 16.6
	H	C2 = 6.2 C3 = 0.8 C4 = 4.2 C5 = 27.8 C6 = 28.2 C7 = 11.4 C8 = 8.5 C9 = 7.5 ≥C10 = 5.4	C3 = 26.2 C4 = 10.7 C5 = 11.4 C6 = 5.4 C7 = 5.5 C8 = 1.9 C9 = 1.2 C11 = 10.9 C13 = 10.5 C15 = 16.5
28L$_S$	L	C2 = 6.0 C3 = tr C4 = 5.5 C5 = 10.7 C6 = 10.5 C7 = 16.3 C8 = 28.6 C9 = 16.4 ≥C10 = 5.0	C3 = 20.1 C4 = 5.0 C5 = 6.1 C6 = 9.7 C7 = 10.5 C8 = 10.6 C9 = 1.0 C11 = 10.4 C13 = 10.3 C15 = 16.3

Keshava Prasad and Bhat, 1987a

		Dienoic fatty acid variation during storage at 50°C	Trienoic fatty acid variation during storage at 50°C	Tetraenoic fatty acid variation during storage at 50°C	Pentaenoic fatty acid variation during storage at 50°C
23S$_S$ control	U	0 d = 0.992 15 d = 1.011 30 d = 1.027 45 d = 1.083 % inc = 9.57	0 d = 0.043 15 d = 0.044 30 d = 0.046 45 d = 4.047 % inc = 9.30	0 d = 0.006 15 d = 0.007 30 d = 0.007 45 d = 0.008 % inc = 33.0	0 d = 0.003 15 d = 0.003 30 d = 0.004 45 d = 0.003 % inc = 33.0

Continued

TABLE 5.201 (Continued)

Fractionation method Author	Fraction designation[1]	General fraction property[2]	Selected chemical species formed during oxidation[3]		
	23S$_S$ oxygen-treated	U	0 d = 1.001 15 d = 1.016 30 d = 1.036 45 d = 1.093 % inc = 9.22	0 d = 1.043 15 d = 0.045 30 d = 0.048 45 d = 0.047 % inc = 2.32	0 d = 0.006 15 d = 0.008 30 d = 0.009 45 d = 0.008 % inc = 50.0	0 d = 0.003 15 d = 0.003 30 d = 0.003 45 d = 0.002 % inc = 33.3
	23S$_S$ copper-treated	U	0 d = 1.022 15 d = 1.037 30 d = 1.057 45 d = 1.116 % inc = 9.20	0 d = 0.043 15 d = 0.046 30 d = 0.048 45 d = 0.046 % inc = 13.95	0 d = 0.006 15 d = 0.007 30 d = 0.310 45 d = 0.008 % inc = 66.0	0 d = 0.003 15 d = 0.003 30 d = 0.004 45 d = 0.003 % inc = 33.0
	23L$_S$ control	U	0 d = 1.258 15 d = 1.277 30 d = 1.303 45 d = 1.374 % inc = 9.22	0 d = 0.052 15 d = 0.053 30 d = 0.055 45 d = 0.056 % inc = 7.69	0 d = 0.007 15 d = 0.007 30 d = 0.008 45 d = 0.009 % inc = 28.5	0 d = 0.004 15 d = 0.005 30 d = 0.005 45 d = 0.004 % inc = 25.0
	23L$_S$ oxygen-treated	U	0 d = 1.341 15 d = 1.361 30 d = 1.388 45 d = 1.464 % inc = 9.17	0 d = 0.052 15 d = 0.054 30 d = 0.057 45 d = 0.056 % inc = 9.60	0 d = 0.007 15 d = 0.009 30 d = 0.010 45 d = 0.009 % inc = 50.0	0 d = 0.004 15 d = 0.005 30 d = 0.006 45 d = 0.005 % inc = 30.0
	23L$_S$ copper-treated	U	0 d = 1.381 15 d = 1.401 30 d = 1.429 45 d = 1.508 % inc = 9.20	0 d = 0.052 15 d = 0.052 30 d = 0.058 45 d = 0.056 % inc = 13.75	0 d = 0.007 15 d = 0.008 30 d = 0.009 45 d = 0.007 % inc = 28.5	0 d = 0.004 15 d = 0.005 30 d = 0.005 45 d = 0.004 % inc = 25.0
	28S$_S$ control	U	0 d = 0.991 15 d = 1.012 30 d = 1.029 45 d = 1.092 % inc = 10.20	0 d = 0.030 15 d = 0.030 30 d = 0.032 45 d = 0.032 % inc = 6.66	0 d = 0.006 15 d = 0.007 30 d = 0.008 45 d = 0.009 % inc = 50.0	0 d = 0.003 15 d = 3.003 30 d = 0.004 45 d = 0.004 % inc = 33.0

Technical Data for Experimental and Commercial Milkfat Fractions 465

28S$_S$ oxygen-treated	U	0 d = 1.020 15 d = 1.030 30 d = 1.038 45 d = 1.101 % inc = 7.94	0 d = 0.030 15 d = 0.031 30 d = 0.032 45 d = 0.034 % inc = 6.66	0 d = 0.006 15 d = 0.007 30 d = 0.008 45 d = 0.007 % inc = 33.3	0 d = 0.003 15 d = 0.003 30 d = 0.004 45 d = 0.004 % inc = 33.0
28S$_S$ copper-treated	U	0 d = 1.030 15 d = 1.010 30 d = 1.070 45 d = 1.120 % inc = 8.83	0 d = 0.030 15 d = 0.031 30 d = 0.032 45 d = 0.034 % inc = 13.33	0 d = 0.006 15 d = 0.007 30 d = 0.008 45 d = 0.016 % inc = 60.5	0 d = 0.003 15 d = 0.004 30 d = 0.004 45 d = 0.003 % inc = 33.0
28L$_S$ control	U	0 d = 1.257 15 d = 1.300 30 d = 1.322 45 d = 1.372 % inc = 9.15	0 d = 0.054 15 d = 0.056 30 d = 0.056 45 d = 0.055 % inc = 3.70	0 d = 0.007 15 d = 0.008 30 d = 0.009 45 d = 0.008 % inc = 28.5	0 d = 0.004 15 d = 0.004 30 d = 0.005 45 d = 0.004 % inc = 25.0
28L$_S$ oxygen-treated	U	0 d = 1.320 15 d = 1.351 30 d = 1.421 45 d = 1.440 % inc = 9.10	0 d = 0.054 15 d = 0.056 30 d = 0.058 45 d = 0.058 % inc = 7.41	0 d = 0.007 15 d = 0.008 30 d = 0.009 45 d = 0.007 % inc = 28.5	0 d = 0.004 15 d = 0.004 30 d = 0.005 45 d = 0.005 % inc = 25.0
28L$_S$ copper-treated	U	0 d = 1.360 15 d = 1.380 30 d = 1.410 45 d = 1.190 % inc = 9.56	0 d = 3.054 15 d = 0.056 30 d = 0.059 45 d = 0.058 % inc = 9.26	0 d = 0.007 15 d = 0.008 30 d = 4.009 45 d = 0.010 % inc = 50.0	0 d = 0.004 15 d = 0.005 30 d = 0.005 45 d = 0.004 % inc = 25.0
Keshava Prasad and Bhat, 1987b		*Production of total carbonyls during storage at 50°C (mg/g fat)*	*Production of monocarbonyls during storage at 50°C (μg/g fat)*		
23S$_S$ control	U	0 d = 3.2 15 d = 3.4 30 d = 5.4 45 d = 8.3	0 d = 0.32 15 d = 0.37 30 d = 0.66 45 d = 0.85		

Continued

TABLE 5.201 (Continued)

Fractionation method Author	Fraction designation[1]	General fraction property[2]	Selected chemical species formed during oxidation[3]	
	$23S_S$ oxygen-treated	U	0 d = 3.1 15 d = 5.3 30 d = 6.0 45 d = 8.6	0 d = 0.31 15 d = 0.56 30 d = 0.69 45 d = 0.98
	$23S_S$ copper-treated	U	0 d = 3.1 15 d = 5.0 30 d = 5.5 45 d = 8.5	0 d = 0.31 15 d = 0.48 30 d = 0.70 45 d = 0.90
	$23L_S$ control	U	0 d = 3.2 15 d = 5.3 30 d = 6.5 45 d = 8.5	0 d = 0.32 15 d = 0.69 30 d = 0.76 45 d = 1.01
	$23L_S$ oxygen-treated	U	0 d = 3.2 15 d = 5.5 30 d = 6.4 45 d = 10.6	0 d = 0.32 15 d = 0.75 30 d = 0.50 45 d = 1.28
	$23L_S$ copper-treated	U	0 d = 3.2 15 d = 5.2 30 d = 6.8 45 d = 9.6	0 d = 0.32 15 d = 0.70 30 d = 0.80 45 d = 1.16
	$28S_S$ control	U	0 d = 2.6 15 d = 5.3 30 d = 5.4 45 d = 8.1	0 d = 0.27 15 d = 0.58 30 d = 0.60 45 d = 0.75
	$28S_S$ oxygen-treated	U	0 d = 2.6 15 d = 5.0 30 d = 5.2 45 d = 9.2	0 d = 0.27 15 d = 0.62 30 d = 0.70 45 d = 0.80
	$28S_S$ copper-treated	U	0 d = 2.6 15 d = 5.4 30 d = 5.7 45 d = 8.2	0 d = 0.27 15 d = 0.60 30 d = 0.60 45 d = 0.80

Technical Data for Experimental and Commercial Milkfat Fractions

28L$_S$ control	U	0 d = 3.7 15 d = 5.4 30 d = 5.6 45 d = 9.8	0 d = 0.36 15 d = 0.70 30 d = 0.70 45 d = 0.85
28L$_S$ oxygen-treated	U	0 d = 3.7 15 d = 6.0 30 d = 6.2 45 d = 11.2	0 d = 0.36 15 d = 0.80 30 d = 0.90 45 d = 1.12
28L$_S$ copper-treated	U	0 d = 3.7 15 d = 5.7 30 d = 6.2 45 d = 10.6	0 d = 0.36 15 d = 0.75 30 d = 0.82 45 d = 0.95

[1] AMF = anhydrous milkfat. The number indicates the fractionation temperature, the letter indicates physical state (L = liquid, S = solid), and the subscript indicates fractionation procedure used (S = single-step). Other designations are those of the original author. Designations in brackets refer to the intact milkfat (S = summer, W = winter).

[2] H = high-melting milkfat fraction (35 < MP < 45°C), L = low-melting milkfat fraction (MP between 10 and 25°C), U = unknown-melting milkfat fractions.

[3] DG = diglyceride, FAME = fatty acid methyl ester, FFA = free fatty acid, inc = increase, MF = milkfat, TG = triglyceride.

[4] nd = not determined (terminology employed by original author).

[5] tr = trace.

TABLE 5.202 Nutritional Composition of Intact Milkfat and Milkfat Fractions

Fractionation method Author	Fraction designation[1]	General fraction property[2]	Cholesterol content	Carotene content	Vitamin A content	Vitamin E content
Crystallization from melted milkfat						
Vacuum filtration						
Arul et al., 1988b	AMF[3]		2.5 (mg/g fat)	—[4]	—	—
	Solid (S.1 + S.2)	VH	2.0	—	—	—
	Intermediate (I.1 + I.2)	M + L	2.3	—	—	—
	Liquid (F.L.)	L	3.8	—	—	—
Centrifugation						
Antila and Antila, 1970	AMF		—	—	2282 (IE/100 g)	1.075 (mg/100 g)
	A	L	—	—	3555	1.705
	B	L	—	—	2335	1.274
	C	M	—	—	856	0.607
	D	VH	—	—	265	0.128
Branger, 1993	AMF pilot-scale		2.32 (mg/g fat)	—	—	—
	32S_S pilot-scale	H	1.76	—	—	—
	32L_S pilot-scale	M	2.40	—	—	—
	25S_S pilot-scale	H	1.90	—	—	—
	25L_S pilot-scale	M	2.56	—	—	—
	20S_S pilot-scale	M	2.17	—	—	—
	20L_S pilot-scale	L	2.51	—	—	—
	17S_S pilot-scale	M	2.14	—	—	—
	17L_S pilot-scale	L	2.62	—	—	—
Separation with the aid of an aqueous detergent solution						
Dolby, 1970a	AMF		—	7.8 (μg/g)	5.2 (μg/g)	—
	Solid fraction	H	—	7.3	4.8	—
	Liquid fraction	L	—	8.3	5.6	—
Norris et al., 1971	AMF [A]		240 (mg/100g fat)	6.8 (μg/g fat)	8.4 (μg/g fat)	—
	25S_S [A]	H	220	5.7	6.6	—
	25L_S [A]	L	250	7.5	9.8	—
	AMF [B]		230	7.0	9.2	—
	26S_S [B]	H	230	6.7	8.2	—
	26L_S [B]	L	250	7.6	10.0	—

Continued

TABLE 5.202 (Continued)

Fractionation method Author	Fraction designation[1]	General fraction property[2]	Cholesterol content	Carotene content	Vitamin A content	Vitamin E content
	AMF [C]		250	7.7	8.6	—
	$26S_S$ [C]	H	220	7.2	7.7	—
	$26L_S$ [C]	L	260	8.5	9.6	—
Crystallization from solvent solution						
Acetone						
Chen and deMan, 1966	AMF		Total = 236 (mg/100 g) Free = 209 Ester = 17	—	—	—
	$15S_M$	U	Total = tr[5] Free = tr Ester = —	—	—	—
	$5S_M$	U	Total = tr Free = tr Ester = —	—	—	—
	$-5S_M$	U	Total = 60 Free = 20 Ester = 40	—	—	—
	$-15S_M$	U	Total = 93 Free = 33 Ester = 60	—	—	—
	$-25S_M$	U	Total = 156 Free = 125 Ester = 31	—	—	—
	$-35S_M$	U	Total = 171 Free = 136 Ester = 35	—	—	—
	$-45S_M$	U	Total = 158 Free = 130 Ester = 28	—	—	—
	$-45L_M$	U	Total = 1020 Free = 972 Ester = 48	—	—	—
Hexane						
Muuse and van der Kamp, 1985	AMF		0.278 (% m/m)	—	—	—
	$12.5S_S$	U	0.231	—	—	—
	$12.5L_S$	U	0.291	—	—	—
Supercritical fluid extraction						
Carbon dioxide						
Bhaskar et al., 1993b	AMF		2.4 (mg/g)	314 (IU/100g)	—	—
	S1	VH	1.18	768	—	—
	S2	H	2.35	—	—	—
	S3	M	2.52	—	—	—
	S4	M	3.64	—	—	—
	S5	M	3.54	—	—	—

Continued

TABLE 5.202 (Continued)

Fractionation method Author	Fraction designation[1]	General fraction property[2]	Cholesterol content	Carotene content	Vitamin A content	Vitamin E content
Bradley, 1989	AMF WM-14		2.02 (mg/g)	—	—	—
	WM-14,1 (80°C, 2300 psig)	U	0.0	—	—	—
	WM-14,2 (80°C, 2500 psig)	U	11.66	—	—	—
	WM-14,3 (80°C, 2800 psig)	U	10.62	—	—	—
	WM-14,4 (80°C, 3050 psig)	U	5.65	—	—	—
	WM-14,5 (80°C, 3350 psig)	U	2.68	—	—	—
	WM-14,6 (80°C, 3750 psig)	U	1.43	—	—	—
	WM-14,7 (80°C, 6000 psig)	U	0.25	—	—	—
	AMF WM-21		2.14	—	—	—
	WM-21,1 (80°C, 2300 psig)	U	0.16	—	—	—
	WM-21,2 (80°C, 2500 psig)	U	12.42	—	—	—
	WM-21,3 (80°C, 2800 psig)	U	4.93	—	—	—
	WM-21,4 (80°C< 3050 psig)	U	1.44	—	—	—
	WM-21,5 (80°C, 3400 psig)	U	0.31	—	—	—
	WM-21,6 (80°C, 6000 psig)	U	0.0	—	—	—
Chen et al., 1992	AMF		2.28 (mg/g)	—	—	—
	10.3 (40°C, 10.3 MPa)	L	2.96	—	—	—
	13.8 (40°C, 13.8 MPa)	H	2.80	—	—	—
	17.2 (40°C, 17.2 MPa)	M	1.09	—	—	—
	20.7 (40°C, 20.7 MPa)	H	1.76	—	—	—
	24.1 (40°C, 24.1 MPa)	H	2.46	—	—	—
	27.6 (40°C, 27.6 MPa)	H	2.41	—	—	—
Chen and Schwartz, 1991	Unknown (1500 psi)	U	2.96 (mg/g)	—	—	—
	Unknown (2000 psi)	U	2.78	—	—	—
	Unknown (2500 psi)	U	1.07	—	—	—

Continued

TABLE 5.202 (Continued)

Fractionation method Author	Fraction designation[1]	General fraction property[2]	Cholesterol content	Carotene content	Vitamin A content	Vitamin E content
	Unknown (3000 psi)	U	1.76	—	—	—
Ensign, 1989	AMF WM-38		2.61 (mg/g fat)	—	—	—
	WM-38,1 (60°C, 2150 psi)	U	7.81	—	—	—
	WM-38,2 (60°C, 2300 psi)	U	6.83	—	—	—
	WM-38,3 (60°C, 2400 psi)	U	6.32	—	—	—
	WM-38,4 (60°C, 2400 psi)	U	5.84	—	—	—
	WM-38,5 (60°C, 2500 psi)	U	5.36	—	—	—
	WM-38,6 (60°C, 2500 psi)	U	4.73	—	—	—
	WM-38,7 (60°C, 2650 psi)	U	4.17	—	—	—
	WM-38,8 (60°C, 2850 psi)	U	3.49	—	—	—
	WM-38,9 (60°C, 2900 psi)	U	3.27	—	—	—
	WM-38,10 (60°C, 2900 psi)	U	2.81	—	—	—
	WM-38,11 (60°C, 2900 psi)	U	1.57	—	—	—
	WM-38,12 (60°C, 2900 psi)	U	0.60	—	—	—
	WM-38,13 (60°C, 2900 psi)	U	0.20	—	—	—
	WM-38,14 (60°C, 2900 psi)	U	0.0	—	—	—
	WM-38,15 (60°C, —)	U	0.0	—	—	—
	AMF WM-39		2.34	—	—	—
	WM-39,1 (60°C, 2000 psi)	U	10.21	—	—	—
	WM-39,2 (60°C, 6000 psi)	U	1.92	—	—	—
	AMF WM-43		1.81	—	—	—
	WM-43,1 (40°C, 1700 psi)	U	1.10	—	—	—
	WM-43,2 (40°C, 6000 psi)	U	1.22	—	—	—
Kankare et al., 1989	AMF2		2.27 (mg/g)	—	—	—
	2E1 (48°C, 120 bar (3 h 12 min) + 150 bar (4h))	U	3.22	—	—	—

TABLE 5.202 (Continued)

Fractionation method Author	Fraction designation[1]	General fraction property[2]	Cholesterol content	Carotene content	Vitamin A content	Vitamin E content
	2E2 (48°C, 200 bar)	U	2.57	—	—	—
	2R 48°C 200 bar	U	1.61	—	—	—
Kaufmann et al., 1982	AMF		0.26 (%)	—	—	—
	I (residue) (80°C, 200 bar)	H	0.23	—	—	—
	II (extract) (80°C, 200 bar)	L	0.55	—	—	—
Rizvi et al., 1990	AMF		232.55[6] (mg/100 g fat)	—	—	—
	Extract-1 (40°C, 241.3 bar)	M	4.70	—	—	—
	Extract-2 (48°C, 275.8 bar)	H	8.02	—	—	—
	Residue	H	145.08	—	—	—
Shukla et al., 1994	Raffinate	H	1.18 (mg/g)	768 (IU/100g)	—	—
Timmen et al., 1984	AMF		0.26 (%)	—	—	—
	Long-chain fraction (80°C, 200 bar)	H	0.23	—	—	—
	Short-chain fraction (80°C, 200 bar)	L	0.55	—	—	—
Short-path distillation						
Arul et al., 1988b	AMF[7]		2.6 (mg/g fat)	—	—	—
	Solid (SF Residue)	H	0.2	—	—	—
	Intermediate (IF Distillate III) (265°C, 100 μm Hg)	M	1.4	—	—	—
	Liquid (LF1 Distillate I + LF2 Distillate II) (245°C, 220 μm Hg + 245°C, 100 μm Hg)	L	16.9	—	—	—

[1] AMF = anhydrous milkfat; MF = milkfat. The number indicates the fractionation temperature, the letter indicates physical state (L = liquid, S = solid), and the subscript indicates fractionation procedure used (M = multiple-step, S = single-step). Other primary designations and designations in parentheses are those of the original author. Designations in brackets refer to the intact milkfat. Fractionation conditions have been specified in parentheses for fractions obtained by supercritical fluid extraction and short-path distillation.

[2] VH = very-high-melting milkfat fraction (MP > 45°C), H = high-melting milkfat fraction (MP between 35 and 45°C), L = low-melting milkfat fraction (MP between 10 and 25°C), U = unknown-melting milkfat fractions (MP not available).

[3] Data not available.

[4] Fractions obtained from Makhlouf et al. (1987). Fraction designations in parentheses correspond to the fraction designations of Makhlouf et al. (1987).

[5] tr = trace.

[6] Cholesterol removal was accomplished simultaneously during extraction in conjunction with in-line adsorption by magnesium silicate.

[7] Fractions obtained from Arul et al. (1988a). Fraction designations in parentheses correspond to the fraction designations of Arul et al. (1988a).

TABLE 5.203 Formulas for Mixtures of Milkfat Fractions with Other Fats and Oils

Fat blend type Author Fractionation method	Milkfat fractions and other fats and oils used[1]	General fraction property[2]	Blend formula Ingredient	%	Technical data available for blends[3]
Milkfat fractions–intact milkfat Milkfat fractions–milkfat fractions					
Kankare and Antila, 1986 Crystallization from melted milkfat (centrifugation)	AMF $24S_M$ fraction $12S_M$ fraction $12L_M$ fraction	VH M L	*Mixture 1* AMF $24S_M$ fraction	55 45	DSC curves
			Mixture 2 $12S_M$ fraction $24S_M$ fraction	75 25	
			Mixture 3 $12L_M$ fraction $24S_M$ fraction	75 25	
Milkfat fractions–cocoa butter					
Timms, 1980b Crystallization from solvent solution (acetone)	AMF $3S_M$ fraction $3L_M$ fraction Cocoa butter	H L	*Mixture 1* AMF Cocoa butter	0→100 100→0	Isosolid diagrams
			Mixture 2 $3S_M$ fraction Cocoa butter	0→100 100→0	
			Mixture 3 $3L_M$ fraction Cocoa butter	0→100 100→0	
Timms and Parekh, 1980 Crystallization from melted milkfat (vacuum filtration)	AMF [S] $25S_S$ fraction $25L_S$ fraction Cocoa butter	H L	*Mixture 1* AMF [S] Cocoa butter	10 90	Solid fat content
			Mixture 2 AMF [S] Cocoa butter	20 80	
			Mixture 3 AMF [S] Cocoa butter	30 70	
			Mixture 4 AMF [S] Cocoa butter	40 60	
			Mixture 5 AMF [S] Cocoa butter	50 50	
			Mixture 6 $25S_S$ fraction Cocoa butter	10 90	
			Mixture 7 $25S_S$ fraction Cocoa butter	20 80	

Continued

TABLE 5.203 (Continued)

Fat blend type Author Fractionation method	Milkfat fractions and other fats and oils used[1]	General fraction property[2]	Blend formula Ingredient	%	Technical data available for blends[3]
			Mixture 8		
			$25S_S$ fraction	30	
			Cocoa butter	70	
			Mixture 9		
			$25S_S$ fraction	40	
			Cocoa butter	60	
			Mixture 10		
			$25S_S$ fraction	50	
			Cocoa butter	50	
			Mixture 11		
			$25L_S$ fraction	10	
			Cocoa butter	90	
			Mixture 12		
			$25L_S$ fraction	20	
			Cocoa butter	80	
			Mixture 13		
			$25L_S$ fraction	30	
			Cocoa butter	70	
			Mixture 14		
			$25L_S$ fraction	40	
			Cocoa butter	60	
			Mixture 15		
			$25S_S$ fraction	50	
			Cocoa butter	50	
			Mixture 16		
			AMF [S]	30	
			Cocoa butter	70	
			Mixture 17		
			AMF [S]	35	
			Cocoa butter	65	
			Mixture 18		
			AMF [S]	40	
			Cocoa butter	60	
			Mixture 19		
			AMF [S]	25	
			$25S_S$ fraction	5	
			Cocoa butter	70	
			Mixture 20		
			AMF [S]	25	
			$25S_S$ fraction	10	
			Cocoa butter	65	
			Mixture 21		
			AMF [S]	25	
			$25S_S$ fraction	15	
			Cocoa butter	60	
Barna et al., 1992 Crystallization from melted milkfat (vacuum filtration)	$30S_M$ fraction $20S_M$ fraction $20L_M$ fraction Cocoa butter	 H M L	*Mixture 1* $30S_M$ fraction Cocoa butter	 20 80	DSC curves Isosolid diagrams

Continued

TABLE 5.203 (Continued)

Fat blend type Author Fractionation method	Milkfat fractions and other fats and oils used[1]	General fraction property[2]	Blend formula Ingredient	%	Technical data available for blends[3]
			Mixture 2		
			20S$_M$ fraction	20	
			Cocoa butter	80	
			Mixture 3		
			20L$_M$ fraction	20	
			Cocoa butter	80	
			Mixture 4		
			30S$_M$ fraction	0→100	
			Cocoa butter	100→0	
			Mixture 5		
			20S$_M$ fraction	0→100	
			Cocoa butter	100→0	
			Mixture 6		
			20L$_M$ fraction	0→100	
			Cocoa butter	100→0	
Yi, 1993 Crystallization from solvent solution (acetone)	AMF [S] 25S$_M$ fraction 20S$_M$ fraction 15S$_M$ fraction 10S$_M$ fraction 5S$_M$ fraction 0S$_M$ fraction 0L$_M$ fraction Cocoa butter	VH VH VH H M M L	*Mixture 1* AMF [S] Cocoa butter *Mixture 2* 25S$_M$ fraction Cocoa butter *Mixture 3* 20S$_M$ fraction Cocoa butter *Mixture 4* 15S$_M$ fraction Cocoa butter *Mixture 5* 10S$_M$ fraction Cocoa butter *Mixture 6* 5S$_M$ fraction Cocoa butter *Mixture 7* 0S$_M$ fraction Cocoa butter *Mixture 8* 0L$_M$ fraction Cocoa butter	6.4 93.6 6.4 93.6 6.4 93.6 6.4 93.6 6.4 93.6 6.4 93.6 6.4 93.6 6.4 93.6	Solid fat content

Milkfat fraction–other fats and oils

Verhagen and Bodor, 1984 Crystallization from melted milkfat (separation technique not reported)	AMF 25S$_S$ fraction 15S$_S$ fraction Sunflower oil Soybean oil	H M	*Mixture 1* AMF 25S$_S$ fraction Sunflower oil *Mixture 2* 15S$_M$ fraction Sunflower oil	40 20 40 60 40	Solid fat content

Continued

TABLE 5.203 (Continued)

Fat blend type Author Fractionation method	Milkfat fractions and other fats and oils used[1]	General fraction property[2]	Blend formula Ingredient	%	Technical data available for blends[3]
			Mixture 3		
			25S$_S$ fraction	40	
			Soybean oil	60	
Kankare and Antila, 1988b Crystallization from melted milkfat (centrifugation)	24S$_M$ fraction Rapeseed oil	VH	*Mixture 1* 24S$_M$ fraction Rapeseed oil	40 60	DSC curves Solid fat content
			Mixture 2		
			24S$_M$ fraction	50	
			Rapeseed oil	50	
			Mixture 3		
			24S$_M$ fraction	60	
			Rapeseed oil	40	
Versteeg, 1991 Crystallization from melted milkfat (vacuum filtration)	13L$_M$ fraction Tallow 0021 (soft fraction obtained in a two-step fractionation of tallow at 21°C)	U	*Mixture 1* 13L$_M$ fraction Tallow 0021	100→0 0→100	Isosolid diagrams

[1] AMF = anhydrous milkfat. The number indicates the fractionation temperature, the letter indicates physical state (L = liquid, S = solid), and the subscript indicates fractionation procedure used (M = multiple-step, S = single-step). Designations in brackets refer to the intact milkfat (S = summer).

[2] VH = very-high-melting milkfat fractions (MP > 45°C), H = high–melting milkfat fraction (MP between 35 and 45°C), M = middle-melting milkfat fractions (MP between 25 and 35°C), L = low-melting milkfat fraction (MP between 10 and 25°C), U = unknown-melting milkfat fractions (MP not available).

[3] See List of Tables and List of Figures on pp. xxiii–xxiv for location of technical data.

TABLE 5.204 Solid Fat Content Profiles for Mixtures of Milkfat Fractions with Other Fats and Oils

Fat blend type Author Fractionation method	Blend formula Ingredient[1]	%	\multicolumn{9}{c}{Solid fat content (%) at temperature (°C)}								
			0	5	10	15	20	25	30	35	40
Milkfat fractions–cocoa butter											
Timms and Parekh, 1980	Mixture 1		—[2]	—	78.5	72.6	65.4	55.4	21.8	0.7	0.0
	AMF [S]	10									
Crystallization from melted milkfat	Cocoa butter	90									
(vacuum filtration)	Mixture 2		—	—	73.0	65.8	56.3	45.9	14.9	0.8	0.0
	AMF [S]	20									
	Cocoa butter	80									
	Mixture 3		—	—	67.6	57.9	46.0	35.0	10.6	0.5	0.0
	AMF [S]	30									
	Cocoa butter	70									
	Mixture 4		—	—	62.5	47.1	—	20.2	6.8	0.5	0.0
	AMF [S]	40									
	Cocoa butter	60									
	Mixture 5		—	—	57.6	37.5	11.7	9.2	4.3	0.6	0.0
	AMF [S]	50									
	Cocoa butter	50									
	Mixture 6		—	—	77.3	69.4	62.1	53.4	23.7	0.3	0.0
	25S$_S$ fraction	10									
	Cocoa butter	90									
	Mixture 7		—	—	73.4	62.6	53.0	45.0	18.8	0.7	0.0
	25S$_S$ fraction	20									
	Cocoa butter	80									
	Mixture 8		—	—	70.7	56.1	37.2	30.8	14.9	1.7	0.2
	25S$_S$ fraction	30									
	Cocoa butter	70									
	Mixture 9		—	—	68.2	51.4	24.7	17.6	9.2	2.8	0.4
	25S$_S$ fraction	40									
	Cocoa butter	60									
	Mixture 10		—	—	63.1	43.8	19.9	17.2	10.9	4.0	0.0
	25S$_S$ fraction	50									
	Cocoa butter	50									
	Mixture 11		—	—	77.1	71.0	63.1	52.8	20.7	0.3	0.0
	25L$_S$ fraction	10									
	Cocoa butter	90									
	Mixture 12		—	—	71.4	63.0	51.8	39.9	12.5	0.0	0.0
	25L$_S$ fraction	20									
	Cocoa butter	80									
	Mixture 13		—	—	64.1	54.9	40.6	28.5	7.4	0.0	0.0
	25L$_S$ fraction	30									
	Cocoa butter	70									
	Mixture 14		—	—	58.0	46.4	30.8	19.5	2.9	0.0	0.0
	25L$_S$ fraction	40									
	Cocoa butter	60									
	Mixture 15		—	—	51.1	26.4	4.7	3.0	1.0	0.0	0.0
	25S$_S$ fraction	50									
	Cocoa butter	50									

Continued

TABLE 5.204 (Continued)

Fat blend type Author Fractionation method	Blend formula Ingredient[1]	%	Solid fat content (%) at temperature (°C)								
			0	5	10	15	20	25	30	35	40
	Mixture 16		—	—	67.4	57.6	48.1	36.4	10.0	0.6	0.0
	AMF [S]	30									
	Cocoa butter	70									
	Mixture 17		—	—	64.5	51.1	40.5	31.1	9.2	0.4	0.0
	AMF [S]	35									
	Cocoa butter	65									
	Mixture 18		—	—	62.9	45.4	26.9	21.6	7.8	0.9	0.0
	AMF [S]	40									
	Cocoa butter	60									
	Mixture 19		—	—	67.3	56.8	47.5	37.1	10.8	1.0	0.0
	AMF [S]	25									
	25S$_S$ fraction	5									
	Cocoa butter	70									
	Mixture 20		—	—	66.1	51.1	37.5	29.9	10.8	1.0	0.0
	AMF [S]	25									
	25S$_S$ fraction	10									
	Cocoa butter	65									
	Mixture 21		—	—	65.1	47.8	26.1	19.9	8.9	1.6	0.0
	AMF [S]	25									
	25S$_S$ fraction	15									
	Cocoa butter	60									
Yi, 1993 Crystallization from solvent solution (acetone)	*Mixture 1*		83.7	81.2	74.3	70.4	62.4	54.9	32.4	0.0	—
	AMF [S]	6.4									
	Cocoa butter	93.6									
	Mixture 2		91.8	88.4	85.4	82.6	79.9	74.3	53.5	0.0	—
	25S$_M$ fraction	6.4									
	Cocoa butter	93.6									
	Mixture 3		91.1	87.5	85.5	82.1	78.5	72.4	49.8	0.0	—
	20S$_M$ fraction	6.4									
	Cocoa butter	93.6									
	Mixture 4		91.0	88.3	85.3	80.9	77.7	70.1	43.0	0.0	—
	15S$_M$ fraction	6.4									
	Cocoa butter	93.6									
	Mixture 5		78.3	71.7	66.7	61.8	59.1	47.8	19.1	0.0	—
	10S$_M$ fraction	6.4									
	Cocoa butter	93.6									
	Mixture 6		80.1	71.0	66.9	63.9	58.5	47.9	22.3	0.0	—
	5S$_M$ fraction	6.4									
	Cocoa butter	93.6									
	Mixture 7		80.6	71.0	69.0	62.9	59.2	45.4	22.6	0.0	—
	0S$_M$ fraction	6.4									
	Cocoa butter	93.6									
	Mixture 8		80.9	71.6	66.1	63.5	59.3	49.4	21.8	0.0	—
	0L$_M$ fraction	6.4									
	Cocoa butter	93.6									

Continued

TABLE 5.204 (Continued)

Fat blend type / Author / Fractionation method	Blend formula Ingredient[1]	%	Solid fat content (%) at temperature (°C)								
			0	5	10	15	20	25	30	35	40
Milkfat fraction–other fats and oils											
Verhagen and Bodor, 1984 Crystallization from melted milkfat (separation technique not reported)	*Mixture 1* AMF 25S$_S$ fraction Sunflower oil	40 20 40	—	—	25	—	12	—	—	0.5	—
	Mixture 2 15S$_M$ fraction Sunflower oil	60 40	—	—	30	—	14	—	—	0.7	—
	Mixture 3 25S$_S$ fraction Soybean oil	40 60	—	—	20	—	11	—	—	1.0	—
Kankare and Antila, 1988b Crystallization from melted milkfat (centrifugation)	*Mixture 1* 24S$_M$ fraction Rapeseed oil	40 60	51.4	—	40.3	—	32.4	—	20.2	—	1.5
	Mixture 2 24S$_M$ fraction Rapeseed oil	50 50	63.2	—	50.9	—	42.4	—	27.6	—	4.1
	Mixture 3 24S$_M$ fraction Rapeseed oil	60 40	74.6	—	61.6	—	52.5	—	35.8	—	8.7

[1] AMF = anhydrous milkfat. The number indicates the fractionation temperature, the letter indicates physical state (L = liquid, S = solid), and the subscript indicates fractionation procedure used (M = multiple-step, S = single-step). Designations in brackets refer to the intact milkfat (S = summer).

[2] Data not available.

Milkfat Fractionation

TABLE 5.205 List of Some Commercially Available Milkfat Fraction–Based Products

Company Location data available[1]	Product name	Melting behavior(°C)[2]	General fraction property[3]	Suggested applications
Alaco New Zealand Melting behavior	Bakery Butter	—[4]	U	General-purpose bakery
	Butter Shortening	—	U	General-purpose bakery
	Butter Sheets	—	U	Croissants Danish pastry Puff pastry
	Confectionery Butterfat 42	42 (MP)	H	Chocolate Fudge Caramel
	Soft Butteroil 21	21	L	Ice cream Recombined dairy products
	Soft Butteroil 28	28	M	Ice cream Recombined dairy products
	Pastry Butter	37	H	Croissants Danish pastry Puff pastry
	Cookie Butter	—	U	Shortbread High-butter biscuits High-butter cookies

Aveve
Belgium
  Selected chemical characteristics
  Melting behavior
	Anhydrous Milk Fat	—	U	General-purpose food and bakery
				Recombined milk products
				Ice cream

	Pure Cow Anhydrous Milk Fat 13–15°C Melting Point	13–15 (MP)	L	—
	Concentrated Butter Formula B Concentrated Butter	28–32	M	Ice cream
	"Creme au Beurre"	28	M	Recombined cream Butter cookies
	Concentrated Butter "4/4"	26-28	M	Doughs Biscuits Chocolate
	Concentrated Butter "Patissier"	28–32	M	Doughs Biscuits Chocolate
	Concentrated Butter "Croissant"	38	H	Doughs
	Concentrated Butter "Millefeuille"	40–42	H	All puff pastry applications
	Concentrated Butter Feuilletage 2000	24–26	L/M	All puff pastry applications

S.A.N. Corman
Belgium
  Fatty acid composition
  Selected chemical characteristics
  Melting behavior
  Solid fat content
	Standard Anhydrous Milk Fat	32 (MP)	M	General-purpose bakery
				Recombined milk products
				Ice cream
				Culinary dressings

Continued

TABLE 5.205 (Continued)

Company Location data available[1]	Product name	Melting behavior(°C)[2]	General fraction property[3]	Suggested applications
	Crème au Beurre	—	U	Butter icings and fillings Petits fours Soft centers for chocolates
	4/4 Concentrated Butter	—	U	Doughs requiring the use of melted butter Liquid doughs (sponge cake)
	Danish 4/4	30	M	Danish butter cookies Shortbread Finger biscuits
	Patissier Concentrated Butter	—	U	Self-raising pastries (Viennese shortcakes, bread, rolls)
	Croissant Concentrated Butter	38	H	Short pastry dough Raised puff pastry (croissants, Danish) Puff pastry
	Millefeuille Concentrated Butter	41	H	Any puff pastry application
	Glacier Concentrated Butter	30–32	M	Ice cream
	Glacier Extra Concentrated Butter	30–32	M	Ice cream
	Extra White Anhydrous Milk Fat	32	M	Soft cheeses (goat, Feta)
	Milk Extra Anhydrous Milk Fat	28	M	Recombined milk products Yogurt Ice cream
Lambelet, 1983 Triglyceride composition Melting behavior DSC curves Solid fat content	Corman fraction (S2)[5]	36–38 (MP)	H	—
	Corman fraction (S3)	41	H	
Büning-Pfaue et al., 1989 Fatty acid composition Triglyceride composition DSC curves	Corman fraction MF-35[6]	35 (MP)	H	—
	Corman fraction MF-38	38	H	
	Corman fraction MF-42	42	H	
	Saumweber fraction MF-28	28	M	
	Saumweber fraction MF-36	36	H	
	Saumweber fraction MF-42	42	H	

[1]Refer to the List of Tables and List of Figures for the exact location of data available.

[2]MP = melting point.

[3]H = high-melting milkfat fractions (MP between 35 and 45°C), M = middle-melting milkfat fractions (MP between 25 and 35°C), L = low-melting milkfat fractions (MP between 10 and 25°C), U = unknown-melting milkfat fractions (MP not available).

[4]Data not reported.

[5]Fractions characterized by Lambelet (1983) were obtained from Corman, Belgium. The S2 and S3 fraction designation is that of Lambelet.

[6]Fractions characterized by Büning-Pfaue et al. (1989) were obtained from Corman, Belgium and from Saumweber, Germany.

TABLE 5.206 Fatty Acid Composition of Commercially Available Milkfat Fraction Products

Company	Product name	Fatty acid (%)[1]												Selected saturated n-chain groupings (%)			Selected ratio (%)
		4:0	6:0	8:0	10:0	12:0	14:0	16:0	18:0	18:1	18:2	18:3	C4–C10	C12–C15	C12, C14, C16	C16–C18	C18:1/C18:0
S.A.N. Corman	Standard Anhydrous Milk Fat	3.0–4.5	1.9–2.7	1.1–1.5	2.1–3.2	2.5–4.0	9.0–11.5	22.0–32.0	9.3–13.0	21.5–30.0	1.5–2.7	0.5–2.5	11.90	15.50	47.50	45.00	2.31
	Danish 4/4	2.9–4.6	1.9–2.9	1.1–1.6	2.3–3.3	2.7–4.0	9.2–11.4	22.7–29.8	9.0–12.6	21.9–29.1	1.5–3.1	0.6–2.4	12.40	15.40	45.20	42.40	2.31
	Croissant Concentrated Butter	2.4–3.7	1.6–2.3	1.0–1.4	2.1–3.0	2.7–4.0	9.7–12.1	25.5–33.1	10.0–15.5	19.5–25.5	1.2–2.7	0.5–2.0	10.40	16.10	49.20	48.60	1.65
	Millefeuille Concentrated Butter	2.0–3.4	1.4–2.1	0.9–1.3	2.1–2.9	2.8–4.0	10.2–12.2	28.0–35.4	10.6–15.8	17.5–22.8	1.0–2.4	0.2–1.8	9.70	16.20	51.60	51.20	1.44
	Glacier Concentrated Butter	3.0–4.6	1.9–2.9	1.1–1.7	2.1–3.3	2.7–4.0	9.1–11.4	22.0–30.0	9.0–12.6	22.0–30.0	1.5–2.9	0.6–2.4	12.50	15.40	45.40	42.60	2.38
	Glacier Extra Concentrated Butter	3.0–4.6	1.9–2.9	1.1–1.7	2.1–3.3	2.7–4.0	9.1–11.4	22.0–30.0	9.0–12.6	22.0–30.0	1.5–2.9	0.6–2.4	12.50	15.40	45.40	42.60	2.38
	Extra White Anhydrous Milk Fat	3.0–4.5	1.9–2.7	1.1–1.5	2.1–3.2	2.5–4.0	9.0–11.5	22.0–32.0	9.3–13.0	21.5–30.0	1.5–2.7	0.5–2.5	11.90	15.50	47.50	45.00	2.31
	Milk Extra Anhydrous Milk Fat	3.1–4.7	2.0–2.9	1.2–1.7	2.4–3.3	2.9–4.0	9.4–11.3	23.0–30.0	8.7–11.7	22.8–29.5	1.8–2.9	0.6–2.3	12.60	15.30	45.30	41.70	2.52
Büning-Pfaue, et al., 1989	Corman fraction MF-35[1]	2.8	1.8	1.1	2.7	3.8	12.2	32.8	10.6	23.1	2.4	–[2]	8.40	16.00	48.80	43.40	2.18
	Corman fraction MF-38	2.0	1.3	0.8	2.3	3.5	11.9	33.7	12.2	22.3	2.4	—	6.40	15.40	49.10	45.90	1.83
	Corman fraction MF-42	1.4	1.0	0.6	1.8	2.9	10.4	31.0	18.0	19.9	2.8	—	4.80	13.30	44.30	49.00	1.11

Saumweber fraction MF-28	2.9	2.0	1.1	2.8	3.9	12.4	32.4	8.3	25.1	2.0	—	8.80	16.30	48.70	40.70	3.02
Saumweber fraction MF-36	2.6	1.8	0.9	2.5	3.7	12.2	34.6	9.8	22.1	2.0	—	7.80	15.90	50.50	44.40	2.26
Saumweber fraction MF-42	2.1	1.5	0.9	2.5	4.0	13.6	38.6	11.4	18.2	1.7	—	7.00	17.60	56.20	50.00	1.60

[1]Fractions characterized by Büning-Pfaue et al. (1989) were obtained from Corman, Belgium and from Saumweber, Germany.

[2]Data not reported.

TABLE 5.207 Triglyceride Composition of Commercially Available Milkfat Fraction Products

Company Author	Product name	Triglyceride (%)[1]															Selected triglyceride groupings (%)				
		C26	C28	C30	C32	C34	C36	C38	C40	C42	C44	C46	C48	C50	C52	C54	C24–C34	C36–C40	≤C40	C42–C5
Lambelet, 1983	Corman fraction (S2)[1]	0.2	0.3	0.7	1.5	3.2	6.5	—[2]	—	—	—	—	—	—	—	—	5.90	6.50	12.40	—
	Corman fraction (S3)	0.2	0.3	0.6	1.3	3.0	5.8	—	—	—	—	—	—	—	—	—	5.40	5.80	11.20	—
Büning-Pfaue et al., 1989	Corman fraction MF-35[3]	0.33[4]	0.8	1.5	2.9	6.4	11.5	12.6	9.3	6.2	5.9	6.3	7.6	9.5	9.0	2.0	11.93	33.40	45.33	46.50
		—	—	—	4.5[5] <C34	2.6	14.9	19.8	13.4	7.7	7.5	9.9	11.7	6.5	1.5	—	36.60	48.10	84.70	44.80
	Corman fraction MF-38	0.3[4]	0.7	1.3	2.6	5.7	10.4	11.4	8.7	6.3	6.2	7.2	8.7	10.6	9.3	1.7	10.63	30.50	41.13	50.00
		—	—	—	6.5[5] <C34	2.5	14.5	18.5	11.3	5.9	7.1	10.3	13.4	7.8	2.3	—	36.50	44.30	80.80	46.80
	Corman fraction MF-42	0.2[4]	0.5	1.0	1.8	3.8	7.1	8.8	7.5	5.4	6.0	7.5	9.5	13.8	12.1	5.8	7.33	23.40	30.73	60.10
		—	—	—	4.8[5] <C34	0.3	11.3	14.9	10.5	7.3	7.9	12.1	15.9	11.2	3.8	—	34.30	36.70	71.00	58.20
	Saumweber fraction MF-28	0.3[4]	0.8	1.5	3.2	7.8	13.6	13.7	9.6	6.3	5.5	5.6	6.9	9.5	9.3	—	13.63	36.90	50.53	43.10
		—	—	—	7.3[5] <C34	2.8	16.4	21.1	13.8	7.6	6.2	8.6	11.3	4.7	0.2	—	36.80	51.30	88.10	38.60
	Saumweber fraction MF-36	0.3[4]	0.7	1.3	2.6	4.8	10.5	11.8	9.5	7.4	6.2	7.1	8.2	9.7	8.4	—	9.73	31.80	41.53	48.70
		—	—	—	5.9[5] <C34	5.6	12.3	16.6	8.7	6.6	7.3	12.1	14.8	1.9	—	—	39.60	37.60	77.20	51.10

Saumweber fraction MF-42	0.2[4]	0.6	1.1	2.1	4.9	8.7	9.6	7.3	6.0	6.8	8.5	10.2	12.2	9.2	4.0	8.93	25.60	34.53	56.90
	—	—	—	3.0[5] <C34	2.7	9.9	15.0	11.0	6.4	8.7	13.2	16.2	9.9	4.0	—	36.70	35.90	72.60	58.40

[1] Fractions characterized by Lambelet (1983) were obtained from Corman, Belgium. The S2 and S3 fraction designation is that of Lambelet.

[2] Data not available.

[3] Fractions characterized by Büning-Pfaue et al.(1989) were obtained from Corman, Belgium and from Saumweber, Germany.

[4] Triglyceride composition determined by GC (wt %).

[5] Triglyceride composition determined by HPLC (area %).

TABLE 5.208 Selected Chemical Characteristics of Commercially Available Milkfat Fraction Products

Company Location	Product name	Iodine value	Refractive index	Reichert–Meissl number	Polenske number	Density (g/mL)
Aveve Belgium	Pure Cow Anhydrous Milk Fat 13–15°C Melting Point	40–45	—[1]	—	—	—
	Concentrated Butter Formula B	29–40	—	—	—	—
	Concentrated Butter "Creme au Beurre"	28–42	—	—	—	—
	Concentrated Butter "4/4"	28–42	—	—	—	—
	Concentrated Butter "Patissier"	28–40	—	—	—	—
	Concentrated Butter "Croissant"	28–42	—	—	—	—
	Concentrated Butter "Millefeuille"	28–42	—	—	—	—
S.A.N. Corman Belgium	Standard Anhydrous Milk Fat	28–42	1.453–1.456 (40°C)	24–32	1–4	0.918 (30°C) 0.908 (40°C)
	Danish 4/4	28–42	1.453–1.456 (40°C)	24–32	1–4	0.918 (30°C) 0.908 (40°C)
	Croissant Concentrated Butter	—	—	—	—	0.918 (30°C) 0.908 (40°C)
	Millefeuille Concentrated Butter	—	—	—	—	0.918 (30°C) 0.908 (40°C)
	Glacier Concentrated Butter	28–42	1.453–1.456 (40°C)	24–32	1–4	0.918 (30°C) 0.908 (40°C)
	Glacier Extra Concentrated Butter	28–42	1.453–1.456 (40°C)	24–32	1–4	0.918 (30°C) 0.908 (40°C)
	Extra White Anhydrous Milk Fat	28–42	1.453–1.456 (40°C)	24–32	1–4	0.918 (30°C) 0.908 (40°C)
	Milk Extra Anhydrous Milk Fat	28–42	1.453–1.456 (40°C)	24–32	1–4	0.918 (30°C) 0.908 (40°C)

[1] Data not available.

TABLE 5.209 Solid Fat Content of Commercially Available Milkfat Fraction Products

Company/author	Product name	Solid fat content (%) at temperature (°C)									
		0	5	10	15	20	25	30	35	40	45
S.A.N. Corman	Standard Anhydrous Milk Fat	—[1]	49	39	28	16	9	5	0	—	—
	Danish 4/4	—	48	38	25	13	8	3	0	—	—
	Croissant Concentrated Butter	—	59	53	42	29	21	13	6	0	—
	Millefeuille Concentrated Butter	—	67	63	55	44	35	25	16	7	0
	Glacier Concentrated Butter	—	48	38	25	13	7	3	0	—	—
	Glacier Extra Concentrated Butter	—	48	38	25	13	7	3	0	—	—
	Extra White Anhydrous Milk Fat	—	49	39	28	16	9	5	0	—	—
	Milk Extra Anhydrous Milk Fat	—	46	37	26	13	4	0	—	—	—
Lambelet, 1983	Corman fraction (S2)[2]	57.1[3]	49.7	44.6	36.0	32.8	27.9	20.9	12.5	3.7	—
		88.6[4]	79.4	67.0	53.9	49.6	42.7	33.2	20.2	4.0	—
		85.8[5]	74.2	58.7	42.3	36.9	30.4	23.6	14.4	2.8	—
	Corman fraction (S3)	64.7[3]	55.7	51.4	41.7	37.9	30.3	26.0	19.8	10.8	—
		90.1[4]	82.7	72.0	60.9	56.4	48.7	39.7	28.4	9.0	—
		87.1[5]	77.3	63.2	48.7	42.9	36.3	29.6	21.2	6.7	—

[1]Data not available.

[2]Fractions characterized by Lambelet (1983) were obtained from Corman, Belgium. The S2 and S3 fraction designation is that of Lambelet.

[3]Solid fat content determined by NMR.

[4]Solid fat content determined by DSC, noncorrected values.

[5]Solid fat content determined by DSC, corrected values.

Chapter 6
Overview of Milkfat Fractionation

This chapter is intended as a general overview of fractionation and the properties of milkfat fractions for those readers who have neither the time nor the desire to study the detailed data presented in Chapter 5. Because there are many variables that affect the fractionation process and the properties of the fractions, it is often difficult to report narrowly defined ranges for fraction characteristics, such as fatty acid composition and solid fat content, and to recommend specific fractionation conditions to produce milkfat fractions with specific properties. One of the most influential variables is the chemical nature of the starting milkfat, which can sometimes have more influence on the resulting fractions than the processing conditions used. For this reason, the effects of fractionation are generally discussed with respect to intact-milkfat, and specific literature citations are not given.

Conventions Used to Describe Milkfat Fractions

Conventions used herein to facilitate the interpretation and discussion of the physical and chemical characteristics of milkfat fractions include the fraction designation and the general fraction property. These conventions also form a framework within which to link the physical and chemical characteristics of milkfat fractions with the functional properties of milkfat fractions in food products.

Fraction Designation

The fraction designation refers to individual milkfat fractions. Milkfat fractions obtained by crystallization from melted milkfat and from solvent solutions have been designated by a number indicating their fractionation temperature, followed by a letter indicating physical state at separation (L = liquid, S = solid) and a subscript that indicates the fractionation procedure employed (M = multiple-step, S = single-step). For example, a $15L_S$ fraction was the liquid fraction obtained at 15°C using single-step fractionation, whereas a $25S_M$ fraction was the solid fraction obtained at 25°C using a multiple-step fractionation procedure. However, for fractions obtained using multiple-step fractionation, the fraction designation does not indicate the step at which the fraction was produced. The overall fractionation schemes (i.e., numbers of steps and temperatures employed) of individual investigators are presented in Tables 5.1, 5.2, 5.3, and 5.4 for milkfat fractions obtained by crystallization from melted milkfat, crystallization from solvent solution, supercritical fluid extraction, and by short-path distillation, respectively.

Fraction designations provided by the original author were used if these fractionation data were not available for fractions obtained by crystallization methods. Designations provided by the original author were used for all fractions obtained by supercritical fluid extraction and short-path distillation.

General Fraction Property

The general fraction property reflects the overall functionality of broad categories of milkfat fractions. These categories are based on the relative melting point of milkfat fractions:

- Very-high-melting milkfat fractions: MP > 45°C
- High-melting milkfat fractions: MP between 35 and 45°C
- Middle-melting milkfat fractions: MP between 25 and 35°C
- Low-melting milkfat fractions: MP between 10 and 25°C

- Very-low-melting milkfat fractions: MP < 10°C
- Unknown-melting milkfat fractions: fractions that could not be categorized by melting behavior because data were not reported

Generalizations of the physical and chemical characteristics of milkfat fractions provide a perspective on the functionality of milkfat fractions in foods, but specific characteristics of individual milkfat fractions within the categories usually vary.

Technologies Used to Fractionate Milkfat

Crystallization from melted milkfat, crystallization from solvent solution, and supercritical fluid extraction are the primary technologies that have been used experimentally to fractionate milkfat; these have been discussed in more detail in Chapter 3. Wide ranges of fractionation conditions have been employed experimentally for each of these technologies, including separation methods, type of solvents, number of fractionation steps, and fractionation temperatures and pressures.

The only technology that is currently employed commercially for milkfat fractionation is crystallization from melted milkfat. Solid and liquid fractions can be separated by vacuum filtration, by pressure filtration, or with the aid of an aqueous detergent solution.

Crystallization from Melted Milkfat

Crystallization from melted milkfat is a relatively simple, purely physical process, and generally, additives are not employed. The exception to this is the separation of the solid and liquid fractions with the aid of an aqueous detergent solution. Solid fractions obtained by crystallization from melted milkfat tend to have some liquid milkfat contained within the crystal structure, and additional liquid milkfat adheres to the surface of the crystals and the spaces between agglomerated crystals. The presence of liquid milkfat in the crystalline solid fraction results in increased yields, decreased melting points, more gradual melting over a broad temperature range compared with more purely solid fractions. Solid milkfat fractions separated with the aid of an aqueous detergent solution tend to contain less liquid milkfat; consequently, these fractions exhibit decreased yields, increased melting points, elevated solid fat content profiles, and a more rapid melting profile than milkfat fractions obtained with other separation methods. Separation of solid and liquid milkfat fractions with the use of pressure filtration also may produce solid fractions of higher purity, especially where the pressure is sufficient to force the liquid milkfat physically from the solid crystal matrix.

Crystallization from Solvent Solution

Crystallization from solvent solution produces solid fractions purer than those obtained from melted milkfat using similar fractionation conditions (e.g., temperature, number of fractionation steps). Crystal formation of milkfat dissolved in a solvent solution occurs without the interference of liquid milkfat because the liquid milkfat remains dissolved in the solvent, resulting in the formation of relatively pure crystal nuclei that grow without entrapping liquid milkfat. Consequently, solid milkfat fractions obtained from solvent solution exhibit decreased yields, increased melting points, and elevated solid fat contents compared with solid fractions obtained from melted milkfat. Solid fractions obtained from solvent solution exhibit solid fat content profiles that show little melting over a broad temperature range, followed by sharp melting over a narrow temperature range.

Concerns associated with fractionation from solvent solution include potential residual solvent, flavor changes, and the hazards and costs involved with operating a solvent recovery system.

Supercritical Fluid Extraction

Supercritical fluid (i.e., carbon dioxide) extraction separates fractions based on molecular weight rather than on melting point, which is the driving force for crystallization processes. The first extracted fraction obtained with increasing temperature and pressure generally exhibits the lowest melting behavior, and melting points of subsequent fractions gradually increase as the extraction process progresses. Generally, the highest-melting fraction is the residual fraction remaining in the extraction vessel after the extraction is complete. The yields for milkfat fractions obtained by supercritical fluid extraction have appeared to vary depending on the fractionation conditions employed (e.g., fractionation temperature and pressure, number of fractionation steps), and trends in yields were not established. The extraction time may also have influenced the yields, but these data were generally not reported.

The extraction fluid most often employed is carbon dioxide, which is nontoxic. The supercritcal fluid extraction process does not employ other additives. Although it requires a high capital investment, supercritical fluid extraction has relatively low operating expenses.

Experimental Milkfat Fractions

Over 850 milkfat fractions have been produced experimentally, and 98% of these fractions were produced by crystallization from melted milkfat, crystallization from solvent solution, or supercritical fluid extraction. Two-thirds of these fractions have been classified based on melting behavior. To facilitate discussions on fractionation conditions and on the effects of fractionation on selected properties of milkfat fractions, the trends reported here have been concluded only from data reported for fractions that have been characterized for melting behavior.

Fractionation Conditions and Yields

Very-High-Melting Milkfat Fractions

Crystallization from melted milkfat   The very-high-melting milkfat fractions produced by crystallization from melted milkfat were solid fractions obtained either from a single-step fractionation or at the first step of a multiple-step fractionation at temperatures between 23 and 34°C. Very-high-melting milkfat fractions obtained from melted milkfat have been separated by vacuum filtration, by pressure filtration, by centrifugation, with the aid of an aqueous detergent solution, and by techniques not reported. Yields for these fractions ranged from 3 to 36%.

Very-high-melting milkfat fractions have also been produced by the refractionation of very-high-melting milkfat fractions, and this has resulted in fractions with increased melting points.

Crystallization from solvent solution   Very-high-melting milkfat fractions produced by crystallization from solvent solution were solid fractions obtained either from single-step fractionation or at the first through fourth steps of multiple-step fractionation procedures. However, the temperature range employed for crystallization from solvent solution was 0 to 32°C, which was a more broad range and lower temperatures than those employed for crystallization from melted milkfat. Very-high-melting milkfat fractions have been obtained from acetone and pentane solutions. Yields for these fractions ranged from 1 to 33%, but the majority of the fractions exhibited yields between 3 and 10%.

Very-high-melting milkfat fractions obtained from solvent solution that have been refractionated at the same temperature from solvent solution exhibited decreased melting points; this behavior was the opposite of that of refractionated fractions obtained from melted milkfat, which exhibited increased melting points.

Supercritical fluid extraction Very-high-melting milkfat fractions obtained by supercritical fluid extraction were the residual fractions that remained after the extraction process had been completed. The yield of one very-high-melting milkfat fraction obtained by supercritical carbon dioxide was reported as 21%.

High-Melting Milkfat Fractions

Crystallization from melted milkfat High-melting milkfat fractions were obtained from melted milkfat, either using single-step fractionation or at the first through fourth steps of multiple-step fractionation procedures. High-melting milkfat fractions generally were solid fractions obtained at fractionation temperatures between 12 and 38°C. High-melting milkfat fractions have been separated using vacuum filtration, pressure filtration, or centrifugation; with the aid of an aqueous detergent solution; by filtration through muslin cloth, a milk filter, a cheese press and a casein-dewatering press; and using separation techniques not reported. Yields for high-melting milkfat fractions obtained from melted milkfat ranged from 7 to 60%. Yields for fractions obtained by single-step fractionation were generally greater than for fractions obtained by multiple-step fractionation at similar temperatures.

Crystallization from solvent solution High-melting milkfat fractions have been obtained from solvent solution using single-step fractionation or at the first through fourth steps of multiple-step fractionation procedures. High-melting milkfat fractions were solid fractions obtained from acetone solution between 0 and 25°C and from ethanol solution at 18°C. Yields for high-melting milkfat fractions obtained from acetone solution ranged from 3 to 48%, and the fraction obtained from ethanol solution exhibited a yield of 13%.

Supercritical fluid extraction High-melting milkfat fractions were the second through eighth extracted fractions collected or the residual fractions that remained after the extraction process had been completed. High-melting milkfat fractions have been obtained by supercritical carbon dioxide extraction using a wide range of fractionation temperatures and pressures. Yields for high-melting milkfat fractions obtained by supercritical carbon dioxide extraction ranged from 1.5 to 80%.

Middle-Melting Milkfat Fractions

Crystallization from melted milkfat Middle-melting milkfat fractions obtained by crystallization from melted milkfat were both solid and liquid fractions obtained using single-step and multiple-step fractionation procedures. The liquid middle-melting fractions were obtained at fractionation temperatures above 20°C, whereas solid middle-melting milkfat fractions were obtained between 11 and 25°C. Middle-melting milkfat fractions obtained from melted milkfat were separated by vacuum filtration, by pressure filtration, by centrifugation; with the aid of an aqueous detergent solution; by filtration through muslin cloth, a milk filter, a cheese press, or a casein-dewatering press; and by techniques not reported. The yields for middle-melting milkfat fractions obtained from melted milkfat ranged from 12 to 98%. Liquid middle-melting milkfat fractions generally exhibited greater yields (> 60%) than solid middle-melting fractions (< 50%).

Crystallization from solvent solution Middle-melting milkfat fractions have been obtained from acetone and pentane solutions. The fractions obtained from acetone solution were either solid fractions, obtained using multiple-step fractionation between 0 and 10°C, or liquid fractions, obtained using single-step or multiple-step fractionation between 21 and 40°C. The middle-melting milkfat fractions obtained from pentane solutions were liquid fractions obtained using single-step fractionation. Yields for middle-melting milkfat fractions obtained from solvent solution exhibited a wide range of yields (5 to 98%), and the yields for liquid middle-melting fractions (> 90%) were generally greater than yields for solid middle-melting milkfat fractions (< 25%).

Supercritical fluid extraction Middle-melting milkfat fractions obtained from supercritical fluid extraction were all extracted fractions and were the first through sixth fractions collected during the extraction process. Middle-melting milkfat fractions have been produced by supercritical fluid extraction using a range of temperatures and pressures. Yields for middle-melting milkfat fractions obtained by supercritical carbon dioxide extraction ranged from 3 to 48%.

Low-Melting Milkfat Fractions

Crystallization from melted milkfat Low-melting milkfat fractions were primarily liquid fractions obtained using single-step and multiple-step fractionation procedures, although several solid low-melting milkfat fractions have also been produced. Liquid low-melting milkfat fractions have been obtained between 4 and 28°C, whereas solid low-melting milkfat fractions were obtained between 4 and 15°C. Low-melting milkfat fractions obtained from melted milkfat were separated by vacuum filtration; by pressure filtration; by centrifugation; with the aid of an aqueous detergent solution; by filtration through muslin cloth, a milk filter, a cheese press, or a casein-dewatering press; and by techniques not reported. Yields for low-melting milkfat fractions obtained from melted milkfat ranged from 10 to 90%.

Crystallization from solvent solution Low-melting milkfat fractions have been produced primarily from acetone solution but also have been obtained from ethanol, isopropanol, and pentane solutions. Liquid low-melting milkfat fractions have been obtained from acetone solution using single-step and multiple-step fractionation procedures at temperatures between 0 and 24°C, whereas solid low-melting milkfat fractions have been produced only with multiple-step fractionation at temperatures below 0°C. Low-melting milkfat fractions obtained from solvent solution exhibited yields of 10 to 89%.

Supercritical fluid extraction All low-melting milkfat fractions produced by supercritical fluid extraction were either the first or second extracted fraction obtained. Yields for low-melting milkfat fractions obtained by supercritical fluid extraction ranged from 0.2 to 20%.

Very-Low-Melting Milkfat Fractions

Crystallization from melted milkfat Very-low-melting milkfat fractions produced by crystallization from melted milkfat were liquid fractions obtained at the second through fifth steps in multiple-step fractionation procedures between 2 and 8°C. Very-low-melting milkfat fractions were separated using vacuum filtration and centrifugation. Yields for very-low-melting milkfat fractions obtained from melted milkfat ranged from 12 to 36%.

Crystallization from solvent solution Very-low-melting milkfat fractions were liquid fractions obtained using single-step or multiple-step fractionation from acetone solution generally between -15 and 0°C. Very-low-melting milkfat fractions also were obtained using multiple-step fractionation from ethanol and isopropanol solutions at -15 and -5°C, respectively. Yields of very-low-melting milkfat fractions obtained from solvent solution ranged from 8 to 64%.

Effects of Fractionation on Selected Characteristics of Milkfat Fractions

Experimental milkfat fractions have been characterized for the following properties:

1. Chemical characteristics
   - Fatty acid composition
   - Triglyceride composition
   - Selected chemical characteristics
   - Flavor profile
2. Physical properties
   - Crystal morphology
   - Crystal size
   - Melting point
   - Thermal profiles (DSC curves)
   - Solid fat content
   - Textural characteristics
3. Oxidative stability
4. Nutritional properties

These characteristics are highly dependent on the chemical composition of the parent intact milkfat and the processing conditions employed during fractionation:

1. Fractionation method
   - Crystallization from melted milkfat
   - Crystallization from solvent solution
   - Supercritical fluid extraction
2. Fractionation procedure
   - Single-step fractionation
   - Multiple-step fractionation
3. Fractionation processing conditions
   - Separation temperature
   - Cooling rate
   - Agitation rate

The characteristics of milkfat fractions obtained by crystallization methods were highly influenced by the physical state of the fraction at the time of separation (i.e., liquid or solid). Similarly, the characteristics for milkfat fractions obtained by supercritical fluid extraction were highly influenced by whether the fraction was an extracted fraction or the residual fraction that remained after the extraction process had been completed. In general:

- Very-high-melting and high-melting milkfat fractions obtained by crystallization methods were solid fractions at the time of separation.
- Very-high-melting milkfat fractions obtained by supercritical fluid extraction were residual fractions.

- Middle-melting and low-melting fractions obtained by crystallization methods were both solid and liquid fractions.
- Middle-melting milkfat fractions obtained by supercritical fluid extraction were both residual and extracted fractions.
- Very-low-melting milkfat fractions obtained by crystallization methods were liquid fractions.
- Very-low-melting milkfat fractions obtained by supercritical fluid extraction were the first extracted fractions collected.

Effects of Fractionation on the Chemical Composition of Milkfat

Fatty acid composition of milkfat fractions General trends observed for the fatty acid compositions of milkfat fractions compared with the intact parent AMFs include (Table 6.1):

- High-melting and very-high-melting milkfat fractions exhibited decreased C4–C10 and C16–C18 unsaturated fatty acid contents, with a concurrent increase in the C16–C18 saturated fatty acid contents, as the fractionation temperature increased.
- The fatty acid composition of middle-melting milkfat fractions was similar to that of intact milkfat.
- Low-melting and very-low-melting fractions exhibited increased C4–C10 and C16–C18 unsaturated fatty acid contents, with a concurrent decrease in C16–C18 saturated fatty acid contents, as the fractionation temperature decreased.

Trends observed for the fatty acid compositions of milkfat fractions obtained by different fractionation methods include the following:

- Milkfat fractions obtained by crystallization from melted milkfat and solvent solutions exhibited gradual increases in the C4–C10 and C16–C18 unsaturated fatty acid contents and gradual decreases in the C16–C18 saturated fatty acid contents as fractionation temperatures decreased.
- Solid milkfat fractions obtained by crystallization from melted milkfat and solvent solutions exhibited lower concentrations of C4–C10, C12–C15 and C16–C18 unsaturated fatty acids, and higher concentrations of C16–C18 saturated fatty acids, compared with liquid fractions obtained using similar conditions.
- Milkfat fractions obtained by supercritical carbon dioxide extraction exhibited gradual decreases in C4–C10 fatty acid contents and gradual increases in the C16–C18 saturated and unsaturated fatty acid contents as fractionation temperatures and pressures increased.
- The residual fractions obtained by supercritical carbon dioxide extraction exhibited decreased C4–C10 fatty acid contents and increased C16–C18 saturated and unsaturated fatty acid contents compared with the extracted fractions.
- The C18:1 fatty acid contents were greater in the residual, higher-melting fractions obtained by supercritical fluid extraction, whereas the C18:1 fatty acid contents were greater in the lower-melting fractions obtained by crystallization from melted milkfat and solvent solution.

Triglyceride composition of milkfat fractions General trends observed for the triglyceride compositions of milkfat fractions compared with the intact parent AMFs include (Table 6.1):

- High-melting and very-high-melting milkfat fractions exhibited decreased C24–C34 and C36–C40 triglyceride contents, with concurrent increases in C42–C54 triglyceride contents, as fractionation temperatures increased.

TABLE 6.1 Effects of Fractionation on the Chemical Composition of Milkfat Fractions Relative to Intact Milkfat[1]

Chemical composition	Very-high-melting (MP > 45°C)			High-melting (MP between 35 and 45°C)			Middle-melting (MP between 25 and 35°C)			Low-Melting (MP between 10 and 25°C)			Very-low-melting (MP < 10°C)		
	Melt	Solvent	SFE	Melt	Solvent	SFE	Melt	Solvent	SFE	Melt	Solvent	SFE	Melt	Solvent	SFE
Fatty Acid															
C4–C10	↓[2]	→	→	→	→	↑	≈/↑	≈	↑	↑	≈/↑	lg↑	↑	↑	—[3]
C12–C15	≈/↑	↑	→	→	→	slt ↑/↓	≈	≈	slt ↑/↓	↑	↑	↑	≈/↑	≈/↑	—
C16–C18 unsaturated	→	→	←	slt ↑/↓	slt ↑/↓	slt ↑/↓	≈/slt →	→	→	→	→	→	→	→	—
C16–C18 saturated	↑	↑	↑	↑	↑	↑	≈/↑	≈/↑	↑	→	→	→	→	→	—
Triglyceride															
C24–C34	→	lg↓	—	→	→	→	≈/↑	≈/slt ↑/↓	↑/↓	↑	↑	lg↑	—	↑	—
C36–C40	→	→	—	→	→	←	≈/↓	→	↑/↓	←	←	←	—	←	—
C42–C54	←	←	—	←	←	←	≈/↓	→	↑/↓	→	→	→	—	→	—

[1] melt = milkfat fractions obtained by crystallization from melted milkfat; solvent = milkfat fractions obtained by crystallization from solvent solution; SFE = milkfat fractions obtained by supercritical fluid extraction.

[2] ↓ = decrease in content; ≈ = approximately equal in content; ↑ = increase in content; lg = large; slt = slight; a combination of these symbols means fractions in this category exhibited several trends.

[3] data not reported.

- The triglyceride compositions of middle-melting milkfat fractions were similar to or exhibited only slight changes in composition compared with intact milkfat.
- Low-melting and very-low-melting fractions exhibited increased C24–C34 and C36–C40 triglyceride contents with concurrent decreases in C42–C54 triglyceride contents as fractionation temperatures decreased.

Trends observed for the triglyceride compositions of milkfat fractions obtained by different fractionation methods include the following:

- Milkfat fractions obtained by crystallization from melted milkfat and from solvent solutions exhibited gradual increases in the C24–C34 and C36–C40 triglyceride contents and gradual decreases in the C42–C54 triglyceride contents as fractionation temperatures decreased.
- Solid milkfat fractions obtained by crystallization from melted milkfat and from solvent solution exhibited lower concentrations of C24–C34 and C36–C40 triglycerides and higher concentrations of C42–C54 triglycerides than did liquid fractions obtained at similar conditions.
- Milkfat fractions obtained by supercritical fluid extraction exhibited more pronounced changes in triglyceride composition as fractionation conditions changed, compared with their corresponding parent milkfat, than did fractions obtained by crystallization from melted milkfat or from solvent solutions.
- The low-melting milkfat fractions obtained by supercritical fluid extraction had C24–C34 concentrations that were markedly greater than intact milkfat, and the very-high-melting milkfat fractions had markedly lower C24–C34 triglyceride concentrations than intact milkfat.
- The residual fractions obtained by supercritical carbon dioxide extraction exhibited decreased concentrations of C24–C34 and increased concentrations of C42–C54 triglycerides compared with the extracted fractions.

Selected chemical characteristics of milkfat fractions Although data for selected chemical characteristics of milkfat fractions were somewhat limited, the following general trends for iodine values of milkfat fractions were observed:

- Iodine values were lowest for very-high-melting and high-melting milkfat fractions and highest for the low-melting and very-low milkfat fraction, obtained by crystallization from melted milkfat and from solvent solutions.
- Iodine values were lower for solid fractions than for liquid fractions obtained by crystallization methods using similar fractionation conditions.
- Iodine values for very-high-melting and high-melting milkfat fractions obtained by supercritical fluid extraction were greater than values for very-high-melting and high-melting fractions obtained by crystallization methods, because of the increased C18:1 fatty acid contents observed in the higher-melting fractions obtained by supercritical fluid extraction.

Flavor of milkfat fractions The flavor of milkfat fractions has been described by free and potential (free + bound) lactone and methyl ketone concentrations and aldehyde concentrations. Decreases in potential lactone and methyl ketone concentrations have been reported for commercial anhydrous milkfat prepared from fresh cream compared with fresh, unheated milkfat, and consequently, there was an overall decrease in flavor potential of the raw materials used for fractionation compared with fresh cream. General trends observed for flavor components of milkfat fractions, compared with intact milkfat, include the following:

- The free and potential (free + bound) lactone and methyl ketone concentrations of very-high-melting and high-melting milkfat fractions were lower than the concentrations reported for intact milkfat.

- The free and potential (free + bound) lactone and methyl ketone concentrations of middle-melting milkfat fractions were slightly greater than concentrations reported for intact milkfat.
- The free and potential (free + bound) lactone and methyl ketone concentrations of low-melting milkfat fractions were greater than concentrations reported for intact milkfat.
- The aldehyde concentrations were elevated for all milkfat fractions compared with intact milkfat, but this was attributed by the original investigators to slight oxidation of the milkfat, which had occurred during AMF production and fractionation.

Data for the concentrations of flavor components of milkfat fractions obtained by different fractionation methods were limited. However, trends observed for flavor components with respect to fractionation conditions, for milkfat fractions classified by melting behavior and for unknown-melting milkfat fractions, include the following:

- The free and potential (free + bound) lactone and methyl ketone concentrations were lowest in the solid fractions and greatest in the liquid fractions produced by crystallization from melted milkfat and from solvent solutions.
- The free and potential (free + bound) lactone and methyl ketone concentrations were lowest in the first fraction obtained by crystallization methods using multiple-step fractionation, and then increased gradually for successive fractions as the fractionation temperature decreased.
- The free lactone concentration of milkfat fractions obtained by supercritical fluid extraction was highest in the first extracted fraction collected, gradually decreased with subsequent fractions collected, and was lowest in the residual fraction that remained after the extraction process had been completed.

Milkfat fractions have also been evaluated by sensory panels, and significant differences in the flavor of milkfat fractions, compared with intact AMF, were not perceived.

Effects of Fractionation on the Physical Properties of Milkfat

Crystallization behavior of milkfat fractions   The crystallization behavior of milkfat fractions includes effects of the crystallization process (nucleation and growth) and crystal characteristics (morphology and size). The crystallization process is the driving force employed for the fractionation of milkfat from melted milkfat and from solvent solution.

The *crystallization process* influences yields, chemical characteristics, and physical properties for milkfat fractions obtained from melted milkfat and solvent solution. The crystallization conditions employed for milkfat fractionation influence the initial number and size distribution of nuclei that are formed, greatly influencing crystal distribution and growth. Crystal nuclei generally are composed of the highest-melting triglycerides in the milkfat sample, and crystal growth occurs by the deposition of successively lower-melting triglycerides onto an already ordered crystal surface.

Factors affecting the crystallization process include chemical composition of the starting material, the cooling rate, and the agitation rate. The presence or absence of solvents also influences the crystallization process and the properties of the milkfat fractions; these differences were discussed earlier in this chapter in the section on technologies used to fractionate milkfat. Trends observed for the effects of selected variables on the crystallization process include the following:

- Increased concentrations of C42–C54 triglycerides in the milkfat fractions resulted in the formation of increased numbers of large spherulite crystals.
- Milkfat fractions with increased concentrations of C42–C54 triglycerides and C16–C18 saturated fatty acids required shorter crystallization periods.

- Milkfat fractions with decreased concentrations of C42–C54 triglycerides and C16–C18 saturated fatty acids require increased crystallization periods because of increased supercooling required to effect nucleation.
- Milkfat fractions obtained from the second and subsequent steps of multiple-step fractionation procedures were influenced by the changes in triglyceride composition of the milkfat that occurred as fractionation temperatures decreased and higher-melting fractions were removed; thus, subsequent fractions had decreased amounts of C42–C54 triglycerides, and these steps required longer crystallization periods.
- Slow cooling favored the formation of larger, more pure spherulite crystals.
- Rapid cooling favored the formation of crystals that contained more liquid milkfat, resulting in decreased yields and increased melting points for the solid fractions.
- Increased agitation rates increased the cooling rate and resulted in shorter crystallization periods.

Crystal morphology data were reported for milkfat fractions obtained by crystallization from melted milkfat and from solvent solution, and by supercritical fluid extraction.

The crystal morphologies were influenced by the relative chemical composition of the milkfat fractions. Milkfat fractions that contained increased concentrations of higher-melting triglycerides or a more homogenous triglyceride profile exhibited stable β and β' crystal forms. Milkfat fractions that contained more lower-melting triglycerides or a more heterogenous triglyceride profile exhibited the β' form as the most stable form. General trends observed for the crystal morphologies of milkfat fractions include the following:

- Crystals from very-high melting and high-melting milkfat fractions exhibited β and β' forms.
- Crystals from middle-melting and low-melting milkfat fractions exhibited the β' form.
- Crystals from milkfat fractions obtained by crystallization from melted milkfat existed in the β' form.
- Crystals from milkfat fractions obtained by crystallization from solvent solution existed in the β or β' form.
- Crystals from milkfat fractions obtained by supercritical fluid extraction existed in the β or β' form.
- α crystals were observed for all milkfat fractions obtained by crystallization from melted milkfat and from solvent solution, but not for milkfat fractions obtained by supercritical fluid extraction.
- Elevated separation temperatures for milkfat fractions obtained by crystallization methods resulted in the formation of a smaller number of stable crystals (β' or β form), because of the increased concentration of higher-melting triglycerides in the crystals.
- Lower separation temperatures for milkfat fractions obtained by crystallization methods resulted in the formation of greater numbers of unstable crystals (α form) because of the increased concentration of liquid milkfat in the crystals.
- Slow cooling resulted in the formation of β' crystals.
- Rapid cooling resulted in the formation of α crystals, which transformed into more stable crystals (β') with storage.

Crystal size data were reported only for high-melting and unknown-melting milkfat fractions obtained by crystallization from melted milkfat. Consequently, comparisons for the crystal size of milkfat fractions obtained by different fractionation methods and for different melting classes of milkfat fractions were not possible. However, the effects of selected crystallization parameters on the crystal size of milkfat fractions were studied, and the following conclusions have been drawn:

- Crystal size was an important parameter for the efficient separation of milkfat fractions obtained from melted milkfat using vacuum filtration (optimal crystal size = 200 to 350 μm) or centrifugation (optimal crystal size = 200 μm).
- Increased agitation rates resulted in the formation of larger numbers of smaller crystals.
- Slow agitation rates resulted in the formation of smaller numbers of larger crystals.
- Rapid cooling resulted in the formation of larger numbers of small crystals (1 to 2 μm).
- Slow cooling resulted in the formation of smaller numbers of larger crystals (50 μm), which eventually aggregated to form even larger crystals (100 to 1000 μm).

Melting behavior of milkfat fractions The melting behavior of milkfat fractions has been described by final melting point, DSC thermal profile, and solid fat content data over a range of temperatures.

Melting points were used as the basis for the classification of the general properties of milkfat fractions. The melting point ranges for each of the major melting classes of milkfat fractions have been defined as

- Very-high-melting milkfat fractions: MP > 45°C
- High-melting milkfat fractions: MP between 35 and 45°C
- Middle-melting milkfat fractions: MP between 25 and 35°C
- Low-melting milkfat fractions: MP between 10 and 25°C
- Very-low-melting milkfat fractions: MP < 10°C

The general effect of chemical composition on melting points of milkfat fractions was that milkfat fractions with higher contents of saturated C4–C10 and unsaturated C16–C18 fatty acids and C24–C34 triglycerides generally exhibited lower melting points than milkfat fractions with lower concentrations of these fatty acids and triglycerides.

The effects of crystallization conditions on milkfat fractions obtained by crystallization from melted milkfat and from solvent solution include the following trends:

- Milkfat fractions generally exhibited increased melting points as fractionation temperatures increased.
- Solid fractions exhibited higher melting points than the liquid fractions obtained under similar conditions.
- Milkfat fractions obtained from solvent solutions exhibited higher melting points than fractions obtained from melted milkfat under similar conditions.
- Solid milkfat fractions obtained from melted milkfat exhibited increases in melting points from 5 to 25°C compared with fractionation temperatures, whereas liquid milkfat fractions exhibited melting points that ranged from -5 to +5°C compared with fractionation temperatures.
- Solid milkfat fractions obtained from solvent solution exhibited increases in melting points of 15 to 40°C compared with fractionation temperatures, whereas liquid fractions exhibited increases in melting points of 3 to 27°C compared with fractionation temperatures.
- Solid milkfat fractions obtained from melted milkfat that were refractionated at similar or higher temperatures yielded fractions with increased melting points.
- Solid milkfat fractions obtained from solvent solution that were refractionated at similar or higher temperatures yielded fractions with decreased melting points.

General trends observed for melting points of milkfat fractions obtained by supercritical fluid extraction include the following:

- The first extracted fractions were the lowest-melting fractions obtained, and melting points increased as the extraction process progressed.
- The melting points of the residual fractions were generally higher than the melting points of extracted fractions.

Thermal profiles (DSC curves) generally exhibited gradual changes in the presence and size of the peaks for the three major melting triglyceride species (high, middle, low) as the melting behavior of the milkfat fractions ranged from very-high-melting to very-low-melting. These changes can be readily observed in Figures 6.1 and 6.2 for milkfat fractions obtained by multiple-step fractionation from melted milkfat and from acetone solution, respectively. Trends observed in the DSC curves of milkfat fractions include the following:

- Very-high-melting milkfat fractions obtained from solvent solution exhibited DSC curves that consisted of one relatively narrow peak for the high-melting triglyceride species.
- Very-high-melting milkfat fractions obtained from melted milkfat and by supercritical fluid extraction exhibited a broader peak for the high-melting triglyceride species than did very-high-melting fractions obtained from solvent solution, indicating a more complex melting behavior in this range.
- Very-high-melting milkfat fractions obtained from melted milkfat exhibited a small peak for the low-melting triglyceride species, but this peak was absent for very-high-melting milkfat fractions obtained from solvent solution and by supercritical fluid extraction.
- High-melting milkfat fractions obtained from melted milkfat and solvent solution exhibited DSC curves that showed major peaks for the high-melting triglyceride species and small peaks for the middle-melting and low-melting triglyceride species.
- High-melting milkfat fractions obtained by supercritical fluid extractions exhibited a major peak for the high-melting triglyceride species and small peaks for the middle-melting triglyceride species, but the peak for the low-melting species was absent.
- Middle-melting milkfat fractions obtained by crystallization from melted milkfat and by supercritical fluid extraction exhibited larger peaks for the middle-melting triglyceride species and smaller peaks for the high-melting and low-melting triglyceride species.
- Middle-melting milkfat fractions obtained by crystallization from solvent solution exhibited larger peaks for the middle-melting triglyceride species and smaller peaks for the low-melting triglyceride species; the peak for the high-melting triglyceride species was present only in some middle-melting milkfat fractions.
- All low-melting milkfat fractions exhibited major peaks for the low-melting triglyceride species and smaller peaks for the middle-melting triglyceride species; some low-melting milkfat fractions exhibited small peaks for the high-melting triglyceride species.
- Very-low-melting milkfat fractions exhibited DSC curves that consisted of one distinct peak for the low-melting triglyceride species.

The fractionation conditions employed also influenced the thermal profiles of milkfat fractions; following are some of these effects:

- Milkfat fractions obtained from melted milkfat (Figure 6.1) contained more liquid milkfat, resulting in DSC curves that exhibited broader peaks for the three major melting triglyceride species than for most milkfat fractions obtained by crystallization from solvent solution (Figure 6.2) and by supercritical fluid extraction.

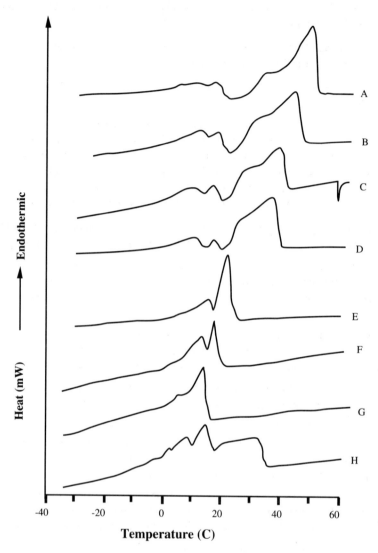

Figure 6.1 Thermal profiles (DSC curves) of milkfat fractions obtained by multiple-step crystallization from melted milkfat (Kaylegian, 1991): A, $34S_M$ [W][1] fraction, VH^2; B, $30S_M$ [W] fraction, H; C, $25S_M$ [W] fraction, H; D, $20S_M$ fraction, H; E, $16S_M$ fraction, M; F, $13S_M$ fraction, L; G, $13L_M$ fraction, L; H, AMF [W]. [1]The number indicates fractionation temperature, the letter indicates the physical state (L = liquid, S = solid), and the subscript indicates the fractionation procedure used (M = multiple-step). Designations in brackets refer to the anhydrous milkfat source (W = winter). AMF = anhydrous milkfat. [2]VH = very-high-melting milkfat fraction, H = high-melting milkfat fraction, M = middle-melting milkfat fraction, L = low-melting milkfat fraction.

- Milkfat fractions obtained by single-step fractionation from melted milkfat and from solvent solutions exhibited peaks for the high-melting triglyceride species that gradually shifted to lower temperatures, became more broad, and developed leading shoulders as the fractionation temperature decreased.

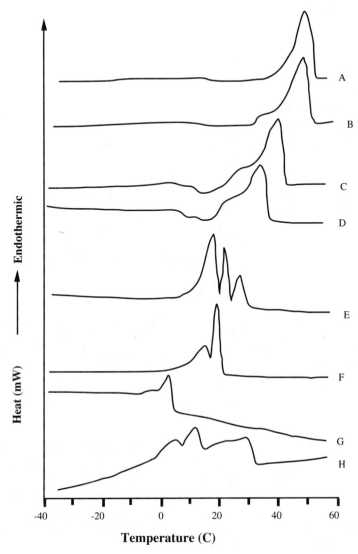

Figure 6.2 Thermal profiles (DSC curves) of milkfat fractions obtained by multiple-step crystallization from acetone solution (Kaylegian, 1991): A, $25S_M$ [W][1] fraction, VH[2]; B, $21S_M$ [W] fraction, VH; C, $15S_M$ [W] fraction, VH; D, $11S_M$ fraction, H; E, $5S_M$ fraction, M; F, $0S_M$ fraction, M; G, $0L_M$ fraction, VL; H, AMF [W]. [1]The number indicates fractionation temperature, the letter indicates the physical state (L = liquid, S = solid), and the subscript indicates the fractionation procedure used (M = multiple-step). Designations in brackets refer to the anhydrous milkfat source (W = winter). AMF = anhydrous milkfat. [2]VH = very-high melting milkfat fraction, H = high-melting milkfat fraction, M = middle-melting milkfat fraction, L = low-melting milkfat fraction.

- Milkfat fractions obtained by multiple-step fractionation exhibited DSC curves that showed greater changes in the peaks for the major melting triglyceride species for successive fractions as fractionation temperatures decreased than did milkfat fractions obtained using single-step fractionation.

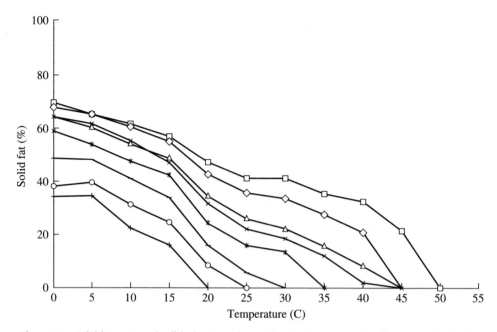

Figure 6.3 Solid fat content of milkfat fractions obtained by single-step crystallization from melted milkfat (Kaylegian, 1991): ▫, 34S$_S$ [W][1] fraction, VH[2]; X 34L$_S$ [W] fraction, M; ◇ 30S$_S$ [W] fraction, H; -, 30L$_S$ [W] fraction, M; △, 25S$_S$ [W] fraction, H; ○, 25L$_S$ [W] fraction, L; ∗, 20S$_S$ fraction, H; +, 20L$_S$ fraction, L.
[1]The number indicates fractionation temperature, the letter indicates the physical state (L = liquid, S = solid), and the subscript indicates the fractionation procedure used (S = single-step). Designations in brackets refer to the anhydrous milkfat source (W = winter). [2]VH = very-high-melting milkfat fraction, H = high-melting milkfat fraction, M = middle-melting milkfat fraction, L = low-melting milkfat fraction.

The *solid fat content* profiles of the major melting classes of milkfat fractions followed expected trends where very-high-melting milkfat fractions exhibited the highest solid fat content profiles, and very-low-melting milkfat fractions exhibited the lowest solid fat content profiles. A wide range of solid fat content profiles have been observed for milkfat fractions that exhibit similar melting points, and this has been illustrated graphically for very-high-melting milkfat fractions (Figure 5.22), high-melting milkfat fractions (Figure 5.23), and middle-melting milkfat fractions (Figure 5.24).

Solid fat content profiles are influenced by the chemical composition of the milkfat fraction, the fractionation conditions employed (Figures 6.3, 6.4, and 6.5), and the types of fractions obtained at separation (i.e., solid, liquid, extract, or residual). The effects of some of these variables on the solid fat content profiles of milkfat fractions include the following:

- Milkfat fractions with higher concentrations of C4–C10 saturated fatty acids, C16–C18 unsaturated fatty acids, and C24–C34 triglycerides generally exhibited lower solid fat content profiles than milkfat fractions that have lower concentrations of these fatty acids and triglycerides.

- Solid fat content profiles of milkfat fractions obtained from melted milkfat exhibited gradual melting throughout the melting range.

- Solid fat content profiles of milkfat fractions obtained from solvent solution exhibited slow melting over a broad temperature range, followed by sharp melting over a narrow temperature range.

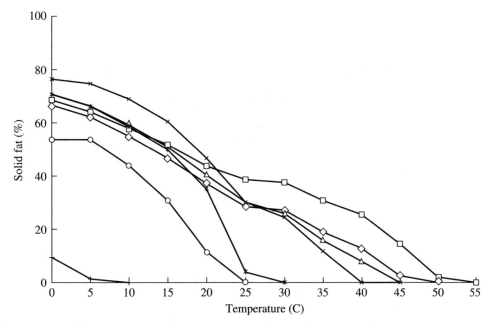

Figure 6.4 Solid fat content of milkfat fractions obtained by multiple-step crystallization from melted milkfat (Kaylegian, 1991): ▫, 34S_M [W][1] fraction, VH[2]; ◊, 30S_M [W] fraction, H; △, 25S_M [W] fraction, H; X, 20S_M fraction, H; ∗, 16S_M [W] fraction, M; o, 13S_M [W] fraction, L; +, 13L_M [W] fraction, L. [1]The number indicates fractionation temperature, the letter indicates the physical state (L = liquid, S = solid), and the subscript indicates the fractionation procedure used (M = multiple-step). Designations in brackets refer to the anhydrous milkfat source (W = winter). [2]VH = very-high-melting milkfat fraction, H = high-melting milkfat fraction, M = middle-melting milkfat fraction, L = low-melting milkfat fraction.

- Milkfat fractions obtained from solvent solution exhibited elevated solid fat content profiles compared with milkfat fractions obtained from melted milkfat using similar fractionation conditions.
- High-melting and low-melting milkfat fractions obtained by supercritical fluid extraction exhibited lower solid fat contents at temperatures below 20°C than did fractions obtained by crystallization from melted milkfat or solvent solution.
- Middle-melting milkfat fractions obtained by supercritical fluid extraction exhibited solid fat content profiles that fell between the solid fat content profiles of middle-melting milkfat fractions obtained from melted milkfat and the profiles of those obtained from solvent solution.
- Liquid and solid milkfat fractions obtained by crystallization methods generally exhibited lower solid fat content profiles as fractionation temperatures decreased.
- Solid fractions obtained by crystallization methods exhibited elevated solid fat content profiles compared with liquid fractions obtained using similar fractionation conditions.
- Residual fractions obtained by supercritical fluid extraction exhibited elevated solid fat content profiles compared with the corresponding extracted fractions.
- Solid high-melting milkfat fractions obtained by crystallization methods using multiple-step fractionation procedures exhibited elevated solid fat content profiles compared with high-melting fractions obtained using single-step fractionation procedures, whereas solid low-melting milkfat fractions obtained by multiple-step fractionation procedures exhibited lower solid fat content profiles than did low-melting fractions obtained using single-step fractionation procedures.

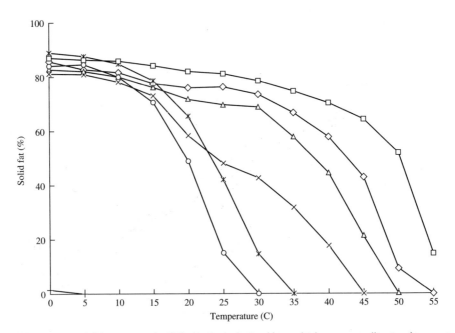

Figure 6.5 Solid fat content of milkfat fractions obtained by multiple-step crystallization from acetone solution (Kaylegian, 1991): ▫, $25S_M$ [W][1] fraction, VH[2]; ◇ $21S_M$ [W] fraction, VH; △, $15S_M$ [W] fraction, VH; X, $11S_M$ fraction, H; ∗, $5S_M$ [W] fraction, M; o, $0S_M$ [W] fraction, M; +, $0L_M$ [W] fraction, VL. [1]The number indicates fractionation temperature, the letter indicates the physical state (L = liquid, S = solid), and the subscript indicates the fractionation procedure used (M = multiple-step). Designations in brackets refer to the anhydrous milkfat source (W = winter). [2]VH = very-high-melting milkfat fraction, H = high-melting milkfat fraction, M = middle-melting milkfat fraction, VL = very-low-melting milkfat fraction.

The effects of the type of fraction obtained at separation (i.e., solid, liquid, extract or residual) were most apparent in the middle-melting milkfat fractions. Liquid middle-melting milkfat fractions exhibited solid fat content profiles that were similar to those of intact milkfat, whereas solid middle-melting milkfat fractions exhibited solid fat content profiles that were higher at temperatures below 20°C than that of intact milkfat, but melted in the same temperature range (25 to 35°C).

Textural characteristics of milkfat fractions The textural characteristics of high-melting, middle-melting, and low-melting milkfat fractions have been determined using cone penetrometry, and the following trends were observed:

- High-melting milkfat fractions exhibited hardness values that were higher than hardness values for the parent intact milkfat.
- Middle-melting and low-melting milkfat fractions exhibited hardness values that were lower than hardness values for the parent intact milkfat.

Effects of Fractionation on the Oxidative Stability of Milkfat

The effects of fractionation on the oxidative stability of milkfat fractions generally has been measured by peroxide value. Trends observed for the peroxide values of milkfat fractions include the following:

- Liquid milkfat fractions exhibited higher peroxide values than did the corresponding solid fractions obtained using single-step fractionation procedures.
- Peroxide values were lowest for the most solid fractions obtained by multiple-step fractions and increased for the fractions obtained at successive steps.

The oxidative stability of milkfat fractions has been studied under a variety of conditions, and the following conclusions have been drawn from these studies:

- The oxidation rate of milkfat fractions increased as the unsaturated fatty acid content increased.
- Low-melting milkfat fractions oxidized at a faster rate than high-melting fractions during storage under all conditions investigated.
- Peroxide values of milkfat fractions increased as storage temperature increased from –40 to 60°C.
- Milkfat fractions stored under lights oxidized faster than the control samples stored in the dark.
- Milkfat fractions that had been treated with copper and iron exhibited greater peroxide values than untreated control samples.
- Production of carbonyl compounds during the oxidation process was greater in the low-melting milkfat fractions than in the high-melting milkfat fractions.
- Solid and liquid milkfat fractions exhibited more stability toward intermolecular polymerization than intact milkfat after 8 h of thermal oxidation.

Effects of Fractionation on the Nutritional Properties of Milkfat

General trends observed for the nutritional composition of milkfat fractions include the following:

- Liquid milkfat fractions obtained by crystallization from melted milkfat and solvent solution using single-step fractionation procedures exhibited greater concentrations of cholesterol, carotene, and vitamin A and E contents than the corresponding solid fractions.
- The first solid fractions obtained by multiple-step fractionation procedures from melted milkfat and from acetone solution exhibited the lowest concentrations of cholesterol, carotene, vitamin A and E, and the concentrations of these compounds increased in fractions obtained at successive steps, and the highest concentration was found in the liquid fraction obtained at the last step.
- Cholesterol concentrations in milkfat fractions obtained by supercritical fluid extraction were greatest in the first extracted fractions, decreased with successive extracted fractions, and were lowest in the residual fractions.
- Approximately 15% of the cholesterol in milkfat occurs in the esterified form. A greater proportion of the cholesterol found in the high-melting and residual fractions was esterified, whereas a greater proportion of the cholesterol found in the lower-melting and extracted fractions was free cholesterol.

The effects of milkfat fractions on different aspects of rat metabolism have also been studied, and conclusions from these studies include the following:

- Thermally polymerized intact milkfat and solid milkfat fractions did not cause differences in the growth of rats compared with fresh milkfats, but the corresponding thermally polymerized liquid milkfat fractions caused marked depressions in growth.
- The iodine values and oleic acid contents of the animal's depot fats increased for rats fed low-melting milkfat fractions and decreased for rats fed very-high-melting milkfat fractions, compared with rats fed intact milkfat.

- Fasting plasma cholesterol levels were greater for rats fed intact milkfat and solid milkfat fractions than for rats fed liquid milkfat fractions.

- Postprandial lipemia data indicated delayed plasma triglyceride clearance or increased hepatic triglyceride secretion with the ingestion of intact milkfat and solid milkfat fractions, but not with liquid milkfat fractions.

Commercially Prepared Milkfat Ingredients

Although commercial milkfat ingredients made from milkfat fractions are available in several countries throughout the world, the availability of technical data for the production of these products and for their chemical and physical characteristics is limited. However, the following generalizations about commercially prepared milkfat ingredients can be made:

- Commercially prepared milkfat ingredients are produced by crystallization from melted milkfat.

- Commercially prepared milkfat ingredients have been classified as high-melting, middle-melting, and low-melting milkfat fractions.

- Chemical and physical characteristics of commercially prepared milkfat fractions were similar to the characteristics for the corresponding experimental high-melting, middle-melting, and low-melting milkfat fractions.

- Commercially prepared milkfat ingredients are designed for specific uses, and primary applications include bakery (pastry and general purpose), dairy products, and chocolate and confections.

Chapter 7

The Functionality of Milkfat and Milkfat Fractions in Foods

The inherent functionality of food ingredients entails chemical and physical interactions between food components. Food products are generally systems that involve complex chemical interactions between the fat, carbohydrate, protein, and water components. The physical form (solid, liquid, gas) of these components affects the way in which they interact. Lipids that are solid at room temperature are commonly referred to as "fats," and lipids that are liquid at room temperature are commonly referred to as "oils".

To facilitate discussion of the functionalities of milkfat ingredients in food systems, food component interactions have been broadly classified:

- Milkfat–fat interactions
- Milkfat–(nonfat) solid interactions
- Milkfat–gas interactions
- Milkfat–water interactions

Although milkfat functionality in foods involves both chemical and physical interactions, functionality is often described by attributes relating only to physical properties, such as firmness and plasticity. Milkfat functionalities can be described from a number of perspectives, and proposed categories of milkfat functionality are presented in Table 7.1. Milkfat functionalities have been categorized by basic functionalities (chemical and physical), by expressed functionality attributes (e.g., flavor, firmness), and by measurements of functionality (e.g., chemical composition, melting behavior).

The functional properties of milkfat ingredients are interrelated and are often difficult to describe in terms of a single functionality (Table 7.1). For example, the firmness of a milkfat ingredient is an expression of the solid phase functionality and can be measured by solid fat content. However, firmness also is influenced by the chemical composition and the crystallization behavior of milkfat. The functionalities of milkfat and milkfat fractions also depend on the application. For example, intact milkfat is too firm to spread easily at refrigerator temperature, but it is not firm enough for optimal functionality in pastries. Because functionality attributes are difficult to isolate and define, the final test of functionality for a milkfat ingredient is performance in the finished product.

Interactions of Milkfat With Other Food Components

Food component interactions involve both the chemical and the physical functionalities of milkfat. The basic chemical functionality of milkfat is expressed as flavor, and interactions between food components affect flavor compound release and the perception of flavor compounds. For example, the relative amounts of milkfat and other lipids in foods are important in determining the release of flavors because of the favorable partitioning of nonpolar flavor compounds into the lipid phase over the aqueous phase. Effects of food components on milkfat flavor generally have not been studied to a great extent. However, factors such as pH have been shown to affect the flavor contributions from volatile fatty acids (Baldwin et al., 1973).

The basic physical functionalities of milkfat are a combination of solid and liquid phase functionalities and are expressed by attributes such as structure formation and lubricity. Temperature is an important factor in functionality because it mediates the amount of solid milkfat and liquid milkfat present in a milkfat ingredient. The contribution of the solid phase and the liquid phase components to the performance of a milkfat ingredient in food systems is complex because both

TABLE 7.1 Functionality Categories of Milkfat Ingredients

Basic functionality	Expressed functionality attributes	Measurements of functionality
I. Chemical		
Flavor	Flavor compounds and precursors	Chemical composition
		Fatty acids
		Triglycerides
II. Physical		
Solid phase	Firmness/hardness	Chemical composition
	Structure formation	Fatty acids
	Plasticity	Triglycerides
	Spreadability	Crystallization behavior
	Viscosity	Crystal morphology
	Aeration properties	Crystal size
	Layering properties	Melting behavior
	Shortening properties	Melting point
	Heat transfer	Solid fat content
		Textural characteristics
Liquid phase	Softness	Chemical composition
	Lubricity	Fatty acids
	Solution properties	Triglycerides
	Dispersion properties	Melting behavior
	Viscosity	Melting point
	Coating properties	Solid fat content
	Flow properties	Textural characteristics
	Surface properties	
	Heat transfer	

phases are present at most temperatures encountered during food processing and consumption. Temperatures often fluctuate during food processing and consumption, and this results in changes in the amounts of solid milkfat and liquid milkfat present in the system over time.

The physical state (solid or liquid) of milkfat greatly affects the way in which milkfat interacts with other food components (e.g., coating properties). In addition, the solid phase and liquid phase functionalities of milkfat can be influenced by other food components present in the system. For example, the structure formation functionality of milkfat in a conventional butter-type (80% milkfat) spread exhibits different behavior than the structure formation functionality of the same milkfat in a low-fat (40%) spread, because the latter has a greater water content.

Food component interactions and the physical functionalities required of milkfat ingredients depend on the application. For example, milkfat ingredients designed for laminated pastries must exhibit the appropriate firmness so that the fat remains solid and does not interact with the flour particles. On the other hand, milkfat ingredients designed for cookies must exhibit lower melting points so that more liquid fat is available to coat the flour particles and allow the cookie to spread during baking. Many food applications require fat ingredients that exhibit different functionalities at different stages of processing, e.g., mixing and baking.

The interactions of milkfat with other food components are not well understood and can be described only in general terms. Although concepts and general characteristics of such interactions are recognized, there is a need for further research in these areas. Further understanding of the basic mechanisms of these interactions will allow more accurate prediction of the functionality of modified milkfats in food products.

Milkfat–Fat Interactions

Fat–fat interactions occur between triglyceride molecules in lipids. Milkfat–fat interactions occur among the triglycerides of milkfat and between the triglycerides of milkfat and those of other fats and oils. Interactions between triglycerides are influenced by the fatty acid composition of individual triglycerides and by the position of the fatty acids on the triglyceride molecules. Triglyceride–triglyceride interactions vary because of different properties provided by short-chain, long-chain, saturated, and unsaturated fatty acids.

Milkfat–fat interactions also occur between the solid and liquid phases of lipids and include solid–solid, solid–liquid, and liquid–liquid interactions. Fat–fat interactions are highly influenced by temperature changes, which mediate changes in the physical state between the solid and liquid phases, and by the time period over which these changes occur.

Because milkfat–fat interactions occur within milkfat, these interactions exert an influence on all functionality attributes of milkfat ingredients (Table 7.1). In food systems that incorporate blends of milkfat and other fats, the milkfat–fat interactions that occur are generally expressed by functionalities such as crystallization behavior and solubility effects.

An example of milkfat–other fat interactions is the incorporation of milkfat into chocolate, which is largely composed of cocoa butter. Milkfat is quite compatible with cocoa butter at usage levels below 30%, although softening may occur to some extent (Timms, 1980b). Use of low-melting milkfat fractions causes softening of chocolate through dissolution of cocoa butter crystals in the low-melting milkfat (Timms, 1980b). On the other hand, middle-melting milkfat fractions form eutectic mixtures with cocoa butter at substitution levels greater than 30%. The formation of eutectic mixtures causes an inhibition of crystallization and results in a softening of the chocolate (Timms, 1980b). Still, limited incorporation of high-melting milkfat fractions is beneficial to cocoa butter crystallization, because it decreases the rate of transformation of β (V) to β (VI) crystals in cocoa butter and thereby inhibits fat bloom in chocolate (Timms, 1980).

Milkfat–(Nonfat) Solid Interactions

Milkfat–(nonfat) solid interactions occur between milkfat and solid food particles, such as carbohydrates and proteins. The interactions between milkfat and nonfat solid particles are highly dependent on the physical state of the milkfat, and mechanisms of functionality are often different for the liquid and solid phases of milkfat.

Milkfat–(nonfat) solid interactions are encountered in most food applications. Functional attributes of milkfat (Table 7.1) that are influenced by milkfat–(nonfat) solid interactions include:

1. Solid phase functionalities
   - Structure formation
   - Layering properties
   - Shortening properties

2. Liquid phase functionalities
   - Coating properties
   - Flow properties
   - Lubricity
   - Dispersion properties
   - Viscosity
   - Surface properties

An example of milkfat–(nonfat) solid interactions is the interaction of milkfat with flour particles to disrupt the gluten network in pastries. Both solid milkfat and liquid milkfat can disrupt the gluten network, but different mechanisms are involved. The disruption of the gluten network in short pastry occurs when liquid milkfat coats the flour particles, which results in a tender pastry structure (Jordan, 1986; Pyler, 1988). The disruption of the gluten network in a laminated pastry occurs when the solid milkfat forms a barrier layer between dough layers, which results in a flaky pastry (Pedersen, 1988).

Milkfat–Gas Interactions

Milkfat–gas interactions result in foams and include the stabilization of gas bubbles (air, carbon dioxide) by solid milkfat particles and the coating of gas bubbles by thin films of liquid milkfat (Pyler, 1988; Shepherd and Yoell, 1976). The stability of fat-containing foams may depend on the physical state, size, and composition of the dispersed fat particle (Pilhofer et al., 1994).

Milkfat–gas interactions are encountered in food products that are mixed or whipped to improve their structure or other functionalities (e.g., spreadability, texture). Examples of these types of food products include cakes, whipped creams, and whipped butters. Functional attributes of milkfat ingredients (Table 7.1) that are influenced by milkfat–gas interactions include :

- Aeration properties
- Structure formation
- Spreadability

An example of milkfat–gas interactions is the use of milkfat in cakes to provide structure and volume. In a cake batter, butter or shortening is first mixed (creamed) to incorporate air before the remainder of the ingredients are added to the batter. The milkfat–air interactions provide an evenly distributed structure for the expansion of leavening gases and water vapor that are released during baking. Milkfats with poor aeration properties incorporate small volumes of air into the cake batter and yield finished cakes with low volumes and poor textures (Baldwin et al., 1972; Shepherd and Yoell, 1976).

Pilhofer et al. (1994) evaluated the functionality of milkfat and milkfat fractions in foams, and reported distinct functional differences in foam behavior between intact milkfat, a solid milkfat fraction, and a liquid milkfat fraction. During foam formation, the solid milkfat fraction resulted in a more collapsed foam than intact milkfat; however, the solid fraction improved the overall foam stability (Pilhofer et al., 1994).

Milkfat–Water Interactions

Milkfat–water interactions result in emulsions when liquid milkfat is involved, and dispersions when solid milkfat is involved. Emulsions and dispersions can occur in the form of water-in-oil systems (e.g., butter) or in the form of oil-in-water systems (e.g., cream). Because of the inherent hydrophobic nature of fats and oils, the formation of stable aqueous emulsions usually requires the use of emulsifiers or phase inversion processes. Both emulsifiers and phase inversion processes are commonly used in the food industry (Chrysam, 1985).

Milkfat–water interactions are encountered in many foods but are of particular importance in foods where the milkfat and water phases are the primary components (e.g., butter, cream). Functionality attributes that are influenced by milkfat–water interactions include:

- Emulsification effects
- Dispersion effects
- Solution effects

- Viscosity effects
- Flow properties
- Surface properties
- Structure formation

Basic Functionality of Milkfat Ingredients

The basic functionality of milkfat is a result of the fundamental chemical and physical interactions that occur with milkfat. Because the functionality categories of milkfat (Table 7.1) are interrelated, it is often difficult to discuss a single functionality as a distinct property. However, to provide a basic framework for the discussion of milkfat functionality, an attempt is made here to address the three major categories of functionality (basic functionality, expressed functionality attributes, and measurements of functionality) as individual properties.

Chemical Functionality

Inherent Chemical Composition

The inherent chemical composition is the most fundamental property associated with milkfat. The chemical composition of milkfat encompasses fatty acid composition, triglyceride composition, and the structure of the fatty acids in the triglyceride molecule. The chemical composition of milkfat is influenced by many factors (e.g., feed, stage of lactation, etc.) that have been discussed in Chapter 2, and the chemical composition is also influenced by processing conditions (e.g., fractionation, hydrogenation). The chemical composition directly affects the physical functionalities of milkfat, and these effects are discussed later in this chapter.

Milkfat Flavor

The flavor of milkfat (active free compounds and precursors) is the only expressed functionality that is solely attributed to the chemical functionality of milkfat. This functionality results from the presence and generation of flavor-active substances that are capable of stimulating chemoreceptors in the nasal or oral cavities.

The unique flavor of milkfat results from free compounds and compounds that are released from bound precursor forms, which are constituents of milkfat triglycerides. Thus, milkfat serves as a reservoir for flavor compounds and flavor precursors. Compounds responsible for milkfat flavor include volatile free fatty acids, lactones, and methyl ketones. The precursors for lactones, and methyl ketones are hydroxy fatty acids and β-keto acids, respectively. Flavor of intact milkfat and milkfat fractions has been discussed in Chapters 2 and 5.

Differences in flavor functionality of milkfat ingredients are illustrated by the flavor of fresh butter compared with that of heated milkfat. Conversion of flavor precursors to active flavor compounds in milkfat during heating yields flavors that are unique to milkfat. Consequently, milkfat provides a flavor functionality that is not available in competing fats and oils.

Physical Functionality

The physical functionalities of milkfat result from complex interactions between the solid and liquid phases and are mediated by temperature. Milkfat melts over the range of -40 to +40°C, and therefore the solid and liquid phases are both present over this temperature range.

Solid Phase Functionality

Although the solid and liquid phases of milkfat each have distinctive functional properties, the dominant functionality is generally dictated by the presence and behavior of the solid phase.

Solid Phase Functionality

Solid milkfat usually exists in the form of crystals, and the solid phase functionality of milkfat is a result of crystallization and melting behaviors. The solid phase functionality of milkfat depends on the physical form and size of the crystals, the solid fat content, the point at which the solid crystal becomes liquid, and how rapidly crystal melting occurs. Functionality attributes associated with solid phase functionality include firmness, plasticity, and structure formation (Table 7.1).

Functionality requirements of milkfat ingredients depend on the specific application, and some food applications require different properties of fats during various stages of processing. For example, cake shortenings need to possess good aeration properties during the mixing stage, but they also should melt quickly during baking to allow the gas bubbles to expand and provide volume to the finished baked goods.

Crystallization behavior Crystallization behavior encompasses the crystallization process (nucleation and growth) and the crystal characteristics (morphology and size). Milkfat crystals are usually present as spherulite-shaped crystals in the β′ form, which is the most stable polymorphic form observed for fats with a heterogeneous chemical composition. Milkfat crystals in the α form have been identified, but they are unstable and convert to the β′ form on tempering or storage. β crystals in milkfat also have been observed, but they are not common. Fractionation parameters employed to produce β crystals must provide the appropriate conditions to allow the high-melting glycerides to crystallize into more pure needle-shaped crystals without the interference and cocrystallization of the lower-melting glycerides. Consequently, the β form of milkfat crystals is associated with very-high-melting milkfat fractions, and generally the β form is found in milkfat fractions that have been obtained from solvent solution.

The sizes of milkfat crystals vary and depend on the processing conditions used to produce the milkfat ingredient. Milkfat crystals range in size from 1–2 μm for fast cooling rates to 40 μm for slow cooling rates and to 100–1000 μm for aggregated spherulite crystals. The effects of processing on crystal size were discussed in Chapter 3, and the crystal characteristics of milkfat fractions were discussed in Chapter 5.

Crystal size can have a large influence on the functional attributes of milkfat ingredients, and this can be illustrated by the aeration properties of milkfats used for cake shortenings. Butter has traditionally been viewed by the bakery industry as having poor aeration properties, and this has been related to the large crystal size normally present in conventional butter. Vegetable-based shortenings designed for cake applications have been plasticized, which reduces the crystal size to less than 5 μm. The smaller crystal size provides improved aerating properties because the small crystals are better able to surround and entrap many small air bubbles into the crystal network. More air can be incorporated into the batter if the fat crystals and air bubbles are small than if the fat crystals and air bubbles are larger. The increased amount of small air bubbles in the cake batter foam results in a better volume and structure of the finished cake. Buchanan et al. (1970) plasticized anhydrous milkfat in a manner similar to that used for vegetable shortenings and reported improved aerating properties of the plasticized milkfat and an increased volume of the finished cakes compared with cakes made from conventional butter. Buchanan et al. (1970) improved milkfat functionality by changing only the crystal size without altering other functionalities, such as solid fat content.

Melting behavior The melting behavior encompasses the transformation of solid milkfat to liquid milkfat and the rate at which this process occurs. Melting behavior is characterized by measurements such as melting point and solid fat content. The melting point is defined as the temperature at which all the milkfat is liquid and the solid phase is no longer present in the system. The solid fat content profile indicates the amount of solid milkfat that is present in the system at any given temperature and also indicates how rapidly the solid phase melts over the temperature range under consideration.

The melting characteristics of milkfat ingredients are closely related to their functionality in food products. This relationship provides a basis for the categorization of milkfat fractions into broad melting groups based on their melting points:

- Very-high-melting milkfat fractions (MP > 45°C)
- High-melting milkfat fractions (MP between 35 and 45°C)
- Middle-melting milkfat fractions (MP between 25 and 35°C)
- Low-melting milkfat fractions (MP between 10 and 25°C)
- Very-low-melting milkfat fractions (MP < 10°C)

Melting behavior data for milkfat fractions have been presented in Chapter 5. Milkfat fractions that had similar melting points did not necessarily exhibit similar solid fat content profiles, and consequently such fractions may exhibit very different functionalities in a selected application. This is dramatically illustrated by the difference in solid fat content profiles between very-high-melting milkfat fractions obtained from melted milkfat and those obtained from solvent solution (Figure 5.22). The very-high-melting milkfat fraction obtained from melted milkfat had a lower solid fat content profile and melted gradually throughout the melting range from 0 to 50°C, whereas the very-high-melting milkfat fraction obtained from solvent solution was more solid and melted slowly from 0 to 40°C, followed by sharp melting prior to the final melting point at 55°C (Kaylegian, 1991).

The importance of the solid fat content profile to milkfat functionality can be illustrated by cold-spreadable butter. Conventional butter is too solid to spread at refrigerator temperature (50% solid at 5°C), but it spreads easily as it warms up to room temperature, where some of the solid fat melts (20% solid at 20°C). The blending of selected milkfat fractions to achieve a target solid fat content profile similar to that of margarine has been a popular approach for the production of cold-spreadable butters (Chapter 8). Butter prepared with selected milkfat fractions has been found to exhibit good spreadability at refrigerator temperature if the solid fat content is between 25 and 40% at 5°C.

Several cold-spreadable butters reported in the literature had good spreadability at refrigerator temperature but often became excessively soft on standing at room temperature. Kaylegian and Lindsay (1992) added a very-high-melting milkfat fraction component (MP > 45°C) to one of these basic cold-spreadable butter formulations and obtained a butter with good spreadability at refrigerator temperature and good structural properties at room temperature. The improved structural functionality of this butter at room temperature was a direct result of the increase in solid fat content profile of the entire milkfat blend. On the other hand, Yi (1993) produced a milkfat fraction ($5S_M$ fraction) from acetone solution that had a solid fat content similar to that of cocoa butter, but it was not very compatible with cocoa butter. The addition of milkfat to chocolate generally improves its resistance to bloom, but when this fraction was blended with cocoa butter in dark chocolate, it enhanced bloom. Although solid fat content is closely related to milkfat functionality in some foods, it may not always provide a good indication of functionality. Thus, such results further stress the importance of performance testing of milkfat ingredients in food applications.

Liquid Phase Functionality

The liquid phase of milkfat can be present in several conditions: free liquid, liquid milkfat that is adsorbed onto spherulite crystals, and liquid milkfat that has been incorporated in solid crystals, as in the formation of mixed crystals. However, the availability of the liquid phase affects milkfat functionality, and the liquid phase functionality of milkfat generally refers to the properties of free, unbound liquid milkfat that is available for interactions.

The liquid fat content is not a commonly reported physical property of fats. However, solid fat contents of milkfat ingredients are commonly determined, and of course, such measurements imply the presence of liquid milkfat. The liquid fat content of milkfat ingredients, therefore, bears an inverse relationship to the amount of solid fat present at any given temperature. Liquid phase functionality is associated with attributes that are unique to liquid fat, such as coating and flow properties. However, it is also associated with the properties and attributes that are influenced by the melting behavior of the solid phase, such as firmness or softness (Table 7.1).

In determining the spreadability of butter, the liquid milkfat content provides functionality along with the solid milkfat phase. The solid milkfat phase in butter functions as a structural matrix. The liquid milkfat phase functions as a filler in the matrix, and it provides lubrication for the spreading or deformation of the solid structure that occurs when the butter is spread with a knife. The data for cold-spreadable butters discussed in Chapter 8 showed that butter made with sufficiently high solid fat content profiles retained good integrity at room temperature. As the proportion of the solid fat content was decreased, with a concurrent increase in liquid fat content, the butters became softer and exhibited poor structural integrity at room temperature. The increase in liquid fat content resulted in an increase in the flow properties of the butter at room temperature and resulted in a very soft butter. The production of successful cold-spreadable butter requires a balance between the liquid fat content needed to provide good spreadability and the solid fat content needed to provide suitable structural properties.

The aeration properties of milkfat are generally discussed from the perspective of crystal size (optimum aeration for crystals is 5 μm), but the liquid fat also provides functionality in aeration properties. Smaller crystal sizes are required for fats to have good aeration properties because of the improved ability of small crystals over larger crystals to encapsulate many small air bubbles and form a dense structural network. However, the liquid phase surrounding the solid phase allows the milkfat crystals to move freely in the system and form the networks necessary to entrap gas bubbles. Liquid milkfat also stabilizes the foam by coating the gas bubbles at the milkfat–gas interface. If the milkfat were too solid, it would not be able to form a foam, because the crystals would not be able to flow easily to surround and stabilize gas bubbles.

Functionality Attributes of Milkfat Ingredients

The functionality attributes of milkfat ingredients, listed in Table 7.1, often have different meanings in various applications. For example, plasticity in puff pastry milkfat implies the ability of milkfat to form thin layers, while plasticity in cake milkfat is related to its aeration properties; firmness in cold-spreadable butter relates to spreadability at refrigerator temperature, while firmness in chocolate relates to the texture or snap of the chocolate, and the firmness of milkfat–cocoa butter mixtures is generally viewed in regard to the softening effects of milkfat on cocoa butter.

The interrelationship of functionality attributes combined with their relationship to basic physical and chemical functionality and interactions with other food components leads to very complicated situations. To provide a basis for the interpretation of milkfat functionality, a general description of functionality attributes is presented here, and factors affecting these functionality attributes are discussed in the next section.

Firmness, Hardness, and Softness

Firmness, hardness, and softness are descriptors encountered in the application of milkfat fractions in many food products. Thus, the firmness, hardness, and softness attributes of milkfat fractions may be used to describe the textural properties of milkfat under various conditions. Firmness and hardness may be used to describe these properties from the perspective of solid phase functionality, whereas softness may be used to describe the liquid phase functionality. Yet, although the firmness, hardness, or softness of milkfat may be heavily dependent on the presence of the solid phase, it is important to realize that the liquid phase is also functional. Firmness of milkfat is a result of fat–fat interactions between solid milkfat and liquid milkfat and, if considered over a range of temperatures, the rate of transition from solid to liquid. The firmness of milkfat fractions is expressed by their melting behavior and textural properties. Measurements that predict firmness of milkfat include melting point, solid fat content profile, penetrometry (cone, disc), and cutting resistance.

Milkfat fractions are often selected for use in food products based on their melting characteristics, but further processing, such as texturization, can also influence firmness. Expressions of firmness in selected applications of milkfat fractions include:

- *Cold-spreadable butter*. Firmness relates to butter spreadability at refrigerator temperature and stand-up properties at room temperature.

- *Bakery applications*. Firmness relates to milkfat functionality during incorporation into the dough or batter and its baking properties. Pastries require high-melting milkfat fractions to promote layering, and cake fats require middle-melting milkfat fractions for good aeration and volume development.

- *Chocolate*. The firmness of milkfat fractions in chocolate relates to their compatibility with cocoa butter. Elevated usage levels of milkfat fractions in chocolate results in softening, which is caused by the solubility effects of the low-melting milkfat fractions and the crystallization effects of the middle-melting milkfat fractions.

Structure Formation

The structure formation functionality of milkfat refers to the ability to form or influence the structure of a fat ingredient or a finished product. Structure formation is a function of the relative amounts and types of solid and liquid phases present as well as the texturization processes employed, but the specifics of structure formation depend on the application. Structure formation can involve milkfat–fat interactions between solid milkfat and liquid milkfat or between milkfat and other fats, milkfat–water interactions in dairy-based spreads, milkfat–gas interactions in bakery products and whipped creams, and milkfat–(nonfat) solid interactions in bakery products and chocolate.

Structure formation is often assessed empirically to determine whether the finished product displays a characteristic structure or feature. Measurements that have been used to assess structure formation also include measurements of firmness or hardness, as well as volume, height, and the diameter of foods, such as baked goods. Recent advances in the use of microscopy in lipid studies have provided a new technology with which to evaluate structure formation within milkfat fractions and finished products.

Milkfat is important to the structure of most foods in which it is used, and examples include:

- *Cold-spreadable butter*. The solid fat forms a network that is filled with liquid fat. If there is too much solid fat present, the butter will not spread, and if there is too much liquid fat present, the butter will oil off and not maintain characteristic butter structure.

- *Laminated pastries*. The milkfat is rolled in between dough layers, which causes a disruption of the gluten network and keeps the layers separate during baking, resulting in the characteristic flaky structure. If the milkfat ingredients contain water (e.g., butter), the water will vaporize during baking, and the milkfat–gas interactions will also affect the structure (i.e., height) of the pastry.

- *Cakes.* Milkfat is creamed in the presence of sugar to incorporate air into the batter, and good aeration properties of a fat will lead to a large volume of air being incorporated in the form of tiny, well-dispersed bubbles, which expand upon baking and form the characteristic structure of cakes.

Viscosity and Flow Properties

Viscosity is a basic rheological property of food products, and is defined as the internal friction of a liquid, which is a function of shear stress and shear rate. The viscosity of a liquid affects the way in which it flows. Many foods possess elastic properties as well as viscosity properties, and these foods have been described as exhibiting viscoelastic behaviors (Bourne, 1982). The rheological properties of fats and other foods are complex and are beyond the scope of this book. However, information on this topic is readily available in books (Bourne, 1982) and in the research literature.

Viscosity is primarily a function of the liquid phase of milkfat, but it is influenced by the solid content. The addition of liquid milkfat to foods generally decreases viscosity, and the addition of solid milkfat tends to increase viscosity. Viscosity properties result from complex interactions between all food components, including milkfat–fat, milkfat–(nonfat) solid, milkfat–gas, and milkfat–water interactions. Viscosity properties are observed in the softening of chocolate that occurs when milkfat is added at relatively high levels.

Viscosity and flow properties can be described in terms of stress and strain, and deformation and yield. Measurements of these properties include yield value, Bingham viscosity, Casson plastic viscosity, Brookfield viscosity, and measurements of shear forces.

Solution Properties

The solution properties of milkfat refer to its ability to dissolve substances, and these properties are largely a function of the liquid phase. Solution properties generally involve liquid fat–liquid fat interactions. Solution properties are also expressed in the ability of milkfat to serve as a carrier and reservoir for fat-soluble flavors, vitamins, and other substances. The solution properties of milkfat influence viscosity, structure formation, and firmness.

Dispersion Properties

The dispersion properties of milkfat include the ability of liquid milkfat to disperse solid particles and other liquids, and the ability of solid and liquid milkfat to be dispersed in other solutions. Over the range of -40 to +40°C, milkfat can be described as a dispersion of solid milkfat crystals in the liquid milkfat phase. Dispersion properties involve milkfat–fat (solid–liquid and liquid–liquid) interactions, milkfat–(nonfat) solid interactions, and milkfat–water interactions. Stable fat–water interactions generally require the presence of emulsifiers. Dispersion properties of milkfat can influence functionalities of milkfat that depend on both the liquid and solid phases.

Spreadability

The spreadability of milkfat is a measure of how easily the product deforms under stress and then flows. Spreadability is a function of the formation and stability of the solid crystal network and the melting properties of the lower-melting milkfat components. Spreadability primarily involves milkfat–fat interactions, but it may also include milkfat–water and milkfat–gas interactions in the cases of butter, dairy-based spreads, and whipped butters.

Spreadability can be characterized by melting behavior and textural properties, and measurements of spreadability include melting point, solid fat content, penetrometry (cone, disc), cutting resistance, and subjective measurements.

Spreadability is important in cold-spreadable butter, low-fat milkfat spreads, and dairy-based spreads. These products are typically found in retail or food service markets and are generally not used as ingredients of other foods. The spreadability properties of butter are also important in the manufacture of baked goods, but bakers often view these functionalities as plasticity instead of spreadability.

Plasticity

The plasticity of milkfat refers to its ability to be workable or pliable over a selected temperature range, and it is a functionality that is generally associated with baking fats. The functional expression of plasticity is similar to that of spreadability, but the functional requirements of plastic fats are often more rigorous than for spreadable fats. Plastic fats generally need to maintain their structural integrity during repeated deformations, such as in the lamination of doughs, whereas spreadable fats may lose some of their structural integrity following the initial deformation, such as spreading butter on bread.

The plasticity of milkfat is a function of the solid and liquid phases, particularly the solid fat content and crystal size. Like spreadability, plasticity primarily involves milkfat–fat interactions, but it may also include milkfat–water interactions (e.g., butter). Plasticity of milkfat ingredients involve the crystallization and melting behaviors of milkfat and can be characterized by solid fat content profile and crystal characteristics.

Plasticity is a functionality that is generally associated with baking fats, and can involve various food component interactions depending upon the application. Examples of the plasticity of milkfat ingredients used in bakery products include:

- *Laminated pastries.* Milkfat is used in laminated pastries to separate dough layers to promote the formation of a flaky finished pastry. The milkfat must be plastic to allow rolling of the milkfat into thin layers between the dough without cracking or oiling off, which results in the absorption of milkfat into the dough;
- *Cakes.* Milkfat is used in cakes to incorporate air into the batter to form a primary structure, which results in the proper texture of the baked cake. Milkfat used in cakes must be plastic to allow the crystals to mediate the incorporation of many tiny air bubbles into a uniform foam.

Aeration Properties

The aeration properties of milkfat refer to its ability to incorporate air into the crystal network and form a stable foam. Aeration properties are closely associated with plasticity. Aeration properties depend on both the solid and liquid phases of milkfat, particularly on crystal characteristics. Fats in the β' form incorporate numerous small bubbles, whereas β fats tend to produce few, relatively large bubbles (Hoerr, 1960). The optimal crystal size for maximum aeration properties is < 1 µm, which allows the crystals to form tight networks around small air bubbles. The liquid phase is also very important because it allows the crystals to flow and form networks around the air bubbles, and it also serves to stabilize the bubbles by coating them. Aeration properties involve milkfat–fat (solid milkfat–liquid milkfat) and milkfat–gas interactions.

Aeration properties of milkfat are often evaluated by the characteristics of the finished product, such as volume and density measurements. Food applications that depend on good aeration properties include:

- *Cakes.* The formation of a cake with good structure and texture depends on the formation of a uniform foam of tiny air bubbles in the batter that expand and set upon baking. Requirements of milkfat for use in cakes include small crystal size and a plastic nature to permit the even distribution of many tiny air bubbles in the cake batter.

- *Whipping creams.* Aeration properties important in whipped creams include the amount of air that can be incorporated (overrun) and the stability of the finished whipped cream.

Layering Properties

The layering properties of milkfat generally refer to their function in laminated pastries. The milkfat forms a thin layer between dough layers and inhibits the formation of a three-dimensional gluten network, resulting in a flaky structure in the finished pastry. The layering properties of milkfat involve milkfat–fat and milkfat–(nonfat) solid interactions. The layering properties of milkfat are highly dependent on the plastic properties, because the milkfat must be plastic to allow rolling. The milkfat must remain in a continuous layer and not be absorbed by the dough, and this can be viewed as a lack of milkfat–(nonfat) solid interaction. The layering properties of milkfat are generally evaluated in terms of height and uniformity of the finished pastry.

Shortening Properties

The shortening properties of milkfat also refer to their ability to influence the texture of baked products by inhibiting the formation of the gluten network. Shortening properties differ from layering properties in that milkfat coats individual flour particles and "shortens" the gluten network, instead of disrupting the gluten network by the formation of a milkfat barrier layer between dough layers. Shortening properties are influenced by both solid and liquid phase functionalities of milkfat. Milkfat needs to be liquid enough to coat the flour particles, but solid enough so that it is not absorbed by the flour particles. Shortening effects are generally evaluated in terms of the sensory properties of the finished pastry, including flakiness, crumbliness, and tenderness.

Coating Properties and Lubricity

The coating properties of milkfat refer to their ability to cover solid particles in the food. Lubricity of a food system results from the coating of nonfat particles by liquid fat, which provides a layer on which the solid particles can slide past each other. Coating and lubricity are functions of the liquid phase and primarily involve milkfat–(nonfat) solid interactions, although milkfat–milkfat (solid–liquid) interactions are also encountered.

Coating and lubricity properties are generally evaluated in terms of structure formation, viscosity, shortening, and sensory properties of the milkfat ingredients and the finished products.

Heat Transfer

Heat transfer properties are inherent physical properties of all food components. Fats are poor thermal conductors compared with water and this greatly affects the processing and handling of food products that contain various levels of fat (e.g., milk, cream, butter). The heat transfer (thermal conductivity) properties of milkfat vary with fatty acid composition, are quite complex, and have been discussed elsewhere (Formo, 1979; Mulder and Walstra, 1974; Sherbon, 1988).

Heat transfer is generally considered a function of the liquid milkfat and involves interactions with all food components. Heat transfer properties within milkfat are affected by the amounts of liquid and solid milkfat present. Thus, heat transfer properties play an important role in the melting properties of solid milkfat.

The heat transfer functionality of milkfat is often viewed as the ability to transfer heat to other food components. Although fats are considered relatively poor thermal conductors, they are commonly used as heating mediums for cooking. Fats are used in cooking because of the desirable flavor and textural characteristics they impart on the cooked food. Foods cooked in fat (pan-fried or deep-fat fried) have much different eating qualities than foods cooked in water (boiled) or air (baked).

Factors Affecting the Functionality of Milkfat Fractions

The factors affecting milkfat functionality include chemical composition, crystallization behavior, and melting behavior (Table 7.1). Chemical composition is an inherent property of milkfat. The crystallization and melting behaviors of milkfat are influenced by the chemical composition of the milkfat and by processing variables used to fractionate milkfat and in the manufacture of finished milkfat ingredients.

Chemical Composition

The chemical composition of milkfat is important to both its flavor and physical properties. The chemical composition of milkfat is a result of milk synthesis by the cow and is related to feed and other factors (Chapter 2). Changes in chemical composition of milkfat at the biosynthesis level may involve specialized feeding programs or genetic modification. Changes in chemical composition can also be accomplished easily after the milk is secreted using processing techniques, such as fractionation and interesterification. The effects of fractionation on the fatty acid and triglyceride composition and flavor potential of milkfat fractions have been discussed in Chapters 5 and 6.

The relationship between the chemical composition and the crystallization behavior of milkfat is not well understood, and it warrants additional research. It is known that fats with a heterogeneous composition tend towards β′crystal formation, and this has been consistently observed for milkfat. The formation of β crystals is generally associated with more pure fat systems, and such crystals have been observed in high-melting milkfat fractions obtained from solvent solution. However, the effects of changes in chemical composition on the crystallization behavior of milkfat and milkfat fractions are largely unknown.

The general relationship that has been observed between chemical composition and the melting behavior of milkfats is that as the concentrations of C4–C10 and C16–C18 unsaturated fatty acids and C24–C34 triglycerides increase, the melting point and solid fat content of the milkfat decrease. Conversely, as the concentrations of C16–C18 saturated fatty acids and C42–C54 triglycerides increase, the melting point and solid fat content of the milkfat increase.

Crystallization Behavior

The crystallization behavior of milkfat is a function of the solid phase and affects all solid phase functionality attributes (Table 7.1). The solid crystals are formed from liquid milkfat, and liquid milkfat is often incorporated into the crystal matrix by the formation of mixed crystals or adhesion to the surface of the crystal. Crystallization behavior encompasses the phenomena of nucleation and crystal growth, and the resulting crystals are characterized by polymorphic form and crystal size. Crystals of the same polymorphic form, but different size, can have different functionalities. For instance, good aeration properties are observed for β′ crystals that are < 1 μm, but poor aeration properties are observed for β′ crystals that are > 5 μm. The effects of crystal size on functionality can also be observed in the firmness of two milkfats with the same solid fat content. In this case the milkfat with smaller crystal sizes will be firmer because the crystals can pack together more tightly than larger crystals, which form a looser network that results in a softer milkfat.

The crystallization behavior of milkfat is influenced by its chemical composition and processing-controlled variables. The effects of fractionation on the crystallization behavior of milkfat fractions have been discussed in Chapters 5 and 6. Effects of processing conditions on crystallization behavior during the manufacture of milkfat ingredients are discussed by application in Chapter 8.

Melting Behavior

The melting behavior of milkfat results from the interaction of the liquid and solid phases combined with the effects of temperature, which mediates the transition between the solid and liquid phases. Melting behavior affects all aspects of milkfat functionality in foods (Table 7.1), but specific effects of functionality depend on the application. For example, a firm milkfat used for pastries requires a solid fat content of 50 to 60% at 5°C, but the required firmness for a milkfat blend used in cold-spreadable butter should be around 30 to 40% solid at 5°C.

The melting behavior of milkfat fractions is influenced by chemical composition and processing variables employed during fractionation. Changes in melting behavior depend on the specific conditions employed and are greatly influenced by:

1. Fractionation method
   - Crystallization from melted milkfat
   - Crystallization from solvent solution
   - Supercritical fluid extraction
   - Short-path distillation
2. Fractionation procedure used
   - Single-step fractionation
   - Multiple-step fractionation
3. Fractionation processing conditions employed
   - Final temperature
   - Cooling rate
   - Agitation rate
4. Type of milkfat fraction obtained at separation
   - Crystallization processes: solid fraction, liquid fraction
   - Extraction and distillation processes: extracted fractions and distillates, residual fraction
5. Texturization processes
   - Use and sequence of stage units: swept-surface chillers, scraped-surface crystallizers, pin-working units, resting tubes
   - Processing conditions: entrance and exit temperatures, throughput rate, agitation rate

The effects of fractionation on the melting behavior of milkfat fractions were discussed in Chapters 5 and 6. Effects of processing conditions employed during the manufacture of milkfat ingredients are discussed by application in Chapter 8. The melting behavior of milkfat can also be modified by selective blending of milkfat fractions and intact milkfat to create a milkfat ingredient with a specific melting profile (Chapter 8).

Summary of the Functionality of Milkfat Ingredients in Food Products

The functionality of a milkfat ingredient can be defined in terms of its basic chemical and physical properties. The chemical functionality of milkfat is expressed as flavor. The physical properties are expressed as solid phase and liquid phase functionalities and are described by many functionality attributes, such as firmness and viscosity (Table 7.1). The functionality of milkfat ingredients is also a function of its interaction with other food components (fat, carbohydrate, protein, water) and the physical state of the food components (solid, liquid, gas). These interactions have been broadly described as milkfat–fat, milkfat–(nonfat) solid, milkfat–gas and milkfat–water. The specific requirements and expressions of milkfat functionality depend on the application.

The interactions between milkfat and other food components are complex and are not well understood. Further research on the relationship between chemical composition and crystallization behavior, as well as the mechanisms behind the interactions of milkfat with other food components, will lead to an improved ability to predict the functionality of milkfat fractions in food products and assist in identifying new uses for milkfat.

Chapter 8
Application of Milkfat Fractions in Foods

Considerations in the selection of fat and oil ingredients include functionality, economic factors, and consumer acceptance. Fats and oils function in many capacities in foods, and the choice of fats for use in a given application depends on why the fat is being used, how it is being used, and what qualities it imparts to the end product. For example, a fat designed for use in pastries will have different requirements from those of a fat designed for deep-fat frying. Functionality attributes important in a pastry fat are strongly related to its physical (solid phase) functionality and include firmness, plasticity, and flavor, whereas functionality attributes important in deep-fat frying are related to physical (liquid phase) and chemical functionalities that include heat transfer properties and flavor qualities. Fats and oils are used in foods to provide functionality:

1. During manufacturing and cooking:

- Structure formation
- Firmness
- Plasticity
- Viscosity
- Solution properties
- Dispersion properties
- Heat transfer

- Lubrication
- Aeration
- Shortening
- Layering
- Flow properties
- Coating properties
- Surface properties

2. In the finished product:

- Structure formation
- Texture
- Firmness
- Softness
- Spreadability
- Pseudomoistness

- Sensory attributes
- Flavor
- Texture
- Mouthfeel
- Appearance

The inherent chemical and physical functionalities of milkfat and milkfat fractions were discussed in Chapter 7.

The price of a fat ingredient is also important relative to its use in a product. A fat may have ideal functionality, but it may not be competitive in price and therefore may not be considered a viable ingredient for a selected application. Milkfat is a relatively expensive fat when compared with vegetable oils, but it is inexpensive compared with other specialty fats, such as cocoa butter. Because milkfat is one of the most expensive edible oils, it is often used in foods that benefit from its flavor. Milkfat has a unique flavor that is inherently desirable to humans and is used in certain foods, such as shortbread cookies and hollandaise sauce, primarily for its flavor contribution. On the other hand, since milkfat is less expensive than cocoa butter and legally allowed in chocolate in the U.S., it is an economically desirable ingredient in chocolates.

Consumer acceptance of food ingredients has become important and influential in the food industry. Consumers are concerned with the influences of diet on health, and this has resulted in trends toward diets with lower fat and cholesterol contents. These trends have brought large numbers of low-fat and low-cholesterol products to the marketplace, but interestingly, there has been a concurrent increase in popularity of premium desserts. This indicates that people are more conservative in their overall diets, but when they indulge themselves, they prefer high-quality foods (Podmore, 1991). This trend provides many opportunities for milkfat ingredients, because milkfat possesses a desirable flavor and is perceived by consumers as a natural and premium ingredient.

Milkfat Ingredients

Milkfat ingredients include conventional products, such as butter and anhydrous milkfat, as well as specialty milkfats that have been modified to change their functionality. Experimental milkfat ingredients based on milkfat fractions have been described in the literature, and some are commercially available in certain countries. Milkfat ingredients made from blends of milkfat fractions are generally designed with specific applications in mind and are produced using five general steps:

1. Fractionating milkfat to create stock fractions
2. Selective blending of fractions to the desired melting characteristics of the base milkfat
3. Blending with any additional functional ingredients
4. Texturizing the final blend
5. Packaging the product

Fractionation of Milkfat to Produce Stock Fractions

Milkfat fraction ingredients are produced using one or more stock milkfat fractions. The nature of the fractionation process provides many opportunities to create milkfat fractions with a wide range of melting characteristics, and fractionation conditions can be easily manipulated to optimize the properties of stock fractions. The exact properties required in a stock fraction will depend on the final application and on how many stock fractions will be used by a manufacturer as the basis for its product line. Stock fractions should exhibit a range of physical and chemical properties that will complement each other when blended. The more stock fractions available, the more flexibility there is in tailoring milkfat ingredients to food products. However, increasing the number of stock fractions increases the production costs, because more fractionation steps are necessary. The balance between flexibility and cost is a decision to be addressed by each manufacturer.

To provide a good range of functionality, at least five stock fractions should be used, and there should be a representative fraction from each of the major melting categories: very-high-melting, high-melting, middle-melting, low-melting, and very-low-melting. As discussed in Chapter 5, the solid fat content profiles of milkfat fractions within a major melting category can be quite different, so it may be advantageous to increase the number of stock fractions to provide increased flexibility during formulation. Intact anhydrous milkfat is also considered a stock fraction and exhibits properties similar to middle-melting milkfat fractions.

Technologies and methodologies available to fractionate milkfat and characterize stock milkfat fractions were discussed in Chapters 3 and 4, respectively. The melting behavior and other functional properties of intact milkfat and milkfat fractions were presented in Chapter 5.

Blending of Milkfat and Milkfat Fractions

The range of melting properties available from milkfat fractions provides an opportunity to blend intact milkfat and milkfat fractions systematically to create ingredients with selected melting profiles. The blending of fractions gives a manufacturer greater flexibility to tailor milkfat ingredients to specific functional requirements than could be accomplished with fractionation alone. Thus, blending of milkfat fractions allows some flexibility to overcome normal compositional changes in the milkfat supply and to compensate for any other processing changes that may change the functionality of a stock fraction. Blending also makes it easier to provide a consistent end product (Deffense, 1989).

However, the literature on milkfat fraction blends is not abundant, and information on commercial milkfat ingredients is limited to finished milkfat ingredients (Chapter 5). As expected, detailed information on the constituent milkfat fractions used to create commercial milkfat ingredients is proprietary. Further, the complex interactions that occur when milkfats are blended with each other and with other fats, such as cocoa butter or vegetable oils, are not fully understood. Therefore, an increased knowledge of the effects of blending on the melting properties and crystallization behavior of milkfat fractions would greatly enhance the ability to accurately predict the functionality of milkfat ingredients made from milkfat fractions with known physical and chemical characteristics.

Incorporation of Functional Additives to Milkfat Ingredients

Many fat systems contain additional ingredients to improve their functionality in selected applications. Functional additives include

- Aqueous phase
- Salt
- Surfactants
- Emulsifiers
- Dough conditioners
- Crumb softeners
- Antifoaming agents
- Antioxidants
- Metal scavengers
- Antimicrobial agents

The use of these functional additives is specific for various applications, such as the use of emulsifiers to improve the aeration functionality of cake shortenings or the use of antimicrobial agents in margarine.

The addition of functional ingredients to fat systems is commonplace in the edible oils industry but not in the dairy industry. The composition of butter has been defined in the Food and Drug Act of June 30, 1906, as the product that is made exclusively from milk or cream, or both, with or without salt, with or without coloring matter, and containing not less than 80 percent milkfat by weight (Newlander and Atherton, 1964). Therefore, additives are not allowed in the product if it is to be labeled "butter". The addition of functional additives to milkfat ingredients may be desirable on a commercial scale but has not been pursued at this time because of the legal implications. Milkfat ingredients available commercially in Europe and New Zealand are made from pure anhydrous milkfat with no additives, other than legally allowed tracer compounds, and can be labeled as butter or concentrated butter in food products (Alaco, undated; Aveve, undated; Corman, undated).

The experimental addition of functional ingredients to milkfat systems has been shown to improve their functionality in selected applications and will be discussed in this chapter in relation to specific applications.

Texturization of Milkfat Ingredients

After the constituent milkfat fractions and any functional ingredients have been properly blended, they are generally crystallized to create a solid finished product. Controlled crystallization of the ingredient blend, followed by physical manipulation of the crystals, results in the formation of the appropriate crystal form and size for optimal physical functionality in the final application. The continuous crystal network formed during crystallization also stabilizes the components of the ingredient blend (e.g., milkfat and aqueous phases) and yields a uniform product with the desired appearance characteristics (Weiss, 1970).

The texturization or plasticization of fats involves controlled crystallization of the molten ingredient blend, followed by physical reduction of crystal size by pin-rotor processing units. Commercial texturizers are available, and include the Votator (Cherry-Burrell, USA), the Perfector (Gerstenberg and Agger, Denmark), and the Kombinator (Schroeder, Germany). Texturization is achieved through the use of four processing units:

1. Swept-surface chillers

   - These are used primarily as a precooler but can also be used as an intermediate cooler.
   - Precoolers ensure that the material entering the crystallizer is at a constant temperature; reduce the heat load on the crystallizer; and cool the fat as close to its melting point as possible to ensure the formation of the largest number of crystal nuclei possible (Black, 1975a; Joyner, 1953).
   - Precoolers provide a way to cool the fat slowly before the initiation of nucleation, which results in a softer fat (Black, 1975a).

2. Scraped-surface heat exchangers

   - These are used to crystallize the fat with agitation to maintain small crystal size.
   - Rapid initial cooling of the fat blend with agitation causes rapid nucleation and further emulsification of fat–aqueous blends (Chrysam, 1985; Joyner, 1953; Weiss, 1970).

3. Pin-rotors or pin-working units

   - These are used to crystallize the fat under high-shear conditions.
   - Most of the crystallization occurs during working and results in a rearrangement of α crystals, formed in the chiller, to more stable β' crystals (Chrysam, 1985; Podmore, 1991).
   - Mechanical work increases the plasticity of shortenings (Podmore, 1991).
   - Working causes a decrease in crystal size (Chrysam, 1985; Weiss, 1970).
   - Working causes vigorous aeration of fat and can incorporate 10 to 15% air into shortening (Chrysam, 1985; Joyner, 1953; Lawson, 1985).
   - The temperature of the fat generally rises 10 to 15°F because of the latent heat of crystallization, and the pin workers or rotors help to quickly dissipate the excess heat (Joyner, 1953).

4. Resting tubes (setting tubes)

   - These are used for final crystal growth.
   - The use of these tubes promotes bonding of crystals, which results in the desired opaque appearance of shortenings (Keogh and Morrissey, 1990).
   - These are often used if product requires a stiffer consistency for packaging (Chrysam, 1985).

In addition to altering the sequence of the processing units, processing conditions, such as temperature and throughput, can be manipulated to provide a range of physical properties in the final product. Because of the complexity of the processing conditions and the crystallization processes that occur, the texturizing of fat blends is considered part art and part science (Weiss, 1970). Processing conditions and some of their effects include the following:

1. Cooling rate/supercooling
   - Slow crystallization, such as when a precooler is used, will result in the formation of large, soft crystals that do not crystallize sufficiently and will continue to crystallize after leaving the scraped-surface crystallizer (Haighton, 1976; Keogh and Morrissey, 1990).
   - Rapid supercooling results in the formation of many small nuclei, which yields a fine crystal structure (Podmore, 1991).
   - The temperature at which the supercooled product is allowed to reach crystal equilibrium is directly related to the temperature range over which the shortening is workable (Joyner, 1953).

2. Refrigerant temperature
   - Decreased refrigerant temperature used in the crystallizers results in increased cold-shearing (Keogh and Morrissey, 1990).

3. Agitation/shear rate
   - Crystallization that occurs with agitation/cold-shearing will result in the formation of very small crystals and less crystal network formation, yielding a softer product (Foley and Brady, 1984; Joyner, 1953; Keogh and Morrissey, 1990).
   - Increases in agitation rate enhance cold-shearing and the formation of smaller crystals (Keogh and Morrissey, 1990).
   - Crystallization that occurs without agitation results in the formation of a very firm crystal network that is of far greater strength than the same proportion of solids in the form of discrete crystals, and this condition adversely affects the plastic range of the fat (Haighton, 1976; Joyner, 1953; Keogh and Morrissey, 1990).

4. Throughput
   - A lower throughput means a higher residence time and results in more cold-shearing of the crystals (Keogh and Morrissey, 1990).
   - Lower throughput decreases the yield value and results in smaller crystals in madeira milkfats (Keogh and Morrissey, 1990).
   - High throughput increases the crystal size, yield value, and brittleness of croissant milkfat (Keogh and Morrissey, 1990).

5. Crystal size
   - The fat becomes firmer as the average crystal size decreases and softer as the average crystal size increases (Joyner, 1953).

Although texturization and tempering protocols have been well-established in vegetable fats and oils industry, these details are not well understood for milkfat, and they warrant further study. There are many options available for processing of fat ingredients to meet a variety of functional demands. To ensure that processing conditions are appropriate for the desired application, the final evaluation should always include performance testing of the finished fat ingredient (Weiss, 1970).

Packaging of Milkfat Ingredients

A variety of sizes and materials are available for packaging milkfat ingredients. Considerations in the choice of packaging material should include shelf life and stability concerns, convenience to the end user, and materials and processing costs. Recycling and disposability of packaging materials are also considerations in the choice of packaging materials. Metal drums are often collected, cleaned, and refilled, and cardboard cartons can be broken down for easy disposal (Weiss, 1970).

Optimizing packaging sizes can help improve the usability of milkfat ingredients. In the U.S., butter and anhydrous milkfat are commonly packaged in 1 lb prints or 50 lb blocks, and the bakery industry has stated that these options are sometimes inconvenient to their mixing operations. An example of optimized packaging is the sheet packaging used for confectioner's margarines for easy incorporation into pastry doughs (Lawson, 1985; Weiss, 1970). Milkfat ingredients commercially available in Europe and New Zealand use the following packaging sizes:

- Bakery products: 454 g sheets, 500 g sheets, 1 kg sheets, 2 kg sheets, 1 kg prints, 10 kg blocks, 12.5 kg blocks, 25 kg blocks
- Confectionery products: 25 kg blocks
- Ice cream and dairy products: 17 kg tins, 25 kg blocks, 200 kg drums, tankers for immediate use.

The sheets are generally foil-wrapped; blocks are contained in polyethylene-lined cartons; tins are varnished on the inside; and drums are made of either metal or fiberboard lined with polyethylene (Alaco, undated; Aveve, undated; Corman, undated). Liquid oils are supplied in tankers for immediate use and are only available within a certain geographical range of the manufacturing facility.

Standard packaging sizes in the edible oils industry in the U.S. are 50 lb corrugated cartons lined with polyethylene bags; 110 lb metal cans; metal drums (400 lb), whereas confectioner's margarines are packaged in 5 or 10 lb slabs (Lawson, 1985; Weiss, 1970).

Tempering of Milkfat Ingredients

Tempering processes affect the crystal form by allowing less stable crystals to melt and recrystallize into more stable forms. Texturization processing of milkfat manipulates the solid crystal matrix to modify characteristics of milkfat, such as spreadability and plasticity. Tempering can be performed after texturization and results in additional crystallization (liquid-to-solid transitions), polymorphic transitions (solid-to-solid), and recrystallization (solid-to-liquid-to-solid transitions) (van Beresteyn, 1972).

Tempering processes employed for fat ingredients depend on the application. For example, vegetable shortenings used in the baking industry are commonly tempered for 1 to 4 days at 26 to 29°C after packaging to ensure proper crystal growth and appropriate consistency (Chrysam, 1985; Joyner, 1953; Lawson, 1985; Weiss, 1970). However, Black (1975a) tempered milkfat at 25–35°C for 1 to 5 days and reported an increase in firmness and an unexpected decrease in plasticity of the milkfat, which was in contrast to the improvement of plasticity observed for vegetable shortenings. There is some uncertainty as to what happens to the structure of plastic fats during the tempering period, and the unexpected results reported by Black (1975a) illustrate the need for further studies of the effects of tempering on milkfat ingredients.

The Use of Milkfat Ingredients in Food Products

This section describes the production, characterization, and performance evaluation of experimental milkfat ingredients used in food products. The first two products considered, cold-spreadable butter and dairy-based spreads, are food products for retail and food service marketing, and the remaining sections describe food products that incorporate milkfat ingredients at the manufacturing level. Information on commercially available milkfat ingredients designed for specific food applications has been included where possible.

Cold-Spreadable Butter

One of the early goals of milkfat research was to increase the spreadability of butter at refrigerator temperature. Research efforts in the 1930s and 1940s concentrated on manipulating traditional dairy processing conditions to produce a more spreadable butter. Cream tempering, churning conditions, butter tempering, and reworking can have a marked effect on the hardness of butter. However, the improved spreadability gained by these physical processes can be lost under home usage conditions (Dolby, 1959; McGillivray, 1972; Mogensen, 1984).

Other methods available for increasing the spreadability of butter include aeration; homogenization, which improves the distribution of water droplets and results in an increase in spreadability; addition of vegetable oils; and the use of additives (soya lecithin or buttermilk powder) (El-Nimr, 1980; Mickle, 1960; Rajah, 1992). Technological advances in interesterification and fractionation have also greatly increased the opportunities to manipulate the physical properties of milkfat and increase the spreadability of butter.

Strategies that rely on purely physical modifications of butter are not always successful. In order to ensure consistent spreadability, McGillivray (1972) suggested that the chemical composition of the cold-spreadable butter should be different from that of the intact milkfat. Interesterification directly modifies the chemical composition of milkfat and yields a more random distribution of the fatty acids on the triglyceride molecule, resulting in an improvement in butter spreadability (Mickle, 1960). Milkfat fractionation is a physical process that results in a change in chemical composition of the fractions compared with the intact milkfat. Considerable efforts have been devoted to using fractionation technology to produce milkfat fractions for incorporation into cold-spreadable butter (Chapter 5). However, only a small proportion of investigators have actually applied these fractions in cold-spreadable butters (Table 8.1), including three U.S. patents for cold-spreadable butter (Bumbalough, 1989; MacCollom, 1970; Norris, 1977).

Desired Functionality Attributes of Cold-Spreadable Butter

A cold-spreadable butter should be easily spreadable when taken out of the refrigerator, should maintain physical integrity at room temperature, and should possess a characteristic butter flavor.

The functionality of cold-spreadable butter is a direct result of the functionality of the milkfat ingredients used, and this functionality can be described in terms of its physical and inherent chemical properties (Chapter 7). Functionality attributes important to cold-spreadable butter include the following:

- Spreadability
- Firmness/hardness
- Softness
- Viscosity
- Flow properties
- Structure formation
- Flavor potential

The physical characteristics of cold-spreadable butters are the result of complex interactions between the solid milkfat phase, the liquid milkfat phase, and the aqueous phase.

Spreadability can be described by the melting behavior of the milkfat and can be correlated to firmness, hardness, or softness values. Development of an appropriate butter structure is related to the melting and crystallization behaviors of the milkfat fractions employed and to the crystallization and texturization processes employed.

Characterization of functionality can be performed for the milkfat fractions or for the finished cold-spreadable butter. Characteristics of milkfat fractions are generally used to predict the spreadability of the butter, and they are useful in determining the blending ratio for the fractions. For example, solid fat content profiles can be used to define a target profile, and then milkfat fractions can be selectively blended to achieve this target (Kaylegian et al., 1993). Melting behavior has been characterized by solid fat content and melting point. Firmness, hardness, and softness have been characterized by cone or disc penetrometry, sectility, and cutting resistance. Viscosity and flow properties have been described by yield values and apparent viscosity. The flavor of the cold-spreadable butters has been evaluated by sensory panels, and the flavor potential of spreadable butters has been determined based on the lactone content.

The spreadability of margarine has often served as a model for the spreadability of butter. Target values for ideal spreadability of butter have been identified in the literature based on the characteristics of spreadable margarine, including the following:

- Solid fat content

  (Deffense, 1987):   (Jamotte and Guyot, 1980):
  5°C = 40%;   4°C = 37%;
  10°C = 30%;   14°C = 25%;
  15°C = 20%;   18°C = 19%
  20°C = 15%;
  25°C = 7%;

- Dropping point = 30°C (Deffense, 1987)
- Penetration value = 100–150 (Jamotte and Guyot, 1980)
- Hardness value = 200 (McGillivray, 1972)
- Yield value = 20–50 kPa (Makhlouf et al., 1987).

The finished butter structure must yield to spreading, or deformation, at refrigerator temperature, and it must not leak oil at room temperature.

Production of Experimental Cold-Spreadable Butters

Experimental cold-spreadable butters have been produced from milkfat fractions obtained from melted milkfat and from solvent solution using three general approaches:

- Addition of a low-melting milkfat fraction to cream or whole butter during the conventional churning process
- Use of a single milkfat fraction as the milkfat source, which is emulsified with other ingredients and texturized appropriately, and
- Selective blending of milkfat fractions to achieve a desired melting profile, followed by emulsification with other ingredients and appropriate texturization

Formulations and processing information for experimental cold-spreadable butters are presented in Table 8.1.

Stock milkfat fractions used to produce experimental cold-spreadable butters Many of the stock milkfat fractions that were added directly to cream or whole butter during churning were categorized as unknown-melting milkfat fractions, because melting behavior data were not available (El-Nimr, 1980; Frede et al., 1980; Kankare and Antila, 1974b). These were liquid fractions obtained from melted milkfat at 12, 25, and 28°C and these should have melting points just below their fractionation temperature (Chapter 5). Therefore, such fractions would be categorized as low-melting or very-low-melting. Dixon and Black (1974) added a low-melting milkfat fraction to cream to produce a cold-spreadable butter.

Stock milkfat fractions comprising 100% of the milkfat source in cold-spreadable butters have been obtained from melted milkfat using a variety of separation methods (Dolby, 1970b; MacCollom, 1970; Vovan and Riel, 1973). The melting properties of these fractions ranged from low-melting to high-melting. MacCollom (1970) produced a spreadable butter with an unknown-melting liquid fraction obtained at 22 to 24°C, and this fraction probably could be categorized as a low-melting milkfat fraction. Fractionation temperatures were not reported for the high-melting, middle-melting, and low-melting milkfat fractions produced by Dolby (1970b). Vovan and Riel (1973) removed a portion (5, 10, or 15%) of the unwanted fraction from the total milkfat, either the fraction that was solid at temperatures above 21°C or the fraction of milkfat that melted between 10 and 21°C.

In cases were the milkfat portion was composed of several milkfat fractions, generally very-high- or high-melting milkfat fractions were combined with low-melting, very-low-melting, or unknown-melting milkfat fractions (Bumbalough, 1989; Deffense, 1987, 1989; Guyot, 1982; Jamotte and Guyot, 1980; Kaylegian and Lindsay, 1992; MacCollom, 1970; Makhlouf et al., 1987; Munro et al., 1978; Norris, 1977). The very-high-melting milkfat fractions obtained from melted milkfat were solid fractions obtained using single-step fractionation or fractions obtained at the first step of a multiple-step fractionation procedure at fractionation temperatures above 26°C (Guyot, 1982; Kaylegian and Lindsay, 1992). The high-melting fractions obtained from melted milkfat were solid fractions obtained between 22 and 30°C using both single-step and multiple-step fractionation procedures. The low-melting and very-low-melting milkfat fractions obtained from melted milkfat were liquid fractions produced by multiple-step fractionation at temperatures below 22°C. The very-high-melting and high-melting milkfat fractions obtained from acetone solution were solid fractions obtained between 10 and 15°C. The unknown-melting milkfat fractions were liquid fractions obtained below 0°C, and they would be categorized as very-low-melting milkfat fractions.

Available physical and chemical characteristics of the stock milkfat fractions used in experimental cold-spreadable butters were discussed in Chapter 5.

Blending of milkfat fractions to create the desired melting profile of experimental cold-spreadable butters Milkfat fractions can be blended with cream, butter, anhydrous milkfat, and other milkfat fractions to modify the melting profile of butter. Butter spreadability can be improved by increasing the total unsaturated content of the fat, generally accomplished by the addition of low-melting milkfat fractions to cream, butter, or anhydrous milkfat. Low-melting milkfat fractions that were added to cream or butter at a level of 20% or more of the total fat content were found to improve the spreadability of butter at refrigerator temperature (Dixon and Black, 1974; El-Nimr, 1980; Frede et al., 1980; Kankare and Antila, 1974b).

Jamotte and Guyot (1980) have suggested that to soften a hard butter, or to harden a soft butter, the simple addition of suitable fractions to the butter oil will suffice. However, to obtain a truly cold-spreadable butter (spreadable at 5°C while maintaining adequate consistency at 20°C), it is essential to recombine an olein fraction that is liquid around 0°C and a solid fraction that has a dropping point of at least 45°C (Jamotte and Guyot, 1980). The cold-spreadability of margarine has been attributed to the fact that there are almost no triglycerides present that melt between 5 and 20°C (Timmen, 1975a). Thus, the goal of many cold-spreadable butter experiments has been the removal of this middle-melting milkfat fraction.

Vovan and Riel removed 5, 10, and 15% of the milkfat that melted between 10 and 21°C and found that as more of this fraction was removed, the more spreadable the butter became. However, Kaylegian and Lindsay (1992) removed the fraction that melted between 20 and 25°C (7% of the milkfat, Table 5.1) and reported no improvement in spreadability compared to the control butter. These results suggested that Kaylegian and Lindsay (1992) did not remove a middle-melting fraction with a sufficiently wide melting range to improve spreadability.

Many cold-spreadable butters (Table 8.1) were produced by combining two milkfat fractions, i.e., a very-high-melting or high-melting milkfat fraction (10 to 50%) and a low-melting or very-low-melting milkfat fraction (50 to 90%) (Bumbalough, 1989; Deffense, 1987, 1989; Guyot, 1982; Kaylegian and Lindsay, 1992; Jamotte and Guyot, 1980; Makhlouf et al., 1987; Munro et al., 1978; Norris, 1977). The blends with the best spreadability performance were reported to be in the range of 20 to 30% very-high-melting or high-melting milkfat fraction (MP > 43°C) combined with 70 to 80% very-low-melting milkfat fraction (MP < 10°C). As the percentage of high-melting fraction increased, the butter became less spreadable at refrigerator temperature. Conversely, as the percentage of low-melting fraction increased, cold-spreadability improved, but the butter became excessively soft at room temperature.

Kaylegian and Lindsay (1992) produced two cold-spreadable butters using three stock milkfat fractions, which were selected for incorporation into butter based on their solid fat content profiles. A low-melting:high-melting butter was made from a low-melting milkfat fraction (90% $13L_M$ fraction) and a high-melting milkfat fraction (10% $30S_S$ fraction) obtained from melted milkfat. This butter exhibited good spreadability at refrigerator temperature, but it had poor physical characteristics at room temperature. To improve the physical integrity of the butter at room temperature, a high-melting (10% of $11S_M$) or very-high-melting (20% of $15S_M$) fraction obtained from acetone solution was added to the low-melting:high-melting butter, with a concurrent reduction in the amount of low-melting milkfat fraction used. The milkfat fractions obtained from acetone solution melted very slowly until just prior to the melting point and contributed to improved structural integrity of the butters at room temperature, compared with the initial low-melting:high-melting butter.

Addition of functional ingredients to experimental cold-spreadable butters   A goal of producing cold-spreadable butter is to keep the formulation as close to the normal composition of butter as possible, and therefore, only essential ingredients are added to the blend (Table 8.1). In cases where the cold-spreadable butter was made by adding a low-melting milkfat fraction to cream or whole butter during the conventional butter process, no additional ingredients were added. Ingredients added to the milkfat blend to produce a recombined cold-spreadable butter have included

- Water
- Skim milk
- Buttermilk powder
- Casein
- Salt

The aqueous phase of recombined butter is usually obtained from water or skim milk, which is then emulsified with the milkfat portion. Generally a protein or phospholipid source, such as buttermilk powder or casein, is added to stabilize the emulsion (Guyot, 1982; Jamotte and Guyot, 1980; Kaylegian and Lindsay, 1992; Makhlouf et al., 1987; Vovan and Riel, 1973). Salt is a common ingredient in conventional butter and is generally added to the recombined butters for flavor. MacCollom (1970) also suggested the addition of diacetyl distillate for flavor and butter color, if desired.

Texturization processes employed in the production of experimental cold-spreadable butters Production methods employed for cold-spreadable butter depended on the approach used to produce the butter (Table 8.1):

- Conventional butter churning processes were used when milkfat fractions were added to cream or to whole butter.
- Recombined butter processes were used when the milkfat portion was combined with other ingredients.

Recombined butter processes were adapted from processes developed for margarine by the edible oils industry.

Recombined butters were made by emulsifying the ingredients (40 to 55°C) with agitation and then pumping the mixture to a scraped-surface heat exchanger. Processing conditions, such as throughput and agitation, employed in the scraped-surface heat exchanger varied depending on the unit used, such as a Votator (Bumbalough, 1989), Perfector (Deffense, 1987, 1989), Kombinator (Guyot, 1982; Jamotte and Guyot, 1980), or laboratory-scale unit (Kaylegian and Lindsay, 1992; Makhlouf et al., 1987). The exit temperatures of the butters were usually between 6 and 13°C, regardless of the texturizing unit employed. Several investigators tempered the butters at 18 to 20°C for several hours and then reworked the butters prior to packaging to improve consistency (Dixon and Black, 1974; Frede et al., 1980; Kaylegian and Lindsay, 1992).

Packaging and tempering of experimental cold-spreadable butters Experimental cold-spreadable butters have been generally formed into small blocks or packaged in plastic tubs and stored in the refrigerator at 4°C or in the freezer at –20°C (Table 8.1). Some butters were tempered (18 to 30°C) for several hours after packaging to improve consistency (Bumbalough, 1989; El-Nimr, 1980; Vovan and Riel, 1973).

Characterization of Experimental Cold-Spreadable Butter Functionality

Physical characteristics of experimental cold-speadable butters The spreadability of the experimental cold-spreadable butters has been evaluated relative to solid fat content profiles (Table 8.2), melting point (Table 8.3), and firmness, hardness, and softness values (Table 8.3).

Almost all of the cold-spreadable butters produced showed a decrease in *solid fat content* at 5°C when compared with the control butters (Table 8.2). Solid fat content values at 5°C decreased as the percentage of middle-melting fraction or high-melting fraction was removed (Vovan and Riel, 1973), as the percentage of low-melting milkfat fractions was increased (Deffense, 1987, 1989; Jamotte and Guyot, 1980; Kaylegian and Lindsay, 1992), or as the melting point of the low-melting fraction decreased (Deffense, 1987, 1989).

Cold-spreadable butters that had solid fat content profiles similar to spreadable margarine at 5°C (35 to 40% solid) and at 20°C (15 to 20% solid) were produced by Vovan and Riel (1973) by removal of 5, 10, or 15% of either the solid fraction or the intermediate fraction. Deffense (1987, 1989) produced two butters with the desired solid fat content at 5°C that were slightly softer than desired at 20°C (12% solid) but resulted in acceptable butters. Kaylegian and Lindsay (1992) produced butters that had the desired solids content at 20°C (15 to 20%) and were slightly harder (44 to 47% solid) than desired at 5°C, but were still spreadable. Butters that had less than 10% solid fat at 20°C generally did not exhibit good physical integrity at room temperature.

The objective of many cold-spreadable butter experiments has been to remove the middle-melting milkfat fraction, and this strategy would result in a relatively flat solid fat content profile curve between 5 and 20°C. However, most spreadable butters produced according to this strategy exhibited a gradual decrease in solid fat content between 5 and 20°C, indicating that there were triglycerides in the blend that melted over this range. Jamotte and Guyot (1980) produced two spreadable butters with 80 and 85% low-melting fraction that exhibited relatively flat melting profiles that ranged from 15% to 8.5% solid fat between 5°C and 20°C.

Jamotte and Guyot (1980) reported *dropping points* of 19.5 to 29.8°C for cold-spreadable butters made with an increasing proportion of liquid fraction (65 to 85%, respectively), compared with a dropping point of 33.5°C for winter butter (Table 8.3). Deffense (1987) reported a dropping point of 32°C for a butter (Butter A) that was close to the desired dropping point of 30°C recommended by Deffense (1987) for spreadable butters.

Firmness, hardness, and *softness* functionalities of experimental cold-spreadable butters have been measured by different methods and at different temperatures and expressed in a variety of units (Table 8.3). Cold-spreadability of butter should be evaluated near refrigerator temperature (4 to 6°C), whereas stand-up properties of butter should be evaluated at room temperature (18 to 20°C). Similar methods have been used by different authors to measure these various functionality attributes. For example, Dolby (1970b) and Norris (1977) used sectility to measure hardness (arbitrary units), and Vovan and Riel (1973) used sectility to measure spreadability (torque measured in g). Guyot (1982) measured spreadability with a cone penetrometer, Dixon and Black (1974) measured firmness with disc penetrometry, El-Nimr (1980) measured softness, and Deffense (1987, 1989) measured hardness using an unspecified type of penetrometry. Consequently, it is difficult to compare the data directly between authors, but the general effects of different formulations on these attributes can be observed in data presented by individual authors (Table 8.3).

Cold-spreadable butters produced by adding a low-melting milkfat fraction to cream or whole butter showed a decrease in cutting resistance (Frede et al., 1980; Kankare and Antila, 1974b) or an increase in softness (El-Nimr, 1980) as the percentage of low-melting milkfat fraction increased. Frede et al. (1980) evaluated the effect of the different surface-active agents in skim milk (casein and whey proteins), whey (whey proteins), and buttermilk (phospholipids) on the cutting resistance of finished butters. Frede et al. (1980) reported little difference between the three aqueous dispersions (resistance = 1.24 to 1.29 N) but reported a decrease in resistance of the experimental butters compared with winter butter (resistance = 1.72 N).

When cold-spreadable butters were produced using single milkfat fractions as the milkfat source, they exhibited a decrease in hardness as the melting point of the milkfat fractions decreased (Dolby, 1970b). Vovan and Riel (1973) reported an increase in spreadability as the amount of solid fraction or intermediate fraction removed increased from 5 to 15%. Interestingly, the spreadability increased more for the butters from which the intermediate fraction was removed than for butters that had the solid fraction removed, further supporting the position that removal of the middle-melting milkfat fraction from butter is essential for improved spreadability.

Cold-spreadable butters made from combinations of milkfat fractions exhibited increases in penetrometry values at 4 and 18°C as the percentage of very-low-melting milkfat fraction increased (Jamotte and Guyot, 1980). Target penetration values for spreadable butter fall within the range of 100 to 150, and these values have been observed for butters that were made with 75 to 85% low-melting milkfat fraction (Guyot, 1982; Jamotte and Guyot, 1980). At a concentration of 65% very-low-melting milkfat fraction, the butter produced by Jamotte and Guyot (Butter 5) had a penetrometer value of 38, which was similar to the control butter at 4°C, but butters produced by Deffense (Butter B; 1987, 1989) had penetrometer values of 105, which were in the target range for spreadable butter. The dropping points of the high-melting and very-low-melting stock milkfat fractions used by Jamotte and Guyot were 42.3 and 5.0°C, compared with melting points of 43.1 and 8.5°C for the fractions used by Deffense (1987, 1989). The final melting point of the high-melting fraction used by Jamotte and Guyot may have been greater than 45°C, in which case it would have been classified as a very-high-melting milkfat fraction; this might have contributed to the differences in penetrometer values observed between the two investigators.

Kaylegian and Lindsay (1992) determined spreadability by an informal assessment procedure and reported that the low-melting:high-melting butter (90% 13L_M fraction) was spreadable at 4°C but too soft at 21°C. Good cold-spreadability at 4°C and physical properties at 21°C were reported for the low-melting:high-melting + 10% very-high-melting butter and the low-melting:high-melting + 20% very-high-melting butter. The content of the 13L_M fraction in these butters was 70 to 80%, and the results were in agreement with the trends observed for other spreadable butters.

Viscosity and flow properties of experimental cold-spreadable butters Yield values, apparent (Bingham) viscosity, and shear forces for experimental cold-spreadable butters were reported by Makhlouf et al. (1987; Table 8.3). Makhlouf et al. (1987) reported that a yield value of less than 50 kPa should be considered acceptable for spreadable butter, and a value of 20 kPa should be similar to margarine. Yield values for experimental cold-spreadable butter ranged from 22.2 to 53.1 kPa at 7°C, compared with a yield value of 108.0 kPa for the control butter. However, yield values ranged from 1.8 to 5.4 kPa at 20°C for the experimental butters and the control butter had a yield value of 28.9 at 20°C, which indicated that the experimental butters were markedly softer than the control butter at room temperature.

Makhlouf et al. (1987) reported that the apparent viscosity did not necessarily correspond to yield value, and suggested that mixed crystallization during the cooling of the butter explained certain aspects of this behavior. A much greater variation in yield values was reported for the experimental cold-spreadable butters compared with the variation observed in the apparent viscosity of these butters. They reported that spreadability of butter correlated well with the inverse of the shear force on the knife during evaluation (Table 8.3).

Overall viscosity trends showed an increase in yield value, a decrease in apparent viscosity, and a decrease in shear force as the fractionation temperature of the low-melting milkfat fraction decreased or as the concentration of the low-melting milkfat fraction increased.

Flavor and flavor potential of experimental cold-spreadable butters Flavor of the experimental cold-spreadable butters has been generally assessed by sensory evaluation. Cold-spreadable butters produced from stock milkfat fractions obtained from melted milkfat generally had a characteristic butter flavor and aroma (Kankare and Antila, 1974b; Kaylegian and Lindsay, 1992). Jamotte and Guyot (1980) reported that their spreadable butters had a greasy taste, but they noted that all recombined butters have a tendency to exhibit this defect because the milkfat component is not sufficiently protected by membrane material. Jamotte and Guyot suggested that an emulsifier such as buttermilk powder be added to the formulation to minimize this defect. Dolby (1970b) noted that the flavoring substances in the butters tended to be concentrated in the lower-melting milkfat fractions. Therefore, spreadable butters made from high-melting milkfat fractions had rather flat flavors, whereas those made from low-melting milkfat fractions exhibited more intense flavors. However, both desirable and undesirable (oxidized or rancid) flavors were included (Dolby, 1970b).

Munro et al. (1978) reported an acceptable flavor in cold-spreadable butter made from milkfat fractions obtained from acetone solution. However, Kaylegian and Lindsay (1992) reported an off-flavor in butters that contained a stock fraction obtained from acetone solution, compared with butter produced only from fractions obtained from melted milkfat. Amer et al. (1985) also noted that milkfat fractions produced by acetone solution were prone to off-flavor formation.

Walker et al. (1977, 1978) determined the lactone potential of the milkfat fractions (Tables 5.146 to 5.150) and for butters made from different milkfat sources (Table 8.4). Differences in lactone potential between cold-spreadable butters made from different milkfat sources were related to the differences in lactone potential of the intact anhydrous milkfat (Table 5.145). Walker et al. (1977, 1978) reported an increase of 150 to 200% in lactone potential of low-melting milkfat fractions and a decrease of 26 to 39% in lactone potential of the high-melting milkfat fractions obtained from acetone solution compared with intact anhydrous milkfat. Spreadable butters prepared from milkfat fractions contained 1.3 to 1.6 times greater potential for lactone formation than butters produced directly from the intact anhydrous milkfats. This increase in lactone potential was caused by the enriched potential of the low-melting milkfat fraction and because of the formation of further quantities of free lactones from precursors during fractionation and processing. Walker et al. (1977) reported that the flavor of recombined butter did not appear to be affected by the variation in lactone content, and all of the spreadable butters produced were considered to be of good flavor quality. However, several of these exhibited a flat flavor and lacked the brightness of traditional butter, and this was attributed to the loss of some of the more volatile butter flavor compounds during the acetone removal process during fractionation.

Summary of Technical Data for Experimental Cold-Spreadable Butters

Technical data for experimental cold-spreadable butters can be summarized:

1. Cold-spreadable butters were produced using three approaches:
   - Addition of a low-melting milkfat fraction to cream or whole butter
   - Use of a single milkfat fraction as the milkfat source, which was then recombined with an aqueous phase
   - Selective blending of milkfat fractions to achieve desired melting profiles, with the blend being then recombined with an aqueous phase

2. Stock milkfat fractions used in cold-spreadable butters were obtained from melted milkfat and from acetone solution and ranged from very-low-melting to very-high-melting milkfat fractions.

3. The preferred method of improving the spreadability of butter at refrigerator temperature while maintaining physical integrity at room temperature was the combination of 20 to 30% high-melting milkfat fraction with 70 to 80% very-low-melting milkfat fraction.

4. As the concentration of the low-melting or very-low-melting fraction increased, butter became more spreadable at refrigerator temperature, but it did not maintain physical integrity at room temperature.

5. Increased concentrations of high-melting or very-high-melting fractions decreased spreadability at refrigerator temperature.

6. The spreadability and firmness characteristics of the butters were measured primarily by solid fat content profile, penetrometry, and sectility.

7. The flavor of cold-spreadable butters made from fractions obtained from melted milkfat was characteristic of butter, but butters produced using fractions obtained from acetone solution tended to contain off-flavors.

Technical Data for Commercial Cold-Spreadable Butters

Although there are several cold-spreadable butters available commercially in Europe and New Zealand, technical data for the manufacture of these products is not available.

Dairy-Based Spreads

Milkfat fractions have been used in three types of dairy-based spreads:

- Low-milkfat spreads
- High-melting butter
- Butter–margarine spreads

Low-milkfat spreads contain milkfat as the only fat source, but there is a reduction in total fat and caloric contents compared with conventional butters. High-melting butter contains milkfat as the only fat source, and this is a full-fat (80%) product. High-melting butter has been produced for potential evaluation in the bakery, chocolate, and confectionery industries (Shukla et al., 1994). Butter–margarine spreads are produced to reduce the cost of the spread and give it spreading properties similar to those of margarine while adding the flavor of butter (Verhagen and Bodor, 1984; Lansbergen and Kemps, 1984). These spreads have a fat content of 80%, which is similar to conventional butter. A number of low-fat butters and butter–margarine blends that are made from intact milkfat are available in the market, but data are not available for the commercial use of milkfat fractions in these products. Several experimental dairy-based spreads produced from milkfat fractions have been reported in the literature (Table 8.5).

Desired Functionality Attributes of Dairy-Based Spreads

The functionality attributes desired in most dairy-based spreads are similar to those for cold-spreadable butter:

- Spreadability
- Firmness
- Softness
- Viscosity
- Flow properties
- Structure formation
- Flavor

Some improvements in spreadability of low-fat butter is achieved by the inherent reduction in fat content. The structure-forming functionality of fats is more complex in low-fat products than in full-fat spreads. The increased water content in the butter requires the use of functional additives to stabilize the liquid milkfat–water emulsion and the solid fat–water dispersion. Factors affecting the functionality of milkfat fractions in foods were discussed in Chapter 7.

Target values for functionality attributes of most dairy-based spreads are similar to those reported for cold-spreadable butter; in particular, hardness evaluated by the method of Haighton (1959), is reported in C-values by Verhagen and Warnaar (1984) as 5°C = 1800 g/cm^2; 10°C = 1300 g/cm^2; and 20°C = 80 g/cm^2.

The desired functionalities of the high-melting butter (Shukla et al., 1994) are similar to those required of fats for pastries and chocolates, and include the following:

- Firmness
- Plasticity (pastries)
- Layering properties (pastries)
- Compatibility with other fats (chocolate)
- Flavor

Production of Experimental Dairy-Based Spreads

Formulation and processing information for low-milkfat spreads, high-melting butter, and butter–margarine spreads made from milkfat fractions are presented in Table 8.5.

Milkfat fractions used in experimental dairy-based spreads Two low-milkfat butters (41%) were produced, one with a middle-melting milkfat fraction produced from melted milkfat (Verhagen and Warnaar, 1984) and one with an unknown-melting milkfat fraction that was designated as a soft milkfat fraction (Keogh, 1988). Deffense (1989) reported that the use of liquid middle-melting milkfat fractions (MP = 26 to 28°C) in reduced-fat butters provided good spreadability. Jamotte and Guyot (1980) have suggested that the preparation of low-fat (40%) butters is facilitated if the creams are enriched in stearins prior to churning.

One high-melting butter was produced using a high-melting milkfat fraction obtained by supercritical carbon dioxide extraction (Shukla et al., 1994). This fraction was the residual (raffinate) fraction that remained after the extraction process had been completed.

Six butter–margarine blends (84% fat) were produced using unknown-melting milkfat fractions obtained from melted milkfat and from acetone solution. The milkfat fractions obtained from melted milkfat were both solid fractions obtained from single-step fractionation at either 15 or 25°C (Verhagen and Bodor, 1984). Three spreads were produced with the solid fraction ($25S_M$ fraction) in the manner of recombined butter, and one spread was produced by recombining the ingredients into cream (40%), followed by conventional churning. Lansbergen and Kemps (1984) first hardened milkfat by hydrogenation to achieve a milkfat triglyceride with two C16 fatty acids and one C2 to C8 fatty acid; they then fractionated this milkfat using multiple-step fractionation from acetone solution at 12, 6, and 0°C. The solid and liquid milkfat fractions were used to produce butter–margarine spreads. Based on the fractionation temperatures, the liquid fractions would probably have been categorized as very-low-melting milkfat fractions (Chapter 5).

Blending of milkfat fractions in experimental dairy-based spreads Low-milkfat spreads and high-melting butter were made using a single milkfat fraction as the only fat source (Table 8.5).

Verhagen and Bodor (1984) produced four butter–margarine spreads from a combination of intact milkfat, the $25S_S$ fraction, $15S_S$ fraction, and sunflower oil. Lansbergen and Kemps (1984) produced one spread from the $12L_M$ fraction and one from a combination of the $0L_M$ and $6L_M$ fractions as the only fat sources. These two spreads could have been considered butters because of the fat content, but they contained added colors (potassium sorbate and lactic acid). The third blend produced by Lansbergen and Kemps was a combination of the $12S_M$ fraction and sunflower oil.

Addition of functional ingredients to experimental dairy-based spreads Functional ingredients are commonly added to margarine and low-fat spreads (Chrysam, 1985). The spread formulations in Table 8.5 contain a combination of the following ingredients:

1. Fat phase:
   - Vegetable oil (sunflower, safflower)
   - Emulsifier (monoglycerides)

2. Aqueous phase:
   - Water
   - Skim milk
   - Salts (NaCl, $NaHPO_4$)
   - Acid (citric, lactic)
   - Emulsifier
   - Thickener or gelling agents (gelatin, hydrocolloids, gums)

- Protein (milk protein, skim milk powder, sodium caseinate)
- Preservatives (potassium sorbate)

The addition of emulsifiers to dairy-based spreads is particularly important in low-fat spreads, where the emulsion is the oil-in-water type instead of the traditional water-in-oil emulsion found in full-fat products. Thickeners or gelling agents are sometimes used in low-fat spreads to provide aqueous phase viscosity to promote the appropriate structure. The function of some salts, acids, and preservatives in spreads is to retard microbial spoilage, which becomes more prevalent as the water content of the finished product increases.

Texturization processes employed in the production of experimental dairy-based spreads The low-milkfat spreads and butter–margarine spreads have been produced using a Votator unit under standard margarine processing conditions (Table 8.5). Keogh (1988) found it necessary to use high stirring speeds in the emulsion tank to achieve emulsions with high stability, and found that the optimal processing treatments were a low throughput rate, high refrigerant temperature, and high agitation rates. The high-melting butter components were blended, pasteurized, and then formed into butter using a single-stage homogenizer (Table 8.5).

Packaging and tempering of experimental dairy-based spreads Detailed information on the packaging and tempering of low-milkfat spreads and butter–margarine spreads was not available. However, in the case of high-melting butter, it was stored overnight to achieve temperature equilibrium prior to physical and chemical analyses (Table 8.5).

Characteristics of Experimental Dairy-Based Spreads

Physical properties of experimental dairy-based spreads *Thermal profiles* (DSC curves) for the high-melting and control butters produced by Shukla et al. (1994) are shown in Figure 8.1. The high-melting butter exhibited a large peak for the high-melting triglyceride species, and the peaks for the low-melting and middle-melting triglyceride species were greatly diminished compared with the control market butter. Thermal profiles for low-milkfat spreads and butter–margarine spreads were not available.

The high-melting butter exhibited a *solid fat content* profile that was elevated compared with the control market butter (Table 8.6). The solid fat content profile of the high-melting butter was also elevated compared with milkfat ingredients for use in pastries (Table 8.10) and chocolate (Table 8.14).

Three butter–margarine spreads, produced from intact milkfat, solid milkfat fractions, and sunflower oil, had solid fat content profiles (Table 8.6) similar to the target values identified for spreadable margarine. These fractions exhibited good cold-spreadability and good physical properties at room temperature (Verhagen and Bodor, 1984). All three spreads produced by Lansbergen and Kemps (1984) were reported to have plasticity and elasticity scores that were comparable to those of margarine, but supporting data were not given.

The *hardness values* of a low-milkfat spread made from a middle-melting milkfat fraction ($25L_S$ fraction, Verhagen and Warnaar, 1984) was slightly lower at 5°C, but similar at 20°C, to the target values for spreadable margarine (Table 8.7). Verhagen and Bodor (1984) produced two butter–margarine spreads using a combination of AMF, $25S_S$ fraction, and sunflower oil, and reported increased hardness as the percentage of solid milkfat fraction increased. They also reported a marked decrease in hardness value at 5°C as the percentage of sunflower oil was increased from 40 to 60%, but this spread had a hardness value similar to margarine at 20°C.

Shukla et al. (1994) evaluated the viscoelastic properties of butter using dynamic mechanical analysis and reported higher complex viscosities and power law consistency indices for the high-melting butter compared with control market butter (Table 8.7). They noted that the high-melting butter exhibited viscoelastic properties at 32°C that were similar to the properties of control market butter at 22°C (Shukla et al., 1994).

Microbial stability of experimental dairy-based spreads   Keogh (1988) reported that as the numbers of large moisture droplets in the spread structure decreased, so did the rate of microbial growth at 4°C. The use of preservatives and antioxidants in the blends also helped to decrease microbial growth.

Flavor of experimental dairy-based spreads   Flavor evaluations of most spreads were not available, although Keogh (1988) reported an increase in rancid and oxidized off-flavors in spreads made without preservatives and antioxidants.

Summary of the Technical Data for Experimental Dairy-Based Spreads

The technical data for dairy-based spreads can be summarized:

- Low-milkfat spreads (40% milkfat) were produced from middle-melting or unknown-melting milkfat fractions using procedures for manufacture of recombined butter.
- Functional ingredients added to low-milkfat spreads included monoglycerides and gelling agents to stabilize the emulsion.
- High-melting butter (80% milkfat) has been produced from a high-melting milkfat fraction obtained by supercritical fluid extraction.
- Butter–margarine spreads have been produced using unknown-melting milkfat fractions obtained from melted milkfat or from solvent solution and combined with sunflower oil.
- Butter–margarine spreads were prepared using procedures for manufacture of recombined butter and by conventional churning.
- Improved spreadability was observed for low-milkfat spreads and butter–margarine spreads compared with conventional butter.

Bakery Products

Milkfat imparts flavor, richness, tenderness, and other desirable attributes to baked products (Baldwin et al., 1972; Keogh and Morrissey, 1990; McGillivray, 1972; Pyler, 1988). Historically, butter and lard were the primary fats used in bakery products. The complex composition of milkfat provides physical properties that exhibit poor functionality in some applications, and seasonal variability often also contributes to inconsistent butter. This has prompted the bakery industry to search for more suitable and consistent products. The vegetable fats and oils industry has responded by tailoring fats to specific baking requirements with the use of fractionation, hydrogenation, and interesterification of fats, followed by a plasticizing process to create more preferred consistencies of finished ingredients. Specialized vegetable shortenings and margarines offer better functionality at a lower price than does butter, and the availability of these has resulted in a decrease in the use of butter in many commercial bakeries. However, the use of milkfat in baked products is still highly desirable because of the unique flavors that are developed on heating (Keogh and Morrissey, 1990).

Despite the success of specialized vegetable shortening and margarines in the baking industry, the U.S. dairy industry has not responded with specialized shortenings based on milkfat. However, specialized products made from fractionated milkfat for bakery applications are available commercially in Europe (Aveve, undated; Corman, undated) and New Zealand (Alaco, undated), and they appear to be successful commercial products.

The required functional attributes for bakery fats depend on the specific application. For example, a milkfat produced for puff pastry usage should be a firm, plastic fat that provides good layering properties. On the other hand, a cake fat generally should be softer and contribute good aeration properties. Experimental and commercial data for bakery fats based on milkfat fractions are presented in this section, and they are grouped by applications that have similar fat requirements. Bakery groups for which there is information available on the use of milkfat fractions include the following:

- Laminated pastries (puff, Danish, croissant)
- Short pastries and cookies
- Cakes
- Miscellaneous products (icings, ice cream cones)

Laminated Pastries

Pastries represent one of the most frequently suggested uses for milkfat fractions, because they are high-quality products that significantly benefit in flavor from the use of milkfat, and they generally can accommodate the increased price of milkfat ingredients compared with vegetable fats. Laminated pastries (i.e., puff pastry, croissants, Danish) made with plasticized high-melting milkfat fractions exhibit improved functionality in the handling of the dough and in the appearance of the finished pastry, compared to those made with conventional butter. Furthermore, they exhibit a more desirable butter flavor and color when compared with pastries made from vegetable shortenings (Baker, 1970b; Cocup and Sanderson, 1987; Deffense, 1987; Humphries, 1971; Kuzdzal-Savoie and Saada, 1978b; Seibel, 1987).

Desired functionality attributes of milkfat ingredients for laminated pastries Fats used in laminated pastries provide flakiness, tenderness, and flavor. A laminated pastry fat must be able to form a thin film (70 µm to 3 mm) between the dough layers during rolling to prevent the formation of an extensive gluten network (Larsen, 1984; Keogh and Morrissey, 1990). During the baking of puff pastry and croissants, vaporizing water in the dough layers gives rise to a gas phase, which causes the pastry to rise before the fat melts, resulting in a flaky, tender pastry (Baldwin et al., 1972; Chrysam, 1985; Keogh, 1988; Keogh and Morrissey, 1990; Larsen, 1984; Pedersen, 1988; Pyler, 1988). During the baking of Danish pastries, the fat melts and is absorbed into the crumb structure of the pastry, and this requires a fat ingredient with a slightly lower melting point than for puff pastry and croissant fats (Chrysam, 1985; Tolboe, 1984).

Functionality attributes (Chapter 7) important to milkfat ingredients for laminated pastries include the following:

- Firmness
- Plasticity
- Layering properties
- Flavor and flavor potential

The layering properties of a pastry milkfat are a function of its firmness and plasticity. If the milkfat is too soft, it will oil-off during lamination and be absorbed by the dough. If the milk fat is too brittle, it will crack and fail to form a continuous fat layer between dough layers. Good layering properties require that the milkfat be plastic over typical manufacturing temperatures and resist work-softening (Baker, 1970b; Deffense, 1987; Frium and Fritsvold, 1984; Podmore, 1991).

Target physical characteristics for pastry milkfats and finished products include the following:

1. Puff pastry
   - Dropping point = 42°C (Deffense, 1987).
   - Slipping point = 41 to 42°C (Keogh, 1988), maximum SLP = 42°C (Keogh and Morrissey, 1990).
   - Baked puff pastry height > 30 mm, shrinkage < 24%, and eccentricity < 15% (Keogh, 1988).

2. Croissant
   - Slipping point = 38°C (Keogh, 1988), maximum SLP = 42°C (Keogh and Morrissey, 1990).
   - Baked croissant volume > 250 mL (Keogh, 1988).

3. Danish pastry
   - Melting point = 31.4°C, IV = 30 (Tolboe, 1984).
   - Solid fat content: 0°C = 73, 10°C = 51, 20°C = 22, 30°C = 9, 40°C = 0% (Tolboe, 1984).

4. General or unspecified pastries
   - Melting point = 35.0°C, IV = 22 (Tolboe, 1984).
   - Softening point = 38°C (Humphries, 1971).
   - Slipping point = 37 to 38°C (Larsen, 1984).
   - Solid fat content: 0°C = 71, 10°C = 56, 20°C = 41, 30°C = 24, 40°C = 2% (Tolboe, 1984).

Maximum slip points (SLP) have been identified to ensure full melting of the fat in the mouth, because incomplete melting of the milkfat results in undesirable sensory properties for finished pastries (Keogh and Morrissey, 1990; Larsen, 1984).

Production of experimental milkfat ingredients for laminated pastries Formulation and processing information for experimental milkfat ingredients used in laminated pastries is presented in Table 8.8. However, information on the production of commercially available milkfat ingredients for use in laminated pastries is proprietary.

Stock milkfat fractions used to produce experimental milkfat ingredients for laminated pastries have been high-melting, solid fractions obtained by fractionation from melted milkfat (Table 8.8). Schaap (1982) used a solid fraction obtained at 24°C by single-step fractionation for puff pastries. Keogh and Morrissey (1990) produced three solid high-melting fractions using single-step fractionation at 35, 38, and 41°C but did not specify which of these fractions was used for their puff pastry experiments. Commercially available milkfat ingredients are also produced by fractionation from melted milkfat (Alaco, undated; Aveve, undated; Corman, undated).

All reported experimental milkfat ingredients used for laminated pastries have been made using a single milkfat fraction as the milkfat source. Barts (1991) has recommended a combination of anhydrous milkfat plus stearin fractions for use in both puff pastries and fermented pastries (croissants and Danish).

Milkfat ingredients used in laminated pastries can be anhydrous products or those in an emulsified form. Experimental puff pastry milkfats have generally contained an aqueous phase, which contributed to the flakiness and volume of the finished pastry by vaporization of water in the milkfat ingredients during baking (Chrysam, 1985; Pedersen, 1988). Commercial milkfat ingredients for laminated pastries have all been anhydrous products, which are marketed as concentrated butters.

Functional ingredients added to emulsified milkfat ingredients used for laminated pastries include (Table 8.8):

- Skim milk (neutral and acidified)
- Emulsifiers (including lecithin)
- Buttermilk powder

The use of skim milk provides an aqueous phase, and the use of emulsifiers and buttermilk powder aid in the stabilization of the emulsion. Pedersen (1988) has used both neutral and acidified skim milk in puff pastries and reported better pastry volume with the milkfat ingredients made from acidified skim milk. Keogh and Morrissey (1990) used lecithin as an emulsifier, and Pedersen (1988) used a combination of lecithin and a commercial emulsifier (Homodan RD, Grinsted Products). Cocup and Sanderson (1987) have recommended the use of monoglycerides in pastry milkfat for improved functionality, and also noted that the absence of salt in the blend helps prevent retardation of fermentation during lamination.

Keogh and Morrissey (1990) studied the effects of lecithin and buttermilk powder on the crystal size and yield value of puff pastry milkfats (Table 8.9). They reported that the addition of lecithin and buttermilk powder increased the crystal size and yield values of the puff pastry milkfats.

All puff pastry and croissant milkfats have been processed through a scraped-surface heat exchanger (Table 8.8). Pedersen (1988) and Keogh and Morrissey (1990) used a Gerstenberg and Agger Perfector, which consisted of scraped-surface coolers (A units), intermediate crystallizers (C units), pin-workers (W units), and setting tubes (S units). Neither Pedersen (1988) nor Schaap (1982) reported processing conditions employed in the production of the milkfat ingredients. Keogh and Morrissey (1990) studied the effect of processing unit sequence on Madeira cake milkfats, and found that only the ACAS processing sequence gave a milkfat ingredient with the proper appearance. They used an ACAS sequence, the lowest refrigerant temperature, and the highest agitation rate possible in the production of all of their croissant and puff pastry milkfats. Keogh and Morrissey (1990) have suggested that the pin-worker unit not be used because it reduces plasticity, but the pin-worker was used successfully by Pedersen (1988) in the production of puff pastry milkfats.

Keogh and Morrissey (1990) also studied the effect of throughput rates on crystal size and yield values of croissant milkfats (Table 8.9), and found that a low throughput (30 kg/h) gave the milkfat ingredient with the best physical properties. They reported an increase in crystal size, yield value, and degree of brittleness as throughput rate increased from 30 to 120 kg/h. Milkfat ingredients processed at 120 kg/h were too brittle to be included in the bakery trials.

Schaap (1982) stored experimental puff pastry milkfats for one week at the dough preparation temperature prior to application but did not report the effects of tempering on the finished products. Pedersen (1988) stored half of the puff pastry milkfats at 5°C after production and half the samples at 18°C for 4 d prior to moving them to cold storage. Pedersen (1988) reported that penetrometer measurements showed that the decrease in firmness due to work softening was smaller for tempered samples than for untempered samples (numerical data not reported; 1988).

Commercial milkfat ingredients outside of the U.S. for laminated pastries are generally packaged in 10, 12.5, or 25 kg blocks, or 1 kg sheets for easy incorporation into the lamination process (Alaco, undated; Aveve, undated; Corman, undated).

Characteristics of experimental and commercial milkfat ingredients for laminated pastries
The firmness of milkfat ingredients for laminated pastries is expressed by melting behavior and solid fat content profiles. Most pastry milkfats contain high-melting milkfat fractions, but some commercially available products for use in general and puff pastry applications contain middle-melting milkfat fractions. Melting behavior ranges reported for pastry milkfats are (Table 8.9):

- Puff pastry: 36 to 42°C, and 26 to 28°C
- Croissant: 37 to 38°C
- Danish pastry: 37 to 38°C
- General pastry: 28 to 37°C.

Milkfats used for pastries have exhibited gradually melting solid fat content profiles (Table 8.10). The puff pastry milkfats were the most solid, followed by the croissant and Danish pastry milkfats. Pastry milkfats have ranged from 45 to 65% solid at 10°C and have melted between 30 and 45°C.

Pedersen (1988) reported that sensory finger judgments of puff pastry milkfat samples made with skim milk had better plasticity than anhydrous puff pastry milkfats.

Yield values for croissant milkfats increased from 1068 to 2429 g/cm^2 as the throughput rate during texturization increased from 30 to 120 kg/h (Table 8.9; Keogh and Morrissey, 1990). The yield value of puff pastry milkfats increased with the addition of lecithin and buttermilk to the aqueous phase (Keogh and Morrissey, 1990). Viscosity decreased as temperature increased for all pastry milkfats (Table 8.9; Corman, undated).

Characteristics of laminated pastries made with milkfat ingredients   Bakery formulas, production methods, and product evaluations of experimental milkfat ingredients in puff pastries and croissants are presented in Table 8.11. Barts (1991) listed the general composition of laminated pastries as follows:

- Puff pastry: about 55% flour, 30 to 35% milkfat, water, sugar, and salt
- Croissant: About 50% flour, about 20% milkfat, water, sugar, yeast, and salt
- Danish pastry: about 40% flour, about 40% fat, water, sugar, eggs, yeast, and salt

Handling properties of tempered puff pastry milkfats during lamination were better and the milkfat was firmer for tempered samples compared with untempered ones, but there were no noticeable effects of tempering on the height of the finished pastry (Pedersen, 1988).

All baked puff pastries prepared with milkfat ingredients exhibited good appearance characteristics that compared favorably with puff pastries made from margarine (Pedersen, 1988; Schaap, 1982). Pedersen reported that a low pH of the aqueous phase and the addition of an emulsifier increased the height of the baked pastry. Keogh and Morrissey (1990) also reported increased pastry heights with the addition of emulsifiers to puff pastry milkfats.

Keogh and Morrissey (1990) studied the effect of texturizing throughput rate on the texture and volume of croissants (Table 8.11). They reported that throughput of 30 to 90 kg/h gave typical croissant textures, and the croissant volume increased as throughput increased.

Short Pastries and Cookies

This category includes short pastries used for tart crusts, pie crusts, shortbread biscuits, and butter cookies. Short pastries differ from the laminated pastries in their formulations, the way in which the milkfat is incorporated into the pastry dough, and the desired functionality of the milkfat ingredients. Short pastries and cookies generally contain more water and sugar than laminated pastries, and eggs are usually added to cookies. Fat is generally incorporated into the dough by mixing it with the other dry ingredients until well-blended, followed by the addition of the liquid ingredients.

Desired functionality attributes of milkfat ingredients for short pastries   Desired functionality attributes of milkfat ingredients for short pastries include the following:

- Shortening properties
- Plasticity

- Lubricity

- Flavor and flavor potential

Fats are used in short pastries and biscuits primarily for the tenderness and shortening qualities they impart. Tenderizing and shortening effects are due to the ability of the fat to lubricate the structure of the product by being dispersed in films and globules in the dough during mixing and preventing the starch and protein components from forming a continuous, three-dimensional gluten network (Jordan, 1986; Podmore, 1991; Pyler, 1988). In places where the fat coats the flour, the gluten development is interrupted, and the result is a baked pastry that is more inclined to melt in the mouth (Podmore, 1991). The shortening effects of milkfat ingredients depend on the firmness and plasticity of the fats. Soft fats are more effective than hard fats because they are easier to disperse in the dough (Chrysam, 1985; Jordan, 1986; Podmore, 1991).

Target physical characteristics for short pastry milkfat include the following:

- Melting point = 32.1°C; IV = 34 (Tolboe, 1984)
- Slipping point = 27.5°C (Jordan, 1986)
- Solid fat content: 0°C = 69, 10°C = 47, 20°C = 28, 30°C = 10, 40°C = 0% (Tolboe, 1984)
- Crystal characteristics = 20μm, β' form (Pyler, 1988).

Production of experimental milkfat ingredients for short pastries   Jordan (1986) has evaluated the use of the solid, high-melting milkfat fraction and the liquid, low-melting milkfat fraction obtained at 28°C from melted milkfat in experimental shortbread biscuits (Table 8.8). Peck (1990) has evaluated a hard milkfat fraction and a soft milkfat fraction obtained from the New Zealand Dairy Board (Confectionery Butterfat 42 and Soft Butteroil 21, respectively) in chocolate chip cookies. Commercially available milkfat ingredients for short pastries are produced from melted milkfat (Alaco, undated; Aveve, undated; Corman, undated).

The experimental milkfat ingredients used by Jordan (1986) were produced from a single milkfat fraction. Barts (1991) has recommended a blend of anhydrous milkfat plus 20% oleic milkfat fractions for optimum functionality in short pastries.

Jordan (1986) did not specify whether the milkfat ingredients used for short pastries were anhydrous shortenings or emulsified butter-type products. Chrysam (1985) has advised against the use of emulsifiers in short pastry fats, because they may be detrimental to flakiness; they create a uniform dispersion of fat instead of the layers that are desired in short pastries. Commercial milkfat ingredients for short pastries all have been anhydrous products.

Processing information for experimental or commercial milkfat ingredients for short pastries was not available.

Information on packaging and tempering of experimental milkfat ingredients for short pastries was not available. Commercial milkfat ingredients for short pastries are generally packaged in 25 kg blocks (Alaco, undated; Aveve, undated; Corman, undated).

Characteristics of milkfat ingredients for short pastries   The firmness of experimental and commercial milkfat ingredients for short pastries is determined by their melting behavior (Table 8.9) and solid fat content profiles (Table 8.10). Commercial short pastry milkfats have been middle-melting and high-melting milkfats that have a melting point range from 28 to 38°C. Solid fat content profiles of these milkfats exhibit gradual melting, and these products are softer than the milkfats used for laminated pastries. Milkfat ingredients for short pastries have ranged from 38 to 56% solid at 10°C and generally have melted between 30 and 35°C.

Characteristics of short pastries made with milkfat ingredients   The general composition of short pastries is about 50% flour, about 25% sugar, and about 25% milkfat (Barts, 1991). Bakery formulas, production methods, and data from evaluations of shortbread biscuits (cookies) and chocolate chip cookies are presented in Table 8.11. Jordan (1986) produced two shortbread biscuits and observed better eating qualities in the biscuit (cookie) made with the low-melting milkfat fraction ($28L_S$ fraction). However, the shortbread made with the $28L_S$ fraction was a slightly too short and had too loose and crumbly a texture, but the shortbread made with the $28S_S$ fraction was too hard and not sufficiently short. Therefore, the ideal shortbread milkfat should exhibit properties that lie somewhere between these two fractions.

Peck (1990) reported that the hard milkfat fraction was difficult to cream into the sugar due to its extreme hardness and low plasticity. Cookies baked with Confectionery Butterfat 42 (Peck, 1990) had a waxy appearance, decreased diameter, and increased height compared with the control cookies, probably due to unequal dispersion and melting of the fat. Soft Butteroil 21 was easier to cream into the sugar than Confectionery Butterfat 42 but became almost liquid on removal from the cooler. Cookies made with the Soft Butteroil 21 had increased diameters and heights compared to control cookies but were very greasy. Both milkfat ingredients increased the hardness of the cookies compared to the control cookies.

The use of low-melting milkfat fractions has been suggested for the prevention of fat bloom that may occur in shortbread biscuits (cookies) during normal shelf life periods (Cocup and Sanderson, 1987).

Cakes

Pastries contain large amounts of flour and little liquid, whereas cakes contain a substantial amount of liquid. Consequently, the functionality of fats in cake systems is quite different than for pastries. Milkfat is traditionally known for providing poor aerating properties and poor plasticity, both of which are required for cakes of good texture and volume (Cocup and Sanderson, 1987; Knightbridge, 1978). The functional requirements for cake fats have been well studied for anhydrous vegetable shortenings.

Cake batters are air-in-fat-in-water emulsions that are made by creaming air into the fat, generally in the presence of sugar, followed by the addition of liquid ingredients to the fat–air mixture. During baking the fat melts (37 to 40°C), and the entrapped air passes into the aqueous phase and causes the batter to expand, creating the desired foam structure. Finally, the foam structure is heat-set by the coagulation of egg whites and starch gelatinization (Baldwin et al., 1972; Carlin, 1944; Shepherd and Yoell, 1976).

If the fat is not completely melted before the cake begins to rise, it will have an adverse effect on cake volume, and this is the case when intact milkfat is used. If a softer milkfat is used, the fat is more completely melted and improved cake volume and texture result (McGillivray, 1972).

Desired functionality attributes of milkfat ingredients for cakes Fats provide various functionalities to cakes (Baldwin et al., 1972; Chrysam, 1985; Keogh, 1988; Keogh and Morrissey, 1990; Pyler, 1988), and these include the following:

- To provide aeration, which results in proper leavening of the product leading to a desirable structure and texture
- To emulsify and stabilize the batter
- To impart tenderness and shortness to the crumb
- To contribute flavor
- To extend the shelf life of the finished product

Shortening plays an important role in determining the structure of cakes. Even though chemical leavening agents are used in cakes, incorporation of considerable amounts (about 270%) of air in the form of minute cells and bubbles during the mixing process is required to produce a cake with proper volume, grain, and texture (Carlin, 1944; Chrysam, 1985; Frium and Fritsvold, 1984; Keogh and Morrissey, 1990; Podmore, 1991; Pyler, 1988).

The functional attributes (described in Chapter 7) that are important to cakes include:

- Aeration properties
- Structure formation
- Plasticity
- Firmness
- Flavor and flavor potential

The melting properties and crystal characteristics of cake milkfats are of critical importance to aeration performance. Target physical characteristics for cake milkfats and baked cakes include the following:

- Softening point = 28°C (Humphries, 1971).
- Slipping point = 36°C (Keogh, 1988).
- Solid fat content: 20°C = 15 to 24%, 40°C = 0 to 5% (Keogh, 1988; Keogh and Morrissey, 1990; Shepherd and Yoell, 1976).
- Crystal characteristics: β' crystals, < 1 μm (Keogh and Morrissey, 1990; Pyler, 1988).
- Cake batter specific gravity < 0.76 (Keogh, 1988).
- Cake batter relative density for plain vegetable shortening = 0.74, for emulsified vegetable shortening = 0.72 (Cocup and Sanderson, 1987).
- Time to reach maximum aeration for plain vegetable shortening = 10 min, for emulsified vegetable shortening = 9 min (Cocup and Sanderson, 1987).
- Baked cake specific volume > 2.0, penetration > 100 (Keogh, 1988).

Pyler (1988) noted that maximum aeration properties of fats have been associated with optimal crystal sizes of less than 1 μm, and that aeration properties significantly decrease as crystal size becomes greater than 5 μm.

Production of experimental milkfat ingredients for cakes Milkfat fractions used in experimental milkfat ingredients for cakes (Table 8.8) have been solid, high-melting milkfat fractions obtained by single-step fractionation at 35, 38, and 41°C (Keogh and Morrissey, 1990), or unknown-melting milkfat fractions obtained by single-step fractionation at 24°C (Schaap, 1982). Interestingly, McGillivray (1972) suggested the use of low-melting or middle-melting milkfats for improved cake volume.

Experimental milkfat ingredients used for cakes have been produced using a single milkfat fraction as the milkfat source. Barts (1991) has recommended the use of intact anhydrous milkfats for most cakes and olein fractions for butter sponges.

Milkfat ingredients used for cakes generally have been anhydrous shortenings, some of which contained emulsifiers that provided positive effects on the functionality of milkfats in cakes (Black, 1975a; Buchanan et al., 1970; Cocup and Sanderson, 1987; Humphries, 1971; Keogh and Morrissey, 1990). Addition of an emulsifier to a cake shortening aids in the dispersion of fat in the batter and results in an improvement in air bubble dispersion, which increases the strength of the batter. Thus, incorporation of an emulsifier enables the use of a higher ratio of sugar and liquid ingredients relative to flour than could be used otherwise (Carlin, 1944; Chrysam, 1985). Emulsifiers used in cake milkfats include mono- and diglycerides (up to 4% w/w based on fat content), lecithin, polysorbates, and sodium stearoyl lactylate (Chrysam, 1985; Cocup and Sanderson, 1987; Keogh and Morrissey, 1990). Chrysam

(1985) has suggested the use of saturated monoglycerides in cake shortenings because they form complexes with the amylose fraction of starch, leading to a softer crumb with better keeping quality than control shortenings.

Buchanan et al. (1970) and Black (1975a) texturized intact milkfat using a swept-surface heat exchanger and produced milkfat shortenings that functioned well in high-ratio cakes. The plasticization process decreased the crystal size of the milkfat, which improved its aeration properties and resulted in improved textural characteristics in the finished cakes. Texturization processes that have been employed in the production of milkfat ingredients for cakes are shown in Table 8.8. Schaap (1982) cooled milkfat to 14°C using a scraped-surface heat exchanger, but no further processing conditions were reported.

Keogh and Morrissey (1990) studied the effects of processing sequence on Madeira cake milkfats using a Gerstenberg and Agger Perfector. The processing units consisted of scraped-surface coolers (A units), intermediate crystallizers (C units), pin-workers (W units), and setting tubes (S units). They evaluated the effects of four processing unit sequences (AC, ACAC, ACAS, and ACAW) on the appearance, crystal size and yield value of the milkfats (Table 8.9). Refrigerant temperature (low) and agitation rate (high) were held constant. The milkfats processed without the S unit had a glazed, petroleum jelly–like appearance, but milkfat processed with the setting tube had a normal opaque appearance, which was caused by additional bonding and crystal growth that took place in the setting tube. The use of the setting tube (S unit) was considered essential for puff pastry and croissant milkfats but optional for Madeira cake milkfats. The hardness of the milkfat was not increased by the setting tube, but crystal size was increased and baking performance marginally improved. The pin-worker decreased the hardness and plasticity of the milkfat, but it did not significantly alter baking performance. Post-hardening was apparent after the AC processing sequence, and this was due to continued crystallization and network formation after packaging.

Once an adequate processing sequence was determined (ACAS), Keogh and Morrissey (1990) evaluated the effect of throughput (30, 60, 90, and 120 kg/h) on the physical and chemical properties of the milkfats (Table 8.9). The effect of throughput rate on the largest crystal size was significant for all fats. Lower throughput rates decreased the largest crystal size and yield value of the Madeira cake fats because of the increased crystal breakdown and cold-shearing of the fats that occurred with longer residence times.

Vegetable shortenings for cakes are commonly tempered for 1 to 4 days at 26 to 29°C after packaging to ensure proper crystal growth and uniform consistency (Chrysam, 1985; Joyner, 1953; Lawson, 1985; Weiss, 1970). Tempering results in the melting of some crystals and the formation of firmer crystals, which produces a structure that is more homogeneous and stronger (Joyner, 1953).

Black (1975a) tempered plasticized, intact milkfat at 25 to 35°C for 1 to 5 days and reported the formation of finer crystals resulting from the melting and cooling of crystals, along with the formation of a more homogenous crystal structure of greater mechanical strength. This caused an increase in firmness and an unexpected decrease in plasticity of the milkfat, which was in contrast to the improvement of plasticity normally observed in vegetable shortenings. There is some uncertainty as to what happens to the structure of tempered plastic fats during the tempering period, and the unexpected results of Black (1975a) illustrate a need for further studies of the effects of tempering on milkfat.

Characteristics of experimental and commercial milkfat ingredients for cakes Milkfat ingredients used for cakes have ranged from low-melting milkfats to high-melting milkfats, and commercial milkfat ingredients produced by Alaco were categorized as unknown-melting because melting behavior data were not available. Melting, drop, and slip points have ranged from 26 to 41°C for cake shortenings (Table 8.9). Solid fat content profiles of cake milkfats have exhibited gradual melting, and they were 36 to 56% solid at 10°C. Generally these cake milkfats were melted at 30°C (Table 8.10).

Characteristics of cakes made with milkfat ingredients Bakery formulas, production methods, and evaluations of cakes made with milkfat ingredients are presented in Table 8.11. Schaap (1982) and Cocup and Sanderson (1987) observed favorable effects on cake batters for milkfat ingredients made from fractionated milkfat compared with unfractionated milkfat or vegetable shortenings. Cocup and Sanderson reported a decrease in the time required for fractionated milkfat to reach maximum aeration compared with intact milkfats, and this indicated improved plasticity of the milkfat ingredients. This effect was further improved by the addition of emulsifiers to the milkfat.

Keogh and Morrissey (1990) reported that the throughput rate employed during the production of the milkfat ingredients was not significantly related to the batter specific gravity or firmness of the finished cake. The specific volume of the finished cakes was inversely related to the specific gravity of the batters.

Miscellaneous Bakery Products

Other bakery products in which milkfat can contribute to sensory properties include icings, breads, and ice cream cones, although very little technical information is available for these products. Selected chemical and physical properties of commercial milkfat ingredients for use in miscellaneous bakery products are presented in Tables 8.9 and 8.10 (Corman, undated).

Ice cream cones and waffles Information on ice cream cones and waffles is limited to the recommendation to use oleic milkfat fractions in ice cream cones and waffles (Barts, 1991).

Breads and brioches Corman (undated) has suggested the use of standard anhydrous milkfat in breads and brioche.

Icings Buttercream icings are water-in-oil emulsions that contain up to 40% fat, sugar, and water (Chrysam, 1985). Icings are dependent on the solid fat in the shortening for their consistency, and it is important that the fat be in the stable β' form for smoothness and aeration. High-melting fats are used to impart body to the icing over an extended temperature range (Chrysam, 1985).

Target physical characteristics for buttercream icing milkfats include the following:

- Melting point = 31.5°C, IV = 31 (Tolboe, 1984).
- Solid fat content: 0°C = 68, 10°C = 50, 20°C = 28, 30°C = 10, 40°C = 0% (Tolboe, 1984).

No experimental data were available for the use of milkfat fractions in icings. However, Barts (1991) has recommended a blend of anhydrous milkfat and oleic milkfat fractions for use in buttercream icings.

Emulsifiers are often added to icing shortenings (2 to 4% based on shortening weight) to improve their aeration properties, and these include mono- and diglycerides and polysorbates (Chrysam, 1985). Chrysam (1985) has recommended the use of soft, unsaturated monoglycerides in icing systems to provide good aeration properties and to contribute dryness, smoothness, and stability to the finished icing.

Summary of Technical Data for the Use of Milkfat Ingredients in Bakery Products

Technical data for milkfat ingredients used in bakery products have been summarized as follows:

- Experimental milkfat ingredients were suitable for use in laminated pastries, short pastries, cooking, and cakes, and commercial milkfat ingredients are available for many bakery applications.
- Solid, high-melting milkfat fractions were recommended for use in pastries, middle-melting milkfat fractions for use in cakes, and middle-melting and low-melting milkfat fractions for use in cookies and shortbread biscuits.

- The texturization of milkfat ingredients to create the desired plasticity and crystal size is crucial for laminated pastries, where the fat needs to rolled to a very thin layer without cracking, and also for cakes, where the crystals size must be < 1 μm for good aerating properties.

Chocolate

Milkfat is an attractive ingredient in chocolate for several reasons (Barna et al., 1992; Bunting, 1991; Jebson, 1974b; Jeffrey, 1991; Jordan, 1986; Kuzdzal-Savoie and Saada, 1988b; McGillivray, 1972; Petersson, 1986; Timms and Parekh, 1980; Yi, 1993), and these include the following:

- Contributions of desirable flavor and textural qualities to milk chocolate
- Inhibition of fat bloom in chocolate
- Compatibility with cocoa butter
- Lower cost than cocoa butter
- Legality in chocolate in the U.S.A

Milkfat is usually incorporated into milk chocolate in the form of a product known as crumb, and the inclusion of milkfat gives milk chocolate its characteristic flavor (Jordan, 1986; Timms and Parekh, 1980). However, it is well known that excessive addition of milkfat to chocolate softens chocolate; thus, milkfat contributes to the soft textures of milk chocolate.

Fat bloom is a common visual and textural defect in chocolate. Mild cases of fat bloom are observed only as a visual defect and do not affect the eating quality of chocolate. In these cases, fat bloom appears as a grayish-white film on the surface of the chocolate which is caused by the scattering of light by clusters of large fat crystals of 5 μm or greater extending from the surface of the chocolate (Petersson, 1988; Yi, 1993). Chocolate with extensive fat bloom also exhibits a grayish-white appearance, but under these conditions fat bloom significantly alters the texture of the chocolate and results in a crumbly and undesirable product.

Fat bloom may result from poor tempering, from incompatible fats added to cocoa butter, or from incorrect cooling of tempered chocolate. Extensive fat bloom may result from temperature fluctuations over long periods of storage. The mechanism by which fat bloom occurs is generally attributed to the migration of lower-melting crystals to the surface and to the polymorphic transition of cocoa butter from the β(V) crystal form to the β(VI) form (Campbell et al., 1969; Petersson, 1988; Yi, 1993). Milkfat is a bloom inhibitor in chocolate, and it is commonly added to dark chocolate solely for this functionality.

Cocoa butter is the most expensive ingredient in chocolate, and many cocoa butter extenders, replacers, and substitutes have been developed to lower the cost of chocolate (Bunting, 1991). Milkfat is less expensive than cocoa butter, can be legally added to chocolate, and improves bloom stability. Thus, there is considerable incentive to increase the amount of milkfat used in chocolate. Unfortunately, excessive addition of intact milkfat to chocolate results in softening of the chocolate, which limits its use. This is of particular concern in dark chocolates, which are known for their characteristic firmness and snap (Yi, 1993). Bunting (1991) has suggested that chocolates with lower melting points may be desirable in colder climates or for use in ice cream and ice cream coatings.

Addition of greater than 30% milkfat to chocolates has been shown to produce unsatisfactory chocolates (Barna et al., 1992; Jordan, 1986; Timms, 1980b). The softening effect of milkfat on chocolate has been attributed to both low-melting and middle-melting fractions (Jordan, 1986; Timms, 1980; Timms and Parekh, 1980). The low-melting fractions cause softening by dissolving the cocoa butter crystals in chocolate. The middle-melting fractions appear to form eutectic mixtures with cocoa butter, which impedes crystallization and results in lower melting points and solid fat content profiles. Timms (1980b) has reported that the high-melting fraction of milkfat did not cause softening of chocolate, and this fraction contributed to the antiblooming effects of milkfat in chocolate.

Research on the use of modified milkfats in chocolate has been focused on the use of harder milkfats, usually produced by hydrogenation or fractionation, to allow for increased usage of milkfat in chocolate without softening (Baker, 1970b; Barts, 1991; Campbell et al., 1969; Hendrickx et al., 1971; Jamotte and Guyot, 1980; Timms and Parekh, 1980). Hydrogenated milkfats were found to provide good bloom inhibition, but these ingredients still softened chocolate at ambient temperatures. Milkfat ingredients made from milkfat fractions with a range of melting points have been used experimentally in chocolates (Table 8.12), and several milkfat ingredients are commercially available for incorporation into chocolates (Table 8.13).

Desired Functionality Attributes of Milkfat Ingredients for Chocolate

Chocolate contains 28–36% fat, with sugar and cocoa dispersed in the fat phase (Jeffrey, 1991). In the case of milk chocolate, milk solids also are dispersed in the fat phase (Jeffrey, 1991). Consequently, the physical, rheological, and sensory properties of the chocolate are determined by the properties of the fat phase (Jeffrey, 1991; Timms, 1980b; Timms and Parekh, 1980). Milkfat ingredients used in chocolate must produce a finished product that is temperable. Tempering of chocolates promotes the formation of the correct crystal form of cocoa butter, results in good appearance and sensory characteristics (gloss, snap, melting characteristics, and flavor), and produces a chocolate that is reasonably resistant to blooming (Campbell, et al., 1969; Chapman, 1971; Jeffrey, 1991; Yi, 1993).

Functionality attributes (Chapter 7) that are important to the use of milkfat fractions in chocolate include the following:

- Compatibility with other fats (crystallization and melting behaviors)
- Firmness
- Flavor

Milkfat is compatible with cocoa butter to a certain extent (approximately 30% replacement level); then additional amounts of milkfat begin to soften the chocolate. Low-melting milkfat fractions soften chocolate by dissolving cocoa butter crystals, whereas middle-melting milkfat fractions form eutectic mixtures with cocoa butter, which impedes crystallization. The softening and eutectic effects of milkfat fraction–cocoa butter mixtures can be observed in solid fat content profiles and isosolid diagrams.

The prediction of milkfat functionality in chocolate is more complex than in cold-spreadable butter, because of fat–fat interactions between the solid and liquid milkfat and cocoa butter and because of the fat–(nonfat) solid interactions between milkfat and cocoa and sugar particles. Although solid fat content profiles have proven useful in predicting milkfat functionality in cold-spreadable butter (Kaylegian and Lindsay, 1992), they are generally not adequate to predict the functionality of milkfat in chocolate. However, Timms and Parekh (1980) stated that the solid fat content of chocolate at 20 to 25°C is a good measure of hardness, and at 30 to 35°C it is a good assessment of sensory or eating properties. The use of milkfat ingredients in chocolate should be based on evaluation of their performance in the finished product.

Production of Experimental Milkfat Ingredients for Chocolate

Stock milkfat fractions used for experimental milkfat ingredients for chocolate Stock milkfat fractions used in experimental chocolates are shown in Table 8.12. Yi (1993) evaluated low-melting, middle-melting, high-melting, and very-high-melting milkfat fractions in dark chocolate. These milkfat fractions were obtained by multiple-step fractionation from acetone solution at temperatures between 0 and 25°C (Yi, 1993). Jebson (1974b) added an unknown-melting milkfat fraction that was designated a hard milkfat fraction to dark chocolates, but did not specify fractionation conditions.

Milkfat fractions added to milk chocolate have included low-melting, middle-melting, and high-melting fractions obtained from melted milkfat. The high-melting milkfat fractions were solid fractions obtained by single-step fractionation or at the first step of a multiple-step fractionation procedure at temperatures above 25°C (Barna et al., 1992; Bunting, 1991; Jordan, 1986). The middle-melting milkfat fraction used was the solid fraction obtained at the second step of a multiple-step fractionation at 20°C, and the low-melting fractions were the liquid fractions obtained at 20°C (Barna et al., 1992; Bunting, 1991).

Timms and Parekh (1980) used the solid, high-melting and the liquid, low-melting fractions obtained by single-step fractionation from melted milkfat at 25°C but did not specify which type of chocolate was evaluated. Baker (1970b) noted that the $30S_M$ fraction was investigated for use in chocolate, but gave no additional information.

Commercially available milkfat ingredients for use in chocolates are shown in Table 8.13. These fractions are high-melting or middle-melting fractions obtained from melted milkfat. Corman has provided a milkfat ingredient for use in chocolate buttercreams and fillings, but melting behavior data have not been available.

Blending of milkfat fractions for experimental milkfat ingredients for chocolate
Experimental milkfat ingredients used for dark chocolates have been produced using a single milkfat fraction as the milkfat source (Jebson, 1974b; Yi, 1993). Milkfat fractions were added to experimental dark chocolates at levels of 2 to 3%. To be an effective antiblooming agent, milkfat must be used at a minimum level of 2% (6% of the total fat), and a 4% addition of milkfat is considered optimum (Timms and Parekh, 1980).

Milkfat fractions added to milk chocolate have been used to replace the cocoa butter portion, while the amount of milkfat normally added to milk chocolate was held constant (Barna et al., 1992; Bunting, 1991). Consequently, the fat phase of experimental milk chocolates contained a milkfat fraction, intact milkfat, and cocoa butter. Barts (1991) has recommended the use of stearic milkfat fractions in milk chocolates for coatings or bars and intact milkfat or oleic milkfat fractions in chocolate creams for filling or decoration.

Addition of functional ingredients to experimental milkfat ingredients for chocolate
Ingredients added to experimental chocolates were those normally found in chocolate and include chocolate liquor, cocoa butter, sugar, lecithin, and vanillin (Table 8.12). Milkfat is generally added to milk chocolate in the form of crumb, which is a blend of cocoa nib and sweetened condensed milk that is dried to a low moisture content (Jordan, 1986). Desirable flavors are produced by Maillard reactions that occur during the production of the crumb; these would not be present if a blend of milkfat fractions and skim milk were used (Jordan, 1986). It is likely that high-melting milkfat fractions could be used to produce a recombined sweetened condensed milk for use in the production of cocoa crumb for milk chocolates.

Texturization processes employed in the production of milkfat ingredients for chocolate
The milkfat ingredients are melted prior to incorporation into the molten chocolate, and therefore the texture of these ingredients is not as crucial as it is for bakery applications. No information has been given for the texturization of milkfat ingredients for use in chocolate.

Packaging and tempering of milkfat ingredients for chocolate Information for the tempering and packaging of experimental milkfat ingredients used in chocolates was not available. The decreased importance of texture in the milkfat ingredient may make the use of tempering milkfat ingredients unnecessary. However, for the development of an acceptable finished chocolate, the full chocolate blend must be tempered prior to use. Tempering protocols used in experimental chocolates were presented in Table 8.12, and will be discussed under characteristics of the finished chocolates.

Commercial milkfat ingredients for use in chocolates are available in 1, 10, and 25 kg blocks.

Characteristics of Milkfat Ingredients for Chocolate

The melting points of milkfat fractions used in experimental chocolates are presented in Table 8.12, and solid fat content profiles are presented in Table 8.14. High-melting milkfat fractions obtained from melted milkfat were 45 to 55% solid at 10°C and melted gradually (Jordan, 1986; Timms and Parekh, 1980). Very-high-melting and high-melting milkfat fractions obtained from acetone solution were more solid at lower temperatures and melted slower compared with fractions obtained from melted milkfat, which was consistent with trends observed in Chapter 5. The very-high-melting and high-melting milkfat fractions were 80 to 90% solid at 10°C and melted very slowly until just before their final melting point. Middle-melting and low-melting fractions exhibited similar behavior to the high-melting fractions, but the fractions melted at lower temperatures. The $5S_M$ fraction obtained from solvent solution exhibited a melting profile similar to cocoa butter (Yi, 1993).

Milkfat Ingredient and Cocoa Butter Mixtures

The interaction of milkfat and cocoa butter is very important to the properties of the finished product. The effects of blending milkfat fractions with cocoa butter were discussed in more detail in Chapter 5. The following data are for milkfat–cocoa butter mixtures that were used to prepare experimental chocolates.

Solid fat content of milkfat–cocoa butter mixtures The solid fat content profiles of the fat phase of the experimental chocolates are shown in Table 8.15. Milkfat fractions obtained from acetone solution were used in experimental chocolates at a level of 2% addition (6% of the total fat). Very-high-melting milkfat fractions obtained from acetone solution ($25S_M$, $20S_M$, and $15S_M$ fractions) increased the solid fat content profile compared with pure cocoa butter, but the high-melting ($10S_M$ fraction), the middle-melting ($5S_M$ and $0S_M$ fractions), and the low-melting ($0L_M$ fraction) fractions decreased the solid fat contents of the mixtures (Yi, 1993). Although the $5S_M$ fraction had a solid fat content profile similar to cocoa butter, it still softened the chocolate and thereby illustrated that solid fat content profiles alone were not sufficient to predict the functionality of milkfat fractions in chocolate.

Solid fat content profiles for milkfat–cocoa butter mixtures used for experimental milk chocolates showed a decrease in solid fat content as the content of milkfat increased and as the melting point of the fractions decreased (Jordan, 1986). Jordan reported a slight increase in solid fat content for the mixture of cocoa butter and the HM40 fraction, and a further increase in solid fat content profile was observed when the HM41 fraction was used. These effects were a result of the increased proportion of high-melting triglycerides in the HM41 fraction compared with the HM40 fraction. Jordan (1986) found that the solid fat content profiles for the milkfat fraction–cocoa butter mixtures agreed closely for similar blends when present in chocolate.

Timms and Parekh (1980) also reported a decrease in solid fat content of the milkfat fraction–cocoa butter mixtures as the melting point of the fractions decreased ($25S_S$ and $25L_S$ fractions) and as the amount of milkfat in the mixture was increased.

Eutectic formation in milkfat cocoa butter mixtures for experimental chocolates In a study on pure milkfat fraction–cocoa butter mixtures, Timms (1980b, Chapter 5) reported that at levels of greater than 30% milkfat addition, eutectic interactions from the middle-melting fraction makes some contribution to the softness of the mixtures. However, this effect was considered small relative to the solution effects of the liquid milkfat fractions. Timms (1980b) has suggested that the high compatibility of milkfat with cocoa butter in comparisons with other cocoa butter replacers is due to the low content of solid fat in milkfat at normal temperatures. Only solid glycerides can show incompatibility (eutectic interaction) of the type observed with cocoa butter and cocoa butter replacer mixtures. Therefore, Timms (1980b) has stated that it should be possible to make satisfactory, although very soft, chocolate containing up to 50% milkfat in the fat phase, without encountering the type of fat bloom caused by solid–solid transformations. Fractionation to remove the low-melting fractions, which are mainly responsible for the softening effect of milkfat, concentrates the middle-melting fractions and increases the softening effects caused by eutectic formation (Timms and Parekh, 1980).

Isosolid diagrams represent the phase behavior of fat mixtures and can be used as an indicator of fat compatibility and eutectic formation. A eutectic is observed as a depression in the isosolid lines where the combined fat system results in an inhibition of crystallization. The isosolid diagrams for several cocoa butter and milkfat mixtures that were used in experimental milk chocolates are presented in Figure 8.1. Cocoa butter formed a strong eutectic with the low-melting milkfat fraction, but mostly at high milkfat contents. This indicates that cocoa butter inhibits the crystallization of the milkfat, but the low-melting milkfat fraction had very little effect on the crystallization of cocoa butter. A slight eutectic was formed between cocoa butter and the middle-melting milkfat fraction at all levels evaluated. A eutectic was formed between cocoa butter and the high-melting milkfat fraction at low replacement levels, and this indicated significant inhibition of cocoa butter crystallization by the milkfat fraction (Barna et al., 1992). These results were in agreement with Timms (1980b) for cocoa butter and milkfat mixtures.

Characteristics of Chocolates Made with Milkfat Ingredients

Milkfat fractions have been incorporated into chocolates using two basic strategies:

- Blending the milkfat fractions with the remaining ingredients at the initial stage of chocolate production, followed by conventional processes (refining, conching) (Jordan, 1986).
- Addition of the milkfat fractions to a low-fat chocolate base (Barna et al., 1992; Bunting, 1991; Yi, 1993).

Tempering properties of experimental chocolates made with milkfat ingredients   All experimental chocolates have been tempered by hand, and the procedures employed are presented in Table 8.12. Tempering chocolate is somewhat of an art, and trial and error was required to determine the best tempering profile for each chocolate as determined by visual observation and subjective evaluation of gloss and mold contraction after setting (Barna et al., 1992). Tempering procedures were altered to produce good chocolate, and these alterations were based more on the replacement level than on the type of fraction used. As the milkfat content in the chocolate increased, tempering temperatures were decreased because of the inhibition of crystallization. Tempering temperatures were decreased up to 1°C for each 5% milkfat fraction in excess of the milkfat content of the control chocolate (Barna et al., 1992).

Tempering difficulties were encountered at the 20% replacement level for the high-melting milkfat fraction and at the 30% replacement level for all milkfat fractions evaluated in milk chocolates (Barna et al., 1992; Bunting, 1991).

Textural characteristics of experimental chocolates made with milkfat ingredients Textural characteristics evaluated for experimental chocolates include melting behavior, hardness, and snap (Table 8.16). Jebson (1974b) reported an increase in the softening point of chocolate with an increase in the softening point of the fractions.

Thermal profiles (DSC curves) for milkfat fraction–cocoa butter blends are shown in Figure 8.2. Thermal profiles obtained by differential scanning calorimetry showed a decrease in the melting points of chocolates as milkfat fraction replacement level increased, compared with the control chocolates (Barna et al., 1992; Bunting, 1991). The shape of the melting curve remained the same regardless of the type or amount of milkfat fraction added. This suggested that the stable polymorphic crystal form had not been altered, and these results were consistent with those observed for isosolid diagrams (Barna et al., 1992).

Hardness or snap of milkfat fractions has been measured using an Instron Universal Testing Machine (Barna et al., 1992; Bunting, 1991; Yi, 1993) or subjective assessments (Jordan, 1986). Hardness values generally increased as the milkfat ingredient content increased. Barna et al. (1992) reported a decrease in hardness as milkfat content increased, regardless of the type of fraction used. Yi (1993) reported that chocolates made with the higher-melting fractions were harder than the control chocolate and chocolates made with the lower-melting fractions and milkfat control, and hardness followed the same trends as the solid fat content profiles for the milkfat–cocoa butter mixtures.

Barna et al. (1992) showed that as the level of milkfat replacement in the chocolate was increased, the plastic viscosity was decreased slightly (Table 8.16). A control chocolate had a Casson plastic viscosity of 20.5 poise, and experimental chocolates had viscosities less than 18 poise. Yield values for these chocolates showed little variation and generally fell within the range of 110 to 130 dyne/cm^2, which were within acceptable ranges for milk chocolates (Barna et al., 1992).

Fat bloom in experimental chocolates made with milkfat ingredients In general, dark chocolate samples made with milkfat fractions showed increased resistance to blooming compared with control samples at all conditions evaluated (Table 8.16). Jebson (1974b) reported that samples stored at 22 and 30°C showed little or no bloom formation after 9 months of storage. In this study (Jebson, 1974b), samples that were temperature-cycled showed bloom after 6 weeks of storage, but the samples made from fractionated milkfat could be cycled twice as long before bloom developed to the same degree as the control chocolates. Yi (1993) reported that dark chocolates made with very-high-melting and high-melting milkfat fractions did not exhibit bloom, but chocolates made with middle-melting and low-melting milkfat fractions exhibited bloom at a 2% replacement level.

Bloom formation has not been evaluated in experimental milk chocolates.

Sensory evaluation of experimental chocolates made with milkfat ingredients The sensory properties of experimental chocolates have been described by sensory panels and objective gloss measurements (Table 8.16). Chocolates made with higher levels of milkfat fractions melted more easily in the hand and in the mouth, and no waxy mouthfeel was detected with any of the fractions (Barna et al., 1992; Bunting, 1991).

High gloss results from smooth, even surfaces that are free from cracks, fissures, and protruding crystals. The two hardest chocolates, the control chocolate and the chocolate made with a 5% replacement of intact milkfat, exhibited the highest gloss (Table 8.16; Barna et al., 1992; Bunting, 1991). In contrast, Jordan (1986) reported that the highest gloss properties were observed for chocolates made with the softest fraction (HM38 fraction).

Summary of the Use of Milkfat Ingredients in Chocolate

Technical data for the use of milkfat ingredients in chocolate have been summarized as follows:

- Milkfat fractions have been used in dark chocolates at levels of 2 to 3% cocoa butter replacement and in milk chocolates at levels up to 25% cocoa butter replacement.
- Milkfat fractions used in experimental chocolates were produced by crystallization from melted milkfat and from acetone solution.

- Very-high-melting and high-melting milkfat fractions gave the hardest chocolates but, at levels of 20 and 30% replacement in milk chocolate, resulted in tempering difficulties.
- Eutectic formation was evident for the 20% replacement level for the high-melting milkfat fractions and at the 30% replacement level for all milkfat fractions evaluated.
- The occurrence of fat bloom on dark chocolates was inhibited by very-high-melting and high-melting milkfat fractions and induced by middle-melting and low-melting milkfat fractions.

Dairy Products

The use of modified milkfat in dairy products has been evaluated to improve the functionality of the finished product (e.g., whipping cream) and, to some extent, for identifying additional outlets for the less marketable milkfat fractions. Milkfat fractions have been used experimentally in recombined butter and milk, butter powder and whole milk powder, whipping cream, several cheeses and cheese spreads, and ice cream (Table 8.17). Milkfat ingredients are also commercially available for use in dairy products (Table 8.18).

Recombined Dairy Products

Tucker (1974) reported the use of a hard milkfat fraction in recombined milk or cream that was then made into butter by conventional churning, but no additional information was given.

Mayhill and Newstead (1992) evaluated the suitability of milkfat fractions in UHT-recombined milk to reduce the formation of a visible cream layer, which normally occurs during storage. Intact anhydrous milkfat, and a very-high-melting and a low-melting milkfat fraction were recombined with water, nonfat milk solids, and emulsifiers and UHT-processed (Table 8.17). The creaming stability index of UHT-recombined milks at storage temperatures of 22, 30, and 40°C was measured monthly for 6 months (Table 8.20). The milk made from the low-melting milkfat fraction showed the most rapid creaming at all storage temperatures. At 22 and 30°C the high-melting milkfat fractions showed the least separation of the milks, but at 40°C had creaming stability indices similar to the milks made from the low-melting milkfat fraction and intact milkfat. Mayhill and Newstead noted that the rates of creaming of the different milkfats were highest at storage temperatures near or above their dropping points.

Commercially available milkfat ingredients for use in recombined dairy products are middle-melting and low-melting milkfat fractions obtained from melted milkfat (Tables 8.18 and 8.19).

Dairy Powders

Black and Dixon (1974) used an unknown-melting milkfat fraction that they described as a hard milkfat fraction for the production of butter powder (Table 8.17). The milkfat fraction was recombined with water, sodium caseinate, sodium citrate, and glycerol monostearate prior to homogenization and spray drying. They reported that the experimental butter powder was similar to the control, but it lacked flavor in baking trials.

Experimental whole milk powders have been made using low-melting and middle-melting milkfat fractions obtained from melted milkfat and high-melting milkfat fractions obtained from acetone solution (Table 8.17). Ingredients added to milkfat fractions for reconstitution into whole milk prior to drying have included skim milk, water, skim milk powder, and glycerol monostearate (Table 8.17).

Baker et al. (1959) prepared agglomerated milk powders with milkfat fractions obtained from melted milkfat and from acetone solution. The wettability and dispersibility of the milk powders improved as the melting points of the milkfat fractions decreased (Table 8.20). The milk powder made with the milkfat fraction obtained from acetone solution had the same wettability and dispersibility as commercial instant skim milk powder, and this may have been due to the sharper separation of fractions normally associated with solvent fractionation. Deffense (1987) also has suggested the use of a low-melting milkfat fraction to improve the reconstitutability of milk powders.

Black and Dixon (1974) used an unknown-melting milkfat fraction that was described as a hard milkfat fraction to produce milk powder. The milkfat fraction was blended with fresh whole milk prior to drying. They reported improved flowability and solubility of the experimental milk powder compared with the control, but the experimental powder lacked flavor and aroma. Tucker (1974) and Malkamäki and Antila (1974) also used high-melting milkfat fractions in whole milk powder and reported experimental milk powders with comparable or improved properties compared with the control. However, all of these products lacked flavor.

Whipping Cream

The use of high-melting milkfat fractions to improve aeration properties, such as whippability and stability, in full-fat and low-fat whipped creams is well known (Bratland, 1983; Jamotte and Guyot, 1980; Tucker, 1974). Experimental whipping creams have been prepared from high-melting milkfat fractions or from solid unknown-melting milkfat fractions obtained at fractionation temperatures above 20°C (Table 8.17). Tucker (1974) reported that physical properties were maintained, whiteness increased, and flavor decreased as the content of the high-melting milkfat fraction in whipping cream was increased (Table 8.20).

Bratland (1983) reported eight examples of whipping creams made with solid unknown-melting milkfat fractions in a U.S. patent (Table 8.17). Milkfat was blended with a combination of serum, water, buttermilk powder, coconut oil, and/or sugar and then homogenized, pasteurized, and homogenized a second time. All experimental whipping creams produced finished products with 100 to 174% overrun at temperatures between 10 and 12°C (Table 8.20). The addition of sugar to the cream increased the freeze-thaw stability of the whipped cream.

Cheese

Very-high-melting and high-melting milkfat fractions obtained from melted milkfat have been used in the manufacture of experimental Edam, Colby, Cheddar, processed cheese, and cheese spreads according to traditional manufacturing processes (Table 8.17). Edam and Colby cheeses produced with high-melting milkfat fractions were comparable in quality to control cheeses, but no information was given on the evaluation of milkfat fractions in Cheddar cheese (Table 8.20; Malkamäki and Antila, 1974; Tucker, 1974).

Thomas (1973a, 1973b) reported the successful replacement of up to 40% of the total milkfat by the very-high-melting milkfat fractions in processed Cheddar cheese and cheese spreads. The experimental cheeses showed no significant differences in sensory evaluation, pH, oil separation, and storage stability compared with the control cheeses (Table 8.20). Penetrometer tests showed an increase in firmness as the amount of very-high-melting milkfat fraction was increased in cheese spreads (Thomas, 1973b).

Ice Cream

The availability of technical data on the use of milkfat fractions in experimental ice cream is limited to a single abstract (Goff et al., 1988). Goff et al. (1988) used unknown-melting milkfat fractions, described as hard and soft fractions, to produce ice creams of standard composition (Table 8.17). They indicated that the soft milkfat fraction produced significantly more fat destabilization than did the hard fractions.

Interestingly, 80% of the milkfat ingredients that are commercially available for use in dairy products are middle-melting milkfat fractions that have been recommended for use in ice cream. However, ice cream in Europe is generally made from anhydrous milkfat ingredients instead of cream, and this would account for the abundance of products available for use in ice cream. The use of milkfat fractions in ice cream may not be as attractive in the U.S., where it would replace the current market for fluid cream.

Summary of the Use of Milkfat Ingredients in Dairy Products

The technical data for milkfat ingredients used in dairy products have been summarized as follows:

- The use of high-melting milkfat fractions in UHT-recombined creams helped to increase the creaming stability during storage.
- The use of high-melting milkfat fractions in butter powder offered no advantages compared with intact milkfat, and low-melting milkfat fractions were detrimental to the butter powder and reconstituted spread.
- Whole milk powders manufactured with milkfat fractions showed improved wettability, dispersibility, and solubility as the melting point of the milkfat fractions decreased.
- High-melting milkfat fractions have been used to produce whipped creams that had good physical and stability properties in full-fat and reduced-fat whipping creams.
- Promising results have been observed for the use of high-melting milkfat fractions in Edam and Colby cheeses.
- High-melting milkfat fractions have been successfully used to replace 40% of the milkfat in processed Cheddar cheese and cheese spreads.
- Decreased flavor has been noted for many dairy products as the amount of high-melting fraction increased.
- Milkfat ingredients are commercially available for use in recombined dairy products and ice cream.

Frying Oils

Intact milkfat has been shown to impart desirable flavors and textures and to increase the storage stability of some fried foods when used as a deep-fat frying medium (Baker, 1970b; Tangel et al., 1977).

The deep-fat frying process is complex and involves fat–(nonfat) solid, fat–water, and fat–fat interactions between the frying medium and the components in the bulk food, as well as interactions with food particles that become suspended in the frying oil. Oxidation, hydrolysis, and polymerization reactions, initiated by the heat of frying (180 to 190°C) and moisture from the food, are common in deep-fat frying. These reactions result in formation of off-flavors, a decrease in the shelf life of the fat, and in an increase in the viscosity of the oil (Keogh, 1989).

The functionality attributes of milkfat ingredients that are important to deep-fat frying include the following:

- Heat transfer properties
- Oxidative stability
- Flavor

A typical target smoke point for a frying oil is 200°C (Keogh, 1989).

Antioxidants are generally not used in deep-fat frying oils, because they are lost by steam distillation during the frying process (Keogh, 1989). Silicone is commonly added to commercial frying shortenings as an antifoaming agent. The addition of silicone (2 to 10 ppm) to intact milkfat increased its smoke point, decreased the total free fatty acid content, decreased foaming during frying, and resulted in improved flavor stability in the fried potato chips (Keogh, 1988; Tangel et al., 1977).

Technical Data for the Use of Milkfat Fractions as Deep-Frying Fats

High-melting and low-melting milkfat fractions were evaluated for suitability as deep-frying fats (Table 8.21). Malkamäki and Antila (1974) reported favorable results for doughnuts fried in a hard milkfat fraction. Low-melting milkfat fractions have often been suggested for use as a frying oil (with the addition of silicone) because they remain liquid at room temperature. Low-melting milkfat fractions may be more prone to oxidation than the high-melting milkfat fractions, and they may benefit from the addition of antioxidants (Baker, 1970b; Jamotte and Guyot, 1980; Keogh, 1989).

Keogh (1989) fried potato chips in intact milkfat, a low-melting milkfat fraction, and corn oil (Table 8.21). Keogh reported that the difference in smoke point between intact milkfat and the low-melting fraction was negligible, and both milkfats had a lower smoke point than the corn oil (187°C and 220°C, respectively). To raise the smoke point of the milkfats, 2 mg/kg silicone was added (7°C increase), the milkfat was partially neutralization to decrease the free fatty acid content (6°C increase), and the milkfats were blended 1:3 with corn oil (12°C increase). These changes raised the smoke point to 192.5°C, which was closer to the recommended smoke point for frying oils. The addition of antioxidants to the milkfat did not result in any significant difference in oxidative stability but did reduce the color loss that occurred during the first day of frying. Oxidized and rancid flavors were not detected in any of the potato chips fried in milkfat, even after the fat had been used for 4 days of frying.

Summary of Technical Data for the Use of Milkfat Fractions as Deep-Frying Fats

The technical data for deep-fat frying with milkfat fractions have been summarized as follows:

- Intact milkfat and low-melting fractions imparted a desirable flavor and texture and increased flavor stability to fried foods (potato chips and doughnuts).
- The low-melting milkfat fraction exhibited properties similar to the intact milkfat.
- Milkfat exhibited good oxidative stability over 4 d of frying.
- The addition of silicone raised the smoke point of the milkfat and reduced oxidation and hydrolysis.
- Partial neutralization of the milkfat also increased the smoke point and oxidative stability of the milkfats.

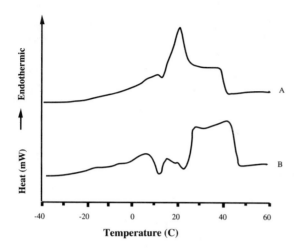

Figure 8.1. Thermal profiles (DSC curves) of dairy-based spreads. A, Market butter (Shukla et al., 1994); B, High-melting butter (Shukla et al., 1994).

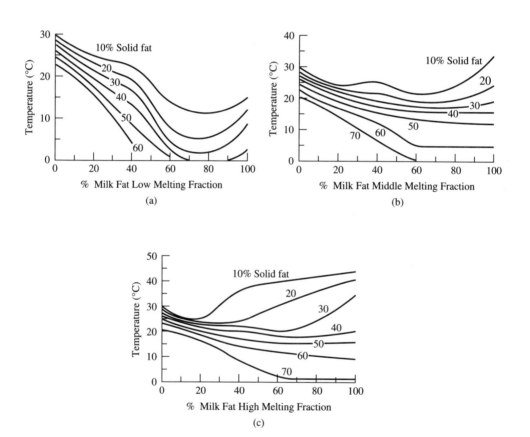

Figure 8.2 Isosolid diagrams of milkfat-cocoa butter mixtures used for experimental chocolates. Milk chocolate: a, low-melting milkfat fraction-cocoa butter (Barna et al., 1992); b, middle-melting milkfat fraction-cocoa butter (Barna et al., 1992); c, high-melting milkfat fraction-cocoa butter (Barna et al., 1992).

Application of Milkfat Fractions in Foods 563

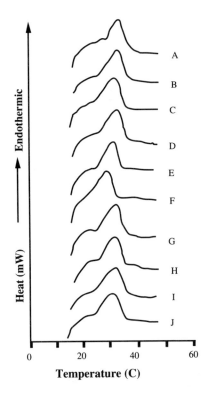 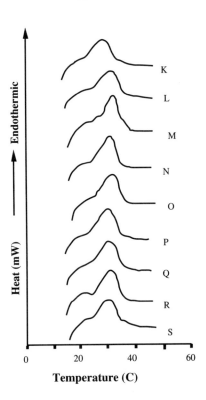

Figure 8.3 Thermal profiles (DSC curves) for experimental chocolates. Milk chocolate: A, Control chocolate (Bunting, 1991); B, Chocolate 1 (5% AMF) (Bunting, 1991); C, Chocolate 2 (5% LMF) (Bunting, 1991); D, Chocolate 3 (5% MMF) (Bunting, 1991); E, Chocolate 4 (5% HMF) (Bunting, 1991); F, Chocolate 5 (10% AMF) (Bunting, 1991); G, Chocolate G (10% LMF) (Bunting, 1991); H, Chocolate 7 (10% MMF) (Bunting, 1991); I, Chocolate 8 (10% HMF) (Bunting, 1991); J, Chocolate 9 (20% AMF) (Bunting, 1991); K, Chocolate 10 (20% LMF) (Bunting, 1991); L Chocolate 11(20% MMF) (Bunting, 1991); M, Control chocolate (Barna et al., 1992); N, Chocolate 1 (5% AMF) (Barna et al., 1992); O, Chocolate 2 (5% LMF) (Barna et al., 1992); P, Chocolate 3 (5% MMF) (Barna et al., 1992); Q, Chocolate 4 (5% HMF) (Barna et al., 1992); R, Chocolate 6 (10% LMF) (Barna et al., 1992); S, Chocolate 10 (20% LMF) (Barna et al., 1992).

TABLE 8.1 Formulations and Processing Information for Experimental Cold-Spreadable Butters

Fractionation method Author Separation method	Stock fractions employed[1]	Butter designation	Milkfat fraction category[2]	Milkfat fraction blend Fraction	%	Butter blend Ingredient	%	Method of production[3]	Packaging and tempering[3]
Dolby, 1970b									
Milkfat fractions obtained from melted milkfat									
Separation with the aid of an aqueous detergent solution	Fraction #1 [1]	Butter 1	H	Fraction #1 [1]	100	Milkfat	40	Emulsified with skim milk to 40% cream, homogenized, and churned as for conventional butter.	—[4]
	Fraction #2 [1]	Butter 2	H	Fraction #2 [1]	100	Skim milk	60		
	Fraction #3 [1]	Butter 3	M	Fraction #3 [1]	100				
	Fraction #4 [1]	Butter 4	L	Fraction #4 [1]	100				
	Fraction #1 [2]	Butter 5	H	Fraction #1 [2]	100				
	Fraction #2 [2]	Butter 6	M	Fraction #2 [2]	100				
MacCollom, 1970									
Liquid fraction collected by scraping from the top of the crystal mass	22–24S_S Fraction 22–24L_S Fraction	Butter A	U	22–24L_S fraction	100	Milkfat Cultured skim milk Water	unknown unknown unknown	Processed as for continuous butter.	—
		Butter B	U	22–24S_S fraction 22–24L_S fraction	unknown unknown	Diacetyl distillate NaCl Color, if desired	unknown unknown unknown		
Vovan and Riel, 1973									
Vacuum filtration	AMF with solid fraction (MP > 21°C) removed	Butter 1	U	5% of solid fraction removed	100	Milkfat Water NaCl Casein	80 16 2 2	Mixture is warmed to 40°C and then pumped to a butter-worker-crystallizer; exit temperature = 13°C.	Molded into blocks (6 × 2.5 × 15 cm), and held at 18°C.
		Butter 2	U	10% of solid fraction removed	100				
		Butter 3	U	15% of solid fraction removed	100				
	AMF with intermediate fraction (10 < MP < 21°C) removed	Butter 4	U	5% of intermediate fraction removed	100				
		Butter 5	U	10% of intermediate fraction removed	100				
		Butter 6	U	15% of intermediate fraction removed	100				

Application of Milkfat Fractions in Foods

Dixon and Black, 1974 Separation method not reported	HMF MMF LMF	15 Butters produced	H, M, L	Specific formulas not reported		Recombined in a swept-surface cooler plus post-worker, and vacuum reworked 1 wk later.
Kankare and Antila, 1974b Centrifugation	$12L_M$ fraction	Produced several butters	U	$12L_M$ fraction	100	250 g boxes.
				Soured churning cream	26.6–33.3 66.7–73.4	Liquid fraction pumped into churn and then mixture churned as for conventional butter.
El-Nimr, 1980 Separation method not reported	$25I_S$ fraction Whole butter	Butter 1	U	$25L_S$ fraction Whole butter	100 	20 80	Liquid fraction added directly to whole butter and worked adequately with wooden augers.
		Butter 2	U	$25L_S$ fraction Whole butter	100	40 60	
		Butter 3	U	$25L_S$ fraction Whole butter	100	60 40	
Frede et al., 1980 Filtration through an 80 μm sieve	$28L_S$ [S] fraction Winter cream Whole butter	Butter 1	U	Cream A: 25% fat made from $28L_S$ [S] emulsified with skim milk Cream B: 25% fat made from winter cream	35 65	—	Cream A is blended with Cream B and pasteurized (90°C), then cooled in swept-surface cooler to 13.5°C, ripened (13.5°C) 20 h, buttered in finisher, washed 3 times (10°C), and kneaded 30 min (15°C).	Tempered at 18–30°C.
		Butter 2	U	Cream A: 25% fat made from $28L_S$ [S] emulsified with whey Cream B: 25% fat made from winter cream	35 65	35 65	
		Butter 3	U	Cream A 25% fat made from $28L_S$ [S] emulsified with buttermilk Cream B 25% fat made from winter cream	35 65	35 65	
		Butter 4	U	$28L_S$ [S] fraction Whole butter	100	35 65	Liquid fraction added directly to whole butter grain during buttering.

Continued

TABLE 8.1 (Continued)

Fractionation method			Milkfat			Butter blend				
Author Separation method	Stock fractions employed[1]	Butter designation	fraction category[2]	Milkfat fraction blend Fraction	%	Ingredient	%	Method of production[3]	Packaging and tempering[3]

Author Separation method	Stock fractions employed[1]	Butter designation	Milkfat fraction category[2]	Milkfat fraction blend Fraction	%	Butter blend Ingredient	%	Method of production[3]	Packaging and tempering[3]
Jamotte and Guyot, 1980									
Vacuum filtration	Liquid fraction = L_4 Solid fraction = Commercial solid fraction B fractionated at 25°C	Butter 1	VL H	Liquid fraction Solid fraction	85 15	Milkfat Defatted milk Buttermilk powder	82.0 17.5 0.5	Recombination carried out in a Schroeder Kombinator type VUK 01 pilot plant: Throughput = 25–28 kg/hr Temperature of chiller = –10 to –16°C Shaft rotated at 840 rpm Exit temperature = 4–8°C.	—
		Butter 2	VL H	Liquid fraction Solid fraction	80 20				
		Butter 3	VL H	Liquid fraction Solid fraction	75 25				
		Butter 4	VL H	Liquid fraction Solid fraction	70 30				
		Butter 5	VL H	Liquid fraction Solid fraction	65 35				
Guyot, 1982									
Vacuum filtration	$5L_M$ fraction Solid fraction	Produced several butters	VL VH	$5L_M$ fraction Solid fraction	65–85 15–35	Milkfat Skim milk Buttermilk powder	unknown unknown 1.5	Ingredients combined in the manner of reconstituted butter and texturized in a Schroeder pilot unit.	—
		Butter 1	VL VH	$5L_M$ fraction Solid fraction	80 20				
		Butter 2	VL VH	$5L_M$ fraction Solid fraction	75 25				
Bratland, 1983									
Centrifugation	$20L_S$ fraction	Butter 1	U	$20L_S$ fraction	100	$20L_S$ fraction Serum	35 65	The milkfat was emulsified with serum from the phase inversion process to produce a 35% fat cream. The cream was pasteurized, cooled and conventionally churned at 10°C.	—
Deffense, 1987									
Vacuum filtration	Stearin #1_S Stearin #3_S 2nd olein$_M$ filtered at 10°C 2nd olein$_M$ filtered at 6°C	Butter A	H	Stearin #3_S 2nd olein$_M$ filtered at 10°C	35 65	—		Made as for recombined butter and texturized in a Gerstenberg and Agger Perfector.	—
		Butter B	H VL	Stearin #1_S 2nd olein filtered at 6°C	30 70				

Reference / Method	Sample		Fraction	%	Ingredient	%	Processing	Packaging/Storage
Makhlouf et al., 1987 Vacuum filtration			$26S_M$ (S.I) fraction	80	Milkfat	80	Blend emulsified with agitation at 55°C and pumped to laboratory butter-worker-crystallizer: four scrapers chilled water 1–2°C, 150–240 L/h throughput = 8 to 19 kg/h agitation = 60 rpm exit temperature = 7 – 9°C.	Packed in plastic containers (8 × 2.5 cm) and stored at –20°C.
	Butter 1	U	$15L_M$ fraction	20	Water	16		
		VH	F.S (> 21) fraction	70	Buttermilk powder	2		
	Butter 2	U	$13L_M$ fraction	30	NaCl	2		
		VH	F.S (> 21) fraction	70				
	Butter 3	L	$9L_M$ (F.L) fraction	30				
		VH	F.S (> 21) fraction	75				
	Butter 4	L	$9L_M$ (F.L) fraction	25				
		VH	F.S (> 21) fraction	75				
	Butter 5	L	$9L_M$ (F.L) fraction	25				
		VH	$26S_M$ (S.I) fraction	80				
	Butter 6	L	$9L_M$ (F.L) fraction	20				
		VH	$26S_M$ (S.I) fraction					
Bumbalough, 1989 Vacuum filtration	Butter A	H	$29.5S_M$ fraction	10–15	Milkfat	80.1	Blend emulsified with agitation at 42–45°C, then pumped to Votator and texturized in a manner similar to margarine.	Packaged in tubs and tempered 4 h at 21°C.
		L	$12.5L_M$ fraction	85–90	Skim milk	18.3–18.6		
					NaCl	1.2–1.6		
Deffense, 1989 Vacuum filtration	Butter A	H	Stearin #3_S	35	—		Made in as for recombined butter and texturized in a Gerstenberg and Agger Perfector.	
		L	2nd olein$_M$ filtered at 10°C	65				
	Butter B	H	Stearin #1_S	30				
		VL	2nd olein$_M$ filtered at 6°C	70				
Kaylegian and Lindsay, 1992 Vacuum filtration	Recombined butter	VH	$34S_M$ [W] fraction	4.0	Milkfat	80.0	Blend emulsified in a Waring blender (40°C), and crystallized in a lab scale scraped-surface heat exchanger: agitation = 40 rpm exit temperature = 6 – 12°C. Tempered 4 h at 21°C and texturized by hand using metal spatulas.	Formed into blocks (10 × 10 × 2 cm) and stored at 4°C.
		H	$30S_M$ [W] fraction	19.0	Skim milk	18.5		
		H	$30L_S$ [W] fraction	12.0	NaCl	1.2		
		M	$25S_M$ [W] fraction	15.3	Buttermilk powder	0.3		
		L	$25L_S$ [W] fraction	14.3				
		L	$16S_M$ [W] fraction	35.4				
	Low-melting butter	M	$13S_M$ [W] fraction	50.0				
		L	$30L_S$ fraction	50.0				
	Low-melting:high-melting butter	L	$25L_S$ fraction	50.0				
		L	$13L_M$ fraction	90.0				
		H	$30S_S$ fraction	10.0				

Continued

TABLE 8.1 (Continued)

Fractionation method			Milkfat			Butter blend				
Author Separation method	Stock fractions employed[1]	Butter designation	fraction category[2]	Milkfat fraction blend Fraction	%	Ingredient	%	Method of production[3]	Packaging and tempering[3]

Milkfat fractions obtained from solvent solution

Kaylegian and Lindsay, 1992

| Acetone | 30S$_S$ [W] fraction from melted milkfat | Low–melting:high-melting + 10% very high-melting butter | L H H | 13L$_M$ fraction from melted milkfat 30S$_S$ fraction from melted milkfat 11S$_M$ fraction from acetone | 80.0 10.0 10.0 | Milkfat Skim milk NaCl Buttermilk powder | 80.0 18.5 1.2 0.3 | Blend emulsified in a Waring blender (40°C), and crystallized in a lab scale scraped-surface heat exchanger: agitation = 40 rpm exit temperature = 6 – 12°C. Tempered 4 h at 21°C and texturized by hand using metal spatulas | Formed into blocks (10 × 10 × 2 cm) and stored at 4°C. |
| | 13L$_M$[W] fraction from melted milkfat 15S$_M$ [W] fraction from acetone 11S$_M$ [W] fraction from acetone | Low–melting:high-melting + 20% very-high-melting butter | L H VH | 13L$_M$ fraction from melted milkfat 30S$_S$ fraction from melted milkfat 15S$_M$ fraction from acetone | 70.0 10.0 20.0 | | | | |

Munro et al., 1978

| Acetone | 10S$_M$ fraction –15L$_M$ fraction | Butter 1 | U U | –15L$_M$ fraction 10S$_M$ fraction | 75 25 | — | | Made using a recombining plant equipped with a scraped-surface heat exchanger and working unit. | — |

Norris, 1977

| Acetone Isopropanol | 11.5S$_M$ fraction –15L$_M$ fraction 20S$_M$ fraction –5L$_M$ fraction | Butter 1 Butter 2 | U U U U | 11.5S$_M$ fraction –15L$_M$ fraction 20S$_M$ fraction –5L$_M$ fraction | 25 75 25 75 | Milkfat Moisture NaCl | 83.1 15.6 1.3 | Made as for recombined butter. | — |

Walker et al., 1977

| Acetone | 14S$_M$ (A) fraction –13L$_M$ (A) fraction 14S$_M$ (C) fraction –13L$_M$ (C) fraction | Butter A Butter C | U U U U | 14S$_M$ (A) fraction –13L$_M$ (A) fraction 14S$_M$ (C) fraction –13L$_M$ (C) fraction | 25 75 25 75 | Milkfat Water Nonfat milk solids NaCl | unknown unknown unknown unknown | Ingredients emulsified and cooled in a scraped-surface heat exchanger to form butter of normal composition. | — |

Walker et al., 1978					
Acetone	14S$_M$ fraction –13L$_M$ fraction	Butter 1	U U	14S$_M$ fraction –13L$_M$ fraction	25 75	—
Fractionation method not reported					
Dixon and Black, 1974					
	Soft-melting fraction	Butter 1	L	Soft-melting fraction Cream	20 80	Soft-melting fraction added to cream before pasteurization and conventionally churned and reworked.
	Cream	Butter 2	L	Soft-melting fraction Cream	40 60	
	Skim milk	Butter 3	L	Soft-melting fraction Cream	50 50	
		Butter 4	L	Soft-melting fraction	100	

[1] Physical and chemical characteristics of stock milkfat fractions are presented in Chapter 5.

[2] VH = very-high-melting milkfat fraction (MP > 45°C), H = high-melting milkfat fraction (MP between 35 and 45°C), M = middle-melting milkfat fraction (MP between 25 and 35°C), L = low-melting milkfat fraction (MP between 10 and 25°C), and VL = very-low-melting milkfat fractions (MP < 10°C), U = unknown-melting milkfat fraction.

[3] Method and data presented as described by the original author.

[4] No methods or data reported.

TABLE 8.2 Solid Fat Content of Experimental Cold-Spreadable Butters

Fractionation method Author	Butter designation	Milkfat fraction category[1]	Solid fat content (%)							
			5°C	10°C	15°C	20°C	25°C	30°C	35°C	40°C
Milkfat fractions obtained from melted milkfat										
Vovan and Riel, 1973										
	Control butter		47.0[2]	44.3	35.9	21.6	13.0	7.7	3.2	1.1
	Butter 1	U	37.9	32.9	23.2	14.1	8.6	3.9	1.3	0.0
	Butter 2	U	37.1	31.4	22.3	14.1	8.5	3.0	0.7	0.4
	Butter 3	U	37.5	32.5	—[3]	15.1	9.3	4.1	1.1	0.0
	Butter 4	U	42.0	37.9	26.8	15.4	10.2	5.0	1.8	0.0
	Butter 5	U	41.6	35.9	23.9	12.7	9.1	3.8	1.3	0.0
	Butter 6	U	40.9	35.9	23.4	12.1	7.9	3.8	—	—
Jamotte and Guyot, 1980										
	Regular butter		34.0 (4°C)	27.5 (8°C)	29.8 (14°C)	21.0 (18°C)	—	—	—	—
	Butter 1	VL H	15.2	14.0	12.0	8.5	—	—	—	—
	Butter 2	VL H	15.8	14.5	13.8	8.5	—	—	—	—
	Butter 3	VL H	23.7	20.1	15.7	12.0	—	—	—	—
	Butter 4	VL H	28.5	24.8	18.5	15.0	—	—	—	—
	Butter 5	VL H	34.0	27.5	25.0	18.5	—	—	—	—
Deffense, 1987										
	Butter A	H L	41.0[2]	30.8	20.0	12.8	8.0	4.8	1.0	0.0
	Butter B	H VL	35.3	25.3	18.0	12.0	8.0	4.8	1.5	0.0
Makhlouf et al., 1987										
	Control butter		44.8	41.0	30.2	16.0	—	—	—	—
	Butter 1	U VH	32.2	26.2	12.4	6.3	—	—	—	—
	Butter 2	U VH	30.0	24.1	13.8	9.6	—	—	—	—
	Butter 3	L VH	29.4	23.1	13.3	9.9	—	—	—	—
	Butter 4	L VH	27.6	21.5	11.3	6.8	—	—	—	—
	Butter 5	L VH	28.2	22.8	12.6	8.0	—	—	—	—
	Butter 6	L VH	27.3	21.1	10.2	6.3	—	—	—	—
Deffense, 1989										
	Butter A	H L	41.0[2]	30.8	20.0	12.8	8.0	4.8	1.0	0.0
	Butter B	H VL	35.3	25.3	18.0	12.0	8.0	4.8	1.5	0.0

Continued

TABLE 8.2 (Continued)

Fractionation method Author	Butter designation	Milkfat fraction category[1]	Solid fat content (%)							
			5°C	10°C	15°C	20°C	25°C	30°C	35°C	40°C
Kaylegian and Lindsay, 1992										
	Control butter		57.9	51.2	37.0	19.9	15.7	7.4	0.0	—
	Recombined butter	VH	61.2	50.4	37.8	21.4	16.4	11.8	1.6	0.0
		H								
		H								
		M								
		L								
		L								
	Low-melting butter	M L	51.3	40.9	33.7	14.0	8.5	0.0	—	—
	Low-melting: high-melting butter	L H	21.9	16.3	11.1	4.3	0.0	—	—	—
Milkfat fractions obtained from solvent solution										
Kaylegian and Lindsay, 1992										
	Control butter		57.9	51.2	37.0	19.9	15.7	7.4	0.0	—
	Low-melting: high-melting + 10% very-high-melting butter	L H H	44.1	33.8	21.3	16.7	10.6	8.8	0.7	0.0
	Low-melting: high-melting + 20% very-high-melting butter	L H VH	47.4	38.6	27.7	20.1	13.8	11.9	7.8	0.0

[1]VH = very-high-melting milkfat fraction (MP > 45°C), H = high-melting milkfat fraction (MP between 35 and 45°C), M = middle-melting milkfat fraction (MP between 25 and 35°C), L = low-melting milkfat fraction (MP between 10 and 25°C), VL = very-low-melting milkfat fractions (MP < 10°C), U = unknown-melting milkfat fraction.

[2]Solid fat content values determined from graphs provided by original author.

[3]Data not available.

TABLE 8.3 Textural Characteristics of Experimental Cold-Spreadable Butters

Fractionation method Author	Butter designation	Milkfat fraction category[1]	Selected textural characteristics[2]
Milkfat fractions obtained from melted milkfat			
Dolby, 1970b			*Hardness determined by sectility after 30 days (arbitrary units)*
	Butter 1	H	12.5°C = 1740
	Butter 2	H	12.5°C = 1100
	Butter 3	M	12.5°C = 565
			7.5°C = 740
	Butter 4	L	12.5°C = 405
			7.5°C = 675
	Butter 5	H	12.5°C = 1790
	Butter 6	M	12.5°C = 820
Vovan and Riel, 1973			*Spreadability determined by sectility (torque in g)*[3]
	Control Butter		4°C = 1600
			13°C = 783
			21°C = 142
	Butter 1	U	4°C = 1517
			13°C = 738
			21°C = 100
	Butter 2	U	4°C = 1417
			13°C = 733
			21°C = 58
	Butter 3	U	4°C = 1367
			13°C = 730
			21°C = 47
	Butter 4	U	4°C = 1500
			13°C = 725
			21°C = 47
	Butter 5	U	4°C = 1358
			13°C = 642
			21°C = 42
Dixon and Black, 1974			Tested firmness with disc penetrometer at 5, 13, and 20°C.
	15 Butters produced	U	Firmness at 5 and 13°C was mainly affected by the proportion of HMF, and at 20°C all butters were too soft for ideal spreadability.
Kankare and Antila, 1974b			*Mohr cutting resistance method after 14 d (resistance in g)*
	Control butter		Resistance = 195–215 g
	Produced several butters	U	Resistance = 73–101 g
El-Nimr, 1980			Softness measured with penetrometry at 5 and 12.5°C.
	Butter 1	U	Showed an increase in softness compared with whole butter as percent low-melting fraction increased.

Continued

TABLE 8.3 (Continued)

Author / Fractionation method	Butter designation	Milkfat fraction category[1]	Selected textural characteristics[2]	
	Butter 2	U		
	Butter 3	U		
Frede et al., 1980			*Cutting Resistance (N) performed at 15°C, after 10 days*	
	Winter butter		Resistance = 1.72 N	
	Butter 1	U	Resistance = 1.24 N	
	Butter 2	U	Resistance = 1.29 N	
	Butter 3	U	Resistance = 1.28 N	
	Butter 4	U	Resistance = 1.21 N	
Jamotte and Guyot, 1980			*Melting behavior (°C)[4]*	*Penetrometry measured after 1 wk (arbitrary units)*
	Winter butter		33.5 (DP)	4°C = 38
				8°C = 65
				14°C = 100
				18°C = 158
	Butter 1	VL / H	19.5	4°C = 106
				8°C = 140
				14°C = 260
				18°C = 300
	Butter 2	VL / H	20.0	4°C = 108
				8°C = 135
				14°C = 260
				18°C = 295
	Butter 3	VL / H	24.0	4°C = 92
				8°C = 122
				14°C = 195
				18°C = 270
	Butter 4	VL / H	29.8	4°C = 80
				8°C = 100
				14°C = 162
				18°C = 225
	Butter 5	VL / H	33.5	4°C = 38
				8°C = 65
				14°C = 100
				18°C = 158
Guyot, 1982			Spreadability measured with cone penetrometer at 4 and 18°C.	
	Produced several butters	VL / VH		
	Butter 1	VL / VH	Results similar to reference margarine.	
	Butter 2	VL / VH	Results similar to reference margarine.	
Bratland, 1983				
	Butter 1	U	Butter was spreadable at refrigerator temperature (4–6°C).	
Deffense, 1987			*Melting behavior (°C)[4]*	*Hardness measured by penetrometry after 1 wk (arbitrary units)*
	Butter A	H / L	32 (DP)	5°C = 96
	Butter B	H / VL		5°C = 105

Continued

TABLE 8.3 (Continued)

Fractionation method Author	Butter designation	Milkfat fraction category[1]	Selected textural characteristics[2]		
Makhlouf et al., 1987			*Yield value (kPa) measured by cone penetrometry using Instron*	*Apparent viscosity (Pa.s.)*	*Bingham viscosity (kPa) × 10^2*
	Regular butter		7°C = 108.0 15°C = 58.9 20°C = 28.9	7°C = 5.25 × 10^5 15°C = 1.25 × 10^5 20°C = 8.7 × 10^4	7°C = 53.58 15°C = 13.09 20°C = 8.89
	Butter 1	U VH	7°C = 52.2 15°C = 13.3 20°C = 1.3	7°C = 2.41 × 10^5 15°C = 3.0 × 10^4 20°C = 7.2 × 10^3	7°C = 24.62 15°C = 3.13 20°C = 0.73
	Butter 2	U VH	7°C = 52.6 15°C = 19.4 20°C = 5.4	7°C = 2.27 × 10^5 15°C = 3.18 × 10^4 20°C = 7.8 × 10^3	7°C = 23.23 15°C = 3.37 20°C = 0.83
	Butter 3	L VH	7°C = 28.4 15°C = 16.0 20°C = 3.2	7°C = 2.02 × 10^5 15°C = 2.76 × 10^4 20°C = 9.9 × 10^3	7°C = 20.48 15°C = 2.92 20°C = 1.02
	Butter 4	L VH	7°C = 22.2 15°C = 11.2 20°C = 1.8	7°C = 2.05 × 10^5 15°C = 2.85 × 10^4 20°C = 7.48 × 10^3	7°C = 20.72 15°C = 2.96 20°C = 0.77
	Butter 5	L VH	7°C = 53.1 15°C = 14.2 20°C = 1.9	7°C = 1.24 × 10^5 15°C = 1.71 × 10^4 20°C = 9.4 × 10^3	7°C = 12.93 15°C = 1.85 20°C = 0.98
	Butter 6	L VH	7°C = 33.7 15°C = 9.0 20°C = 1.3	7°C = 1.4 × 10^5 15°C = 1.58 × 10^4 20°C = 9.0 × 10^3	7°C = 14.34 15°C = 1.67 20°C = 0.91
Deffense, 1989			*Hardness measured by penetrometry after 1 wk (arbitrary units)*		
	Butter A	H L	5°C = 96		
	Butter B	H VL	5°C = 105		
Kaylegian and Lindsay, 1992			*Informal assessment of spreadability performed by spreading sample on a cracker with a plastic knife*		
	Control butter		4°C = very firm and difficult to spread 21°C = easy to spread and retained physical form		
	Recombined butter	VH H H M L L	4°C = very firm and difficult to spread 21°C = easy to spread and retained physical form		
	Low-melting butter	M L	4°C = firm and somewhat spreadable 21°C = almost melted and did not retain physical form		
	Low-melting: high-melting butter	L H	4°C = moderately soft and spreadable 21°C = amost melted		

Continued

TABLE 8.3 (Continued)

Fractionation method Author	Butter designation	Milkfat fraction category[1]	Selected textural characteristics[2]	
Milkfat fractions obtained from solvent solution				
Kaylegian and Lindsay, 1992			*Informal assessment of spreadability performed by spreading sample on a cracker with a plastic knife*	
	Control butter		4°C = very firm and difficult to spread 21°C = easy to spread and retained physical form	
	Low-melting: high-melting + 10% very-high-melting butter	L H H	4°C = moderately soft and spreadable 21°C = easy to spread and retained physical form	
	Low-melting: high-melting + 20% very-high-melting butter	L H VH	4°C = moderately firm and spreadable 21°C = easy to spread and retained physical form	
Munro et al., 1978			*Hardness measured (arbitrary units)*	
	Regular butter		5°C = 1500	
	Butter 1	U U	5°C = 220	
Norris, 1977			*Hardness measured by sectility (arbitrary units)*	*Stand-up value (arbitrary units), should have value < 5 for satisfactory spreadability*
	Butter 1	U	5°C = 135	Stand-up = 1.3
	Butter 2	U	5°C = 180	Stand-up = 2.0
Fractionation method not reported				
Dixon and Black, 1974			*Disc penetrometry (5°C)*	
	Butter 1	L	Noted a reduction in firmness at > 20% soft melting fraction.	
	Butter 2	L		
	Butter 3	L		
	Butter 4	L		

[1]VH = very-high-melting milkfat fraction (MP > 45°C), H = high-melting milkfat fraction (MP between 35 and 45°C), M = middle-melting milkfat fraction (MP between 25 and 35°C), L = low-melting milkfat fraction (MP between 10 and 25°C), and VL = very-low-melting milkfat fractions (MP < 10°C), U = unknown-melting milkfat fraction.

[2]Methods and data presented are those described by the original authors.

[3]Data calculated from original graphs.

[4]DP = dropping point.

TABLE 8.4 Lactone Concentration of Cold-Spreadable Butters

Author	Butter	δ-C10		γ-C12		δ-C12		δ-C14		δ-C16		Total	
Fractionation method	Designation	Free	Potential	Free	Potential	Free	Potential	Free	Potential	Free	Potential	Free	Potential
Walker et al., 1977													
Crystallization from acetone solution	Butter A	6.3	15.0	—[1]	1.1	16.8	32.8	24.2	39.7	11.1	20.6	58.4	109.2
	Butter C$_3$	18.2	25.2	3.3	4.6	45.9	56.6	55.8	73.5	30.8	41.2	1547.0	201.1

Lactone concentration (μg/g fat)

[1] Data not available.

TABLE 8.5 Formulations and Processing Information for Dairy-Based Spreads

Type of Spread / Author / Fractionation method	Stock milkfat fractions employed[1]	Spread designation	Milkfat fraction category[2]	Fat phase blend Fraction	%	Spread blend Ingredient	%	Method of production[3]	Packaging and tempering[3]
Low-milkfat spreads									
Verhagen and Wanaar, 1984									
Fractionation from melted milkfat	25L$_S$ fraction	Spread 1	M	25L$_S$ fraction	100	*Fat phase* Milkfat Monoglyceride *Aqueous phase* Water Na-caseinate NaCl NaHPO$_4$•2H$_2$O Emulsifier Citric acid	39.5 0.2 50.1 8.4 1.0 0.15 0.1 0.1	Aqueous phase was emulsified in the fat phase and the mixture was cooled and worked in a Votator unit.	—[4]
Keogh, 1988									
Fractionation method not reported	Soft milkfat fraction	Spread 1	U	Soft milkfat fraction	100	*Fat phase* Milkfat Distilled monoglyceride *Aqueous phase* Water Sodium caseinate Gelatin NaCl	39.5 0.5 52.0 5.0 2.0 1.0	Aqueous phase was blended and added slowly to the fat phase with agitation (40–42°C), then processed in a single scraped-surface cooler (Votator A-unit).	—
High-melting butter									
Shukla et al., 1994									
Fractionation by supercritical carbon dioxide extraction	Raffinate	HMT butter	H	Raffinate	100	*Fat phase* Raffinate *Aqueous phase* Water NaCl Skimmed milk powder	80.5 16.0 1.5 2.0	The fraction was mixed with the aqueous solution, pasteurized; and formed into butter using a single-stage homogenizer operated at 700 kPa.	The homogenized samples were rapidly cooled to test temperatures and stored overnight for temperature equilibrium.

Continued

TABLE 8.5 (Continued)

Type of Spread / Author / Fractionation method	Stock milkfat fractions employed[1]	Spread designation	Milkfat fraction category[2]	Fat phase blend		Spread blend		Method of production[3]	Packaging and tempering[3]
				Fraction	%	Ingredient	%		
Butter-margarine spreads									
Bratland, 1983									
Fractionation from melted milkfat	$20L_S$ fraction	Spread 1	U	$20L_S$ fraction Groundnut oil	60 40	—		The fat blend was emulsified with serum from the phase inversion process to produce a 35% fat cream. The cream was pasteurized and cooled, then churned at 5°C.	—
Verhagen and Bodor, 1984									
Fractionated from melted milkfat	$25S_S$ fraction $15S_S$ fraction	Spread 1	U	Intact milkfat $25S_S$ fraction Sunflower oil	40 20 40	*Fat phase* Fat blend Monoglyceride *Aqueous phase (pH 4.7)* Acidified skim milk	 83.9 0.1 16.0	Aqueous phase was emulsified with the fat phase and then prepared in a Votator unit with the following sequence: AACA (A = cooling unit, C = crystallizing unit), Exit temperature: 1st A unit = 15°C 2nd A unit = 10°C C unit = 13°C 3rd A unit = 10°C	—
		Spread 2	U	$15S_S$ fraction Sunflower oil	60 40	*Fat phase* Fat blend Monoglyceride *Aqueous phase (pH 4.7)* Acidified skim milk	 83.9 0.1 16.0		
		Spread 3	U	$25S_S$ fraction Sunflower oil	40 60	*Fat phase* Fat blend Monoglyceride *Aqueous phase (pH 4.7)* Acidified skim milk	 83.9 0.1 16.0		
		Spread 4	U	Intact milkfat $25S_S$ fraction Sunflower oil	50 25 25	Fat blend Skim milk Water	40 30 30	Artificial cream was prepared by emulsifying the fat phase with the skim milk and water, ripening overnight at 10°C, and subsequently churned until the desired texture and plasticity was obtained.	
Lansbergen and Kemps, 1984									
Fractionation of hardened milkfat	$12S_M$ [1]fraction $6L_M$ [1] fraction	Spread 1	U	$12L_M$ fraction	100	*Fat phase* $12L_M$ fraction	 83.64	Margarine prepared in a Votator apparatus.	—

	from acetone solution		$12L_M$ [2] fraction $0L_M$ [2] fraction		Color 0.2 *Aqueous phase (pH 4.2)* Water 16.0 Potassium sorbate 0.1 Lactic acid 0.06
Spread 2	U	$12S_M$ fraction Sunflower oil	17.5 82.5		*Fat Phase* $12S_M$ fraction 14.64 Sunflower oil 69.0 Color 0.2 *Aqueous Phase (pH 4.2)* Water 16.0 Potassium sorbate 0.1 Lactic acid 0.06
Spread 3	U	$6L_M$ fraction $0L_M$ fraction	unknown unknown		*Fat Phase* $6L_M+0L_M$ fraction 83.64 Color 0.2 *Aqueous Phase (pH 4.2)* Water 16.0 Potassium sorbate 0.1 Lactic acid 0.06

[1] Physical and chemical characteristics of stock milkfat fractions are presented in Chapter 5.

[2] H = high-melting milkfat fraction (MP between 35 and 45°C); M = middle-melting milkfat fraction (MP between 25 and 35°C); U = unknown-melting milkfat fraction.

[3] Methods and data presented are those of the original author.

[4] Data not available.

TABLE 8.6 Solid Fat Content Profiles of Dairy-Based Spreads

Type of Spread Author	Spread designation	Milkfat fraction category[1]	Solid fat content (%)					
			10°C	15°C	20°C	25°C	30°C	35°C
High-melting butter								
Shukla et al., 1994	Market butter	U	72.8[2] 7°C	37.2 17°C	—[3]	13.3 27°C	—	0
	HMT butter	H	76.8	71.6	—	55.6	—	18.0
Butter–margarine spreads								
Verhagen and Bodor, 1984								
	Spread 1	U	25	—	12	—	0.5	—
	Spread 2	U	30	—	14	—	—	0.7
	Spread 3	U	20	—	11	—	—	1.0

[1] H = high-melting milkfat fraction (MP between 35 and 45°C), U = unknown-melting milkfat fraction.

[2] Data estimated from graphs provided by original authors.

[3] Data not available.

TABLE 8.7 Textural Characteristics of Dairy-Based Spreads

Type of Spread Author	Spread designation	Milkfat fraction category[1]	Selected textural characteristics[2]			
Low-milkfat spreads						
Verhagen and Warnaar, 1984			Hardness values (g/cm^2)			
	Spread 1	M	5°C	10°C	15°C	20°C
			1600	1200	600	80
High-melting butters						
Shukla et al., 1994			*Storage modulus G' (MPa•$s^{n'}$)*	*Loss modulus G'' (MPa•$s^{n''}$)*	*Complex viscosity $\eta*$ (MPa•s^{n*})*	
	Market butter	U	12°C	12°C	12°C	
			$G' = 11.76$	$G'' = .47$	$\eta* = 1.77$	
			$n' = .035$	$n'' = -.58$	$n* = -.967$	
			17°C	17°C	17°C	
			$G' = 6.08$	$G'' = .41$	$\eta* = .28$	
			$n' = .074$	$n'' = -.337$	$n* = -.929$	
			22°C	22°C	22°C	
			$G' = .42$	$G'' = .06$	$\eta* = .068$	
			$n' = .122$	$n'' = -.206$	$n* = -.887$	
			27°C	27°C	27°C	
			$G' = .02$	$G'' = .004$	$\eta* = .003$	
			$n' = .128$	$n'' = -.130$	$n* = -.886$	
	HMT butter	H	17°C	17°C	17°C	
			$G' = 16.76$	$G'' = 1.16$	$\eta* = 2.69$	
			$n' = .038$	$n'' = -.582$	$n* = -.968$	
			22°C	22°C	22°C	
			$G' = 7.8$	$G'' = .34$	$\eta* = 1.24$	
			$n' = .049$	$n'' = -.365$	$n* = -.952$	
			27°C	27°C	27°C	
			$G' = 2.89$	$G'' = .21$	$\eta* = .46$	
			$n' = .047$	$n'' = -.216$	$n* = -.953$	
			32°C	32°C	32°C	
			$G' = .24$	$G'' = .04$	$\eta* = .039$	
			$n' = .121$	$n'' = -.212$	$n* = -.894$	
Butter–margarine spreads						
Verhagen and Bodor, 1984[2]			*Hardness values (g/cm^2)*			
	Spread 1	U	5°C	10°C	20°C	
			1400	800	100	
	Spread 2	U	2100	1300	150	
	Spread 3	U	700	400	90	

[1]H = high-melting milkfat fraction (MP between 35 and 45°C), M = middle-melting milkfat fraction (MP between 25 and 35°C), U = unknown-melting milkfat fraction.

[2]Methods and data presented are those of the original authors.

TABLE 8.8 Formulations and Processing Information for Experimental Bakery Milkfat Ingredients

Bakery category / Author / Milkfat fractionation method	Stock milkfat fractions employed[1]	Milkfat ingredient designation	Milkfat fraction category[2]	Milkfat fraction blend — Fraction	%	Milkfat ingredient blend — Ingredient	%	Method of production[3]	Packaging and tempering[3]
Pastry milkfats									
Schaap, 1982									
Fractionated from melted milkfat	$24S_S$ fraction	Puff pastry milkfat 1	U	$24S_S$ fraction	100	—[4]		Milkfat was melted and cooled to 14°C with a scraped-surface heat exchanger.	Stored for about one week at the dough-preparing temperature prior to application in the product.
Pedersen, 1988									
Fractionation method not reported	Stearin fraction	Puff pastry milkfat 1	H	Stearin fraction	100	Milkfat	100	Crystallization was carried out on Gerstenber and Agger Perfector consisting of a high-pressure piston pump, 3 chilling tubes, and 2 pin-rotor machines.	Half samples were stored at 5°C immediately after production, and half stored at 18°C for 4 d before moving to cold storage.
		Puff pastry milkfat 2	H	Stearin fraction	100	Milkfat Skim milk	84 16		
		Puff pastry milkfat 3	H	Stearin fraction	100	Milkfat Skim milk Emulsifier (Homodan RD, Grinsted Products) Soya lecithin	82.75 16.0 0.75 0.5		
		Puff pastry milkfat 4	H	Stearin fraction	100	Milkfat Acidified skim milk (pH 3.5)	84 16		
		Puff pastry milkfat 5	H	Stearin fraction	100	Milkfat Acidified skim milk (pH 3.5) Emulsifier (Homodan RD, Grinsted Products) Soya lecithin	82.75 16 0.75 0.5		
Keogh and Morrissey, 1990									
Fractionated from melted milkfat	Milkfat fraction (38)[5]	Croissant milkfat 1	H	Milkfat fraction (38)	100	—		Processed in Gerstenberg and Agger Perfector with two scraped-surface	—

	Croissant milkfat 2	H	Milkfat fraction (38)	100	surface coolers (type A), two intermediate crystallizers (type C), pin-working unit (W), and setting tube (S). ACAS sequence: refrigerant T = −20°C agitation rate = 900 rpm throughput = 20 kg/h throughput = 30 kg/h
	Croissant milkfat 3	H	Milkfat fraction (38)	100	throughput = 60 kg/h
	Croissant milkfat 4	H	Milkfat fraction (38)	100	throughput = 90 kg/h
	Croissant milkfat 5	H	Milkfat fraction (38)	100	throughput = 120 kg/h
	Puff pastry milkfat 1[6]	H	unknown	Milkfat 100	ACAS sequence throughput = 30 kg/h
	Puff pastry milkfat 2	H	unknown	Milkfat Lecithin 0.5	throughput = 30 kg/h
	Puff pastry milkfat 3	H	unknown	Milkfat Buttermilk powder 1.6	throughput = 30 kg/h
	Puff pastry milkfat 4	H	unknown	Milkfat Lecithin 0.5 Buttermilk powder 1.6	throughput = 30 kg/h

Short pastry milkfats

Jordan, 1986
Fractionation from melted milkfat

$28S_S$ fraction $28L_S$ fraction	Shortbread biscuit milkfat 1	H	$28S_S$ fraction	100	—
	Shortbread biscuit milkfat 2	L	$28L_S$ fraction	100	

Continued

TABLE 8.8 (Continued)

Bakery category Author Milkfat fractionation method	Stock milkfat fractions employed[1]	Milkfat ingredient designation	Milkfat fraction category[2]	Milkfat fraction blend		Milkfat ingredient blend		Method of production[3]	Packaging and tempering[3]
				Fraction	%	Ingredient	%		
Cake milkfats									
Schaap, 1982									
Fractionation from melted milkfat	$24S_S$ fraction	Short cake milkfat 1	U	$24S_S$ fraction	100	—		Milkfat was melted and cooled to 14°C with a scraped surface heat exchanger	Stored for about one week at the dough preparation temperature prior to application in the product.
Keogh and Morrissey, 1990									
Fractionation from melted milkfat	Milkfat fraction (35) Milkfat fraction (38) Milkfat fraction (41)	Madeira cake milkfat 1	H	Milkfat fraction (35)	100	*Madeira Milkfat* Milkfat Lecithin Buttermilk powder	unknown 0.5 1.6	Processed in Gerstenberg and Agger Perfector with two scraped-surface coolers (type A), two intermediate crystallizers (type C), pin-working unit (W), and setting tube (S): throughput = 60 kg/h refrigerant T = −20°C agitation rate = 900 rpm AC sequence used	Aluminum foil wrapped samples of 0.5 kg were placed in storage at 4, 15, and 21°C.
		Madeira cake milkfat 2	H	Milkfat fraction (35)	100			ACAC sequence used	
		Madeira cake milkfat 3	H	Milkfat fraction (35)	100			ACAS sequence used	
		Madeira cake milkfat 4	H	Milkfat fraction (35)	100			ACAW sequence used	
		Madeira cake milkfat 5	H	Milkfat fraction (35)	100			ACAS sequence throughput = 30 kg/h	
		Madeira cake milkfat 6	H	Milkfat fraction (35)	100			ACAS sequence throughput = 60 kg/h	
		Madeira cake	H	Milkfat fraction (35)	100			ACAS sequence throughput = 90 kg/h	

Madeira cake milkfat 8	H	Milkfat fraction (35) 100	ACAS sequence throughput = 120 kg/h
Madeira cake milkfat 9	H	Milkfat fraction (38) 100	ACAS sequence throughput = 30 kg/h
Madeira cake milkfat 10	H	Milkfat fraction (38) 100	ACAS sequence throughput = 60 kg/h
Madeira cake milkfat 11	H	Milkfat fraction (38) 100	ACAS sequence throughput = 90 kg/h
Madeira cake milkfat 12	H	Milkfat fraction (38) 100	ACAS sequence throughput = 120 kg/h
Madeira cake milkfat 13	H	Milkfat fraction (41) 100	ACAS sequence throughput = 30 kg/h
Madeira cake milkfat 14	H	Milkfat fraction (41) 100	ACAS sequence throughput = 60 kg/h
Madeira cake milkfat 15	H	Milkfat fraction (41) 100	ACAS sequence throughput = 90 kg/h
Madeira cake milkfat 16	H	Milkfat fraction (41) 100	ACAS sequence throughput = 120 kg/h

[1]Physical and chemical characteristics of stock milkfat fractions are presented in Chapter 5.

[2]H = high-melting milkfat fraction (MP between 35 and 45°C), L = low-melting milkfat fraction (MP between 10 and 25°C), and U = unknown-melting milkfat fraction.

[3]Method and data presented are described by the original author.

[4]Data not available.

[5]The number in parentheses refers to the slip point as reported by the author.

[6]Three solid, high-melting milkfat fractions were obtained at 35, 38 and 41°C, but the specific fraction used for puff pastries was not specified.

TABLE 8.9 Chemical and Physical Properties of the Milkfat Phase of Experimental and Commercial Milkfat Ingredients for Bakery Products

Bakery category Source Bakery product Milkfat origin	Milkfat ingredient designation	Milkfat fraction category[1]	Melting behavior[2] (°C)	Iodine value	Other physical properties[3]	Textural properties[3]	Other properties[3]
Pastry milkfats							
Alaco Company[4]							*Recommended uses*
All Pastry Commercial milkfats	Pastry Butter	H	37 (MP)	—[5]	—	—	Croissants, puff and Danish pastry
	Buttersheets	H	37	—	—	—	Croissants, puff and Danish pastry
Aveve Company[4]							*Recommended uses*
All Pastry Commercial milkfats	Concentrated butter feuilletage 2000	M	24–26 (MP)	—	—	—	Puff pastry
	Concentrated butter patissier	M	28–32	—	—	—	General pastry doughs
	Concentrated butter croissant	H	38	—	—	—	Croissants, doughs
	Concentrated butter millefeuille	H	40–42	—	—	—	Puff pastry
Corman Company[4]							*Recommended uses*
Commercial milkfat	Millefeuille concentrated butter		41 (MP)	—	—	—	Puff pastry
	Croissant concentrated butter		38	—	—	—	Croissant, Danish, puff pastry
	Standard anhydrous milkfat		32	—	—	—	General pastry
Tolboe, 1984							
Danish pastry Commercial fraction obtained from Corman	Pastry milkfat	H	35.0 (MP)	22	—	—	—
	Cookie milkfat	M	32.1	34	—	—	—
Pedersen, 1988[3]						*Plasticity*		
Fractionation method not reported	Puff pastry milkfat 1	H	36 (SP)	—	—	Finger judgments of plasticity have shown that all samples made with skim milk were better than those made with anhydrous stearin Measurements of penetration before and after work-softening show	—
	Puff pastry milkfat 2	H	36	—	—		—
	Puff pastry milkfat 3	H	36	—	—		—

Continued

TABLE 8.9 (Continued)

Bakery category Source Bakery product Milkfat origin	Milkfat ingredient designation	Milkfat fraction category[1]	Melting behavior[2] (°C)	Iodine value	Other physical properties[3]	Textural properties[3]	Other properties[3]
	Puff pastry milkfat 4	H	36	—	—	that decrease in firmness due to work-softening was smaller for tempered samples than for untempered ones	—
	Puff pastry milkfat 5	H	36	—	—		—
Deffense, 1989							*Recommended uses*
Puff, croissant and Danish pastry	Feuilletage	H	42 (DP)	—	—	—	Puff pastry
Commercial milkfats obtained from Corman	Feuilletage 2000	M	26	—	—	—	Puff pastry except flaky yeast dough
	Croissant	H	39	—	—	—	Croissant, Danish pastry
Keogh and Morrissey, 1990					*Largest crystal size (µm)*	*Yield value (g/cm²)*	*Fat appearance*
Fractionated from melted milkfat	Croissant milkfat 1	H	38 (SLP)	—	64	1068	Smooth
	Croissant milkfat 2	H	38	—	60	1997	Smooth
	Croissant milkfat 3	H	38	—	72	2277	Slightly brittle
	Croissant milkfat 4	H	38	—	101	2429	Brittle
	Croissant milkfat 5	H	38	—	—	—	Very brittle
	Puff pastry milkfat 1	H	—	—	67	2252	—
	Puff pastry milkfat 2	H	—	—	74	2425	—
	Puff pastry milkfat 3	H	—	—	125	2708	—
	Puff pastry milkfat 4	H	—	—	104	2615	—
Short pastry milkfats							
Alaco Company							*Recommended uses*
Short pastries, general bakery applications Commercial milkfat	Bakery Butter	U	—	—	—	—	Cakes, muffins, cookies, short pastries and yeast goods
	Butter Shortening	U	—	—	—	—	Anhydrous form of Bakery Butter
Aveve Company							*Recommended uses*
Commercial milkfat	Concentrated butter creme au beurre	M	28 (MP)	—	—	—	Butter cookies

Continued

TABLE 8.9 (Continued)

Bakery category
  Source
    Bakery product
      Milkfat origin
	Milkfat ingredient designation	Milkfat fraction category[1]	Melting behavior[2] (°C)	Iodine value	Other physical properties[3]	Textural properties[3]	Other properties[3]

Corman Company							*Recommended uses*
Commercial milkfat	Croissant concentrated butter	H	38 (MP)	—	—	—	Short pastry
	Anhydrous milkfat	M	32	—	—	—	Biscuits, pastry
	Danish 4/4 concentrated butter	M	30	—	—	—	Danish butter cookies, shortbread, biscuits
Tolboe, 1984							
Vanilla rings Commercial fractions obtained from Corman	Pastry milkfat	H	35.0 (MP)	22	—	—	—
	Cookie milkfat	M	32.1	34	—	—	—
Jordan, 1986							
Fractionation from melted milkfat	Shortbread biscuit milkfat 1	H	39.0 (MP)	—	—	—	—
	Shortbread biscuit milkfat 2	L	23.9	—	—	—	—
Peck, 1990							
Commercial fractions from New Zealand Dairy Board	Confectionery butterfat 42	H	42 (MP)	—	—	—	—
	Soft butteroil 21	L	21	—	—	—	—
Cake milkfats							
Alaco Company							*Recommended uses*
Short pastries, general bakery applications Commercial milkfat	Bakery butter	U	—	—	—	—	Cakes, muffins, cookies, short pastries and yeast goods
	Butter shortening	U	—	—	—	—	Anhydrous form of Bakery Butter
Aveve Company							*Recommended uses*
Doughs Commercial milkfat	Concentrated butter 4/4	M	26–28 (MP)	—	—	—	Doughs, biscuits

Continued

TABLE 8.9 (Continued)

Bakery category Source Bakery product Milkfat origin	Milkfat ingredient designation	Milkfat fraction category[1]	Melting behavior[2] (°C)	Iodine value	Other physical properties[3]	Textural properties[3]	Other properties[3]
Tolboe, 1984							
Pound cake Commercial milkfat from Corman	Pastry milkfat	H	35.0 (MP)	22	—	—	—
	Cookie milkfat	L	32.1	34	—	—	—
Deffense, 1989							*Recommended uses*
Pound cake Commercial milkfat from Corman	Quatre-quart	M	28 (DP)	—	—	—	Pound cake, choux pastry, madeleines
Keogh and Morrissey, 1990 Fractionation from melted milkfat					*Largest crystal (μm)*	*Yield value (g/cm²)*	*Fat appearance*
	Madeira cake milkfat 1	H	35 (SLP)	—	35	2056	Glazed
	Madeira cake milkfat 2	H	35	—	45	1701	Glazed
	Madeira cake milkfat 3	H	35	—	88	1698	Slighty brittle
	Madeira cake milkfat 4	H	35	—	57	1290	Glazed
	Madeira cake milkfat 5	H	35	—	61.0	467	—
	Madeira cake milkfat 6	H	35	—	73.5	803	—
	Madeira cake milkfat 7	H	35	—	89.0	1444	—
	Madeira cake milkfat 8	H	35	—	88.5	1708	—
	Madeira cake milkfat 9	H	38	—	67.0	1292	—
	Madeira cake milkfat 10	H	38	—	82.0	1647	—
	Madeira cake milkfat 11	H	38	—	33.5	1899	—
	Madeira cake milkfat 12	H	38	—	88.5	2429	—
	Maderia cake milkfat 13	H	41	—	72.0	627	—
	Madeira cake milkfat 14	H	41	—	83.0	1475	—
	Madeira cake milkfat 15	H	41	—	147.5	2445	—
	Madeira cake milkfat 16	H	41	—	99.0	3786	—

Continued

TABLE 8.9 (Continued)

Bakery category Source Bakery product Milkfat origin	Milkfat ingredient designation	Milkfat fraction category[1]	Melting behavior[2] (°C)	Iodine value	Other physical properties[3]	Textural properties[3]	Other properties[3]
Other bakery applications							
Corman Company							*Recommended uses*
Commercial milkfat	Anhydrous milkfat	M	32 (MP)	—	—	—	Bread making

[1] H = high-melting milkfat fraction (MP between 35 and 45°C), M = middle-melting milkfat fraction (MP 25 and 35°C), L = low-melting milkfat fraction (MP 10 and 25°C), and U = unknown-melting milkfat fraction.

[2] MP = melting point, DP = dropping point, SLP = slip point, SP = softening point.

[3] Method and data presented are described by the original author.

[4] Milkfat ingredients supplied by commercial companies have standard sensory (flavor, color), oxidative stability (peroxide value, free fatty acid value) and microbiological (total plate count, coliform, yeast-mold, salmonella) specifications.

[5] Data not available.

TABLE 8.10 Solid Fat Content Profiles of the Milkfat Portion of Experimental and Commercial Milkfat Ingredients Used in Bakery Products

Bakery category Source Milkfat origin	Milkfat ingredient designation	Milkfat fraction category[1]	Solid fat content (%)								
			5°C	10°C	15°C	20°C	25°C	30°C	35°C	40°C	45°C
Pastry milkfats											
Corman Company											
Commercial milkfat	Millefeuille concentrated butter	H	67	63	55	44	35	25	16	7	0
	Croissant concentrated butter	H	59	53	42	29	21	13	6	0	—[2]
	Standard anhydrous milkfat	M	49	39	28	16	9	5	0	—	—
Tolboe, 1984											
Commercial milkfat from Corman	Pastry milkfat	H	—	56	—	41	—	24	—	2	—
	Cookie milkfat	M	—	47	—	28	—	10	—	0	—
Pedersen, 1988											
Fractionation method not reported	Puff pastry milkfat 1	H	66.1	63.6	—	39.4	—	21.2	11.5	2.6	—
	Puff pastry milkfat 2	H	66.1	63.6	—	39.4	—	21.2	11.5	2.6	—
	Puff pastry milkfat 3	H	66.1	63.6	—	39.4	—	21.2	11.5	2.6	—
	Puff pastry milkfat 4	H	66.1	63.6	—	39.4	—	21.2	11.5	2.6	—
	Puff pastry milkfat 5	H	66.1	63.6	—	39.4	—	21.2	11.5	2.6	—
Deffense, 1989											
Commercial milkfats from Corman	Feuilletage	H	72	66	56	45	35	25	15	5	0
	Feuilletage 2000	M	66	56	45	27	3	0	—	—	—
	Croissant	H	63	57	47	33	23	14	7	0	—
Keogh and Morrissey, 1990											
Fractionated from melted milkfat	Croissant milkfat 1	H	—	44.5	33.4	16.9	9.3	—	—	—	—
	Croissant milkfat 2	H	—	44.5	33.4	16.9	9.3	—	—	—	—
	Croissant milkfat 3	H	—	44.5	33.4	16.9	9.3	—	—	—	—
	Croissant milkfat 4	H	—	44.5	33.4	16.9	9.3	—	—	—	—
	Croissant milkfat 5	H	—	44.5	33.4	16.9	9.3	—	—	—	—
Short pastry milkfats											
Corman Company											
Commercial milkfat	Croissant concentrated butter	H	59	53	42	29	21	13	6	0	—

Continued

TABLE 8.10 (Continued)

Bakery category Source Milkfat origin	Milkfat ingredient designation	Milkfat fraction category[1]	Solid fat content (%)								
			5°C	10°C	15°C	20°C	25°C	30°C	35°C	40°C	45°C
	Anhydrous milkfat	M	49	39	28	16	9	5	0	—	—
	Danish 4/4 concentrated butter	M	48	38	25	13	8	3	0	—	—
Tolboe, 1984 Commercial milkfat from Corman	Pastry milkfat	H	—	56	—	41	—	24	—	2	—
	Cookie milkfat	M	—	47	—	28	—	10	—	0	—
Jordan, 1986 Fractionation from melted milkfat	Shortbread biscuit milkfat 1	H	—	—	—	29.3	—	—	—	—	—
	Shortbread biscuit milkfat 2	L	—	—	—	6.5	—	—	—	—	—
Cake milkfats											
Tolboe, 1984 Commercial milkfat from Corman	Pastry milkfat	H	—	56	—	41	—	24	—	2	—
	Cookie milkfat	L	—	47	—	28	—	10	—	0	—
Deffense, 1989[5] Commercial milkfat from Corman	Quatre-quart	M	46	36	23	6	1	0	—	—	—
Keogh and Morrissey, 1990 Fractionation from melted milkfat	Madeira cake milkfat 1	H	—	44.7	30.3	15.5	8.5	—	—	—	—
	Madeira cake milkfat 2	H	—	44.7	30.3	15.5	8.5	—	—	—	—
	Madeira cake milkfat 3	H	—	44.7	30.3	15.5	8.5	—	—	—	—
	Madeira cake milkfat 4	H	—	44.7	30.3	15.5	8.5	—	—	—	—
	Madeira cake milkfat 5	H	—	44.7	30.3	15.5	8.5	—	—	—	—
	Madeira cake milkfat 6	H	—	44.7	30.3	15.5	8.5	—	—	—	—
	Madeira cake milkfat 7	H	—	44.7	30.3	15.5	8.5	—	—	—	—
	Madeira cake milkfat 8	H	—	44.7	30.3	15.5	8.5	—	—	—	—
	Madeira cake milkfat 9	H	—	45.2	33.4	16.9	9.3	—	—	—	—
	Madeira cake milkfat 10	H	—	45.2	33.4	16.9	9.3	—	—	—	—
	Madeira cake milkfat 11	H	—	45.2	33.4	16.9	9.3	—	—	—	—
	Madeira cake milkfat 12	H	—	45.2	33.4	16.9	9.3	—	—	—	—

Continued

TABLE 8.10 (Continued)

Bakery category Source Milkfat origin	Milkfat ingredient designation	Milkfat fraction category[1]	Solid fat content (%)								
			5°C	10°C	15°C	20°C	25°C	30°C	35°C	40°C	45°C
	Maderia cake milkfat 13	H	—	47.7	34.4	21.1	13.0	—	—	—	—
	Madeira cake milkfat 14	H	—	47.7	34.4	21.1	13.0	—	—	—	—
	Madeira cake milkfat 15	H	—	47.7	34.4	21.1	13.0	—	—	—	—
	Madeira cake milkfat 16	H	—	47.7	34.4	21.1	13.0	—	—	—	—
Other bakery applications											
Corman Company Commercial milkfat	Anhydrous milkfat	M	49	39	28	16	9	5	0	—	—

[1]H = high-melting milkfat fraction (MP between 35 and 45°C), M = middle-melting milkfat fraction (MP between 25 and 35°C), L = low-melting milkfat fraction (MP between 10 and unknown-melting milkfat fraction).

[2]Data not available.

TABLE 8.11 Formulations and Evaluation of Milkfat Ingredients in Bakery Products

Bakery category / Author	Milkfat ingredient designation	Milkfat fraction category[1]	Bakery Formula Ingredient	wt (g)	Method of production[2]	Sensory Evaluation	Textural properties	Physical properties	Other properties
Pastry milkfats									
Schaap, 1982	Puff pastry milkfat 1	U	*Puff pastry* —[3]		—	Puff pastry appeared to be of excellent quality.	—	—	—
Tolboe, 1984	Pastry milkfat	H	*Danish pastry* Flour	1200	The dough is mixed just to homogeneity, T = 17°C. The pastry is made by rolling 50% or 40% butter fat in the dough, which is rolled 3 × 3 and to a final thickness of 4 mm. The sheeted dough is cut into triangles (40 g) and wound-formed to Danish horns (croissants). Proving 45 min at 32°C and 80% R.H. Baking 12 min at 220°C.	The pastry milkfat produced Danish pastry with a flaked structure just as good as a typical Danish pastry margarine.	—	—	—
			Water	500					
			Egg	250					
			Yeast	100					
			Sugar	80					
			Salt	15					
	Cookie milkfat	M				The quality of pastry containing cookie butter is inferior because the dough layers have not separated into the desired flake structure, but have stuck together in thick layers.			
Pedersen, 1988	Puff pastry milkfat 1	H	*Puff pastry* —		—	All pastry butters containing skim milk gave considerably higher pastries than anhydrous stearin. Low pH of the aqueous phase and addition of emulsifier had a positive effect on the size of the finished pastry.	—	—	—
	Puff pastry milkfat 2	H							
	Puff pastry milkfat 3	H							
	Puff pastry milkfat 4	H							
	Puff pastry milkfat 5	H							
Keogh and Morrissey, 1990	Croissant fat 1	H	*Croissants* Flour	500	Croissant ingredients were mixed to a smooth dough. The dough was pinned out to a thickness of about 20 mm and	—	*Texture* Uneven	*Volume (mL)* 268	—
			Water	245					
			Egg	45					
			Yeast	18					

Application of Milkfat Fractions in Foods

Sample		Ingredient	Amount	Type	Value	Notes
Croissant fat 2	H	Gluten flour	9			chilled in a refrigerator at 16°C in about 45 min. The folding-in milkfat was pinned out to 2/3 size of the dough sheet and chilled with the dough. The butter was placed on the dough sheet and folded in. The dough was rested and chilled for 10 min after the first half turn. After the second turn the dough was rested and chilled for 30 min before the third half turn. A 10 min resting/chilling was given before the fourth half turn. The dough was placed in a refrigerator for 1.5 h before pinning out to a thickness of 4 mm and cut into croissants with base 125 mm wide and 250 mm high at the apex. The croissants were rolled into shape and proved at 27°C for 40 min and baked for 17 min at 220°C.
		Salt	9			
		Sugar	9			
		Milkfat	8	Typical	236	
		Milkfat for folding in	154			
Croissant fat 3	H			Typical	284	
Croissant fat 4	H			Typical	282	
Puff pastry fat 1	H	*Puff pastry*			—	The English method for making puff pastry was used by mixing dough from the flour, cake margarine, salt and water. The milkfat was sheeted and laminated into the dough, giving six half-turns before final sheeting and cutting. During laminating and before baking, appropriate recovery periods were given. Vol-au-vent rings were baked at 215°C for 20 min.
		Flour	500			
		Cake margarine	62			
		Salt	5			
		Water	275			
		Milkfat	359			
Puff pastry fat 2	H				—	The degree of pastry expansion was acceptable in all samples as measured by pastry height. The sample containing buttermilk powder was the most uniform as measured by eccentricity.
Puff pastry fat 3	H					
Puff pastry fat 4	H					

Continued

TABLE 8.11 (Continued)

Bakery category Author	Milkfat ingredient designation	Milkfat fraction category[1]	Bakery Formula Ingredient	wt (g)	Method of production[2]	Sensory Evaluation	Textural properties	Physical properties	Other properties
Short pastry milkfats									
Tolboe, 1984			*Vanilla rings*						
	Pastry milkfat	H	Milkfat	200	After mixing, the dough is formed to vanilla rings and baked 9 min at 180°C.	Vanilla rings made with pastry milkfat were too hard.	—	—	—
			Water	40					
			Egg	50					
			Sugar	175					
			Flour	325					
			Salt	2.5					
			Vanilla	unknown					
	Cookie milkfat	M				Vanilla rings made with cookie milkfat produced larger cookies with a nice crispness.			
Jordan, 1986			*Shortbread biscuits*						
	Biscuit milkfat 1	H	Milkfat	100	Milkfat, water, and sugar were creamed together before mixing in the flour. The dough was rolled out and baked at 200°C.	Harder, compact structure, less layered structure, less dry, less crumbly, coarse breakdown.	—	—	—
			Water	25					
			Castor sugar	50					
			Plain flour	175					
	Biscuit milkfat 2	L				Softer, open structure, more layered structure, drier, more crumbly, fine breakdown.			
Peck, 1990			*Chocolate chip cookies*				Hardness (N)	Height (cm)	Diameter (cm)
	Confectionery butterfat 42	H	Milkfat	280	In a large Cuisinart: cream butter, granular sucrose, and brown sugar for 50 s; scrape sides. Add whole egg, water, vanilla, and invert sugar, blend 15 s and scrape sides. Add flour, nonfat dry milk, baking powder, and baking soda, blend 15 s, scrape side, blend another 15 s and scrape sides. Add chocolate chips and blend manually with a spoon. Cover and refrigerate dough for 24 h. Shape 50 g samples with a #20 ice cream	The waxy appearance of these cookies suggest that the milkfat did not melt uniformly.	1 h = 22.4	1.73	8.24
			Granular sucrose	150			72 h = 43.6		
			Brown sugar	150					
			Ivert sugar	50					
			Egg	60					
			Water	48					
			Vanilla	6					
			Nonfat dry milk	3.5					
			Flour	454					
	Soft butteroil 21	L	Chocolate chips	350		This milkfat was easier to cream into the sugar than the Confectionery Butterfat 42, but became almost liquid after	1 h = 30.7	1.82	8.5
			Baking powder	2 tsp			72 h = 38.32		
			Baking soda	1/4 tsp					

Cake milkfats

Schaap, 1982	Short cake milkfat 1	U	Short cakes	—	Short cakes appeared to be of excellent quality.
			scoop and place on cookie sheet. Bake in a preheated oven for 22 min at 135°C.		
			removal from the cooler.		
				A favorable effect on the dough viscosity was observed.	

Tolboe, 1984			Pound cake		No difference in cake volume or cake texture was observed between the two milkfats. The creaming speed of the fat-sugar mixture was lower for pastry milkfat than for cookie milkfat, which was lower than for a cake shortening.
	Pastry milkfat	H	Milkfat (19°C) 400	Mixing in a Hobart N 50, gear 2 in 3 steps. Milkfat and sugar are creamed for 5 min, eggs and water are added and mixed for 5 min, flour and baking powder are added and mixed for 5 min. Baking 50 min at 170°C.	
			Sugar 500		
			Egg 500		
			Water 80		
			Flour 500		
	Cookie milkfat	M	Baking powder 10		

Cocup and Sanderson, 1987   *Cake batter*

				Relative density of cake batter	Time to reach maximum aeration (min)
Standard butter				0.86	13
Fractionated milkfat	U			0.79	11
Emulsified fractionated milkfat	U			0.73	9
Plain shortening				0.74	10
Emulsified cake shortening				0.72	9

Keogh and Morrissey, 1990   *Madeira cake*

				Batter specific gravity	Cake specific volume	Cake penetration (mm)
Madeira cake milkfat 1	H	High-ratio flour 300.0	The Madeira cake batter was made on a Hobart 3-speed mixer. The dry ingredients were mixed to a crumble for two min	0.72	2.20	9.9
		Castor sugar 240.0				
		Baking powder 3.8				
Madeira cake milkfat 2	H	Salt 8.8		0.76	2.10	9.8

Continued

TABLE 8.11 (Continued)

Bakery category Author	Milkfat ingredient designation	Milkfat fraction category[1]	Bakery Formula Ingredient	wt (g)	Method of production[2]	Sensory Evaluation	Textural properties	Physical properties	Other properties
	Madeira cake milkfat 3	H	Milkfat	197.0	at speed 1. The liquid ingredients were mixed and added gradually at speed 1 and then mixed for 5 min at speed 2. The cake batter (360 g) was weighed immediately into a tin, lined with greaseproof paper; two cakes were obtained from each mix. The cakes were baked for 45 min at 193°C.		0.71	2.10	10.0
	Madeira cake milkfat 4	H	Eggs	240.0			0.75	2.10	9.6
	Madeira cake milkfat 5	H	Milk	60.0			0.76	1.92	11.1
	Madeira cake milkfat 6	H	Water	38.0			0.76	1.89	9.9
	Madeira cake milkfat 7	H					0.71	1.91	10.8
	Madeira cake milkfat 8	H					0.80	1.84	10.0
	Madeira cake milkfat 9	H					0.71	2.25	12.2
	Madeira cake milkfat 10	H					0.70	2.25	12.5
	Madeira cake milkfat 11	H					0.67	2.45	16.0
	Madeira cake milkfat 12	H					0.69	2.40	15.8
	Madeira cake milkfat 13	H					0.68	2.20	10.4
	Madeira cake milkfat 14	H					0.74	2.15	10.0
	Madeira cake milkfat 15	H					0.71	2.10	9.7
	Madeira cake milkfat 16	H					0.72	2.15	10.0

[1]VH = very-high-melting milkfat fraction (MP > 45°C), H = high-melting milkfat fraction (MP between 35 and 45°C), M = middle-melting milkfat fraction (MP between 25 and 35°C), L = low-melting milkfat fraction (MP between 10 and 25°C), and VL = very-low-melting milkfat fractions (MP < 10°C), U = unknown-melting milkfat fraction.

[2]Method and data presented are by the original author.

[3]Data not available.

TABLE 8.12 Formulations and Processing Information for Milkfat Fractions in Chocolate

Type of chocolate Author Milkfat fractionation method	Stock fractions employed[1]	Chocolate designation	Milkfat fraction category[2]	Fat phase blend Fraction	%	Chocolate formula Ingredient	%	Method of chocolate production[3]	Chocolate tempering[3]
Dark chocolate									
Jebson, 1974b Fractionation method not reported	Hard milkfat fraction	Chocolate 1	U	Hard milkfat fraction	3	—[4]		—	—
				Cocoa butter	97				
Yi, 1993 Fractionation from acetone solution	Intact AMF	Chocolate 1		Cocoa butter (control)	100	Sugar	51.2	Chocolate samples were prepared by adding the fat blend to unfinished semi-sweet chocolate base.	Chocolate was melted to 60°C and held for 30 min with constant stirring, then cooled to 27°C and stirred until constant viscosity was reached. The chocolate was then heated to 33°C and held until viscosity was constant. Then the chocolate was cooled to 29°C and held until viscosity was constant. Samples were moulded and held at 15°C for 24 h prior to analysis.
				Added Fat Blend		Chocolate liquor	37.8		
	$25S_M$ fraction	Chocolate 2		Intact milkfat	100	Cocoa butter	8.7		
	$20S_M$ fraction	Chocolate 3	VH	$25S_M$ fraction	100	Added fat blend	2.0		
	$15S_M$ fraction	Chocolate 4	VH	$20S_M$ fraction	100	Lecithin	0.4		
	$10S_M$ fraction	Chocolate 5	VH	$15S_M$ fraction	100	Vanillin	0.1		
	$5S_M$ fraction	Chocolate 6	VH	$10S_M$ fraction	100				
	$0S_M$ fraction	Chocolate 7	H	$5S_M$ fraction	100				
	$0L_M$ fraction	Chocolate 8	M	$0S_M$ fraction	100				
		Chocolate 9	M	$0L_M$ fraction	100				
			L						
Milk chocolate									
Jordan, 1986 Commercially obtained fractions from two Belgian companies	HM41	Chocolate 1		Butteroil	23	Cocoa liquor	13.0	The cocoa liquor, cocoa butter and milkfat were melted and the sugar, milk powder and lecithin added, to form a stiff paste. The paste was milled with a Pascall small-scale three-roller refiner and remelted at	The refined and conched paste was tempered by hand in a stainless steel beaker. The samples were temperated by being cooled to 25°C and reheated to 29°C.
	HM40			Cocoa butter	77	(Cocoa butter 7%)			
	HM38	Chocolate 2	H	HM41 fraction	23	Cocoa butter	21.2		
				Cocoa butter	77	Milkfat	8.4		
		Chocolate 3	H	HM40 fraction	23	Icing sugar	43.4		
				Cocoa butter	77	Skimmed milk powder	13.6		

Continued

TABLE 8.12 (Continued)

Type of chocolate Author Milkfat fractionation method	Stock fractions employed[1]	Chocolate designation	Milkfat fraction category[2]	Fat phase blend Fraction	%	Chocolate formula Ingredient	%	Method of chocolate production[3]	Chocolate tempering[3]
		Chocolate 4	H	HM38 fraction	23	Lecithin	0.4	50°C.	
				Cocoa butter	77				
		Chocolate 5		Butteroil	25				
				Cocoa butter	75				
		Chocolate 6	H	HM41	25				
				Cocoa butter	75				
		Chocolate 7	H	HM40	25				
				Cocoa butter	75				
		Chocolate 8	H	HM38	25				
				Cocoa butter	75				
Bunting, 1991 Fractionation from melted milkfat	Milkfat from chocolate base Intact AMF $30S_M$ fraction (HMF) $20S_M$ fraction (MMF) $20L_M$ fraction (LMF)	Chocolate 1 (5% AMF)		Intact AMF Milkfat from base Cocoa butter	4.1 16.6 79.3	Sugar Whole milk powder Chocolate liquor Cocoa butter Lecithin Added fat blend	47.61 18.88 13.76 11.55 0.32 7.88	A low-fat milk chocolate base was obtained from Guittard Chocolates Co. This base contained 26.5% milkfat from milk powder and was refined and conched at the plant. Additional fat at a level of 7.9% was added to this chocolate base to prepare a standard 32.5% milk chocolate.	Molten chocolate at 60°C was cooled with agitation to 25°C (T_1) until max. viscosity was reached. The chocolate was reheated to 33.3°C (T_2) until a viscosity reached a min. The chocolate was then reheated to 33°C (T_3) for further processing. Chocolates were molded and stored at 15°C < 50% relative humidity for 24 h prior to analysis. $T_1 = 25.0°C$ $T_2 = 33.0°C$ $T_3 = 32.0°C$ $T_1 = 25.0°C$ $T_2 = 33.0°C$ $T_3 = 32.0°C$ $T_1 = 25.0°C$ $T_2 = 33.0°C$ $T_3 = 32.0°C$
		Chocolate 2 (5% LMF)	L	$20L_M$ fraction Milkfat from base Cocoa butter	4.1 16.6 79.3				
		Chocolate 3 (5% MMF)	M	$20S_M$ fraction Milkfat from base Cocoa butter	4.1 16.6 79.3				

Chocolate 4 (5% HMF)	H	30S$_M$ fraction Milkfat from base Cocoa butter	4.1 16.6 79.3	$T_1 = 25.0°C$ $T_2 = 33.0°C$ $T_3 = 32.0°C$
Chocolate 5 (10% AMF)		Intact AMF Milkfat from base Cocoa butter	8.3 16.6 75.1	$T_1 = 25.0°C$ $T_2 = 33.0°C$ $T_3 = 32.0°C$
Chocolate 6 (10% LMF)	L	20L$_M$ fraction Milkfat from base Cocoa butter	8.3 16.6 75.1	$T_1 = 25.0°C$ $T_2 = 33.0°C$ $T_3 = 31.0°C$
Chocolate 7 (10% MMF)	M	20S$_M$ fraction Milkfat from base Cocoa butter	8.3 16.6 75.1	$T_1 = 25.0°C$ $T_2 = 33.0°C$ $T_3 = 31.0°C$
Chocolate 8 (10% HMF)	H	30S$_M$ fraction Milkfat from base Cocoa butter	8.3 16.6 75.1	$T_1 = 25.0°C$ $T_2 = 33.0°C$ $T_3 = 31.0°C$
Chocolate 9 (20% AMF)		Intact AMF Milkfat from base Cocoa butter	16.5 16.6 66.9	$T_1 = 24.5°C$ $T_2 = 31.0°C$ $T_3 = 31.0°C$
Chocolate 10 (20% LMF)	L	20L$_M$ fraction Milkfat from base Cocoa butter	16.5 16.6 66.9	$T_1 = 24.0°C$ $T_2 = 31.0°C$ $T_3 = 31.0°C$
Chocolate 11 (20% MMF)	M	20S$_M$ fraction Milkfat from base Cocoa butter	16.5 16.6 66.9	$T_1 = 24.5°C$ $T_2 = 30.5°C$ $T_3 = 31.0°C$
Chocolate 12 (20% HMF)	H	30S$_M$ fraction Milkfat from base Cocoa butter	16.5 16.6 66.9	untempered
Chocolate 13 (30% AMF)		Intact AMF Milkfat from base Cocoa butter	24.5 16.6 58.9	untempered
Chocolate 14 (30% LMF)	L	20L$_M$ fraction Milkfat from base Cocoa butter	24.5 16.6 58.9	untempered
Chocolate 15 (30% MMF)	M	20S$_M$ fraction Milkfat from base Cocoa butter	24.5 16.6 58.9	untempered
Chocolate 16 (30% HMF)	H	30S$_M$ fraction Milkfat from base Cocoa butter	24.5 16.6 58.9	

Continued

TABLE 8.12 (Continued)

Type of chocolate Author Milkfat fractionation method	Stock fractions employed[1]	Chocolate designation	Milkfat fraction category[2]	Fat phase blend		Chocolate formula		Method of chocolate production[3]	Chocolate tempering[3]
				Fraction	%	Ingredient	%		
Barna et al., 1992 Fractionation from melted milkfat	Milkfat from chocolate base Intact AMF $30S_M$ fraction (HMF) $20S_M$ fraction (MMF) $20L_M$ fraction (LMF)	Chocolate 1 (5% AMF)		Intact AMF Milkfat from base Cocoa butter	4.1 16.6 79.3	Sugar Whole milk powder Chocolate liquor Cocoa butter Lecithin Added fat blend	47.61 18.88 13.76 11.55 0.32 7.88	A low-fat milk chocolate base was obtained from Guittard Chocolates Co. This base contained 26.5% milkfat from milk powder and was refined and conched at the plant. Additional fat at a level of 7.9% was added to this chocolate base to prepare a standard 32.5% milk chocolate.	Molten chocolate at 60°C was cooled with agitation to 25°C (T_1) until maximum viscosity was reached. The chocolate was reheated to 33.3°C (T_2) until a viscosity reached a minimum. The chocolate was then reheated to 33°C (T_3) for further processing. Chocolates were moled and stored at 15°C < 50% relative humidity for 24 h prior to analysis.
		Chocolate 2 (5% LMF)	L	$20L_M$ fraction Milkfat from base Cocoa butter	4.1 16.6 79.3				T_1 = 25.0°C T_2 = 33.0°C T_3 = 32.0°C
		Chocolate 3 (5% MMF)	M	$20S_M$ fraction Milkfat from base Cocoa butter	4.1 16.6 79.3				T_1 = 25.0°C T_2 = 33.0°C T_3 = 32.0°C
		Chocolate 4 (5% HMF)	H	$30S_M$ fraction Milkfat from base Cocoa butter	4.1 16.6 79.3				T_1 = 25.0°C T_2 = 33.0°C T_3 = 32.0°C
		Chocolate 5 (10% AMF)		Intact AMF Milkfat from base Cocoa butter	8.3 16.6 75.1				T_1 = 25.0°C T_2 = 33.0°C T_3 = 32.0°C
		Chocolate 6 (10% LMF)	L	$20L_M$ fraction Milkfat from base Cocoa butter	8.3 16.6 75.1				T_1 = 25.0°C T_2 = 33.0°C T_3 = 31.0°C
		Chocolate 7 (10% MMF)	M	$20S_M$ fraction Milkfat from base Cocoa butter	8.3 16.6 75.1				T_1 = 25.0°C T_2 = 33.0°C T_3 = 31.0°C

Chocolate		Component	%	Tempering
Chocolate 8 (10% HMF)	H	$30S_M$ fraction	8.3	$T_1 = 25.0°C$
		Milkfat from base	16.6	$T_2 = 33.0°C$
		Cocoa butter	75.1	$T_3 = 31.0°C$
Chocolate 9 (20% AMF)		Intact AMF	16.5	$T_1 = 24.5°C$
		Milkfat from base	16.6	$T_2 = 31.0°C$
		Cocoa butter	66.9	$T_3 = 31.0°C$
Chocolate 10 (20% LMF)	L	$20L_M$ fraction	16.5	$T_1 = 24.0°C$
		Milkfat from base	16.6	$T_2 = 31.0°C$
		Cocoa butter	66.9	$T_3 = 31.0°C$
Chocolate 11 (20% MMF)	M	$20S_M$ fraction	16.5	$T_1 = 24.5°C$
		Milkfat from base	16.6	$T_2 = 30.5°C$
		Cocoa butter	66.9	$T_3 = 31.0°C$
Chocolate 12 (20% HMF)	H	$30S_M$ fraction	16.5	untempered
		Milkfat from base	16.6	
		Cocoa butter	66.9	
Chocolate 13 (30% AMF)		Intact AMF	24.5	untempered
		Milkfat from base	16.6	
		Cocoa butter	58.9	
Chocolate 14 (30% LMF)	L	$20L_M$ fraction	24.5	untempered
		Milkfat from base	16.6	
		Cocoa butter	58.9	
Chocolate 15 (30% MMF)	M	$20S_M$ fraction	24.5	untempered
		Milkfat from base	16.6	
		Cocoa butter	58.9	
Chocolate 16 (30% HMF)	H	$30S_M$ fraction	24.5	untempered
		Milkfat from base	16.6	
		Cocoa butter	58.9	

Type of chocolate not identified

Timms and Parekh, 1980

Fractionation from melted milkfat, vacuum filtration	$25S_S$ [S] fraction $25L_S$ [S] fraction	Chocolate 1	H	$25S_S$ fraction	10	—
				Cocoa butter	90	
		Chocolate 2	H	$25S_S$ fraction	20	
				Cocoa butter	80	
		Chocolate 3	H	$25S_S$ fraction	30	
				Cocoa butter	70	

Continued

TABLE 8.12 (Continued)

Type of chocolate Author Milkfat fractionation method	Stock fractions employed[1]	Chocolate designation	Milkfat fraction category[2]	Fat phase blend Fraction	%	Chocolate formula Ingredient	%	Method of chocolate production[3]	Chocolate tempering[3]
		Chocolate 4	H	25S$_S$ fraction	40				
				Cocoa butter	60				
		Chocolate 5	H	25S$_S$ fraction	50				
				Cocoa butter	50				
		Chocolate 6	L	25L$_S$ fraction	10				
				Cocoa butter	90				
		Chocolate 7	L	25L$_S$ fraction	20				
				Cocoa butter	80				
		Chocolate 8	L	25L$_S$ fraction	30				
				Cocoa butter	70				
		Chocolate 9	L	25L$_S$ fraction	40				
				Cocoa butter	60				
		Chocolate 10	L	25L$_S$ fraction	50				
				Cocoa butter	50				

[1]Physical and chemical characteristics of stock milkfat fractions are presented in Chapter 5.

[2]VH = very-high-melting milkfat fraction (MP > 45°C), H = high-melting milkfat fraction (MP between 35 and 45°C), M = middle-melting milkfat fraction (MP between 25 and 35°C), L = low-melting milkfat fraction (MP between 10 and 25°C), and U = unknown-melting milkfat fraction.

[3]Method and data presented are by the original author.

[4]No methods or data reported.

TABLE 8.13 Commercial Milkfat Ingredients for Use in Chocolate

Company	Milkfat ingredient	Melting behavior (°C)[1]	Milkfat fraction category[2]	Recommended uses	Packaging
Alaco	Confectionery butterfat 42	42 (MP)	H	Confectionery and chocolate applications	25 kg block
Aveve	Concentrated butter 4/4	26–28 (MP)	M	Chocolates	10 or 25 kg blocks
	Concentrated butter patissier	28–32	M	Chocolates	10 or 25 kg blocks
Corman	Concentrated butter creme au beurre	—[3]	U	Chocolate buttercream, buttercream filling for fine chocolates	1 kg blocks

[1] MP = melting point.

[2] H = high-melting milkfat fraction (MP between 35 and 45°C), M = middle-melting milkfat fraction (MP between 25 and 35°C), U = unknown-melting milkfat fraction.

[3] Data not available.

TABLE 8.14 Solid Fat Content of the Experimental Milkfat Ingredients Used in Chocolates

Type of chocolate Author	Milkfat fraction category[1]	Milkfat ingredient	Solid fat content (%)							
			5°C	10°C	15°C	20°C	25°C	30°C	35°C	40°C
Dark chocolate										
Yi, 1993										
		Cocoa butter control	84.0	79.7	75.3	72.3	63.0	35.7	0.0	—[2]
		Intact milkfat	47.8	38.2	29.5	16.7	7.0	3.7	0.7	0.0
	VH	25S_M fraction	78.3	77.2	76.2	74.6	74.7	71.5	64.4	59.7
	VH	20S_M fraction	67.4	64.3	61.1	59.8	60.9	54.2	48.9	46.4
	VH	15S_M fraction	77.4	83.8	70.1	64.3	61.3	59.6	49.8	36.7
	H	10S_M fraction	69.1	64.4	59.1	48.9	36.9	32.0	16.6	0.0
	M	5S_M fraction	85.9	82.0	77.8	70.2	50.3	21.1	0.0	0.0
	M	0S_M fraction	88.9	83.5	76.3	61.1	21.1	0.0	—	—
	L	0L_M fraction	0.0	—	—	—	—	—	—	—
Milk chocolate										
Jordan, 1986										
	H	HM41 fraction	—	55.7	46.2	36.9	34.4	27.4	21.6	—
	H	HM40 fraction	—	58.4	48.3	37.1	34.1	28.1	21.6	—
	H	HM38 fraction	—	46.7	35.2	24.3	21.6	16.1	9.1	—
Type of chocolate not identified										
Timms and Parekh, 1980										
	H	25S_S fraction	—	52.8	41.8	31.4	26.9	20.0	11.0	2.0
	L	25L_S fraction	—	41.9	24.6	7.6	2.2	0.9	0.0	0.0

[1] VH = very-high-melting milkfat fraction (MP > 45°C), H = high-melting milkfat fraction (MP between 35 and 45°C), M = middle-melting milkfat fraction (MP between 25 and 35°C), L = low-melting milkfat fraction (MP between 10 and 25°C).

[2] Data not available.

TABLE 8.15 Solid Fat Content of the Fat Phase Used in Experimental Chocolates

Type of chocolate Author	Chocolate designation	Milkfat fraction category[1]	Fat phase blend		5°C	10°C	15°C	20°C	25°C	30°C	35°C	40°C
Dark chocolate												
Yi, 1993			*Added Fat Blend*									
	Chocolate 1		Cocoa butter control	100	84.0	79.8	75.3	72.3	63.1	35.7	0.0	—[2]
	Chocolate 2		Intact milkfat	100	81.2	74.3	70.4	62.4	54.9	32.4	0.0	—
	Chocolate 3	VH	25S_M fraction	100	88.4	85.4	82.6	79.9	74.3	53.5	0.0	—
	Chocolate 4	VH	20S_M fraction	100	87.5	85.5	82.1	78.5	72.4	49.8	0.0	—
	Chocolate 5	VH	15S_M fraction	100	88.3	85.3	80.9	77.7	70.1	43.0	0.0	—
	Chocolate 6	H	10S_M fraction	100	71.8	66.7	61.8	59.1	47.8	19.1	0.0	—
	Chocolate 7	M	5S_M fraction	100	71.0	66.9	63.9	58.5	47.9	22.3	0.0	—
	Chocolate 8	M	0S_M fraction	100	71.0	69.0	62.9	59.2	45.4	22.6	0.0	—
	Chocolate 9	L	0L_M fraction	100	71.6	66.1	63.5	59.3	49.4	21.8	0.0	—
Milk chocolate												
Jordan, 1986												
	Chocolate 1		Butteroil	23	—	—	—	46.0	40.8	26.9	1.2	—
			Cocoa butter	77								
	Chocolate 2	H	HM41 fraction	23	—	—	—	51.1	42.2	26.9	2.5	—
			Cocoa butter	77								
	Chocolate 3	H	HM40 fraction	23	—	—	—	47.5	38.3	26.9	4.5	—
			Cocoa butter	77								
	Chocolate 4	H	HM38 fraction	23	—	—	—	46.0	35.8	20.8	1.6	—
			Cocoa butter	77								
	Chocolate 5		Butteroil	25	—	—	—	43.3	36.1	22.5	1.1	—
			Cocoa butter	75								
	Chocolate 6	H	HM41 fraction	25	—	—	—	49.1	39.0	27.2	3.2	—
			Cocoa butter	75								
	Chocolate 7	H	HM40 fraction	25	—	—	—	44.9	33.6	23.5	4.8	—
			Cocoa butter	75								
	Chocolate 8	H	HM38 fraction	25	—	—	—	43.8	32.5	18.8	1.9	—
			Cocoa butter	75								
Type of chocolate not identified												
Timms and Parekh, 1980												
	Chocolate 1	H	25S_S fraction	10	—	77.3	69.4	62.1	53.4	23.7	0.3	0.0
			Cocoa butter	90								
	Chocolate 2	H	25S_S fraction	20	—	73.4	62.6	53.0	45.0	18.8	0.7	0.0
			Cocoa butter	80								
	Chocolate 3	H	25S_S fraction	30	—	70.7	56.1	37.2	30.8	14.9	1.7	0.2
			Cocoa butter	70								
	Chocolate 4	H	25S_S fraction	40	—	68.2	51.4	24.7	17.6	9.2	2.8	0.4
			Cocoa butter	60								
	Chocolate 5	H	25S_S fraction	50	—	63.1	43.8	19.9	17.2	10.9	4.0	0.0
			Cocoa butter	50								

Continued

TABLE 8.15 (Continued)

Type of chocolate Author	Chocolate designation	Milkfat fraction category[1]	Fat phase blend		Solid fat content (%)							
					5°C	10°C	15°C	20°C	25°C	30°C	35°C	40°C
	Chocolate 6	L	$25L_S$ fraction Cocoa butter	10 90	—	77.1	71.0	63.1	52.8	20.7	0.3	0.0
	Chocolate 7	L	$25L_S$ fraction Cocoa butter	20 80	—	71.4	63.0	51.8	39.9	12.5	0.0	0.0
	Chocolate 8	L	$25L_S$ fraction Cocoa butter	30 70	—	64.1	54.9	40.6	28.5	7.4	0.0	0.0
	Chocolate 9	L	$25L_S$ fraction Cocoa butter	40 60	—	58.0	46.4	30.8	19.5	2.9	0.0	0.0
	Chocolate 10	L	$25L_S$ fraction Cocoa butter	50 50	—	51.1	26.4	4.7	3.0	1.0	0.0	0.0

[1]VH = very-high-melting milkfat fraction (MP > 45°C), H = high-melting milkfat fraction (MP between 35 and 45°C), M = middle-melting milkfat fraction (MP between 25 and 35°C), L = low-melting milkfat fraction (MP between 10 and 25°C).

[2]Data not available.

TABLE 8.16 Evaluations of Milkfat Ingredients in Experimental Chocolate

Type of chocolate Author Milkfat origin	Chocolate designation	Milkfat fraction category[1]	Fat phase blend		Bloom[2]	Gloss[2]	Snap[2]	Other properties[2]
			Fraction	%				

Dark chocolate

Jebson, 1974b
Fractionation method not reported

| | Chocolate 1 | U | Hard milkfat fraction | 3 | Samples stored at 22°C showed no bloom > 9 mo. Samples stored at 30°C showed some bloom > 9 mo, but less bloom was apparent on the samples made with hard fraction. Samples that were temperature cycled (16 h at 22°C, 8 h at 30°C) showed bloom after 6 wk storage, but samples made with hard fraction could be stored twice as long before bloom developed to same degree as control. | —[3] | — | — |
| | | | Cocoa butter | 97 | | | | |

Yi, 1993
Fractionation from acetone solution

			Added Fat Blend		Bloom rate (ΔWI/days of cycling)		Instron hardness (lbs/mm)	
	Chocolate 1		Cocoa butter control	100	0.54	—	3.9	—
	Chocolate 2	VH	Intact milkfat	100	0.45		3.0	
	Chocolate 3	VH	25S$_M$ fraction	100	0.07		4.2	
	Chocolate 4	VH	20S$_M$ fraction	100	0.03		3.6	
	Chocolate 5	VH	15S$_M$ fraction	100	0.05		4.0	
	Chocolate 6	H	10S$_M$ fraction	100	0.05		3.1	

Continued

TABLE 8.16 (Continued)

Type of chocolate Author Milkfat origin	Chocolate designation	Milkfat fraction category[1]	Fat phase blend Fraction	%	Bloom[2]	Gloss[2]	Snap[2]	Other properties[2]
	Chocolate 7	M	5S$_M$ fraction	100	0.78		3.3	
	Chocolate 8	M	0S$_M$ fraction	100	1.06		3.4	
	Chocolate 9	L	0L$_M$ fraction	100	1.06		2.8	
Milk chocolate								
Jordan, 1986						*Subjective assessment*	*Subjective assessment*	
Commercially obtained fractions from two Belgian companies	Chocolate 1		Butteroil	23	—	55	8	—
			Cocoa butter	77				
	Chocolate 2	H	HM41 fraction	23		45	6	
			Cocoa butter	77				
	Chocolate 3	H	HM40 fraction	23		45	5	
			Cocoa butter	77				
	Chocolate 4	H	HM38 fraction	23		55	6	
			Cocoa butter	77				
	Chocolate 5		Butteroil	25		55	4	
			Cocoa butter	75				
	Chocolate 6	H	HM41	25		15	3	
			Cocoa butter	75				
	Chocolate 7	H	HM40	25		15	3	
			Cocoa butter	75				
	Chocolate 8	H	HM38	25		25	4	
			Cocoa butter	75				
					Yield value (dynes/cm²)	*Gloss (% light reflected)*	*Instron hardness (force per inch thickness)[4]*	*Casson plastic viscosity (poise)[4]*
Bunting, 1991 Fractionation from melted milkfat	Control chocolate		Milkfat from base	16.6	—	29.6	14.1	20.4
			Cocoa butter	83.4				
	Chocolate 1 (5% AMF)		Intact AMF	4.1	131.5	32.1	14.3	18.0
			Milkfat from base	16.6				
			Cocoa butter	79.3				
	Chocolate 2 (5% LMF)	L	20L$_M$ fraction	4.1	161.1	21.3	10.7	14.7
			Milkfat from base	16.6				
			Cocoa butter	79.3				

Application of Milkfat Fractions in Foods 611

Sample		Component					
Chocolate 3 (5% MMF)	M	20S$_M$ fraction Milkfat from base Cocoa butter	4.1 16.6 79.3	116.7	—	9.9	16.3
Chocolate 4 (5% HMF)	H	30S$_M$ fraction Milkfat from base Cocoa butter	4.1 16.6 79.3	139.4	18.6	7.4	16.7
Chocolate 5 (10% AMF)		Intact AMF Milkfat from base Cocoa butter	8.3 16.6 75.1	118.3	16.5	9.7	13.7
Chocolate 6 (10% LMF)	L	20L$_M$ fraction Milkfat from base Cocoa butter	8.3 16.6 75.1	123.2	21.3	9.7	15.9
Chocolate 7 (10% MMF)	M	20S$_M$ fraction Milkfat from base Cocoa butter	8.3 16.6 75.1	130.5	13.2	7.6	16.6
Chocolate 8 (10% HMF)	H	30S$_M$ fraction Milkfat from base Cocoa butter	8.3 16.6 75.1	128.5	12.3	6.9	15.3
Chocolate 9 (20% AMF)		Intact AMF Milkfat from base Cocoa butter	16.5 16.6 66.9	116.6	—	9.4	14.9
Chocolate 10 (20% LMF)	L	20L$_M$ fraction Milkfat from base Cocoa butter	16.5 16.6 66.9	121.9	24.1	5.0	—
Chocolate 11 (20% MMF)	M	20S$_M$ fraction Milkfat from base Cocoa butter	16.5 16.6 66.9	129.3	—	7.9	—
Chocolate 12 (20% HMF)	H	30S$_M$ fraction Milkfat from base Cocoa butter	16.5 16.6 66.9	122.9	—	—	—
Chocolate 13 (30% AMF)		Intact AMF Milkfat from base Cocoa butter	24.5 16.6 58.9	131.9	—	—	13.1
Chocolate 14 (30% LMF)	L	20L$_M$ fraction Milkfat from base Cocoa butter	24.5 16.6 58.9	107.5	—	—	14.0
Chocolate 15 (30% MMF)	M	20S$_M$ fraction Milkfat from base Cocoa butter	24.5 16.6 58.9	111.3	—	—	15.9

Continued

TABLE 8.16 (Continued)

Type of chocolate Author Milkfat origin	Chocolate designation	Milkfat fraction category[1]	Fat phase blend Fraction	%	Bloom[2]	Gloss[2]	Snap[2]	Other properties[2]
	Chocolate 16 (30% HMF)	H	30S$_M$ fraction Milkfat from base Cocoa butter	24.5 16.6 58.9	110.4	—	—	15.6
Barna et al., 1992								*Casson plastic viscosity (poise)[4]*
Fractionation from melted milkfat	Control chocolate		Milkfat from base Cocoa butter	16.6 83.4	—	—	*Instron hardness (force per inch thickness)[4]* 14.1	20.4
	Chocolate 1 (5% AMF)		Intact AMF Milkfat from base Cocoa butter	4.1 16.6 79.3			14.3	18.0
	Chocolate 2 (5% LMF)	L	20L$_M$ fraction Milkfat from base Cocoa butter	4.1 16.6 79.3			10.7	14.7
	Chocolate 3 (5% MMF)	M	20S$_M$ fraction Milkfat from base Cocoa butter	4.1 16.6 79.3			9.9	16.3
	Chocolate 4 (5% HMF)	H	30S$_M$ fraction Milkfat from base Cocoa butter	4.1 16.6 79.3			7.4	16.7
	Chocolate 5 (10% AMF)		Intact AMF Milkfat from base Cocoa butter	8.3 16.6 75.1			9.7	13.7
	Chocolate 6 (10% LMF)	L	20L$_M$ fraction Milkfat from base Cocoa butter	8.3 16.6 75.1			9.7	15.9
	Chocolate 7 (10% MMF)	M	20S$_M$ fraction Milkfat from base Cocoa butter	8.3 16.6 75.1			7.6	16.6
	Chocolate 8 (10% HMF)	H	30S$_M$ fraction Milkfat from base Cocoa butter	8.3 16.6 75.1			6.9	15.3
	Chocolate 9 (20% AMF)		Intact AMF Milkfat from base Cocoa butter	16.5 16.6 66.9			9.4	14.9

Chocolate	Type	Fraction	Value	Value	Value
Chocolate 10 (20% LMF)	L	20L$_M$ fraction	16.5		
		Milkfat from base	16.6		5.0
		Cocoa butter	66.9		
Chocolate 11 (20% MMF)	M	20S$_M$ fraction	16.5		
		Milkfat from base	16.6		7.9
		Cocoa butter	66.9		
Chocolate 12 (20% HMF)	H	30S$_M$ fraction	16.5		
		Milkfat from base	16.6		—
		Cocoa butter	66.9		
Chocolate 13 (30% AMF)		Intact AMF	24.5	13.1	
		Milkfat from base	16.6		—
		Cocoa butter	58.9		
Chocolate 14 (30% LMF)	L	20L$_M$ fraction	24.5	14.0	
		Milkfat from base	16.6		—
		Cocoa butter	58.9		
Chocolate 15 (30% MMF)	M	20S$_M$ fraction	24.5	15.9	
		Milkfat from base	16.6		—
		Cocoa butter	58.9		
Chocolate 16 (30% HMF)	H	30S$_M$ fraction	24.5	15.6	
		Milkfat from base	16.6		—
		Cocoa butter	58.9		

[1]VH = very-high-melting milkfat fraction (MP > 45°C), H = high-melting milkfat fraction (MP between 35 and 45°C), M = middle-melting milkfat fraction (MP between 25 and 35°C), L = low-melting milkfat fraction (MP between 10 and 25°C), and U = unknown-melting milkfat fraction.

[2]Methods and data presented are those of the original author.

[3]No methods or data reported.

[4]Data are interpreted from original figure.

TABLE 8.17 Formulations and Processing Information for Milkfat Ingredients Used in Experimental Dairy Products

Dairy category / Author / Milkfat fractionation method	Stock milkfat fractions employed[1]	Product designation	Milkfat fraction category[2]	Milkfat fraction blend — Fraction	%	Finished product blend — Ingredient	%	Method of production[3]
Recombined dairy products								
Tucker, 1974 Fractionation from melted milkfat	Hard milkfat fraction	Recombined butter	H	Hard milkfat fraction	100	Hard milkfat fraction / Other ingredients	unknown / unknown	Low-pressure homogenization prior to pasteurization was used to incorporate the hard fat in recombined milk or cream for product preparation. Specific product preparation methods not reported.
Mayhill and Newstead, 1992 Fractionation method not reported	Intact AMF / Soft milkfat fraction / Hard milkfat fraction	Recombined UHT–milk 1		Intact AMF	100	Intact AMF / Nonfat milk solids / Water	3.2 / 8.5 / 88.3	Skim milk powder was reconstituted in water at 55°C. The emulsifier was dissolved in the melted fat and the fat added to the mixing tank and agitated for 10 min prior to homogenization (50–55°C, 200 bar). Two-stage homogenizers were used, but the pressure was applied on the second stage only. The milk was cooled to 18–20°C by a plate heat exchanger and held at 4°C overnight before UHT processing. The milks were UHT processed in an Alfa-Laval type-D UHT pilot plant. Preheat temperature = 75°C, UHT = 145°C for 5 s, and homogenization at 75°C, 250 bar stage and 50 bar second stage. Milks were cooled to 25°C for aseptic filling.
		Recombined UHT–milk 2	L	Soft milkfat fraction	100	Soft milkfat fraction / Nonfat milk solids / Water	3.2 / 8.5 / 88.3	
		Recombined UHT–milk 3	VH	Hard milkfat fraction	100	Hard milkfat fraction / Nonfat milk solids / Water	3.2 / 8.5 / 88.3	
		Recombined UHT–milk 4		Intact AMF	100	Intact AMF / Nonfat milk solids / Span 65 (sorbitan tristearate) / Water	3.2 / 8.5 / 0.1 / 88.2	
		Recombined UHT–milk 5	L	Soft milkfat fraction	100	Soft milkfat fraction / Nonfat milk solids / Span 65 (sorbitan tristearate) / Water	3.2 / 8.5 / 0.1 / 88.2	
		Recombined UHT–milk 6	VH	Hard milkfat fraction	100	Hard milkfat fraction / Nonfat milk solids / Span 65 (sorbitan tristearate) / Water	3.2 / 8.5 / 0.1 / 88.2	

Recombined UHT–milk 7		Intact AMF	100	Intact AMF 3.2
				Nonfat milk solids 8.5
				Span 40 (sorbitan monopalmitate) 0.1
				Water 88.2
Recombined UHT–milk 8	L	Soft milkfat fraction	100	Soft milkfat fraction 3.2
				Nonfat milk solids 8.5
				Span 40 (sorbitan monopalmitate) 0.1
				Water 88.2
Recombined UHT–milk 9	VH	Hard milkfat fraction	100	Hard milkfat fraction 3.2
				Nonfat milk solids 8.5
				Span 40 (sorbitan monopalmitate) 0.1
				Water 88.2
Recombined UHT–milk 10		Intact AMF	100	Intact AMF 3.2
				Nonfat milk solids 8.5
				Tween 21 (polyoxyethylene-4-sorbitan monolaurate) 0.1
				Water 88.2
Recombined UHT–milk 11	L	Soft milkfat fraction	100	Soft milkfat fraction 3.2
				Nonfat milk solids 8.5
				Tween 21 (polyoxyethylene-4-sorbitan monolaurate) 0.1
				Water 88.2
Recombined UHT–milk 12	VH	Hard milkfat fraction	100	Soft milkfat fraction 3.2
				Nonfat milk solids 8.5
				Tween 21 (polyoxyethylene-4-sorbitan monolaurate) 0.1
				Water 88.2
Recombined UHT–milk 13		Intact AMF	100	Intact AMF 3.2
				Nonfat milk solids 8.5
				Hymono 8903 (90% monoglycerides) 0.1
				Water 88.2

Continued

TABLE 8.17 (Continued)

Dairy category / Author / Milkfat fractionation method	Stock milkfat fractions employed[1]	Product designation	Milkfat fraction category[2]	Milkfat fraction blend — Fraction	%	Finished product blend — Ingredient	%	Method of production[3]
		Recombined UHT–milk 14	L	Soft milkfat fraction	100	Soft milkfat fraction Nonfat milk solids Hymono 8903 (90% monoglycerides) Water	3.2 8.5 0.1 88.2	Skim milk was heated to 145°F, and butter oil was added to give a final dried product containing approximately 25% fat, and the mixture was homogenized at 3000 psi. The mixture was evaporated in a vacuum pan to 42% solids and then dried in a Scott spray drier: inlet temperature = 318–330°F, outlet temperature = 160–170°F.
		Recombined UHT–milk 15	VH	Hard milkfat fraction	100	Soft milkfat fraction Nonfat milk solids Hymono 8903 (90% monoglycerides) Water	3.2 8.5 0.1 88.2	
Dairy powders								
Baker et al., 1959 Fractionation from melted milk and from acetone solution	Intact AMF $5S_S$ [S] fraction from acetone solution $22S_M$ [W] fraction from melted milkfat $28S_S$ [W] fraction from melted milkfat	Milk powder 1	L	$5S_S$ [W] fraction	100	—[4]		
		Milk powder 2	L	$22S_M$ [W] fraction	100			
		Milk powder 3	M	$28S_S$ [W] fraction	100			
		Milk powder 4		Intact milkfat	100			
Black and Dixon, 1974 Fractionation from melted milkfat	Hard milkfat fraction	Milk powder 1	U	Hard milkfat fraction Intact milkfat in whole milk	50	Hard milkfat fraction reconstituted to whole milk Fresh whole milk	50 50	The two milks were blended, homogenized, and dried in a pilot-scale spray drier
		Butter powder 1	U	Hard milkfat fraction	100	Hard milkfat fraction Sodium caseinate Glycerol monostearate Sodium citrate Water	unknown unknown unknown unknown unknown	The ingredients were mixed hot, homogenized, and dried in a pilot-scale dryer and cooled in a fluid bed cooler.

Malkamäki and Anila, 1974						
Fractionation from melted milkfat	Hard milkfat fraction	Milk powder 1	U	Hard milkfat fraction	100	Hard milkfat fraction Skim milk	unknown unknown	The milkfat fraction was homogenized in skim milk, and the milk powder was prepared using ordinary technology.
Tucker, 1974						
Fractionation from melted milkfat	Hard milkfat fraction	Milk powder 1	H	Hard milkfat fraction	100	Hard milkfat fraction Other ingredients	unknown unknown	Low pressure homogenization prior to pasteurization was used to incorporate the hard fat in recombined milk or cream for product preparation. Specific product preparation methods not reported.
Whipping cream						
Tucker, 1974						
Fractionation from melted milkfat	Hard milkfat fraction	Whipping cream 1	H	Hard milkfat fraction	100	Hard milkfat fraction Other ingredients	unknown unknown	Low pressure homogenization prior to pasteurization was used to incorporate the hard fat in recombined milk or cream for product preparation. Specific product preparation methods not reported.
Bratland, 1983						
Fractionation from melted milkfat	28S$_S$ fraction 25S$_S$ fraction 24S$_S$ fraction 20S$_S$ fraction	Whipping cream 1	U	20S$_S$ fraction	100	20S$_S$ fraction Serum	unknown unknown	The milkfat was emulsified with sufficient serum from the phase inversion process to produce a 35% fat cream, the pasteurized, packed, and held for 24 h at 4°C.
		Whipping cream 2	U	24S$_S$ fraction	100	24S$_S$ fraction Serum	unknown unknown	Milkfat was emulsified with serum from the phase inversion process to produce a 22% fat cream, then homogenized (single-stage, 110 kg/cm^2), pasteurized (90°C, 15 s) and further homogenized (single-stage, 40 kg/cm^2), cooled, and held for 24 h at 4°C.

Continued

TABLE 8.17 (Continued)

Dairy category Author Milkfat fractionation method	Stock milkfat fractions employed[1]	Product designation	Milkfat fraction category[2]	Milkfat fraction blend		Finished product blend		Method of production[3]
				Fraction	%	Ingredient	%	
		Whipping cream 3	U	24S$_S$ fraction	100	24S$_S$ fraction Serum	unknown unknown	Milkfat was emulsified with serum from the phase inversion process to produce a 20% fat cream, then homogenized (single-stage, 130 kg/cm^2), UHT sterilized (142°C, 5 s) and further homogenized (single-stage, 45 kg/cm^2), aseptically packaged and held for 24 h at 4°C.
		Whipping cream 4	U	24S$_S$ fraction	100	24S$_S$ fraction Coconut oil (1:1 milkfat:coconut oil) Serum	unknown unknown unknown	Milkfat and coconut oil were emulsified with serum from the phase inversion process to produce a 25% fat cream, then homogenized (single-stage, 110 kg/cm^2), pasteurized (90°C, 15 s) and further homogenized (single-stage, 40 kg/cm^2), cooled and held for 24 h at 4°C.
		Whipping cream 5	U	25S$_S$ fraction	100	25S$_S$ fraction Buttermilk powder Water	20.0 8.7 71.3	The milkfat was emulsified with water and buttermilk powder at 55°C to a cream of 20% fat and 7.8% nonfat milk solids. The pH was adjusted to 6.64–6.46 and the cream was homogenized (300 kg/cm^2), pasteurized (109°C), and then homogenized (40 kg/cm^2). The cream was slowly stirred and cooled to 4°C, and then heated to 28°C and cooled again to 4°C.

Whipping cream 6	U	28S$_S$ fraction	100	28S$_S$ fraction Buttermilk powder Water	11.4 11.2 77.4	The milkfat was emulsified with water and buttermilk powder at 55°C to a cream of 12% fat and 10% nonfat milk solids. The pH was adjusted to 6.35, and the cream was homogenized (60 kg/cm²), pasteurized (120°C). The cream was slowly stirred and cooled to 4°C, and then heated to 28°C and cooled again to 4°C.
Whipping cream 7	U	28S$_S$ fraction	100	28S$_S$ fraction Buttermilk powder Water	14.0 10.4 75.6	The milkfat was emulsified with water and buttermilk powder at 55°C to a cream of 15% fat and 9.3% nonfat milk solids. The pH was adjusted to 6.20–6.25 with lactic acid, and the cream was homogenized (424 kg/cm²), pasteurized (265–270°C), and homogenized (60 kg/cm²). The cream was slowly stirred and cooled to 4°C, and then heated to 28°C and cooled again to 4°C.
Whipping cream 8	U	28S$_S$ fraction	100	28S$_S$ fraction Buttermilk powder Water Sugar	14.4 10.0 55.1 20.5	The milkfat was emulsified with water, buttermilk powder and sugar at 55°C. The pH was adjusted to 6.20–6.25 with lactic acid and the cream was homogenized (424 kg/cm²), pasteurized (265–270°C), and homogenized (60 kg/cm²). The cream was slowly stirred and cooled to 4°C, and then heated to 32°C and cooled again to 4°C.

Continued

TABLE 8.17 (Continued)

Dairy category Author Milkfat fractionation method	Stock milkfat fractions employed[1]	Product designation	Milkfat fraction category[2]	Milkfat fraction blend		Finished product blend		Method of production[3]
				Fraction	%	Ingredient	%	

Cheese

Malkamäki and Antila, 1974
Fractionation from melted milkfat

| | Hard milkfat fraction | Edam cheese | U | Hard milkfat fraction | 100 | Hard milkfat fraction
Skim milk
Other ingredients | unknown
unknown
unknown | The hard milkfat fraction was homogenized with skim milk, and the Edam cheese was prepared with ordinary technology. |

Tucker, 1974
Fractionation from melted milkfat

| | Hard milkfat fraction | Colby cheese | H | Hard milkfat fraction | 100 | Hard milkfat fraction
Other ingredients | unknown
unknown | Low-pressure homogenization prior to pasteurization was used to incorporate the hard fat in recombined milk or cream for product preparation. Specific product preparation methods not reported. |
| | | Cheddar cheese | H | Hard milkfat fraction | 100 | Hard milkfat fraction
Other ingredients | unknown
unknown | |

Thomas, 1973a
Fractionation from melted milkfat

	$20S_S$ fraction Intact AMF	Processed cheese 1 (control)				Low-fat cheese Full-fat cheese Old cheese Water Salt	0 unknown 30 unknown unknown	The milkfat fractions or intact milkfat were added to the low-fat cheese to a normal fat level. Milkfat, water and cooking salt were added so that the blends approximated normal composition. Processed cheese were prepared in the manner of Thomas (1970).
		Processed cheese 2	VH	$20S_S$ fraction	100	Low-fat cheese Full-fat cheese Old cheese Water Salt	5 unknown 30 unknown unknown	
		Processed cheese 3	VH	$20S_S$ fraction	100	Low-fat cheese Full-fat cheese Old cheese Water Salt	10 unknown 30 unknown unknown	

Application of Milkfat Fractions in Foods   621

Processed cheese 4	VH	20S$_S$ fraction	100	Low-fat cheese	20
				Full-fat cheese	unknown
				Old cheese	30
				Water	unknown
				Salt	unknown
Processed cheese 5	VH	20S$_S$ fraction	100	Low-fat cheese	30
				Full-fat cheese	unknown
				Old cheese	30
				Water	unknown
				Salt	unknown
Processed cheese 6	VH	20S$_S$ fraction	100	Low-fat cheese	40
				Full-fat cheese	unknown
				Old cheese	30
				Water	unknown
				Salt	unknown
Processed cheese 7		Intact AMF		Low-fat cheese	20
				Full-fat cheese	unknown
				Old cheese	30
				Water	unknown
				Salt	unknown

Thomas, 1973b
Fractionation from
melted milkfat   20S$_S$ fraction

Cheese spread 1 (control)		Butter		Mature cheese (9 mo)	30.0
				Young cheese (2 mo)	10.0
				Butter	11.6
				Skim milk powder	12.0
				Emulsifying salt	3.0
				Condensate	5.0
				Added water	28.4
Cheese spread 2 (50% substitution)	VH	20S$_S$ fraction		Mature cheese (9 mo)	30.0
				Young cheese (2 mo)	10.0
				Butter	5.8
				20S$_S$ fraction	4.7
				Skim milk powder	12.0
				Emulsifying salt	3.0
				Cooking salt	0.1
				Condensate	5.0
				Added water	29.4

Cheese spreads were prepared by the method of Thomas (1970). The ingredients were heated to 88°C for 7 min. Two thirds of the water was added initially and the other third after 5 min of processing. The cheese melt was poured into 340 g plastic cups and transferred to a cold room at 5°C.

Continued

TABLE 8.17 (Continued)

Dairy category / Author / Milkfat fractionation method	Stock milkfat fractions employed[1]	Product designation	Milkfat fraction category[2]	Milkfat fraction blend — Fraction	%	Finished product blend — Ingredient	%	Method of production[3]
		Cheese spread 3 (100% substitution)	VH	$20S_S$ fraction		Mature cheese (9 mo)	30.0	
						Young cheese (2 mo)	10.0	
						$20S_S$ fraction	9.5	
						Skim milk powder	12.1	
						Emulsifying salt	3.0	
						Cooking salt	0.2	
						Condensate	5.0	
						Added water	30.3	
Ice cream								
Goff et al., 1988								
Fractionation from melted milkfat	Hard milkfat fraction	Ice cream 1	U	Hard milkfat fraction	100	Hard milkfat fraction	unknown	Ice cream mixes of standard composition were prepared and pasteurized, homogenized (17.2/3/2 MPa), cooled to 4°C, and aged 24 h. The mixes were frozen to −5°C in a batch freezer for 15 min.
	Soft milkfat fraction					Other ingredients	unknown	
		Ice cream 2	U	Soft milkfat fraction	100	Soft milkfat fraction	unknown	
						Other ingredients	unknown	

[1]Physical and chemical characteristics of stock milkfat fractions are presented in Chapter 5.

[2]VH = very-high-melting milkfat fraction (MP > 45°C), H = high-melting milkfat fraction (MP between 35 and 45°C), M = middle-melting milkfat fraction (MP between 25 and 35°C), L = low-melting milkfat fraction (MP between 10 and 25°C), and U = unknown-melting milkfat fraction.

[3]Methods and data presented are those of the original author.

[4]Data not available.

TABLE 8.18 Information for Commercial Milkfat Ingredients for Dairy Products

Company	Milkfat ingredient	Melting behavior (°C)[1]	Milkfat fraction category[2]	Recommended uses	Packaging	Comments
Alaco						
	Soft butteroil 21	21 (MP)	L	Recombined dairy products Ice cream	200 kg drum 950 kg Optipak	
	Soft butteroil 28	28	M	Recombined dairy products Ice cream	25 kg block 200 kg drum 950 kg Optipak	
Aveve						
	Anhydrous milkfat	—[3]	U	Recombined dairy products Ice cream	1 lb, 2 lb, 2 kg, 5 lb 5 kg and 15 kg tins 100-200 kg drums	
	Concentrated butter Formula B	28–32 (MP)	M	Ice cream	20 or 25 kg blocks	
	Concentrated butter creme au beurre	28	M	Recombined cream	10 or 25 kg blocks	
Corman						
	Standard anhydrous milkfat	32 (MP)	M	Recombined dairy products Ice cream	17 kg tins 25 kg blocks 200 kg drums Tankers	
	Extra white anhydrous milkfat	32	M	Goat, Feta, Kefalotyri, and Regatello cheeses	17 kg tins 25 kg blocks 200 kg drums Tankers	White color is suitable for white cheeses
	Milk extra anhydrous milkfat	28	M	Recombined dairy products Ice cream	17 kg tins 25 kg blocks 200 kg drums Tankers	
	Glacier concentrated butter	30–32	M	Ice cream	10 and 25 kg blocks Tankers	
	Glacier extra concentrated butter	30–32	M	Ice cream	25 kg blocks Tankers	Has neutral flavor

[1]MP = melting point.

[2]M = middle-melting milkfat fraction (MP between 25 and 35°C), L = low-melting milkfat fractions (MP between 10 and 25°C), U = unknown-melting milkfat fraction.

[3]Data not available.

TABLE 8.19 Solid Fat Content Profiles of Commercial Milkfat Ingredients for Use in Dairy Products

Company	Milkfat ingredient	Milkfat fraction category[1]	Solid fat content (%)						
			5°C	10°C	15°C	20°C	25°C	30°C	35°C
Corman									
	Standard anhydrous milkfat	M	49	39	28	16	9	5	0
	Extra white anhydrous milkfat	M	49	39	28	16	9	5	0
	Milk extra anhydrous milkfat	M	46	37	26	13	4	0	0
	Glacier concentrated butter	M	48	38	25	13	7	3	0
	Glacier extra concentrated butter	M	48	38	25	13	7	3	0

[1]M = middle-melting milkfat fraction (MP between 25 and 35°C).

TABLE 8.20 Evaluation of Experimental Milkfat Ingredients in Dairy Products

Dairy category Author	Product designation	Milkfat fraction category[1]	Selected evaluations[2]					
				Stability index[3,4]				
			1 mo	2 mo	3 mo	4 mo	5 mo	6 mo
Recombined dairy products								
Mayhill and Newstead, 1992								
	Recombined UHT–milk 1		22°C = 0.81	0.67	0.63	0.68	0.66	0.67
			30°C = 0.80	0.62	0.67	0.61	0.61	0.56
			40°C = 0.74	0.60	0.63	0.50	0.55	–[5]
	Recombined UHT–milk 2	L	22°C = 0.86	0.68	0.65	0.61	0.57	0.56
			30°C = 0.79	0.62	0.61	0.56	0.53	0.51
			40°C = 0.75	0.57	0.55	0.53	0.50	0.51
	Recombined UHT–milk 3	VH	22°C = 0.90	0.80	0.77	0.72	0.63	0.75
			30°C = 0.86	0.69	0.63	0.62	0.73	0.61
			40°C = 0.81	0.67	0.66	0.52	–	–
	Recombined UHT–milk 4		22°C = 0.87	0.72	0.67	0.63	0.64	0.61
			30°C = 0.81	0.64	0.66	0.57	0.64	0.63
			40°C = 0.75	0.61	0.58	0.55	0.54	0.50
	Recombined UHT–milk 5	L	22°C = 0.85	0.72	0.71	0.66	0.67	0.62
			30°C = 0.81	0.63	0.67	0.60	0.62	0.53
			40°C = 0.75	0.62	0.60	0.55	0.54	0.51
	Recombined UHT–milk 6	VH	22°C = 0.93	0.85	0.81	0.76	0.73	0.80
			30°C = 0.90	0.73	0.74	0.67	0.70	0.65
			40°C = 0.82	0.63	0.65	0.55	0.57	0.51
	Recombined UHT–milk 7		22°C = 0.88	0.72	0.73	0.65	0.62	0.62
			30°C = 0.82	0.71	0.68	0.60	0.62	0.61
			40°C = 0.78	0.62	0.67	0.55	0.62	0.57
	Recombined UHT–milk 8	L	22°C = 0.87	0.75	0.75	0.66	0.65	0.64
			30°C = 0.86	0.71	0.75	0.62	0.62	0.57
			40°C = 0.80	0.67	0.68	0.65	0.61	0.55
	Recombined UHT–milk 9	VH	22°C = 0.92	0.79	0.77	0.72	0.65	0.73
			30°C = 0.87	0.70	0.69	0.60	0.62	0.58
			40°C = 0.81	0.67	0.65	0.50	0.50	–
	Recombined UHT–milk 10		22°C = 0.89	0.80	0.75	0.77	0.67	0.69
			30°C = 0.86	0.73	0.75	0.66	0.68	0.71
			40°C = 0.80	0.75	0.62	0.61	0.58	0.57

Continued

TABLE 8.20 (Continued)

Dairy category Author	Product designation	Milkfat fraction category[1]	Selected evaluations[2]			
	Recombined UHT-milk 11	L	22°C = 0.90	0.78	0.75	0.72	0.72	0.68
			30°C = 0.88	0.73	0.73	0.67	0.62	0.57
			40°C = 0.82	0.67	0.68	0.65	0.57	0.57
	Recombined UHT-milk 12	VH	22°C = 0.92	0.82	0.81	0.78	0.78	0.77
			30°C = 0.87	0.72	0.65	0.63	0.68	0.60
			40°C = 0.84	0.61	0.67	0.58	0.53	0.55
	Recombined UHT-milk 13	L	22°C = 0.85	0.72	0.75	0.58	0.52	0.50
			30°C = 0.78	0.63	0.71	0.50	0.55	—
			40°C = 0.73	0.56	0.56	—	0.51	—
	Recombined UHT-milk 14	L	22°C = 0.86	0.68	0.67	0.57	0.55	0.56
			30°C = 0.81	0.62	0.65	0.54	0.58	—
			40°C = 0.76	0.59	0.58	0.60	0.59	—
	Recombined UHT-milk 15	VH	22°C = 0.92	0.78	0.70	0.71	0.63	0.63
			30°C = 0.88	0.72	0.72	0.58	0.67	0.60
			40°C = 0.80	0.55	0.54	—	—	—

Dairy powders

			Wettability (%) Ungraded powder 25°C	Particle size 0.5–1.0 mm 24°C	Dispersibility (%) Ungraded powder 25°C	Particle size 0.50–1.0 mm 24°C
Baker et al., 1959	Milk powder 1	L	87	88	80	80
	Milk powder 2	L	60	62	63	45
	Milk powder 3	M	40	43	46	37
	Milk powder 4		26	20	54	20
	Commercial skim milk powder		51	—	82	—
	Commercial instant skim milk powder		89	90	16	81
	Commercial whole milk powder		13	—	4.5	—
	Commercial agglomerated whole milk powder		2.9	4.7		4.9
Black and Dixon, 1974	Milk powder 1	U	Experimental product was similar to control. Flowability was improved, although desirable flavor and aroma were lacking.			

Tucker, 1974	Butter powder 1	U	Experimental product was similar to control. Baking trials indicated a lack of flavor.
	Milk powder 1	H	Whole milk powders were similar to controls, except there was a slight loss of flavor that was evident with increased proportions of hard milkfat fraction.
Whipping cream			
Tucker, 1974	Whipping cream 1	H	Whiteness increased and physical characterstics were maintained, but flavor decreased with increased proportions of hard milkfat fraction.
Bratland, 1983	Whipping cream 1	U	This cream gave a good stable whip with a volume increase of approximately 100% at 12°C.
	Whipping cream 2	U	This cream gave a firm and stable whip with a volume increase of 140–160% at 10°C. This product was suitable for piping and decoration.
	Whipping cream 3	U	This cream produced a very stable whip with a volume increase of 140–160% at 12°C. This product was suitable for piping and decoration.
	Whipping cream 4	U	This cream gave a firm and stable whip with a volume increase of 130–150% at 10°C.
	Whipping cream 5	U	On whipping for 107 s at a starting temperature = 9.4°C and a final temperature = 14.4°C, the tempered cream showed a stable overrun of 121% and the whipped product was firm and of excellent taste.
	Whipping cream 6	U	The tempered cream when whipped for 210 s gave a stable, firm product of good taste with an overrun of 174%.
	Whipping cream 7	U	Whipping characteristics of this product were in line with others.
	Whipping cream 8	U	Whipping characteristics of this product were in line with the others, the sugar content imparting good freeze–thaw stability.
Cheese			
Malkamäki and Antila, 1974	Edam cheese	U	Promising results were obtained with the use of hard milkfat fraction in Edam cheese.

Continued

TABLE 8.20 (Continued)

Dairy category Author	Product designation	Milkfat fraction category[1]	Selected evaluations[2]		
Tucker, 1974	Colby cheese	H	Colby cheese with < 37% moisture and > 50% in dry matter exhibited equal grade quality and comparable acceptability to control cheese after 6 mo storage at 15°C.		
	Cheddar cheese	H	Evaluation not reported.		
Thomas, 1973a	Processed cheese 1 (control)	VH	No change in body or texture was observed for samples at 35 or 45°C.		
	Processed cheese 2	VH	None of the samples showed any oil separation or sweating, indicating that the fat was completely emulsified and stable.		
	Processed cheese 3	VH			
	Processed cheese 4	VH	No significant differences were observed in color, appearance, flavor or texture by sensory panelists.		
	Processed cheese 5	VH			
	Processed cheese 6	VH			
	Processed cheese 7	VH			
Thomas, 1973b			Firmness (Penetration mm)	Stability tests	Sensory evaluation
	Cheese spread 1 (control)	VH	23.5	None of the samples showed any fat separation, indicating that the emulsion was stable.	There were no significant differences between the spreads for color, flavor, mouthfeel, or spreadability.
	Cheese spread 2 (50% substitution)	VH	19.4		
	Cheese spread 3 (100% substitution)	VH	16.1		

[1]VH = very-high-melting milkfat fraction (MP > 45°C), H = high-melting milkfat fraction (MP between 35 and 45°C), M = middle-melting milkfat fraction (MP between 25 and 35°C), L = low-melting milkfat fraction (MP between 10 and 25°C), and U = unknown-melting milkfat fraction.

[2]Methods and data presented are described by the original author.

[3]Data estimated from graphs provided by the original author.

[4]Stability index: 1.0–0.85 = no observed cream layer, 0.85–0.70 = discernable pale cream layer, 0.70–0.65 = cream layer with cream adhesion to container walls, <0.65 = thick yellowing cream layer, with cream clumps or surface skin.

[5]Data not available.

TABLE 8.21 Evaluation of Milkfat Fractions in Deep-Fat Frying

Author Fractionation method	Stock milkfat fractions [1]	Product designation	Milkfat fraction category [2]	Finished product Ingredient	%	Method of frying [3]	Selected evaluations [3]					
Baker, 1970b Fractionation method not reported	Liquid milkfat fraction	Frying milkfat 1	U	— [4]	—		Liquid milkfat fraction had the advantage of remaining liquid when cool. Deep fried products were not greasy. Gradual decolorization and limited resistance to oxidation was observed.					
Malkamäki and Antila, 1974 Fractionation from melted milkfat	Hard milkfat fraction	Frying milkfat 1	U	—	—		The hard milkfat fraction was suitable for frying doughnuts at 160–180°C.					
Keogh, 1989 Fractionation from melted milkfat	Soft milkfat fraction	Butter oil frying fat	L	Butter oil Silicone	100 2 mg/kg	2.5 L of each oil was placed in separate Teflon-coated deep fat fryers. The temperature of each oil was brought to 180 ± 7°C, and maintained for 7 h each day for 4 consecutive days. A commercial brand of deep-frozen potato chips was used.	*Smoke point (°C)* Day 0 = 187.0 Day 4 = 171.5	*Absorbance* 2.03 0.15	*Viscosity* 34.9 41.5	*Peroxide value* 0.20 3.50	*TBA value* 0.58 1.70	*FFA value* 0.25 0.51
		Soft butter oil fraction frying fat	L	Soft butter oil fraction Silicone	100 2 mg/kg		Day 0 = 187.0 Day 4 = 171.5	2.09 0.15	34.7 36.8	0.20 3.50	0.24 1.31	0.14 0.66
		Corn oil frying fat		Corn oil Silicone	100 2 mg/kg		Day 0 = 220.0 Day 4 = 205.5	0.15 0.18	33.2 33.8	0.20 2.80	0.29 0.69	0.06 0.17

[1] Physical and chemical characteristics of stock milkfat fractions was presented in Chapter 5.

[2] L = low-melting milkfat fraction (MP between 10 and 25°C).

[3] Methods presented are those of the original author.

[4] Data not available.

Chapter 9

Future Directions for Milkfat Fractionation

Over 850 milkfat fractions have been reported in the literature and have been produced using a variety of fractionation methods and processing conditions. These fractions exhibit a wide range of chemical and physical properties, and consequently, they exhibit a wide range of functionalities. Thus, milkfat is a versatile starting material for the production of specialty fats that are tailored for specific applications.

Over two-thirds of the milkfat fractions reported in the literature have been characterized for melting behavior. Milkfat fractions have been categorized based on melting behavior to facilitate the discussion of the functionality of milkfat and milkfat fractions in food products. Melting categories for milkfat fractions include

- Very-high-melting milkfat fractions (MP > 45°C)
- High-melting milkfat fractions (MP between 35 and 45°C)
- Middle-melting milkfat fractions (MP between 25 and 35°C)
- Low-melting milkfat fractions (MP between 10 and 25°C)
- Very-low-melting milkfat fractions (MP < 10°C)
- Unknown-melting milkfat fractions (MP data not available)

Generalizations of the physical and chemical characteristics of milkfat fractions provide a perspective of the functionality of milkfat fractions in foods, but specific characteristics of individual milkfat fractions within the categories can vary.

Milkfat Ingredients

Milkfat ingredients possess a desirable flavor and are perceived by consumers as premium, natural ingredients. Milkfat exhibits a unique flavor functionality that cannot be mimicked by other fats or oils. Milkfat flavor is an expression of the chemical functionality of milkfat and involves free flavor compounds and precursor compounds that are bound to the triglycerides and are released on heating.

The functionalities of milkfat ingredients are often viewed in terms of physical properties (crystallization and melting behaviors) that depend on the chemical composition (fatty acid and triglyceride compositions) of the milkfat. The chemical composition and physical properties of milkfat can be influenced by many factors during milkfat biosynthesis and processing. These factors include feed and stage of lactation during milkfat synthesis as well as processing conditions employed during fractionation and texturization, such as fractionation and separation methods, number of fractionation steps, cooling rate, and agitation rate.

The functionality of milkfat ingredients in food products involves interactions with other food components (protein, carbohydrates, water, air, carbon dioxide), and these interactions are influenced by the physical states (solid, liquid, gas) of these components. Interactions between milkfat and other food components have been broadly categorized as milkfat–milkfat, milkfat–other fat, milkfat–(nonfat) solid, milkfat–water, and milkfat-gas interactions. The functionalities required of milkfat ingredients and the food component interactions encountered depend on the application.

Milkfat functionality can be easily modified through the use of fractionation, selective blending, and appropriate texturization. These processes generally do not involve additives, and thus milkfat retains a natural, premium image. Blending of fractions gives a manufacturer greater flexibility to tailor milkfat ingredients to specific functional requirements than could be accomplished with fractionation alone. Blending of milkfat fractions allows some flexibility to overcome normal compositional changes in the milkfat supply and to compensate for any other processing changes that may change the functionality of a stock fraction. Therefore, it is relatively easy to provide a consistent end product to the customer.

Applications for Specialty Milkfat Ingredients

Current dietary trends towards low-fat and low-cholesterol foods have created some limitations relative to the usage of milkfat in some food products. However, current trends also suggest an increase in the consumption of indulgence foods, such as premium baked goods, confections, and frozen desserts. As a result, these trends have created many opportunities for specialized milkfat ingredients.

Milkfat ingredients can impart desirable flavor, mouthfeel, structure, texture, moistness, and appearance characteristics to food products. The functionality attributes required of milkfat are highly dependent on the application, and some applications require different functionalities at different stages of processing (e.g., mixing, baking) and handling (e.g., shipping, consumption).

Current Applications for Specialty Milkfat Ingredients

Based on the data reported in experimental literature for milkfat fractions and information obtained from commercial literature for milkfat ingredients, the most common uses for milkfat fractions are in foods that traditionally contain milkfat. Although the use of milkfat fractions in these foods often replaces butter or anhydrous milkfat, milkfat fractions provide greater functionality and result in improved handling properties during processing and in improved qualities in the finished products.

Important applications for specialty milkfat ingredients are bakery products (pastries, cakes, cookies) and chocolate confections. Less important applications for specialty milkfat ingredients include a variety of dairy products, such as cold-spreadable butter and other dairy-based spreads, cheese, and ice cream. Other applications for milkfat fractions include sauces and condiments in which butter flavors are considered desirable.

Potential Applications for Specialty Milkfat Ingredients

Many of the potential applications suggested here require additional research to determine the suitability of selected milkfat fractions in these applications. Some applications require product development–type research, such as the use of milkfat fractions in sauces, but some applications will require considerable basic research. Included in the latter group are the uses of milkfat fractions in the fabrication of medium-chain triglyceride ingredients and certain nonfood applications, such as ingredients for cosmetics.

In general, increased knowledge of milkfat interactions will allow more accurate prediction of the functionality of milkfat ingredients in food products and facilitate the evaluation of milkfat ingredients for suitability in potential food and nonfood applications. An increased knowledge of the effects of blending milkfat fractions will be critical to the success of specialty milkfats.

Potential applications for milkfat fractions in food and nonfood products include the following:

- *Specialized dairy products.* High-melting milkfat fractions might be used in the production of recombined sweetened condensed milk to allow the production of crumb for use in milk chocolate. This would allow the incorporation of higher levels of milkfat in the chocolate while maintaining the characteristic flavors derived from the crumb.

- *Confections.* Fats are important in the structure of candies and affect viscosity and tenderness of the finished product (Kinsella, 1970). Fat also acts as a moisture barrier, imparts an overall richness of texture, and provides flavor to candies, such as caramels and toffees. The use of milkfat fractions in chewy confections and hard toffees may be beneficial to the flavors and textures of these products.

- *Salad dressings.* Jamotte and Guyot (1980) have suggested that milkfat fractions with a melting point less than 8°C may find use in providing flavors and other functionality to salad oils or mayonnaise.

- *Sauces and other convenience foods.* Dairy-based sauces, such as hollandaise, may benefit from the use of low-melting milkfat fractions, which could maintain the sauce in a liquid condition at room temperature. Other potential uses for milkfat fractions in convenience foods include frozen foods, microwaveable entreés, and microwaveable popcorn, where flavor and physical properties may provide needed functionality.

- *Flavoring agents.* Fractionation of milkfat has been shown to enhance the flavor potential of the lower-melting milkfat fractions, and these fractions may provide good starting materials for concentrated and lipolyzed butter flavors.

- *Nutritional foods.* Schultz and Timmen (1966) have suggested the use of milkfat fractions with increased concentrations of unsaturated fatty acids in the production of dietetic dairy products and sauces, such as butter, yogurt, mayonnaise, and butter sauce. The improved nutritional properties of these products result from the decreased saturated fat content compared to intact milkfat.

- *Medium-chain triglycerides and structured lipids.* Medium-chain triglycerides and structured lipids have proven beneficial in the treatment of hospital patients with special nutritional requirements, particularly for high-trauma patients. These specialized fats are metabolized quickly to provide a high energy source to the patient instead of following metabolic processes similar to long-chain triglycerides, where they are likely to be deposited in body fat reserves (Kennedy, 1991). Medium-chain triglycerides and structured lipids for clinical use are generally comprised of C8 and C10 fatty acids and are currently made by interesterification of vegetable oils, such as coconut and palm kernel oil (Babayan et al., 1990; Kennedy, 1991).

    Apparently, the use of milkfat as a commercial source of C8 and C10 fatty acids has not been seriously considered, because of the concerns about cholesterol associated with milkfat (Babayan et al., 1990). However, in a U.S. patent Babayan and coworkers (1990) reported that medium-chain triglycerides produced using milkfat improved the overall nutrition and dietary support in hypercatabolic mammals without elevating cholesterol levels. Babayan et al. (1990) have suggested that the evaluation of middle-melting and low-melting milkfat fractions with increased concentrations of C8 and C10 fatty acids in medium-chain triglycerides for nutritional use is warranted.

- *Sucrose polyesters.* The use of milkfat in the production of sucrose polyesters has been investigated to a limited extent, and the initial results have suggested that the use of milkfat to produce fat substitutes are feasible. However, further research is needed to develop this application for milkfat fractions (Connolly et al., 1991; Drake et al., 1992). Drake et al. (1992) reported that the hardness of the sucrose polyester depended on the hardness of the milkfat.

- *Edible films.* Waxes and high-melting lipids are often added to edible films to improve their water vapor properties. There are no legal limits to the use of milkfat in edible films, and finished products may benefit from the flavor functionality of the milkfat components (Shellhammer et al., 1993). The use of milkfat fractions as components in edible films is currently under investigation (Shellhammer et al., 1993).

- *Cosmetics.* The use of milkfat fractions in cosmetic and other nonfood applications has been proposed, but not specifically documented in the literature. It is possible that emollients and waxes traditionally used in cosmetics could be produced from milkfat, but the feasibility of these types of uses is yet to be reported.
- *Other uses.* Wilson (1975) has reported that a dairy company in New Zealand used an original process to obtain three milkfat fractions, and the very-high-melting milkfat fraction was used for candles.

Research Needs Identified for Milkfat Fractions

There is a pressing need for continued basic research on the chemical and physical characterization of the myriad component triglycerides of milkfat. This information is critical to understanding the nutritional consequences of milkfat consumption as well as understanding the functionality of milkfat in food applications. Only after this information is assembled will it be possible to truly reap the benefits of milkfat fractionations.

The interactions among the triglycerides of milkfat and between milkfat and other food components are indeed complex and not well understood. Further research on the basic chemical and physical properties of milkfat fractions and on mechanisms behind the interactions of milkfat fractions with other food components will lead to an improved ability to predict the functionality of milkfat fractions in food products and also assist in identifying new uses for milkfat.

Basic Research Needs Identified for Milkfat Fractions

Basic research needs identified in this book have focused on the inherent chemical and physical functionalities of milkfat fractions and their interactions with other food components, and include the following:

1. A need for increased knowledge of the fundamental melting and crystallization properties of milkfat fractions:
   - To allow more accurate prediction of the functionality of milkfat fractions and blends of milkfat fractions based on known physical and chemical characteristics
   - To understand thoroughly and predict the effects of processing variables on the texturizing and tempering of milkfat ingredients and blends of milkfat ingredients and other fats (e.g., cocoa butter)
   - To learn more about the relationship between fatty acid and triglyceride compositions and structure and the crystallization characteristics of milkfat and milkfat fractions
2. A need for increased knowledge of the mechanisms responsible and the factors affecting interactions between milkfat fractions and other food constituents:
   - Solid milkfat–liquid milkfat (mixed crystallization effects, solubility effects)
   - Milkfat–other fat (mixed crystallization effects, solubility effects, eutectic formation)
   - Milkfat–water (emulsion properties, viscosity effects)
   - Milkfat–solid particles (dispersion properties, lubricity effects, viscosity effects)
   - Milkfat–air (aeration properties, foam properties)

   Advances in technology have allowed more precise determination of chemical and physical characteristics (e.g., triglyceride profile, solid fat content profile), and these data continually lead to a better understanding of the relationship between milkfat characteristics and functionality.

The availability of techniques such as microscopy (Heathcock, 1993; Heertje et al., 1988; Juriaanse and Heertje, 1988; Kawanari et al., 1992) greatly enhances the ability to study the behavior and microstructure of milkfat fractions in foods.

Applications-Based Research Needs Identified for Milkfat Fractions

Research needs also have been identified for selected applications of milkfat fractions, and these are discussed in the following paragraphs.

Milkfat Ingredients for Bakery Products

There is a notable lack of experimental data relating to specialized bakery milkfat production as well as of technical data relating the functionality of these milkfats and their performance in bakery products. Information relating to the formulation and production of commercially produced products is unavailable. Thus, research on these topics is needed.

The effects of texturization and functional ingredients on the behavior of vegetable shortenings and margarines are well known in the edible oil industry. However, these effects are only sparsely known for milkfat, and the complex composition of milkfat makes these effects even more elusive. The changes that occur during texturization and tempering have not yet been fully clarified and warrant further research (Black, 1975a; Keogh and Morrissey, 1990).

Bakery products represent complex food systems that involve solid milkfat–liquid milkfat interactions, milkfat–(nonfat) solid interactions (sugar, flour), milkfat–liquid (water, milk, eggs), and milkfat–gas (air, carbon dioxide) interactions. A greater understanding of the mechanisms and factors affecting these interactions would be beneficial in predicting the functionality of milkfat ingredients in bakery products, and research is warranted to enhance applications of milkfat in foods.

Recent advances in the determination of the microstructure of butter and its influence on pastry quality offer a new vantage point to study the effects of the milkfat structure on the functionality of the milkfat ingredients and on the finished pastry (Kawanari et al., 1992). Furthermore, Heathcock (1993) has recommended the use of microscopy in determining the structure of many fat-containing products.

Milkfat Ingredients for Chocolates

Chocolates are complex systems that represent solid milkfat–liquid milkfat interactions, milkfat–other fat (cocoa butter–milkfat) interactions, and milkfat–(nonfat) solid (sugar, cocoa) interactions. One difficulty with studying the incorporation of milkfat into chocolates is that crystallization characteristics of mixed fat systems are not well understood (Barna et al., 1992; Hogenbirk, 1984). Yi (1993) indicated that solid fat content profiles of milkfat and cocoa butter mixtures could not be predicted by simple mixing of the profiles for each component, indicating the limitations of using solid fat content profiles curves to predict characteristics of fat mixture. An increased understanding of the crystallization of mixed fats and the effects of tempering on these systems would greatly enhance our ability to predict the compatibility of milkfat and milkfat fractions with cocoa butter systems. Additionally, an understanding of the influences of fatty acid positioning in triglycerides on milkfat crystallization and compatibility with other fats should provide a means to overcome barriers in the use of milkfat fractions in cocoa butter applications.

Another potential area of research that should be pursued relates to the use of milkfat fractions in the production of recombined sweetened condensed milk to allow the production of milk crumb from a high-melting milkfat fraction. An ingredient of this type could be incorporated into milk chocolate without sacrificing the flavors derived from the crumb, and it should allow the use of higher levels of milkfat in chocolates.

Milkfat Ingredients for Cold-Spreadable Butters

Cold-spreadable butter is a good illustration of the use of analytical information, such as solid fat content profiles, to predict the functionality and behavior of milkfat in food products. However, it has been shown that interactions between milkfat fractions cannot always be predicted using solid fat content, and that the final test of suitability is still actual performance evaluation of the milkfat ingredient in the finished product. These observations suggested a need for further research on fat–fat interactions of milkfat fractions, to increase the understanding of this phenomenon and allow more accurate prediction of the behavior and functionality of milkfat fractions.

The use of solvents in the preparation of dairy food ingredients has not been permitted, but solvents are used in the manufacture of other food ingredients, including vegetable oils (Thomas, 1985) and flavoring agents (Heath and Reineccius, 1986). However, because of the concerns surrounding the use of solvents, it would be desirable to explore alternative processing technologies, such as fractionation from melted milkfat and pressure filtration (Deffense, 1993b) to obtain suitable very-high-melting milkfat fractions for chocolate and cold-spreadable butter formulations.

Milkfat Ingredients for Dairy-Based Spreads

Further needed research on dairy-based spreads is similar to that suggested for cold-spreadable butter regarding fat–fat interactions between milkfat fractions. Additional research to further the understanding of fat–fat interactions between milkfat and other fats (e.g., sunflower oil) and of liquid milkfat–water and solid milkfat–water interactions also seems warranted.

Commercialization of Milkfat Fractionation in the United States

Commercialization of milkfat fractionation has been successful in several countries, including Belgium, France, and New Zealand, and these examples should serve as a model for the fractionation of milkfat in the United States. At this time, the most likely method for the commercial production of milkfat fractions in the United States is crystallization from melted milkfat. This method involves a relatively simple physical process that does not employ additives, and thus the milkfat fractions retain their natural image and designation. Fractionation by supercritical fluid extraction does not employ additives either, and this process may provide a future means for the production of certain specialty milkfat ingredients. Singh and Rizvi (1994) have suggested that continuous supercritical fluid extraction processes are economically attractive compared with batch processes, and therefore continuous processes are competitive with other methods available for milkfat fractionation.

Successful commercialization of milkfat fractionation in the United States also depends to some extent on the manufacturer's ability to tailor milkfat ingredients to customer specifications and provide application support in a variety of products.

References

Abu-Lehia, I.H. (1989) Physical and chemical characteristics of camel milkfat and its fractions. *Food Chem. 34,* 261.

Alaco, Undated. Product information brochure. New Zealand Dairy Board, Wellington, New Zealand.

Alfa-Laval. (1987) *Dairy handbook,* Alfa-Laval, Lund, Sweden, p. 191.

Amer, M.A., Kupranycz, D.B., and Baker, B.E. (1985) Physical and chemical characteristics of butterfat fractions obtained by crystallization from molten fat. *J. Amer. Oil Chem. Soc. 62,* 1551.

Andrianov, Y.P., and Kornelyuk, B.V. (1974) Crystallization of butterfat under isothermal conditions. In *Proc. XIX Int. Dairy Congr.,* New Delhi, India, vol. E, p. 223.

Anonymous. (1982) Dairy Market Trends. *Dairy Field 165,* 39.

Antila, V. (1966) Fatty acid composition, solidfication and melting of Finnish butter fat. *Meijeritieteellinen Aikakauskirja 27,* 1. (cited by Antila and Antila, 1970)

Antila, V. (1979) The fractionation of milkfat. *Milk Ind. 81,* 17.

Antila, V. (1984). Fractionation of milkfat. In *Milkfat and Its Modification, Contributions at a LIPID-FORUM symposium,* Marcuse, R., Scandinavian Forum for Lipid Research and Technology, Göteborg, Sweden, p. 85.

Antila, V., and Antila, M. (1970) Der Nährwert verschiedener Milchfett-Fraktionen [The nutritional value of various fractions of milk fat]. *Fette Seifen Anstrichm. 72,* 285.

Antila, V., and Kankare, V. (1983) The fatty acid composition of milk lipids. *Milchwissenschaft 38,* 478. (cited by Kankare and Antila, 1986; 1988a; Kankare et al., 1989)

AOAC (Association of Official Analytical Chemists). (1984, 1990) *Official Methods of Analysis of the Association of Official Analytical Chemists,* 15th ed., Association of Official Analytical Chemists, Inc., Arlington, VA.

AOCS (American Oil Chemists' Society). (1962, 1973, 1989) *Official Methods of the American Oil Chemists' Society,* American Oil Chemists' Society, Illinois.

Apps, P.J., and Willemse, C. (1991) Fast, high precision fingerprinting of fatty acids in milk. *J. High Resolution Chrom. 14,* 802.

Arul, J., Boudreau, A., Makhlouf, J., Tardif, R., and Bellavia, T. (1988a) Fractionation of anhydrous milk fat by short-path distillation. *J. Amer. Oil Chem. Soc. 65,* 1642.

Arul, J., Boudreau, A., Makhlouf, J., Tardif, R., and Grenier, B. (1988b) Distribution of cholesterol in milk fat fractions. *J. Dairy Res. 55,* 361.

Arul, J., Boudreau, A., Makhlouf, J., Tardif, R., and Sahasrabudhe, M.R. (1987) Fractionation of anhydrous milk fat by superficial carbon dioxide. *J. Food Sci. 52,* 1231.

Aveve, undated. Product information brochure. Brussels, Belgium.

Avvakumov, A.K. (1974) Dilatometric study of mixtures of milkfat and its fractions. In *Proc. XIX Int. Dairy Congr.,* New Delhi, India, vol. E, p. 243.

Babayan, V.K., Blackburn, G.L., and Bistrian, B.R. (1990) *Structured Lipid Containing Dairy Fat.* U.S. Patent 4,952,606.

Badings, H.T., Schaap, J.E., DeJong, C., and Hagedoorn, H.G. (1983a) An analytical study of fractions obtained by stepwise cooling of melted milk fat. 1. Methodology. *Milchwissenschaft 38,* 95.

Badings, H.T., Schaap, J.E., DeJong, C., and Hagedoorn, H.G. (1983b) An analytical study of fractions obtained by stepwise cooling of melted milk fat. 2. Results. *Milchwissenschaft 38,* 150.

Badings, H.T., van der Pol, J.J.G., and Wassink, J.G. (1975) In *Proc. 1st Int. Symp. Glass Capillary Chrom.,* p. 175. (cited by Schaap et al., 1975)

Badings, H.T., and Wassink, J.G. (1963) *Neth. Milk Dairy J.* 17, 133. (cited by Bhat and Rama Murthy, 1983)

Baer, R.J. (1991) Alteration of the fatty acid content of milk fat. *J. Food Protection 54*, 383.

Baker, B.C. (1970a) The fractionation of butterfat, and the properties of selected fractions. In *Proc. XVIII Int. Dairy Congr.*, Sydney, Australia, vol. 1E, p. 241.

Baker, B.C. (1970b) Use of butterfat fractions for special purposes. In *Proc. XVIII Int. Dairy Congr.*, Sydney, Australia, vol. 1E, p. 244.

Baker, B.E., Bertok, E., and Samuels, E.R. (1959) Studies on milk powders. III. The preparation and properties of milk powders containing low-melting butter oil. *J. Dairy Sci. 42*, 1038.

Baldwin, R.E., Cloninger, M., and Lindsay, R.C. (1973) Thresholds for fatty acids in buffered solutions. *J. Food Sci. 38*, 528.

Baldwin, R.R., Baldry, R.P., and Johansen, R. (1972) Fat systems for bakery products. *J. Am. Oil Chem. Soc. 49*, 473.

Banks, W. (1991a) Chemical and physical properties of milkfat. In *Utilizations of Milkfat*, Bull. No. 260, Int. Dairy Fed., p. 4.

Banks, W. (1991b) Possibilities of changing the composition of milkfat by feeding. In *Utilizations of Milkfat*, Bull. No. 260, Int. Dairy Fed., p. 5.

Banks, W. (1991c) Hydrogenation and Dehydrogenation. In *Utilizations of Milkfat*, Bull. No. 260, Int. Dairy Fed., p. 13.

Banks, W. (1991d) Milkfat production. In *Milk Fat Production, Technology and Utilization*, Rajah, K.K., and Burgess, K.J., Society for Dairy Technology, Huntingdon, Cambridgeshire, England, p 1.

Banks, W., Christie, W.W., Clapperton, J.L., and Girdler, A.K. (1987) The trisaturated glycerides of bovine milk fat. *J. Sci. Food Agric. 39*, 303.

Banks, W., Clapperton, J.L., and Ferrie, M.E. (1976) The physical properties of milk fats of different chemical compositions. *J. Soc. Dairy Technol. 29*, 86. (cited by Banks et al., 1985)

Banks, W., Clapperton, J.L., and Girdler, A.K. (1985) On the fractional melting of milk fat and the properties of the fractions. *J. Sci. Food Agric. 36*, 421.

Banks, W., Clapperton, J.L., and Girdler, A.K. (1989a) Fractional melting of hydrogenated milk fat. *J. Dairy Res. 56*, 265.

Banks, W., Clapperton, J.L., Girdler, A.K., and Steele, W. (1989b) Fractionation of hydrogenated milk fat using thin layer chromatography. *J. Sci. Food Agric. 48*, 495.

Barbano, D.M., and Sherbon, J.W. (1975) Stereospecific Analysis of High Melting Triglycerides of Bovine Milk Fat and Their Biosynthetic Origin. *J. Dairy Sci. 51*, 1.

Barna, C.M., Hartel, R.W., and Metin, S. (1992) Incorporation of milkfat fractions into milk chocolate. *Manufacturing Confectioner 72*, 107.

Barnicoat, C.R. (1944) *Analyst, Lond. 69*, 176. (cited by Norris et al., 1971; Sherbon et al., 1972)

Barts, R. (1991) The use of milkfat and milkfat fractions in the food industry. In *Utilizations of Milkfat*, Bull. No. 260, Int. Dairy Fed., p. 19.

Bartsch, A., Schuff, P., and Büning-Pfaue, H. (1990) Untersuchungen zur Mischbarkeit von Fetten am Beispiel von Mischungen eines laurischen Fettes mit Milchfettfraktionen [Investigations on the compatibility of fats—introduced by the example of blends of a lauric fat and milk fat fractions]. *Fat Sci. Technol. 92*, 213.

Bassette, R., and Acosta, J.S. (1988) Composition of milk products. In *Fundamentals of Dairy Chemistry*, 4th ed., Wong, N.P. Van Nostrand Reinhold, New York, p. 39.

Belousov, A.P., and Vergelesov, V.M. (1962) Polymorphism in butterfat. In *Proc. 16th Int. Dairy Congr.*, København, sec. III:1, p. 122.

Beyerlein, U., and Voss, E. (1973) Probleme der technischen Fraktionierung von Butterfett mit Hilfe des Fettkörnchen-Filtrationsverfahrens. II. Zur Bildung und Morphologie der Sphärolithen in temperieretem Butterfett [Problems related to technological fractionation of butterfat by means of filtration of fat granules. II. Formation and morphology of spherulites in tempered butterfat]. *Kieler Milchwirtschaftliche Forschungsberichte 25,* 49.

Bhalerao, V.R., Johnson, O.C., and Kummerow, F.A. (1959) Effect of thermal oxidative polymerization on the growth-promoting value of some fractions of butterfat. *J. Dairy Sci. 42,* 1057.

Bhaskar, A.R., Rizvi, S.S.H., and Harriott, P. (1993a) Performance of a packed column for continuous supercritical carbon dioxide processing of anhydrous milk fat. *Biotechnol. Prog. 9,* 70.

Bhaskar, A.R., Rizvi, S.S.H., and Sherbon, J.W. (1993b) Anhydrous milk fat fractionation with continuous countercurrent supercritical carbon dioxide. *J. Food Sci. 58,* 748.

Bhat, G.S., and Rama Murthy, M.K. (1983) Distribution and production of carbonyls during autoxidation in low and high melting fraction of cow and buffalo milk fats. *Indian J. Dairy Sci. 36,* 308.

Biernoth, G., and Merk, W. (1985) *Fractionation of Butterfat Using a Liquefied Gas or a Gas in the Supercritical State,* U.S. patent 4,504,503.

Black, R.G. (1972) Milkfat fractions - definitions, terminology and description. *Aust. J. Dairy Technol. 27,* 46. (cited by Black, 1973, 1975b; Keogh and Higgins, 1986b)

Black, R.G. (1973) Pilot-scale studies of milk fat fractionation. *J. Dairy Technol. 28,* 116.

Black, R.G. (1974a) A rapid softening-point test. *Aust. J. Dairy Technol. 29,* 23.

Black, R.G. (1974b) Mechanical separation techniques in the fractionation of milkfat. In *Proc. XIX Int. Dairy Congr.,* New Delhi, India, vol. E, p. 654.

Black, R.G. (1975a) The plasticizing of milkfat: 1. Process variables. *Aust. J. Dairy Technol. 30,* 60.

Black, R.G. (1975b) Partial crystallization of milkfat and separation of fractions by vacuum filtration. *Aust. J. Dairy Technol. 30,* 153.

Black, R.G., and Dixon, B.D. (1974) Use of milkfat fractions in standard dairy products. In *Proc. XIX Int. Dairy Congr.,* New Delhi, India, vol. E, p. 627.

Boldingh, J., and Taylor, R.J. (1962) Trace constituents of butterfat. *Nature 194,* 909.

Bornaz, S., Fanni, J., and Parmentier, M. (1993) Butter Texture: The Prevalent Triglycerides. *J. Am. Oil Chem. Soc. 70,* 1075.

Bornaz, S., Novak, G., and Parmentier, M. (1992) Seasonal and regional variation in triglyceride composition of french butterfat. *J. Am. Oil Chem. Soc. 69,* 1131.

Boudreau, A., and Arul, J. (1993) Cholesterol reduction and fat fractionation technologies for milk fat: an overview. *J. Dairy Sci. 76,* 1772.

Bourne, M.C. (1982) *Food Texture and Viscosity: Concept and Measurement,* Academic Press, New York.

Bradley, R.L., Jr. (1989) Removal of cholesterol from milk fat using supercritical carbon dioxide. *J. Dairy Sci. 72,* 2834.

Branger, A.I. (1993) Separation of Fractionally Crystallized Milk Fat by Centrifugal Methods, M.S. Thesis, University of Wisconsin.

Bratland, A. (1983) *Production of a Dairy Emulsion,* European Patent 0095001.

Breckenridge, W.C., and Kuksis, A. (1968a) Specific distribution of short-chain fatty acids in molecular distillates of bovine milk fat. *J. Lipid Res. 9,* 388.

Breckenridge, W.C., and Kuksis, A. (1968b) Structure of bovine milk fat triglycerides. 1. Short and medium chain lengths. *Lipids 3,* 291.

Brennand, C.P., Ha, J.K., and Lindsay, R.C. (1989) Aroma properties and thresholds of some branched-chain and other minor volatile fatty acids occurring in milkfat and meat lipids. *J. Sensory Studies 4,* 105.

British Standards Institution. (1961) *British Standard 769.* (cited by Norris et al., 1971; Sherbon et al., 1972)

Bruker. (1993) *Minispec Newsletter, Special SFC Issue.* Bruker Spectrospin, Milton, Ontario, Canada.

Buchanan, R.A., Townsend, F.R., and Black, R.G. (1970) Plasticized butterfat shortening for cakes. In *Proc. XVIII Int. Dairy Congr.*, Sydney, Australia, vol. 1E, p. 250.

Buege, J.A., and Aust, S.D. (1978) Microsomal lipid peroxidation. In *Methods in Enzymology*, vol. LII, Fleischer, S., and Packer, L., Academic Press, New York, p. 302.

Bumbalough, J.E. (1989) *Spreadable Product Having an Anhydrous Milk Fat Component,* U.S. Patent 4,839,190.

Büning-Pfaue, H., Eggers, R., and Bartsch, A. (1989) Vergleich von Milchfetten aus der Kristallisationsfraktionierung und einem kontinuierlichen Fraktionierungsverfahren mittels überkritischem Kohlendioxid [Comparison of milk fat fractions obtained by crystallization with those obtained in a continuous fractionation process using supercritical carbon dioxide]. *Fat Sci. Technol. 91,* 92.

Bunting, C.M. (1991) Effects of the Addition of Milkfat and Milkfat Fractions on the Quality of Milk Chocolate, M.S. Thesis, University of Wisconsin-Madison.

Campbell, L.B., Andersen, D.A., and Keely, P.G. (1969) Hydrogenated milk fat as an inhibitor of the fat bloom defect in dark chocolate. *J. Dairy Sci. 52,* 976.

Carlin, G.T. (1944) A microscopic study of the behavior of fats in cake batters. *Cereal Chem. 21,* 189.

Cebula, D.J., and Smith, K.W. (1991) Differential scanning calorimetry of confectionery fats. Pure triglycerides: effects of cooling and heating rate variation. *J. Am. Oil Chem. Soc. 68,* 91.

Chang, J.M., Boylston, T.D., Lin, H., Luedecke, L.O., and Shultz, T.D. (1994) Survey of the conjugated linoleic acid (CLA) content of dairy products. *IFT Annual Meeting/Book of Abstracts.* Abstract 12F-1, p. 35.

Chapman, G.M. (1971) Cocoa butter and confectionery fats. Studies using programmed temperature X-ray diffraction and differential scanning calorimetry. *J. Am. Oil Chem. Soc. 48,* 824.

Chen, H., and Schwartz, S.J. (1991) Fractionation of butteroil with supercritical carbon dioxide. *J. Dairy Sci. 74 (Supp. 1),* 130.

Chen, H., Schwartz, S.J., and Spanos, G.A. (1992) Fractionation of butter oil by supercritical carbon dioxide. *J. Dairy Sci. 75,* 2659.

Chen, P.C., and deMan, J.M. (1966) Composition of milk fat fractions obtained by fractional crystallization from acetone. *J. Dairy Sci. 49,* 612.

Chen, Z.Y., and Nawar, W.W. (1991) Prooxidative and antioxidative effects of phospholipids on milk fat. *J. Am. Oil Chem. Soc. 68,* 938.

Chin, S.F., Liu, W., Storkson, J.M., Ha, Y.L., and Pariza, M.W. (1992) Dietary sources of conjugated dieonic isomers of linoleic acid, a newly recognized class of anticarcinogens. *J. Food Comp. and Anal. 5,* 185.

Christie, W.W. (1979) The effects of diet and other factors on the lipid composition of ruminant tissues and milk. *Prog. Lipid Res. 17,* 245.

Christie, W.W. (1983) The composition and structure of lipids. In *Developments in Dairy Chemistry— 2 Lipids,* Fox, P., Applied Science Publishers, New York, p. 1.

Christie, W.W. (1987) In *High Performance Liquid Chromatography and Lipids,* Pergamon Press, New York, p. 134. (cited by Chen et al., 1992; Fouad et al., 1990)

Christie, W.W. (1987) A stable silver-loaded column for the separation of lipids by high performance liquid chromatography. *J. High Resolut. Chromatogr., Chromatogr. Commun. 10,* 148. (cited by Laakso et al., 1992)

Christie, W.W. (1988) Separation of molecular species of triacylglycerols by high-performance liquid chromatography with a silver ion column. *J. Chromatogr. 454,* 273. (cited by Laakso et al., 1992)

Christopherson, S.W., and Glass, R.L. (1969) *J. Dairy Sci. 52,* 1289. (cited by Kaufmann et al., 1982; Muuse and van der Kamp, 1985)

Chrysam, M.M. (1985) Table spreads and shortenings. In *Bailey's Industrial Oil and Fat Products,* vol. 3, Applewhite, T.H., John Wiley and Sons, New York, p. 41.

Cocup, R.O., and Sanderson, W.B. (1987) Functionality of dairy ingredients in bakery products. *Food Technol. 41*, 86.

Connolly, J.F., Canton, M., Keogh, K., and Charteris, B. (1991) Low calorie fat from dairy ingredients. In *Dairy Levy Reporting Symposium Poster Session*, The Agriculture and Food Development Authority, Moorepark Research Centre, Fermoy, Co. Cork, Ireland.

Conte, J.A., Jr., and Johnson, B.R. (1992) *Production of Low Cholesterol Butter Oil by Vapor Sparging*, U.S. Patent 5,092,964.

Corman, undated. Product information brochure. Goé, Belgium.

Davis, J.G., and Macdonald, F.J. (1953) *Richmond's Dairy Chemistry*, Charles Grrifin and Co. Ltd. (cited by Lakshminarayana and Rama Murthy, 1985)

Deeth, H.C., and Fitz-Gerald, C.H. (1983) Lipolytic enzymes and hydrolytic rancidity in milk and milk products. In *Developments in Dairy Chemistry—2 Lipids*, Fox, P., Applied Science Publishers, New York, p. 195.

Deffense, E. (1987) Multi-step butteroil fractionation and spreadable butter. *Sonderdruck aus Fett Wiss. Technol. - Fat Sci. Technol. 13*, 1.

Deffense, E. (1989) Fractionated milk fat products in bakery products. In *Proc. New Uses for Milk*, The Dairy Sciences Research Center, Laval University, Quebec, Canada, p. 79.

Deffense, E. (1993a) The turbo deodorizer. *Intl. News on Fats Oils and Related Materials (INFORM) 4*, 498 Abstract E6-2.

Deffense, E. (1993b) Milk fat fractionation today. *J. Am. Oil Chem. Soc. 70*, 1193.

deMan, J.M. (1961a) Physical properties of milk fat. I. Influence of chemical modification. *J. Dairy Res. 28*, 81.

deMan, J.M. (1961b) Physical properties of milk fat. II. Some factors influencing crystallization. *J. Dairy Res. 28*, 117.

deMan, J.M. (1963) Polymorphism in milk fat. *Dairy Sci. Abst. 25*, 219.

deMan, J.M. (1964) Determination of the fatty acid composition of milk fat by dual-column temperature programmed gas-liquid chromatography. *J. Dairy Sci. 47*, 546. (cited by Chen and deMan, 1966; deMan, 1968; Schaap et al., 1975)

deMan, J.M. (1968) Modification of milk fat by removal of a high melting glyceride fraction. *J. Inst. Can. Technol. Alliment. 1*, 90.

deMan, J. M., and Finoro, M. (1980) Characteristics of milk fat fractionated by crystallization from the melt. *Can. Inst. Food Technol. J. 13*, 167.

deMan, J.M., and Wood, F.W. (1959) Polarized light microscopy of butterfat crystallization. In *Proc. XV Int. Dairy Cong.*, vol. 2. p. 1010. (cited by El-Ghandour et al., 1976)

Deroanne, C. (1976) La cristallisation et le fractionnement naturel de la matière grass butyrique. *Le Lait 56*, 39.

Deroanne, C., and Guyot, A. (1974) Classical and differential thermal analysis of milk fats modified by fractional crystallisation from the melted fat and by partial hydrogenation. *Bull. Res. Agron. 2*, 261. Gembloux, Belgium.

De Smet Engineering, undated. Product information brochure. Antwerp, Belgium.

Dimick, P.S., and Walker, N.J. (1967) *J. Dairy Sci. 50*, 97. (cited by Walker 1972)

Dimick, P.S., Walker, N.J., and Patton, S. (1969) Occurrence and biochemical origin of aliphatic lactones in milk fat—a review. *J. Agr. Food Chem. 17*, 649.

Dixon, B.D., and Black, R.G. (1974) The manufacture of butter of controlled firmness using fractionated milkfat. In *Proc. XIX Int. Dairy Congr.*, New Delhi, India, vol E., p. 650.

Dolby, R.M. (1959) Methods of controlling the consistency of butter. *Aust. J. Dairy Technol. 14*, 103.

Dolby, R.M. (1961) Determination of the softening point of butterfat. *Aust. J. Dairy Technol. 16,* 89.

Dolby, R.M. (1970a) Chemical composition of fractions of milkfat separated by a commercial process. In *Proc. XVIII Int. Dairy Congr.*, Sydney, Australia, vol. 1E, p. 242.

Dolby, R.M. (1970b) Properties of recombined butter made from fractionated fats. In *Proc. XVIII Int. Dairy Congr.*, Sydney, Australia, vol. 1E, p. 243.

Doležálek, J., Forman, L., and Vodičková, M. (1976) Fractionation of milk fat by centrifugation in dispersed state. In *Scientific Papers of the Prague Institute of Chemical Technology,* E 45 Food, p. 37.

Drake, M.A., Younce, F., Cleary, D., and Swanson, B.G. (1992) Melting onset and hardness of milkfat blend sucrose polyesters. In *1992 IFT Annual Meeting/Book of Abstracts*, Institute of Food Technologists, Chicago, IL, Abstract 705 p. 179.

El-Ghandour, M.A., and El-Nimr, A.A. (1982) Dilatation properties of local cow's and buffalo's milk fat and their fractions. II. Regime 6-16-13. *Annals Agric. Sci. 27,* 125.

El-Ghandour, M.A., Helal, F.R., and Hofi, A.A. (1976) Comparative study for microscopic crystal structure of buffalos' and cows' butterfat. *Egyptian J. Dairy Sci. 4,* 161.

Elliot, J.M. (1990) Modification of Butteroil by Lipase-Catalyzed Acyl-Exchange Reactions in Anhydrous Media, M.S. Thesis, University of Wisconsin-Madison.

Ellis, R., and Wong, N.P. (1975) Lactones in butter, butteroil, and margarine. *J. Am. Oil Chem. Soc. 52,* 252.

El-Nimr, A.A. (1980) How to increase butter softening. *Dairy Ind. Int. 7,* 45.

El-Nimr, A.A., and Bass, A. (1979) *Milchwissenschaft*, accepted for publication. (cited by El-Nimr, 1980)

El-Nimr, A.A., and El-Ghandour, M.A. (1980) Dilatation properties of local cow's and buffalo's milk fat and their fractions. I. Regime 16-6-13, *Res. Bull. No. 1325,* Faculty of Ag., Ain Shams Univ., Cairo, Egypt. (cited by El-Ghandour and El-Nimr, 1982)

Ensign, K.K. (1989) Supercritical Carbon Dioxide Extraction of Beef Tallow and Milkfat, M.S. Thesis, University of Wisconsin-Madison.

Erickson, D.R., and Dunkley, W.L. (1964) *Anal. Chem. 36,* 1055. (cited by Antila and Antila, 1970)

Esdaile, J.D., and Wilson, B.W. (1974) Butter oil fractionation by liquation. In *Proc. XIX Int. Dairy Congr.*, New Delhi, India, vol. E, p. 678.

Evans, A.A. (1976) Colour measurements in milkfat fractionation. *N.Z. J. Dairy Sci. Technol. 11,* 73.

Fairley, P., German, J.B., and Krochta, J.M. (1994) Phase behavior and mechanical properties of tripalmitin/butterfat mixtures. *J. Food Sci. 59,* 321.

Ferry, J.D. (1980) *Viscoelastic Properties of Polymers*, 3rd ed. John Wiley & Sons, New York. (cited by Shukla et al., 1994)

Feuge, R.O., and Bailey, A.E. (1944) Measurement of the consistency of plastic vegetable fats. A standard micropenetration technique. *J. Am. Oil Chem. Soc. 21,* 78. (cited by Richardson, 1968)

Fjaervoll, A. (1970a) Anhydrous milk fat fractionation offers new applications for milk fat. *Dairy Ind. Int. 35,* 502.

Fjaervoll, A. (1970b) Fractionation of milkfat. In *Proc. XVIII Int. Dairy Congr.*, Sydney, Australia, vol. 1E, p. 239.

Foley, J., and Brady, J.P. (1984) Temperature-induced effects on crystallization behaviour, solid fat content and the firmness values of milk fat. *J. Dairy Res. 51,* 579.

Formo, M.W. (1979) Physical properties of fats and fatty acids. In *Bailey's Industrial Oil and Fat Products*, vol. 1., 4th ed., Swern, D., John Wiley and Sons, New York, p. 177.

Forss, D.A. (1971) The flavors of dairy fats—a review. *J. Am. Oil. Chem. Soc. 48,* 702.

Fouad, F.M., Van de Voort, F.R., Marshall, W.D., and Farrel, P.G. (1990) A critical evaluation of thermal fractionation of butter oil. *J. Amer. Oil Chem. Soc. 67,* 981.

Frankel, E.N. (1985) Chemistry of autoxidation: mechanism, products and flavor significance. In *Flavor Chemistry of Fats and Oils*, Min, D.B., and Smouse, T.H., American Oil Chemists' Society, Champaign, IL, p. 1.

Frede, E. (1991) Interesterification. In *Utilizations of Milkfat*, Bull. No. 260, Int. Dairy Fed. p. 12.

Frede, E., Peters, K.H., and Precht, D. (1980) Verbesserungder Konsistenz der Butter mit Hilfe der Fettfraktionierung sowie einer besonderen Temperaturbehandlung des Rahms [Improvement of the consistency of butter by means of fat fractionation and of a special temperature treatment of the cream]. *Milchwissenschaft 35*, 287.

Frede, E., and Thiele, H. (1987) Analysis of milkfat by HPLC. *J. Am. Oil Chem. Soc. 64*, 521.

Frium, N., and Fritsvold, S. (1984) Hydrogenated marine oils for baking. In *Milkfat and Its Modifications Contributions at a LIPIDFORUM symposium*, Marcuse, R., Scandinavian Forum for Lipid Research and Technology, Göteborg, Sweden, p. 107.

Fujikawa, T., Hamashima, M., and Yasuda, K. (1971) *Yukagaku* 20, 138. (cited by Shishikura et al., 1986)

Gibson, J.P. (1991) The potential for genetic change in milk fat composition. *J. Dairy Sci. 74*, 3258.

Goff, H.D., Sherbon, J.W., and Jordan, W.K. (1988) The influence of milkfat crystallization on emulsion stability during ice cream manufacture. *J. Dairy Sci. 71 (Supp 1)*, 80 Abstract D55.

Grall, D.S., and Hartel, R.W. (1992) Kinetics of butterfat crystallization. *J. Amer. Oil Chem. Soc. 69*, 741.

Gresti, J., Bugaut, M., Maniongui, C., and Bezard, J. (1993) Composition of molecular species of triacylglycerols in bovine milk fat. *J. Dairy Sci. 76*, 1850.

Grishchenko, A.D., and Ivantsova, L.S. (1974) Effect of seasonal variations in the contents of high-melting glycerides in milkfat on its crystallization. In *Proc. XIX Int. Dairy Congr.*, New Delhi, India, vol. E, p. 653.

Grummer, R.R. (1991) Effect of feed on the composition of milk fat. *J. Dairy Sci. 74*, 3244.

Gurr, M. (1992) Saturated fatty acids and hypercholesterolemia: are all saturated acids equal? *Lipid Technol. 4*, 93.

Guth, G., and Grosch, W. (1992) Furan fatty acids in butter and butter oil. *Z. Lebensm. Unters Forsch. 194*, 360.

Guyot, A.L. (1982) Cristallisation fractionée sans solvant et beurres frigotartinables [Fractionated crystallization without solvents and butter spreadability]. In *Proc. XXI Int. Dairy Congr.*, Moscow, Russia, vol. 1, p. 329.

Ha, J.K., and Lindsay, R.C. (1991a) Contributions of cow, sheep, and goat milks to characterizing branched-chain fatty acid and phenolic flavors in varietal cheeses. *J. Dairy Sci. 74*, 3267.

Ha, J.K., and Lindsay, R.C. (1991b) Volatile branched-chain fatty acids and phenolic compounds in aged Italian cheese flavors. *J. Food Sci. 56*, 1241.

Haighton, A.J. (1959) The measurement of the hardness of margarine and fats with cone penetrometers. *J. Amer. Oil Chem. Soc. 36*, 345. (cited by Keogh and Morrissey, 1990; Verhagen and Bodor, 1984; Verhagen and Warnaar, 1984)

Haighton, A.J. (1976) Blending, chilling, and tempering of margarines and shortenings. *J. Am. Oil Chem. Soc. 53*, 397.

Hamm, D.L., Hammond, E.G., and Hotchkiss, D.K. (1968) Effect of temperature on rate of autoxidation of milk fat. *J. Dairy Sci. 51*, 483.

Hamman, H., and Sivik, B. (1991) Fractionation of gluten lipids with supercritical carbon dioxide. *Fat Sci. Technol. 93*, 104.

Hamman, H., Söderberg, I., and Sivik, B. (1991) Physical properties of butter oil fractions obtained by supercritical carbon dioxide extraction. *Fat Sci. Technol. 93*, 374.

Hammond, E.G. (1989) The flavors of dairy products. In *Flavor Chemistry of Lipid Foods*, Min, D.B., and Smouse, T.H., American Oil Chemists' Society, Champaign, IL, p. 222.

Hawke, J.C., and Taylor, M.W. (1983) Influence of nutritional factors on the yield, composition and physical properties of milkfat. In *Developments in Dairy Chemistry—2 Lipids*, Fox, P., Applied Science Publishers, New York, p. 37.

Hayakawa, M., and deMan, J.M. (1982) Consistency of fractionated milk fat as measured by two penetration methods. *J. Dairy Sci. 65*, 1095.

Heath, H.B., and Reineccius, G. (1986) *Flavor Chemistry and Technology*, Van Nostrand Reinhold, New York.

Heathcock, J.F. (1993) Microscopy of fats. *Lipid Technol. 5*, 4.

Heertje, I., Leunis, M., van Zeyl, W.J.M., and Berends, E. (1987) Product morphology of fatty products. *Food Microstructure 6*, 1.

Heertje, I., van Eendenburg, J., Cornelissen, J.M., and Juriaanse, A.C. (1988) The effect of processing on some microstructural characteristics of fat spreads. *Food Microstructure 7*, 189.

Helal, F.R., El-Ghandour, M.A., and Hofi, A.A. (1977) Polymorphism in milk fat and its fractions. *Egyptian J. Dairy Sci. 5*, 37.

Helal, F.R., El-Ghandour, M.A., and Khader, A.E. (1984) Polymorphism in intermediate glycerides of Egyptian buffalo's and cow's milk fat during winter and summer. *Minufiya J. of Agric. Res. 8*, 279.

Hendrickx, H., De Moor, H., Huyghebaert, A., and Janssen, G. (1971) Manufacture of chocolate containing hydrogenated butterfat. *Rev. Int. Choc. 26*, 190.

Heuvel, H.M., and Lind, K.C.J.B. (1970) *Anal. Chem. 42*, 1044. (cited by Sherbon et al., 1972)

Higgins, A.C., and Keogh, M.K. (1986) Anhydrous milk fat 2. Hydrolytic aspects. *Irish J. Food Sci. and Technol. 10*, 23.

Hinrichs, J., Heinemann, U., and Kessler, H.G. (1992) Differences in the composition of triglycerides in summer and winter milk fat. *Milchwissenschaft 47*, 495.

Hoerr, C.W. (1960) Morphology of fats, oils, and shortenings. *J. Am. Oil Chem. Soc. 37*, 539.

Hogenbirk, G. (1984) Compatibility of specialty fats with cocoa butter. *Manufacturing Confectioner 63*, 59.

Huebner, V.R., and Thomsen, L.C. (1957) Spreadability and hardness of butter. II. Some factors affecting spreadability and hardness. *J. Dairy Sci. 40*, 839.

Humphries, M.A. (1971) Use of fractionated milkfat in bakery products. *N.Z. J. Dairy Sci. Technol. 6*, 28.

International Dairy Federation. (1984) *The World Market for Butter, Bull. 170*, p. 4.

Ip, C., and Sciemeca, J.A. (1994) Mammary cancer prevention by conjugated linoleic acid. *IFT Annual Meeting/ Book of Abstracts*. Abstract 4-4 p. 7.

ISI. (1966) IS: 3508 *Methods of Sampling and Test for Ghee (Butterfat)*, Indian Standards Institution, Manak Bhavan, New Delhi, India. (cited by Sreebhashyam et al., 1981)

Itabashi, Y., Myher, J.J., and Kuksis, A. (1993) Determination of positional distribution of short-chain fatty acids in bovine milk fat on chiral columns. *J. Am. Oil Chem. Soc. 70*, 1177.

Iverson, J.L., and Sheppard, A.J. (1986) Determination of fatty acids in butter fat using temperature-programmed gas chromatography of the butyl esters. *Food Chem. 21*, 223.

Iyer, M., Richardson, T., Amundson, C.H., and Boudreau, A. (1967) Improved technique for analysis of free fatty acids in butteroil and Provolone cheese. *J. Dairy Sci. 50*, 285. (cited by Richardson, 1968)

Jacobs, M.B. (1964) *The Chemical Analysis of Foods and Food Products*, Van Nostrand Company, Inc., Princeton, New Jersey. (cited by Youssef et al., 1977)

Jamotte, P., and Guyot, A. (1980) *Final report on research dealing with the preparation and use of new types of butter fats modified by fractional crystallization in the absence of solvents*, Ministry of Agric., Dairy Stn., Chaussee de Namur, Belgium.

Jasperson, H., and McKerrigan, A.A. (1957) The use of differential curves in the dilatometry of fats. *J. Sci. Food Agric. 8*, 46.

Jebson, R.S. (1970) Fractionation of milkfat into high and low melting point components. In *Proc. XVIII Int. Dairy Congr.*, Sydney, Australia, vol. E, p. 240.

Jebson, R.S. (1974a) Further studies on the fractionation of milkfat. In *Proc. XIX Int. Dairy Congr.*, New Delhi, India, vol. E, p. 751.

Jebson, R.S. (1974b) The use of fractions of milkfat in chocolate. In *Proc. XIX Int. Dairy Congr.*, New Delhi, India, vol. E, p. 761.

Jeffrey, M.S. (1991) The effect of cocoa butter origin, milk fat and lecithin levels on the temperability of cocoa butter systems. *Manufacturing Confectioner 71*, 76.

Jenness, R. (1988) Composition of milk. In *Fundamentals of Dairy Chemistry*, 4th ed., Wong, N.P., Van Nostrand Reinhold, New York, p. 1.

Jensen, R.G., and Clark, R.M. (1988) Lipid composition and properties. In *Fundamentals of Dairy Chemistry*, 4th ed., Wong, N.P., Van Nostrand Reinhold, New York, p. 171.

Jensen, R.G., Ferris, A.M., and Lammi-Keefe, C.J. (1991) The composition of milk fat. *J. Dairy Sci. 74*, 3228.

Jensen, R.G., Rubano Galluzzo, D., and Bush, V.J. (1990) Selectivity is an important characteristics of lipases (acylglycerol hydrolases). *Biocatalysis 3*, 307.

Jensen, R.G., Sampugna, J., Carpenter, D.L., and Pitas, R.E. (1967) Structural analysis of triglyceride classes obtained from cow's milk fat by fractional crystallization. *J. Dairy Sci. 50*, 231.

Jesse, E.V. (1989) World butterfat situation and outlook. In *Proc. Milkfat—trends and utilization*, Dairy Res. Conf. Center for Dairy Res., Univ. of Wisconsin-Madison, p. 9.

Johnson, B.R., and J.A. Conte, Jr. (1991) *Production of Low-Cholesterol Milk Fat by Solvent Extraction*, U.S. Patent 4,997,668.

Jordan, M.A. (1986) *Studies on butter oil*, Leatherhead Food R.A. Research Report, No. 568.

Joyner, N.T. (1953) The plasticizing of edible fats. *J. Am. Oil Chem. Soc. 30*, 526.

Juriaanse, A.C., and Heertje, I. (1988) Microstructure of shortenings, margarine and butter — a review. *Food Microstructure 7*, 181.

Kabara, J.J., McLaughlin, J.T., and Riegel, C.A. (1961) Quantitative microdetermination of cholesterol using tomatine as precipitating agent. *Anal. Chem. 33*, 305. (cited by Chen and deMan, 1966)

Kalo, P., and Kemppinen, A. (1993) Mass Spectrometric Identification of Triacylglycerols of Enzymatically Modified Butterfat Separated on a Polarizable Phenylmethylsilicone Column. *J. Am. Oil Chem. Soc. 70*, 1209.

Kalo, P., Parviainen, P., Kemppinen, A., and Antila, M. (1986) Triglyceride composition and melting and crystallisation properties of inter-esterified butter fat/tallow mixtures. *Meijeritieteellinen Aikakauskirja 44*, 37.

Kaneda, T., Nakajima, A., Fujimoto, K., Kobayashi, T., Kiriyama, S., Ebihara, K., Innami, T., Tsuji, K., Tsuji, E., Kinumaki, T., Shimada, H., and Yoneyama, S. (1980) *J. Nutri. Sci. Vitaminol. 26*, 497. (cited by Shishikura et al., 1986)

Kankare, V., and Antila, V. (1974a) Fractionation of milkfat. In *Proc. XIX Int. Dairy Congr.*, New Delhi, India, vol. E, p. 225.

Kankare, V., and Antila, V. (1974b) Use of low-melting milkfat fractions for improvement of the spreadability of butter. In *Proc. XIX Int. Dairy Congr.*, New Delhi, India, vol. E, p. 671.

Kankare, V., and Antila, A. (1986) Melting characteristics of milk fat and milk fat fractions. *Meijeritieteellinen Aikakauskirja 44*, 67.

Kankare, V., and Antila, A. (1988a) Über die physikaliscen und chemichen Eigenschaften der Milchfettenfraktionen [Physical and chemical characteristics of milk fat fractions]. *Fat Sci. Technol. 90*, 171.

Kankare, V., and Antila, V. (1988b) Melting characteristics of milk fat, milk fat fractions and fat mixtures. *Meijeritieteellinen Aikakauskirja 46*, 25.

Kankare, V., Antila, V., Harvala, T., and Komppa, V. (1989) Extraction of milk fat with supercritical carbon dioxide. *Milchwissenschaft 44,* 407.

Kaufmann, H.P. (1958) *Analyse der Fette und Fettproduckte,* BD. I, S. Springer-Verlag, Berlin-Göttingen-Heidelberg, p. 957. (cited by Antila and Antila, 1970)

Kaufmann, W., Biernoth, G., Frede, E., Merk, W., Precht, D., and Timmen, H. (1982) Fraktionierung von Butterfett durch Extraktion mit überkritischem CO_2 [Fractionation of butterfat by extraction with supercritical CO_2]. *Milchwissenschaft 37,* 92.

Kawanari, M., Okamoto, K., Matsumoto, K., Tanimoto, M., Kimura, T., and Takeya, K. (1992) Micro-structure of butter and its influence on the pastry quality. *Nippon Shokuhin Kogyo Gakkaishi 39,* 36.

Kaylegian, K.E. (1991) Physical and Chemical Properties of Milkfat Fractions, M.S. Thesis, University of Wisconsin-Madison.

Kaylegian, K.E., Hartel, R.W., and Lindsay, R.C. (1993) Application of modified milkfat in food products. *J. Dairy Sci. 76,* 1782.

Kaylegian, K.E., and Lindsay, R.C. (1992) Performance of selected milkfat fractions in cold-spreadable butter. *J. Dairy Sci. 75,* 3307.

Keenan, T.W., Dylewski, D.P., Woodford, T.A., and Ford, R.H. (1983) Origin of milk fat globules and the nature of the milk fat globule membrane. In *Developments in Dairy Chemistry—2 Lipids,* Fox, P., Applied Science Publishers, New York, p. 83.

Keenan, T.W., Mather, I.H., and Dylewski, D.P. (1988) Physical equilibria: lipid phase. In *Fundamentals of Dairy Chemistry,* 4th ed., Wong, N.P., Van Nostrand Reinhold, New York, p. 511.

Kehagias, C., and Radema, L. (1973) Storage of butter oil under various conditions. *Neth. Milk Dairy J. 27,* 379.

Kemppinen, A., and Kalo, P. (1993) Fractionation of the Triacylglycerols of Lipase-Modified Butter Oil. *J. Am. Oil Chem. Soc. 70,* 1203.

Kennedy, J.P. (1991) Structured lipids: fats of the future. *Food Technol. 45,* 76.

Keogh, M.K. (1988) Opportunities for fats: low fat spreads, deep fat frying, baking. In *Proc. Food and Industrial Uses of Fats Including Milkfat: The Potential of New Technology,* The Agriculture and Food Development Authority, Moorepark Research Centre, Fermoy, Co. Cork, Ireland.

Keogh, M.K. (1989) Anhydrous milk fat. 5. Deep fat frying. *Irish J. Food Sci. Technol. 13,* 129.

Keogh, M.K., and Higgins, A.C. (1986a) Anhydrous milk fat 1. Oxidative stability aspects. *Irish J. Food Sci. and Technol. 10,* 11.

Keogh, M.K., and Higgins, A.C. (1986b) Anhydrous milk fat 3. Fractionation aspects. *Irish J. Food Sci. and Technol. 10,* 35.

Keogh, M.K., and Morrissey, A. (1990) Anhydrous milk fat. 6. Baked foods. *Irish J. Food Sci. Technol. 14,* 69.

Kerkhoven, E., and deMan, J.M. (1966) Composition of the trisaturated glycerides of milk fat. *J. Dairy Sci. 49,* 1086.

Kermasha, S., Kubow, S., Safari, M., and Reid, A. (1993) Determination of the positional distribution of fatty acids in butterfat triacylglycerols. *J. Am. Oil Chem. Soc. 70,* 169.

Keshava Prasad, P.K., and Bhat, G.S. (1987a) Variation in conjugated polyunsaturated fatty acids content during autoxidation of milk-fat fraction. *Indian J. Anim. Sci. 57,* 1000.

Keshava Prasad, P.K., and Bhat, G.S. (1987b) Influence of oxygen and copper on autoxidation of milk fat fractions. *Indian J. Dairy Sci. 40,* 467.

Khalifa, M.Y., and Mansour, A.A. (1988) Physical, chemical and organoleptic properties of butter oil fractions. *Egyptian J. Dairy Sci. 16,* 47.

King, N. (1955) *The Milk Fat Globule Membrane and Some Associated Phenomena,* Commonwealth Agricultural Bureaux, Farnham Royal, Bucks, England.

Kinsella, J.E. (1970) Functional chemistry of milk products in candy and chocolate manufacture. *Manufacturing Confectioner 19*, 45.

Kinsella, J.E. (1975) Butter flavor. *Food Technol. 29*, 82.

Kinsella, J.E., Patton, S., and Dimick, P.S. (1967) The flavor potential of milk fat. A review of its chemical nature and biochemical origin. *J. Am. Oil Chem. Soc.* 449.

Kirkpatrick, K.J. (1982) Utilization of natural and modified milk fat in the food industry. In *Dairy Ingredients in Foods*, Int. Dairy Fed., Doc. 147, p. 31.

Kishonti, E. and Sjöström, G. (1965) *Svenska Mejertidn. 57*, 637. (cited by Mattsson et al., 1969)

Kleyn, D.H. (1992) Textural aspects of butter. *Food Technol. 46*, 118.

Knightbridge, P. (1978) Milkfat blends in bakery products. In *Proc. XX Int. Dairy Congr.*, Congrilait, France, p. 986.

Koops, J. (1960) *Off. Org. FNZ 52*, 333 (1960). *Alg. Zuivelbl.* 53, 127 (1960), *Missets Zuivel 66*, 304 (1960). (cited by Kehagias and Radema, 1973)

Kuksis, A., and Breckenridge, W.C. (1968) Triglyceride composition of milk fats. In *Dairy Lipids and Lipid Metabolism*, Brink, M.F., and Kritchevsky, D., AVI Publishing Co., Westport, CT.

Kuksis, A., Marai, L., and Myher, J.J. (1972) Triglyceride structure of milk fats. *J. Am. Oil Chem. Soc. 50*, 193.

Kuksis, A., Marai, L., and Myher, J.J. (1991) Reversed-phase liquid chromotography-mass spectrometry of complex mixtures of natural triacylglycerols with chloride-attachment negative chemical ionization. *J. of Chrom. 588*, 73.

Kuksis, A., and McCarthy, M.J. (1962) Gas-liquid chromatographic fractionation of natural triglyceride mixtures by carbon number. *Can. J. Biochem. Physiol. 40*, 679.

Kulkarni, S., and Rama Murthy, M.K. (1987) Relation between properties of fat and rheological characteristics of butter. *Indian J. Dairy Sci. 40*, 232.

Kupranycz, D.B., Amer, M.A., and Baker, B.E. (1986) Effects of thermal oxidation on the constitution of butterfat, butterfat fractions and certain vegetable oils. *J. Am. Oil Chem. Soc. 63*, 332.

Kuwabara, Y., Hidaka, H., Asahara, K., and Sagi, N. (1991) *Method for Dry Fractionation of Fats and Oils*, U.S. Pat. 5,045,243.

Kuzdzal-Savoie, S., and Kuzdzal, W. (1968) Utilisation des sels de baryum en vue du dosage des acides volatils du beurre. *Le Lait 155*, 475. (cited by Lechat et al., 1975)

Kuzdzal-Savoie, S., and Saada, M. (1978a) Cristallisation et fractionnement de la matière grasse laitière anhydre. 1. Modalités de la cristallisation et du fractionnement. *La Technique Laitière 920*, 8.

Kuzdzal-Savoie, S., and Saada, M. (1978b) Cristallisation et fractionnement de la matière grasse laitière anhydre. 2. Résultats. *La Technique Laitière 921*, 28.

Laakso, P.H., Nurmela, K.V.V., and Homer, D.R. (1992) Composition of the triacylglycerols of butterfat and its fractions obtained by an industrial melt crystallization process. *J. Agric. Food Chem. 40*, 2472.

Laakso, P., and Kallio, H. (1993a) Triacylglycerols of winter butterfat containing configurational isomers of monoenoic fatty acyl residues. I. disaturated monoenoic triacylglycerols. *J. Am. Oil Chem. Soc. 70*, 1161.

Laakso, P., and Kallio, H. (1993b) Triacylglycerols of winter butterfat containing configurational isomers of monoenoic fatty acyl residues. II. saturated dimonoenoic triacylglycerols. *J. Am. Oil Chem. Soc. 70*, 1173.

Lai, H-C., and Ney, D.M. (1992) Defined milkfat fractions alter postprandial lipemia in meal-fed rats *FASEB J. 6*, A1496 Abstract 3245.

Lakshminarayana, M., and Rama Murthy, M.K. (1985) Cow and buffalo milk fat fractions. Part I: Yield, physico chemical characteristics and fatty acid composition. *Indian J. Dairy Sci. 38*, 256.

Lakshminarayana, M., and Rama Murthy, M.K. (1986) Cow and buffalo milk fat fractions. Part III: Hydrolytic and autoxidative properties of milk fat fractions. *Indian J. Dairy Sci. 39,* 251.

Lambelet, P. (1983) Comparison of N.M.R. and D.S.C. methods for determining solid contents of fats application to milk fat and its fractions. *Lebensm.-Wiss. u.-Technol. 16,* 90.

Lambelet, P., and Raemy, A.J. (1983) *J. Am. Oil Chem. Soc.* in press. (cited by Lambelet, 1983)

Lansbergen, G.J.T., and Kemps, J.M.A. (1983) *Wet Fractionation of Hardened Butterfat,* European Patent 0.074,146.

Lansbergen, G.J.T., and Kemps, J.M.A. (1984) *Hardened Butterfat in Margarine Fat Blends,* U.S. Patent 4,479,976.

Lanzani, A., Bondioli, P., Mariani, C., Folegatti, L., Venturini, S., Fedeli, E., and Barreteau, P. (1994) A new short-path distillation system applied to the reduction of cholesterol in butter and lard. *J. Am. Oil Chem. Soc. 71,* 609.

Larsen, N.E., and Samuelsson, E.G. (1979) Some technological aspects on fractionation of anhydrous butterfat. *Milchwissenschaft 34,* 663.

Larsen, P. (1984) Roll-in fat for use in Danish pastry the effect of fat composition on quality of Danish pastry, fresh-baked and baked after a deep-freezing period. In *Milkfat and Its Modifications Contributions at a LIPIDFORUM symposium,* Marcuse, R., Scandinavian Forum for Lipid Research and Technology, Göteborg, Sweden, p. 83.

Lawson, H.W. (1985) *Standards for Fats and Oils,* The L.J. Minor Food Service Standards Series vol. 5., AVI Publishing Co., Inc., Westport, CT.

Lea, C.H. (1931) Effect of light on the oxidation of fats. In *Proc. of Royal Society of London 1028,* 175. (cited by Keshava Prasad and Bhat, 1987b; Lakshminaryana and Rama Murthy, 1985)

Lea, C.H. (1939) *Rancidity in Edible Fats,* Chemical Publ. Co., New York. (cited by Bhat and Rama Murthy, 1983)

Lechat, G., Varchon, P., Kuzdzal-Savoie, S., Langlois, D., and Kuzdzal, W. (1975) Cristallisation fractionnée de la matière grasse laitière anhydre. *Le Lait 55,* 295.

Lees, T.M., and DeMuria, P.J. (1962) *J. Chrom. 8,* 108. (cited by Norris et al., 1971)

Loftus Hills, G., and Thiel, C.C. (1945–46) *J. Dairy Res. 14,* 340. (cited by Kehagias and Radema, 1973)

Luddy, F.E., Barford, R.A., Herb, S.F., and Magidman, P. (1968) A rapid and quantitative procedure for the preparation of methyl esters of butteroil and other fats. *J. Am. Oil Chem. Soc. 45,* 549. (cited Arul et al., 1987; Makhlouf et al., 1987)

Lund, P. (1981) Isometric fatty Acids in milk fat. In *Lipidforum 11th Scandinavian Symposium on Lipids,* Lipidforum Scandinavian Forum for Lipid Research and Technology, Bergen, Norway, p. 117.

Lund, P. (1988) Analysis of butterfat triglycerides by capillary gas chromatography. *Milchwissenschaft 43,* 159.

Lund, P., and Danmark, H. (1987) Chemical and physical characteristics of butterfat fractions. In *Proc. 14th Scandinavian Symposium on Lipids,* Marcuse, R., Lipidforum Scandinavian Forum for Lipid Research and Technology, Göteborg, Sweden, p. 53.

Lutton, E.S. (1945) *J. Am. Chem. Soc. 67,* 524. (cited by Timms, 1980b)

Lutton, E.S. (1955) The polymorphism of glycerides. An application of X-ray diffractions. *J. Soc. Cosmetic Chem. 6,* 26. (cited by Helal et al., 1977, 1984)

Lynch, J., and Barbano, D.M. (1988) *Lab Manual, Institute of Food Science,* Cornell Univ., Ithaca, NY. (cited by Bhaskar et al., 1993b)

Lynch, J., and Barbano, D.M. (1991) *Laboratory Manual: Preparation of Quantitative GLC Fatty Acid Methyl Ester Standards,* Cornell Univ., Ithaca, NY. (cited by Shukla et al., 1994)

Lynch, J.M., Senyk, G.F., Barbano, D.M., Bauman, D.E., and Hartnell, G.F. (1988) Influence of sometribove (recombinant methionyl bovine somatotropin) on milk lipase and protease activity. *J. Dairy Sci. 71 (Supp 1),* 100, Abstract D115.

MacCollom, M.S. (1970) *Fractionating 99 to 100% Milk Fat and Making Butter from the Separated Fats,* U.S. Patent 3,519,435.

Macrae, A.R. (1983) Lipase-catalyzed interesterification of fats and oils. *J. Am. Oil Chem. Soc. 60,* 291.

Makhlouf, J., Arul, J., Boudreau, A., Verret, P., and Sahasrabudhe, M.R. (1987) Fractionnement de la matière grass laitière par cristallisation simple et son utilisation dans la fabrication de beurres mous. *Can. Inst. Food Sci. Technol. 20,* 236.

Malkamäki, J., and Antila, M. (1974) Use of hard milkfat fraction in foodstuffs. In *Proc. XIX Int. Dairy Congr.,* New Delhi, India, vol. E, p. 628.

Mangold, H.K., and Mallins, D.C. (1960) *J. Am. Oil Chem. Soc. 37,* 383. (cited by Norris et al., 1971)

Maniongui, C., Gresti, J., Bugaut, M., Gauthier, S., and Bezard, J. (1991) Determination of bovine butterfat triacylglycerols by reversed-phase liquid chromatography and gas chromatography. *J. Chrom. 543,* 81.

Marai, L., Myher, J.J., and Kuksis, A. (1983) Analysis of triacylglycerols by reversed-phase high pressure liquid chromatography with direct liquid inlet mass spectrometry. *Can. J. Biochem. Cell Biol. 61,* 840.

Marchis-Mouren, G., Sards, L., and Desnuelle, P. (1959) *Arch. Biochem. Bio Phys. 83,* 309. (cited by Lakshminarayana and Rama Murthy, 1985)

Martine, F. (1982) Fractionnement et hydrogenation de la matière grasse butyrique. *La Technique Laitière 967,* 17.

Mattsson, S., Swartling, P., and Nilsson, R. (1969) The major fatty acids in whole milk fat and in a fraction obtained by crystallization from acetone. *J. Dairy Res. 36,* 169.

Mayhill, P.G., and Newstead, D.F. (1992) The effect of milkfat fractions and emulsifier type on creaming in normal-solids UHT recombined milk. *Milchwissenschaft 47,* 75.

McBean, L.D., and Speckmann, E.W. (1988) Nutritive value of dairy foods. In *Fundamentals of Dairy Chemistry,* 4th ed., Wong, N.P., Van Nostrand Reinhold, New York, p. 343.

McCarthy, M.J., Kuksis, A., and Beveridge, J.M.R. (1962) Gas-liquid chromatographic analysis of the triglyceride composition of molecular distillates of butter oil. *Can. J. Biochem. Physiol. 40,* 693.

McGillivray, W.A. (1972) Softer butter from fractionated fat or by modified processing. *N.Z. J. Dairy Sci. Technol. 7,* 111.

McNeill, G.P. (1988) Modification of the physical properties of fats and oils by enzymatic and chemical processes. In *Proc. Food and Industrial Uses of Fats Including Milkfat: the Potential of New Technology,* The Agriculture and Food Development Authority, Moorepark Research Centre, Fermoy, Co. Cork, Ireland, p. 133.

McNeill, G.P., O'Donoghue, A., and Connolly, J.F. (1986) Quantification and identification of flavour components leading to lipolytic rancidity in stored butter. *Irish J. Food Sci. Technol. 10,* 1.

Mertens, W.G., and deMan, J.M. (1972) Automatic melting point determination of fats. *J. Am. Oil Chem. Soc. 49,* 366. (cited by deMan and Finoro, 1980)

Metcalfe, L.D., Schmitz, A.A., and Pelka, J.R. (1966) *Anal. Chem. 38,* 514. (cited by Norris et al., 1971)

Micich, T.J. (1990) Behavior of polymer-support digitonin with cholesterol in the absence and presence of butter oil. *J. Agric. Food Chem. 38,* 1839.

Micich, T.J. (1991) Behavior of polymer-supported tomatine toward cholesterol in the presence or absence of butter oil. *J. Agric. Food Chem. 39,* 1610.

Micich, T.J., Fogalia, T.A., and Hosinger, V.H. (1992) Polymer-supported saponins: an approach to cholesterol removal from butteroil. *J. Agric. Food Chem. 40,* 1321.

Mickle, J.B. (1960) *Changing the Molecular Structure of Milk Fat as a Means of Improving the Spreadability of Butter*, Oklahoma State University Experimental Station Tech. Bull. T-83.

Milk Industry Foundation. (1991) *Milk Facts*, 1991 ed., Washington, D.C.

Mogensen, G. (1984) Technological and quality aspects in milk fat modification. In *Milkfat and Its Modification Contributions at a LIPIDFORUM symposium*, Marcuse, R., Scandinavian Forum for Lipid Research and Technology, Göteborg, Sweden, p. 54.

Moore, J.L., Richardson, T., and Amundson, C.H. (1965) The chemical and physical properties of interesterified milk fat fractions. *J. Am. Oil Chem. Soc. 42*, 796. (cited by Richardson, 1968)

Mortensen, B.K. (1983) Physical properties and modification of milkfat. In *Developments in Dairy Chemistry—2 Lipids*, Fox, P., Applied Science Publishers, New York, p. 159.

Mukherjea, R.N., Leeder, J.G., and Chang, S.S. (1966) Improvement of keeping quality of butteroil by selective trace-hydrogenation and winterization. *J. Dairy Sci. 49*, 1381.

Mulder, H., and Walstra, P. (1974) *The Milk Fat Globule Emulsion Science as Applied to Milk Products and Comparable Foods*, Commonwealth Agricultural Bureaux, Farnham Royal, Bucks., England, and Centre for Agricultural Publishing and Documentation, Wageningen, the Netherlands.

Munro, S.D., Bissell, T.G., Jebson, R.S., Norris, R., and Taylor, M.W. (1978) fat. In *Proc. XX Int. Dairy Congr.*, Paris, France, p. 862.

Munro, D.S., and Illingworth, D. (1986) Milk fat based food ingredients: present and potential products. *Food Technol. Aust. 38*, 335.

Muuse, B.G., and van der Kamp, H.J. (1985) Detection of the presence of fractionated butterfat by crystallization of the high-melting saturated triglycerides. *Neth. Milk Dairy J. 39*, 1.

Myher, J.J., Kuksis, A., and Marai, L. (1988) Identification of the more complex triacylglycerols in bovine milk fat by gas chromatography-mass spectrometry using polar capillary columns. *J. Chrom. 452*, 93.

Myher, J.J., Kuksis, A., and Marai, L. (1993) Identification of the less common isologous short-chain triacylglycerols in the most volatile 2.5% molecular distillate of butter oil. *J. Am. Oil Chem. Soc. 70*, 1183.

Nawar, W.W. (1985a) Lipids. In *Food Chemistry*, 2nd ed., Fennema, O.R., Marcel Dekker, Inc., New York, p. 176.

Nawar, W.W. (1985b) Chemistry of thermal oxidation of lipids. In *Flavor Chemistry of Fats and Oils*, Min, D.B., and Smouse, T.H., American Oil Chemists' Society, Champaign, IL, p. 39.

NEN 6350. (1977) *Determination of the total sterol content*, NNI, Delft, the Netherlands. (cited by Muuse and van der Kamp, 1985)

Newlander, J.A., and Atherton, H.V. (1964) *The Chemistry and Testing of Dairy Products*, 3rd ed., Olsen Publishing Co., Milwaukee, WI.

Ney, D.M. (1991) Symposium: The role of the nutritional and health benefits in the marketing of dairy products. *J. Dairy Sci. 74*, 4002.

Ney, D.M., and Lai, H.C. (1992) Hepatic secretion and plasma clearance of butterfat triacylglycerols. *Int. News on Fats, Oils and Rel. Mat. (INFORM) 3*, 551 Abstract III5.

Norris, F.A. (1982) Extraction of fats and oils. In *Bailey's Industrial Oil and Fat Products*, 4th ed., Swern, D., John Wiley and Sons, New York, vol. 2, p. 175.

Norris, G.E., Gray, I.K., and Dolby, R.M. (1973) Seasonal variations in the composition and thermal properties of New Zealand milk fat. II. Thermal properties of milk fat and their relation to composition. *J. Dairy Res. 40*, 311.

Norris, R. (1977) *Milk Fat Containing Products and Processes Therefor*, U.S. Patent 4,005,228.

Norris, R., Gray, I.K., McDowell, A.K.R., and Dolby, R.M. (1971) The chemical composition and physical properties of fractions of milk fat obtained by a commercial fractionation process. *J. Dairy Res. 38*, 179.

Oakenfull, D., and Sidhu, G.S. (1991) Processing technology for cholesterol extraction. In *Proc. March 1991 Conference on Fat and Cholesterol Reduced Foods*, International Business Communications, South Natick, MA.

Padley, F.B., and Timms, R.E. (1978) *Lebensm. Wiss. Technol. 11*, 319. (cited by Timms, 1980b)

Palmquist, D.L., Beaulieu, A.D., and Barbano, D.M. (1993) Feed and Animal Factors Influencing Milk Fat Composition. *J. Dairy Sci. 76*, 1753.

Pariza, M.W., Albright, K., Benjamin, H., Storkson, J., Liu, W., and Chin, S. (1993) Biological significance of conjugated dienoic derivatives of linoleic acid. In *University of Wisconsin Center for Dairy Research Annual Report*, Madison, WI, p. 147.

Parks, O.W., Keeney, M., and Schwartz, D.P. (1961) Bound aldehydes in butteroil. *J. Dairy Sci. 44*, 1940.

Parodi, P.W. (1970) *Aust. J. Dairy Technol. 25*, 200. (cited by Black, 1973; Parodi, 1974b, 1979)

Parodi, P.W. (1972) *Aust. J. Dairy Technol. 27*, 90. (cited by Parodi, 1974a)

Parodi, P.W. (1974a) The composition of a high melting glyceride fraction from milkfat. *Aust. J. Dairy Technol. 29*, 20.

Parodi, P.W. (1974b) Variation in the fatty acid composition of milkfat: Effect of stage of lactation. *Aust. J. Dairy Technol. 29*, 145.

Parodi, P.W. (1979a) Stereospecific distribution of fatty acids in bovine milk fat triglycerides. *J. Dairy Res. 46*, 75.

Parodi, P.W. (1979b) Relationship between trisaturated glyceride compostion and the softening point of milk fat. *J. Dairy Res. 46*, 633.

Parodi, P.W. (1981) Relationship between triglyceride structure and softening point of milk fat. *J. Dairy Res. 48*, 131.

Parodi, P.W. (1982) Positional distribution of fatty acids in the triglyceride classes of milk fat. *J. Dairy Res. 49*, 73.

Parodi, P.W., and Dunstan, R.J. (1971) The relationship between the fatty acid composition, the softening point and the refractive index of milkfat. *Austr. J. Dairy Technol. 26*, 29.

Peck, S.M. (1990) Response Surface Analysis of Ingredient Contribution to Quality Characteristics of Cookies, M.S. Thesis, Ohio State University.

Pedersen, A. (1988) Puff pastry butter—a new product in the dairy industry. *Special Issue of: Danish Dairy and Food Ind.*

Petersson, B. (1986) Pulsed NMR method for solid fat content determination in tempering fats Part II. Cocoa butters and equivalents in blends with milk fat. *Fette Seifen Anstrichm. 88*, 128.

Pilhofer, G.M., Lee, H.C., McCarthy, M.J., Tong, P.S., and German, J.B. (1994) Functionality of milk fat in foam formation and stability. *J. Dairy Sci. 77*, 55.

Podmore, J. (1991) Bakery applications of milkfat and margarine. In *Milk Fat: Production, Technology and Utilization*, Rajah, K.K., and Burgess, K.J., Society of Dairy Technology, Huntingdon, Cambridgeshire, England, p. 102.

Prasad, S., and Gupta, S.K. (1983) Use of fractionated butterfat on the quality of butter powder and its spread. *Asian J. Dairy Res. 2*, 196.

Precht, D. (1988) Fat crystal structure in cream and butter. In C*rystallization and Polymorphism of Fats and Fatty Acids*, Garti, N., and Sato, K., Marcel Dekker, Inc., New York, p. 305.

Precht, D., Frede, E., and Timmen, H. (1984) Effect of feeding on the milk fat composition. In *Milkfat and Its Modification Contributions at a LIPIDFORUM Symposium*, Marcuse, R., Scandinavian Forum for Lipid Resesarch and Technology, Göteborg, Sweden, p. 69.

Pyler, E.J. (1988) *Baking Science and Technology*, 3rd ed., Sosland Publishing Co., Merriam, KS, vol., 1, p. 443.

Rajah, K.K. (1991) Anhydrous milkfat and fractionated products. In *Milk Fat: Production, Technology and Utilization*, Rajah, K.K., and Burgess, K.J., Society of Dairy Technology, Huntingdon, Cambridgeshire, England, p. 37.

Rajah, K. (1992) Recent Technology for the manufacture of butter, margarine and spreads. *Lipid Technol. 4,* 129.

Rama Murthy, M.K., and Narayanan, K.M. (1971) *Milchwissenchaft 26,* 693. (cited by Kulkarni and Rama Murthy, 1987; Lakshminarayana and Rama Murthy, 1985)

Ramesh, B., and Bindal, M.P. (1987a) *Indian J. Dairy Sci. 40,* 94. (cited by Ramesh and Bindal, 1987b)

Ramesh, B., and Bindal, M.P. (1987b) Softening point, melting point and fatty acid composition of solid and liquid fractions of cow, buffalo and goat ghee. *Indian J. Dairy Sci. 40,* 303.

Reimer, L., and Pfefferkorn, G. (1973) In *Rasterelektronenmikroskopie*, Springer Verlag, Berlin, Ch. 7. (cited by Schaap et al., 1975).

Richardson, T. (1968) Studies on milk fat and milk fat fractions. In *Dairy Lipids and Lipid Metabolism*, Brink, M.F., and Kritchevsky, D., AVI Publishing Co., Westport, CT, p. 4.

Richardson, T., and JiminezFlores, R. (1989) *Extraction du Cholestérol des Produits Laitiers à l'Aide de Saponine,* University of California-Davis. U.S. Pat. No. 421,153. (cited by Boudreau and Arul, 1993)

Richardson, T., and Korycka-Dahl, M. (1983) Lipid oxidation. In *Developments in Dairy Chemistry—2 Lipids*, Fox, P., Applied Science Publishers, New York, p. 241.

Riel, R.R. (1966) Study of the dilatometric and melting properties of modified milk fat. In *Proc. XVII Int. Dairy Congr.*, München, Germany, sec. C:2, p. 295.

Riel, R.R., and Paquet, R. (1972) Procédé continu de fractionnement des graisses par fusion et filtration. *J. Inst. Can. Sci. Technol. Aliment. 5,* 210.

Rifaat, I.D., Farag, M.S., Hilal, F.R., and El-Sadek, G.M. (1973) X-ray diffraction and infrared studies on crystal formation in high melting fraction of buffalo butter fat. *Egyptian J. Dairy Sci. 1,* 85.

Rizvi, S.S.H. (1991) Supercritical fluid processing of milk fat. In *Dairy Center News 3,* 1. Northeast Dairy Foods Research Center, Ithaca, NY.

Rizvi, S.S.H., Lim, S., Nikoopour, H., Singh, M., and Yu, Z. (1990) Supercritical fluid processing of milk fat. In *Engineering and Food, vol. 3. Advanced Processes*, Spiess, W.E.L., and Schubert, H., Elsevier Applied Science, New York, p. 145.

Rizvi, S.S., Yu, Z-R., Bhaskar, A., and Rosenberry, L. (1993) Phase equilibria and distribution coefficients of δ-lactones in supercritical carbon dioxide. *J. Food Sci. 58,* 996.

Rolland, J.R., and Riel, R.R. (1966) Separation of milk fat fractions by centrifugation. *J. Dairy Sci. 49,* 608.

Sadler, A.M., and Wong, N.P. (1970) Milk fat utilization in foods. In *Byproducts from Milk*, 2nd ed, Webb, B.H., AVI Publishing Company, Inc., Westport, CT, p. 274.

Sampugna, J., Carpenter, D.L., Marks, T.A., Quinn, J.G., Pereira, R.L., and Jensen, R.G. (1966) Interpretation of pancreatic lipase analyses of cow's milk fat triglyceride structure. *J. Dairy Sci. 49,* 163. (cited by Jensen et al., 1967)

Schaap, J.E. (1982) Application of fractionated milkfat as shortening in flour confectionery products. In *Dairy Ingredients in Foods*, Int. Dairy Fed., Doc. 147, p. 51.

Schaap, J.E., Badings, H.T., Schmidt, D.G., and Frede, E. (1975) Differences in butterfat crystals, crystallized from acetone and from the melt. *Neth. Milk Dairy J. 29,* 242.

Schaap, J.E., and Rutten, G.A.M. (1976) Effect of technological factors on the crystallization of bulk milk fat. *Neth. Milk Dairy J. 30,* 197.

Schaap, J.E., and van Beresteyn, E.C.H. (1970) Fractioneren van melkvet. *Voedingsmiddelentechnologie 1,* 553.

Schroder, B.G., and Baer, R.J. (1990) Utilization of cholesterol-reduced milk fat in fluid milks. *Food Technol. 44,* 145.

Schulz, M.E., and Timmen, H. (1966) Versuche zur Milchfettfraktionierung und Möglichkeiten ihrer Anwendung [Experiments on milk fat fractionation and its possible application]. In *Proc. XVII. Int. Dairy Congr.*, München, Germany, sec. C:2, p. 155.

Schwartz, D.P., Haller, H.S., and Keeney, M. (1963) Direct quantitative isolation of monocarbonyl compounds from fats and oils. *Anal. Chem. 35,* 2191. (cited by Bhat and Murthy, 1983; Keshava Prasad and Bhat, 1987b)

Schwartz, D.P., and Parks, O.W. (1961) *Anal. Chem. 33,* 1396. (cited by Walker, 1972)

Schwartz, D.P., Parks, P.W., and Yoncoskie, R.A. (1966) *J. Am. Oil Chem. Soc. 43,* 128. (cited by Walker, 1972)

Seibel, W. (1987) Verwendung von fraktioniertem Butterreinfett (Butterziehfett). *Zuckerund Susswarenwirtschaft 40,* 122.

Shehata, A.A.Y., deMan, J.M., and Alexander, J.C. (1970) A simple and rapid method for the preparation of methyl esters of fats in milligram amounts for gas-chromatography. *Can. Inst. Food Technol. J., 3,* 85.

Shehata, A.A.Y., deMan, J.M., and Alexander, J.C. (1971) Triglyceride composition of milk fat. I. Separation of triglycerides by column and gas-liquid chromatography. *J. Inst. Can. Technol. Aliment. 4,* 61.

Shehata, A.A.Y., deMan, J.M., and Alexander, J.C. (1972) Triglyceride composition of milk fat: II. Separation of triglycerides by argentation-TLC and GLC. *Can. Inst. Food Sci. Technol. J. 5,* 13.

Shellhammer, T.H., Fairley, P., German, J.B., and Krochta, J.M. (1993) Water vapor barrier properties of simulated milk fat fractions. In *1993 IFT Annual Meeting Technical Program: Book of Abstracts*, Institute of Food Technologists, Chicago, IL, Abstract 922.

Shepherd, I.S., and Yoell, R.W. (1976) Cake emulsions. In *Food Emulsions*, Friberg, S., Marcel Dekker Inc., New York, p. 215.

Sherbon, J.W. (1963) The Physical State and Thermodynamic Properties of Selected Milk Fat Systems, Ph.D. Thesis, University of Minnesota.

Sherbon, J.W. (1974) Crystallization and fractionation of milk fat. *J. Amer. Oil Chem. Soc. 51,* 22.

Sherbon, J.W. (1988) Physical properties of milk. In *Fundamentals of Dairy Chemistry*, 4th ed., Wong, N.P., Van Nostrand Reinhold, New York, p. 409.

Sherbon, J.W., and Dolby, R.M. (1973) Preparation and fractionation of the high melting glyceride fractions of milk fat. *J. Dairy Sci. 56,* 52.

Sherbon, J.W., Dolby, R.M., and Russell, R.W. (1972) The melting properties of milk fat fractions obtained by double fractionation using a commercial process. *J. Dairy Res. 39,* 325.

Shishikura, A., Fujimoto, K., Kaneda, T., Arai, K., and Saito, S. (1986) Modification of butter oil by extraction with supercritical carbon dioxide. *Agric. Biol. Chem. 50,* 1209.

Shukla, A., Bhaskar, A.R., Rizvi, S.S.H., and Mulvaney, S.J. (1994) Physiochemical and rheological properties of butter made from supercritically fractionated milk fat. *J. Dairy Sci. 77,* 45.

Simoneau, C., Fairley, P., Krochta, J.M., and German, J.B. (1994) Thermal behavior of butterfat fractions and mixtures of tripalmitin and butterfat. *J. Am. Oil Chem. Soc. 71,* 795.

Singh, B., and Rizvi, S.S.H. (1994) Design and economic analysis for continuous countercurrent processing of milk fat with supercritical carbon dioxide. *J. Dairy Sci. 77,* 1731.

Smith, L.M. (1961) *J. Dairy Sci. 44,* 607. (cited by Mattsson et al., 1969)

Smith, L.M., and Vasconcellos, A. (1974) Some factors affecting hydrogenation of milk fat. *J. Am. Oil Chem. Soc. 51,* 26.

Sokolov, F.S. (1974) Increasing the stability of milkfat by the use of antioxidants. In *Proc. XIX Int. Dairy Congr.*, New Delhi, India, vol. E, p. 796.

Sonntag, N.O.V. (1979a) Structure and composition of fats and oils. In *Bailey's Industrial Oil and Fat Products*, vol 1, 4th ed., Swern, D., John Wiley and Sons, New York, p. 1.

Sonntag, N.O.V. (1979b) Composition and characteristics of individual fats and oils. In *Bailey's Industrial Oil and Fat Products*, vol 1, 4th ed., Swern, D., John Wiley and Sons, New York, p. 289.

Sonntag, N.O.V. (1982) Analytical methods. In *Bailey's Industrial Oil and Fat Products*, vol 2, 4th ed., Swern, D., John Wiley and Sons, New York, p. 407.

Sperber, R.M. (1989) New technologies for cholesterol reduction. *Food Proc. 50,* 154.

Sreebhashyam, S.K., Gupta, S.K., and Patel, A.A. (1981) A comparative study of buffalo and cow butterfat fractions. *Indian J. Dairy Sci. 34,* 310.

Stark, W., Urbach, G., and Hamilton, J.S. (1976) Volatile compounds in butter oil V. The quantitative estimation of phenol, o-methoxyphenol, m-and p-cresol, indole, and skatole by cold-finger molecular distillation. *J. Dairy Res. 43,* 479.

Stepanenko, T.A., and Tverdokhleb, G.V. (1974) Chemical composition and physico-chemical properties of milkfat fractions obtained without use of solvents. In *Proc. XIX Int. Dairy Congr.*, New Delhi, India, vol E., p. 206.

Tangel, F.P., Jr., Leeder, J.G., and Chang, S.S. (1977) Deep fat frying characteristics of butteroil. *J. Food Sci. 42,* 1110.

Taylor, M.W., and Hawke, J.C. (1975a) The triacylglycerol compositions of bovine milkfats. *N. Z. J. Dairy Sci. Technol. 10,* 40.

Taylor, M.W., and Hawke, J.C. (1975b) Structural analysis of the triacylglycerols of bovine milkfats. *N.Z. J. Dairy Sci. Technol. 10,* 49.

Thomas, A.E., III. (1985) Fractionation and winterization: processes and products. In *Bailey's Industrial Fat and Oil Products*, vol. 3, Applewhite, T.H., John Wiley and Sons, New York, p. 1.

Thomas, M.A. (1973a) The use of a hard milkfat fraction in processed cheese. *Aust. J. Dairy Technol. 28,* 77.

Thomas, M.A. (1973b) Cheese spreads using hard milkfat fraction. *Aust. J. Dairy Technol. 28,* 151.

Thompson, S.Y., Ganguly, J., and Kon, S.K. (1949) *Br. J. Nutr. 3,* 50. (cited by Norris et al., 1971)

Timmen, H. (1974) Gas Chromatographic detection of milkfat fractionation. In *Proc. XIX Int. Dairy Congr.*, New Delhi, India, vol. E, p. 491.

Timmen, H. (1975a) Die Verwendung von Milchfettfraktionen zur Verbesserung der Streichfähigkeit der Butter. *Die Molkerei.-Zeitung der Milch 29,* 1259.

Timmen, H. (1975b) *Milchwissenschaft 30,* 329. (cited by Kaufmann et al., 1982)

Timmen, H. (1978) Improvement of oxidation stability of pure butterfat by antioxidants. In *Proc. XX Int. Dairy Congr.*, Congrilait, Paris, France, p. 865.

Timmen, H., Frede, E., and Precht, D. (1984) Characterization of short and long chain butterfat fractions obtained with supercritical carbon dioxide. In *Proc. Milkfat and Its Modification Contributions at a LIPIDFORUM Symposium*, Marcuse, R., Scandinavian Forum for Lipid Research and Technology, Göteborg, Sweden, p. 92.

Timms, R.E. (1978a) *Aust. J. Dairy Technol. 33,* 4. (cited by Timms, 1980b; Timms and Parekh, 1980)

Timms, R.E. (1978b) The solubility of milk fat, fully hardened milk fat and milk fat hard fraction in liquid oils. *Aust. J. Dairy Technol. 33,* 130.

Timms, R.E. (1978c) Automatic determination of the softening point of milk fat, milk fat fractions and butter. *Aust. J. Dairy Technol. 33,* 143.

Timms, R.E. (1979) The physical properties of blends of milk fat with beef tallow and beef tallow fractions. *Aust. J. Dairy Technol. 34,* 60.

Timms, R.E. (1980a) The phase behaviour and polymorphism of milk fat, milk fat fractions and fully hardened milk fat. *Aust. J. Dairy Technol. 35,* 47.

Timms, R.E. (1980b) The phase behavior of mixtures of cocoa butter and milk fat. *Lebensm.-Wiss. u.-Technol. 13,* 61.

Timms, R.E., and Parekh, J.V. (1980) The possibilities for using hydrogenated, fractionated or interesterified milk fat in chocolate. *Lebensm.-Wiss. u.-Technol. 13,* 177.

Tirtiaux, A. (1983) Tirtiaux fractionation: industrial applications. *J. Amer. Oil Chem. Soc. 60,* 473.

Tirtiaux, F. (1976) Le fractionnement industriel des corps gras par cristallisation dirigée — procédé Tirtiaux. *Oléagineux 31,* 279.

Tirtiaux, s.a. Fractionnement, undated. Product information brochure. Fleurus, Belgium.

Tolboe, O. (1984) Physical characteristics of butter fat and their influence on the quality of Danish pastry and cookies. In *Milkfat and Its Modifications Contributions at a LIPIDFORUM Symposium,* Marcuse, R., Scandinavian Forum for Lipid Research and Technology, Göteborg, Sweden, p. 43.

Tucker, V.C. (1974) Uses for hard milkfat fraction. In *Proc. XIX Int. Dairy Congr.,* New Delhi, India, vol. E, p. 762.

Tverdokhleb, G.V., and Avvakumov, A.K. (1978) Phase changes in butterfat during secondary structure formation of butter during storage. In *Proc. XX Int. Dairy Congr.,* Congrilait, Paris, France, p. 868.

Tverdokhleb, G.V., Stepanenko, T.A., and Nesterov, V.N. (1974) Formation of milkfat crystals as a function of chemical composition and cooling conditions. In *Proc. XIX Int. Dairy Congr.,* New Delhi, India, vol. E, p. 214.

Tverdokhleb, G.V., and Vergelesov, V.M. (1970) Factors controlling crystallization of milkfat. In *Proc. XVIII Int. Dairy Congr.,* Sydney, Australia, vol. 1E, p. 211.

Urbach, G. (1963) *J. Chrom. 12,* 196. (cited by Bhat and Rama Murthy, 1983)

Urbach, G. (1990) Effect of feed on flavor in dairy foods. *J. Dairy Sci. 73,* 3639.

Urbach, G. (1991) Butter flavor in food systems. *CSIRO Food Res. Quarterly 51,* 50.

Urbach, G., Stark, W., and Hamilton, J.S. (1970) Isolation and estimation of phenol, o-methoxyphenol, and m- and p-cresols in butter. In *Proc. XVIII Int. Dairy Congr.,* Sydney, Australia, vol. 1E, p. 233.

USDA Specifications. (1975) Federal Register, Oct. 10.

van Beresteyn, E.C.H. (1972) Polymorphism in milk fat in relation to the solid/liquid ratio. *Neth. Milk Dairy J. 26,* 117.

van Beresteyn, E.C.H., and Walstra, P. (1971) De invloed van de dispersiteit van melkvet op het kristallisatiegedrag [Effect of the dispersion of milkfat on the crystallization behavior]. *Voedingsmiddelentechnologie 2,* 14.

van den Embden, J.C., Haighton, A.J., van Putte, K., Vermaas, L.F., and Waddington, D. (1978) *Fette Seifen Anstrichmittel 80,* 180. (cited by Kaufmann et al., 1982)

van Wijngaarden, D. (1967) Modified rapid preparation of fatty acid esters from lipids for gas chromatographic analysis. *Anal. Chem. 39,* 849. (cited by Norris et al., 1971; Sherbon and Dolby, 1972)

Vasishtha, A.K., Leeder, J.G., and Chang, S.S. (1970) Trace hydrogenation of butteroil at low temperatures. *J. Food Sci. 35,* 395.

Verhagen, L.A.M., and Bodor, J. (1984) *Spreadable Water-in-Oil Emulsion Based on a High-Melting Butterfat Fraction and a Liquid Oil,* U.S. Patent 4,438,149.

Verhagen, L.A.M., and Warnaar, L.G. (1984) *Low-calorie Spread Based on a Low-Melting Butterfat Fraction,* U.S. Patent 4,436,760.

Versteeg, C. (1991) Milkfat fractionation and cholesterol removal. *Food Res. Quarterly 51,* 32.

Voss, E., Beyerlein, U., and Schmanke, E. (1971) Probleme der technischen Fraktionierung von Butterfett mit Hilfe des Fettkörnchen-Filtrationsverfahrens. *Milchwissenschaft 26,* 605.

Vovan, X.V., and Riel, R.R. (1973) Amélioration de la tartinabilté du beurre par soustraction de la fraction intermédiaire. *Can. Inst. Food Sci. Technol. J. 6,* 254.

Walker, N.J. (1972) Distribution of flavour precursors in fractionated milkfat. *N.Z. J. Dairy Sci. Technol. 7,* 135.

Walker, N.J. (1974) Flavour potential of fractionated milkfat. In *Proc. XIX Int. Dairy Congr.*, New Delhi, India, vol. E, p. 218.

Walker, N.J., and Bosin, W.A. (1971) Comparison of SFI, DSC and NMR methods for determining solid-liquid ratios in fats. *J. Am. Oil Chem. Soc. 48*, 50.

Walker, N.J., Cant, P.A.E., and Keen, A.R. (1977) Lactones in fractionated milkfat and spreadable butter. *N.Z. J. Dairy Sci. Technol. 12*, 94.

Walker, N.J., Cant, P.A.E., and Keen, A.R. (1978) Lactones in fractionated milkfat and spreadable butter. In *Proc. XX Int. Dairy Congr.*, Congrilait, Paris, France, p. 700.

Walstra, P. (1983) Physical chemistry of milk fat globules. In *Developments in Dairy Chemistry—2 Lipids*, Fox, P., Applied Science Publishers, New York, p. 119.

Weihrauch, J.L. (1988) Lipids of milk: deterioration. In *Fundamentals of Dairy Chemistry*, 4th ed., Wong, N.P., Van Nostrand Reinhold, New York, p. 215.

Weiss, T.J. (1970) *Food Oils and Their Uses,* AVI Publishing Co., Inc., Westport, CT.

Wilson, B.W. (1975) Techniques of fractionation of milk fat. *Aust. J. Dairy Technol. 30*, 10.

Wolff, J.P. (1968) *Manuel d'Analyse des Corps Gras*, Azoulay, Paris, France, p. 115. (cited by Deroanne, 1976)

Wong, D.W.S. (1989) *Mechanism and Theory in Food Chemistry*, Van Nostrand Reinhold, New York, p. 1.

Woo, A.H., and Lindsay, R.C. (1983) Statistical correlation of quantitative flavor intensity assessments and individual free fatty acid measurements for routine detection and prediction of hydrolytic rancidity off-flavors in butter. *J. Food Sci. 48*, 1761.

Woodrow, I.L., and deMan, J.M. (1968) Polymorphism in milk fat shown by X-ray diffraction and infrared spectroscopy. *J. Dairy Sci. 51*, 996.

Yeh, A., Liang, J.H., and Hwang, L.S. (1991) Separation of fatty acid esters from cholesterol in esterified natural and synthetic mixtures by supercritical carbon dioxide. *J. Am. Oil Chem. Soc. 68*, 224.

Yi, M. (1993) Effects of Milk Fat on Chocolate Bloom, M.S. Thesis, University of Wisconsin-Madison.

Yom, H.C., and Bremel, R.D. (1993) Genetic engineering of milk composition: modification of milk components in lactating transgenic animals. *Am. J. Clin. Nutr. 58 (suppl)*, 299S.

Youssef, A.M., Salama, F.A., and El-Ghanam, M.S. (1977) Fractional crystallization of cow and buffalo milk fats from acetone. *Alex. J. Agric. Res. 25*, 459.

Index

Acid degree value, *see* Lipolysis
Aeration properties, general functionality, 519
Alfa Laval fractionation process, 46
Anhydrous milkfat
    manufacture of, 21
    used to produce milkfat fractions, 145
Antioxidants, 35

Bakery products, 542
    cakes, 548
        characteristics of cakes made with experimental milkfat ingredients, 594
        characteristics of experimental and commercial milkfat ingredients for cakes, 550, 586, 591
        desired functionality attributes of milkfat ingredients for cakes, 548
        productions of experimental milkfat ingredients for cakes, 549, 582
    cookies, *see* short pastries
    laminated pastries, 543
        characteristics of experimental and commercial milkfat ingredients for laminated pastries, 545, 586
        characteristics for laminated pastries made with milkfat ingredients, 546, 586
        desired functionality attributes of milkfat ingredients for laminated pastries, 543
        production of experimental milkfat ingredients for laminated pastries, 544, 582
    short pastries, 546
        characteristics of milkfat ingredients for short pastries, 547, 586
        characteristics of short pastries made with milkfat ingredients, 547, 594
        desired functionality attributes of milkfat ingredients for short pastries, 546
        production of experimental milkfat ingredients for short pastries, 547, 582
Biological processes to modify milkfat, 14
    diet, 14
    genetic modification, 15
Blends of milkfat fractions, 17, 209
    with cocoa butter, 17, 209, 473
    with milkfat and other milkfat fractions, 209
    with other edible oils, 17, 210, 473
Butter, *see* Dairy products

Cakes, *see* Bakery products
Cheese, *see* Dairy products

Chemical composition of milkfat, 22
    factors affecting, 20, 521
    functionality, 513
    see Fatty acid composition
    see Flavor
    see Triglyceride composition
Chemical processes available to modify milkfat, 13
    hydrogenation, 13
    interesterification, 13
Chocolate, 552
    antiblooming effects of milkfat, 552
    characteristics of chocolate made with experimental milkfat ingredients, 556, 599, 610
    characteristics of experimental milkfat ingredients for chocolate, 555, 607, 608
    commercial milkfat ingredients used in chocolate, 606
    desired functionality attributes of milkfat ingredients for chocolate, 553
    production of experimental milkfat ingredients for chocolate, 554
    softening effects of milkfat, 552
Cholesterol
    content of intact milkfat and milkfat fractions, 207, 468, 507
    methodology, 90, 137
    milkfat and hypercholesterolemia, 36
    removal methods, 15
Cloud point, *see* Melting point
Coating properties, general functionality, 520
Cocoa butter, blends with milkfat, 17, 209, 473, 555
Confections, 633
Convenience foods, 633
Cookies, *see* Bakery products
Cosmetics, 633
Cream, *see* Dairy products
Crystal
    characteristics, 30, 84
        methodology, 84, 117
        of intact milkfat, 30, 146, 430
        of milkfat fractions, 499
            very-high-melting, 155, 431
            high-melting, 168, 432, 434
            middle-melting, 180, 432
            low-melting, 194, 433
            unknown-melting, 205, 433, 435
    growth, 30
    nucleation, 29
Crystallization behavior
    factors affecting, 521
    general functionality, 514

Crystallization behavior *(continued)*
　of intact milkfat, 29
Crystallization
　fractionation, 39
　　from melted milkfat, 8, 43
　　　methodology, 8, 43, 55, 490
　　　milkfat fractions produced, 228
　　　　very-high-melting, 148, 491
　　　　high-melting, 159, 492
　　　　middle-melting, 172, 492
　　　　low-melting, 185, 493
　　　　very-low-melting, 199, 493
　　　　unknown-melting, 203
　　from solvent solution, 9, 49
　　　methodology, 8, 49, 67, 490
　　　milkfat fractions produced, 247
　　　　very-high-melting, 150, 491
　　　　high-melting, 161, 492
　　　　middle-melting, 173, 492
　　　　low-melting, 188, 493
　　　　very-low-melting, 199, 493
　　　　unknown-melting, 203
　　　solvents employed, 50
　kinetics, 42
　process, 29, 39, 498
　　growth, 30
　　nucleation, 29
　process variables, 29, 40, 529
　　agitation rate, 41, 529
　　cooling rate, 40, 529
　　crystallization temperature, 41, 529
　　crystallization time, 42
　　seeding, 40
　　supercooling, 40
　texturization of milkfat ingredients, 528

Dairy-based spreads, *see* Dairy products
Dairy powders, *see* Dairy products
Dairy products
　butter
　　cold-spreadable, 531
　　　characterization of experimental cold-spreadable butters, 535
　　　commercial cold-spreadable butters, 538
　　　desired functionality attributes, 531
　　　production of experimental cold-spreadable butters, 532, 564
　　reworking and tempering, 13
　cheese, 559, 615, 626
　commercial milkfat ingredients for use in dairy products, 624
　cream, tempering, 12
　dairy-based spreads, 529
　　characteristics of experimental dairy-based spreads, 541, 580–581
　　desired functionality attributes, 539
　　production of experimental dairy-based spreads, 540, 577
　dairy powders, 558, 615, 626
　ice cream, 559
　recombined dairy products, 558, 615, 626
　specialized dairy products, 632
　whipping cream, 559, 615, 626
Density, *see* Specific gravity
De Smet fractionation process, 45
Deterioration of milkfat, *see* Lipolysis and Oxidation
Diet manipulation of the dairy cow, 14, 21
Differential scanning calorimetry, *see* Thermal profile and Solid fat content
Dilatometry, *see* Solid fat content
Dispersion properties, general functionality, 518
Dropping point, *see* Melting point

Edible films, 633

Fatty acid
　composition
　　of commercial milkfat ingredients, 211, 482
　　of intact milkfat, 22, 145, 302–309, 345–348
　　of milkfat fractions, 495
　　　very-high-melting, 152, 310–313, 349–352
　　　high-melting, 165, 314–320, 353–356
　　　middle-melting, 177, 320–324, 357–359
　　　low-melting, 190, 325–333, 360–364
　　　very-low-melting, 200, 333–334, 364
　　　unknown-melting, 204, 334–344, 365–369
　　methodology, 81, 91
　synthesis, 19
Firmness
　general functionality, 516
　see Texture
Flavor
　compounds, 26
　　aldehydes, 28, 429–430
　　branched chain fatty acids, 28
　　lactones, 27, 422–426
　　low molecular weight compounds, 28
　　methyl ketones, 27, 427–428
　　other compounds, 28
　flavor agents, 633
　general functionality, 513
　measurement of flavor compounds, 84, 111
　of experimental cold-spreadable butters, 537, 576
　of experimental dairy-based spreads, 542
　of intact milkfat, 26, 146, 422, 427, 429
　of milkfat fractions, 497
　　very-high-melting, 155, 423, 427, 429
　　high-melting, 167, 423, 428, 429
　　middle-melting, 180, 424, 428, 429
　　low-melting, 194, 424, 428, 430

unknown-melting, 204, 425
off-flavors, 34–35
Flow properties, general functionality, 518
Fraction nomenclature, 141, 489
    fraction designation, 141, 489
    general properties, 141, 489
Fractionation
    commercial fractionation of milkfat, 10, 211
    conditions used to produce milkfat fractions
        very-high-melting, 148, 491
        high-melting, 159, 492
        middle-melting, 172, 492
        low-melting, 185, 493
        very-low-melting, 199, 493
        unknown-melting, 203
    literature available, 4
    effects of fractionation on milkfat fractions, 494
    experimental fractionation of milkfat, 142
    methods, 8, 39
        crystallization from melted milkfat, 8, 43, 55, 490
        crystallization from solvent solution, 9, 49, 67, 490
        short-path distillation, 9, 52, 79
        supercritical fluid extraction, 9, 51, 74, 491
    procedures, 42
        multiple-step, 42
        single-step, 42
        refractionation, 43
    separation techniques, 56, 60
        aqueous detergent solutions, 48
        centrifugation, 48
        other techniques, 48
        pressure filtration, 47
        vacuum filtration, 47
Free fatty acid value, see Lipolysis
Frying oils, 560, 630
Functionality, 3, 509, 525
    attributes, 516
        aeration properties, 519
        coating properties, 520
        dispersion properties, 518
        firmness, hardness, softness, 516
        heat transfer, 520
        layering properties, 520
        lubricity, 520
        plasticity, 519
        shortening properties, 520
        solution properties, 518
        spreadability, 518
        structure formation, 517
        viscosity and flow properties, 518
    basic, 513
        chemical, 513
        physical, 513
    factors affecting functionality, 4, 521

Genetic manipulation of the dairy cow, 15

Hardness
    general functionality, 516
    see Texture
Heat transfer properties, general functionality, 520
Hydrogenation, 13
Hydrolysis, see Lipolysis

Ice cream, 559
Interactions with other food components, 509
    milkfat-fat, 511
    milkfat-gas, 512
    milkfat-(non-fat)-solid, 511
    milkfat-water, 512
Interesterification, 13
Iodine value
    methodology, 83, 111
    of commercial milkfat ingredients, 211, 486
    of intact milkfat, 26, 146, 408–410
    of milkfat fractions, 497
        very-high-melting, 155, 410–411
        high-melting, 167, 411–414
        middle-melting, 180, 414–416
        low-melting, 194, 416–419
        very-low-melting, 201, 419
        unknown-melting, 204, 420–421
Isosolid diagrams
    of milkfat fraction-cocoa butter mixtures for experimental chocolates, 556, 562
    of mixtures of milkfat fractions and other fats and oils, 225

Laminated pastries, see Bakery products
Layering properties, general functionality, 520
Lipolysis
    lipolytic enzymes, 35
    lipolytic off-flavors, 35
    methodology, 35, 89, 135
        free fatty acid value, 89, 135
        acid degree value, 89
Lubricity, general functionality, 520

Medium-chain triglycerides, 633
Melting behavior
    factors affecting, 521
    general functionality, 31, 514
    influence of chemical composition, 33
    see Melting point
    see Solid fat content
    see Thermal profile
Melting point
    methodology, 85, 121
        capillary melting point, 85, 121
        cloud point, 85, 121

Melting point methodology *(continued)*
    dropping point, 85, 121
    slipping point, 85, 121
    softening point, 85, 121
    of commercial milkfat ingredients, 211
    of experimental cold-spreadable butters, 536
    of experimental milkfat ingredients for bakery products, 545, 547, 550, 586
    of experimental milkfat ingredients for use in chocolate, 555
    of intact milkfat, 31, 147, 263–266
    of milkfat fractions, 500
        very-high-melting, 156, 267–270
        high-melting, 168, 271–277
        middle-melting, 181, 278–282
        low-melting, 195, 283–288
        very-low-melting, 201, 289–290
Milkfat biosynthesis, 19
    fatty acid synthesis, 19
    triglyceride synthesis, 19
    milkfat globule formation, 19
Milkfat fractionation, *see* Fractionation
Milkfat fractions
    fraction designation, 141, 489
    general fraction properties, 141, 489
    experimentally-prepared, 142, 491
        very-high-melting, 148
        high-melting, 158
        middle-melting, 171
        low-melting, 184
        very-low-melting, 198
        unknown-melting, 202
    commercially-prepared, 211, 480, 508
Milkfat ingredients, 526, 631
    blending of milkfat and milkfat fractions, 526
    incorporation of functional additives, 527
    packaging, 530
    stock milkfat fractions, 526
    tempering, 530
    texturization, 528
Milkfat, usage
    trends, 1
    as a food ingredient, 2, 530

Nuclear magnetic resonance, *see* Solid-fat content
Nutritional foods, 633
Nutritional properties
    aspects of metabolism of milkfat and milkfat fractions in rats, 208
    cholesterol, *see* Cholesterol
    conjugated linoleic acid, 37
    methodology, 90
    of commercial milkfat ingredients, 212
    of milkfat and milkfat fractions, 36, 207, 468, 507
    vitamin content, *see* Vitamin content

Oxidation of milkfat, 34
    factors affecting oxidative stability, 34, 206
    methodology, 34, 89, 135
        active oxygen method, 89
        peroxide value, 89, 135
        thiobarbituric acid value, 89, 135
        measurements of aldehyde content, 89, 135
    oxidation mechanisms, 34
    oxidative off-flavors, 34
    oxidation products formed in intact milkfat and milkfat fractions, 207, 461, 507
    oxidative stability characteristics of commercial milkfat ingredients, 212
    oxidative stability characteristics of intact milkfat and milkfat fractions, 206, 452, 506

Packaging of milkfat ingredients, 530
Penetrometry, *see* Texture
Peroxide value, *see* Oxidation of milkfat
Phase diagrams of mixtures of milkfat fractions and cocoa butter, 226
Physical processes to modify milkfat, 12
    butter tempering and reworking, 13
    cream tempering, 12
    fractionation, *see* Fractionation
Plasticity, general functionality, 519
Polenske number
    methodology, 83, 111
    of commercial milkfat ingredients, 486
    of intact milkfat, 26, 408–410
    of milkfat fractions
        high-melting, 167, 411–414
        middle-melting, 180, 414–416
        low-melting, 194, 416–419

Recombined dairy products, *see* Dairy products
Refractive index
    methodology, 83, 111
    of commercial milkfat ingredients, 211, 486
    of intact milkfat, 26, 146, 408–410
    of milkfat fractions
        high-melting, 167, 411–414
        middle-melting, 180, 414–416
        low-melting, 194, 416–419
        very-low-melting, 201, 419
        unknown-melting, 204, 420–421
Reichert-Meissl number
    methodology, 83, 111
    of commercial milkfat ingredients, 486
    of intact milkfat, 26, 408–410
    of milkfat fractions
        high-melting, 167, 411–414
        middle-melting, 180, 414–416
        low-melting, 194, 416–419
        very-low-melting, 201, 419

Salad dressings, 633
Saponification number
  methodology, 84, 111
  of intact milkfat, 26
Sectility, *see* Texture
Shortening properties, general functionality, 520
Short pastries, *see* Bakery products
Short-path distillation
  methodology, 9, 52, 79
  milkfat fractions produced, 258
    high-melting, 163
    middle-melting, 174
    low-melting, 189
Slipping point, *see* Melting point
Softening point, *see* Melting point
Softness
  general functionality, 516
  see Texture
Solid fat content
  methodology, 86, 121
    differential scanning calorimetry, 87, 121
    dilatometry, 87, 121
    nuclear magnetic resonance, 87, 121
  of experimental cold-spreadable butters, 531
  of experimental dairy-based spreads, 539
  of experimental and commercial milkfat ingredients for bakery products, 545, 547, 550, 591
  of experimental milkfat ingredients for chocolates, 555, 608
  of commercial milkfat ingredients, 211, 487
  of commercial milkfat ingredients for dairy products, 625
  of intact milkfat, 31, 148, 436–438
  of milkfat fractions, 504
    very-high-melting, 157, 223, 438–439
    high-melting, 170, 223, 440–442
    middle-melting, 183, 224, 443–444
    low-melting, 196, 445–448
    very-low-melting, 202, 448
  of mixtures of milkfat and other fats and oils, 477, 555
Solid fat index, *see* Solid fat content
Solution properties, general functionality, 518
Solvent fractionation
  see Crystallization, from solvent solution
Specific gravity
  methodology, 83, 111
  of commercial milkfat ingredients, 211, 486
  of intact milkfat, 26, 408–410
  of milkfat fractions
    high-melting, 167, 411–414
    middle-melting, 180, 414–416
    low-melting 416–419
    unknown-melting, 204, 420
Spreadability, general functionality, 518

Structured lipids, 633
Structure formation, general functionality, 517
Sucrose polyesters, 633
Supercritical fluid extraction
  fractionation, 9, 51
    methods, 9, 51, 74
    factors affecting fractionation, 52, 491
  milkfat fractions produced, 253
    very-high-melting, 151, 491
    high-melting, 162, 492
    middle-melting, 174, 493
    low-melting, 188, 493
    unknown-melting, 203

Tempering
  butter, 13
  cream, 12
  milkfat ingredients, 530
Texture
  methodology, 87, 132
    cone penetrometry, 88, 132
    disc penetrometry, 88, 132
    constant speed penetrometry, 88, 132
    sectility, 88, 132
    apparent viscosity, 88, 132
  textural characteristics
    of experimental cold-spreadable butter, 536, 572
    of experimental dairy-based spreads, 541, 581
    of experimental and commercial milkfat ingredients for bakery products, 586
    of experimental chocolates, 556
    of intact milkfat, 148, 449
    of milkfat fractions, 506
      high-melting, 171, 449
      middle-melting, 184, 450
      low-melting, 198, 450
      unknown-melting, 206, 451
Texturization of milkfat ingredients, 528
Thermal profile
  methodology, 86, 121
  of experimental chocolates, 563
  of experimental dairy-based spreads, 541, 562
  of commercial milkfat ingredients, 211, 227
  of intact milkfat, 31, 147, 213–214
  of milkfat fractions, 501
    very-high-melting, 156, 215–216
    high-melting, 169, 216–218
    middle-melting, 181, 218–220
    low-melting, 195, 220–222
    very-low-melting, 201, 222
  of mixtures of milkfat fractions and other fats and oils, 224
Thiobarbituric acid value, *see* Oxidation of milkfat
Tirtiaux fractionation process, 44

Triglyceride
  composition
    of commercial milkfat ingredients, 211, 484
    of intact milkfat, 23, 146, 370–373, 389–392
    of milkfat fractions, 495
      very-high-melting, 154, 373–374, 393–394
      high-melting, 166, 375–379, 395–398
      middle-melting, 179, 379–381, 399–401
      low-melting, 192, 382–384, 402–404
      very-low-melting, 200, 384, 404–405
      unknown-melting, 204, 385–388, 405–408
    methodology, 82, 91
  structure, 25
  synthesis, 19

Vegetable oils, blends with milkfat and milkfat fractions, 17, 210, 473
Viscosity
  general functionality, 518
  of experimental cold-spreadable butters, 531
  of experimental milkfat ingredients for bakery products, 546
  see Texture
Vitamin content
  methodology, 90, 137
  of commercial milkfat ingredients, 212
  of intact milkfat and milkfat fractions, 36, 207, 468

Whipping cream, *see* Dairy products

Yield
  of milkfat fractions
    very-high-melting, 151, 267–270
    high-melting, 163, 271–277
    middle-melting, 174, 278–282
    low-melting, 189, 283–288
    very-low-melting, 199, 289–290
    unknown-melting, 203, 291–301